Transportation and Power Grid in Smart Cities

Transportation and Power Grid in Smart Cities

Communication Networks and Services

Edited By
Hussein T. Mouftah and Melike Erol-Kantarci
University of Ottawa
Ottawa, Canada

Mubashir Husain Rehmani
COMSATS Institute of Information Technology
Wah Cantt, Pakistan

WILEY

This edition first published 2019

Registered Offices
John Wiley & Sons, Inc., 111 River Street, Hoboken, NJ 07030, USA
John Wiley & Sons Ltd, The Atrium, Southern Gate, Chichester, West Sussex, PO19 8SQ, UK

Editorial Office
The Atrium, Southern Gate, Chichester, West Sussex, PO19 8SQ, UK

For details of our global editorial offices, customer services, and more information about Wiley products visit us at www.wiley.com.

Wiley also publishes its books in a variety of electronic formats and by print-on-demand. Some content that appears in standard print versions of this book may not be available in other formats.

Library of Congress Cataloging-in-Publication Data

Names: Mouftah, Hussein T., editor. | Erol-Kantarci, Melike, editor. | Rehmani, Mubashir Husain, 1983- editor.
Title: Transportation and power grid in smart cities : communication networks and services / edited by Hussein T. Mouftah, Melike Erol-Kantarci, Mubashir Husain Rehmani.
Description: Hoboken, NJ : John Wiley & Son, 2019. | Includes bibliographical references and index. | Description based on print version record and CIP data provided by publisher; resource not viewed.
Identifiers: LCCN 2018012509 (print) | LCCN 2018028974 (ebook) | ISBN 9781119360094 (pdf) | ISBN 9781119360117 (epub) | ISBN 9781119360087 (cloth : alk. paper)
Subjects: LCSH: Smart power grids–Communication systems. | Urban transportation.
Classification: LCC TK3105 (ebook) | LCC TK3105 .T73 2018 (print) | DDC 388.3/12–dc23
LC record available at https://lccn.loc.gov/2018012509

Cover image: © chinaface/iStockphoto; © oonal/iStockphoto; © ansonmiao/iStockphoto
Cover design by Wiley

Set in 10/12pt WarnockPro by SPi Global, Chennai, India

MIX
Paper from responsible sources
FSC® C013604

Contents

List of Contributors

Taufik Abrao
Department of Electrical Engineering
Londrina State University (DEEL-UEL)
Brazil

Ibrahim Abualhaol
Carleton University
Ottawa
Canada

José A. Aguado
University of Málaga
Spain

Ozgur B. Akan
Koc University
Turkey

Kemal Akkaya
Florida International University
Miami, FL
USA

Fadi Al-Turjman
Antalya Bilim University
Antalya
Turkey

Muhammmad Amjad
COMSATS Institute of Information Technology
Wah Cantt
Pakistan

Muhammad Rizwan Asghar
Department of Computer Science
The University of Auckland
New Zealand

Ravil Bikmetov
University of North Carolina at Charlotte
NC
USA

Shengxia Cai
Zhou Enlai School of Government
Nankai University
China

Hakki C. Cankaya
Fujitsu Network Communications
Richardson, TX
USA

Oktay Cetinkaya
Next-generation and Wireless Communications Laboratory
Koc University
Turkey

Periklis Chatzimisios
Department of Informatics
Thessaloniki
Greece

Lei Chu
Research Center for Big Data Engineering and Technology
State Energy Smart Grid Research and Development Center, Shanghai, China

and

Department of Electrical Engineering
Shanghai Jiaotong University
Shanghai, China

Melike Erol-Kantarci
School of Electrical Engineering and Computer Science
University of Ottawa
Canada

Mahmoud Gad
Cognitive Labs Inc
Ottawa
Canada

Hervé Ganem
Gemalto
Paris, France

Fabrizio Granelli
Department of Information Engineering and Computer Science
University of Trento
Italy

Halil Gulacar
Department of Computer Engineering
Istanbul Technical University
Turkey

Sanket Gupte
Department of Computer Science and Electrical Engineering
University of Maryland Baltimore County
Baltimore

Guillaume Habault
IMT Atlantique - IRISA
France

Syed Hashim Raza Bukhari
COMSATS Institute of Information Technology
Wah Cantt
Pakistan

and

COMSATS Institute of Information Technology
Attock
Pakistan

Xing He
Research Center for Big Data Engineering and Technology
State Energy Smart Grid Research and Development Center, Shanghai, China

and

Department of Electrical Engineering
Shanghai Jiaotong University
Shanghai, China

Ekram Hossain
Department of Electrical and Computer Engineering
University of Manitoba
Canada

Md Maruf Hossain
Department of Natural and Applied Sciences (Engineering Technology Program)
University of Wisconsin-Green Bay
USA

Muhammad Ismail
Texas A&M University at Qatar
Doha, Qatar

Ljupco Jorguseski
TNO
The Hague
Netherlands

Khurram Kazi
Draper Laboratory
Cambridge MA
USA

Francois Lemercier
Itron and IMT Atlantique - IRISA
France

Zenan Ling
Research Center for Big Data Engineering and Technology
State Energy Smart Grid Research and Development Center, Shanghai, China

and

Department of Electrical Engineering
Shanghai Jiaotong University
Shanghai, China

Haichun Liu
Research Center for Big Data Engineering and Technology
State Energy Smart Grid Research and Development Center, Shanghai, China

and

Department of Electrical Engineering
Shanghai Jiaotong University
Shanghai, China

Qi Liu
Key Laboratory of Smart Grid of Ministry of Education
Tianjin University
China

Mohammad Upal Mahfuz
University of Wisconsin-Green Bay
Wisconsin
USA

Mohamed Mahmoud
Tennessee Technological University
TN, USA

Patrick Maille
IMT Atlantique - IRISA
France

Prodromos-Vasileios Mekikis
Department of Informatics and Telecommunication
University of Athens
Athens
Greece

Nicolas Montavont
IMT Atlantique - IRISA
France

Hussein T. Mouftah
School of Electrical Engineering and Computer Science (EECS)
University of Ottawa
Canada

Seyedamirabbas Mousavian
School of Business
Clarkson University
Potsdam
NY
USA

Ahmed O. Nasif
Department of Engineering Technology
University of Wisconsin-Oshkosh
Wisconsin
USA

Jimmy Jessen Nielsen
Aalborg University
Denmark

Sema F. Oktug
Computer Engineering Department
Istanbul Technical University
Turkey

Leonardo D. Oliveira
Department of Telecommunications and Control Engineering
University of Sao Paulo
Brazil

Mustafa Ozger
Next-generation and Wireless Communications Laboratory
Koc University
Turkey

Georgios Z. Papadopoulos
IMT Atlantique – IRISA
France

Marbin Pazos-Revilla
Tennessee Technological University
TN, USA

Konstantinos Plataniotis
Department of Electrical and Computer Engineering
University of Toronto
Canada

Petar Popovski
Aalborg University
Denmark

Khalid Qaraqe
Texas A&M University at Qatar
Doha, Qatar

Robert Qiu
Research Center for Big Data Engineering and Technology
State Energy Smart Grid Research and Development Center, Shanghai, China

and

Department of Electrical Engineering
Shanghai Jiaotong University
Shanghai, China

and

Tennessee Technological University
Cookeville TN
USA

Md. Abdur Rahman
Department of Electrical and Electronic Engineering
American International University-Bangladesh
Bangladesh

Mubashir Husain Rehmani
COMSATS Institute of Information Technology
Wah Cantt
Pakistan

Dhaou Said
School of Electrical Engineering and Computer Science (EECS)
University of Ottawa
Ottawa
Canada

Erchin Serpedin
Texas A&M University
College Station
USA

Ahmed Sherif
Tennessee Technological University
Cookeville
TN, USA

Sajid Siraj
University of Leeds
United Kingdom

and

COMSATS Institute of Information Technology
Wah Cantt
Pakistan

Petros Spachos
School of Engineering
University of Guelph
Canada

Alicia Triviño-Cabrera
University of Málaga
Spain

Tariq Umer
COMSATS Institute of Information Technology
Wah Cantt
Pakistan

Muhammad Usman
Department of Information Engineering and Computer Science
University of Trento
Italy

Binod Vaidya
School of Electrical Engineering and Computer Science (EECS)
University of Ottawa
Canada

John Vardakas
Iquadrat Informatica
Barcelona
Spain

Christos Verikoukis
Telecommunications Technological Centre of Catalonia (CTTC/CERCA)
Barcelona
Spain

Shouxiang Wang
Key Laboratory of Smart Grid of Ministry of Education
Tianjin University
China

Lei Wu
Electrical and Computer Engineering Department
Clarkson University
Potsdam
NY
USA

M. Yasin Akhtar Raja
Center for Optoelectronics and Optical Communication
University of North Carolina at Charlotte
USA

Yusuf Yaslan
Computer Engineering Department
Istanbul Technical University
Turkey

Mohamed Younis
Department of Computer Science and Electrical Engineering
University of Maryland Baltimore County
USA

Ioannis Zenginis
Iquadrat Informatica
Barcelona
Spain

Haibin Zhang
TNO
The Hague
The Netherlands

Ziming Zhu
Toshiba Research Europe Ltd
Bristol
United Kingdom

Preface

In recent years, there has been an increasing trend in population moving toward urban regions and large cities. It is envisioned that the future cities around the world will be smart cities. Plenty of efforts have been made to improve the quality of inhabitants of smart cities by integrating different technologies in their day-to-day lifestyle. These improvements include advancement in public facilities such as water systems, transportation systems, and the electricity system. In smart cities, information and communication technologies (ICT) will play a vital role for providing services in urban environments. These services include real-time monitoring and control through wireless sensor and actuator networks. Smart grids (SGs), intelligent transportation systems (ITS), the Internet of Things (IoT), electric vehicles (EVs), and wireless sensor networks (WSNs) will be the building blocks of future smart cities. "Smart grid" refers to the modernization of the traditional power grid by incorporating two-way digital communication support at the generation, transmission, and distribution levels. "Intelligent transportation system" refers to making the vehicular traffic smarter by reducing congestion, optimizing fuel consumption, choosing shorter routes, and improving safety, as well as allowing self-driving cars by using communication and sensing technologies. The "Internet of Things" refers to a worldwide network of interconnected objects uniquely addressable, based on standard communication protocols and allows people and things to be connected anytime, anyplace, with anything and anyone, ideally using any path/network and any service. The IoT can be very useful for resource management in the context of smart cities. Wireless sensor networks are composed of sensor nodes capable of performing sensing. The application of WSNs ranges from environmental monitoring to forest fire detection and from power system applications to disaster, security, emergency applications in urban environments. Electric vehicles aim to reduce vehicle emissions and can also be envisaged as mobile power stations, which can introduce the consumer-generated energy to the main electrical grid. All these technologies will somehow help to build a smart city.

This book provides detailed insights on communication networks and services for transportation and power grid for the future smart cities. The book aims to be a complementary reference for the smart city governors, utility operators, telecom operators, communications engineers, power engineers, electric vehicle service providers, university professors, researchers, and students who would like to grasp the advances in smart cities, smart grid, and intelligent transportation. This book accommodates 23 book chapters authored by world-renowned experts, all presenting their views on transportation and power grid in smart cities with a focus on communication networks and services. The chapters are organized in five parts.

Part I: Communication Technologies for Smart Cities focuses on the latest advancements in smart grid communications including cognitive radio based solutions, device-to-device communications, and 5G. Part I consists of five chapters.

Chapter 1 "Energy-Harvesting Cognitive Radios in Smart Cities," authored by Mustafa Ozger, Oktay Cetinkaya and Ozgur B. Akan, discusses the potential use cases of energy harvesting cognitive radios in smart cities along with research challenges that need to be addressed. Cognitive radio is a revolutionary technology that allows for opportunistic use of the unused spectrum frequencies to increase the communication capabilities and improve the overall system performance. On the other hand, energy harvesting brings a new perspective to the operation of cognitive radio, and their use in smart cities bare many opportunities that can lead to remarkable advances.

Chapter 2 "LTE-D2D Communication for Power Distribution Grid: Resource Allocation for Time-Critical Applications," authored by Leonardo Dagui de Oliveira, Taufik Abrao, and Ekram Hossain, focuses on device-to-device communications and LTE integration for applications in time-critical smart grid infrastructure. The authors propose a full duplex LTE-D2D scheduler to improve the capacity of LTE networks to enhance their performance in smart city applications.

Chapter 3 "5G and Cellular Networks in the Smart Grid," authored by Jimmy Jessen Nielsen, Ljupco Jorguseski, Haibin Zhang, Hervé Ganem, Ziming Zhu, and Petar Popovski, describes and analyzes the most relevant wireless cellular communication technologies for supporting the smart grid. Under the umbrella of 3GPP, the authors have looked specifically at releases up to and including Release 13, as well as considering the non-3GPP technologies such as IEEE 802.11ah, Sigfox, and LoRa. The authors provide a solid tutorial on cellular networks and their use in smart grid and smart cities.

Chapter 4 "Machine-to-Machine Communications in the Smart City—a Smart Grid Perspective," authored by Ravil Bikmetov, M. Yasin Akhtar Raja, and Khurram Kazi, presents intelligent Machine-to-Machine Communication techniques that can be used in smart cities. The chapter discusses optimization algorithms and machine learning techniques for efficient management of energy services in smart cities.

Chapter 5 "5G and D2D Communications at the Service of Smart Cities," authored by Muhammad Usman, Muhammad Rizwan Asghar and Fabrizio Granelli, provides an excellent survey on 5G and Device-to-Device (D2D) communications in the context of smart cities. The chapter presents smart city scenarios, their communication requirements, and the potential impact on the life of citizens as well as discussing the impact of big data on smart cities with potential security and privacy concerns.

Part II: Emerging Communication Networks for Smart Cities consists of five chapters that focus on emerging networks for smart cities.

Chapter 6 "Software Defined Networking and Virtualization for Smart Grid," authored by Hakki C. Cankaya, gives a comprehensive review of the state of the art in smart grid and SDN. The chapter then discusses the use cases for SDN for smart grid as well as several smart city scenarios.

Chapter 7 "GHetNet: A Framework Validating Green Mobile Femtocells in Smart-Grids," authored by Fadi Al-Turjman, focuses on energy-efficiency aspects of communications systems and presents energy-based analysis of femtocells in ultra-large scale (ULS) applications such as the smart grid.

Chapter 8 "Communication Architectures and Technologies for Advanced Smart Grid Services," authored by Francois Lemercier, Guillaume Habault, Georgios Z. Papadopoulos, Patrick Maille, Periklis Chatzimisios, and Nicolas Montavont, presents communication architectures and technologies employed in the smart grid and the requirements for next-generation smart grid systems. The chapter compares existing routing families in the constrained-based smart grid environment.

Chapter 9 "Wireless Sensor Networks in Smart Cities: Applications of Channel Bonding to Meet Data Communication Requirements," authored by Syed Hashim Raza Bukhari, Sajid Siraj, and Mubashir Husain Rehmani, motivates the use of WSN-based solutions in smart cities. The authors have introduced a channel-bonding technique for cognitive radios that can meet the requirements of high-bandwidth applications in smart cities. The chapter concludes with interesting future directions that pinpoint the open issues in this very active area of research.

Chapter 10 "A Prediction Module for Smart City IoT Platforms," authored by Sema F. Oktug, Yusuf Yaslan, and Halil Gulacar, brings the IoT perspective to the smart city discussion of our book. The authors emphasize the significance of prediction and present a tool that is used for prediction of traffic jams in populated smart cities.

Part III: Renewable Energy Resources and Microgrid in Smart Cities covers integration of renewable energy sources and the use of microgrids in smart cities. This part consists of four chapters.

Chapter 11 "Integration of Renewable Energy Resources in the Smart Grid: Opportunities and Challenges," authored by Mohammad Upal Mahfuz, Ahmed O. Nasif, Md Maruf Hossain, and Md Abdur Rahman, presents the opportunities and the corresponding challenges of integrating renewable energy resources in the smart grid. This chapter extensively discusses the impact of renewable energy on sustainable smart grid and sustainable cities.

Chapter 12 "Environmental Monitoring for Smart Buildings," authored by Petros Spachos and Konstantinos Plataniotis, focuses on smart buildings, which are an important component of the smart city. Their chapter introduces a wireless sensor network system that monitors the quality of the air in smart buildings. A framework is proposed for real-time remote monitoring of the carbon dioxide in a complex indoor environment showing an excellent real-world implementation of a smart building.

Chapter 13 "Cooperative Energy Management in Microgrids," authored by Ioannis Zenginis, John Vardakas, Prodromos-Vasileios Mekikis, and Christos Verikoukis, presents a cooperative energy management model for buildings that can exchange the energy produced by their PV panels or stored at their energy storage systems (ESSs), in a smart way so that the excess energy of buildings with energy surplus is consumed by buildings of the same microgrid with energy deficit. Energy management of buildings is certainly an important part of smart cities, and this chapter has a special focus on buildings and energy management.

Chapter 14 "Optimal Planning and Performance Assessment of Multi-Microgrid Systems in Future Smart Cities," authored by Shouxiang Wang, Lei Wu, Qi Liu, and Shengxia Cai, introduces optimal planning for multi-microgrids (MG). A microgrid is a small-scale power system containing distributed generation, loads, ESSs, and a control system. MGs have high flexibility so that they can be connected to the distribution network or work in an isolated mode when grid faults occur in the distribution network.

Therefore, they are anticipated to have a significant role in smart cities. This chapter provides valuable results on resilient planning of multi-microgrids.

Part IV: Smart Cities, Intelligent Transportation System and Electric Vehicles includes four chapters focusing on electric vehicles and intelligent transportation solutions that are a part of smart cities.

Chapter 15 "Wireless Charging for Electric Vehicles in the Smart Cities: Technology Review and Impact," authored by Alicia Triviño-Cabrera and José A. Aguado, provides an extensive review on wireless power transfer applied to the charging of electric vehicles and studies the scheduling algorithms that control the timing of the charge process in a group of EVs. The authors provide useful insights and present open issues in this exciting field of research.

Chapter 16 "Channel Access Modelling for EV Charging/Discharging Service through Vehicular ad hoc Networks (VANETs) Communications," authored by Dhaou Said and Hussein T. Mouftah, first presents the scheduling problem for electric vehicle (EV) charging/discharging and then introduces a specific case study of the channel access modelling for the EV charging service based on the IEEE802.11p/DSRC protocol.

Chapter 17 "Intelligent Parking Management in Smart Cities," authored by Sanket Gupte and Mohamed Younis, focuses on a popular smart city application, namely intelligent parking. The authors provide a taxonomy of parking systems, they survey existing solutions, and highlight the most important aspects, advantages, and shortcomings, and then they categorize regular parking systems based on the sensing infrastructure, communication infrastructure, storage infrastructure, application infrastructure, and user interfacing.

Chapter 18 "Electric Vehicle Scheduling and Charging in Smart Cities," authored by Muhammad Amjad, Mubashir Husain Rehmani, and Tariq Umer, provides an overview of different scheduling approaches involved in recharging of EVs in smart cities. The authors highlight various objectives such as frequency regulation, reduction of power losses, minimization of total cost for recharging, and integration of renewable energy resources during the controlled scheduling of EVs recharging. They also discuss communications requirements and resource management for scheduling the EVs recharging in smart cities.

Part V: Security and Privacy Issues and Big Data in Smart Cities is the last part of this book and covers security and privacy issues as well as big data in smart cities. It consists of five chapters.

Chapter 19 "Cybersecurity and Resiliency of Transportation and Power Systems in Smart Cities," authored by Seyedamirabbas Mousavian, Melike Erol-Kantarci, and Hussein T. Mouftah, demonstrate the vulnerability of the phasor measurement unit (PMU) networks and electric vehicle infrastructure (EVI) to cyberattacks. The chapter is an excellent guide to building response models against cyberattacks in the smart grid and smart cities.

Chapter 20 "Protecting the Privacy of Electric Consumers in the Smart City," authored by Binod Vaidya and Hussein T. Mouftah, underlines privacy concerns in the smart grid as well as emphasizing the aspects of privacy principles including privacy-by-design. It provides a basis for better understanding of current state-of-the-art privacy engineering along with privacy impact assessment and privacy-enhancing technologies.

Chapter 21 "Privacy Preserving Power Charging Coordination Scheme in the Smart Grid," authored by Ahmed Sherif, Muhammad Ismail, Marbin Pazos-Revilla, Mohamed

Mahmoud, Kemal Akkaya, Erchin Serpedin and Khalid Qaraqe, discusses the general security requirements for the smart grid and present potential cybersecurity solutions.

Chapter 22 "Securing Smart Cities Systems and Services: A Risk-based Analytics-Driven Approach," authored by Mahmoud Gad and Ibrahim Abualhaol, proposes a risk-based analytics-driven approach to design and operation of smart city critical infrastructure. In their approach, big data aggregated by different smart city systems and users is utilized by a fusion center analytics engine to provide functionalities to minimize the cybersecurity risk during operations.

Chapter 23 "Spatiotemporal Big Data Analysis for Smart Grids Based on Random Matrix Theory," authored by Robert Qiu, Lei Chu, Xing He, Zenan Ling and Haichun Liu, recognizes the fact that data are more and more easily accessible in smart grids. Therefore, big data analytics and data-driven approaches become natural tools for the future grid. In this chapter, the authors provide interesting insights on these tools.

This book is designed to be a handbook for the researchers in the academia and the industry who desire to learn the state of the art and open issues in communication technologies for the transportation and power grid of the future smart cities.

Hussein T. Mouftah, Melike Erol-Kantarci, Mubashir Husain Rehmani

Section I

Communication Technologies for Smart Cities

1

Energy-Harvesting Cognitive Radios in Smart Cities

Mustafa Ozger[1], Oktay Cetinkaya[1] and Ozgur B. Akan[2]

[1] Research & Teaching Assistant, Next-generation and Wireless Communications Laboratory, Koc University, Turkey
[2] Professor of Electrical and Electronics Engineering, Director of Next-generation and Wireless Communications Laboratory, Koc University, Turkey

1.1 Introduction

Wireless communication has been experiencing tremendous advancements. These developments have triggered new wireless networking paradigms and communication services. For instance, 5G is being studied by the research community to provide mobile broadband communications in wireless networks. Recently, the concept of IoT (Atzori et al., 2010), which is one of the key elements of 5G wireless networks, has been proposed to connect every device to the Internet, such as wireless sensor nodes, RFID tags, household appliances, etc.

As these technologies grow, they have been applied to various real-world problems. One of the most important application areas is the management of cities in a more efficient and smarter way. The smart city is a vision that extracts information from systems in the city to take measures for its management. This vision can be realized if information and communication technologies are employed in these systems to observe and manage them. Wireless sensors can be utilized as key elements for observing systems such as hospitals, highways, transportation networks, and power grids (Su et al., 2011). The transmission of the sensor observations about the city needs Internet connection to inform city officials. This fact leads to the utilization of the IoT since it can integrate all the facilities of the city with the Internet. Hence, the realization of the smart city vision becomes possible since the systems can be sensed, analyzed, and integrated with the use of communication technologies (Jalali et al., 2015). This enables the ability of managing the city in a cleverer and more efficient way in terms of city infrastructure, services, communication, business, energy, water, and so forth.

The smart city uses information and communication technologies to manage cities in an integrated manner (Zhang, 2010). The holistic view of the smart city can be seen in Figure 1.1.1, where smart grid, smart Transportation, smart communication, smart building, smart home, and smart infrastructure are bound together. With the use of next-generation information technologies, core systems are sensed, and the extracted information is analyzed for better management of the city and for improving quality of life in cities. To this end, it is envisioned that sensors are being deployed in different

Transportation and Power Grid in Smart Cities: Communication Networks and Services, First Edition.
Edited by Hussein T. Mouftah, Melike Erol-Kantarci and Mubashir Husain Rehmani.

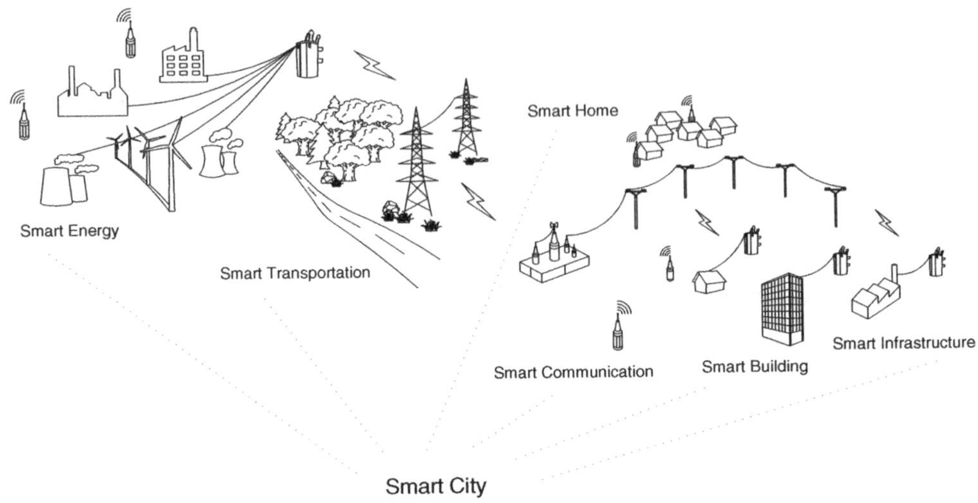

Figure 1.1.1 Smart city architecture.

key areas of the cities, such as power grids, water and underground systems, oil and gas pipelines, railways, roads, schools, hospitals, stations, airports, and so on.

The application areas of the smart city are diverse, and they connect every corner of the city to the Internet. This provides global intelligence over the management, and it paves the road for "Internet + Internet of things = smart planet" (Zhang et al., 2010). The smart city is the application of the concept of the smart planet to a specific region. The surveillance of infrastructure and environments of the city with the help of sensor technology achieves intelligent urban management and services (Su et al., 2010). The features of the smart city provide development of efficient urban strategies such as construction of smart homes, wireless cities, and smart transportation systems (Su et al., 2010).

The use of IoT to realize the smart city vision brings important challenges. With IoT technology, it is estimated that the number of devices connected to Internet will reach to 16 billion in 2020 (EU, 2010). Due to this increase, wireless data traffic that is being fulfilled in cities will reach excessive levels, and the available spectrum will become scarcer. Ever-growing demand in wireless communications has also increased the spectrum scarcity problem. Furthermore, fixed allocation of the spectrum worsens the problem of inefficient spectrum use. While the licensed spectrum bands are underutilized, the unlicensed ones are crowded, and the wireless communication is no longer feasible in these bands. To overcome the spectrum inefficiency and scarcity problems, CR technology is proposed (Mitola and Maguire, 1999). CR-capable wireless devices can access the licensed spectrum bands opportunistically and hence increase the spectrum utilization efficiency (Haykin, 2008). On the other hand, wireless nodes in these systems are resource constrained. Even though the majority of sensor nodes have duty cycling, a conventional battery in a sensor node depletes in less than a year. Therefore, an auxiliary or even a completely distinct source, such as heat, light, motion, and electromagnetic (EM) waves must be exploited to ensure sensors' operation. In this regard, EH technologies come into prominence to build wireless sensor networks (WSNs) that are free from battery constraints (Sudevayalam and Kulkarni, 2011). Hence, these challenges promote novel methodologies for obtaining the spectrum efficiency and delivering the power to the wireless devices. The most promising answers that obviate these problems are CR and EH technologies.

1.1.1 Cognitive Radio

Excessive use of wireless devices has caused spectrum scarcity problem in industrial, scientific, and medical (ISM) bands (FCC, 2002). This problem has revealed a new access technology to the EM spectrum. The enabling technology is CR (Mitola and Maguire, 1999). The most important feature of CR is the ability to adapt its operating frequency to certain frequency bands for wireless communication. Hence, this feature provides a scheme enabling dynamic access to the spectrum. With this feature, wireless devices can access to the licensed spectrum bands opportunistically. Hence, CR nodes could coexist with license holders, which are primary users.

Dynamic spectrum access has one constraint, which is interference to the licensed users. There are two interference schemes. The first one is that CR must cause no interference to the primary users. The second one is that CRs may cause interference within a certain threshold. The interference is avoided with the help of cognitive cycle operations of CR-enabled wireless devices. These operations are spectrum sensing, spectrum decision, and spectrum handoff, which are explained in detail as follows.

- **Spectrum sensing:** Spectrum sensing is the most important feature of CR due to the ability to get information about the usage of the spectrum bands. It reveals unused spectrum bands, i.e., spectrum opportunity for the wireless devices. They share these spectrum bands without causing interference to the licensed users. There are different methods to identify the spectrum opportunities, which are energy detection, waveform detection, cyclo-stationary detection, and matched filter detection.
- **Spectrum decision:** After detecting the vacant spectrum bands, CR nodes should decide on which spectrum band to transmit their data. The spectrum band for the communication is determined according to the requirement of the CR networks. The decision is performed in a centralized or distributed manner. However, the dynamic radio environment poses challenges since a common control channel, which is proposed generally in literature, may not be present most of the time. The exchange of spectrum sensing results may degrade the communication quality of CR nodes since the control packets may overwhelm the CR network.
- **Spectrum handoff:** This functionality prevents the interference caused by secondary transmission to the primary users. It provides the ability to stop the transmission of secondary users when a primary user arrives.

The features of CR increase overall spectrum utilization by using the underutilized licensed spectrum bands without causing any interference to the licensed users. Furthermore, they relieve the traffic jam in ISM bands. This also increases the energy efficiency since they decrease collisions among the secondary nodes, i.e., CR, by using the underutilized licensed spectrum bands.

1.1.2 Cognitive Radio Sensor Networks

WSNs have an event-driven communication nature that results in "bursty" traffic depending on event characteristics. Dynamic spectrum access proposes a spectrum-efficient solution for the problems related to spectrum limitations of WSNs. Dynamic change of operating spectrum to utilize the unused licensed spectrum by the CR-enabled sensor nodes improves the overall spectrum utilization. Hence, a new networking paradigm is proposed in Akan et al. (2009), named CRSNs. This networking paradigm has changed the fixed spectrum utilization adopted by the WSN nodes.

CR capability of the sensor nodes provides coexistence with the licensed users, which enables sensor observations to be conveyed in a multi-hop manner over the available licensed spectrum bands. The advantages of CR capability of sensor nodes are dynamic spectrum access, opportunistic channel usage for bursty traffic, and adaptability for reducing power consumption. Hence, CR decreases the probability of collision and packet loss, which accordingly decreases the energy consumption of the sensor nodes.

The advantages of the CR usage in WSNs reveal a number of possible application areas of CRSNs. They may be listed as indoor sensing applications, multimedia applications, multiclass heterogeneous sensing applications, and real-time surveillance applications.

1.1.3 Energy Harvesting and Energy-Harvesting Sensor Networks

Sensor nodes have an important constraint, which is limited battery power. This restricts the operation duration of sensor networks. Because of the CR capability, the sensor nodes require more energy as the cognitive cycle operations are energy-requiring processes. Hence, they drain the battery sooner, and the lifetime of the sensor network is decreased.

One of the solutions to this problem is EH. It enables to power the wireless devices from natural resources such as light, temperature gradient, and different ambient sources such as radio frequency signals and EM fields (Weddell et al., 2013). The practicality of this method comes from the fact that the sensor nodes no longer need to have their batteries replaced or recharged when they are depleted. The harvested energy from the ambient resources is utilized to run the wireless devices autonomously. This provides more energy for the communication, which results in higher quality of service (QoS). Furthermore, cognitive cycle operations can be performed with higher precision due to increased energy budget thanks to harvesting.

In this chapter, we overview all the mentioned solutions to overcome the challenges related to the realization of the smart city, and therefore we explore EH-CR wireless nodes. To this end, we first explain CRSNs and how to combine the harvesting methods with them. The motivations for spectrum-aware and self-sustaining communications in the smart city are revealed. Furthermore, challenges posed by the use of these methods in the smart city are investigated. In the light of these discussions, we study a networking architecture for the IoT, i.e., Cognitive Energy-Harvesting IoT. By using this architecture, we also overview a general implementation framework for EH-CRs.

1.2 Motivations for Using Energy-Harvesting Cognitive Radios in Smart Cities

The smart city vision necessitates observing the key system elements in cities and extracting the information about these elements for better management of the cities. The most important aspect of this vision is the observation part, which is handled by the sensors that are deployed on different systems of the city such as hospitals, roads, bridges, and so on. This makes the city smarter, which means more efficient, sustainable, and livable (NRDC n. d.).

The advantages of the smart city motivate the utilization of information and communication technologies. However, cities present harsh environments, which poses challenges on the sensor technology. In the context of the smart city, sensor nodes are deployed in different remote geographical areas that make them inaccessible in case of battery depletion or failures. Hence, the utilization of EH-CRs is crucial in the smart city. Motivations of using these radios are explained in detail in the following subsections.

1.2.1 Motivations for Spectrum-Aware Communications

A CR node can sense the spectrum, find vacant bands, and change its transmission parameters to use these vacant bands for communication. This provides opportunistic spectrum access (OSA). Since the licensed spectrum is not fully utilized, these bands are used opportunistically. CRSN is a distributed network that senses the environment and collaboratively conveys their readings dynamically over available spectrum bands in a multi-hop manner to satisfy application-specific requirements (Akan et al., 2009).

The CR capability offers spectrum-aware communications. This type of communication provides licensed bands for the sensors for opportunistic access. Spectrum-aware communications utilize the most favorable channel among the idle spectrum bands after spectrum sensing. This helps reducing transmission errors and the number of retransmissions and increasing reliability of the communication. Furthermore, CR capability fulfills the delay requirement of the applications by utilizing the idle channels. This may reduce the number of hops to the destination in the network, which also decreases the total delay.

The smart city applications require efficient and timely data transfer from the systems of the city. Since the unlicensed bands are heavily used in the cities, the spectrum-aware communications become the most important solution to overcome the spectrum scarcity problem. Furthermore, the advantages of the spectrum-aware communications such as increased reliability and decreased latency further motivate its utilization. Hence, it fulfills the requirements of the smart city.

1.2.2 Motivations for Self-Sustaining Communications

Sensor technology is a key element for different IoT applications. Depending on the application, sensor nodes sense and transmit data either intermittently or in a periodic manner. Data transmission consumes a few milliwatts of power and microwatts of it during sleep mode (Moghe et al., 2009; Sudevalayam and Kulkarni, 2011). However, a typical sensor battery depletes in less than a year. If we consider the number of deployed sensors and their energy requirements in the domain, there is a clear need of an auxiliary or even a distinct source. However, this may or may not be an option for each application mostly due to size constraints, maintenance, and/or deployment costs. Large number of sensor node utilization and their individual energy demands require energy-efficient solutions and sustainable use of resources. Hence, this also promotes energy-efficient solutions by encouraging battery-less systems.

EH resolves the problem of limited lifetime of wireless devices such as sensor nodes (Moghe et al., 2009). This approach is very beneficial for the sensor nodes, which have limited battery power. It exploits the ambient resources, namely radio frequency (RF) signals, heat, and movement variations, vibrations, EM and sound waves, and so forth, and provides utilizable energy to the wireless devices.

The digital skin of the smart city, which is sensor networks deployed in different parts of the city, performs periodical monitoring and requires an excessive amount of energy. The sensor deployment in large numbers and the difficulty in replacing the battery motivate the use of EH approaches for the smart city. Hence, it stands as a way of providing more reliable, durable, and profitable alternative for the proper system operation.

1.3 Challenges Posed by Energy-Harvesting Cognitive Radios in Smart Cities

Sensor nodes have limited energy, memory, and processing power. Furthermore, CR operation increases the energy consumption of these nodes with additional operations such as spectrum sensing and spectrum management. These challenges are intensified by the smart city. In the following subsection, we list these challenges.

- *Bandwidth*: Depending on the application, the sensors in the smart city demand high bandwidth. For example, some applications require multimedia delivery. This poses an important challenge to overcome in a dynamic radio environment. The heterogeneity of spectrum bands and their dynamic availability change the bandwidth of the network, which requires flexibility in bandwidth of the licensed bands.
- *Quality of Service (QoS)*: CR offers flexibility to satisfy QoS requirements of the smart city. The QoS level depends on the application and the observed key elements of the cities. QoS is measured in terms of reliability, throughput, and delay. The most important factor for the change of QoS level is the licensed user activities.
- *Coverage area*: The smart city must provide coverage for very large geographical areas. However, the licensed user activity will vary spatially. This poses challenges for the coverage since the nodes in a region may not have any spectrum opportunity for some period of time due to licensed user activities. Hence, we cannot extract information from those regions, which degrades the vision of the smart city.
- *Scalability*: Monitoring such a large geographical area will be performed by large number of sensors. This requires scalable solutions to enable the smart city. Furthermore, this forces the communication infrastructure to be flexible and adaptive to any change in the network.
- *Reliability and delay*: The sensory information from the elements of the city should be reliably delivered in real time. However, the reliability of the communication may be degraded by the licensed user activities. This may distort the channel conditions and cause network outages and interference caused by the transmissions of primary users. Furthermore, intermittent connectivity of the CR nodes causes an increase in the delay.
- *Computational capability*: The computational capability of the CRSN nodes is limited, and this poses a challenge to perform complex tasks. The sensor nodes are also energy constrained, which exacerbates the limited computational complexity of the nodes.

- *Intermittency in available sources*: Consistency of available sources cannot be guaranteed. For example, in RF EH the energy is harvested from the ambient RF signals. However, these signals are not always present, and hence, harvestable energy is not continuous. This is also the case for different EH resources such as wind and light.
- *Integration of the sensor networks to the Internet*: For the realization of a smart city, the extracted data should be delivered to the city authorities or the citizens. However, in such a large geographical area with varying deployment conditions, the connection of the sensors to the Internet poses a challenge.
- *Service differentiation*: ZigBee, IEEE 802.15.4, 6LoWPAN, and similar technologies are utilized in the IoT domain. Various sensor types must cooperate with each other to achieve a desired goal in the smart city. Different sensor nodes deployed in a system of a city must be compatible with one gateway, which provides the Internet connection to the system.

CR and EH pose some significant challenges. These challenges are intensified by the physical conditions in the cities. Furthermore, the harsh environmental conditions of cities decrease the possibility to transmit the observed data to the cloud for the analysis and the evaluation of the critical data.

1.4 Energy-Harvesting Cognitive Internet of Things

1.4.1 Definition

The advancements in wireless technology make interaction between people and the digital world more powerful and sophisticated. Wireless communications technology plays a vital role for these advancements. In this respect, the IoT is one of the most innovative paradigms for next-generation wireless networks (Gubbi et al., 2013).

Every electronic device will be connected to the Internet as a result of the notion of the IoT (Gubbi et al., 2013). A clear definition has not been determined for the IoT despite the fact that concept of the IoT is obvious. It can be viewed as a union of next-generation wireless devices that sense the surrounding and gather information from the physical world (Castellani et al., 2010).

Connection of the wireless devices with each other and to the Internet results in congestion and a high volume of traffic in the EM spectrum. As these devices in the IoT mainly use already crowded ISM bands, the problem of spectrum scarcity is exacerbated by the introduction of the IoT. Furthermore, the licensed spectrum bands are underutilized. CR technology has been proposed as a solution to the problems of spectrum scarcity in unlicensed spectrum bands and the under-utilization of licensed spectrum bands (EU, 2010). Wireless devices with CR capability can use the spectrum opportunistically. Unlicensed users, i.e., CR nodes, coexist with licensed users, i.e., primary users (PUs). Cognitive cycle operations enable CR nodes to use the licensed channels in an opportunistic manner. These operations provide scanning of the spectrum for an opportunity to transmit and ceasing the transmission if PU activities exist during communication. Hence, cognitive capability of devices increases the efficiency of the spectrum utilization and overcomes the spectrum scarcity problem. For instance, Wu et al. (2014) present the cognitive IoT (CIoT) to enable cognition in the IoT devices in terms of spectrum usage and the interaction with the physical and social worlds.

The wireless devices in IoT are resource constrained. Furthermore, cognitive cycle operations deplete the battery of the wireless devices sooner. Moreover, it is not possible to replenish their batteries. Hence, EH is an efficient method to power wireless devices in the CIoT. Therefore, the networking paradigm of Energy-Harvesting Cognitive Internet of Things, consisting of EH-CRs, is a key enabler of the smart city by satisfying its requirements.

1.4.2 Energy-Harvesting Methods in IoT

In this subsection, we overview the existing EH methods in the IoT domain. Harvesting methods can be separated into two groups depending on the energy demand and availability in harvestable resources, named Harvest-Use and Harvest-Store-Use, respectively (Sudevalayam and Kulkarni, 2011). Furthermore, the harvestable resources can be categorized according to their controllability and predictability. With the help of these categories, leading harvesting techniques are compared and their advantages and disadvantages are discussed in Table 1.1.

EH from light sources is a well-studied method of power provision that gathers energy from ambient lights, either from the sun or from artificial light sources, and is

Table 1.1 Comparison of Energy-Harvesting Techniques.

Technique \ Features	Energy availability	Characteristics	Advantages	Disadvantages
Solar	poor	ambient, uncontrollable, predictable	environmental, independent of grid, high output voltage	depends on sunlight, deployment constraints
Thermal	poor	ambient, uncontrollable, unpredictable	environmental, independent of grid, scalability	depends on thermal gradient, requires efficient heat sinking
Airflow	good	non-ambient, uncontrollable, unpredictable	environmental, independent of grid	fluctuating density, hard to implement, requires construction
Motion	fair	non-ambient, controllable, unpredictable	no external power source, compact configuration, light weight	charge leakage, highly variable output
RF	good	Non-Ambient, uncontrollable, predictable	abundant in urban lands, allows mobility	scarcity in rural areas, low power density, distance dependent
M-field	good	non-ambient, controllable, predictable	no external power source, easy to implement, light weight	requires high current flow, safety vulnerabilities
E-field	excellent	non-ambient, controllable, predictable	no need of current flow, easy to implement, always available	capacitive, mechanical constraints

based on the phenomenon known as the photovoltaic (PV) effect (Sudevalayam and Kulkarni, 2011). Solar EH is performed by solar cell inlaid photovoltaic panels in mostly outdoor applications for the monitoring of overhead power lines (Moghe et al., 2009; Sudevalayam and Kulkarni, 2011). For indoor applications, specialized photovoltaic materials, which are better suited for diffused lights, are employed for taking advantage of the light emitted from ambient lighting elements.

Although the PV modules are getting cheaper, easier to use, and more efficient each passing day, due to the dramatic fluctuations on the output power and ongoing installation and maintenance costs, they have limited applicability in mission critical applications (Akan et al., 2017). Another EH method, kinetic energy harvesting (KEH), is the conversion of ambient movement energy into electrical power. Wind turbines and, on a smaller scale, anemometers are being utilized for exploiting airflow energy to enable wide-scale communications structured in open space; however, their performance is highly threatened by the environmental variables similarly to solar energy related techniques (Moghe et al., 2009; Sudevalayam and Kulkarni, 2011). Piezoelectric materials, similarly, are used often for gathering energy from highly random and mostly unpredictable motion variations driven by external factors (Matiko et al., 2009; Moghe et al., 2009). KEH is an applicable method for both indoor and outdoor domains, as there is a variety of sources that can be conveniently exploited to drive low-power wireless autonomous devices. However, constituting a generalized harvesting system, especially for vibrating sources, is an ongoing issue, because the conversion efficiency highly varies with the resonant frequency of the vibration, which makes necessary a specialized design for each source (Moghe et al., 2009; Sudevalayam and Kulkarni, 2011; Zhao et al., 2013).

Thermal energy harvesting, i.e., thermoelectric generation (TEG), is simply based on converting temperature gradients into utilizable electric power; this is based on the Seebeck effect, which occurs in semiconductor junctions. TEG is an innate power provision technique for Smart Grid communications, in which temperature swings between the grid assets and the environment are used to extract energy (Sudevalayam and Kulkarni, 2011). Similarly to environmental sources such as solar and airflow, it strictly depends on the ambient variables and therefore may fail to satisfactorily provide stable power in some cases. For less power-requiring sensor nodes; Peltier/thermoelectric coolers and thermocouples are widely used for building delay-tolerant wireless networks. Although harnessing power by using temperature gradient between the systems sounds promising, there is a fundamental limit, namely, the Carnot cycle, to the maximum efficiency at which energy can be harvested from a temperature difference (Akan et al., 2017).

Regarding the intensive use of GSM (global system for mobile) networks in urban areas, radio frequency (RF) signals attracted both academia and industry in recent years (Sudevalayam and Kulkarni, 2011). RF EH simply targets RF signals emitted from base stations, network routers, modems, smartphones, tablets, and any other wireless signal sources and collects them via large aperture power-receiving antennae, subsequently converting the gathered waves into utilizable DC power for the sensor nodes. An RF EH circuit includes a transducer for converting EM waves into alternating current (AC) and a rectifier for converting the AC power into the direct current (DC). In addition to them, RF filters, voltage multipliers, and couplers can be also utilized for more advanced harvesting procedures. Even though this method delivers reliable solutions regardless of the environmental variables, deployment inflexibility and low power provision efficiency hamper its utilization in some applications (Cetinkaya and Akan, 2017a; Sudevalayam and Kulkarni, 2011).

Wireless networks can also be powered by exploiting EM fields, i.e., magnetic and electric fields, emitted from current-carrying and/or tension-holding conductive materials (Sudevalayam and Kulkarni, 2011; Zhu et al., 2009). Magnetic field (M-field) EH is based on coupling magnetic fields surrounding the AC current-carrying conductors by simply clamping them with current transformers. This technique provides an adequate rate of power, and its utilization is less complex; however, the availability of energy is affected severely by the current density on the power line. As the M-field occurs due to an AC current, the line must be loaded to allow sufficient current flow. Furthermore, mostly due to space limitations, it might not be an efficient solution for applications in which it is not that practical to clamp the conductor. As the field density decreases quadratically with the distance from the field-emitting source, this approach also requires very close placement for the most effective EH performance. In wide open spaces, such as Smart Grid and smart city applications, this approach performs sufficiently in scavenging power from the targeted/attached assets to operate remote monitoring devices. However, gathering energy in a safe way from a high current–carrying asset that has physical contact with the harvester is still a challenging issue. To mitigate the safety concerns, M-field-based methods need to be equipped with advanced protection mechanisms and components.

Similar to M-field-related approaches, electric field (E-field) EH (EFEH) also targets surrounding, unutilized field flow for operating low-power autonomous devices. E-field EH is simply based on obstructing the free charges that are being emitted by a conductor with voltage potential and draining them via the displacement current. E-field EH was first tried on high and middle voltage overhead power lines by utilizing the E-field in abundance (Moghe et al., 2009; Moghe et al., 2015; Zhangl et al., 2009; Zhao et al., 2013). First results revealed the potential of this method in powering wireless devices placed for condition monitoring procedures and accordingly encouraged further efforts for building advanced Smart Grid services. Later, it was applied to low voltage systems as mounting single-phase AC power cords with metallic sheets (Cetinkaya and Akan, 2017a; Chang et al., 2012; Choi et al., 2014; Honda et al., 2015). These efforts disclosed that it is also possible to constitute an applicable EFEH methodology for applications in which E-field intensity is considerably low. In short, E-field is the only source that is neither intermittent nor dependent on the load. As the voltage and the frequency are firmly regulated and exactingly maintained, the E-field is therefore stable and predictable in its behavior. Thus, it can be regarded as the most promising way to compose long-term and self-sustainable communication systems notwithstanding the ambient factors (Cetinkaya and Akan, 2017b). A detailed comparison of the aforementioned methods is illustrated in Table 1.1.

1.4.3 System Architecture

In the system architecture, the nodes transmit their readings opportunistically to the access point or the sink that has the Internet connection. Furthermore, the sink or the access point has the CR capability.

Network topologies may differ according to the application. These topologies are explained as follows.

Ad hoc CRSN in the IoT: In this architecture, there is no central entity for regulating the spectrum-aware communications. Hence, the nodes exchange control packets among themselves to perform interference free communication. The readings of the sensors are transmitted in multiple hops in an ad hoc manner. Spectrum management is performed cooperatively between the nodes.

Clustered CRSN in the IoT: In a dynamic radio environment, the nodes can only communicate according to their spectrum availability. The nodes located within a neighborhood have similar vacant channels. It is highly possible that they have common channels among them. This provides a common control channel between the cluster members to exchange the control data for spectrum-aware communications. This is the most convenient architecture to support the communication.

Heterogeneous CRSN in the IoT: In this architecture, some special nodes that have more power sources are deployed. These nodes may have special responsibilities such as spectrum bargaining (Haykin 2008).

Mobile CRSN in the IoT: In addition to the dynamic radio environment, the nodes are mobile in this CRSN architecture. This makes the operation of CRSN more dynamic. Solutions for dynamic spectrum access should also consider the mobility in this architecture.

1.4.4 Integration of Energy-Harvesting Cognitive Radios with the Internet

CRSNs are distributed networks having no central entity regulating the communication. Due to the large operation coverage of the smart city, sensor observations are conveyed in a multi-hop manner. Furthermore, some nodes can reach the sink or access point in one hop. The CRSNs are the main components of devices that can extract information from the environment; hence, they are the main pillar of the IoT. The harvested information by the CRSN nodes is important for the actions to be taken and the information processing. The integration of CRSN with the mobile networks or the Internet realizes the IoT and manages communications in CRSNs. To this end, we overview architectures (Alcaraz et al., 2010; Honda et al., 2015) to enable the connection of CRSNs with the Internet.

Stack-based (Christin et al., 2010) and topology-based (Alcaraz et al., 2010) solutions are the two approaches that integrate WSNs to the Internet. The stack-based integration approaches require similarity between the network stacks of the sensor network and the Internet. Front-end solution, gateway solution, and TCP/IP solution are the stack-based integration approaches. There is independency between the Internet and the sensor networks in the front-end solution. In this approach, the sensor network uses its own protocol. On the other hand, the Internet uses TCP/IP. The gateway approach supports tunneling for the exchange of information and the integration between the sensor networks and the Internet. In the TCP/IP approach, sensor networks and the Internet use the same protocol, which is the TCP/IP solution advocating the integration by utilizing the same networking protocols in both sides. On the other hand, the topology-based approaches depend on the deployment of the Internet-connected nodes in the sensor networks.

The spectrum awareness by CR capability makes an important difference for the stack-based integration solutions. The access point or the base station, which provides the Internet connection, must behave as a CR node and operate in an opportunistic manner. They are not resource limited, which makes them capable of complex operations. In a front-end solution, the base station has two interfaces: one for CRSN and one for Internet connectivity. The packets from CRSN nodes are gathered at the base station by spectrum-aware communications. The base station sends the collected information to the cloud or to a user via an Internet connection using TCP/IP. The Internet connection of the CRSN nodes is achieved by the base stations at the edge of the network. This architecture is the classical approach existing in the literature.

The gateway approach supports the addressing of the CRSN nodes such that a user on the Internet communicates with specific nodes in the network. However, the bottleneck in this approach is again the base station since IP tunneling is implemented, and the information must traverse the base station. This approach uses extra bits for the encapsulation of the IP packet. This degrades the benefits of this approach.

The full integration with the Internet is achieved by the TCP/IP solution. A user on the Internet can access any node in CRSN. For this approach, IPv6 is very promising for sensor networks, as it enables sensor nodes to be directly connected to the Internet (Hui and Culler, 2008). The use of IPv6 can realize the TCP/IP solution for CRSNs and increase the possibility of usage CRs in the smart city. It provides interoperability between CRSN and different network systems in the smart city, which will emulate always-on link communication (Hui and Culler, 2008).

In the topology-based integration solutions, location and deployment of the nodes, as well as the base stations, which are directly connected to the Internet, are all important (Christin et al., 2009). Multiple nodes have access to the Internet in hybrid solutions, which are one of the topology-based integration approaches. These nodes are the neighbors of base stations. In an access point solution, sensor nodes have direct access to the access points.

1.5 A General Framework for EH-CRs in the Smart City

1.5.1 Operation Overview

As its definition implies, this concept of smart city emerges from the need of better management of the cities. Hence, many national governments are pushed to adopt information and communications technologies to provide better services in the cities, which realizes the smart city concept (Schaffers, 2011). With the application of the smart city, city resources are efficiently utilized, costs of the city are reduced, and the life quality of the city residents is increased (Zanella et al., 2014).

The adoption of the IoT in the smart city also provides new services for the citizens and the governments. These services may be structural health, waste management, air quality and traffic congestion monitoring, and smart lighting and parking, to name a few (Zanella et al., 2014). IoT networks will find a number of deployment areas in the city to realize the concept of the smart city. These areas include transport systems, bridges, electric grid, hospitals, schools, cultural sites, and so forth. Hence, with the IoT concept, the Internet becomes even more pervasive, which also helps realizing the smart city. Every device will be connected to each other and the Internet.

The business side of the smart city is studied in terms of the public actors and the city governments (Walravens and Ballon, 2013). However, there are some open issues

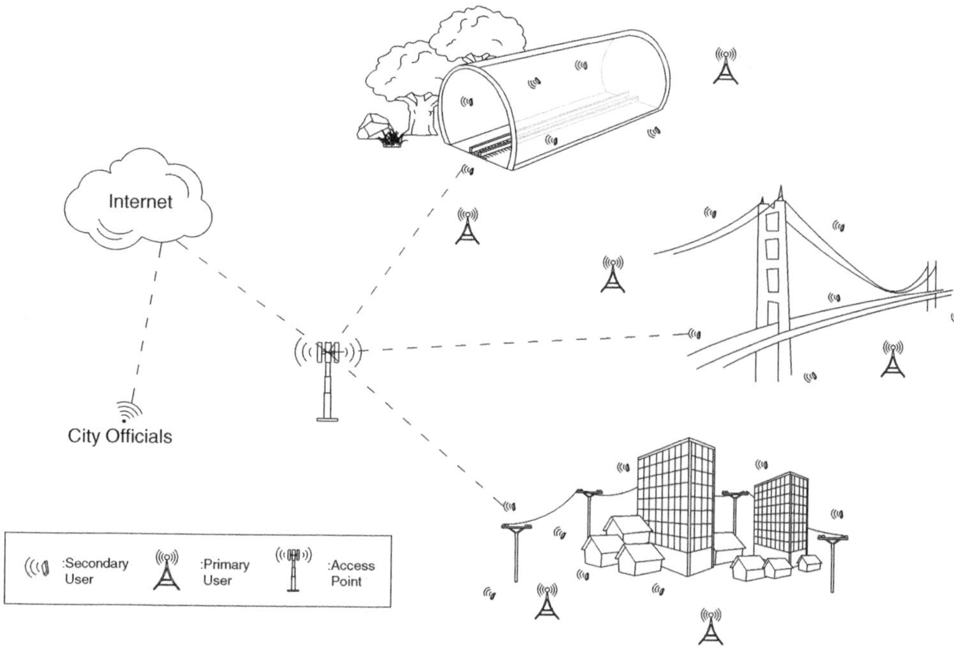

Figure 1.5.1 Operation overview of EH-CR in the smart city.

in the technical side of the smart city. First of all, due to excessive increase in wireless communications, there are some challenges. The smart city concept requires to sense the systems of the city and to report the application-specific data to the city officials. This causes an enormous amount of data transfer from different part of the city, which also increases wireless traffic in the city. Hence, new communication paradigms should be utilized. To this end, CR is an efficient solution to overcome the excessive wireless traffic in the city by opportunistic spectrum access. Secondly, wireless devices deployed in the smart city are resource constrained. The reporting about the systems in the city causes energy depletion of the wireless nodes. Since the maintenance of these nodes is a difficult task, EH methods are perfect candidates to extend the lifetime of these nodes. To this end, the concept of EH-CRs becomes very important to address the technical challenges for enabling the smart city.

From the holistic point of view, operation overview of EH-CRs in the Smart Cities can be explained as in Figure 1.5.1. The EH-CR nodes deployed in the systems of the city report their sensing results to the city officials thanks to the approaches enabling integration of EH-CRs with the Internet. By using the spectrum opportunistically, they send their information in a more reliable manner. With the adoption of EH, they do not face the shortages in battery power. They increase the communication efficiency and power efficiency of the smart city.

1.5.2 Node Architecture

The node architecture of an EH-CR node in the smart city contains four main units (Figure 1.5.2), which are antenna unit, ultra-low power communication unit, ultra-low

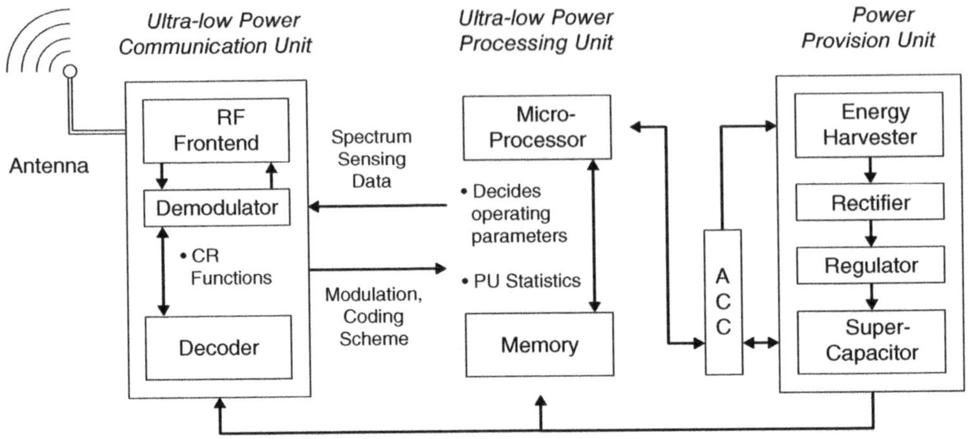

Figure 1.5.2 Node architecture of EH-CR node.

power processing, and power-provisioning unit. The ultra-low power communication unit provides the capability of channel switching, channel sensing, modulation, and power control. The power-provisioning unit has an energy harvester, a rectifier, a regulator and a supercapacitor. It provides energy to the antenna unit, communication unit, and power processing unit. The aim of this unit is to convert an available resource into utilizable electrical power to operate the nodes. The processing unit decides the operating parameters, which consist of a microprocessor and memory. The communication unit has RF frontend, demodulator, and decoder to provide the communication with CR functionalities.

The block stated as ACC in Figure 1.5.2, i.e., autonomous connection circuit, needs to be employed to switch between harvesting and nodal operation stages. This circuit simultaneously observes the voltage level on the storage element, i.e., the supercapacitor, and accordingly enables charge transfer when the harvested energy is sufficient enough for sensory operations and disengages the supercapacitor from the circuit to turn back the harvesting period when the voltage level descends below a certain threshold. This operation, held under the control of the processing unit, not only prevents undesired discharge of the supercapacitor but also allows more frequent transmission cycles by shortening the time exerted for harvesting processes (Cetinkaya and Oktay, 2017b).

1.5.3 Network Architecture

The networks consisting of EH-CRs have ad hoc and cluster-based architectures. The centralized architecture would not be feasible due to limited harvestable energy. The harvestable energy cannot support the energy required for the transmission and the reception of the control data with the central entity regulating the communication.

In ad hoc networks, each EH-CR node detects the vacant channels with the spectrum sensing. The sensing results are exchanged in the neighborhood to decide on which channel to operate for communication. This may lead to high power consumption for the nodes due to failures in the exchange of control data and false spectrum sensing results. Hence, cooperation between nodes is required to increase the reliability of the

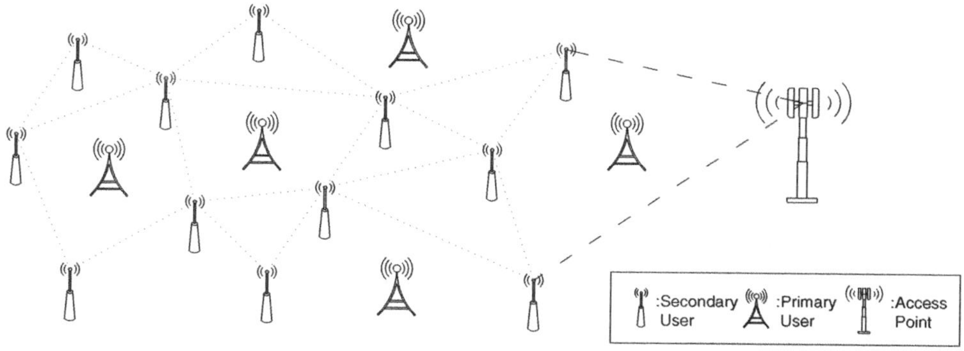

Figure 1.5.3 Ad hoc network architecture of EH-CR in the smart city.

spectrum-aware communications; however, it may increase the amount of the control data. Ad hoc network architecture can be seen in Figure 1.5.3.

The cluster-based architecture supports cooperation in the network by enabling more local coordinators in the network. These local coordinators are called cluster heads, where the spectrum sensing results of the cluster members are gathered for determining the operating frequency among the cluster members. This increases the reliability since the nodes in a neighborhood reach an agreement on the parameters of the spectrum-aware communications.

However, the cluster heads require more power since they are the most active nodes in their corresponding cluster. These issues make the cluster head selection procedure important. The cluster heads should have the highest number of available channels, since they should communicate with as many neighbors as possible to increase their connectivity. Furthermore, the node exchanges information with its members, which increases the power consumption of the cluster heads in comparison to the ordinary nodes. Hence, the selection of the cluster heads in the network should depend on the harvested energy as well as the number of available idle channels. This architecture is illustrated in Figure 1.5.4.

1.5.4 Application Areas

The main application areas of the networks consisting of EH-CRs are the systems of cities. These application areas can be outlined as follows (Hernandez-Munoz et al., 2011).

- *Transport systems*: Provisioning of traffic data by dynamic mapping, monitoring of parking lots, pollution detection in different parts of the city, and monitoring of the transportation network.
- *Hospitals*: Monitoring of hospital conditions such as temperature of vaccines and people with disabilities.
- *Alert Services*: Deployment of EH-CRs in different parts of the city for alerting services such as traffic control and communication services to alert citizen about critical situations.
- *Energy systems*: Decreasing the outages and failures and monitoring of water and gas consumption by smart metering systems.

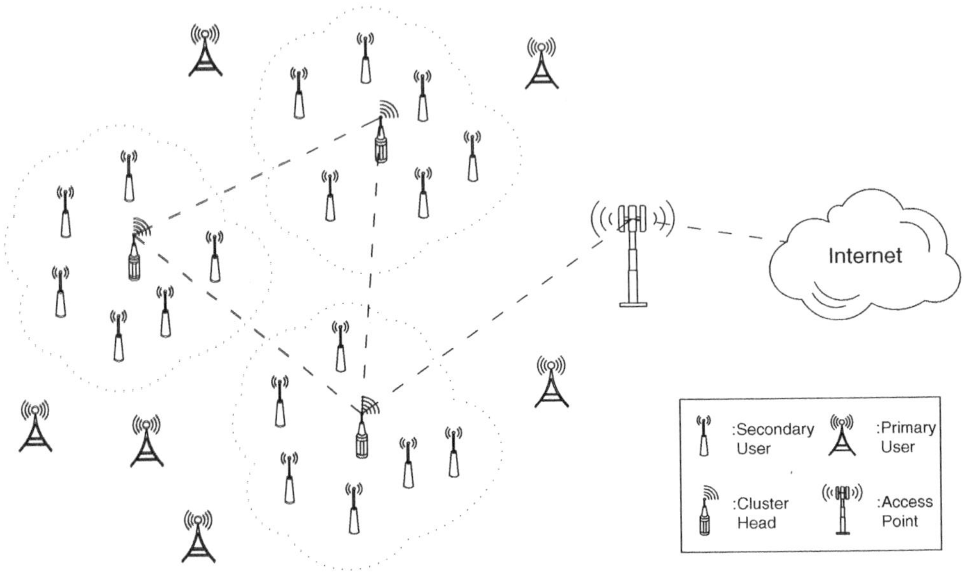

Figure 1.5.4 Clustered network architecture of EH-CR in the smart city.

1.6 Conclusion

In this chapter, we overview the joint use of two important techniques for the realization of the smart city concept, which are CR and EH techniques. We discuss these techniques and their use in the smart city. We explain the motivations for spectrum-aware and self-sustaining communications. The challenges posed by utilization of these techniques in the smart city are studied, and IoT networks with CR and EH capabilities are discussed. According to these discussions, a general framework for EH-CRs is explained for the smart city concept.

References

Akan, O., Karli O., and Ergul, O. 2009, 'Cognitive Radio Sensor Networks', *IEEE Network*, vol. 23, 34–40.

Akan, O.B., Cetinkaya, O., Koca, C., and Ozger, M. 2017, 'Internet of Hybrid Energy Harvesting Things', *IEEE IoT Journal*.

Alcaraz, C., Najera, P., Lopez, J., and Roman, R., 2010, 'Wireless sensor networks and the internet of things: Do we need a complete integration?' *Proceedings 1st International Workshop on the Security of the Internet of Things (SecIoT'10)*.

Atzori, L., Iera, A., and Morabito, G., 2010, 'The Internet of Things: A Survey', *Computer Networks*, vol. 54, no. 15, 2787–2805.

Castellani, A.P., Bui, N., Casari, P., Rossi, M., Shelby, Z., and Zorzi, M., 2010, 'Architecture and Protocols for the Internet of Things: A Case Study', *Proceedings of IEEE PerCom*, 678–683.

Cetinkaya, O., Akan, O.B., 2017a, 'Electric-field Energy Harvesting Wireless Networks', *IEEE Wireless Communications,* vol. 24, no. 2, 34–41.

Cetinkaya, O. and Akan, O.B., 2017b, 'Electric-field Energy Harvesting from Lighting Elements for Battery-less Internet of Things', *IEEE Access,* vol. 5, 7423–7434.

Chang, K.-S., Kang, S.-M., Park, K.-J., Shin, S.-H., Kim, H.-S., and Kim, H.-S., 2012, 'Electric field energy harvesting powered wireless sensors for smart grid', *Journal of Electrical Engineering and Technology,* vol. 7, no. 1, 75–80.

Choi, D., Gong S., and Park, K., 2014, 'Stray electric field energy harvesting technology using MEMS switch from insulated AC power line', *Electronics Letters,* vol. 50, no. 17, 1236–1238.

Christin, D., Reinhardt, A., Mogre, P.S., and Steinmetz, R., 2009, 'Wireless Sensor Networks and the Internet of Things: Selected Challenges', *Proceedings 8th GI/ITG KuVS Fachgesprch, "Drahtlose Sensornetze".*

European Union, 2010, *Vision and Challenges for Realising the Internet of Things,* EU.

Federal Communications Commission, 2002, 'Spectrum Policy Task Force', *Rep. ET Docket* no. 02-135, FCC.

Gubbi, J., Buyya, R., Marusic, S., and Palaniswami, M., 2013, 'Internet of Things (IoT): A vision, architectural elements, and future directions', *Future Generation Computer Systems (Elsevier),* vol. 9, 1645–1660.

Haykin, S., 2005, 'Cognitive radio: brain-empowered wireless communications', *IEEE Journal on Selected Areas,* vol. 23, no. 2, 201–220.

Hernandez-Munoz, J.M., Vercher, J.B., Munoz, L., and Galache, J. A., 2011, 'Smart cities at the Forefront of the Future Internet', *Proceedings of The Future Internet,* 447–462.

Honda, M., Sakurai, T., and Takamiya, M., 2015, 'Wireless temperature and illuminance sensor nodes with energy harvesting from insulating cover of power cords for building energy management system', *Proceedings. IEEE PES Asia-Pacific Power and Energy Engineering Conf.,* 1–5.

Hui, J.W. and Culler, D.E., 2008, 'IP is dead, long live IP for wireless sensor networks', in *Proc. of the SenSys Conf.,* 15–28.

Jalali, R., El-Khatib, K., and McGregor, C., 2015, 'Smart City Architecture for Community Level Services through the Internet of Things', *Proceedings of 18th International Conference on Intelligence in Next Generation Networks,* 108–113.

Matiko, J.W., et al., 2013, 'Review of the application of energy harvesting in buildings', *Measurement Science and Technology,* vol. 25, no. 1.

Mitola, J. and Maguire, G.Q., 1999, 'Cognitive radio: making software radios more personal', *IEEE Personal Communications,* vol. 6, no. 4, 13–18.

Moghe, R., Yang, Y., Lambert, F., and Divan, D., 2009,' A scoping study of electric and magnetic field energy harvesting for wireless sensor networks in power system applications', *in Proceedings IEEE Energy Convers. Congr. and Expo.,* 3550–3557.

Moghe, R., Iyer, A., Lambert, F.C., and Divan, D., 2015, 'A Low-Cost Electric Field Energy Harvester for an MV/HV Asset-Monitoring Smart Sensor', *IEEE Trans. on Ind. App.,* vol. 51, no. 2, 1828–1836.

Natural Resources Defense Council (NRDC) n. d., *What are smarter cities?,* NRDC. Available from http://smartercities.nrdc.org/about.

Schaffers, H., Komninos, N., Pallot, M., Trousse, B., Nilsson, M., and Oliveira, A., 2011, 'Smart Cities and the future Internet: Towards cooperation frameworks for open innovation', *The Future Internet, Lect. Notes Comput. Sci.,*vol. 6656, 431–446.

Su, K., Li, J., and Fu, H., 2011, 'Smart city and the applications', *Proceedings of International Conference Electronics Communications Control (ICECC)*, 1028–1031.

Sudevalayam, S. and Kulkarni, P., 2011, 'Energy harvesting sensor nodes: Survey and implications', *IEEE Comm. Surveys & Tutorials*, vol. 13, no. 3, 443–461.

Walravens, N. and Ballon, P., 2013, 'Platform Business Models for Smart Cities: From Control and Value to Governance and Public Value', *IEEE Communications Magazine*, vol. 51, no. 6, 72–79.

Weddell, A.S., Magno, M., Merrett, G.V., Brunelli, D., Al-Hashimi, B.M., and Benini, L., 2013, 'A survey of Multi-Source Energy Harvesting Systems', in *Proceedings of the Conference on Design, Automation and Test in Europe*, 905–908.

Wu, Q., et al., 2014, 'Cognitive Internet of things: A new network paradigm beyond connection', *IEEE Internet Things Journal*, vol. 1, no. 2, 129–143.

Zanella, A., Bui, N., Castellani, A., Vangelista, L., and Zorzi, M., 2014, 'Internet of Things for Smart Cities', *IEEE Internet of Things Journal*, vol. 1, no. 1.

Zangl, H., Bretterklieber, T., and Brasseur, G., 2009, 'A feasibility study on autonomous online condition monitoring of high-voltage overhead power lines', *IEEE Trans. on Instrum. and Meas.*,vol. 58, no. 5, 1789–1796.

Zhang, Y., 2010, 'Interpretation of Smart Planet and Smart City', *China Information Times*, vol. 10, 38–41.

Zhao, X., Keutel, T., Baldauf, M., and Kanoun, O., 2013, 'Energy harvesting for a wireless-monitoring system of overhead high-voltage power lines,' *IET Gen., Trans. and Distr.*, vol. 7, no. 2, pp. 101–107.

Zhu, M., Baker, P.C., Roscoe, N.M., Judd, M.D., and Fitch, J., 2009, 'Alternative power sources for autonomous sensors in high voltage plant', *Proceedings IEEE Electrical Insulation Conf.*, 36–40.

2

LTE-D2D Communication for Power Distribution Grid: Resource Allocation for Time-Critical Applications

Leonardo D. Oliveira[1], *Taufik Abrao*[2] *and Ekram Hossain*[3]

[1] *Department of Telecommunications and Control Engineering, University of Sao Paulo, Sao Paulo, Brazil*
[2] *Department of Electrical Engineering, Londrina State University (DEEL-UEL), Parana state, Brazil*
[3] *Professor, Department of Electrical and Computer Engineering, University of Manitoba, Winnipeg, Manitoba, Canada*

2.1 Introduction

A brief concept of the smart city consists of using the technology to improve the urban infrastructure, to optimize the use of resources and services to turn the urban center a better place to live. Part of this infrastructure, the smart grid (SG) is the evolution of the power distribution grid to the smart city. There exist bidirectional flows of information and energy, and improving the communication in the distribution grid can directly improve the "intelligence" of the city, saving energy and turning it into a low-carbon zone (Deakin and Reid, 2016). In the power system, there are several communication requirements for different services, such as measurement and automation. The last one demands an independent network for security purposes, and it is the focus of this chapter.

When dealing with automation, the power system communications traditionally used protocols not compatible with rigorous latency and reliability requirements, such as Modbus, DNP3 (Distributed Network Protocol), IEC 60870-104, ICCP (Inter-control center communication protocol), and other proprietary ones, and the time-critical applications such as control and protection were hardwired. However, since the publication of the global standard "IEC 61850 – Communication Networks and Systems in Substations" in 2004, communications for time-critical applications have been feasible. There is a gradual migration of world power grid systems to this standard, and since it has been successfully applied inside, the future trend is to extend it to outside substations (Mohagheghi et al., 2009).

When dealing with inter-substation communication, the demanded services consist of wide-area monitoring systems and distributed control/protection. These involve the collection of real-time information over many substations and types of equipment to provide the required information to the control center to perform the main system control and to keep the system available and stable. The requirements for such applications are strict, e.g., reliability between $99.00 - 99.99\%$ and a maximum latency between $15 - 200$ milliseconds (Gungor et al., 2011; Lu et al., 2013).

The implementation of such standard inside a substation is mostly done over optic fiber and Ethernet cables. Nevertheless, for outside substations, wireless

Transportation and Power Grid in Smart Cities: Communication Networks and Services, First Edition.
Edited by Hussein T. Mouftah, Melike Erol-Kantarci and Mubashir Husain Rehmani.

communication technologies offer potential alternatives due to lower costs and higher flexibility. These options are based on the IEEE (Institute of Electrical and Electronics Engineers) standards such as the 802.15.4, 802.11, and 802.16, 802.22, as well as satellite, 3GPP LTE (Long-Term Evolution), and IoT (Internet of Things). It is particularly attractive to exploit the emerging technologies such as the multi-tier heterogeneous networks, device-to-device (D2D) communications, full-duplex (FD) communications, massive multiple-input multiple-output (massive-MIMO) communications, energy-aware communication, and energy harvesting, among others (Hossain and Hasan, 2015).

The D2D communication underlaying/overlaying LTE cellular networks, henceforth entitled LTE-D2D, has been intensively discussed in literature and standardization committees in recent years, and it is of particular interest due to its potential for reducing delay and boost reliability, while minimizing power consumption (Fodor et al., 2012; Jänis et al., 2009). Various aspects of D2D communication underlaying/overlaying LTE networks as well as the use of LTE for SG have been studied in the literature. However, the investigations of LTE-D2D technologies for SG communications are very limited (Kalalas et al., 2016b).

Starting with a brief introduction to the SG architecture, this chapter first provides a review of the communications technologies and network protocols used in the power distribution grid, e.g., Modbus, DNP 3.0, and IEC 61850. Then the communications requirements of the different applications for distribution automation (DA), i.e., distribution system monitoring, control, and protection outside the substations are reviewed. To this end, a resource allocation strategy for an LTE-D2D system model is presented for a power distribution grid based on an optimization formulation. Moreover, a D2D half/full-duplex approach is suggested to help to avoid overload in the communication system. Preliminary results are presented and discussed aiming to bring into discussion the advantages and disadvantages of half/full-duplex LTE-D2D technology for power distribution grids.

2.2 Communication Technologies for Power Distribution Grid

An SG consists of a huge number of devices distributed over a large geographical region, with different demands and connected through a communication network. The efficiency of SG relies on the reliability of such network, which must also be cost-effective. In the present section, an overview of the SG, potential communication technology solutions, a survey on power system protocols, as well as an alternative for modeling the data traffic for DA purpose are presented.

2.2.1 An Overview of Smart Grid Architecture

A conventional power system consists fundamentally of four agents: generation, transmission, distribution, and consumer. The energy production occurs in power plants (hydro, thermal, nuclear, wind, or solar, among others), while the transmission system, which consists of high-voltage substations and transmission lines, transports the energy to the distribution substations. The last ones deliver the electricity to customers through feeders located near them. In such a power system, the communication is usually unidirectional and located only in the generation/transmission infrastructures,

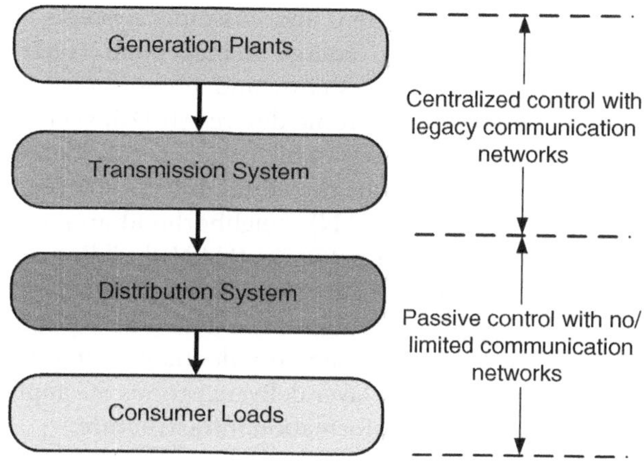

Figure 2.1 Conventional power system with unidirectional communication (Khan and Khan, 2013).

being passive and limited in the distribution and consumers structures. Figure 2.1 depicts the power and communication blocks of a conventional power system.

The evolution of the conventional power grid to the SG arises mainly due to the use of a communication network. The SG concept consists of two-way flows of electricity and information, and with some intelligence in the management system, it is possible to achieve self-monitoring, remote check/test, and self-healing, among others (Fang et al., 2012). Dozens of survey papers have been published in recent years related to the communication architecture for smart grid (of Energy, 2010; Farhangi, 2010; Lo and Ansari, 2012; Khan and Khan, 2013; Gungor et al., 2013; Meng et al., 2014; Kuzlu et al., 2014; Erol-Kantarci and Mouftah, 2015).

The SG comprises the following layers: power system, control, communication, security, and application (Lo and Ansari, 2012; Kuzlu et al., 2014), as can be seen in Figure 2.2. As previously mentioned, the power system involves generation,

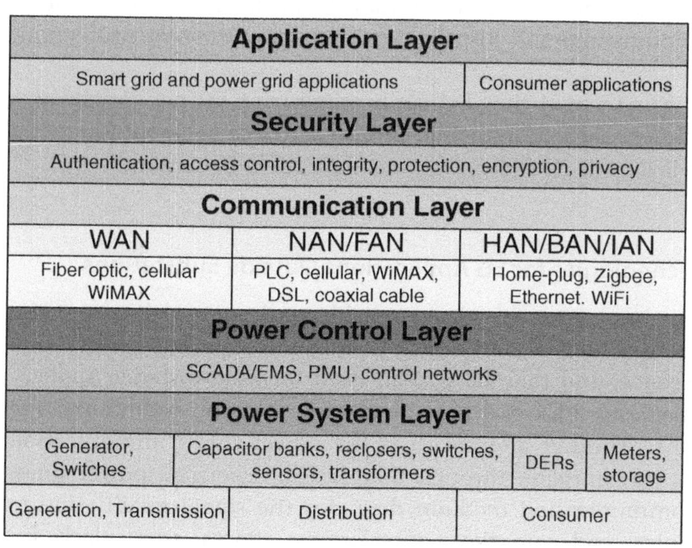

Figure 2.2 Layers of an SG system.

transmission, distribution, and customer systems. The power control layer contains the control networks, energy management (EMS), supervisory control and data acquisition (SCADA) systems, and phasor measurement unit (PMU). The various types of intelligent electronic devices (IEDs), such as breaker IED, measurement unit (MU) IED, and PC IEDs, which can be considered specialized controllers, are also included in this layer. In the communication layer, the networks can be classified as wide area networks (WAN), neighborhood area networks (NAN)/field area network (FAN), and home area networks (HAN)/building area network (BAN)/industrial area network (IAN) (Kuzlu et al., 2014; Ho et al., 2013; Farhangi, 2010; Lo and Ansari, 2012). The technologies adopted in these networks are summarized in Subsection 2.2.2. The security layer provides data confidentiality, integrity, authentication, and availability. Finally, the application layer delivers various SG applications to consumers and utilities based on an existing information infrastructure.

Two main features of communications in SG are the advanced metering infrastructure (AMI) and the DA applications(Zhabelova and Vyatkin, 2012). A typical AMI system uses smart meters to exchange information between consumers and power utilities for monitoring, operating, and billing purposes (Gungor et al., 2011). The AMI systems require uplink transmission of small and intermittent packets, with flexible latency and reliability requirements. Although each user demands a relatively low bandwidth, the expected huge number of devices implies a high total throughput for the whole system (METIS, 2013).

In the SG context, DA deals mainly with time-critical communication (Vyatkin et al., 2010; Zhabelova and Vyatkin, 2012). Its demanded services consist of wide-area monitoring and distributed control/protection, which involve the collection of real-time information over many substations and equipment. The control center requires these data for agile actuation and to keep the system available (Gungor et al., 2011). Such applications are associated with stringent communication requirements in terms of network latency and reliability (IEC, 2015). Examples of DA applications are fault detection, automatic capacitor bank control, fast distribution bus protection scheme, load tap changer control and monitoring, and distribution feeder overcurrent protection, among others.

The focus of this chapter is on the real-time DA applications in an SG network involving: a) the control layer, which enables monitoring, control, and management functions and b) the communication layer, which allows two-way flows of information.

2.2.2 Communication Technologies for SG Applications Outside Substations

Copper and fiber-wired technologies constitute reliable and secure data transfer options, offering increased capacity and low latency. Optical fiber systems are immune to electromagnetic interference, and they have been used for long-distance applications, while Ethernet has been used for real-time protection, control, and monitoring functions inside substations. Another alternative is the power line communication (PLC), which enables data transmission through the existing electrical power lines. However, the nature of communication medium degrades the signal quality due to attenuation, dispersion, noise, and sometimes interference effects, particularly for high frequencies (Galli et al., 2011). Also, PLC/wireless hybrid network solutions have been proposed, aiming to achieve redundancy (Lai and Messier, 2012). Other wired alternatives are digital subscriber line (DSL) and coaxial cable (Kuzlu et al., 2014).

The use of wired communication solutions can be economically and physically prohibitive for most SG applications. Therefore, wireless technologies arise as an alternative for enabling communication almost anywhere with relatively low cost and at the same time being scalable and dynamic. In what follows, the wireless communication standards with potential application in a SG context are briefly discussed.

The IEEE 802.15.4 standard is the foundation of the ZigBee Alliance. In particular, the amendment "g" corresponds to medium access control modifications, so as to cover outdoor low data rate and wireless smart metering utility network, with a data gateway and multi-hop techniques, where devices operate on a vast scale and low-power applications (Meng et al., 2014). Despite the low-cost and low-energy advantages, the low data rate (up to 250 Kbps), short range, security vulnerability, and susceptibility to interference are convincing drawbacks when applying this standard for SG.

The IEEE 802.11 standard, used by the WiFi Alliance, constitutes a mature and widely adopted wireless technology that is suitable for home applications in the SG, with data rates reaching the Gbps (Cena et al., 2010; Parikh et al., 2013). Some advances have been made in amendments: the "s" version allows multi-hop networks, while the "ah" supports large-scale wireless networks, and the "e" provides the quality of service (QoS) support for WiFi networks, with layer 2 traffic class differentiations and priorities. However, limiting factors are the relatively short coverage range (up to 250 meters outdoors) and the use of unlicensed frequency bands, which may become susceptible to interference and vulnerabilities (interference and security issues) in the SG context.

The IEEE 802.16 standard, used for WiMAX technology, has also been proposed to support the bidirectional SG communication. Its design can provide larger and sufficient coverage for the SG system, with low latency (less than 100 ms for round-trip time), high data rates (up to tens of Mbps), coverage range about dozens of kilometers, advanced security protocols, and QoS support (Aalamifar and Lampe, 2016). The main drawback of using WiMAX in such application is the required investments for infrastructure installation and maintenance.

Alternatively, satellite communication has advantages such as the significant broadband and reliability capabilities. Some disadvantages include the small number of public satellites, the increased energy consumption associated with the considerable communication distance, which could be a limitation for autonomously operating smart devices, and communication latency and the installation and maintenance costs, which are relatively larger compared with other wireless technologies (Chang et al., 2014).

In the arena of cellular technology, the preferred one is the LTE because it is more recent than other ones and because of its ubiquity and also its potential/flexibility. It has the following advantages: operation in licensed bands, mature and wide coverage, high data rate, low latency, and high system reliability and availability. As deficiency, it is oriented for human broadband applications (voice and data), and it does not support time-critical applications (Cheng et al., 2011). Also, the total cost becomes comparable to WiMAX technology if the electric power operators prefer deploying their own backhaul due to full control and security perspectives in data traffic.

The modern RF technologies designed for IoT, such as LoRa and SigFox, usually have low data rate (bandwidth < 50 kHz) and also high latency, which make them suitable for metering and non-critical control applications. LTE-based systems, particularly

LTE machine-type communication (LTE-M), are more expensive than other IoT technologies, but their higher data rate and lower latency suit for applications with high-reliability requirements.

2.2.3 Communication Networks for SG

A possible characterization of the power grid distribution system as a function of the communication networks is: legacy wired backhaul network, WAN, NAN/FAN, and HAN/BAN/IAN (Kuzlu et al., 2014; Ho et al., 2013; Farhangi, 2010; Lo and Ansari, 2012).

The HAN/BAN/IAN networks consist of smart devices sending/receiving electrical measurement and control data. These devices are located in a home, building, or industry, and they exchange data to achieve a better consumption management and support home/building/industry automation. The requirements for such networks are low power consumption, low cost, simplicity, secure communication, low data rate, and area coverage up to 200 m^2. The current technologies adopted to support such functionality and requirements are ZigBee, WiFi, ZWave, PLC (HomePlug), Bluetooth, and Ethernet.

The NAN/FAN networks have various applications with distinct requirements, such as smart metering, demand response (DR), distribution automation, electric transportation, firmware updates and program configuration, outage and restoration management, service switch operation, customer information and messaging, and premises network administration (Kuzlu et al., 2014). They also gather information from HAN/BAN/IAN and enable communication to the WANs. The NAN/FAN network is a critical segment of the SG since it is responsible for transporting a huge volume of different types of data and distribution signals between utility companies/control centers and smart field devices. These applications require communication technologies that support higher data rate (up to 100 Mbps) and larger coverage distance (up to 10 km). NAN/FAN applications can be implemented over ZigBee, WiFi, and PLC, as well as over long distance wired and wireless technologies, such as WiMAX, cellular, and DSL.

A WAN aggregates information of widely dispersed structures, and in particular, it uses data aggregation points (DAP). Its primary applications are wide-area protection, control, and monitoring, which require information on the entire power system: generation (power plants and the distributed energy resources, or DER), transmission, distribution, and customers. These wide-area applications require much higher data rate (up to 1 Gbps) combined with a long distance coverage (up to 100 km), provided by communication technologies like cellular and WiMAX. Especially for remotely located power plants and substations, satellite communication is attractive to provide redundancy when other communication technologies become physically or cost prohibitive.

The legacy wired backhaul network involves critical generation and high voltage transmission systems. Power plants and substations are usually interconnected, and a connection with control centers allows essential control functions of the power system, such as power/voltage/frequency control, scheduling, and load dispatch. The communication links are preferably built using optical fiber within the transmission line cables (OPGW, optical ground wire systems), which provides high capacity and reliability. Other alternatives, used mainly for redundancy, are PLC, all-dielectric self-supporting cable, digital radio, and satellite. With the inclusion of DER, it is desirable to implement such communication link with wireless technologies, because of their higher flexibility and lower costs compared to the wired ones. Cellular and

Table 2.1 Comparison among wired and wireless candidate technologies for SG applications (Adapted from (Kuzlu et al., 2014)).

Technology	Max. data rate	Coverage range	Networks
Wired communication			
Fiber optic	40 Gbps	Up to 100 km	FAN/NAN, WAN
DSL	1-100 Mbps	1.5-5.0 km	FAN/NAN
Coaxial cable	172 Mbps	Up to 28 km	FAN/NAN
PLC (Narrowband)	10-500 Kbps	Up to 100 km	FAN/NAN
PLC (HomePlug)	14-200 Mbps	Up to 200 m	HAN
Ethernet	10 Mbps - 10 Gbps	Up to 120 m	HAN
Wireless communication			
Zigbee	250 Kbps	Up to 1600 m	HAN, FAN/NAN
WiFi	2-600 Mbps	Up to 100 m	HAN, FAN/NAN
WiMAX	75 Mbps	Up to 50 km	FAN/NAN, WAN
Cellular	100 Mbps	Up to 50 km	FAN/NAN, WAN
Satellite	1 Mbps	100-6000 km	WAN

WiMAX are among the wireless technology options for DER integration to WAN. Nevertheless, the legacy wired backhaul will remain, because of its importance to the power system stability and its negligible costs for the operational expenditure (OPEX) and capital expenditure (CAPEX) when compared to the total costs of these critical structures. Table 2.1 summarizes a comparison of wired and wireless communication technologies for SG applications (Kuzlu et al., 2014).

In an SG, the DA is carried out in the WAN and NAN/FAN, and the LTE-D2D technology may be viable in such networks. In the next section, there is a review of the main SCADA protocols employed in the control layer of an SG, with a focus on the promising IEC 61850 standard.

2.3 Overview of Communication Protocols Used in Power Distribution Networks

The commonly used protocols in the automation of power systems are Modbus, DNP3, IEC 60870, ICCP, and IEC 61850. Despite being widely used inside substation and generation plants, there is still need to find optimal combinations of communication protocols and network topology in decentralized or agent-based microgrid control systems (Bani-Ahmed et al., 2014). In what follows, these protocols are introduced.

2.3.1 Modbus

Modbus transmission protocol was introduced in 1979 by the company Modicon (now Schneider) for process control systems. It is an application layer messaging protocol used for client/server communication between devices connected to the same network. There are two ways to implement this protocol: serial or via Transmission Control Protocol/Internet Protocol (TCP/IP). The serial implementation is built over wire

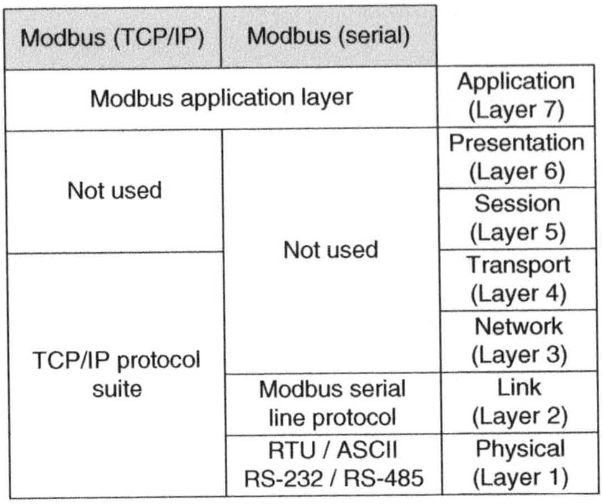

Figure 2.3 Modbus communication stack.

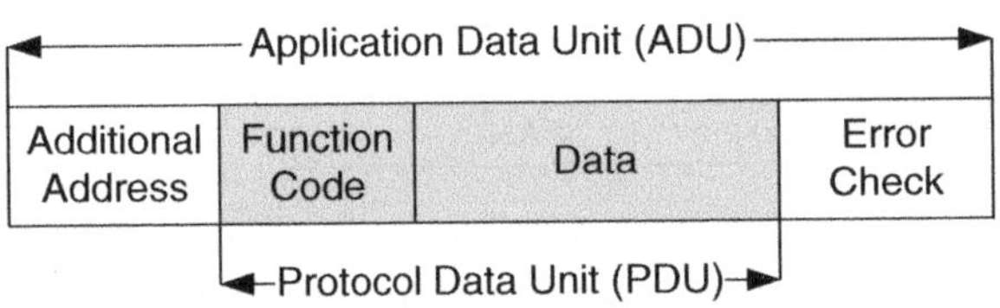

Figure 2.4 General Modbus frame.

RS-232, 422, or 485, fiber, or wireless, while the TCP/IP option is built over Ethernet cable, fiber, or wireless (Mohagheghi et al., 2009; Modbus-ID, 2012). The Modbus communication stack is shown in Figure 2.3.

Usually, the Modbus messages are query/response type, and the communication is initiated by the client, with a single request to a particular device, which answers back. Broadcast/no-response-type messages are also available, and in the TCP/IP version, the server can initiate a transmission via sending an exception message. The frame format is fixed and the same for client and server, with construction as described in what follows (Golshani et al., 2012). Figure 2.4 depicts a general Modbus frame, with its application data unit (ADU) and protocol data unit (PDU). Modbus definitions are itemized in the following sequence.

- Address: mapped over 1 byte, it identifies the required device.
- Function code: carried out over 1 byte field, it informs to the requested server what type of action to perform.
- Error check: expressed in 2 bytes, it adopts a cyclic redundancy check (CRC-16) on the message frame.
- Data: size varies according to the function code of the current message, with the maximum size of ADU being 256 bytes (from the first Modbus implementation on serial line network).

In order to provide reliable communication, the reception of a message must be synchronized with its transmission. Under the Modbus serial, frame synchronization is established by limiting the elapsed time between receipts of characters. If this period expires, then the pending message will be discarded, and the next byte received will be treated as the address field of a new message frame (Golshani et al., 2012).

Usually, communication between high-level equipment (controllers and human-machine interfaces) is established via TCP/IP, while exchange information between high and low level, such as controllers with drivers, is carried out via the serial option. By using gateways, these implementation types can exist in a communication network at the same time (Golshani et al., 2012).

2.3.2 IEC 60870

The "IEC 60870 - Telecontrol Equipment and System" is a standard published initially in 1988 by the International Electrical Commission (IEC) and is mainly used in European countries. Its primary application is for supervisory control and data acquisition in electrical plants and transmission systems. It is an open standard composed of six main parts, where part five, IEC 60870-5, is for communication protocols, and it consists of the following sections:

- IEC 60870-5-1: transmission frame formats;
- IEC 60870-5-2: data link transmission services;
- IEC 60870-5-3: general structure of application data;
- IEC 60870-5-4: definition and coding of information elements;and
- IEC 60870-5-5: basic application functions.

The IEC 60870-5 protocols follow the Enhanced Protocol Architecture (EPA) model, which simplifies the ISO standard (OSI model) in three layers: application (and user process) layer, data-link layer, and physical layer (Medina et al., 2009). The companions IEC 60870-5-101/104 provide functions and objects to SCADA systems, while the IEC 60870-5-102/103 provide support to electrical protection systems.

The "IEC 60870-5-101 - Companion Standard for Basic Telecontrol Tasks" is based on serial communication. It supports V.24/V.28 (RS-232) interface with data rates from 9600 to 115200 bit/s. This protocol has some compelling features, such as priority classes, data classification in 16 groups for interrogation command, on-demand and automatic data updating, and time synchronization. The communication can be balanced or unbalanced, and there are two types of frames: fixed length, restricted to use for frames carrying no user data, and variable length. Each frame can carry a maximum of one application service data unit (ASDU), and the message structure is formed by the data link layer, as described in the following.

- Link address: 1 or 2 octets for unbalanced and 0, 1, or 2 octets for balanced communication.
- Length: 2 octets, repeated twice, and the two values must be equal for the frame to be accepted by the receiver.
- Control field: 1 byte; it is central to the operation of the transmission procedures, and it depends on the transmission mode (balanced or unbalanced) and whether the communication is a primary or secondary message.
- Checksum: 1 byte; it is the modulo 256 sum of the frame user data (link user data + address + control field).
- User data: variable, up to 253 octets of link user data.

Figure 2.5 depicts the data link frame and the application layer ASDU structure carried by it.

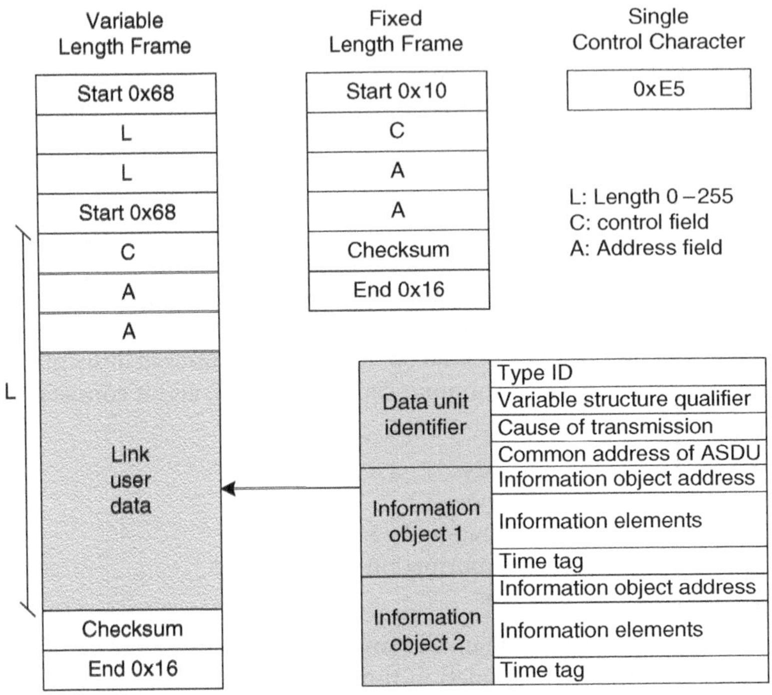

Figure 2.5 Message structure under IEC 60870-5-101 communication.

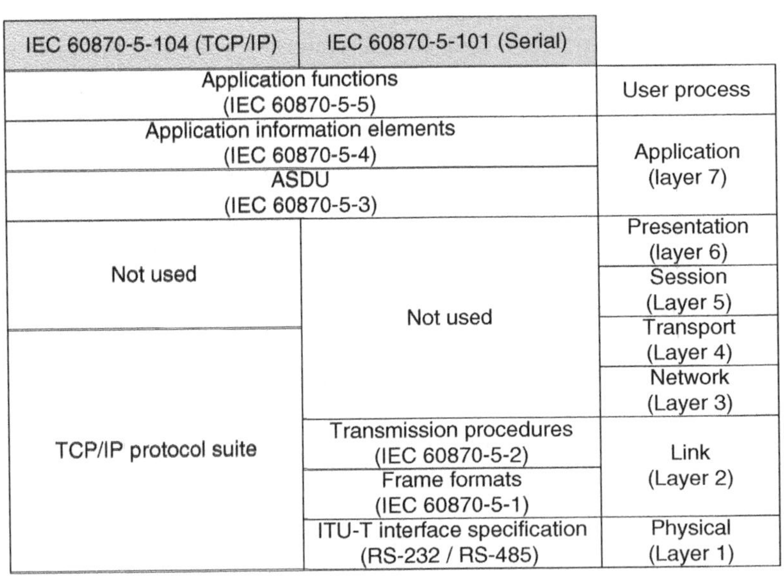

Figure 2.6 IEC 60870 communication stack.

The "IEC 60870-5-104 – Network Access using Standard Transport Profiles" is a combination of the application layer of IEC 60870-5-101 and the transport functions of TCP/IP. It is an evolution in the physical and data transport procedure, with the lower levels of the protocol being completely replaced by the TCP and IP transport and network protocols, respectively. Figure 2.6 illustrates the communication stacks of both companions. More details about this standard can be found in (Clarke et al., 2004).

2.3.3 DNP3

The DNP3 is an open domain IEEE Standard for Electric Power Systems Communications (DNP3, 2012) and it was developed by General Electric Company (GE) based on the early parts of IEC 60870-5. They made it public in 1993, and the ownership was given to the newly formed DNP Users Group (Mohagheghi et al., 2009). It is widely used in power systems around the world and especially in North America, South America, South Africa, Asia, and Australia. Initially designed to operate in the conventional power grid, its application over TCP/IP is a cost-efficient and backward-compatible solution (Wei and Wang, 2013).

Different from Modbus, DNP3 messages can provide time stamp and data quality, and they are not restricted to transferring one data point since multiple data types (e.g., Boolean, floating point) can be encapsulated in a single message to reduce data traffic. Furthermore, DNP3 offers robust features for enabling flexibility and security (Bani-Ahmed et al., 2014; Clarke et al., 2004) including:

- reporting of an exception/event without being polled by the master;
- fragmentation of messages into multiple frames to provide optimum error control and rapid communication flows;
- peer-peer and master-slave topology, also with multiple masters;
- user-definable objects;
- "changed data" only responses;
- broadcast messages;
- secure configuration/file transfers;
- addressing for over 65,000 devices on a single link; and
- data link and application layer confirmation.

Regarding the OSI model, the DNP3 protocol follows the EPA with an additional pseudo-transport layer. Beyond the serial option, this protocol can also be built over the TCP and IP protocols, as shown in Figure 2.7.

The frame structure is divided into header and data, which extends through the layers. At the highest layer of the stack, the application layer breaks down the data into smaller-sized blocks, the ASDUs, and each one has the application header, application protocol control information (APCI), to each chunk and builds the application protocol data unit (APDU). The application layer header is divided into application control code and application function code, where application control code contains information regarding whether the fragment is single-fragment or multi-fragment, whether the application fragment confirmation is requested, and whether fragment was unsolicited. Application layer function code indicates the actual purpose of the message, i.e., what the slave should do. An APDU passed to the pseudo-transport layer is called a transport service data unit (TSDU). TSDUs are broken down into smaller blocks, which are called

<table>
<tr><th>DNP3 (TCP/IP)</th><th>DNP3 (Serial)</th><th></th></tr>
<tr><td colspan="2">DNP3 data object library</td><td rowspan="2">Application
(Layer 7)</td></tr>
<tr><td colspan="2">DNP3 application layer</td></tr>
<tr><td colspan="2" rowspan="2">Not used</td><td>Presentation
(Layer 6)</td></tr>
<tr><td>Session
(Layer 5)</td></tr>
<tr><td>DNP3 transport functions</td><td rowspan="3">DNP3 transport functions</td><td rowspan="3">Transport
(Layer 4)</td></tr>
<tr><td>DNP3 data link functions</td></tr>
<tr><td>TCP / UDP</td></tr>
<tr><td>IP</td><td>Not used</td><td>Network
(Layer 3)</td></tr>
<tr><td rowspan="3">IP</td><td>DNP3 data link layer</td><td rowspan="3">Link
(Layer 2)</td></tr>
<tr><td>Transmission procedures
(IEC 60870-5-2 / balanced)</td></tr>
<tr><td>Frame formats
(IEC 60870-5-1 / FT3)</td></tr>
<tr><td>Ethernet (IEEE 802.3)</td><td>ITU-T interface
specification (RS-232)</td><td>Physical
(Layer 1)</td></tr>
</table>

Figure 2.7 DNP3 communication stack.

transport protocol data units (TPDUs). Each TPDU is made up of one byte of the header and a maximum of 249 bytes of data. The data link layer receives the overall 250 bytes of TPDUs from the pseudo-transport layer (Golshani et al., 2012; Clarke et al., 2004). DNP3 employs an object to let the server device understand what kind of information is required, and all valid data formats are identified by the object number and the variation number. Figure 2.8 depicts the frame structure of the DNP3 message, as presented here.

According to the FT3 frame format (taken from IEC 60870), a 10-byte header is added to each TPDU, including 2 bytes CRC for providing error detection. These 2 bytes CRC will be repeated for each block of data in LPDUs. The maximum number of data blocks in LPDU is 16. Each block consists of 16 bytes of data (except the last block, which may have less than 16 bytes according to the data size) and 2 bytes of CRC. The LPDU maximum size is 292 bytes, of which 250 bytes are data. The physical layer sends data as bitstream over the determined physical media. When DNP3 is transmitted over TCP/IP, the transport, network, and data link layers related to the TCP/UDP (User Datagram Protocol), IP, and Ethernet Local Area Network (LAN) are added at the bottom of the pseudo-transport and data link layers of DNP3 (Golshani et al., 2012; Mohagheghi et al., 2009).

2.3.4 IEC 61850

The "IEC 61850 – Communication Networks and Systems in Substations" is a global standard for substation protection, communication, and control (IEC, 2015). It is divided into ten sections, as presented in Figure 2.9.

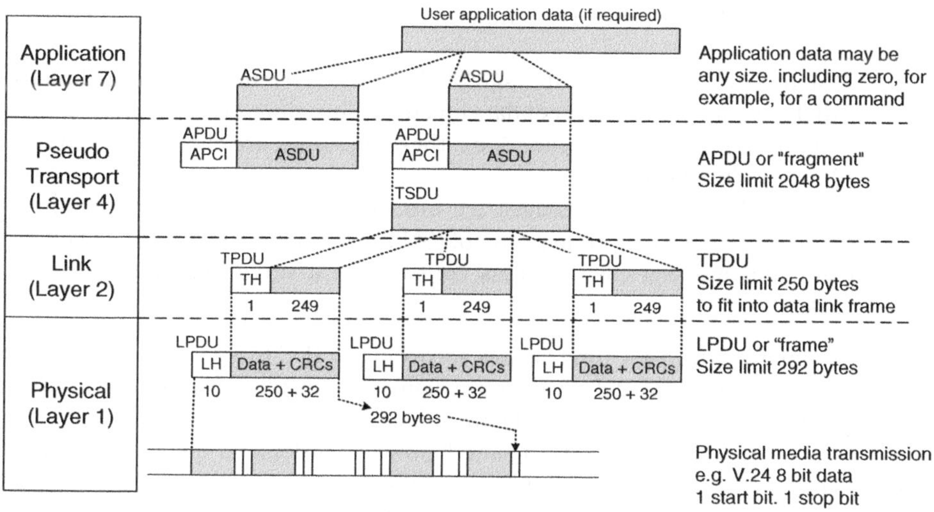

Figure 2.8 Message structure for DNP3 (Clarke et al., 2004).

<table>
<tr><td>Part 1</td><td colspan="3">Introduction and Overview</td></tr>
<tr><td>Part 2</td><td colspan="3">Glossary of Terms</td></tr>
<tr><td>Part 3</td><td colspan="3">General Requirements</td></tr>
<tr><td>Part 4</td><td colspan="3">System and Project Management</td></tr>
<tr><td>Part 5</td><td colspan="3">Communication Requirements</td></tr>
<tr><td>Part 6</td><td colspan="3">Substation Automation System Configuration</td></tr>
<tr><td>Part 7</td><td colspan="3">Basic Communication Architecture</td></tr>
<tr><td rowspan="2">Part 8</td><td rowspan="2">Mapping to Ethernet and MMS</td><td rowspan="2">Part 9</td><td>Sampled Measured Values</td></tr>
<tr><td>Mapping to Ethernet</td></tr>
<tr><td>Part 10</td><td colspan="3">Conformance Testing</td></tr>
</table>

Figure 2.9 Structure of the IEC 61850 standard (IEC, 2015; Mackiewicz, 2006).

The standard works with the concept of a virtual model comprising both physical and logical device models. A physical device is a device that connects to the network, defined by its network address. Within each physical device, there may be one or more logical devices. Figure 2.10 depicts the devices complying with the communication stack. In order to support different automation applications, the standard specifies seven types of messages mapped over the OSI-7 layer stack (IEC, 2015; Mackiewicz, 2006; Kalalas, 2014).

- Type 1: Generic Object Oriented Substation Event (GOOSE) and Generic Substation Event (GSE) are time-critical messages, which require immediate action at receiving IEDs, and they are directly addressed to second-layer Ethernet data link;
- Type 2: with medium transmission speed, they are referred to client/server-based Internet protocol, and they are mapped over the TCP/IP stack;

- Type 3: for low-speed auto-control functions, it is client/server type and mapped over TCP/IP stack;
- Type 4: designed to carry raw data, sampled values (SV) messages are directly addressed to the second-layer Ethernet data link to achieve time-critical performance;
- Type 5: used for large file transfer functions, these messages are client/server-based IP, mapped over TCP/IP stack;
- Type 6: time synchronization messages are broadcast using the UDP/IP, mapped directly on the simple network time protocol (SNTP) (Dehalwar et al., 2015); and
- Type 7: command messages with access control, client/server and mapped over TCP/IP stack.

Table 2.2 summarizes transfer time classes according to IEC 61850, with application examples (IEC, 2015). Note the delay values are defined for inside substation communication, while inter-substation values will be presented further.

In the P1 class, the total transmission time below the order of a quarter of a cycle (5 ms for 50 Hz, 4 ms for 60 Hz), since their application relies on trips and blocking. For the P2 class, the total transmission time shall be in the order of half of a cycle (10 ms

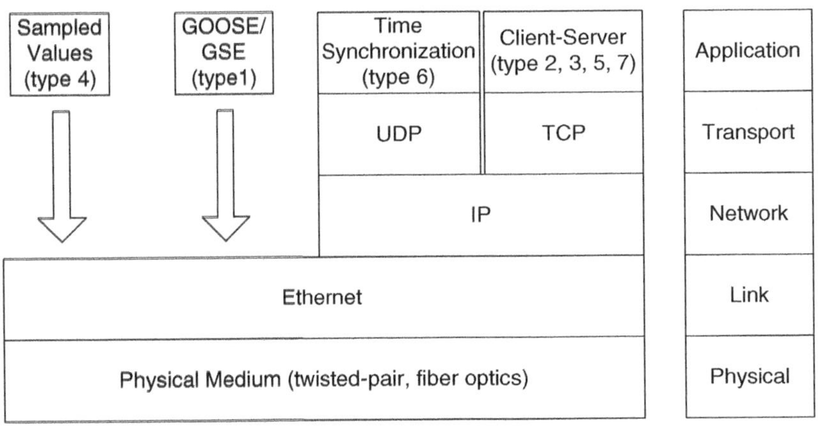

Figure 2.10 Message communication stack of the IEC 61850 standard (Parikh et al., 2013).

Table 2.2 Transfer time classes according to IEC 61850 standard (Adapted from (IEC, 2015)).

Performance/ transfer time class	Transfer time [ms]	Application examples
P1/TT6	≤ 3	trips, blocking
P2/TT5	≤ 10	releases, status changes
P3/TT4	≤ 20	fast automatic interactions
P4/TT3	≤ 100	slow automatic interactions
P5/TT2	≤ 500	operator commands
P6/TT1	≤ 1000	events, alarms
P7/TT0	> 1000	files, events, log contents

for 50 Hz, 8 ms for 60 Hz), and its application regards status changes. The P3 class has a total transmission time in the order of a cycle (20 ms for 50 Hz, 17 ms for 60 Hz), with utilization in fast automatic interactions. P4 transfer time is for slow automation functions, which are less demanding than protection but more demanding than operator actions. For the P5 class, the total transmission time shall be half the operator response time (≤ 1) second regarding event and response (bidirectional). In the P6 class, the total transmission time shall be half the operator response time (≤ 1) second regarding event and response (unidirectional). The P7 class is used for files, events, and log contents and has no strict transfer time requirement (Patel et al., 2016). According to the IEC 61850-5 standard, the overall transfer time is given by:

$$t = t_a + t_b + t_c, \tag{2.1}$$

where t_a and t_c are communication-processing times, internal to the devices, and t_b is the network delay. Note the time requirement presented in Table 2.2 is applicable for the complete transmission chain, but the current work acts only over the network latency t_b. So the challenge is to provide communication that respects the requirements and the distinct IEC 61850 performance classes.

2.3.5 SCADA Protocols for Smart Grid: Existing State-of-the-Art

Although extensively used in industrial and power plants, the simplicity and lack of features of the **Modbus protocol**, jointly with the strict requirements of SG networks, explain the few works with its use for SG. In (Kenner et al., 2015; Gerhard et al., 2013; Schlegl et al., 2014), the authors use Modbus over TCP/IP to communicate the voltage level of a medium-voltage grid. They evaluated two possible architectures for collecting data from energy analyzers, and the performance was analyzed for real-time monitoring, load peaks, and automated regulation of the power grid. Results show that the protocol is suitable for monitoring purposes, but the high dependency between latency and number of devices is critical for SG applications. Other works in literature mainly describe utilization cases for:

- a management system for underground substations, with communication via serial Modbus and DNP3 protocols (de Andrade Barreto and Grimoni, 2013);
- a communication and control system with communication between substation and control center is carried out with IEC 61850 standard and communication to gather data from field done via Modbus TCP/IP protocol (Temiz et al., 2015); and
- a communication network using serial Modbus protocol for measurement equipment (Gandhi and Adhyaru, 2015).

The main issue addressed in the literature applying **DNP3 protocol** to SG is whether its performance is suitable for time-critical purposes. In (Ortega et al., 2013), it is evaluated over a heterogeneous TCP/IP network, and the authors conclude that the DNP3 communication is severely degraded, particularly when the traffic load exceeds 85%, suggesting it is not suitable for time-critical SG communication. Also, for heterogeneous networks, (Manbachi et al., 2015) presents a real-time communication platform for an SG adaptive volt-VAR optimization of distribution networks; authors' finding indicates the use of IEC 61850 standard as an alternative for latency-constrained applications. In a smart microgrid scenario with a control center, (Lu et al., 2011; Wei and Wang,

2013) conclude that DNP3 over TCP/IP or UDP/IP is not suitable to be directly used for time-critical message transmission since both were unable to meet the stringent delay requirement of SG. Along with such approach of minimizing the latency of DNP3 protocol, (Padilla et al., 2014) proposes the use of DNP3 directly over the Ethernet, entitled E-DNP3, and field programmable gate array (FPGA) technology as the first step to the electric power system automation. This approach improves the delivery time and provides a versatile solution for real-time information exchange in electrical applications.

As the most up-to-date option, **IEC 61850 standard** is the first candidate for fulfilling protection, control, and monitoring in WAN and FAN/NAN of distribution grids in SG (Mohagheghi et al., 2009, 2011). The importance of distributed generators in the grid control (power, voltage, and frequency) and the adoption of current standards are highlighted in (Jaloudi et al., 2011), drawing the attention to the adoption of IEC 61850 to ensure interoperability. An overview of the information and communication needs for the application of this standard in the SG context is presented in (Naumann et al., 2014), and a simulation tool for such evaluation can be found in (Yang et al., 2011).

An open communication architecture proposal for the application of IEC 61850 to smart grid automation is discussed in (Han et al., 2014). According to the authors, the proposed mapping method to web services + IEC 60870-5-104 can improve interoperability of distributed remote terminal units and reduce their installation and maintenance costs.

The authors in (Das et al., 2012) discuss the factors affecting end-to-end communication delay inside a substation network, which can be extended to the SG outside the substation. Relevant results for practical applications are the optimization of the size of packets (select small and avoid large ones) and the avoidance of sending a variety of data streams at the same time to ensure real-time and reliable data transmission.

In (Dehalwar et al., 2015), the authors propose a review of the recent IEEE 802.22 standard aiming to support massive data communication in the SG context. The superframe structure, which consists of 16 frames each of 10 ms, is capable of sending data on uplink frequency using upstream burst, fulfilling the real-time requirements of some IEC 61850 protocols, with the performance classes of different message types being used for teleprotection and transmitting synchrophasor over WAN.

Finally, LTE has been presented as a communication preference for SG automation based on IEC 61850. In (Patel et al., 2016), the authors focus on the QoS aspects to guarantee the time-critical messages delivery, while (Kalalas et al., 2016a) focuses on the scheduling aspects.

2.4 Power Distribution System: Distributed Automation Applications and Requirements

2.4.1 Distributed Automation Applications

An SG consists of a power system controlled and monitored through a network with two-way flow of information. In particular, the DA applications provide real-time operation information of grid structure, automation control, data communication, and information management to monitor and control the distribution grid. It is usually built on open standard system and protocols, and typical equipment monitored are fault detectors, reclosers, switches, and voltage regulators, among others. The SG-demanded services include real-time system state determination, real-time congestion management, phase angle monitoring, disturbance propagation monitoring, protection for wide-area disturbances, estimation of the load model parameters, validation of generator models, stability, and control, among others (Mohagheghi et al., 2011). Some specific DA applications include Volt/VAR control (VVC), distribution system demand response (DSDR), distribution system monitoring and maintenance, and fault detection, clearing, isolation and restoration (FDCIR) (Kuzlu et al., 2014). The DA applications can be divided into three broad categories as follows: wide-area protection, control, and monitoring (Terzija et al., 2011).

Wide-area protection provides a fully automatic protection against widespread blackouts, transmission congestion, and stressed conditions, or unexpected events, and its uses involve load shedding and adaptive islanding.

Wide-area control provides automatic self-healing capabilities that exceed functionality delivered by local control and responds faster than conventional manual control by a control center. It also works as an additional layer to prevent possible blackouts and facilitates electrical commerce by providing a real-time instability control.

Wide-area monitoring aims at providing system data in real-time from a group of IEDs and PMUs. IEDs transmit snapshots of device status and measurement data to the SCADA and the EMS over a WAN. PMUs, on the other hand, enable time-synchronized snapshots of a power network including voltage and current phase angles, i.e., wide-area measurements (Kuzlu et al., 2014).

2.4.1.1 Voltage/Var Control (VVC)

The primary objective of VVC is to reduce energy loss in the power network. For this, the adjustment of voltage is carried out along a distribution circuit in order to flatten the voltage profile and also for power factor correction (load compensation). For a distribution system, it is necessary to monitor and control the feeder line devices, with the main equipment being the shunt capacitor (size/status), voltage-regulating transformers (tap positions), and the on-load tap changer (OLTC) at the substation. Moreover, it is necessary a regulation of the voltage of the power generation, and in the distribution case, it is required an active control over the DER sources. This optimization problem is subject to operational constraints of the equipment as well as the network constraints for the power system (Kuzlu et al., 2014; Mohagheghi et al., 2011). Figuring out as an optimization problem, VVC can be formulated by min{power losses, total demand}, subject to one or more of the following constraints:

- load balance equations;
- bus voltage constraints;
- line/transformer current constraints;
- power factor constraints at the main substation;
- reactive power constraints at the main substation;
- maximum allowable number of operations for each transformer tap for a given time duration; and
- maximum allowable number of switching operations for each capacitor for a given time duration.

The transmission network voltage is controlled by a well-known three-step control method (Richardot et al., 2006), each step acting with a different time constant. Primary voltage control (PVC) locally regulates generators voltages at their set-point values, with a response time faster than one second. Secondary voltage control (SVC) is automatically performed on a large-area scale, with a response time about one minute, aiming to realize reactive power production-consumption balance. Tertiary voltage control (TVC) is manually performed with a time constant of 10 to 30 min at a country scale to lower reactive energy transfer between areas. It is based on optimization methods taking into account economic and technical aspects of power system operation. At the distribution level, connected DER usually do not participate in voltage regulation, which is performed by other means, such as tap-changers or capacitor banks.

One practical implementation of an automatic VVC using DNP3 with LTE-based architecture can be found in (Angioni et al., 2015), where it is carried out in a central controller, which regulates the voltage with 1 second periodicity. A coordinated VVC is presented in (Sarmina et al., 2013), where the authors derive a solution for the optimization problem using a heuristic approach, specifically the particle swarm optimization (PSO) algorithm. The VVC optimization problem is the minimization of the power as a function of the voltage, the current, the apparent power, and the power factor, as well as the status of the different voltage controller equipment in the grid (capacitor, voltage regulating transformer, OLTC) and also the DER sources. It is important to inform that these main control elements can be easily modeled using IEC 61850.

2.4.1.2 Fault Detection, Isolation, and Restoration (FDCIR)

The FDCIR algorithms are used for detecting and isolating faults in the distribution system and restoring power to the customers located in the outage area, hence improving the system reliability and availability of service. One criterion of the fault isolation is to ensure that the fault is isolated with a minimum amount of unserved load. This mechanism is primarily dependent on the protection scheme employed at the feeders. In general, two main protection schemes can be considered that are currently being used worldwide (Mohagheghi et al., 2011):

- A distributed protection scheme, mostly used in rural distribution networks or when vast distances have to be covered. In such a scheme, multiple breakers and/or reclosers are located on the same feeder and contribute directly to remove the fault as close to where it occurs as possible (North American topology).
- A centralized protection scheme, where protection devices able to clear the fault are exclusively located in the primary substation. This means that when a fault occurs, only one breaker can trip (or perform a reclosing action). Any other switching equipment on the line does not have fault interruption capabilities. In the current state of the art, no more than 2.5 switches per feeder are remotely controllable (European topology).

Following the occurrence of a fault in the distribution system and its consequent isolation from the rest of the network, the number of customers without electricity is not only those connected to the faulty circuit. Service restoration algorithms are therefore applied in the data management system (DMS) in order to identify the location of the fault and to supply electricity to the customers located in the outage area, while the faulty circuit is being repaired. Traditionally, the selection of the restoration sources and paths is determined only by the capacity margin of the restoration source as well as the rating limits of the equipment along the restoration path. When the additional load exceeds the capacity of these generators, a demand response signal can either be sent to the customers on the restoration path aiming to lower their load or it can be responsible for activating and adjusting the generation set points of the DER units. An example of an FDIR algorithm is presented in (Mohagheghi et al., 2011).

2.4.2 Requirements for Distributed Automation Applications

Table 2.3 summarizes the requirements for wide-area protection, control, and monitoring applications. The data sampling for some NAN/FAN applications depends on the equipment. For distribution system monitoring and maintenance, the capacitor bank

Table 2.3 Requirements for different DA applications.

Application	# bytes	Sampling	Latency	Reliability
NAN/FAN				
DA - distribution system monitoring and maintenance	100-1000	varies	<5 s	>99.5%
DA - Volt/VAR control (command to field device)	150-250	varies	<5 s	>99.5%
DA - DSDR	150-250	varies	<4 s	>99.5%
DA - FDCIR	25	varies	<5 s	>99.5%
WAN				
Wide Area Protection				
Adaptive islanding	4-157	0.1 s	<0.1 s	>99.9%
Predictive under frequency load shedding	4-157	0.1 s	<0.1 s	>99.9%
Wide Area Control				
Closed-loop transient stability control	4-157	0.02-0.1 s	<0.1 s	>99.9%
Wide-area power oscillation damping control	4-157	0.1 s	<0.1 s	>99.9%
Wide-area voltage stability control	4-157	0.5-5 s	<5 s	>99.9%
Cascading failure control	4-157	0.5-5 s	<5 s	>99.9%
FACTS and HVDC control	4-157	30 s - 2 min	<2 min	>99.9%
Precalculation transient stability control	4-157	30 s - 2 min	<2 min	>99.9%
Wide Area Monitoring				
Local power oscillation monitoring	> 52	0.1	<30 s	>99.9%
Wide-area power oscillation monitoring	> 52	0.1	<0.1 s	>99.9%
Local voltage stability monitoring	> 52	0.5-5 s	<30 s	>99.9%
Wide-area voltage stability monitoring	> 52	0.5-5 s	<5 s	>99.9%
PMU-based state estimation	> 52	0.1	<0.1 s	>99.9%
Dynamic state estimation	> 52	0.02-0.1 s	<0.1 s	>99.9%
PMU-assisted state estimation	> 52	30 s - 2 min	<2 min	>99.9%

controller (CBC) sampling is 1 per device per hour (24 × 7); the feeder fault detector is 1 per device per week (24 × 7); recloser is 1 per device per 12 h (24 × 7); switch is 1 per device per 12 h (24 × 7); voltage regulator (VR) is 1 per device per hour (24 × 7). For the VVC (command from DMS to field devices), the open/close CBC sampling is 1 per device per 12 h (24 × 7); the open/close switch is 1 per device per week (24 × 7); step up/down VR is 1 per device per 2 h (24 × 7). For the DSDR (command from DMS to field devices), the open/close CBC sampling is 1 per device per 5 min; the open/close switch is 1 per device per 12 h; the step up/down VR is 1 per device per 5 min (1 – 6 h duration, 4 – 8 times a year). Finally, for FCIR (command from DMS to field devices), 1 per device per isolation/reconfiguration event within < 1.5 min of fault event).

2.5 Analysis of Data Flow in Power Distribution Grid

2.5.1 Model for Power Distribution Grid

A few elements are selected to exchange data for a DA application in an SG network. Figure 2.11 illustrates single-line diagrams for standard generator, transformer, feeder, and transfer bays, which will be adopted for composing the automation grid model.

First of all, the communication inside a typical substation is divided into three layers: process level, bay level, and station level, with the following equipment:

- Merging Unit (MU) IED: processes and combines the signals from field current and voltage transformers (CT and VT), and then it transmits the digital voltage and current output to the process bus.
- Breaker IED: controls the breakers positions and monitors their state and condition. Thereby, it receives the switching commands from the PC IEDs and sends state change event messages to corresponding protection IEDs through the process bus.
- PC IED: it is a multifunctional device that bundles the protection and control functions for the bay unit that it serves within the substation.
- Server: concentrates the substation data and provides data to workstations and external control centers.

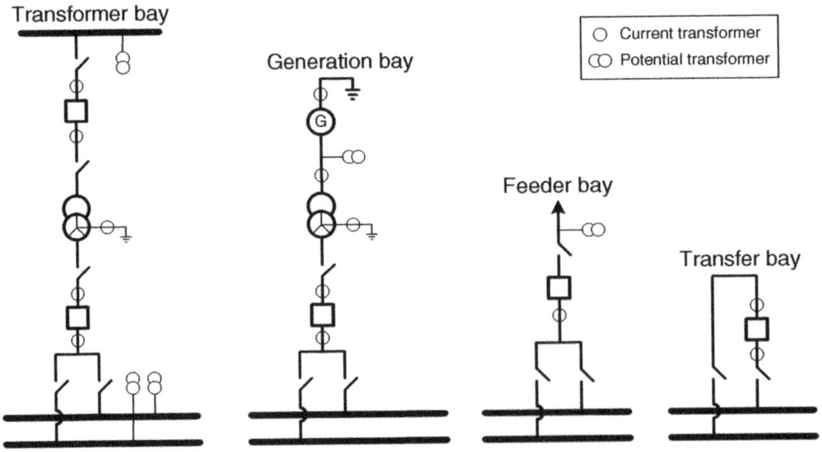

Figure 2.11 Single line diagram of different bays (Adapted from (Zhang et al., 2015)).

According to (Das et al., 2012; Zhang et al., 2015), the communication between these IEDs inside a substation is composed basically of five data flow types. In the SG context, and in particular for outside substations, the equipment and communication are simplified. Therefore, two flows of communication are defined herein for equipment in different locations of the power distribution grid.

- Inside Substation:
  - *MU IED ⇒ PC IED*: MU IEDs in every bay send cyclical and fixed-size SVs to PC IEDs within the same bay at a specific rate. According to IEC 61850-9-2LE, the data set of ASDU comprises information related to four voltages and four currents (three phases and neutral for each). The packet size for SVs with two ASDUs is 219 bytes in total, with a sampling rate for SVs being 4800 Hz.
  - *Breaker/PC IEDs ⇒ Server*: In normal conditions, the breaker and PC IEDs in all bays send meter values and breaker status information (cyclic GOOSE) to the server, with a constant time interval of 20 ms, and the message size is set to 144 bytes.
  - *PC IED ⇒ Breaker IED*: When a fault occurs in a feeder bay, its PC IED sends trip signals to the breaker IED in the feeder bay and the bus bay. These trip signals belong to stochastic data flow with short duration and size set to 204 bytes, with the arriving times of messages obeying a Poisson distribution with $\lambda = 500$. According to IEC 61850-8-1, after the first transmission, this type of message needs to be repeatedly sent in a time sequence of $2^n(n = 1, 2, 3, \ldots)$ ms (IEC, 2015).
  - *Related with Trip Signals*: Once the breaker IED acts based on the received trip message, it will send a GOOSE message to the corresponding PC IED and the server. Meanwhile, the PC IED will send a GOOSE message to the server as well. These messages are the typical case of burst data flow. For the ON-state of burst data, the data size is set to be 144 bytes, and the arriving time interval is 0.1 ms. The performance of ON-state obeys the Pareto distribution with typical parameters of $k = 512\mu s$ and $\alpha = 1.1$, whereas the generation of OFF-state obeys the Poisson distribution with $\lambda = 263.16$, while the average duration time for the ON-state is 6.2 ms, and that for the OFF-state is 3.8 ms.
  - *Server ⇒ Workstations*: The file transfer protocol (FTP) server is used to send a large amount of data to the workstation in a randomly selected time, which indicates that messages from the server to the station PC belong to the stochastic data flow. The total data amount for file transfer is supposed to be hundreds of megabytes. Assume that a single packet size for FTP application is 1000 bytes and the time interval for consecutive packets obeys an exponential distribution with parameter $1/\lambda = 1$ ms. Thus, the average data rate for file transfer is 8 Mb/s.
- Outside Substation:
  - *Server ⇒ Control Center/Server*: The server communicating with the control center is modeled as a cyclic data flow, composed by meter values and equipment status of the entire substation or equipment, with a sampling time of 100 or 500 ms, and also a stochastic FTP data flow with the time interval between packets being modeled by an exponential distribution.
  - *Server ⇔ Breaker/PC IED (different location)*: equipment that is not composed by PC IED or a server is supposed to communicate with adjacent servers. The data flow are composed by similar communication inside substations but with

sampling time of 100 or 500 ms. The GOOSE burst messages transmission obeys the repetitions defined in IEC 61850-8-1.

The inside communication data flows ensure the secure operation of the substation, while the outside ones permit the agents of the power system to perform the wide area protection, control, and monitoring. Figure 2.12 sketches the communication inside and inter-substation for SG and DA purposes.

In the next subsection, the traffic model for these seven data flow types are described based on the classification of cyclic, stochastic, and burst data.

2.5.2 IEC 61850 Traffic Model

According to IEC 61850-5, there are seven types of messages in substations: fast message, medium-speed message, low-speed message, raw data message, file transfer function, time synchronization message, and access control command. The examples and performance requirements for communication inside substation networks is summarized in Table 2.4 (IEC, 2015; Das et al., 2012), while Table 2.5 briefly compares the time delivery requirements for inside and outside substation communication (IEEE, 2005).

In the present work, the MMS messages are adopted in wide-area monitoring and slow control, since they are characterized as more delay tolerant than the GOOSE ones. Inside substations, their functions involve exchange information between control and protection layers with higher levels, i.e., communication between IEDs and control centers. GOOSE messages usually carry critical data information, such as a fault "trip" for breaker actuation. From the perspective of data flow characteristics in the time domain, the model of IEC 61850 messages have been divided into three types: cyclic data, stochastic data, and burst data (Das et al., 2012; Zhang et al., 2015), as described in the sequence.

2.5.2.1 Cyclic Data Flow

In IEC 61850 communication, the cyclic data flow is generated by periodic sampling of CT and PT and also meter values and status of breakers. In the DA context, this information comes from generation plants, substations, DERs (solar and wind), and regulating transformers, among others. Such structures have two types of cyclic data, as detailed in the following list.

- SVs, generated by MU IEDs in the substation process level and then transmitted to PC IEDs in the substation bay level. They represent time-critical information that contains large amounts of data flow.
- Meter values and breaker status information, transmitted from the device in bay level to the server in station level at a certain time interval, which belong to the cyclic type of GOOSE messages (IEC, 2004). This kind of cyclic GOOSE data is comparatively stable and a medium-speed message type.

The cyclic data m_c has fixed length and time interval of transmission, and it can be modeled as a function of three variables: sampling rate f_c of the measurement, i.e., the number of cyclic data type arrived per unit time, message length ℓ_c, which contains all

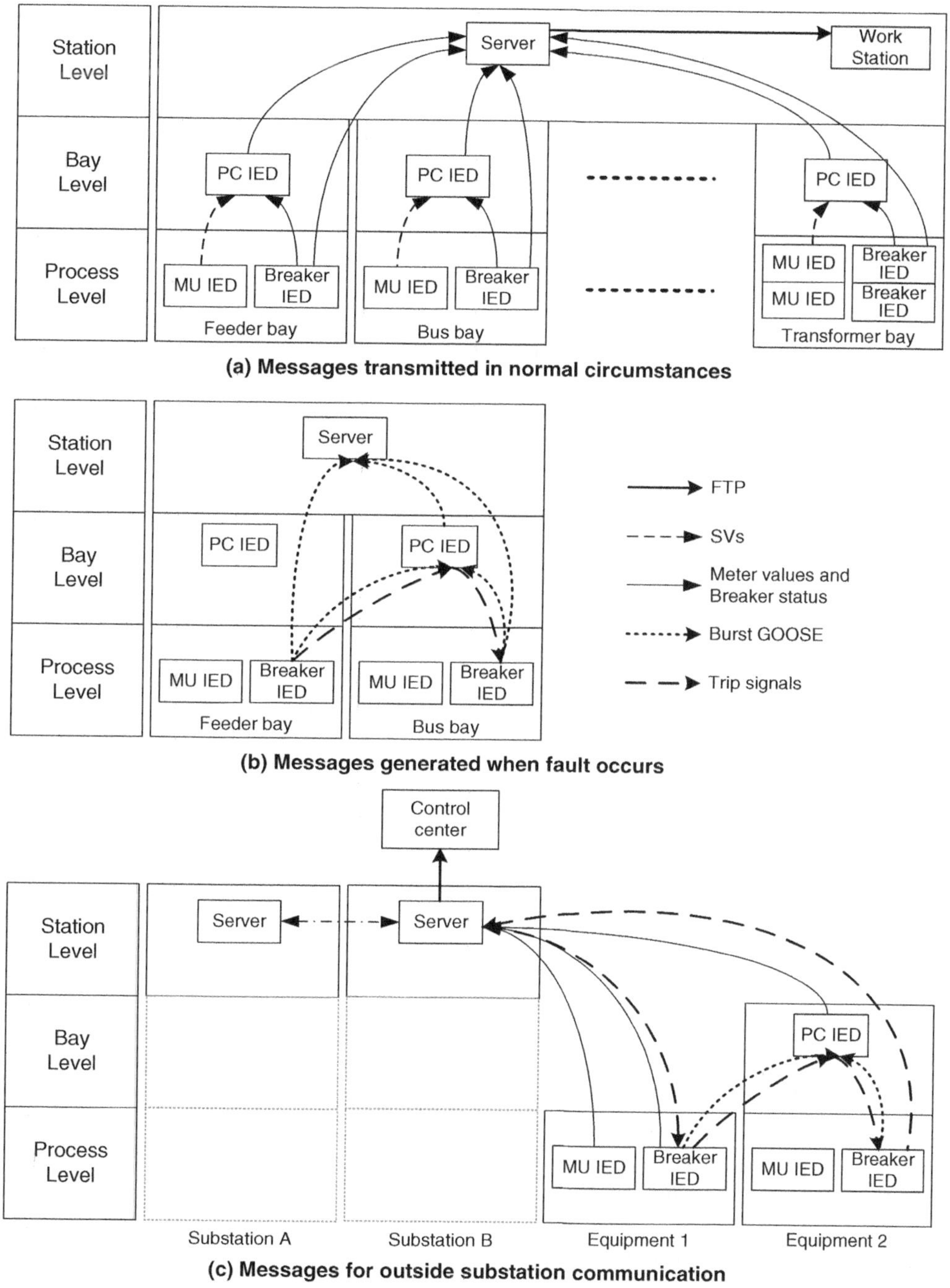

Figure 2.12 Data flow diagram for a substation: (a) in normal circumstances; (b) during fault occurrence; and (c) for communication outside substations (Adapted from (Zhang et al., 2015)).

Table 2.4 Message types and performance requirements in a substation network (Adapted from (IEC, 2015; Das et al., 2012)).

Message type	Example messages	Transfer time [ms]
Fast message	trigger	10-100
	complex block or release	10-100
	fast broadcast message	1
	process state changed	10-100
	trip	1
Medium speed message	process value in R.M.S.	50-1000
	request for synchronism check	10-100
	interlocking	10-100
	process state	1-100
	external state	1-100
Low speed message	measured or meter value	100-1000
	non-electrical process value	1000-5000
	fault value	0.1-5000
	event/alarm	100-1000
	mode of operation	10-100
	set point	100-1000
	acknowledgment	10-1000
	data and time	100-1000
Raw data message	process value	0.1-10
File transfer	report	1000-5000
	mixed fault information	1000-5000
	mixed fault data	5000
	event/alarm list	100-1000
	ID data, setting	1000-5000
	diagnostic data	5000
Time synchronization Message	synchronization pulse	0.1-10
Command message with access control	command	1-1000

Table 2.5 Simplified time delivery requirements for communication inside and outside substations (SS) (Adapted from (IEEE, 2005)).

Message type	Example messages	Inside SS	Outside SS
Fast	protection	4 – 6 ms	8 – 12 ms
Medium speed	monitoring and control	16 ms	1 s
Low speed	operations and maintenance	1 s	10 s
File transfer	program files	60 s	10 min

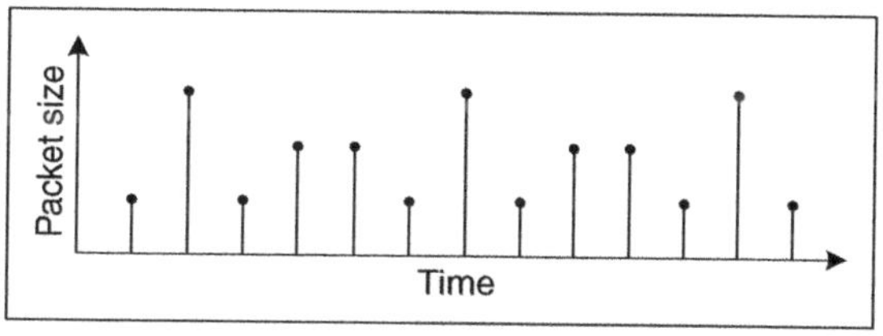

Figure 2.13 Example of the generation of cyclic messages for outside substation communication with IEC 61850.

the layers frame header, address field, and so on, and the delay of transmission d_c, which is the sum of the transmitter, receiver, and network delays:

$$m_c = f(f_c, \ell_c, d_c). \tag{2.2}$$

According to IEC 61850-5, inside a substation, the SV messages should be transmitted and processed within a maximum delay $d_c < d_s = 3$ ms for performance classes P2 and P3; otherwise, the reliability and electrical efficiency of the system could deteriorate. Meanwhile, the preferred sampling rate for SVs is 4800 Hz. Similarly, the end-to-end time delay for breaker and equipment status information, which belongs to the medium-speed message type, has to satisfy a maximum delay $d_c < d_s = 100$ ms.

For the communication in the DA context, it is adopted in this work SV with a $d_c < d_s = 100$ ms, with a sampling rate for the measurements of 10 Hz. With such frequency, it is supposed to perform wide-area control and monitoring, but no distributed protection is expected. The first actuation of equipment will be only local, and a further FDCIR procedures should be carried on. Equivalently, considering the time requirement for medium messages outside a substation in the order of 1 second (IEEE, 2005), the maximum delay for breaker and equipment status information is set to $d_s = 500$ ms. An example of the cyclic data flow outside a substation is shown in Figure 2.13.

2.5.2.2 Stochastic Data Flow

Stochastic data are typical event-driven data, which are usually triggered by command or fault events and also by configuration of equipment. Stochastic data in substations can be mainly classified into (Zhang et al., 2015):

- Type 1: Transformer tap modulation, switch operation message, trip message, protection function interlocking, time synchronization, capacitor switching, and so on. It usually has the features of small size and short duration, while the transmission time should meet the requirements of the fast message type.
- Type 2: Protection setting modification, event log checking, recording data transmission, file transfer, and so on. It is larger in size and will usually cause a sudden increase of network flow, while the real-time requirement of transmission is not strictly specified. According to (IEEE, 2005), there is a time delivery requirement of minutes while downloading, and it is the parameter configuration of an IED for uploading.

The arrival of stochastic data can be modeled by the Poisson process. For a period τ, the probability that k packets arrive is

$$p\{n(\tau) = k\} = \frac{(\lambda t)^k e^{-\lambda t}}{k!}, \tag{2.3}$$

where $n(\tau)$ is the number of packets arrived in the interval τ and λ is the Poisson distribution parameter (average arrival rate of packets). The time interval between two

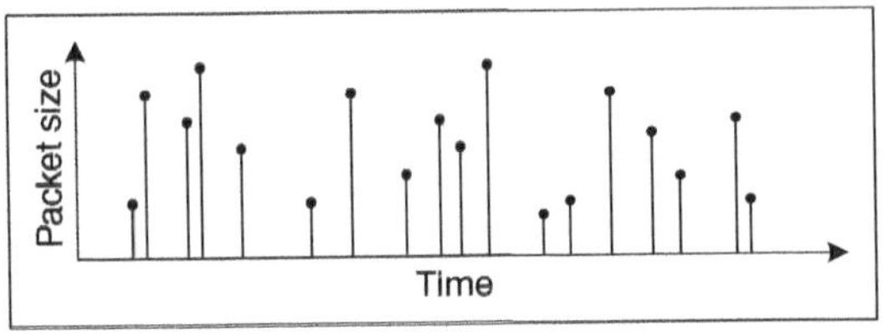

Figure 2.14 Example of stochastic messages generated for communication in the power distribution grid.

consecutive packets is defined by a variable with negative exponential distribution as follows:

$$g(t) = \lambda e^{-\lambda t}, \ \lambda \geq 0. \tag{2.4}$$

Figure 2.14 depicts an example of the stochastic data flow generation for communication outside a substation.

2.5.2.3 Burst Data Flow

During a random time, the IEC 61850 burst data is dependent on previously occurred events. Usually, it contains information about protection actions generated after a fault occurs, and it belongs to a GOOSE message. When a fault occurs, the transmission of a GOOSE message is changed from cyclic mode to burst mode (IEC 61850-8-1). In burst mode, the transmission interval for a GOOSE message changes from τ_0 to $\tau_i = 2^i \times S(i = 1, 2, 3, \ldots)$, where S is set in the substation configuration description file, τ_0 is the constant time interval for a GOOSE message in normal state, and τ_i represents the transmission interval of the ith GOOSE packet since trip message generates and ends when τ_i reaches τ_0.

Due to the characteristics of long-range dependence and self-similarity, which presents the same burstiness at different time scales, it is adopted herein the heavy-tailed distribution and the ON/OFF function (Zhang et al., 2015) to model the burst data flow. Supposing that the time duration of the ON-state for a single data source obeys the Pareto distribution, its cumulative distribution function (cdf) is

$$p(\tau \leq t) = 1 - \left(\frac{k}{t}\right)^{\alpha}, \qquad 0 < k \leq t, \ \ \alpha > 0, \tag{2.5}$$

where k is the minimum possible value of τ, which represents the minimum duration of the ON-state, and α is a positive parameter. The Pareto distribution is characterized by a scale parameter k and a shape parameter α, known as the tail index.

The time duration for the OFF-state obeys a negative exponential distribution of a Poisson process, with the same probability density function shown in Equation (2.4). As long as the duration of the ON- or OFF-state obeys a heavy-tailed distribution, the accumulation of large numbers of these ON- and OFF-states will form self-similar burst data flow, with only one parameter H, so-called Hurst coefficient, given by

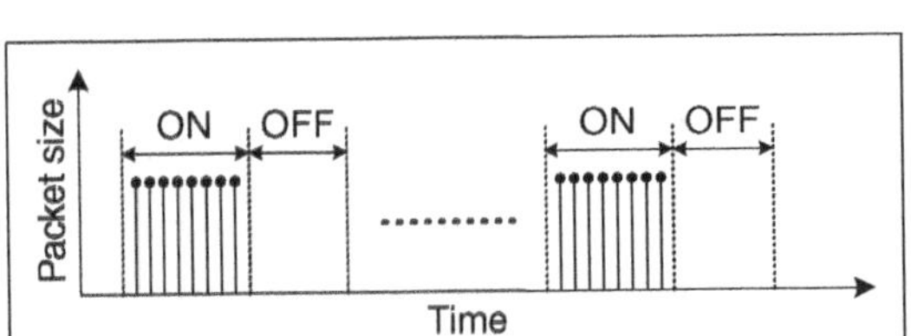

Figure 2.15 Example of burst data generated for communication outside substations in DA.

$H = (3 - \alpha_{min})/2$, with α_{min} being the smaller value assumed by the parameter α for heavy-tailed distribution between ON- and OFF-states. Figure 2.15 illustrates an example of burst data flow generated with the ON-OFF states for communication outside a substation.

2.6 LTE-D2D for DA: Resource Allocation for Time-Critical Applications

In this section, we provide an overview of the LTE standard, followed by the adaptation of IEC 61850 messages (MMS and GOOSE) to LTE and a resource allocation model for time-critical applications

2.6.1 Overview of LTE

The LTE communication networks are designed to support high-speed multimedia unicast and broadcast services, and its primary goal is to provide high data rate, low latency, and packet-optimized radio access technology supporting flexible bandwidth deployments. There are three fundamental technologies that have shaped the LTE radio interface design: multicarrier technology, multiple-antenna technology, and the application of packet-switching to the radio interface (Sesia et al., 2011).

Due to the diversity of wireless users, LTE needs to support a range of classes with different demands. In general, each class attaches different priorities to aspects such as peak data rate, UE size, cost, and battery life. The quality of service (QoS) to support the different user classes, and the approach adopted for LTE is simple, based on the concept of data flows and bearers. Data flows are assigned to bearers, and three individual bearers (Radio, S1, and S5/S8) are successively combined in order to provide a virtual end-to-end QoS connectivity. A bearer can be considered as a set of multiple QoS requirements that are indicated by the QoS Class Identifier (QCI). Each QCI describes the type of service that makes use of the virtual connection (e.g., conversational voice, streaming video, signaling, or best effort). The QCI is a scalar used as a reference to node-specific parameters that control packet forwarding treatment (e.g., scheduling weights, admission thresholds, queue management thresholds, or link layer protocol configuration) and that have been pre-configured by the operator owning the node, e.g., eNodeB (3GPP, 2016). The QCI is defined in terms of the following performance characteristics:

- resource type: Guaranteed Bit Rate (GBR) or not (Non-GBR);
- priority;
- packet delay budget;
- packet error/loss rate.

There are different QoS classes in LTE networks, designed to support various demanded services. However, the classes rely on human-based services, and they do not satisfy the strict requirements of the DA communication. Hence, similar to (Kalalas et al., 2016a), here it is proposed to create two QoS classes to support the IEC 61850 traffic, with the adoption of dedicated bearers. The standardized characteristics for QCI, with the two new proposed classes, are summarized in Table 2.6.

An essential component to achieving an efficient utilization of the available radio resources is the scheduler. It is important in situations of limited network resources,

Table 2.6 Standardized QCI characteristics for LTE (Adapted from (3GPP, 2016)).

QCI	Priority Level	Packet delay Budget [ms]	Packet error Loss rate	Example services
1	2	100	10^{-2}	Convers. Voice
2	4	150	10^{-3}	Convers. Video
3	3	50	10^{-3}	Real-time gaming
4	5	300	10^{-6}	Non-convers. video
65	0.7	75	10^{-2}	Mission critical PTT
66	2	100	10^{-2}	Non-mission critical PTT
5	1	100	10^{-6}	IMS signaling
6	6	100	10^{-6}	Video TCP-based
7	7	300	10^{-3}	Real-time
8	8	100	10^{-6}	Video TCP-based
9	9	300	10^{-6}	Video TCP-based
69	0.5	60	10^{-6}	Mission critical live
70	5.5	200	10^{-6}	Mission critical data
10	0.1	100	10^{-6}	IEC 61850 (GOOSE/SV)
11	0.2	200	10^{-6}	IEC 61850 (MMS)

where different types of users, each one with specific performance requirements, share the network. The scheduling activity needs, simultaneously, to optimize the system capacity and ensure QoS to different types of users. The physical layer at the eNodeB (eNB) collects the channel state information (CSI) and uses it to determine the modulation order (QPSK, 16QAM or 64QAM) and coding scheme through adaptive modulation and coding (AMC) based on the channel conditions.

In the time domain, radio resources are distributed every transmission time interval (TTI), which is considered the minimum scheduling unit in LTE and corresponds to a subframe with a duration of 1 ms. In the frequency domain, the available bandwidth, specifically 1.4, 3, 5, 10, 15, or 20 MHz, is divided into blocks of sub-channels, each including 12 subcarriers with a spacing of 15 KHz. Hence, each block of sub-channel has a bandwidth of 180 KHz, and along with the seven symbols in the time domain, it constitutes a resource block (RB). The physical layer parameters for the uplink transmission of a single carrier frequency division multiple access (SC-FDMA) system are presented in Table 2.7 (Sesia et al., 2011).

The number of RBs varies from 6 to 110, according to the available bandwidth. Two consecutive RBs form a scheduling block, which is the smallest resource unit that a scheduler can allocate to a user. Each scheduling block has two times seven symbols and 12 subcarriers, and each combination of one symbol and one subcarrier is called a resource element (RE).

2.6.2 IEC 61850 Protocols over LTE

The application of IEC 61850 protocols is not inherent to LTE systems. At the present subsection, the LTE within IEC 61850 messages is discussed to allow integration of

Table 2.7 LTE uplink SC-FDMA physical layer parameters (Sesia et al., 2011).

Parameter	Value	Comments
Subframe duration	1 ms	
Slot duration	0.5 ms	
Subcarrier spacing	15 kHz	
SC-FDMA symbol duration	66.67μs	
CP duration	normal CP:	5.2μs first symbol in each slot
		4.69μs all other symbols
	extended CP:	16.67μs all symbols
Number of symbols per slot	normal CP: 7	
	extended CP: 6	
Number of subcarriers per RB	12	

networks in the WAN and LAN/FAN contexts. The selected messages from the IEC 61850 standard is the MMS, SV, and GOOSE. First, the MMS messages deployment is proposed for monitoring and slow control, where a control center and the distributed devices exchange data. Second, the application of SV and GOOSE messages is intended for control and protection purposes, with communication between equipment. In the literature, some works involving the integration of IEC 61850 protocols to the LTE are available in (Pham, 2013), (Karagiannis et al., 2014), (Kalalas, 2014) and (Allam et al., 2015).

2.6.2.1 Mapping MMS over LTE

In order to extend DA outside a substation and to integrate IEC 61850 standard with LTE system, it is necessary to map the MMS messages in the TCP/IP stack. As discussed earlier, MMS is not one protocol, but it is composed of a set of protocols, as presented

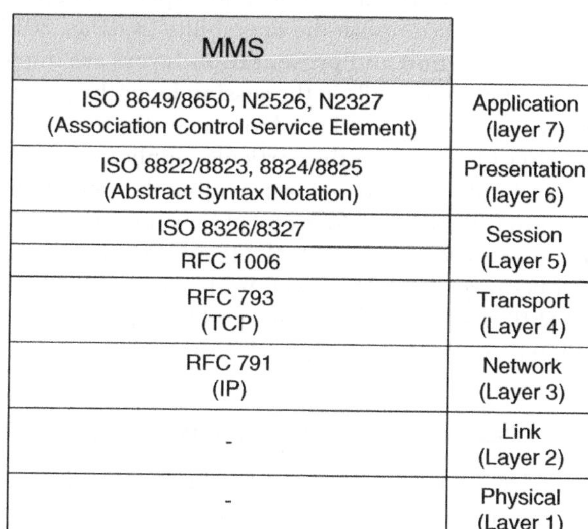

Figure 2.16 MMS message stack (Adapted from (IEC, 2004), (Kalalas, 2014) and (Pham, 2013)).

in Figure 2.16. With such stack, the MMS message will be used for SG communication over LTE systems.

In the **application layer** are placed the MMS services, where the client is connected to the server for specific functions, such as reading or writing local variables. The association control service element (ACSE, ISO 8649/8650) protocol is used to set up and maintain an association between the parties.

The **presentation layer** is used to transmit information between open systems deploying connection-oriented or connectionless mode transmission. It ensures that the information content of presentation data is preserved and adds structure to the units of data that are exchanged. The MMS uses ASN.1 as abstract syntax notation at this layer (Sorensen and Jaatun, 2008).

In the **session layer**, the protocols ISO 8326/8327 and RFC 1006 are implemented for managing the remote procedure calls. MMS is an ISO protocol that requires the transport protocol to exchange information between peers to be in discrete units of information called transport protocol data units (TPDUs). Therefore, RFC 1006 describes that all TPDUs shall be encapsulated in separate units called TPKTs, and the TPKT layer is used to provide these discrete packets to the OSI connection oriented transport protocol (COTP) on top of TCP. With the COTP and TPTK adaptation layers, MMS can run on TCP/IP protocol stack, making it more popular and a widely accepted standard (Pham, 2013).

2.6.2.2 Mapping GOOSE over LTE

The property of being routable over wide-area IP networks is necessary and suited for a protocol to support WAN applications. The IEC 61850-90-5 technical report (IEC, 2012) describes a standardized mechanism to route IEC 61850-8-1 GOOSE packets by forwarding them as inter-substation traffic, exchanged among different geographic locations (endpoints). The report specifies a communication profile that allows GOOSE to be transferred over an IP-based network in a secure and routable manner with the use of UDP (user datagram protocol) as the transport protocol. The UDP is a TCP/IP standard defined in RFC 768 and provides a connectionless datagram service that offers best-effort delivery. Also, application-specific adaptation protocols are added in the session layer to accomplish the operability (Kalalas, 2014).

The **application** and **presentation layers** are implemented in the upper layer of IEC 61850-1 GOOSE, while in the **session layer** the protocol consists of three parts. The first one is a session protocol used to convey the essential parameters required to satisfy the use cases identified for the particular application. The second is the ITU X.234 information technology protocol for providing the OSI connectionless transport services. And the third is the RFC 1240 protocol for supporting OSI connectionless transport services on top of UDP (Kalalas, 2014). Figure 2.17 sketches the protocol stack for transmitting GOOSE over IP in the WAN. With the stacks illustrated in Figures 2.16 and 2.17, the WAN IP network operated over LTE systems and the devices can transmit and route GOOSE and MMS messages.

2.6.3 Resource Allocation in uplink LTE-D2D for DA Applications

Consider initially a single cell LTE system (one eNB), where a conventional cellular user equipment (CUEs) share the spectrum with machine user equipment

Figure 2.17 GOOSE message stack in OSI-7 layer (IEC, 2012; Kalalas, 2014).

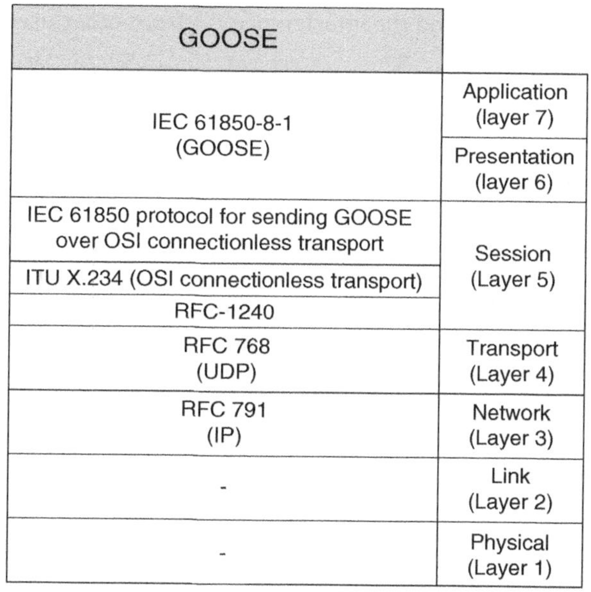

(MUEs) which perform communication for DA applications. Let $\mathcal{C} = \{1, 2, \dots, C\}$ and $\mathcal{M} = \{1, 2, \dots, M\}$ denote the set of CUEs and the set of MUEs, respectively. A CUE is denoted by $u_c \in \mathcal{C}$, while a MUE is denoted by $u_m \in \mathcal{M}$. Moreover, the system bandwidth is divided into N orthogonal RBs with set denoted by $\mathcal{N} = \{1, 2, \dots, N\}$. It is assumed that MUEs can establish D2D and full-duplex (FD) communication links, while CUEs can establish only communication links via eNB.

2.6.3.1 Problem Formulation

In order to start with the resource allocation formulation, the capacity of a communication link of a user u over the nth RB sub-band $B^{(n)}$ is first defined, which is (Dahlman et al., 2007)

$$r_u^{(n)} = B^{(n)} \log_2(1 + \gamma_u^{(n)}), \quad [\text{bits/s}] \tag{2.6}$$

where $\gamma_u^{(n)}$ is the instantaneous signal-to-interference-plus-noise ratio (SINR). The value of the SINR for each user depends on the system configuration, and it must take into account the transmit power, the instantaneous complex channel gains (direct and interference links), link distances, and the noise power. Consider the following notation for the communication link of user $u \in \{\mathcal{C} \cup \mathcal{M}\}$: $P_u^{(n)} \geq 0$ is the transmit power of user u in the nth RB, the CSI for the direct link is expressed as $h_u^{(n)}$; the interference CSI from D2D user u_m or a conventional user u_c is denoted by $g_{u_m,u}^{(n)}$ and $g_{u_c,u}^{(n)}$, respectively, and σ^2 is the noise power. The SINR can be expressed as (Hasan and Hossain, 2015)

$$\gamma_u^{(n)} = P_u^{(n)} \cdot \frac{h_u^{(n)}}{I_u^{(n)} + I_{\mathrm{SI}}^{(n)} + \sigma^2}, \quad \forall u \in \{\mathcal{C} \cup \mathcal{M}\}. \tag{2.7}$$

where $I_{\mathrm{SI}}^{(n)} = \beta$ is the self-interference from imperfect FD nodes, with the coefficient $0 \leq \beta \leq 1$ being used to take the self-interference cancellation into account in the SINR calculation; and the interference $I_u^{(n)}$ from other users is given by

$$I_u^{(n)} = \sum_{\forall u_b \in \mathcal{M}, u_b \neq u} x_{u_b}^{(n)} P_{u_b}^{(n)} g_{u_b,u}^{(n)} + \sum_{\forall u_b \in \mathcal{C} \cup \mathcal{M}, u_b \neq u} q_{u_b}^{(n)} P_{u_b}^{(n)} g_{u_b,u}^{(n)}, \tag{2.8}$$

where $x_u^{(n)} \in \{0, 1\}$ is a binary decision variable for the RB allocation of D2D links, being 1 if the nth RB is assigned to user $u \in \mathcal{M}$ and 0 otherwise. Similarly, $q_u^{(n)} \in \{0, 1\}$ is a binary decision variable for the RB allocation for communication through an eNB link, being 1 if the nth RB is assigned to user u and 0 otherwise.

In LTE, the RB allocation is orthogonal among communication links to eNB, i.e., ideally, there is no interference among users inside the same cell when no D2D communication occurs. Hence, in the present model if $u \in C$ the second parcel of Equation (2.8) is null since the interference comes only from the MUEs, which are supposed to reuse the RBs so as to guarantee the necessary resources to their communication.

Finally, let the total achievable data rate for a user u, considering all subcarriers, be denoted as

$$r_u = \sum_{n=1}^{N} (q_u^{(n)} + x_u^{(n)}) r_u^{(n)}, \qquad \forall u \in \{C \cup \mathcal{M}\}. \tag{2.9}$$

In what follows, the objective function and constraints of the resource allocation optimization problem are described, in order to allow the design of the scheduler of the LTE-D2D system for DA applications.

Objective function The main objective of the resource allocation problem here is to maximize the throughput of the LTE-D2D system. For such aim, it is necessary to specify: 1) what MUEs can establish a D2D link; 2) the assignment of the RBs among all users; and 3) the power allocation so as to fulfill the CUEs and MUEs requirements simultaneously. The variables of the resource allocation problem are $q_u^{(n)}, x_u^{(n)}$, and $P_u^{(n)}$. Based on Equations (2.6) and (2.9), the objective function for the resource allocation problem for LTE-D2D systems can be formulated as:

$$\max_{q_u^{(n)}, x_u^{(n)}, p_u^{(n)}} \sum_{\forall u \in \{C \cup \mathcal{M}\}} r_u \tag{2.10}$$

Note that Equation (2.10) regards both CUEs and MUEs data rate, which implies in the total throughput of the communication system.

Set of constraints The proposed communication model requires a stringent latency and reliability and a flexible data rate for MUE links. In most cases, for the CUEs this constraint is less strict compared to MUEs. Moreover, there are power and RB allocation constraints for both MUEs and CUEs, and they must be considered. Hence, aiming to ensure a minimum data rate and to avoid harmful interference, these requirements are incorporated as constraints in the resource allocation problem.

The reliability of a communication link is a function of the SINR. However, it is usual to express this reliability as a function of more useful parameters, such as the transmission queue length, which is directly related to the delay (Chang, 1994; Aijaz et al., 2014; Kalalas et al., 2016a), or the number of received bits, which describes a probability that during a time span, at least a number of bits has been successfully received (Sun et al.,

2016). The present work adopts queuing theory to establish the reliability and latency constraints for the LTE uplink resource allocation problem, which will be incorporated in the scheduler.

The principle of the scheduler is to adjust the transmit power and to select if the communication switches from the eNB to a D2D and/or FD communication, which, although cause some interference in the system, can exploit the spatial reuse and avoid queue overload. For such approach, the scheduler is divided into priority classes according to QoS adopted in the LTE system and the two proposed for IEC 61850, as presented in Table 2.6.

Delay and queue stability In the eNB, the users are divided into A classes, each one with different QoS. These users share the same resources, and the scheduler is responsible for allocating each user, according to its class. Let $\mathcal{A} = \{1, \dots, A\}$ denote the set of classes, and each class $a \in \mathcal{A}$ is defined by its arrivals, request size ℓ_a bytes/request, the maximum allowed transfer time $\tau_a^{\max}$, and the probability that an arriving request belongs to class a, $p(a)$.

The scheduler consists of A queues, each one for the traffic of the classes that are presented in Table 2.6. According to (Kalalas et al., 2016a), the average delay can be expressed as a sum of the scheduler delay and the core LTE network domain, t_{core}, as:

$$t_{a,\text{total}} = t_a(\mathbf{q}, \mathbf{x}) + t_{\text{core}}, \tag{2.11}$$

where $\mathbf{q}$ and $\mathbf{x}$ are the allocation vectors for eNB and D2D links, respectively. Employing a preemptive-resume queuing discipline, where an ongoing service is interrupted by the arrival of a class with a higher priority, and later the service continues from where it was interrupted (Kalalas, 2014; Tijms, 2003), in a system with A independent queues, its average delay can be modeled as (Kalalas et al., 2016a):

$$t_a(\mathbf{q}, \mathbf{x}) = \frac{\lambda_a \ell_a^2}{R_a(\mathbf{q}, \mathbf{x})[R_a(\mathbf{q}, \mathbf{x}) - \lambda_a \ell_a]} + \frac{\lambda_a}{R_a(\mathbf{q}, \mathbf{x})}, \tag{2.12}$$

where λ_a is the parameter for the Poisson process, and $R_a(\mathbf{q}, \mathbf{x})$ is the data rate for class a. The delay constraints, which also guarantee the queue stability, can be written as

$$t_{a,\text{total}} \leq \tau_a^{\max}, \qquad \forall a \in \{1, \dots, A\}. \tag{2.13}$$

RB assignment For the RB assignment, the values of $q_u^{(n)} \in \{0, 1\}$ and $x_u^{(n)} \in \{0, 1\}$ have to be satisfied. Furthermore, the assignment of RB must be orthogonal for communication with eNB, i.e., only interference from D2D communication is possible. Mathematically, this can be expressed by the following constraint:

$$\sum_{\forall u \in \{\mathcal{C} \cup \mathcal{M}\}} q_u^{(n)} \leq 1, \ \forall n \in \mathcal{N}, \tag{2.14}$$

which guarantees that each RB is assigned to only one user. Moreover, for all MUEs,

$$x_u^{(n)} + q_u^{(n)} \leq 1, \ \forall u \in \{\mathcal{C} \cup \mathcal{M}\}, \ \forall n \in \mathcal{N}, \tag{2.15}$$

which ensures the communication is available only in one mode each time, i.e., either communicate through eNB or through a D2D link.

Transmit power There are requirements for the transmission power of both CUEs and MUEs. The transmit power must satisfy a non-negative condition and also be limited to a maximum transmit power, $P_{\text{eNB}}^{\max}$ for eNB links and $P_{\text{D2D}}^{\max}$ for D2D links, i.e.,

$$p_u^{(n)} \geq 0,$$

$$\sum_{n=1}^{N} q_u^{(n)} p_u^{(n)} \leq P_{\text{eNB}}^{\max} \tag{2.16}$$

$$\text{and} \quad \sum_{n=1}^{N} x_u^{(n)} p_u^{(n)} \leq P_{\text{D2D}}^{\max}. \tag{2.17}$$

Data rate The objective function for the resource allocation problem consists in maximizing the system throughput. Furthermore, it is necessary to guarantee a minimum data rate, $r_u^{\min}$, related to the service required by each user. Using Equation (2.9), this constraint can be described as

$$r_u \geq r_{u_c}^{\min}, \quad \forall u \in \{\mathcal{C} \cup \mathcal{M}\}. \tag{2.18}$$

2.6.3.2 Scheduler Design

The design of the scheduler for the proposed LTE-D2D system can be formulated as a nonlinear programming model; for that, the objective function and the constraints presented above are taken into account. The proposed scheduler is formulated as:

$$\max_{q_u^{(n)}, x_u^{(n)}, p_u^{(n)}} \sum_{\forall u \in \{\mathcal{C} \cup \mathcal{M}\}} r_u \tag{2.19a}$$

subject to

$$q_u^{(n)}, x_u^{(n)} \in \{0,1\}, \sum_{\forall u \in \{\mathcal{C} \cup \mathcal{M}\}} q_u^{(n)} \leq 1, \qquad \forall u \in \{\mathcal{C} \cup \mathcal{M}\}, \tag{2.19b}$$

$$q_u^{(n)} + x_u^{(n)} \leq 1 \qquad \forall u \in \mathcal{M}, \ \forall n \in \mathcal{N}, \tag{2.19c}$$

$$p_u^{(n)} \geq 0, \sum_{n=1}^{N} q_u^{(n)} p_u^{(n)} \leq P_{\text{eNB}}^{\max}, \qquad \forall u \in \{\mathcal{C} \cup \mathcal{M}\}, \tag{2.19d}$$

$$p_u^{(n)} \leq P_{\text{D2D}}^{\max}, \qquad \forall u \in \mathcal{M} \tag{2.19e}$$

$$t_{a,\text{total}} \leq \tau_a^{\max}, \qquad \forall a \in \mathcal{A} \tag{2.19f}$$

$$r_u \geq r_u^{\min}, \qquad \forall u \in \{\mathcal{C} \cup \mathcal{M}\}. \tag{2.19g}$$

The objective function (2.19a) is defined to reach a maximum throughput. Constraints (2.19b) and (2.19c) capture the requirement that each RB is assigned to only one user, and each user performs only one communication mode. Constraint (2.19d) and (2.19e) limit the transmit power of each user. Constraint (2.19f) ensures that the overall transfer time for class a should be at most $\tau_a^{\max}$, which is defined by the QoS of such user. Finally, a minimum data rate for each user is ensured by constraint (2.19g).

The resource allocation problem is described by the objective function in Equation (2.19a) and the set of constraints in Equations (2.19b)-(2.19g). Clearly, this is a mixed-integer nonlinear program (MINLP) with a nonconvex feasible set. Therefore, the resource allocation problem is NP-hard, i.e., there is no polynomial-time algorithm to solve the problem optimally. Among the alternatives to solve such problem, it is possible to apply a relaxation approach and try to turn the problem convex, or it is possible to adopt a heuristic approach, including a genetic algorithm (GA) and the particle swarm optimization (PSO), to deal with the MINLP problem. To solve the proposed optimization problem efficiently, the proposed heuristic algorithm is divided into two main blocks, as described later.

The first block carries out an evaluation of the D2D links for the MUEs. Assuming that the MUEs are fixed (i.e., no mobility), a time scale of hundreds of milliseconds can be established, with only the large scale fading (path-loss and shadowing) being considered, and when available, a D2D FD link is established between MUEs. The MUEs that may perform D2D communication are defined *a priori* using a look-up table (LuT), which are located in the eNB. After executing the resource allocation procedure in a centralized way, all information is sent to the users as an update.

Once established what users perform D2D FD communication, the scheduler acts on each TTI interval for queue enhancement, according to the QoS of each class. Every TTI, the scheduler calculates the SINR values according to the CSI that receives from the user equipment and using the information of the queue, optimizes the throughput, by assigning RBs in a sequential order from the higher-priority classes to lower-priority classes. The pseudo-code for the scheduler and resource allocation scheme is described in the following algorithm.

Algorithm 1 LTE scheduling and resource allocation scheme for D2D FD (2.19a).

```
Load D2D LUT
Load QoS table
Initiate:
for Every TTI do
    if mod(TTI,50 ms)=0 then
        procedure SELECT D2D FD LINKS(q_u^(n), x_u^(n), P_u^(n))
            Calculate SINR for all eNB links
            Calculate SINR for D2D FD links
            Decide whether D2D or not and adjust power.
        end procedure
    end if
    procedure SCHEDULING HIGH PRIORITY(q_u^(n), x_u^(n), P_u^(n))
        Calculate t_a for all classes and r_u for all users
        RB assignment/power adjust for all users
    end procedure
end for
Output: P, q and x for all C and M.
```

Some illustrative numerical simulation results are presented next in order to characterize the performance and complexity of such scheduler and to discuss the viability of IEC 61850 over LTE for real-time communication in SG applications.

2.6.3.3 Numerical Evaluation

To evaluate the performance and complexity of such scheduler, consider the power system equipment communicating MMS and GOOSE messages, as modeled previously,

Table 2.8 Main system parameters adopted in the simulations.

Parameter	value
LTE	
Bandwidth	5 MHz
Transmission Time Interval (TTI)	1 ms
Number of RBs in one TTI	50
Number of subcarriers per RB	12
Number of symbols per RB	84
Transmission mode	SISO
Channel model	suburban
LTE delay	10 ms
# of Users	
Generation plants	2
Substations	3
DER (solar and wind)	5
Transformers	5
Conventional users	9

while conventional users perform the typical LTE applications, including VoIP, Internet, video streaming, real-time gaming, and FTP, modeled according to (Orange et al., 2007).

The simulation results are obtained based on the LTE parameters presented in Table 2.8. It is assumed that the number of MUEs is constant, i.e., the distribution grid equipment is known *a priori*, in order to evaluate the D2D/FD link availability.

The capacity of the uplink on an LTE cell depends on the channel bandwidth, network loading (which impacts the overhead), the configuration of antennas, whether it is MIMO or SISO, and the modulation coding scheme (MCS), which can be QPSK, 16- or 64-QAM, but the typical users can support only 16-QAM. In terms of LTE overhead, there exist the pilot, random access, CP, and guard band overheads, which results in about 25% of the total transmitted bits. Finally, the number of transmitted bits depends on the SINR of the signal, which is a factor for defining the code rate and the modulation order. Table 2.9 shows the SINR range to CQI mapping (Jar and Fettweis, 2012) adopted in the simulation.

The power grid topology adopted for simulation consists of an infrastructure with fixed positions, such that two generation power plants, three substations, ten DERs, and also ten regulation transformers are randomly distributed. The conventional LTE network users are distributed randomly following a uniform distribution inside the cell. One example of such distribution is depicted in Figure 2.18. The number of bytes presented in Table 2.10 includes only the overhead as depicted in Figure 2.16 and Fig. 2.17. These values consider usual plants configurations and also values defined in (Das et al., 2012; Zhang et al., 2015) for IEC 61850 communication aspects inside a substation. For instance, one generation plant usually has generation, transfer, and transmission bays, and in each one there are breakers, SV, and PC IEDs, which results

Table 2.9 SINR range to CQI mapping, based on (Jar and Fettweis, 2012).

SINR range [dB]	CQI	Modulation	Code rate	Bits per RE
$-7 \rightarrow -5$	1	QPSK	0.0761	0.1523
$-5 \rightarrow -3$	2	QPSK	0.1172	0.2344
$-3 \rightarrow -1$	3	QPSK	0.1884	0.3770
$-1 \rightarrow 1$	4	QPSK	0.3008	0.6016
$1 \rightarrow 3$	5	QPSK	0.4384	0.8770
$3 \rightarrow 5$	6	QPSK	0.5879	1.1758
$5 \rightarrow 7$	7	16QAM	0.3691	1.4766
$7 \rightarrow 8.5$	8	16QAM	0.4785	1.9141
$8.5 \rightarrow 10$	9	16QAM	0.6015	2.4063
$10 \rightarrow 11.5$	10	64QAM	0.4550	2.7305
$11.5 \rightarrow 13.5$	11	64QAM	0.5537	3.3223
$13.5 \rightarrow 15$	12	64QAM	0.6504	3.9023
$15 \rightarrow 17$	13	64QAM	0.7539	4.5234
$17 \rightarrow 19.5$	14	64QAM	0.8525	5.1152
≥ 19.5	15	64QAM	0.9258	5.5547

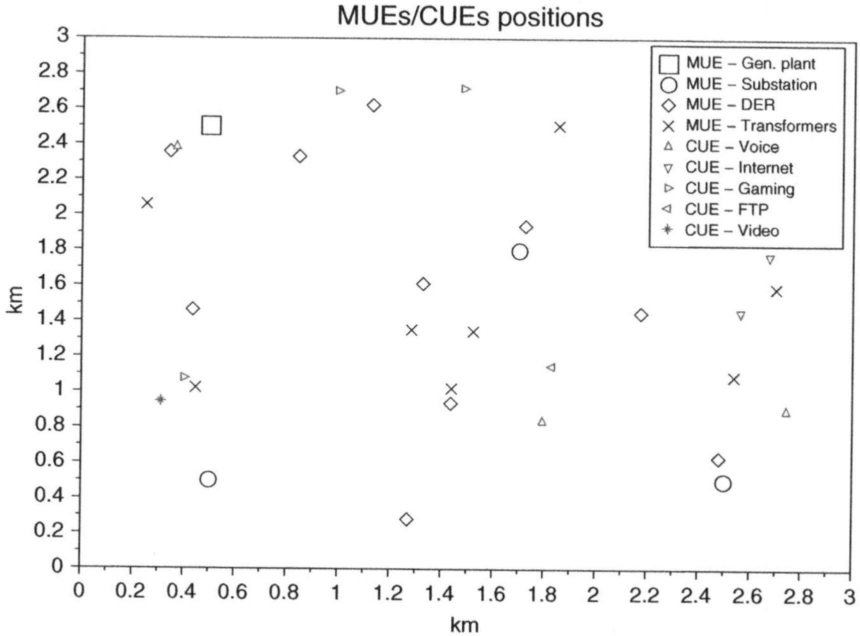

Figure 2.18 Example of MUEs and CUEs distributed in a LTE cell.

Table 2.10 Size of data generation, in [bytes], for different users.

User	Cyclic SV	Cyclic Status	Stochastic Event	Stochastic FTP	Burst
Gen. plant	4 × 219	6 × 144	10 × 204	1 × 1 M	2 × 144
Substation	6 × 219	9 × 144	20 × 204	1 × 1 M	4 × 144
DER	1 × 219	3 × 144	4 × 204	1 × 1 M	1 × 144
Transformer	2 × 219	2 × 144	2 × 204	1 × 10 k	1 × 144

in the multiplying factors in Table 2.10. Burst transmission occurs generated by a previous event, and it happens about 100 ms after the trip signal, being modeled as GOOSE messages for all MUEs.

In such a setup, each plant or equipment has one redundant IED that transmits the signal to outside equipment. Hence, an additional delay $d_i = 10$ ms is included in the total delay equation. Considering the application for wide-area protection and control, as presented in Table 2.3, the latency requirements are 100 milliseconds and 5 seconds, depending on the application. The sampling rate of status and SV values are set to 100 ms, and the background traffic is modeled according to (Orange et al., 2007). The number of available RBs for a 5 MHz bandwidth is 50 for every TTI, and each RB is composed by 12 subcarriers and 7 symbols, which results in a total of 84 symbols in each RB. With such parameters, one can see in Figure 2.19 the cdf of the delay for MMS and GOOSE/SV messages in the proposed scheduling scheme. It is worth mentioning

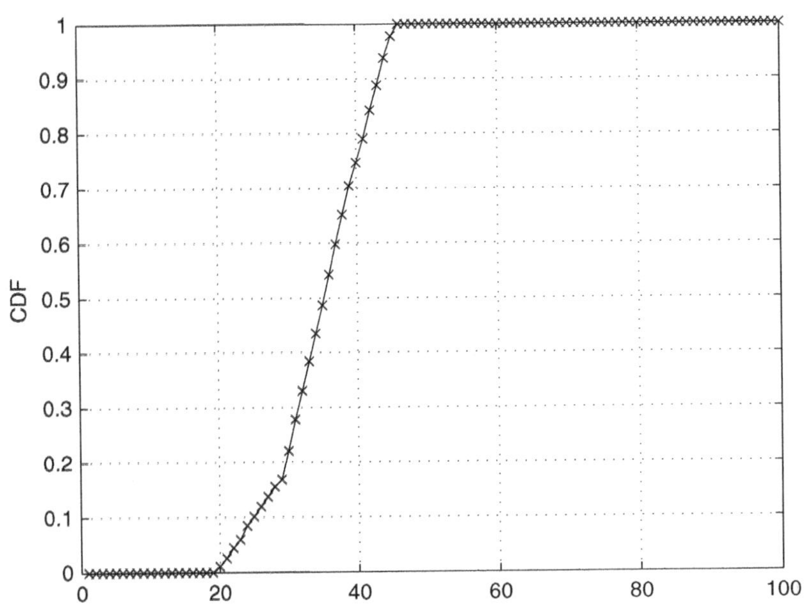

Figure 2.19 Cdf of the delay for IEC 61850 messages.

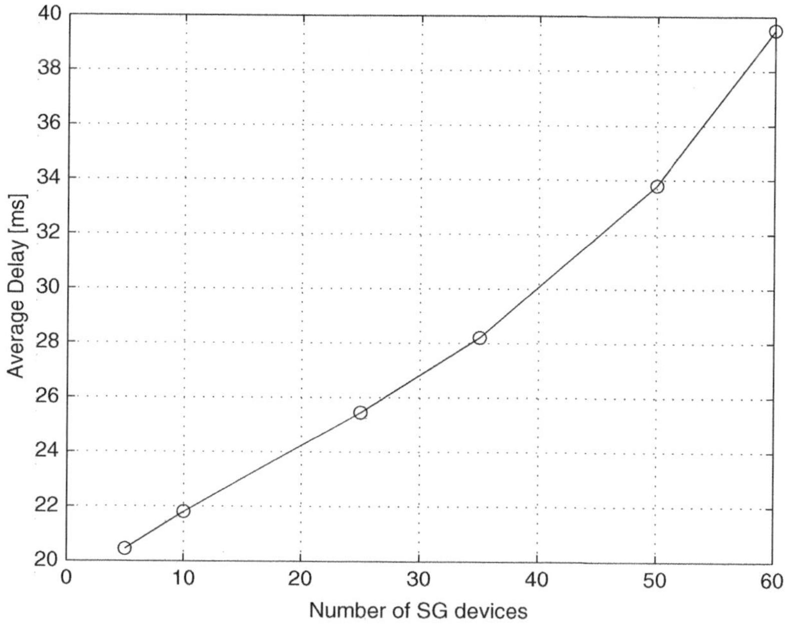

Figure 2.20 Average latency of the conventional users as a function of the number of SG devices.

that this delay is achieved in detriment of the conventional user traffic, which has a higher delay.

When increasing the number of MUEs, the latency of CUEs becomes critical. Figure 2.20 depicts the behavior of average voice/Internet users latency as a function of the number of MUEs in the network. Since the MUEs has higher priority than the CUE ones, the higher the number of devices, the more deteriorated is the latency of conventional users, which constitutes a limitation of such scheduling scheme. In this figure, the increasing number of devices corresponds mainly to the DER and transformers, which are the trend of the SG, and the number of substations and electric power plants is set to a constant value. For a number of 60 MUE devices, the scheduling mechanism does not converge to a stable solution, since the number of IEC 61850 messages becomes a limiting factor to the entire LTE network.

The importance of SV delay constraints is presented in (IEC, 2015) and highlighted in (Zhang et al., 2015) for inside substation network, and it also extends to outside communication and affects the viability of power distribution grid over the LTE infrastructure. For instance, consider a generation power plant that transmits 4 SV messages of 219 bytes, where each message contains a set of 4 current and voltage measurements. Depending on the SINR, MCS, and configuration, a 5 MHz LTE network supports up to 80 Mbps, but usual values are about 20 Mbps. Assuming a sampling rate for SV of $s_{SV} = 50$ ms, one generation plant would have a data rate of

$$r_{SV} = 4 \text{ messages} \cdot 219 \frac{\text{bytes}}{\text{message}} \cdot 8 \frac{\text{bits}}{\text{byte}} \cdot \frac{1}{0.05 \text{ s}} \approx 140\text{Kbps}.$$

When considering also the status and measurements messages, the demanded data rate for cyclic information of a generation plant is about 280 Kbps. When a high number of users are communicating, such demand can cause overload in the system, especially when a trip event occurs, resulting in burst messages. Especially in such time, the scheduler shifts users to D2D/FD mode so as to diminish the delay of transmitting IEC 61850 messages. Note that the power grid configuration adopted for the simulations requires about 3 Mbps of the LTE network resources only for cyclic messages. The MUEs located in the edge cell are susceptible to increasing delay, as detailed in (Louvros and Paraskevas, 2014), and in this particular case, it is desirable that the scheduler shifts communication to D2D/FD, since it improves the overall data rate of such users.

2.7 Conclusion

Due to the importance of the SG in the context of the smart cities, this chapter gave a survey on the automation aspects of a power distribution grid. The discussion involved improving throughput and diminishing transmission delay of the LTE networks, aiming to support time-critical communication for DA. An FD LTE-D2D scheduler was proposed to improve the capacity of LTE networks to supply such demand of communications in smart cities.

First, different wired and wireless communication technologies that are candidates to be applied in the SG were presented. Then, these technologies were linked with automation of power distribution grids, with the types of network (HAN, FAN, and WAN), and services considered inside each one. Among the several technologies presented, it was highlighted the use of LTE networks, because they are reliable and achieve good performance regarding data rate and latency.

Then, the open protocols commonly used in distribution networks, such as Modbus, IEC 60870, DNP3.0 and IEC 61850, were characterized. Their key features were presented and discussed concerning their application to the SG. The IEC 61850 standard, under wireless technology, which allows the encapsulation of its protocols and minimum latency requirements, was presented as a promising candidate to extend the supervision, control, and protection from inside to outside substations, allowing integration of DER in the power distribution grid. In the end, a survey of the application of SCADA systems to the SG was discussed, where it is easy to realize the potential of the IEC 61850 standard.

Among the distribution automation demands, the chapter highlighted the VVC and FDCIR. The requirements of these applications and other ones were presented regarding the network range, the size of the messages, data sampling rate, latency, and reliability. These requirements were defined such that the response of the control system increases the power system availability and keeps it stable.

The IEC 61850 data flow in the distribution network was discussed, and particularly for MMS and GOOSE messages, and their characterization takes place both internally and externally to the substation. The modeling of DA messages considers deterministic, stochastic, and burst random variables generation. The types of messages have been classified as fast, medium speed, low speed, raw data, file transfer, time synchronization, and command.

Finally, a resource allocation scheme for this DA time-critical application was proposed. It is considered an extension of the conventional LTE network, with the possibility of FD and D2D communication for cases of high data rate, especially when burst messages are caused when protection events trigger action in the entire grid. The protocols of the standard IEC 61850 are mapped in the LTE communication, which allows their transmission and integration with an existing LTE network. About the SV messages, which are deterministic with high data rate, they become a limiting factor to increasing the number of DA devices in the SG network.

There are still many many potential future research directions, issues, and opportunities involving the integration of the IEC 61850 standard for communication outside substations, and particularly for performance integration of DERs. In the LTE context, additional studies are needed about the impact of D2D communication for edge users, the potential of FD, densification, and the use of Heterogeneous Networks (HetNet), among others.

References

3GPP. 3rd generation partnership project; technical specification group services and system aspects; policy and charging control architecture (release 13). Technical report, TS 23.203 V13.7.0, Mar. 2016.

F. Aalamifar and L. Lampe. Optimized WiMAX profile configuration for smart grid communications. *IEEE Transactions on Smart Grid*, PP(99):1–10, 2016. ISSN 1949-3053. doi: 10.1109/TSG.2016.2536145.

A. Aijaz, M. Tshangini, M. R. Nakhai, X. Chu, and A. H. Aghvami. Energy-efficient uplink resource allocation in LTE networks with M2M/H2H co-existence under statistical QoS guarantees. *IEEE Transactions on Communications*, 62 (7):2353–2365, Jul. 2014. ISSN 0090-6778. doi: 10.1109/TCOMM.2014.2328338.

Ali M. Allam, Mohamed Saied, and Ihab Aly. Adapting LTE M-2-M protocol standard to fulfil latency requirements for IEC 61850 GOOSE messages. *Int. J. Communications, Network and System Sciences*, 8 (5):154–160, 2015. ISSN 1913-3715. doi: 10.4236/ijcns.2015.85017.

A. Angioni, A. Sadu, F. Ponci, A. Monti, D. Patel, F. Williams, D. della Giustina, and A. Dedé. Coordinated voltage control in distribution grids with LTE based communication infrastructure. In *Environment and Electrical Engineering (EEEIC), 2015 IEEE 15th International Conference on*, pages 2090–2095, Jun. 2015. doi: 10.1109/EEEIC.2015.7165498.

A. Bani-Ahmed, L. Weber, A. Nasiri, and H. Hosseini. Microgrid communications: State of the art and future trends. In *International Conference on Renewable Energy Research and Application (ICRERA)*, pages 780–785, Oct. 2014. doi: 10.1109/ICRERA.2014.7016491.

G. Cena, L. Seno, A. Valenzano, and C. Zunino. On the performance of IEEE 802.11e wireless infrastructures for soft-real-time industrial applications. *IEEE Transactions on Industrial Informatics*, 6 (3):425–437, Aug. 2010. ISSN 1551-3203. doi: 10.1109/TII.2010.2052058.

Cheng-Shang Chang. Stability, queue length, and delay of deterministic and stochastic queueing networks. *IEEE Transactions on Automatic Control*, 39 (5):913–931, May 1994. ISSN 0018-9286. doi: 10.1109/9.284868.

D. Chang, J. Lee, and T. H. Lin. Smart satellites in smart grids. In *IEEE International Conference on Smart Grid Communications (SmartGridComm)*, pages 487–492, Nov. 2014. doi: 10.1109/SmartGridComm.2014.7007694.

Peng Cheng, Li Wang, Bin Zhen, and Shihua Wang. Feasibility study of applying LTE to smart grid. In *IEEE 1st International Workshop on Smart Grid Modeling and Simulation (SGMS)*, pages 108–113, Oct. 2011. doi: 10.1109/SGMS.2011.6089018.

Gordon Clarke, Deon Reynders, and Edwin Wright. *Practical Modern SCADA Protocols*. Newnes, Oxford, 1st edition, 2004. ISBN 978-0-7506-5799-0. doi: http://dx.doi.org/10.1016/B978-075065799-0/50015-2.

Erik Dahlman, Stefan Parkvall, Johan Sköld, and Per Beming. *3G Evolution: HSPA and LTE for Mobile Broadband*. Elsevier, 2007.

N. Das, W. Ma, and S. Islam. Comparison study of various factors affecting end-to-end delay in IEC 61850 substation communications using OPNET. In *22nd Australasian Universities Power Engineering Conference (AUPEC)*, pages 1–5, Sep. 2012.

G. de Andrade Barreto and J. A. Baesso Grimoni. Underground low voltage smarter grid at CEB. In *IEEE PES Conference on Innovative Smart Grid Technologies Latin America (ISGT LA)*, pages 1–5, Apr. 2013. doi: 10.1109/ISGT-LA.2013.6554433.

Mark Deakin and Alasdair Reid. Smart cities: Under-gridding the sustainability of city-districts as energy efficient-low carbon zones. *Journal of Cleaner Production*, 2016. ISSN 0959-6526. doi: http://doi.org/10.1016/j.jclepro.2016.12.054. URL http://www.sciencedirect.com/science/article/pii/S0959652616321096.

V. Dehalwar, A. Kalam, M. L. Kolhe, and A. Zayegh. Review of IEEE 802.22 and IEC 61850 for real-time communication in smart grid. In *International Conference on Computing and Network Communications (CoCoNet)*, pages 571–575, Dec. 2015. doi: 10.1109/CoCoNet.2015.7411245.

DNP3. IEEE standard for electric power systems communications - distributed network protocol (DNP3). *IEEE Std 1815-2012 (Revision of IEEE Std 1815-2010)*, pages 1–821, Oct. 2012. doi: 10.1109/IEEESTD.2012.6327578.

M. Erol-Kantarci and H. T. Mouftah. Energy-efficient information and communication infrastructures in the smart grid: A survey on interactions and open issues. *IEEE Communications Surveys Tutorials*, 17 (1):179–197, Firstquarter 2015. ISSN 1553-877X. doi: 10.1109/COMST.2014.2341600.

X. Fang, S. Misra, G. Xue, and D. Yang. Smart grid - the new and improved power grid: A survey. *IEEE Communications Surveys Tutorials*, 14 (4):944–980, Apr. 2012. ISSN 1553-877X. doi: 10.1109/SURV.2011.101911.00087.

H. Farhangi. The path of the smart grid. *IEEE Power and Energy Magazine*, 8 (1):18–28, Jan. 2010. ISSN 1540-7977. doi: 10.1109/MPE.2009.934876.

G. Fodor, E. Dahlman, G. Mildh, S. Parkvall, N. Reider, G. Miklos, and Z. Turanyi. Design aspects of network assisted device-to-device communications. *IEEE Communications Magazine*, 50 (3):170–177, Mar. 2012. ISSN 0163-6804. doi: 10.1109/MCOM.2012.6163598.

S. Galli, A. Scaglione, and Z. Wang. For the grid and through the grid: The role of power line communications in the smart grid. *Proceedings of the IEEE*, 99 (6):998–1027, Jun. 2011. ISSN 0018-9219. doi: 10.1109/JPROC.2011.2109670.

P. Gandhi and D. Adhyaru. Critical data reliability issues with Modbus protocol in smart grid application. In *International Conference on Electrical, Electronics, Signals, Communication and Optimization (EESCO)*, pages 1–5, Jan. 2015. doi: 10.1109/EESCO.2015.7253788.

Z. Gerhard, K. Susanne, V. Klaus, W. Thomas, and K. Markus. Acquiring energy data from a medium-voltage grid for future smart grid solutions: A practical smart grid application example realized by use of cellular communication networks of the 2nd and 3rd generation. In *11th Workshop on Intelligent Solutions in Embedded Systems (WISES)*, pages 1–8, Sep. 2013.

M. Golshani, G. A. Taylor, I. Pisica, and P. Ashton. Investigation of open standards to enable interoperable wide area monitoring for transmission systems. In *47th International Universities Power Engineering Conference (UPEC)*, pages 1–6, Sep. 2012. doi: 10.1109/UPEC.2012.6398434.

V. C. Gungor, D. Sahin, T. Kocak, S. Ergut, C. Buccella, C. Cecati, and G. P. Hancke. Smart grid technologies: Communication technologies and standards. *IEEE Transactions on Industrial Informatics*, 7 (4):529–539, Nov. 2011. ISSN 1551-3203. doi: 10.1109/TII.2011.2166794.

V. C. Gungor, D. Sahin, T. Kocak, S. Ergut, C. Buccella, C. Cecati, and G. P. Hancke. A survey on smart grid potential applications and communication requirements. *IEEE Transactions on Industrial Informatics*, 9 (1):28–42, Feb. 2013. ISSN 1551-3203. doi: 10.1109/TII.2012.2218253.

Guozheng Han, Bingyin Xu, Kaijun Fan, and Guangxian Lv. An open communication architecture for distribution automation based on IEC 61850. *International Journal of Electrical Power and Energy Systems*, 54:315–324, 2014. ISSN 0142-0615. doi: http://dx.doi.org/10.1016/j.ijepes.2013.07.013. URL http://www.sciencedirect.com/science/article/pii/S0142061513003104.

M. Hasan and E. Hossain. Distributed resource allocation for relay-aided device-to-device communication under channel uncertainties: A stable matching approach. *IEEE Transactions on Communications*, 63 (10): 3882–3897, Oct. 2015. ISSN 0090-6778. doi: 10.1109/TCOMM.2015.2466238.

Q. D. Ho, Y. Gao, and T. Le-Ngoc. Challenges and research opportunities in wireless communication networks for smart grid. *IEEE Wireless Communications*, 20 (3):89–95, Jun. 2013. ISSN 1536-1284. doi: 10.1109/MWC.2013.6549287.

E. Hossain and M. Hasan. 5G cellular: key enabling technologies and research challenges. *IEEE Instrumentation Measurement Magazine*, 18(3):11–21, Jun. 2015. ISSN 1094-6969. doi: 10.1109/MIM.2015.7108393.

IEC. IEC 61850 - communication networks and systems for power utility automation - part 8.1: Specific communication service mapping (SCSM) - mappings to MMS (ISO 9506-1 and ISO 9506-2) and to ISO/IEC 8802-3. Technical report, IEC Technical Committee Number 57 (TC57) Working Group 10 (WG10), 2004.

IEC. IEC 61850 - communication networks and systems for power utility automation - part 9.5: Use of IEC 61850 to transmit synchrophasor information according to IEEE C37.118. Technical report, IEC Technical Committee Number 57 (TC57) Working Group 10 (WG10), 2012.

IEC. IEC 61850 - communication networks and systems for power utility automation. Technical report, IEC Technical Committee Number 57 (TC57) Working Group 10 (WG10), 2015.

IEEE. IEEE standard communication delivery time performance requirements for electric power substation automation. *IEEE Std 1646-2004*, pages 1–24, 2005. doi: 10.1109/IEEESTD.2005.95748.

S. Jaloudi, E. Ortjohann, A. Schmelter, P. Wirasanti, and D. Morton. Communication strategy for grid control and monitoring of distributed generators in smart grids using IEC and IEEE standards. In *2nd IEEE/PES International Conference and Exhibition on*

Innovative Smart Grid Technologies (ISGT Europe), pages 1–6, Dec. 2011. doi: 10.1109/ISGTEurope.2011.6162689.

P. Jänis, C. Yu, K. Doppler, C. Ribeiro, C. Wijting, K Hugl, O. Tirkkonen, and V. Koivunen. Device-to-device communications underlaying cellular communications systems. *International Journal of Communications, Network and System Sciences*, 2 (3):169–247, Jun. 2009.

M. Jar and G. Fettweis. Throughput maximization for LTE uplink via resource allocation. In *International Symposium on Wireless Communication Systems (ISWCS)*, pages 146–150, Aug. 2012. doi: 10.1109/ISWCS.2012.6328347.

C. Kalalas. Enabling LTE for control system applications in a smart grid context. Master's thesis, KTH Royal Institute of Technology, Aug. 2014.

C. Kalalas, L. Gkatzikis, C. Fischione, P. Ljungberg, and J. Zarate. Enabling IEC 61850 communication services over public LTE infrastructure. In *IEEE International Conference on Communications (ICC)*, pages 1–6, May 2016a. doi: 10.1109/ICC.2016.7510937.

C. Kalalas, L. Thrybom, and J. Zarate. Cellular communications for smart grid neighborhood area networks: A survey. *IEEE Access*, 4:1469–1493, 2016b. ISSN 2169-3536. doi: 10.1109/ACCESS.2016.2551978.

Georgios Karagiannis, Giang T. Pham, A. Dung Nguyen, Geert J. Heijenk, Boudewijn R. Haverkort, and Frans Campfens. Performance of LTE for smart grid communications. In Kai Fischbach and Udo R. Krieger, editors, *Measurement, Modelling, and Evaluation of Computing Systems and Dependability and Fault Tolerance: 17th International GI/ITG Conference, MMB & DFT 2014, Bamberg, Germany, March 17-19, 2014. Proceedings*, pages 225–239. Springer International Publishing, Cham, 2014. ISBN 978-3-319-05359-2. doi: 10.1007/978-3-319-05359-2_16.

S. Kenner, R. Thaler, M. Kucera, K. Volbert, and T. Waas. Smart grid architecture for monitoring and analyzing, including Modbus and REST performance comparison. In *12th International Workshop on Intelligent Solutions in Embedded Systems (WISES)*, pages 91–96, Oct. 2015.

Reduan H. Khan and Jamil Y. Khan. A comprehensive review of the application characteristics and traffic requirements of a smart grid communications network. *Computer Networks*, 57 (3):825–845, Feb. 2013. ISSN 1389-1286. doi: http://dx.doi.org/10.1016/j.comnet.2012.11.002. URL http://www.sciencedirect.com/science/article/pii/S1389128612003751.

Murat Kuzlu, Manisa Pipattanasomporn, and Saifur Rahman. Communication network requirements for major smart grid applications in HAN, NAN and WAN. *Computer Networks*, 67:74–88, Jul. 2014. ISSN 1389-1286. doi: http://dx.doi.org/10.1016/j.comnet.2014.03.029. URL http://www.sciencedirect.com/science/article/pii/S1389128614001431.

S. W. Lai and G. G. Messier. Using the wireless and PLC channels for diversity. *IEEE Transactions on Communications*, 60 (12):3865–3875, Dec. 2012. ISSN 0090-6778. doi: 10.1109/TCOMM.2012.081512.110408.

C. H. Lo and N. Ansari. The progressive smart grid system from both power and communications aspects. *IEEE Communications Surveys Tutorials*, 14 (3):799–821, Third Quarter 2012. ISSN 1553-877X. doi: 10.1109/SURV.2011.072811.00089.

Spiros Louvros and Michael Paraskevas. Analytical average throughput and delay estimations for LTE uplink cell edge users. *Computers and Electrical Engineering*,

40 (5):1552–1563, 2014. ISSN 0045-7906. doi: http://dx.doi.org/10.1016/j.compeleceng.2014.03.008. URL http://www.sciencedirect.com/science/article/pii/S0045790614000627.

X. Lu, Z. Lu, W. Wang, and J. Ma. On network performance evaluation toward the smart grid: A case study of DNP3 over TCP/IP. In *IEEE Global Telecommunications Conference (GLOBECOM)*, pages 1–6, Dec. 2011. doi: 10.1109/GLOCOM.2011.6134406.

X. Lu, W. Wang, and J. Ma. An empirical study of communication infrastructures towards the smart grid: Design, implementation, and evaluation. *IEEE Transactions on Smart Grid*, 4 (1):170–183, Mar. 2013. ISSN 1949-3053. doi: 10.1109/TSG.2012.2225453.

R. E. Mackiewicz. Overview of IEC 61850 and benefits. In *IEEE/PES Power Systems Conference and Exposition (PSCE)*, pages 623–630, Oct. 2006. doi: 10.1109/PSCE.2006.296392.

M. Manbachi, A. Sadu, H. Farhangi, A. Monti, A. Palizban, F. Ponci, and S. Arzanpour. Real-time communication platform for smart grid adaptive Volt-VAR optimization of distribution networks. In *IEEE International Conference on Smart Energy Grid Engineering (SEGE)*, pages 1–7, Aug. 2015. doi: 10.1109/SEGE.2015.7324592.

V. Medina, I. Gomez, D. Oviedo, E. Dorronzoro, S. Martin, J. Benjumea, and G. Sanchez. IEC-60870-5 application layer over TCP/IP for an open and flexible remote unit. In *IEEE International Symposium on Industrial Electronics*, pages 420–425, Jul. 2009. doi: 10.1109/ISIE.2009.5217929.

W. Meng, R. Ma, and H. H. Chen. Smart grid neighborhood area networks: a survey. *IEEE Network*, 28 (1):24–32, Jan. 2014. ISSN 0890-8044. doi: 10.1109/MNET.2014.6724103.

METIS. Deliverable D1.1 - scenarios, requirements and KPIs for 5G mobile and wireless system. Technical report, Mobile and wireless communications Enablers for the Twenty-twenty Information Society (METIS), Apr. 2013.

A. Modbus-ID. Modbus application protocol 1.1b. http://www.modbus.org/docs/Modbus_Application_Protocol_V1_1b.pdf, 2012.

S. Mohagheghi, J. Stoupis, and Z. Wang. Communication protocols and networks for power systems - current status and future trends. In *IEEE/PES Power Systems Conference and Exposition (PSCE)*, pages 1–9, Mar. 2009. doi: 10.1109/PSCE.2009.4840174.

S. Mohagheghi, J. C. Tournier, J. Stoupis, L. Guise, T. Coste, C. A. Andersen, and J. Dall. Applications of IEC 61850 in distribution automation. In *IEEE/PES Power Systems Conference and Exposition*, pages 1–9, Mar. 2011. doi: 10.1109/PSCE.2011.5772491.

A. Naumann, I. Bielchev, N. Voropai, and Z. Styczynski. Smart grid automation using IEC 61850 and CIM standards. *Control Engineering Practice*, 25:102–111, 2014. ISSN 0967-0661. doi: http://dx.doi.org/10.1016/j.conengprac.2013.12.001. URL http://www.sciencedirect.com/science/article/pii/S0967066113002335.

U.S. Department of Energy. Communications requirements of smart grid technologies. Technical report, U.S. Department of Energy, 2010.

Orange, China Mobile, KPN, NTT DoCoMo, Sprint, T-Mobile, Vodafone, and Telecom Italia. R1-070674: LTE physical layer framework for performance verification. Technical report, 3GPP TSG RAN WG1, meeting 48, Feb. 2007.

A. Ortega, A. A. Shinoda, and C. M. Schweitzer. Performance analysis of smart grid communication protocol DNP3 over TCP/IP in a heterogeneous traffic environment. In *IEEE Colombian Conference on Communications and Computing (COLCOM)*, pages 1–6, May 2013. doi: 10.1109/ColComCon.2013.6564828.

E. Padilla, K. Agbossou, and A. Cardenas. Towards smart integration of distributed energy resources using distributed network protocol over ethernet. *IEEE Transactions on Smart Grid*, 5 (4):1686–1695, Jul. 2014. ISSN 1949-3053. doi: 10.1109/TSG.2014.2303857.

P. P. Parikh, T. S. Sidhu, and A. Shami. A comprehensive investigation of wireless LAN for IEC 61850 - based smart distribution substation applications. *IEEE Transactions on Industrial Informatics*, 9 (3):1466–1476, Aug. 2013. ISSN 1551-3203. doi: 10.1109/TII.2012.2223225.

D. Patel, M. I. N. Mohamed, S. Z. R. Mehdi, F. Williams, A. Sadu, F. Ponci, and A. Monti. Investigating the performance of QoS enabled LTE networks for IEC 61850 based smart grid applications. In *IEEE International Energy Conference (ENERGYCON)*, pages 1–6, Apr. 2016. doi: 10.1109/ENERGYCON.2016.7513965.

Giang T. Pham. Integration of IEC 61850 MMS and LTE to support smart metering communications. Master's thesis, University of Twente - Faculty of Electrical Engineering, Mathematics and Computer Science (EEMCS), Aug. 2013.

O. Richardot, A. Viciu, Y. Besanger, N. Hadjsaid, and C. Kieny. Coordinated voltage control in distribution networks using distributed generation. In *IEEE/PES Transmission and Distribution Conference and Exhibition*, pages 1196–1201, May 2006. doi: 10.1109/TDC.2006.1668675.

Mohd Khairun Nizam Mohd Sarmina, Worawat Nakawiro, MohdZamriChe Wanik, Mohd FadzilMohd Siam, ZahrulFaizi Hussien, Ahmad Asrul Ibrahim, and Ahmad Kamil Mat Hussin. Coordinated voltage control in distribution network with renewable energy based distributed generation. *Science Research Engineering*, 5 (1B):208–214, Jan. 2013. ISSN 1947-3931. doi: 10.4236/eng.2013.51B038.

Peter Schlegl, Patrick Robatzek, Markus Kucera, Klaus Volbert, and Thomas Waas. Performance analysis of mobile radio for automatic control in smart grids. In *2nd International Conference on Advances in Computing, Communication and Information Technology (CCIT)*, pages 135–141, Nov. 2014. doi: 10.15224/978-1-63248-051-4-89.

Stefania Sesia, Issam Toufik, and Matthew Baker. *LTE - The UMTS Long Term Evolution: From Theory to Practice, 2nd edition.* John Wiley & Sons, 2011. ISBN 978-0-470-66025-6.

Jan Tore Sorensen and Martin Gilje Jaatun. An analysis of the manufacturing messaging specification protocol. In *5th International Conference on Ubiquitous Intelligence and Computing*, Jun. 2008. doi: 10.1007/978-3-540-69293-5_47.

W. Sun, D. Yuan, E. G. Ström, and F. Brännström. Cluster-based radio resource management for D2D-supported safety-critical V2X communications. *IEEE Transactions on Wireless Communications*, 15 (4):2756–2769, Apr. 2016. ISSN 1536-1276. doi: 10.1109/TWC.2015.2509978.

A. Temiz, Ö Kahraman, A. Nadar, M. S. Smiai, S. S. Almutairi, and S. Alshahrani. A smart grid controller for low voltage PV network. In *IEEE 5th International Conference on Power Engineering, Energy and Electrical Drives (POWERENG)*, pages 263–267, May 2015. doi: 10.1109/PowerEng.2015.7266330.

V. Terzija, G. Valverde, D. Cai, P. Regulski, V. Madani, J. Fitch, S. Skok, M. M. Begovic, and A. Phadke. Wide-area monitoring, protection, and control of future electric power networks. *Proceedings of the IEEE*, 99 (1):80–93, Jan. 2011. ISSN 0018-9219. doi: 10.1109/JPROC.2010.2060450.

Henk C. Tijms. *A First Course in Stochastic Models.* Wiley, 2003. ISBN 0-471-49880-7.

V. Vyatkin, G. Zhabelova, N. Higgins, K. Schwarz, and N. K. C. Nair. Towards intelligent smart grid devices with IEC 61850 Interoperability and IEC 61499 open control architecture. In *IEEE/PES Transmission and Distribution Conference and Exposition*, pages 1–8, Apr. 2010. doi: 10.1109/TDC.2010.5484272.

Mingkui Wei and W. Wang. Toward distributed intelligent: A case study of peer to peer communication in smart grid. In *IEEE Global Communications Conference (GLOBECOM)*, pages 2210–2216, Dec. 2013. doi: 10.1109/GLOCOM.2013.6831403.

S. Yang, B. Yoo, H. S. Yang, Y. Ahn, M. Hyun, and H. S. Jang. Feasibility study of various communication networks for smart grid. In *International Conference on Advanced Power System Automation and Protection (APAP)*, volume 2, pages 993–997, Oct. 2011. doi: 10.1109/APAP.2011.6180696.

G. Zhabelova and V. Vyatkin. Multiagent smart grid automation architecture based on IEC 61850/61499 intelligent logical nodes. *IEEE Transactions on Industrial Electronics*, 59 (5):2351–2362, May 2012. ISSN 0278-0046. doi: 10.1109/TIE.2011.2167891.

Z. Zhang, X. Huang, B. Keune, Y. Cao, and Y. Li. Modeling and simulation of data flow for VLAN-based communication in substations. *IEEE Systems Journal*, PP(99):1–12, 2015. ISSN 1932-8184. doi: 10.1109/JSYST.2015.2428058.

3

5G and Cellular Networks in the Smart Grid

Jimmy Jessen Nielsen[1]*, Ljupco Jorguseski*[2]*, Haibin Zhang*[2]*, Hervé Ganem*[3]*,*
Ziming Zhu[4] *and Petar Popovski*[1]

[1]*Aalborg University, Aalborg, Denmark*
[2]*TNO, The Hague, The Netherlands*
[3]*Gemalto, Paris, France*
[4]*Toshiba Research Europe ltd., Bristol, United Kingdom*

3.1 Introduction

The increasing penetration of Distributed Energy Resources (DERs) such as electric vehicles, photovoltaics, and wind turbines in the distribution grid introduces bidirectional power flows and makes the consumption patterns of customers less predictable than it has been previously. As a result, it becomes necessary for the Distribution System Operator (DSO) to achieve observability below substation level, to make stable operation of the grid possible. This observability can be achieved through real-time monitoring of the distribution grid facilitated by communication technologies, thereby making the power grid into a smart grid. A real-time monitoring platform using the cellular network is introduced in Nielsen (2017). The platform relies on the deployment of more advanced phasor measurement devices in the mid/low voltage grid, as well as a secure data exchange through public communication network. The type of traffic that does not originate from a human request, such as in the smart grid, is commonly referred to as machine-type communication (MTC) or machine-to-machine (M2M) communications. A reliable M2M communication is important in order to ensure the integration of machine-type data traffic with the existing communication systems. The reader is referred to Erol-Kantarci (2013) for a comprehensive study of the M2M technologies and standards.

Due to their ability to provide wireless network coverage to large geographical areas, cellular networks are promising candidates to enable MTC in the smart grid. The development of previous generations of cellular networks, from GPRS (2.5G) through LTE (4G), has been mostly concerned with increasing the data rates to support human-oriented services, such as multimedia content on smart phones and mobile devices. The upcoming 5G networks are focused both on higher data rates and efficient support of machine-type communication (MTC), thereby becoming relevant for various smart city applications. Additionally, a number of Internet of Things (IoT) communication protocols are emerging that are expected to be able to support a subset of MTC applications. Typically, two vastly different types of MTC are considered, namely

Transportation and Power Grid in Smart Cities: Communication Networks and Services, First Edition.
Edited by Hussein T. Mouftah, Melike Erol-Kantarci and Mubashir Husain Rehmani.

massive MTC and **mission-critical MTC**, where the latter requires ultra-reliable and often ultra-low-latency connections.

3.1.1 Massive MTC

The massive MTC covers the types of applications that require exchanges with a very large number of devices. Real-time, reliability, and data rate requirements are typically not very high, which means that the challenge in supporting such applications lies in the ability of the network to efficiently manage the access of a vast number of devices. This is poorly addressed by the traditional cellular technologies, since the establishment and maintenance of individual devices' connections has a large overhead compared to the typical small payloads of many MTC applications. Consequently, the resources required for control traffic will for some applications and scenarios grow significantly, thereby turning the control channels into bottlenecks. This is contrary to the typical case in current cellular networks, where usually the bandwidth allocated for data is the limiting factor. Within the smart grid, this type of applications covers, for example, power quality and consumption reporting, where a large number of smart meters transmit status messages periodically in order to facilitate optimal power grid operation.

In the current and upcoming 3GPP specifications, there are several initiatives aiming to extend LTE toward a better support of massive MTC applications, for example through the definition of device categories with lower capability and lighter access modes, where the amount of control signaling is reduced. Besides the 3GPP cellular technologies, there are several other emerging network technologies that are directly targeting the support of a large number of low capability devices. For example, from the IEEE 802.11 family of protocols comes the 802.11ah standard, which operates in sub-GHz ISM bands and supports a large number of devices. Operating in the same sub-GHz ISM bands are the IoT protocols Sigfox and LoRa, which are especially targeting low-throughput networks.

3.1.2 Mission-Critical MTC

A key example of mission-critical MTC in the context of a smart city is related to the smart power grid. One of the strictest requirements is posed by wirelessly connected grid protection relays. Here it is required that the trigger signal arrives at the protective relay within milliseconds and with very high probability, in order to protect the grid in case of, e.g., damage to overhead power lines. That is, the required levels of reliability for mission-critical MTC applications may easily exceed "five nines" (99.999%), which is far beyond what current cellular networks typically deliver. Considering 5G in a narrow sense, i.e., as the next generation of cellular networks after 4G, it will certainly feature a new radio interface specification, not present in 4G, optimized in terms of reliability and low latency (Boccardi, 2014). However, for mission-critical applications with strict latency and reliability requirements, it will be necessary to take advantage of the added redundancy that multiple communication technologies can offer when used simultaneously (Nielsen, 2016). An example of a less strict application is that of real-time monitoring and control, where detailed power measurements are collected periodically from strategic locations in the grid to provide a complete system view. Depending on the type of power network, the update interval is in the order of a few ms up to 1 or more seconds. In this chapter, we will study specifically this use case.

3.1.3 Secure Mission-Critical MTC

Another aspect of mission-critical MTC is that of secure communication. While, from a communications perspective, the reliability is typically defined as the ability to deliver a data packet within a specified deadline, it is implicitly expected that the data packet has not been tampered with along the way. A security architecture is therefore a big part of the mission-critical MTC, and it is often made of onion-like layers, where a security layer is protecting a specific type of access to communication. This layered model results from the distributed responsibility for credential management. In this chapter, we will describe the different types of layer and associated protection, and pitch on the responsibility and liability associated to credential management. For mission-critical MTC, it is necessary to ensure end-to-end security. We will discuss the implications and potential challenges of implementing end-to-end security from the device end point to the application in the cloud. A key aspect of a security architecture is access control and credentials management, and it is challenging to design a solution that scales well, especially for large systems, such as the smart city. We will address these challenges and end up with the notion of role-based access control.

The security architecture will be considered for the specific case of 3GPP communications and the possible role of a Mobile Network Operator (MNO) to help secure power grid applications. On the other hand, the use of non-3GPP communications and non-IP based communication protocols raise new challenges to secure communication. We will provide an overview of the problems and the possible solutions.

The contents of this chapter are organized as follows. In section 3.2 we elaborate on the motivation for the smart grid and its features. Thereafter, in section 3.3 a traffic model for the smart grid is presented. In the sections 3.4 and 3.5 we describe the non-3GPP cellular technologies and current and upcoming 3GPP cellular technologies, respectively, that are relevant for supporting smart grid communication. Additionally, we present the challenges and possible solutions for ensuring end-to-end security in smart grid systems in section 3.6. Finally, the chapter is summarized in section 3.7.

3.2 From Power Grid to Smart Grid

The existing power generation and delivery system is based on many-decades-old vertical broadcast infrastructure (i.e., few-to-many distribution), where a few central power plants provide all the electricity production in a country or region and dispatch this electricity to consumers via a large network of cables and transformers. The electric power flows in one direction from the power generators to the consumers, and the monitoring information is handled only at the operation side, as illustrated in Figure 3.1. Based on load forecasting models developed over time, the utility providers generally over-provision for the demand considering peak load conditions. This approach is costly when the average demand is much lower than the peak because electricity, once produced, has to be consumed since energy storage is normally very expensive. Also, given the increasing and dynamic demand for energy, it may be difficult, perhaps impossible, in the longer run, to match the supply to demand at all times (Goel, 2013).

It is attractive in such a situation to apply two-way communications between the grid and the customers and providing incentives (e.g., through variable pricing) to encour-

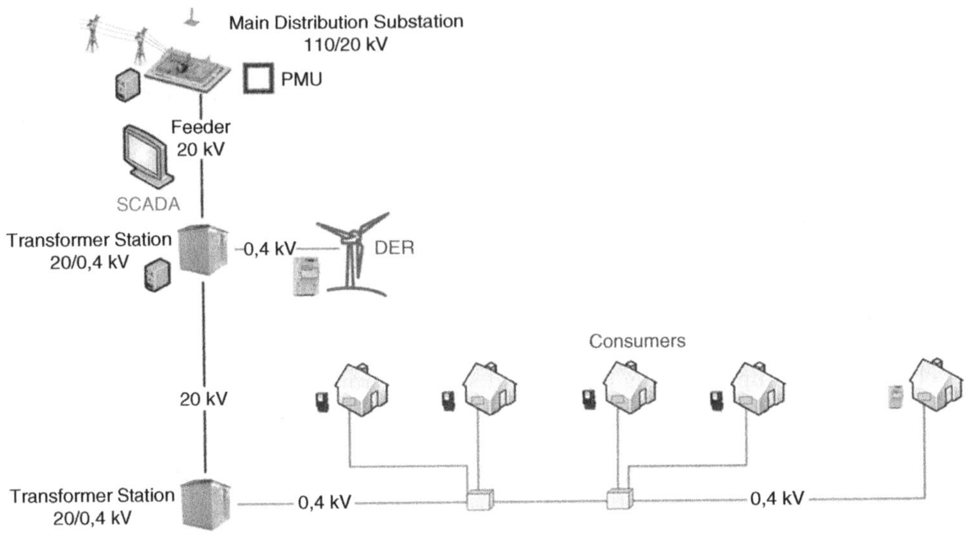

Figure 3.1 Today's power grid where the DSO is operating blindly, i.e., without real-time feedback of costumers' consumption.

age the consumers to reschedule the consumption so as to improve utilization of the available capacity. This necessitates the flow of metering information from the customer premises to the grid management facilities to analyze the demand and the flow of control information in the opposite direction to encourage the customers to manage their demand. The bidirectional flow of information will provide the utility operators a full visibility and hence more effective control of the grids. In particular, the utility operators and consumers can communicate and cooperate in order to achieve more efficient energy consumption management (Zhu, 2012).

Distributed generation using various renewable energy sources offer a key contribution to the environmental problem. It is expected in the future that energy can be generated, stored, and released back to the grids at distribution level, even by households at the low-voltage network. However, their integration into existing power grids comes with a whole new set of barriers, such as the intermittency of generation, and importantly, the lack of real-time monitoring and control capability for such a highly distributed energy system.

In addition, electric vehicles (EVs) are becoming a main component of energy consumption, storage, and supply in the power network. EVs can serve as a distributed and mobile energy source in the electricity market, optimally scheduled and dispatched to meet the dynamic demand of energy and to respond to emergency situations. EVs are a major electricity consumer and draw a significant amount of power in order to retain sufficient battery capacity. For the grid operators, such high loads attached to the grid will have to be managed carefully.

From the DSO's point of view, the introduction of distributed generation makes it more challenging to assure the quality of power delivered to customers, for example, in terms of preventing voltage violations. Since power production may in principle take place anywhere in the grid, it is necessary to obtain detailed real-time measurements of

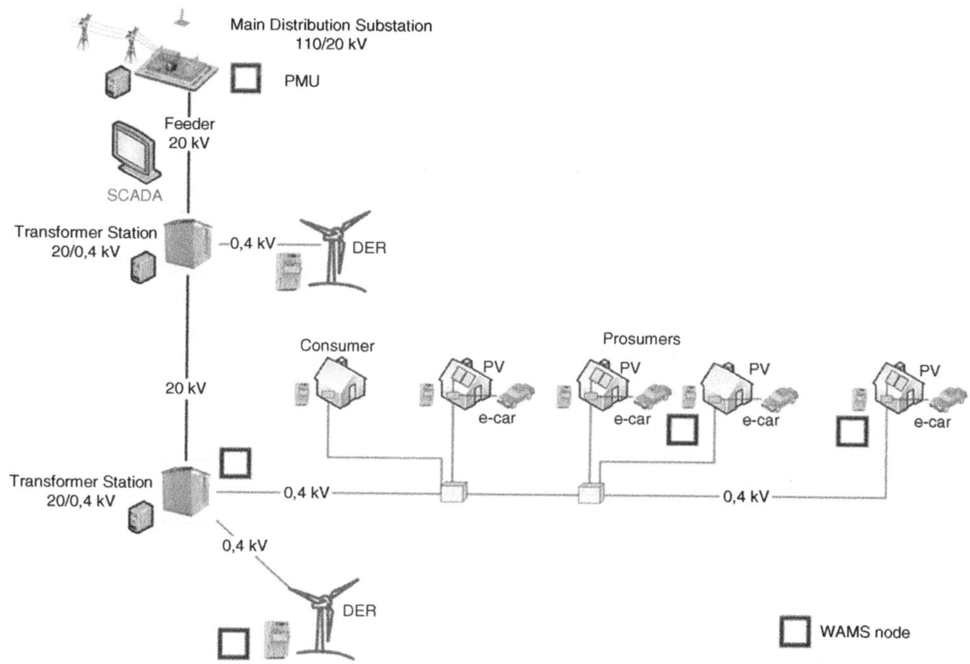

Figure 3.2 Smart distribution grid with real-time measurements of consumption and production at prosumer locations.

the grid state in multiple locations to ensure the required power quality and maintain stable grid operation.

In order to realize all the aforementioned features in the future power network, it is believed that a fundamental evolution, generally recognized as the *smart grid*, is needed. The core of such evolution is the integration of a fast and reliable information and communication technology (ICT) network, in order to provide accurate grid monitoring and interconnection between the power grid and the intelligent information processing systems. Measurements are provided by smart meters and Wide Area Measurement System (WAMS) nodes (Phasor Measurement Unit, PMU-like devices) that are deployed deep into the distribution grid, as shown in Figure 3.2. The control center is able to detect, analyze, and respond fast to dynamic operational status and emerging problems. It is envisaged that the smart grid communication network will be a heterogeneous system of wired and wireless technologies, including fiber and power line communications, cellular networks, and wireless local area networks.

National governments and various relevant stakeholders have already launched massive investments on smart grid research projects and made significant progress. The US Department of Energy (DoE) states that a smart grid uses digital technology to improve reliability, security, and efficiency of the electricity system (DOE, 2012). Its Smart Grid Research and Development Program has set the following performance targets for 2030: 20% reduction in the nation's peak energy demand; 100% availability to serve all critical loads at all times and a range of reliability services for other loads; 40% improvement in system efficiency and asset utilization to achieve a load factor of 70%; and 20% of electricity capacity from distributed and renewable energy sources (200 GW). Europe presents a highly interconnected and more robust system that can intelligently integrate the behavior and actions of all users to ensure sustainable, economic, and secure electricity supply (EC, 2006).

3.3 Smart Grid Communication Requirements

As discussed in the previous section, the increasing production of energy from renewables at the individual household level, through roof-mounted solar panels and small wind turbines, forces the DSO to introduce more advanced measurement feedback systems to ensure stable operation of the distribution grid. Traditionally, the DSO was only collecting consumption information from each household a few times per year, through manual reading of the electricity consumption meter. In many places, those manual meters have been replaced by smart meters, which allow the DSO/utility company to automatically collect measurements from all smart meters on a much more regular basis, typically once per night or every few hours (Hossain, 2012). There exist many different communication solutions for smart meters that use, for example, Power Line Communications (PLC), proprietary wireless solutions, or cellular networks; however, the underlying traffic model is the same.

3.3.1 Traffic Models and Requirements

In industry publications, there are many different examples of traffic models and requirement specifications for the smart grid (Hossain, 2012; Deshpande, 2011; Khan, 2013; Lander, 2013). We have found that the most coherent and comprehensive specification is the OpenSG "Smart Grid Networks System Requirements Specification" (described in Hossain, 2012) from the Utilities Communications Architecture (UCA) user group. While this model specifies in great detail the messages that are communicated between different smart grid entities, it is necessary to make assumptions on a specific system deployment before an actual traffic model can be extracted. An example of this is presented in Nielsen (2015), where the assumed system deployment results in the smart meter traffic model in Table 3.1. The table shows the downlink and uplink traffic rates for default reporting intervals as well as for reduced reporting intervals, which would enable DSO/utility companies to monitor certain aspects of the distribution grid state in close to real time.

At the time of writing, there is no standardized traffic model for describing the reporting behavior of WAMS nodes for the distribution grid. In the transmission grid, PMUs are deployed to obtain synchro-phasor measurements that facilitate monitoring, control, and protection of the grid (Martin, 2008). PMUs are time synchronized, and measurements are sampled and collected up to 50 or 60 times per second, as described in the PMU standards IEEE 1588 (Lee, 2005) and C37.118 (Martin, 2008). Time synchronization is often achieved through the use of GPS receivers. To support the real-time collection of PMU measurements in the transmission grid, the PMUs are typically interconnected through a network based on optical fibre links.

For the use of synchro-phasor measurements in the distribution grid, the real-time requirements depend on the intended application. For monitoring and control purposes, reporting intervals of 1 to 10 s may well be sufficient, whereas protection functions require much faster reaction times in the order of 50 to 250 ms (Goel, 2013). Since

Table 3.1 Derived Smart Meter Traffic Model.

	Downlink	Uplink				
Use case \ reporting interval	**default**	**default**	**5 min**	**1 min**	**30 sec**	**15 sec**
Meter reading	1.25	11K	95K	475K	950K	1.9M
Service switch	3	6	6	6	6	6
Prepay	3.5	8	8	8	8	8
Meter events	0	50	50	50	50	50
Islanded distr. cust. storage	2	5	5	5	5	5
DR-DLC	400	0.5	0.5	0.5	0.5	0.5
Premise network admin	1	1	1	1	1	1
Price	10K	2.4K	2.4K	2.4K	2.4K	2.4K
Firmware / program update	30K	5	5	5	5	5
Total	40.4K	13.4K	97K	477K	952K	1.9M

Table 3.2 Derived Traffic Model of Distribution Grid WAMS Node from SUNSEED Field Trial.

	1 Hz	5 Hz	10 Hz	25 Hz	50 Hz
Goodput (Kbps)	3.2	16	32	81	162
Throughput (Kbps)	3.8	19	38	95	190

the deployment locations may be in prosumer households or small industrial premises, an optical fiber–based communication infrastructure cannot be assumed. For ease of deployment and to ensure reconfigurability of additional distributed generators added to the distribution grid, wireless connectivity is the most obvious choice.

Within the European project SUNSEED[1], a low-cost distribution grid PMU has been developed and deployed in a large-scale field trial. An example of a traffic model for distribution grid WAMS nodes has been derived from network traces. Each synchro-phasor measurement is approximately 405 bytes, since it is encoded with the human-readable data format JSON. Including TCP and IP headers and assuming that each measurement is sent in a separate data packet, the size of the transmitted packet amounts to 470–480 bytes. The resulting goodput and throughput (including overhead) is listed in Table 3.2.

From a throughput perspective, the requirements are quite modest in comparison to other current streaming applications such as video streaming, where several Mbit/s are easily consumed. The key difference is in the real-time requirements. The WAMS synchro-phasor measurements should be delivered within a fraction of a second with a very high reliability, whereas video streaming typically relies on buffering several seconds of data to smoothen out the impact of jitter.

1 www.sunseed-fp7.eu.

An example of a foreseen highly mission-critical MTC application in the smart grid is that of wirelessly connected grid protection relays (METIS, 2013). Here it is required that the trigger signal arrives at the protective relay within milliseconds and with very high reliability (>99.999%), in order to protect the grid in case of, for example, damage to overhead power lines. This is far beyond what current cellular networks typically deliver, even though the message to deliver is very small.

In the following sections, the most promising wireless cellular communication technologies are described and analyzed with respect to their suitability of supporting the smart grid. Initially we consider technologies that operate in the unlicensed spectrum, and hereafter we consider 3GPP technologies.

3.4 Unlicensed Spectrum and Non-3GPP Technologies for the Support of Smart Grid

The main non-3GPP wireless technologies gaining momentum in Europe and North America are IEEE 802.11ah (Adame, 2014), Sigfox (SIGFOX, n.d.), and LoRa™ (LoRa Alliance, n.d.). These technologies are explained in Section 3.4.1, Section 3.4.2, and Section 3.4.3, respectively.

The Sigfox and LoRa™ systems are also the basis for the so-called low-throughput network (LTN) standardization in ETSI (ETSI, LTN FA, 2014; ETSI, LTN P&I, 2014). The main characteristic of these LTN networks is support for low data rates (e.g., up to 50 Kbps for LoRa™) as they are targeting IoT applications that very occasionally transmit very limited amount of data (e.g., some sensor value or alarm/event). Unlike IEEE 802.11ah, the Sigfox and LoRA™ systems are designed for long-range communication links even though they are using limited transmission power (e.g., few tens or hundreds mW). The limited transmission power is constrained by the utilization of industrial, scientific, and medical (ISM) bands, e.g., around 868 MHz in Europe and around 915 MHz in the USA. To increase uplink reception quality in case of unsynchronized uplink transmissions, Sigfox and LoRa™ systems receive the uplink signals via multiple access points (if in range), i.e., the so-called cooperative reception (ETSI, LTN FA, 2014).

3.4.1 IEEE 802.11ah

The IEEE 802.11 wireless local area network (WLAN) standards, generally recognized as the Wi-Fi technologies, have significant advantages in terms of low cost and ease of deployment. However, for MTC traffic, the main limitation of current 802.11 WLAN is the coverage, power consumption, and the limited capacity in terms of number of stations. This is because of the high operating frequency (2.4 GHz and 5 GHz) and the current contention-based medium access control protocols do not work well in large networks. Recent development in the IEEE 802.11 has been focused on the enhanced capability in massive and dense scenarios. In particular, the task group TGah proposes to use a group-based contention scheme in the Media Access Control (MAC) of its proposed amendment 11ah in order to cope with massive power constrained MTC stations (STAs) per access point (AP; IEEE, 2016).

The physical (PHY) layer of IEEE 802.11ah is MIMO OFDM-based transmission operating using sub-1 GHz license-free bands. Some examples of the bands include

863–868 MHz in Europe, 950–958 MHz in Japan, and 902–928 MHz in the US. Owing to the superior propagation feature of lower frequency, the 802.11ah provides longer transmission range, with 6 dB higher link margin compared to that at 2.4 GHz. Channel bandwidths of 1 MHz and 2 MHz are expected to be adopted in Europe, although configurations using 4, 8, and 16 MHz are also allowed in the US. For transmission modes that use at least 2 MHz, the PHY layer is designed based on 10 times down-clocking of 802.11ac's PHY layer, which operates in 20, 40, 80, and 160 MHz channels with 32 or 64 OFDM subcarriers. The 1 MHz channel is used in order to extend the coverage; however the number of data subcarriers per OFDM symbol is reduced to 24. The IEEE 802.11ah supports a small PHY packet length of 256 bytes, which is typical for MTC.

The MAC layer is designed to maximize the number of stations supported while endeavoring to maintain minimum energy consumption. A group-synchronized distributed coordination function is featured in the MAC of the proposed amendment. The idea of grouping network nodes has been widely explored, for example, the clustering strategies in wireless sensor networks and ad hoc networks. It increases energy efficiency, decreases management complexity, and optimizes other network performance metrics. In the proposed 11ah standard, a 13-bit Association IDentifier (AID) is used to label more than 8000 STAs for one AP. The AID classifies stations into 4 pages, each containing 32 blocks. There are further 8 sub-blocks in each block, and up to 8 stations in each sub-block. The structure of the AID is depicted in Figure 3.3. The AIDs are allocated to the STAs according to a grouping mechanism. In addition,

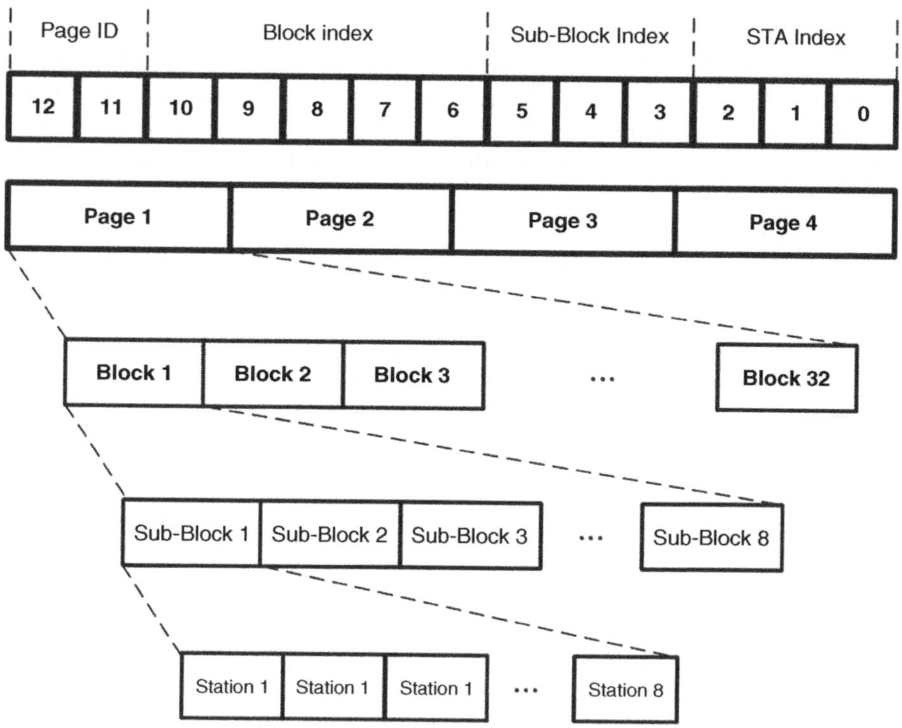

Figure 3.3 AID structure proposed in the IEEE 802.11ah.

three categories of STA are defined. The TIM stations need to listen to the AP's beacon to send or receive data. They are active in the scheduled Restricted Access Windows (RAW). TIMs can be allocated with different RAWs. Only the STAs within the same group will be contending for transmission with each other at once. The non-TIM stations negotiate with the AP to access the channel at the Periodic Restricted Access Window (PRAW) defined in every TIM period. The unscheduled stations are allowed to access the channel in the Any Access window in the TIM period. STAs that are not in permitted RAW can switch to sleep mode and save power. Under such mechanism, collisions and delays caused by the contention-based MAC process can be reduced significantly in a dense network scenario. The detailed STA regrouping technique is described as implementation specific.

The unexpected transmission from the so-called hidden nodes has significant impact on the performance of the IEEE 802.11 WLAN, especially in large cells. It occurs when a STA is visible from the AP but is outside the sensitivity range of certain STA communicating with that AP. It is another major cause of collision for the CSMA/CA-based media access control protocols. The potential hidden node problem can be managed by using the grouping of STAs.

A novel STA regrouping method is proposed in SUNSEED (2016) consisting of a signaling process and a regrouping algorithm. The aim of the regrouping is to minimize the potential transmission collisions caused by the hidden node problem through identifying all potential hidden node pairs and allocate them into different contention groups. Firstly, the AP acquires global knowledge of potential hidden node pairs of the STAs in the network via a signaling technique. Figure 3.4 depicts the time slot allocation in the IEEE 802.11ah MAC protocol. The information collection will happen in the multicast, downlink (DL) and uplink (UL) slots in every TIM. The AP conducts a roll call for all the connected STAs. The STAs respond to the AP with ACK in order, i.e., predefined time slots, and the other STAs listen to the channel. If STA_i cannot hear the transmission from STA_j in the corresponding time slot, then it will record STA_j as a hidden node.

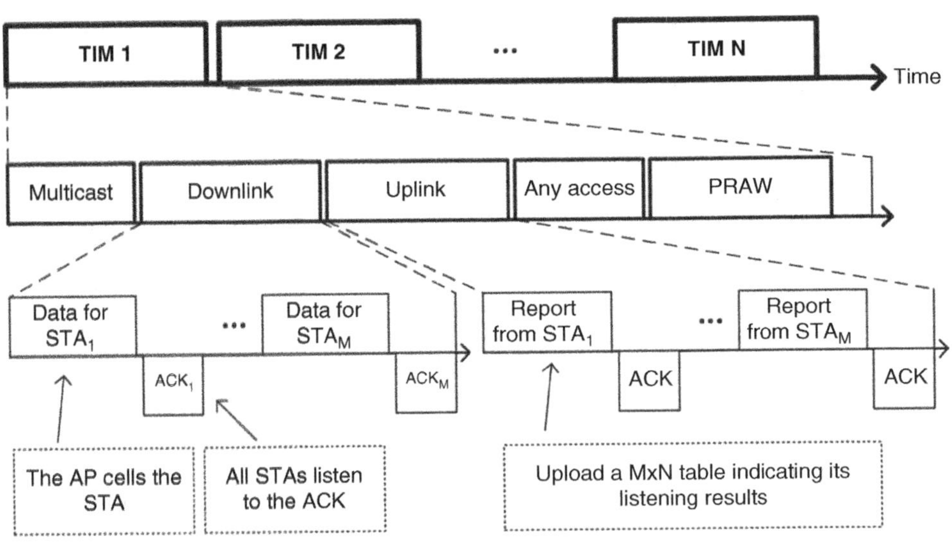

Figure 3.4 Collecting hidden node information using the IEEE 802.11ah MAC.

The STAs report the listening results to the AP in the UL slots. The AP is then able to create a table of the potential hidden nodes for the network and keeps updating the table until no report is collected. In addition to the hidden node information, the AP will consider the active level of each STA based on its type. For example, in smart grids, a WAMS device can have 100 times more measurements and reporting than a smart meter. Such active levels are integrated to form a comprehensive knowledge of potential impact of hidden nodes. The AP then regroups the hidden STAs into different contention groups according to a Viterbi-like algorithm, in order to minimize the potential collisions in the whole network. In a case study considering a distribution grid monitoring scenario, a 50% reduction of collision was achieved using regrouping, compared with random grouping methods (SUNSEED, 2016).

Finally, the IEEE 802.11ah proposes the use of 18-byte MAC layer frame header, instead of 28 bytes in current standards in order to improve the frame utilization efficiency, as the data payload in MTC is often very small. The standard also incorporates bidirectional transmit opportunity (TXOP) and target wake times, which makes it suitable for smart grid applications.

3.4.2 Sigfox's Ultra-Narrow Band (UNB) Approach

The Sigfox radio access link is based on UNB transmissions with, e.g., a channelization mask of 100 Hz in Europe (or 600 Hz in USA). Since the Sigfox technology uses the ISM band, its transmission power is limited to, e.g., 25 mW in uplink and 500 mW in downlink for Europe. The UNB transmission is a suitable technique due to the low noise contribution resulting in low receive sensitivity levels (typically around −140 dBm) for successful reception at low bit rates. The typical link budget is around 160 dB, which can result in rather large cells. In order to have simpler end devices, reduce the system overhead, and consequently, the battery usage, the end devices are not trained by pilot and synchronization signals. The Sigfox version called Cooperative Ultra Narrow Band (C-UNB) was one alternative for clean slate solutions investigated in 3GPP (3GPP, TR45.820) but not selected for standardization in 3GPP Release 13. Note here that due to regulatory aspects for the unlicensed spectrum usage (e.g., maximum transmission duty cycle) the commercial Sigfox systems are limited by the maximum (e.g., 140) number of transmitted messages per day. Consequently, smart grid applications where frequent data messaging is required might not be supported by these systems, even more so if the messages are not small.

On the security aspect, Sigfox messages (up- or downlink) are authenticated both ways, and each message carries a unique ID. Messages sent by Sigfox devices are signed using a key stored in the device. Messages received by the base stations are first forwarded to the Sigfox cloud using a VPN connection to be transmitted from there to application servers via the https protocol. Data ciphering, when used, must be performed at the application level, and the application deployment team is responsible for the distribution of the ciphering/deciphering keys.

The commercial deployment of Sigfox networks is currently ongoing worldwide. This is arranged via Sigfox partnering companies and according to the reports available on the Internet[2], there are already some European countries with nation-wide coverage (e.g., Ireland, France, Belgium, and Netherlands).

---

2 See http://www.Sigfox.com/coverage.

3.4.3 LoRa™ Chirp Spread Spectrum Approach

The LoRa™ system is based on so-called chirp spread spectrum transmission, where the chirp signal continuously varies in frequency with configurable modulation bandwidth between 8 kHz and 500 kHz. The spreading factor can be from 128 (2^7) to 4096 (2^{12}) and together with the modulation bandwidth has to be configured as a trade-off between the desired data rate, transmission range, and power usage (i.e., time on air). Depending on the particular LoRa™ product, the receiver sensitivity is around −140 dBm, which for 20 dBm output power results in a typical link budget of 160 dB. This enables long communication ranges similar to Sigfox systems. The supported data-rates are between 0.3 Kbps and 50 Kbps. The LoRa™ systems define different end devices to suit different needs of uplink versus downlink transmission and battery power classes (LoRa Alliance, n.d.).

The wide area network built from end nodes (e.g. sensors) communicate wirelessly with LoRa™ gateway nodes (i.e., similar as cellular base stations or Wi-Fi access points) that are further connected via a backhaul system with a network server labelled as LoRaWAN. This star architecture for the LoRaWAN is similar as the architecture of the Wi-Fi networks and also comparable with the cellular networks, although without the support for real mobility of the end nodes (i.e., no handover support but nomadic mobility support).

At this moment, the networks deployed based on the LoRa™ technology are providing traditional IoT services that require small data transfer, no stringent delay requirements, and long battery life. Examples include asset tracking, smart meters' support for electricity grids, transport and logistics services, and smart cities (smart street lightning, building management, air quality monitoring, leakage in water distribution systems, etc.). Typically, cellular operators reuse their sites for installing additional LoRa™ gateways and providing transfer of data to a third party via LoRa™ sensors.

The security mechanisms of LoRa are specified in the LoRaWAN networking stack and are at this point relatively basic and likely to evolve in future specifications of the LoRaWAN stack. They address communications occurring between a LoRa device and a LoRa network server via a LoRa gateway.

The security bootstrap of a Lora device using the LoraWAN stack is summarized in Figure 3.5.

As a first step before the deployment, LoRa devices are configured. The configuration includes a device unique ID (DevEUI), an application unique key (AppEUI), and an application-specific key (AppKey).

The parameters DevEUI, AppEUI, and AppKey are also preloaded in the LoRa network server. When the LoRa device is being deployed and attempts to join a LoRa network, it sends an initial join request including a device-generated random number, the unique device ID and the unique application ID. Upon reception of this request, the Lora network server computes two session keys derived from AppKey: a network session key, NwkSKey, and an application session key, AppSKey. The Lora device performs independently the same computations. Then all application data sent to the LoRa server will be ciphered using the application session keys and include a message identification code computed with the network session keys. All MAC messages will be themselves ciphered with the network session key.

In this context, the party operating the network server should also be the one operating the LoRa application, but this assumption is not very realistic. A real-life situation is

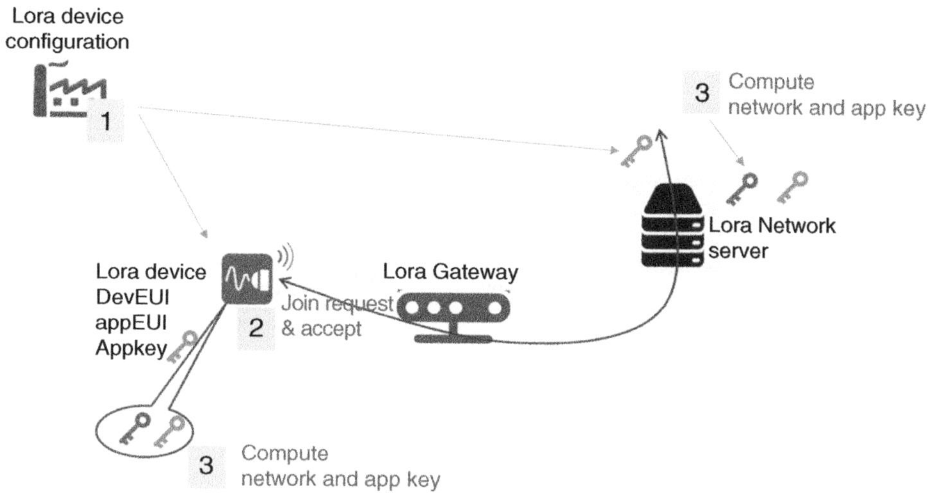

Figure 3.5 Security bootstrap of Lora device (LoraWAN).

the one in which the LoRa network server is operated by one party (a communication service provider such as a cellular operator) and the application operated by another party (the application provider). Therefore, there is a need to separate the credentials so that the AppSKey used to secure application end to end is not known to the communication service provider. Such a need may be answered using the architecture shown on Figure 3.6, also based on the LoraWAN stack, and where a separate key management server is used to manage the credentials. According to this figure, the security bootstrap of a Lora device will involve the following steps. In the first step and before the device deployment, the devices are preconfigured. This step may occur when the LoRa device is manufactured. The configuration includes a device unique ID (DevEUI), an application unique key (AppEUI), and an application-specific key (AppKey). The parameters DevEUI, AppEUI, and AppKey are also preloaded in the LoRa key management server. When the LoRa device is being deployed and attempts to join a LoRa network, it sends an initial join request including a device-generated nonce, the unique device ID, and the unique application ID. Upon reception of this request, the LoRa network server forwards the request to the LoRa key management server. Using the nonce, DevEUI, APPEUI, and AppKey, the key management server computes two derived session keys: a network session key, NwkSKey, and an application session key, AppSKey. It transmits the AppSKey to the application server and the NwkSkey to the LoRa network server. Then, all application data sent from the device will be sent to the application server via the LoRa server. Data will be ciphered using the application session key and include a message identification code computed with the network session key. All MAC messages sent to the Lora network server will be ciphered with the network session key.

Regarding the issue of who operates the LoRa key management server, there are two distinct cases:

(I) The communication service provider operating the network server is sufficiently trusted by the party deploying the LoRa application. In this case the key management server will be operated by the communication service provider, and the benefit

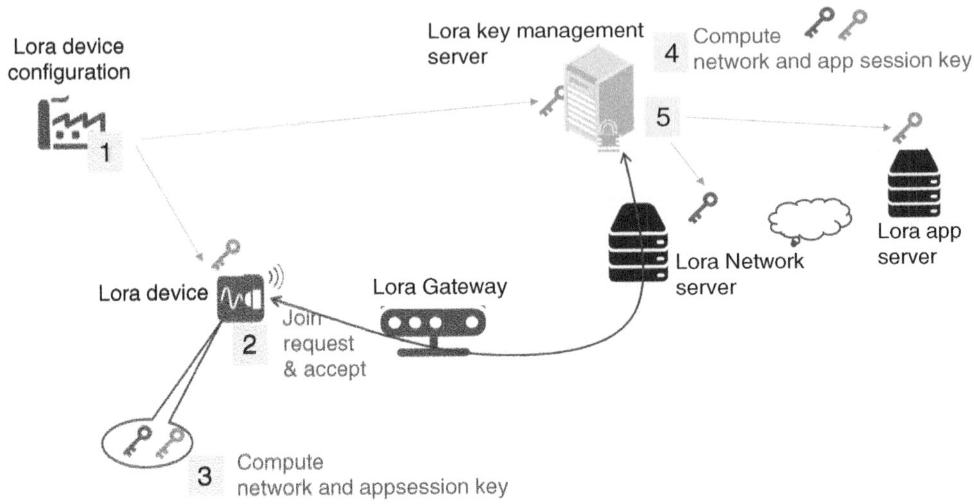

Figure 3.6 LoRa device security bootstrap using separate network and application server.

of the architecture described in Figure 3.6 will lie in the possibility to place key management operation in a separate trusted server, possibly serving several LoRa networks.

(II) The communication service provider is not sufficiently trusted; then an independent neutral party could operate the LoRa key management server. This scheme will ensure complete separation between the application and network credentials.

3.5 Cellular and 3GPP Technologies for the Support of Smart Grid

The advantage of using 3GPP technologies for IoT applications is their almost worldwide coverage and usage of licensed spectrum that, unlike the unlicensed spectrum, provides for higher reliability and interference control. Today's 3GPP technologies are largely optimized for high data rate (human) communications. This makes 3GPP technologies less suitable for wide-area IoT deployments where battery usage and cost might be important requirements. In Section 3.5.1 we provide an overview of the shortcomings of 3GPP technologies up to Release 11 when used for wide-area IoT deployments. The ongoing improvements within the various 3GPP working groups (Releases 12–14) for the domain of wide-area IoT are presented in Section 3.5.2.

3.5.1 Limits of 3GPP Technologies up to Release 11

The 3GPP technologies up to Release 11, especially 3G and 4G technologies, are **optimized for high-rate data transmissions.** This optimization means, among others, increasing the system carrier bandwidth (e.g., 200 KHz carrier in GSM toward an aggregate of up to five 20 MHz carriers in LTE-A), multiple-input and multiple-output

(MIMO) transmission and reception, and higher-order modulation schemes (e.g. 64QAM). The support of these functionalities at the device increases the complexity and performance requirements of its hardware components, **resulting ultimately in higher costs and high power consumption**.

For devices' **power-saving purposes,** 3GPP specifies a so-called discontinuous reception (DRX) process where the device periodically (i.e., the so-called DRX cycle) wakes up to monitor for paging messages and be reachable. The longer the DRX cycle, the lower the device power consumption but also the lower the responsiveness of the communication. The possible values of DRX cycles (e.g., up to 2118 ms and 2560 ms for GSM and LTE, respectively) were defined with communication scenarios involving humans in mind. Wide-area IoT deployments might allow for longer DRX cycles enabling longer **battery life time of the devices** at the cost of a (somewhat) reduced but still acceptable responsiveness of the communication.

3GPP systems have **relatively large signaling overhead** when compared to rather small amounts of user data that is typically transferred in IoT deployments. For example, during one LTE data session setup procedure about 66–72 bytes or 120–193 bytes of control signals need to be transmitted in the uplink or downlink, respectively (3GPP, TR36.822). Similarly, the GPRS and UMTS attach procedure also involves the delivery of non-negligible signaling overhead. **Reducing the measurement and signaling effort for mobility purposes** at the device in terms of regular pilot signal's measurements and reporting as well as location area updates; e.g., IoT deployments with fixed devices can also reduce devices' power consumption.

Deep indoor coverage is an important issue for IoT applications, especially if the majority of the deployed devices are located deep in the buildings and at fixed locations. The development of 3GPP systems up to Release 11 was targeting scenarios with less demanding propagation shadowing and indoor penetration conditions and user speed of, e.g., 3 km/hour emulating human behavior in pedestrian environments.

Most 3GPP technologies operate in **a frequency division duplexing (FDD) mode**, allowing simultaneous transmission and reception at base stations or devices, which has the disadvantage of expensive RF components (i.e., the duplexer). Delay-tolerant IoT applications can **allow a half-duplex frequency division duplexing (HD-FDD) mode** where transmission and reception can occur in turns removing the need for expensive RF components, with the cost of reduced user throughput.

3.5.2 Recent Enhancements of 3GPP Technologies for IoT Applications (Releases 12–13)

A high-level overview of the 3GPP developments for IoT applications is provided in Figure 3.7. The time lines shown in this figure correspond with the (indicative) formal freeze dates of the various 3GPP Releases. Actual implementation time lines may depend on various factors such as the level of complexity of the functionality and/or urgent market demand.

Here, the scope is limited to the following IoT enabling functionalities in 3GPP Releases 12 and 13:

- LTE Cat-0 and Cat-M1 devices; and
- narrow-band Internet of Things (NB-IoT) and Cat-NB1 devices.

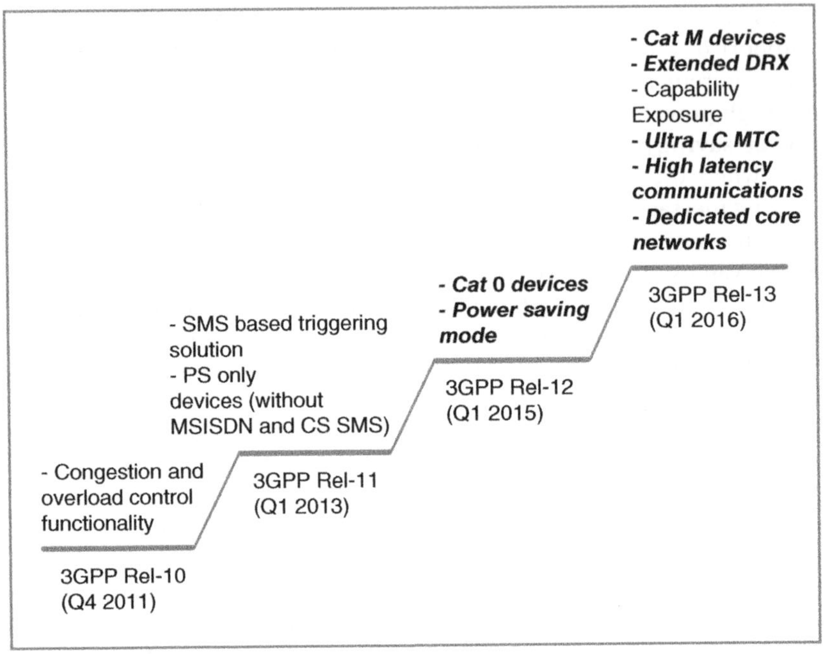

Figure 3.7 A high-level overview of new machine-type communications related functionality with the subsequent 3GPP releases.

3.5.2.1 LTE Cat-0 and Cat-M1 devices

The LTE Cat-0 terminal category was introduced in 3GPP Release 12 with the following major features:

- 1 Rx antenna compared to a minimum of 2 Rx antennas for other device categories. This reduces the number of RF chains and components by the devices, as well as baseband processing, e.g., for channel estimation.
- Peak date rate of 1 Mbps in downlink/uplink (DL/UL), compared to those of 10 Mbps / 5 Mbps for the lowest Cat-1 of non-M2M LTE devices.
- Half-duplex FDD (optional), for delay-tolerant services. This is to reduce RF components, with a side benefit of lower noise figure at the devices.

LTE Cat-0 devices have about 50% lower cost due to reduced hardware complexity in comparison with Cat-1 devices (Nokia, 2014). This might be comparable to a 2G GPRS modem. However, it should be noted that the price of devices is not only determined by the hardware costs but is, e.g., also dependent on the production volume, which is hard to predict at this stage. Although the peak power consumption is reduced, the power consumption in practice will depend on the activity level of devices (e.g., communication pattern and the mobility of the devices) and whether other energy-saving enhancements are introduced.

The Cat-M1 terminals have been specified for 3GPP Release 13, aiming to further reduce the device complexity compared with Cat-0 devices (3GPP, RP141865) and with the following major features:

- Improved coverage, with link budget of 155.7 dB. This is achieved via transmission time interval (TTI) bundling and repetition, i.e., sending the same (or a different version) of a packet multiple times without waiting for feedback.
- Up to ~10 years of device battery life with 5 watt hour battery.
- A lower bandwidth of 1.08 MHz for Cat-M1 devices. It is noted that the LTE network may still have the normal LTE bandwidth of up to 20 MHz (per carrier), but only 1.08 MHz of the total system bandwidth is relevant for the Cat-M1 devices, i.e., signals that are relevant for these Cat-M1 devices are completely sent within this 1.08 MHz bandwidth (3GPP, TS36.300).
- Variable data rates between 10 Kbps and 1 Mbps, depending on coverage needs.
- Implementation of a lower UE maximum transmit power of 20 dBm, instead of conventional 23 dBm (which sacrifices 3 dB on the achievable coverage). This allows integration of a power amplifier in a single chip solution.
- Support of half-duplex operation.

Cat-M1 devices have about 50% lower cost (in the sense of hardware complexity) in comparison with Cat-0 devices (Nokia, 2014).

3.5.2.2 Narrow-Band Internet of Things (NB-IoT) and Cat-NB1 Devices

3GPP has also specified a clean-slate RAN solution for cellular IoT applications, named as **narrow-band Internet of Things (NB-IoT) (3GPP, RP151621)**, which provides improved indoor coverage, support for a massive number of low-throughput devices, low delay sensitivity, ultra-low device cost, low device power consumption, and optimized network architecture. This is largely based on a previous 3GPP study item results (3GPP, TR45.820). NB-IoT devices use bandwidth of 180 kHz for both downlink and uplink. In the downlink, OFDMA is used with subcarrier spacing of 15 kHz. In the uplink, SC-FDMA is used with two possible sub-carrier spacings: 15 kHz and 3.75 kHz.

To further reduce the complexity and cost of Cat-NB1 devices, half-duplex operation and single HARQ process are specified for NB-IoT. The Cat-NB1 devices have been designed to reach the following key performance targets:

- Low device cost, i.e., lower than that of Cat-M1 devices.
- Up to ~10 years of device battery life with 5 watt hour battery.
- Improved coverage, with link budget of 164 dB.
- Support of massive number of devices (around 50,000 devices per cell).
- Deployment flexibility and co-existence with existing technologies.

Next to reusing repetitions (as in Cat-M1) the improved coverage for Cat-NB1 devices is due to the narrower transmission bandwidth.[3]

The battery life improvement is due to longer IDLE mode discontinuous reception (DRX) cycles for Cat-NB1 up to 3 hours (versus ~44 min for Cat-M1) while in connected mode both device classes have same standardized DRX cycle lengths.

Further, NB-IoT reuses the LTE authentication and key agreement (AKA) scheme for security purposes.

The NB-IoT standardization in Release 14 in 3GPP (frozen end of 2017) introduced enhancements of the UE positioning accuracy, multicast transmissions, etc. The further

3 In uplink instead of using the full 12 subcarriers, i.e., the 180 KHz bandwidth, the terminal can use 1, 3, 4, or 12 subcarriers.

IoT enhancements regarding improved coverage (e.g., via multi-hop relaying), scalability, and ultra-reliable low-latency communication (URLLC) are also considered as important topics for 5G standardization in Release 15, which is due end of 2018.

It should be noted that 3GPP is also standardizing the extended-coverage GSM IoT (EC-GSM-IoT). EC-GSM-IoT is a further enhancement of eGPRS to support high-capacity, long-range, low-energy, and low-complexity cellular IoT applications. The performance targets are similar as with Cat-M1 and Cat-NB1 terminals, i.e., extended coverage with link budget of 164 dB (154 dB) for UE with 33 dBm (23 dBm) output power, up to ~10 years battery life, and throughput of up to 240 Kbps.

3.5.3 Performance of Cellular LTE Systems for Smart Grids

The capacity of cellular systems is typically defined as the maximum number of supported users per cell having a desired performance for a particular service. In relation to smart grids, a desired performance from a communication point of view is having reliable transfer of the smart grid data from the source to the destination within a predefined end-to-end delay. The quantification of number of supported users per LTE cell for a given end-to-end delay criteria (e.g., 95% maximum delay value) has been analyzed in Zhang (2016) and SUNSEED (2016). The analysis in Zhang (2016) quantifies the impact of the amount of reserved resources on the LTE carrier as well as the scheduling policy. It can be concluded that a scheduling algorithm that tries to minimize the end-to-end delay is very beneficial and should be deployed for smart grid applications support. Further, the amount of reserved physical resources on the LTE carrier is determined by the upper bound for the delay requirement and the expected number of users per LTE cell. For example, from the LTE simulation results in an urban environment and only a single physical resource block (e.g., 180 kHz in frequency and 1 ms in time domain) reserved for the smart grid traffic corresponds to 250 or 2500 supported smart grid nodes per LTE cell if the desired 95% end-to-end delay is lower than 1 s or 10 s, respectively. When the LTE performance is compared between the regular LTE system and the NB-IoT feature, it can be seen from SUNSEED (2016) that for a single physical resource block (PRB) the delay performance is increased by roughly a factor of three. The main reason for this is that the NB-IoT design was driven from the coverage and efficient battery usage point of view and not from having a minimum delay as the main driver.

An additional relevant comparison is the communication network performance between the non-3GPP systems such as LoRa™ and Sigfox and the 3GPP-based wireless networks for wide-area IoT deployments, which is the relevant deployment scenario for smart grid support. This comparison is done by using four important criteria: coverage, capacity, battery usage, and costs. This is illustrated in Table 3.3. From this comparison table, the following trends can be observed:

- Now that 3GPP Release 13 specification of NB-IoT is finished and commercial products are becoming available, the 3GPP technology can compete on coverage and low power usage with LoRa™ and Sigfox networks for low-rate, battery-powered wide-area IoT deployments. Which wireless technologies will have a predominant role in this IoT deployment segment also depends on market trends and the whole business environment.

- LoRa™ and Sigfox networks are not suitable for wide-area IoT deployments where end devices need higher throughputs (e.g., above 50 Kbps) and more symmetric uplink and downlink communications.

3.5.4 LTE Access Reservation Protocol Limitations

For traditional mobile broadband uses, the capacity of an LTE network is usually limited by the resources available for data payload transmissions. However, when used for machine-to-machine (M2M) or machine-type communications (MTC), such as smart grid data traffic, where a large number of individual data packets are sent from many different devices in the network, the capacity of the different control channels may instead be a limiting factor. The goal of the analysis presented in this section is to identify the potential bottlenecks in LTE when used for MTC and to study different bottleneck scenarios through a proposed mathematical model and protocol simulations. The full analysis is published in the journal article by Madueno (2016), whereas this section summarizes and emphasizes the most relevant concepts and results in relation to smart grid communications.

As already outlined, the traffic profile generated by smart-grid monitoring devices is an example of MTC/M2M traffic, characterized by sporadic transmissions of small amounts of data from a very large number of terminals. This is in sharp contrast with the bursty and high data-rate traffic patterns of the human-centered services. Another important difference is that smart grid services typically require a higher degree of network reliability and availability than the human-centered services (Goel, 2013). So far, cellular access has been optimized to human-centered traffic, and M2M-related standardization efforts came into focus only recently (3GPP, Rel-11 overview).

Due to the sporadic, i.e., intermittent, nature of M2M communications, it is typically assumed that M2M devices will have to establish the connection to the cellular access network every time they perform reporting. Usually, the inactivity timer in LTE is around 10 sec, meaning that after 10 sec of idle connectivity, the UE will have to go through the steps of the LTE access procedure to obtain a new connection. Since the establishment of many connections may pose a significant burden to the operation of the access protocol, it is important to consider the whole procedure associated with the transmission of a data (report) in order to properly estimate the number of M2M devices that can be supported in an LTE access network.

3.5.4.1 LTE Access Procedure

The downlink resources in LTE in the case of frequency division duplexing (FDD) are divided into time-frequency units, where the smallest unit is denoted as a *resource element* (RE). Specifically, the time is divided in frames, where every frame has ten subframes, and each subframe is of duration of 1 ms. An illustration of a subframe is presented in Figure 3.8. Each subframe is composed in time by 14 OFDM-modulated symbols, where the amount of bits of each symbol depends on the modulation used, which could be QPSK, 16QAM, or 64QAM. The system bandwidth determines the number of frequency units available in each subframe, which is typically measured in resource blocks (RBs), where an RB is composed of 12 frequency units and 14 symbols, i.e., a total of 168 REs. The amount of RBs in the system varies from 6 RBs in 1.4 MHz systems to 100 RBs in 20 MHz systems.

Table 3.3 Comparison of IEEE 802.11ah, LoRa™, and Sigfox with 3GPP IoT Developments.

	802.11ah	LoRa™	Sigfox	3GPP <= Rel. 12	3GPP >= Rel. 13
Coverage	Up to 1 km outdoor	160 dB typical link budget with cell range up to few (tens) km. Might be influenced by interference in ISM band.	160 dB typical link budget with cell range up to few (tens) km. Might be influenced by interference in ISM band.	Typically, 140 dB increased up to 155 dB with CE improvements. Probably lower cell range (or indoor penetration) than LoRa™ or Sigfox	160 dB typical link budget with cell range up to few (tens) km.
Capacity	Up to 8191 STAs associated with 1 AP. Throughput: 0.15–4 Mbps (1 MHz), 0.65–7.8 Mbps (2 MHz)	Up to few hundreds bps throughput per sensor and highly asymmetric uplink (more data than downlink). Few (tens) of thousand sensors per cell. Might be influenced by interference in ISM band.	Up to 50 Kbps throughput per sensor and highly asymmetric uplink (more data than downlink). Few (tens) of thousand sensors per cell. Might be influenced by interference in ISM band.	Up to 200 Kbps (Cat-M) throughput per sensor and more symmetrical uplink vs downlink capacity. Up to few thousand sensors per cell depending on throughput and delay requirements. More predictable and planned due to licensed bands.	Up to 50 Kbps (Cat-NB1) throughput per sensor and more symmetrical uplink vs downlink capacity. Up to few tens thousands of sensors per cell depending on throughput and delay requirements. More predictable and planned due to licensed bands.
Battery usage	From months to a few years	~10 years or more	~10 years or more	Few years till 10 years (yet to be proven)	~10 years or more (yet to be proven)
Costs	Low CAPEX due to ISM band and low-cost module, easy deployment, low OPEX due to independence of network operators.	Low CAPEX due to ISM band and low-cost sensors (< $5), low OPEX due to battery lifetime.	Low CAPEX due to ISM band and low-cost sensors (< $5), low OPEX due to battery lifetime.	Depends on business case as provided by Mobile Network Operator	Depends on business case as provided by Mobile Network Operator

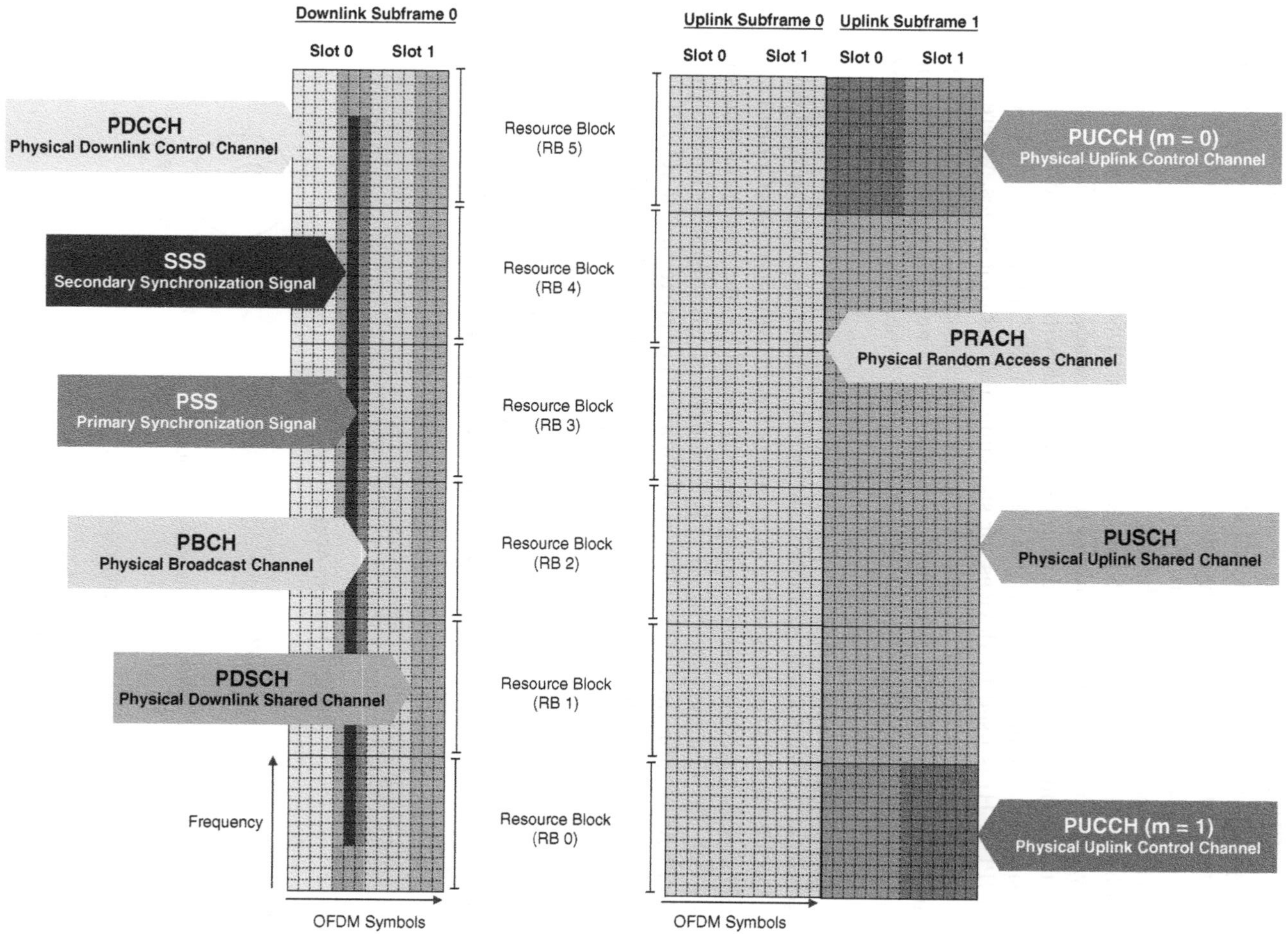

Figure 3.8 Simplified illustration of downlink and uplink subframe organization in a 1.4 MHz system (Nielsen, 2017).

In the downlink, there are two main channels; these are the physical downlink control channel (PDCCH) and the physical downlink shared channel (PDSCH). The PDCCH carries the information about the signaling/data being transmitted on the current PDSCH and the information about the resources that the devices need to use for the physical uplink shared channel (PUSCH), as illustrated in Figure 3.8. Therefore, signaling and data messages consume resources both in the control and shared data channels. The PDCCH is located in the beginning of each subframe. The remaining resources are used for the physical broadcast channel (PBCH), primary and secondary synchronization signals (PSS and SSS respectively), and PDSCH, as shown in Figure 3.8. Obviously, there is a scarcity of resources for MAC messages in the PDSCH.

The uplink resources are organized as in the downlink, with the main difference being that the smallest resource that can be addressed is an RB. The physical uplink shared channel (PUSCH) is used by devices for signaling and data messages, where several devices can be multiplexed in the same subframe. When present, the PRACH occupies 6 RBs and occurs periodically, from once in every two frames (20 subframes) to once in every subframe. A typical PRACH period is once every 5 subframes (3GPP, R2-105212).

3.5.4.2 Connection Establishment

The first part of the connection establishment in LTE is the access reservation procedure (ARP). The ARP in LTE consists of the exchange of four Medium Access Control (MAC) protocol messages between the accessing device denoted UE and the eNodeB, as shown in Figure 3.9. The first message (MSG 1) is a random access preamble sent in the first available PRACH subframe, also called random access opportunity (RAO). The number of subframes between two RAOs varies between 1 and 20. The used preambles are chosen randomly from the set of 64 orthogonal preambles, where usually at most 54 are available for contention purposes and 10 are reserved. The contention scheme is based on slotted ALOHA (3GPP, TS36.321; 3GPP, TS36.213), but unlike in typical ALOHA scenarios, the eNodeB usually cannot detect which preambles have been activated if multiple activations (collisions) have occurred (Sesia, 2011; Thomsen, 2013).

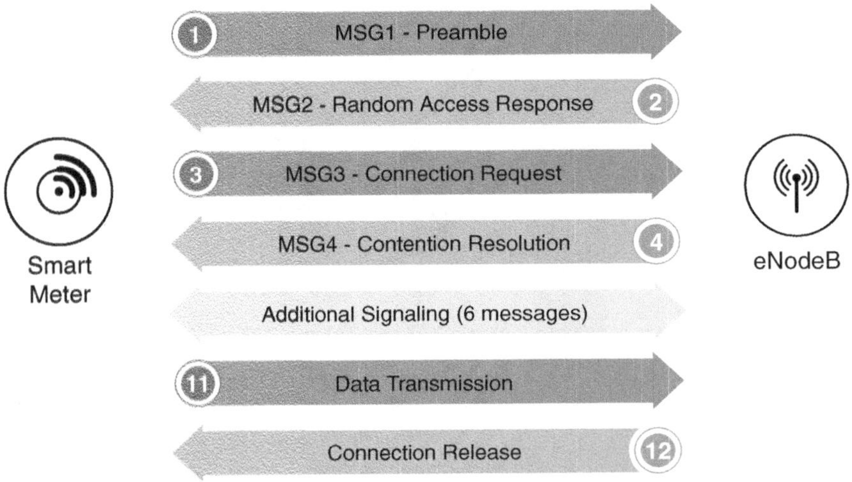

Figure 3.9 Message exchange between a smart meter and the eNodeB (Nielsen, 2017).

In MSG 2, the eNodeB sends a random access response (RAR) to all detected preambles. The contending devices listen for the MSG 2 in the downlink channel. If it is not received within a short time period and the maximum allowed number of MSG 1 transmissions has not been reached, the device backs off and restarts the random access procedure. If received, MSG 2 includes uplink grant information that indicates the RB in which the connection request (MSG 3) should be sent. The connection request specifies the requested service type, e.g., voice call, data transmission, or measurement report. When two devices select the same preamble (MSG 1), they receive the same MSG 2 and experience collision when they send their respective MSG 3 in the same RB.

In contrast to the collisions for MSG 1, the eNodeB is able to detect collisions for MSG 3. The eNodeB only replies to the MSG 3s that did not experience collision, by sending message MSG 4 (i.e., RRC connection setup). The message MSG 4 may lead to two different outcomes: either the required RBs are allocated or the request is denied in case of insufficient network resources. The latter is, however, unlikely in the case of M2M communications, becauase of the small payloads. If the MSG 4 is not received within a period of time since MSG 1 was sent, the random access procedure is restarted. Finally, if a device does not successfully finish all the steps of the random access procedure within the permitted number of transmissions, an outage is declared.

After the ARP exchange finishes, there is an additional exchange of MAC messages between the smart meter and the eNodeB, whose main purposes are to establish security and quality of service for the connection, as well as to indicate the status of the buffer at the device. Those are represented by the "additional signaling" block in Figure 3.9.

Besides MAC messages, there are PHY messages included in the connection establishment (3GPP, TR36.822). A complete list of both PHY and MAC messages exchanged during connection establishment, data report transmission, and connection termination can be seen in, for example, Madueno (2016). For every downlink message, a preceding downlink grant in the PDCCH is required. Similarly, every time a UE wishes to transmit in the uplink after the ARP, it first needs to ask for the uplink resources by transmitting a scheduling request in the PUCCH. This is followed by provision of an uplink grant in the PDCCH by the eNodeB.

3.5.4.3 Numerical Evaluation of LTE Random Access Bottlenecks

We consider a typical 5 MHz (25 RBs) cell configured with one RAO every 5 subframes, 54 available preambles for contention, and a backoff value of 20 ms (METIS, 2013). In addition, we also investigate the performance of the smallest bandwidth cell in LTE, which corresponds to a 1.4 MHz (6 RBs) and RAO interval of 20 subframes.We focus in the lowest modulation in LTE (QPSK). Further, we consider that monitoring devices are reporting every 10 s, which allows for frequent monitoring and control of the smart grid. Report sizes of both 100 and 1000 bytes are considered to illustrate the impact of small and large payload on the system performance. Results are shown both for the analytical model presented in Madueno (2016) and corresponding simulation results. The considered performance metric is the service outage, P_{outage}, as a function of the number of supported users. The outage probability is defined as the probability of a device not being served before reaching the maximum number of failed PRACH transmission attempts (here we allowed 10 transmission attempts).

First, in Figure 3.10 we investigate the performance of traditional LTE 1.4 and 5 MHz systems and compare those results to the case of a simplified access procedure where the

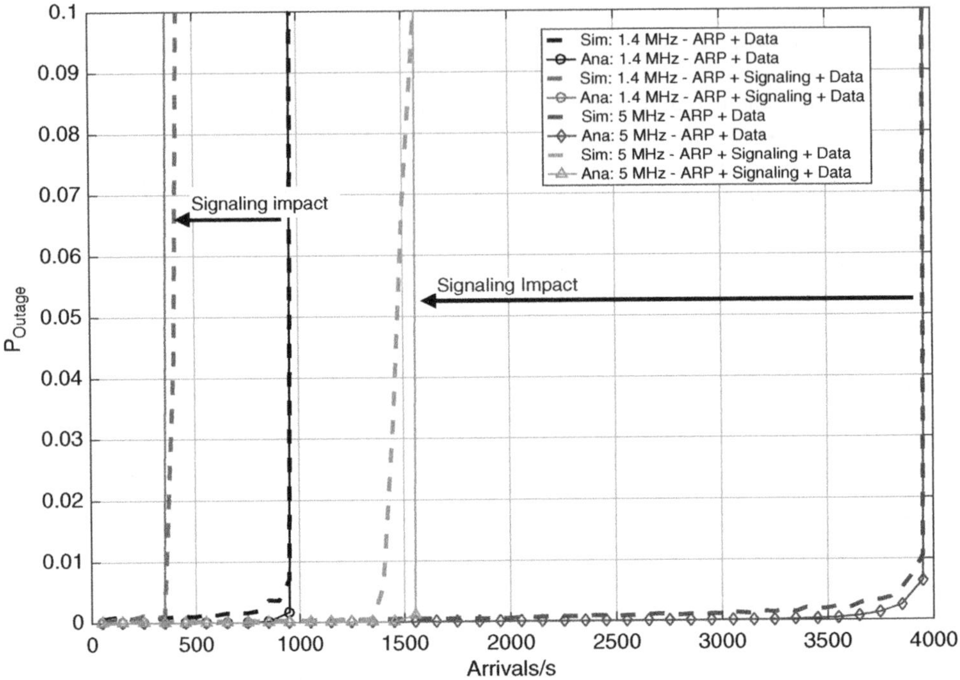

Figure 3.10 Outage comparison for only ARP and data transmission (ARP + Data) and full message exchange (ARP + Signaling + Data; Nielsen, 2017).

"additional signaling" following MSG 4 in Figure 3.9 is omitted. The striking conclusion is that for both the 1.4 MHz and 5 MHz cases, the number of supported arrivals can be increased by almost a factor of 3, going from 400 to 1000 arrivals/s and from 1500 to 4000 arrivals/s, respectively, if the additional signaling could be omitted. Obviously, *the additional signaling must be accounted for as it has a large impact on the system performance.*

Second, we consider only the reduced signaling case where immediately after the ARP (i.e., after MSG 4), the data transmission starts. The impact of the payload (MAC layer limitations) becomes clear in Figure 3.11. A 1.4 MHz system can support a few hundreds (100 arrivals/s) for large eSM payloads (1000 bytes) and up to 1000 arrivals/s for small SM payloads (100 bytes). As expected, increasing the bandwidth does help to increase the capacity of the system, raising the number of supported arrivals to 700 arrivals/s and 4000 arrivals/s, respectively. It should be noted that if the ARP is neglected and the focus is solely on the data capacity, up to 9000 arrivals/s can be supported. When compared to our results, where the different ARP limitations are taken into account, it is clear that for M2M scenarios, data capacity–based analyses are too simplistic and give overly optimistic results.

A key finding of the study of bottlenecks in the LTE access reservation protocol is that especially in M2M scenarios such as smart grids, the signaling channels PRACH and PDCCH may limit the number of supported devices. In case of larger payloads, (e.g., PMU measurements), the PUSCH channel that carries the payload data may instead be the limiting factor. The study also demonstrates how inefficient LTE is, if the used

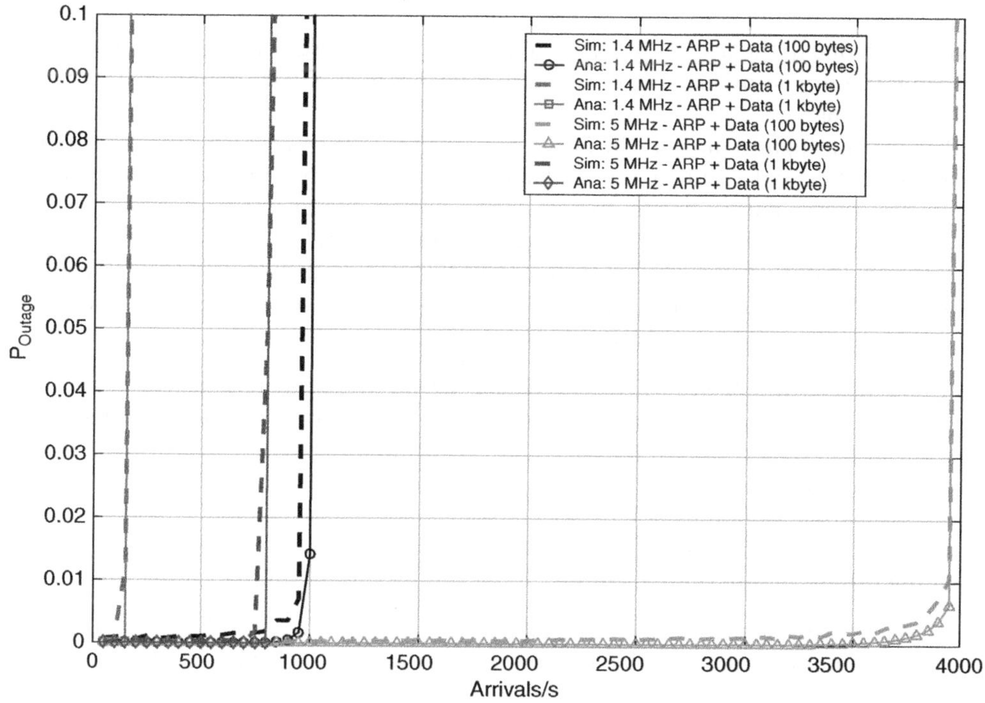

Figure 3.11 Probability of outage in LTE with respect the number of M2M arrivals per second in a 1.4 MHz and 5 MHz system for different models and payload sizes (Nielsen, 2017).

reporting interval causes devices to go through the access reservation protocol for every message transmitted. Specifically, it was shown that if a simplified access procedure with less signaling messages was used, the number of supported devices could be increased by almost three times.

3.5.5 What Can We Expect from 5G?

Many research, standardization, and business-related bodies have arrived to a consensus that 5G wireless networks will support three generic services: enhanced mobile broadband (eMBB), massive machine-type communication (mMTC), and ultra-reliable low latency communications (URLLC). eMBB is about supporting very high data rates and consistent wireless connectivity and coverage, regardless of whether the user is in a crowd or in a rural area. mMTC refers to the support of massive number of simple machine-type devices and, in that sense, it is an evolution of NB-IoT. URLLC are aimed at supporting mission-critical connectivity, with packet error rates going down to 10^{-5} or even 10^{-8} in a controlled industrial environment, while the latency of a packet delivery goes below 5 ms, even below 1 ms in some cases.

Differently from the previous generations of mobile communication systems, 5G will be very much focused on supporting various *vertical industries*, such as energy, transport, manufacturing, and health care. The smart grid communication plays a central role in defining the relationship between 5G and the energy sector. Even more, smart grid communication represents a large business potential for 5G (5GPP-IA, 2015), both due to the use of URLLC in applications for energy protection, control, and distributed automation because of the importance of mMTC for supporting the future smart metering.

Perhaps the highest novelty in 5G is brought by URLLC, which aims to provide extremely robust links with guaranteed latency and reliability. The existence of URLLC links can significantly change the design approach to distributed cyber-physical systems, and one of the main use cases is related to smart grid distribution automation. Specifically, 3GPP is considering the use of URLLC in wide-area monitoring and control systems for smart grids (3GPP, TR22862). Initial performance requirements have been derived from EC FP7 project METIS (METIS, 2013):

- Throughput: from 200 to 1521 bytes reliably (99.999 %) delivered in 8 ms.
- One trip–time latency between any two communicating points should be less than 8 ms for an event-triggered message that may occur anytime.
- Device density:
  - dense urban hundreds of UEs per km^2;
  - urban around 15 UEs per km^2;
  - populated rural max 1 UE per km^2.

As a specific example, the use case of smart grid substation control for power outage avoidance requires end-to-end latency lower than 1 ms and packet loss probability lower than 10^{-4}.

In summary, there is a symbiotic relationship between two of the modes in 5G (mMTC and URLLC) and the smart grid communication, since the former provides the technology and the latter represents one of the most important use cases that encompasses both mMTC and URLLC.

3.6 End-to-End Security in Smart Grid Communications

Security is an important issue for smart grid communications, and this is easily understood when considering the potential disruptions resulting from the unavailability of the power grid. The role of the security framework is to protect the grid against unfriendly attacks. Its definition starts with the definition of the security objectives, which may be stated as follows (EG, 2012):

- avoid international, cross-border, national, or regional power outage (availability objective);
- avoid international, cross-border, national, or regional damage to the infrastructure;
- avoid significant impact to energy market participants; this includes fraud prevention;
- avoid significant impact on operations and maintenance of the grid;
- protect citizen personal data by insuring privacy of communications; and
- protect people from safety issues (including those caused by equipment malfunction or used outside of safety limits).

The definition of the security risks is also depending upon the attackers' profiles. The following profiles are commonly identified (EG, 2012):

- individual hackers or criminals trying to gain some advantage (financial or other) by hacking the grid or individual device end points;
- crime organizations motivated by ideological, political, or financial reasons trying to disrupt the functioning of the grid or target customer data;
- large groups of protesting citizens, manipulated by a few active individuals and acting in reaction to specific issues; and
- foreign government organizations aiming at disorganizing the functioning of another country society.

The protection of the data communications occurring in a smart grid is central to meet the security objectives. Security, however, often appears like an onion, made of several layers, and the main reason for this situation is that several actors coexist in the security chain, and they need to operate without interfering with each other. The superposition of security schemes is the simplest way to achieve separation of roles, even if this benefit is paid by a higher overhead in terms of data transmitted and computing power. This overhead translates in terms of energy consumption, which is sometimes an important issue in IoT use cases.

For smart grid communications, the following security actors may be identified:

- the communication service provider (i.e., telecom operator);
- the IT team of the DSO or associated partners;
- the DSO or associated partners' application management and deployment team; and
- end-point device manufacturers.

Smart grid applications often involve the deployment of three distinct security layers:

- network access security;
- transport level security; and
- application level security.

In the following sections, each of these layers is treated individually.

3.6.1 Network Access Security

This type of security targets the protection of the data transiting through networks. For data transiting via public networks, the communication service provider will generally provide security mechanisms under its control to secure the data transmission. In 3GPP wireless networks, UICC cards are used to secure the communication from the terminal equipment to the core of the MNO network. The use of private access point names (APN) protected by specific credentials is another option to isolate data further from the public network by restricting/limiting access to a private network.

Another security layer, possibly managed by the IT team of the DSO or one of its partners, may be added to the telco-managed data protection. Virtual private networks (VPN) can be used to create this protection layer protecting the transferred data via public networks in a ciphered tunnel. VPNs are usually a corporate-controlled protection. They are used to connect data centers together or to connect communicating devices or human beings to data centers. Data is protected up to the corporate VPN gateway and will need an extra security scheme to extend security beyond that point.

Network access generally includes an access rights–management layer, which is defining network access rules and does not deal with application access control. Finally, network access security provides an overall data protection possibly shared by multiple application streams.

3.6.2 Transport Level Security

This type of security primarily aims at protecting point-to-point data communication between two communicating nodes by securing the transport layer. It provides a good specific protection for the data stream of particular applications. Transport level security is independent from the communication channel (4G, Wi-Fi, PCL, or Ethernet). It does not require any ciphering of the data payload and is, therefore, transparent from the data-handling part of the application. Transport level security is very widely used along with IP communications; in this case TCP links are secured using either TLS protocols and UDP links using the DTLS protocol or adaptation of TLS for datagram transmissions. Transport level security may be used as complement to network access security to protect data in the LAN part of the communication. The credentials used to setup transport level protection are sometimes also used to enforce application access control. Transport level credentials are typically managed by the application deployment or management team.

3.6.3 Application Level Security

This type of security aims at the protection of the application payload. It is not commonly implemented when there a transport exist, but may be necessary when the transport layer is absent (see above the case of LoRa or Sigfox communications). As an analogy, setting up end-to-end application level security may be compared to setting up voice encryption from handset to handset for mobile communications. It involves the setup of a credential distribution mechanism to protect voice communication. Application level security is useful to protect the payload when the application data has to transit via several untrusted communication nodes, requiring an overall data protection from the data source to the data destination with a single set of credentials. Application level credentials are managed by the application deployment team.

Figure 3.12 illustrates the use of different data protection schemes for a communicating WAMS sending data through a 3G/3G communication modem. We see three types of data protections involved: 3G/4G network access, VPN, and transport security. Application level security has been omitted to simplify the figure.

3.6.4 End-to-End Security

From the security standpoint, the need to achieve end-to-end security is now prevailing and becoming widely accepted. End-to-end security involves the protection of the whole communication path, from source to destination. The simplest way to achieve it over a multi-hop communication scheme is to protect each hop of the communication independently from the next. Such security is sometimes referred as "hop-by-hop security," and it is very common to see hop-by-hop security implemented at least at the transport level, where each hop is protected using a hop specific set of credentials. This approach makes it possible to have credentials for each hop managed by different parties.

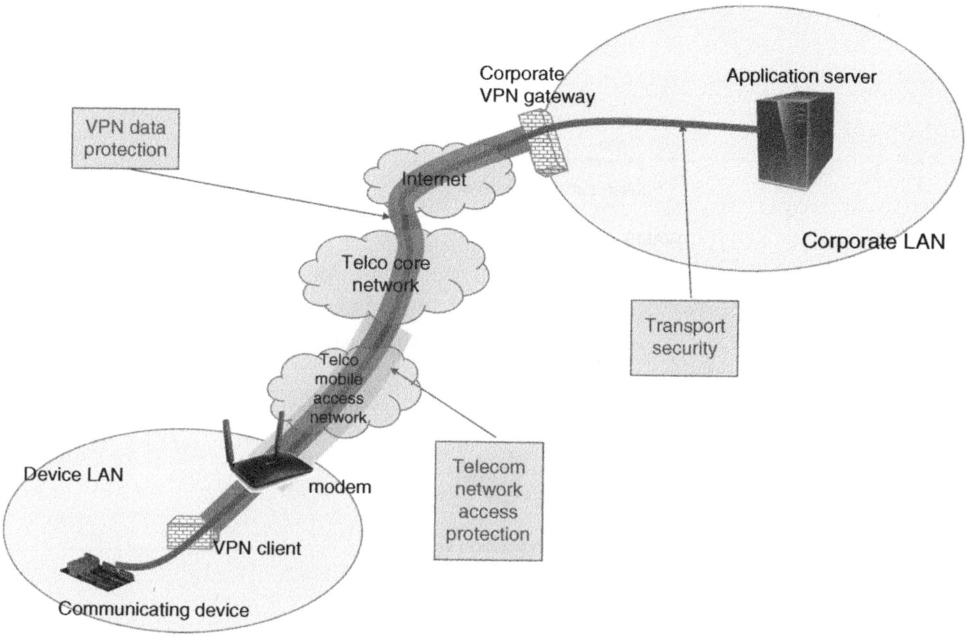

Figure 3.12 Illustration of different types of data protection schemes.

3.6.5 Access Control

Access control is used to regulate access to controlled resources. The meaning of access control depends upon the resource that needs to be protected. Each of the security layers described above may be associated to a specific kind of access control:

- Access control for telecom networks regulate access to telco-operated networks.
- Access control for VPN controls, which may set up a point-to-point VPN connection.
- Access control for IoT or smart-grid applications defines the rules of interaction between different IoT nodes, and such control is required in almost all IoT applications.

The main problem, however, is to manage access control in a heterogeneous environment composed of a number of independent applications. Furthermore, a responsibility and liability problem may arise when communicating IoT nodes are not owned by the same business entity. The questions arising are: Who may manage access control? And what is the liability involved in doing so?

Authorization delegation helps to solve those problems. It consists in delegating access control to an external entity, which manages access control rules. This notion of authorization delegation has emerged during the last decade to solve the problem of controlling access to cloud data from cloud applications. The Oauth2 protocol is commonly used for this purpose. It presents the great advantage of making possible the centralization of access control decisions in a single platform (policy decision point). Access rights granted are typically materialized in the form of digital access tokens, which need to be presented at the access control enforcement point to gain access to protected resources.

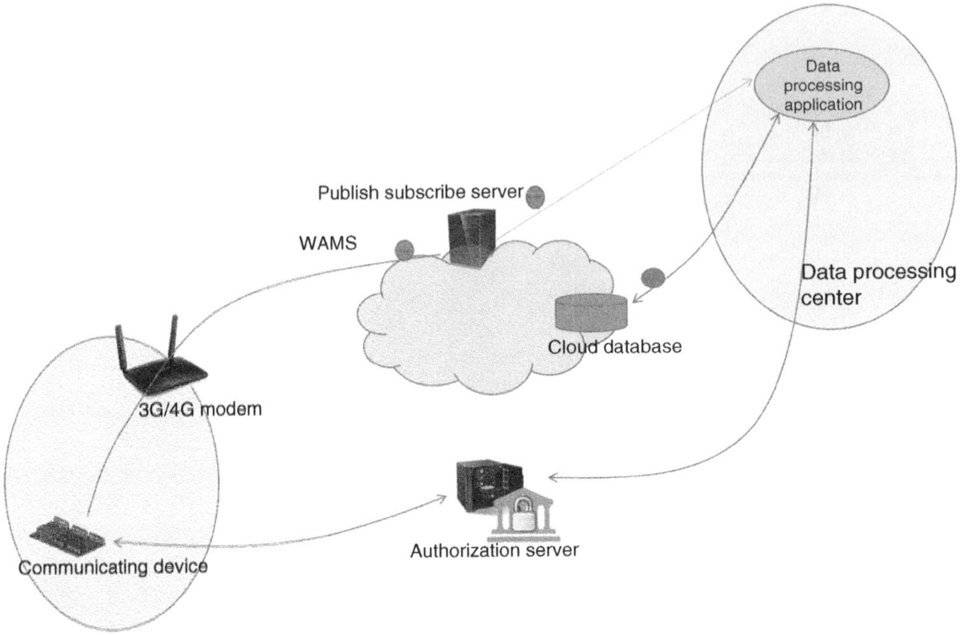

Figure 3.13 Illustration of access control delegation for a smart-grid application.

Figure 3.13 illustrates the benefits of access control delegation, by considering a typical smart grid use case.

In this figure a WAMS node is periodically sending power measurements data to the cloud using a publish/subscribe protocol (such as the popular MQTT protocol). Data published from the WAMS is sent to a publish/subscribe server in the cloud (publishing is typically performed on a specific "topic"). A data-processing application will subscribe to the data sent by the WAMS device, process it, and store the result in a cloud database. Two resources in this scenario need to be access controlled:

- Publish/subscribe: Controlling publish/subscribe involves controlling who may publish or subscribe on a specific topic.
- Database: controlling access to the database involves controlling who can read, query, or write in the cloud database.

A common situation is that the publish/subscribe server and the database will have their own access control system. A first possibility is therefore to manage access rights independently in the database and in the publish/subscribe server. Unfortunately, this approach is time consuming and error prone and can lead to the introduction of security weaknesses because access rights are generally granted in the context of a work flow (as the one described). It is therefore likely that the need will arise to grant or remove simultaneously publish/subscribe and database access rights.

Access control delegation enables the delegation of access control management to an independent authorization server. This server may possibly tie publish/subscribe and database access rights together, allowing for holistic access control management at the level of the work flow. The Oauth2 protocol mentioned above may be used to implement such authorization delegation. In this case authorization decisions made by the authorization server are materialized by digital access tokens to be presented at the access enforcement point (publish/subscribe server or database).

Who is responsible in such a scenario to operate the authorization server? Different possibilities may be envisaged:

- When a single business actor is involved, this actor may operate the authorization server for its own benefit.
- When two or more business actors are involved, one of them may be trusted to operate the authorization server for the mutual benefit of all parties. This is a good solution if there is no or little liability involved.
- When liability is important, the authorization server may be operated as a service by a separate trusted entity.

3.7 Conclusions and Summary

The smart grid presents a wide variety of applications with very different communication requirements, ranging from a potentially huge number of sensors that do not communicate very often, through a moderate number of measurement and control nodes with a moderate data rate, to mission-critical grid protection that requires extremely low latency and ultra-reliable communication. In all cases the communications require end-to-end security to protect privacy, integrity, and confidentiality, manage permissions, and avoid malicious attacks.

In the present chapter, we have described and analyzed the most relevant wireless cellular communication technologies for supporting the smart grid. Specifically, we have considered the non-3GPP technologies IEEE 802.11ah, Sigfox, and LoRa. Under the umbrella of 3GPP, we have looked specifically at releases up to and including Rel. 11 as well as Rel. 12 and Rel. 13.

3GPP's proposed RAN solution for cellular IoT in Rel. 13, denoted NB-IoT, is able to compete with Sigfox and LoRa in terms of coverage and low power for low-rate, battery-powered wide-area IoT deployments. Further, it is able to support higher and more symmetric (uplink/downlink) throughputs.

LTE is capable of supporting real-time smart-grid applications such as real-time monitoring and control, where the latency and reliability requirements are not as extreme as for grid-protection functions. Specifically, LTE is well suited for measurement stream with reporting intervals around 1 to 10 seconds. For longer intervals, the LTE access reservation procedure becomes a bottleneck and effectively reduces the number of supported devices by as much as 3 times, due to signaling overhead. For such applications, with more seldom communication, the lightweight IoT technologies are better suited.

For the strictest smart grid applications related to grid protection, the existing wireless cellular technologies are unable to provide the required performance and reliability. Such applications are, however, covered by the 5G use case ultra-reliable low-latency communications (URLLC). Furthermore, the massive machine-type-communications (mMTC) use case is also highly relevant for the smart grid.

Protecting the smart grid asset is an important issue, and securing smart grid communications is a critical step to achieve this goal. Smart grid communication may be protected using different layers of security the presence of which is essentially motivated by the need to achieve separation of responsibility between different actors involved in security management. Finally, proper management of access control in a smart grid ecosystem is critical; authorization delegation may help to centralize access control to several independent protected resources in a smart grid ecosystem and implement a work flow–based homogeneous access control.

References

3GPP Technical Report TR 36.822, V11.0.0, 'LTE Radio Access Network (RAN) enhancements for diverse data applications', September 2012.

3GPP Work Item Description RP-141865, 'Further LTE Physical Layer Enhancements for MTC', September 2014.

3GPP Work Item Description RP-151621, 'Narrowband IOT', September 2015.

3GPP TR 45.820 v13.1.0, 'Cellular system support for ultra-low complexity and low throughput Internet of Things (CIoT)', Release 13, November 2015.

3GPP TS 36.300 v13.4.0, 'Evolved Universal Terrestrial Radio Access (E-UTRA) and Evolved Universal Terrestrial Radio Access Network (E-UTRAN); Overall Description; Stage 2', Release 13, June 2016.

3GPP, 'Overview of 3GPP release 11,' 3rd Generation Partnership Project (3GPP), Tech. Rep.

3GPP, 'MTC simulation assumptions for RACH performance evaluation', 3rd Generation Partnership Project (3GPP), TR R2-105212, August 2010.

3GPP, 'TS 36.321 E-UTRA medium access control (MAC) protocol specification', Tech. Rep., 2015.

3GPP, 'TS 36.213 E-UTRA physical layer procedures', Tech. Rep., 2015.

3GPP, 'TR 36.822: LTE Radio Access Network (RAN) enhancements for diverse data applications, Rel. 11', Tech. Rep., September 2011.

3GPP TR 22862 v14.1.0, 'Feasibility Study on New Services and Markets Technology Enablers for Critical Communications', Release 14, September 2016.

5GPPP Infrastructure Association, '5G and Energy', version 1.0, September 2015.

Adame, T., Bel, A., Bellalta, B., Barcelo, J., and Oliver, M., 2014, 'IEEE 802.11AH: the WiFi approach for M2M communications', in *IEEE Wireless Communications*, vol. 21, no. 6, December 2014, 144–152.

Boccardi, F., Heath, R.W., Lozano, A., Marzetta, T.L., and Popovski P., 2014, 'Five disruptive technology directions for 5G', in *IEEE Communications Magazine*, vol. 52, no. 2, February 2014, 74–80.

Deshpande, J.G., Kim, E., and Thottan, M., 2011, 'Differentiated services QoS in smart grid communication networks', *Bell Labs Technical Journal*, vol. 16, no. 3, 2011, 61–81.

US Department of Energy, 2012, 'Smart Grid Research & Development: Multi-Year Program Plan (MYPP)', URL: http://www.oe.energy.gov.

European Commission, 2006, 'European Smart Grids Technology Platform—Vision and Strategy for Europe's Electricity Networks of the Future', Directorate-General for Research—Sustainable Energy Systems, URL: http://www.smartgrids.eu.

Expert Group on the Security and Resilience of Communication Networks and Information Systems for Smart grids, 2012, 'Cybersecurity of the smart grids; summary report'.

ETSI GS LTN 002 v1.1.1, 2014, 'Low Throughput Networks (LTN); Functional Architecture', September 2014.

ETSI GS LTN 003 v1.1.1, 2014, 'Low Throughput Networks (LTN); Protocols and Interfaces', September 2014.

Erol-Kantarci, Melike, Mouftah, H.T., 2013, 'Machine-to-machine communications in the smart grid,' *Machine-To-Machine Communications—Architectures, Technology, Standards, and Applications*, Misic, J. and Misic, V. (eds.), 2013, Taylor & Francis.

Goel, S., Bush, S.F., and Bakken, D., 2013, 'IEEE vision for smart grid communications: 2030 and beyond'. New York, NY, USA, Institute of Electrical and Electronic Engineers.

Hossain, E., Han, Z., and Poor, H.V., 2012, *Smart grid communications and networking*. Cambridge University Press.

IEEE P802.11- Task Group AH, 2016, 'P802.11ah draft 8.0', URL: http://grouper.ieee.org/groups/802/11/Reports/tgah update.htm.

Khan, R.H. and Khan, J.Y., 2013, 'A comprehensive review of the application characteristics and traffic requirements of a smart grid communications network', *Computer Networks*, vol. 57, no. 3, 825–845.

Lander, M., Svoboda, P., Nikaein, N., and Rupp, M., 2013, 'Traffic Models for Machine Type Communications', in *Proc. of the International Symposium on Wireless Communication Systems* (ISWCS 2013), Aug. 2013.

Lee, K., Eidson, J.C., Weibel, H., and Mohl, D., 2005, 'IEEE 1588-standard for a precision clock synchronization protocol for networked measurement and control systems', in *Conference on IEEE*, vol. 1588, 2005, 2.

LoRa Alliance, URL: https://www.lora-alliance.org/.

Madueño, G.C., Nielsen, J.J., Kim, D.M., Pratas, N.K., Stefanović, Č., and Popovski, P., 2016, 'Assessment of LTE Wireless Access for Monitoring of Energy Distribution in the Smart Grid', in *IEEE Journal on Selected Areas in Communications*, vol. 34, no. 3, March 2016, 675–688.

Martin, K., Hamai, D., Adamiak, M., Anderson, S., Begovic, M., Benmouyal, G., Brunello, G., Burger, J., Cai, J., Dickerson, B., Gharpure, V., Kennedy, B., Karlsson, D., Phadke, A., Salj, J., Skendzic, V., Sperr, J., Song, Y., Huntley, C., Kasztenny, B., and Price, E., 2008, 'Exploring the IEEE standard c37.118-2005 synchrophasors for power systems', Power Delivery, *IEEE Transactions*, vol. 23, no. 4, Oct 2008, 1805–1811.

METIS D1.1, April 2013, 'Scenarios, requirements and KPIs for 5G mobile and wireless system'.

Nielsen, J.J., Corrales Madueno, G., Pratas, N.K., Sørensen, R. B., Stefanovic, C., and Popovski, P., 'What can wireless cellular technologies do about the upcoming smart metering traffic?' 2015, *IEEE Communications Magazine*, vol. abs/1502.01188.

Nielsen, J.J., and Popovski, P., 2016, 'Latency analysis of systems with multiple interfaces for ultra-reliable M2M communication', in the 17th IEEE International workshop on Signal Processing advances in Wireless Communications, IEEE, 1–6.

Nielsen, J.J., et al., 2014, 'Secure Real-Time Monitoring and Management of Smart Distribution Grid Using Shared Cellular Networks', in *IEEE Wireless Communications*, vol. 24, no. 2, April 2017, 10–17.

Nokia white paper, 2014, 'Optimizing LTE for the Internet of Things'.

Sesia, S., Toufik, I., and Baker, M., 2011, *LTE—The UMTS Long Term Evolution: From Theory to Practice*. Wiley.

SIGFOX white paper, 'M2M and IoT redefined through cost effective and energy optimized connectivity', URL: http://www.Sigfox.com/static/media/Files/Documentation/SIGFOX_Whitepaper.pdf.

SUNSEED D3.2.2, 2016, 'Communication solution design for field trial setup'.

Thomsen, H., Pratas, N., Stefanovic, C., and Popovski, P., 2013, 'Analysis of the LTE Access Reservation Protocol for Real-Time Traffic', *IEEE Commun. Lett.*, vol. 17, no. 8, Aug. 2013, 1616–1619.

Zhang, H., Jorguseski, L., Chrysalos, M., Golinski, M., Toh, Y., 2016, 'LTE Delay Assessment for Real-Time Management of Future Smart Grids', SMARTGIFT conference, 2016.

Zhu, Z., Lambotharan, S., Chin, W.H., and Fan, Z., 2012, 'Overview of Demand Management in Smart Grid and Enabling Wireless Communication Technologies', *IEEE Wireless Communications Magazine*, vol.19, no.3, June 2012, 48–56.

4

Machine-to-Machine Communications in the Smart City—a Smart Grid Perspective

Ravil Bikmetov[1], *M. Yasin Akhtar Raja*[2] *and Khurram Kazi*[3]

[1] *University of North Carolina at Charlotte, NC, USA*
[2] *Center for Optoelectronics and Optical Communication, University of North Carolina at Charlotte, NC, USA*
[3] *Principal Member of Technical Staff, Draper Laboratory, Cambridge MA, USA*

4.1 Introduction

"A Smart City is an urban system that uses information and communication technology (ICT) to make both its infrastructure and its public services more interactive, more accessible and more efficient" (Pellicer at al., 2013). The infrastructure of the smart city (SC) is required to be equipped with state-of-the-art technological solutions to support citizens' interaction with urban elements in the most efficient and convenient way. Such requirements are one of the main factors driving recent advancements in ICT, associated sensing manifolds, and data-handling peripherals to support the SC paradigm (Pellicer et al., 2013; Zanella et al., 2014; Dohler, Boswarthick, and Alonso-Zárate, 2012; Giffinger et al., 2007; Sun et al., 2015).

The emergence of SCs was prompted by three main factors: a growing migration of constantly increasing world population from rural areas to urban centers (forecast to reach 70% by 2050), inevitable shortage of natural resources for the world population, and concerns about the environment and climate change. By 2020, the SC market is projected to exceed $400 billion with 600 cities around the globe (Maddox, 2016).

The SC concept has been developed to address such urban challenges by efficient management of rapidly growing city infrastructure and services, while meeting the needs of its citizens. According to the initial scheme of SC originally proposed by the Center of Regional Science at the Vienna University of Technology (Center of Regional Science, 2007), the foremost aspects of the SC concept are:

- *Smart economy (competitiveness)*—innovative entrepreneurship and productivity, flexibility of labor market, international embeddedness, and ability to transform.
- *Smart governance (citizen participation)*—public participation in decision making, public and social services, transparent control and political strategies.
- *Smart people (social and human capital)*—high level of qualification, commitment to lifelong learning, social and ethnic plurality, flexibility, and creativity.
- *Smart mobility (transport and ICT)*—local and international accessibility of safe transport systems and availability of sustainable ICT infrastructure.

Transportation and Power Grid in Smart Cities: Communication Networks and Services, First Edition.
Edited by Hussein T. Mouftah, Melike Erol-Kantarci and Mubashir Husain Rehmani.

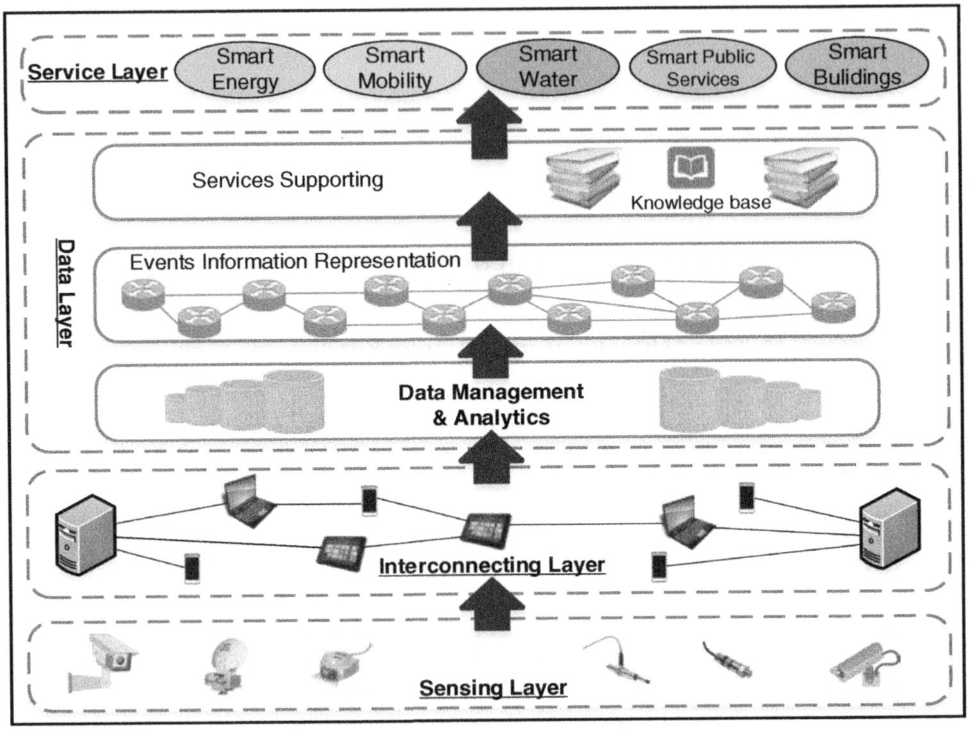

Figure 4.1 A functional architecture of the smart city concept. Sensing, Interconnecting, data, and service are four layers of this infrastructure. Data collection from lower layers to higher layers is shown by arrows.

- *Smart environment (natural resources)*—energy efficiency, sustainable resource management, and environmental protection.
- *Smart living (quality of life)*—cultural facilities, health conditions, individual safety, housing quality, education facilities, and high-level social organization.

Smart city's (SC's) functional architecture is represented in Figure 4.1 from an information and communication technology (ICT) point of view (Pellicer et al., 2013; Sun et al., 2015). The previously described SC's aspects are located at the service layer of this infrastructure. While other interacting layers of a SC (see Figure 4.1) are part of an essential infrastructure for any human-populated clusters: cities, towns and villages, etc., the smartness is afforded by information and awareness enabled by ICT.

Each layer of the SC's functional architecture has its own components and functionality (Sun et al., 2015; Sun et al., 2014a; Zhuge and Sun, 2010; Sun et al., 2016; Sun and Jara, 2014; Sun et al., 2014b; Pouryazdan et al., 2016; Ganti, Ye, and Lei, 2011; Campbell et al., 2016; Campbell et al., 2008) that are briefly described in the following.

- ***Sensing layer***
  Represented by radio-frequency identification (RFID) and utilized for ubiquitous sensing enabled by wireless sensor networks (WSN) that are also becoming popular and essential people-centric urban sensing (Campbell et al., 2016; Campbell et al., 2008). Further, people-centric sensing can be categorized into three classes: personal

sensing, which focuses on personal monitoring and archiving; social sensing, which focuses on sharing information within social and special interest groups; and public sensing, which focuses on sharing data with everyone for the greater public good (such as entertainment or community action; Campbell et al., 2008). With the popularity of smart phones containing various sensors such as camera, audio, accelerometer, GPS (Global Positioning System), gyroscope, compass, proximity, and ambient light, and among others a new community/public sensing paradigm called "mobile crowd-sensing (MCS)" rapidly emerging. This paradigm is defined as a sensing by a large group of mobile devices and sharing mutual information in this group for further data analysis (measuring, mapping, estimating, prediction, etc.) of processes of presenting a common interest for this group (Pouryazdan, 2016; Ganti, Ye, and Lei, 2011).

- ***Interconnecting layer***
  It is utilized for data transmission and information exchange among different devices and different domains in the layer (Sun et al., 2015; Sun et al., 2014a; Zhuge and Sun, 2010; Sun et al., 2016; Sun and Jara, 2014; Sun et al., 2014b; Pouryazdan, 2016; Ganti, Ye, and Lei, 2011; Campbell et al., 2016; Campbell et al., 2008.
- ***Data layer***
  It is utilized for storage of massive and heterogeneous data generated from a variety of monitoring devices in the sensing layer (Sun et al., 2014b), extracting useful information from the large amount of data being sensed and representing the meaningful information in reasonable and efficient ways (Sun et al., 2014a; Sun and Jara, 2014; Sun et al., 2014b), decision making and service supporting (Sun et al., 2016; Sun and Jara, 2014), and knowledge maintenance and management (Zuge and Sun, 2010).
- ***Services (application) layer***
  This layer is utilized to provide services to the communities of a smart city in an efficient and reliable manner. It includes the main aspects of the SC's initial scheme and performs a direct interaction with the citizens of the SC (Sun et al., 2015; Sun et al., 2014a; Zhuge and Sun, 2010; Sun et al., 2016; Sun and Jara, 2014; Sun et al., 2014b; Pouryazdan et al., 2016; Ganti, Ye, and Lei, 2011; Campbell et al., 2016; Campbell et al., 2008).

This chapter focuses on the smart energy concept for a SC, which is presented by the smart grid (SG) paradigm. The first main section is designated for the architecture and common characteristics of SGs. The second major section describes intelligent machine-to-machine interactions in SGs. In the third section, optimization algorithms for energy production, distribution, and consumption are introduced and briefly discussed. The last section finally outlines the machine learning techniques in efficient energy services and from a management perspective. A quick overview of future perspectives of machine-to-machine communications within SGs is given in section five.

4.2 Architecture and Characteristics of Smart Grids for Smart Cities

A pre-introduced concept of a SC was developed to achieve and maintain efficient management in all areas that belong to any city. Those include urban planning, infrastructure, transport, energy, civic services, education, health, public safety, and so on. Within a SC concept, these areas should satisfy both the needs of individual citizens and the communities as a whole. In any city or population cluster, the energy infrastructure is the foremost important function continuously supporting all other domains and civic aspects of the city. A reliable and efficient energy service can significantly improve the performance of other areas and aspects of a SC including its management efficiency, safety, and resources conservation.

A smart grid (SG) can be considered as "an electric system that uses information, two-way, cyber-secure communication technologies, and computational intelligence in an integrated fashion across electricity generation, transmission, substations, distribution and consumption to achieve a system that is clean, safe, secure, reliable, resilient, efficient, and sustainable" (Fang et al., 2013). A smart grid (SG) paradigm was developed to ensure efficient, reliable, and environmentally friendly energy service for the citizens of a population cluster emerging in a SC (Fang et al, 2012; NIST, 2014; NIST, 2010; NIST, 2012). The SG paradigm modernizes the existing power systems through self-healing, automation, and remote monitoring and control. Besides, the SG informs and educates consumers about their energy usage, costs, and alternative options, to enable autonomous decision making about how and when to use electricity and fuels. In addition, the SG provides safe, secure, and reliable integration of distributed and renewable energy sources to support consumers' comfort experience and environment. All these enhancements of an energy infrastructure make it more reliable, sustainable, and resilient by placing a SG in the heart of the SC. To support their functions and civic services, SCs depend on their SGs, which ensure resilient delivery of energy, present opportunities for its conservation, improve efficiency of its consumption, and most importantly, enable coordination between urban infrastructure operators and energy service providers.

4.2.1 Definition of a Smart Grid and Its Conceptual Model

A *traditional energy grid* can be generally regarded as an interconnected network for electricity transmission and its delivery from generation sites to the consumers' domains.

During the conventional development of existing energy grids, the electric power industry had traditionally dedicated more attention and resources to generation and power transmission networks rather than to power distribution networks, as shown in Figure 4.2. This process occurred due to traditional paradigm that the demand for electricity completely dictates its generation (Fang et al, 2012; López et al., 2014; Tuttle et al., 2016; Ardito et al., 2013). As a result, bulk generation plants and power transmission systems have been traditionally monitored and controlled using legacy communication networks, which often allow a certain level of centralized coordination. Moreover, power distribution grids have been traditionally passive systems with limited communication capabilities. This fact significantly reduced flexibility, sustainability, and efficiency of traditional energy grids (Figure 4.2; Fang et al, 2012; López et al., 2014; Tuttle et al., 2016; Ardito et al., 2013).

In comparison with traditional energy grids (Figure 4.2), a smart grid (SG) represents a true revolution in energy distribution and supply including its users' domains involving a wide variety of technologies and numerous standards developed to ensure reliability and interoperability (Fang et al., 2012; NIST, 2014; NIST, 2010; NIST, 2012; Ardito et al.,

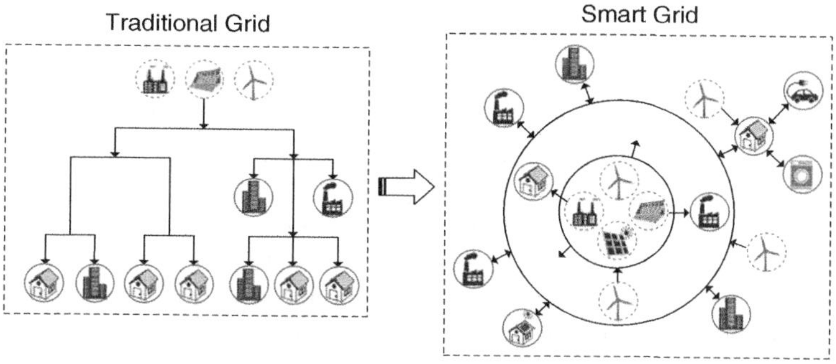

Figure 4.2 Transition from traditional grids to smart grids.

2013). Following a standardization process, the International Electrotechnical Commission (IEC) has created a free "Smart Grid Standards Mapping Tool" (SGSMT) allowing the identification of the standards that are required for any part of the SGs (IEC, 2016). Besides IEC, the standards from other technical organizations were also included in the SGSMT.

During the standardization process, the Smart Grid Interoperability Panel (SGIP) at the National Institute of Standards and Technology (NIST), a private/public partnership funded by different industry stakeholders in cooperation with the United States federal government, focused on the development of a framework for coordinating all SG stakeholders and accelerating standards harmonization and interoperability (NIST, 2010). As a first step of this development, the SGIP established the SG conceptual model, illustrated in Figure 4.3. The first version of this conceptual model was published in January 2010 (NIST, 2010), and it was reviewed and updated in February 2012 (NIST, 2012). The NIST report describing the SG conceptual model identifies about 80 existing standards that can be used to support SG developments. Besides, this report identifies high priority gaps, for which new or revised standards are directly needed (NIST, 2014; NIST, 2010; NIST, 2012).

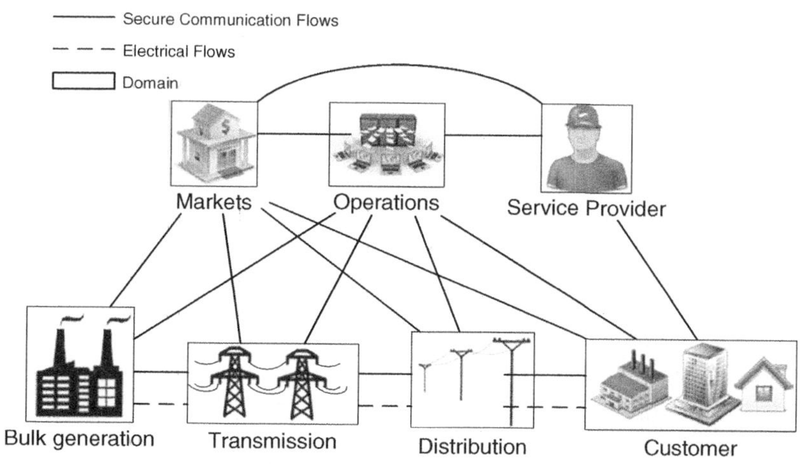

Figure 4.3 NIST Smart Grid Conceptual Model.

As shown in Figure 4.3, the SG conceptual model defines seven domains as well as the electrical and communication flows among them (NIST, 2012). The electrical flows involve the traditional subsystems of the electrical grid: bulk generation, transmission, distribution, and customer domains. Communication and information data flows create a mesh topology between almost every domain illustrating the outstanding importance of communications in the SG (NIST, 2010; NIST, 2012). Based on the terminology developed for this conceptual model, each domain and its subdomains encompass SG "actors" and "applications." Actors include devices such as smart meters, data concentrators, various buffers; systems such as energy consumption measurement and control; programs; and stakeholders that make decisions and exchange information. Applications are tasks performed by one or more actors within a domain (home and building automation, etc.) (NIST, 2010; NIST, 2012).

A specific design of a SG can be obtained by implementation of such a conceptual model for a given domain, a given application, its specific requirements, the actors involved in this application, and the description of interactions between the actors. Table 4.1 summarizes the main actors included in each domain of the SG conceptual model (NIST, 2012). Although other standardization institutions (IEEE, ETSI, and others) have defined their own SG conceptual models, the SG conceptual model developed by National Institute of Standards and Technology (NIST) is the most widely accepted (Fang et al., 2013; López et al., 2014; Smart Grid Coordination Group, 2012; Smarter Grid Solutions UK, 2014).

In the following, each domain is briefly described in reference to its typical functionality and role. The corresponding diagrams are used to provide a quick visual and comprehensive functionality.

The customer domain of the SG conceptual model (Figure 4.4) is an energy consumption domain that consists of stakeholders supported by the entire infrastructure of a SG (NIST, 2012). The actors in this domain are smart meters, Internet of Things (IoT) sensors, and other intelligent electronic devices (IEDs). These actors can manage consumers' energy usage and its generation and control the information flow between

Table 4.1 Domains and Actors in the NIST Smart Grid Conceptual Model (NIST, 2012).

Domains	Actors
Bulk generation	Power plants—generators of electricity in bulk quantities
Transmission	Transmission system operators—carriers of bulk electricity over long distances
Distribution	Distribution system operators—distributors of electricity to and from customers
Customers	End users of electricity able to generate, store, and manage their energy utilization
Operations	Managers of the power flow to and from customers
Markets	Commodity markets that specifically control the trade and supply of energy (actors from independent system operators and regional transmission organizations)
Service providers	Energy services companies—organizations providing services to electrical customers and utilities (aggregators, retailers, etc.)

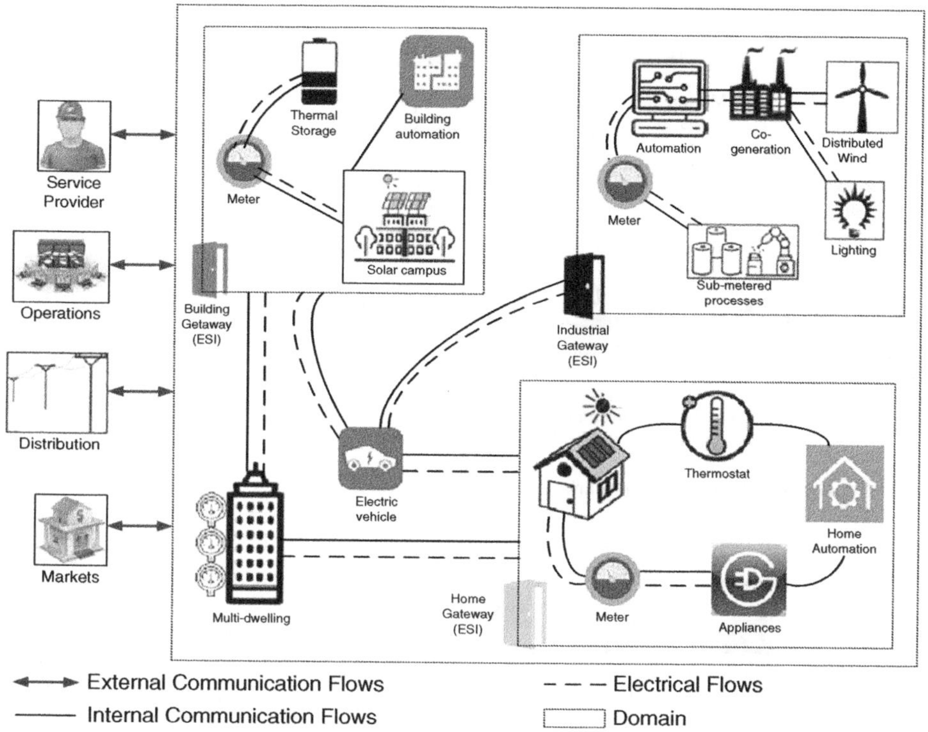

Figure 4.4 General architecture of the customer domain.

the customer domain and other domains of the SG conceptual model (cf. Figure 4.4). To perform these tasks, several actuators are used in the customer domain, such as home and building automation systems. Utility meters and energy service interfaces (ESIs) are the boundaries of the customer domain (Figure 4.4). More details about energy consumers within the SG will be discussed in Subsection 4.1.6.

The markets domain is a part of the Smart Grid (SG) conceptual model (cf. Figure 4.3), where energy grids assets are bought and sold (e.g., Figure 4.5) (NIST, 2010; NIST, 2012). The main functions of this domain are energy price exchange and balance of supply and demand within the power system. The boundaries of the markets domain are at the edge of the operations domain controlling the SG, the domains of supplying assets, such as generation and transmission, and the customer domain (Figure 4.5) (NIST, 2012). Communication flow between the markets domain and all other domains must be reliable, traceable, and auditable to ensure an efficient matching of energy production with its consumption. Besides, these communications must support e-commerce standards for integrity and non-repudiation. The participation of distributed energy resource (DER) in the markets domain is persistently increasing and becomes more interactive (Fang et al., 2012; Ardito et al., 2013; Fadlullah and Kato, 2015).

According to the SG conceptual model (Figure 4.3), the service provider domain shares interfaces with the markets, operations, and customer domains (Figure 4.6). Communication flow through these interfaces is very critical for each domain connected to the service provider: operations by system control and awareness, markets

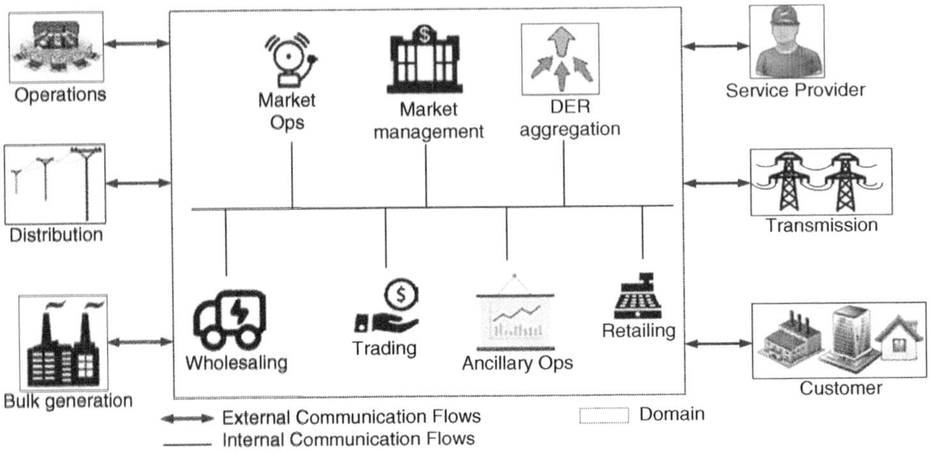

Figure 4.5 General architecture of the markets domain.

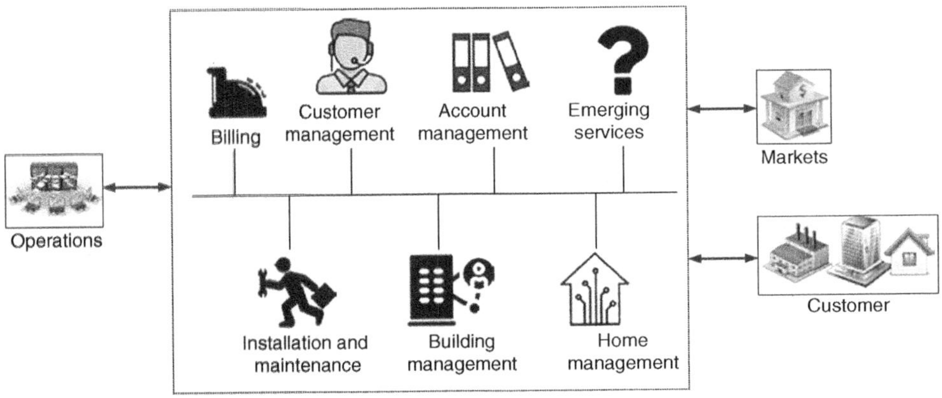

Figure 4.6 General architecture of the service provider domain.

and customer by the grid's efficiency enabling its economic growth and development of "smart" services (NIST, 2010; NIST, 2012). In the service provider domain, illustrated in Figure 4.7, actors perform services to support business processes of power system producers, distributors, and customers. The examples of these processes are billing and customer account management, management of energy use, home energy generation, etc. (CEN, CENELEC, and ETSI, 2012). More details about service providers within the SG will be given in Subsection 4.1.7.

According to the SG conceptual model (Figure 4.3), the operations domain has secure communication connections with all other domains of a Smart Grid (Figure 4.7) (NIST, 2012). Representing the major operated utility, this domain is responsible for the smooth operation of the power system (NIST, 2010; NIST, 2012). Controlling planning and service delivery processes, the operations domain supplements the functionality of the service provider and market domains (NIST, 2010; NIST, 2012).

The generation domain, illustrated in Figure 4.9, is electrically connected to the transmission or, in some cases, to the distribution domain and shares communication

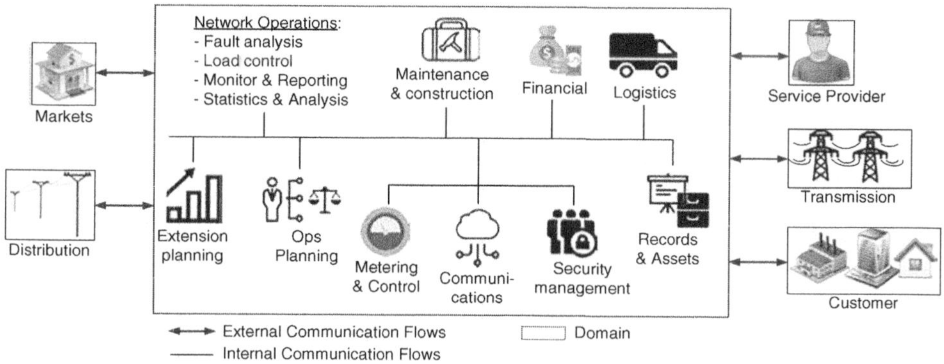

Figure 4.7 General architecture of the operations domain.

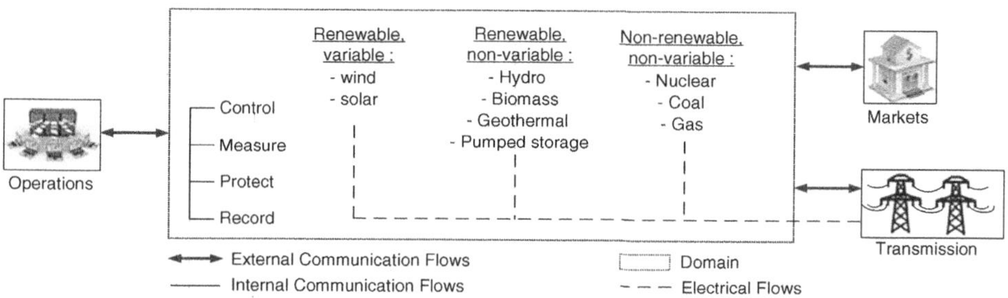

Figure 4.8 General architecture of the generation domain.

interfaces with the operations, markets, transmission and distribution domains (Figure 4.3) (NIST, 2010; NIST, 2012). The generation domain communicates key performance and quality of service for variable energy sources, which can be renewable or non-renewable (e.g., Figure 4.8). In this domain, electricity is produced from other forms of energy, which may include a wide variety of sources: chemical combustion, nuclear fission, water flow, wind, solar radiation, and geothermal heat (Figure 4.8) (NIST, 2010; NIST, 2012). Therefore, applications in the generation domain are the first processes in the electricity delivery to Smart Grid's customers. Besides, various physical actors are presented in this domain: protection relays, remote terminal units, equipment monitors, fault recorders, user interfaces, and programmable logic controllers (Figure 4.8). The boundaries of the generation domain are either the transmission or the distribution domain (NIST, 2010; NIST, 2012).

The transmission domain of the SG conceptual model (cf. Figure 4.3) performs bulk transfer of electrical power from the generation to the distribution domain through multiple substations (see Figure 4.9) (NIST, 2010; NIST, 2012). The transmission domain is typically operated by a transmission-owning utility, which can be represented by a regional transmission operator (RTO) or an independent system operator (ISO). The main responsibility of the RTO or ISO is to maintain stability of the electric grid by balancing energy generation with its load across the transmission network. The physical actors of the transmission domain (Figure 4.9) are remote terminal units, substation meters, protection relays, power quality monitors, phasor measurement units, sag monitors, fault recorders, and substation user interfaces (NIST, 2010; NIST, 2012).

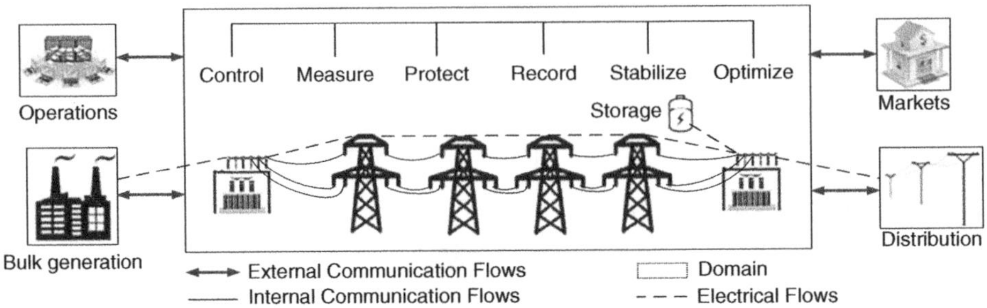

Figure 4.9 General architecture of the transmission domain.

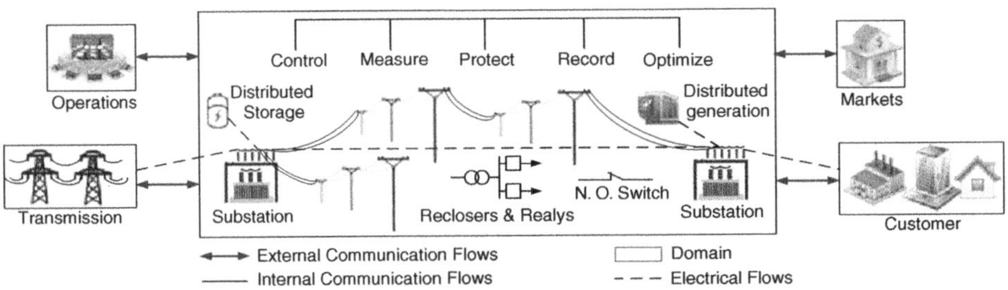

Figure 4.10 General architecture of the distribution domain.

As it was described previously, the Smart Grid concept brought decentralization to the energy distribution component of the electric power system. Because of this, many communications and electrical interfaces are considered to work in both directions supporting a bidirectional flow. Based on these considerations, the general architecture of the distribution domain was developed (Figure 4.10) (NIST, 2010; NIST, 2012). In such architectural design, distribution actors may have local peer-to-peer communication with more centralized communication methodology in several cases (Fang et al., 2012; NIST, 2010; NIST, 2012; Fadlullah and Kato, 2015).

In the Smart Grid, the distribution domain will communicate in real time more closely with the operations domain to manage the power flow associated with a more dynamic markets domain (NIST, 2010; NIST, 2012). The markets domain will communicate with the distribution domain in ways that will affect localized consumption and generation. In turn, these behavioral changes in consumption and generation due to market forces may have electrical and structural impacts on the distribution domain and the larger grid (NIST, 2010; NIST, 2012). In this architecture, service providers may communicate with the customer domain using the infrastructure of the distribution domain. The typical applications within the distribution domain are usually divided into the following categories: substation monitoring and control, management of energy storage unit, management of distributed generation, and control of SG's protection and optimization mechanisms (NIST, 2012).

4.2.2 Standardization Approach in Smart Grids

Beside the SG conceptual model (NIST, 2012), one of the main outcomes of the Smart Grid Interoperability Panel (SGIP) activity is the elaboration of standards, practices, and

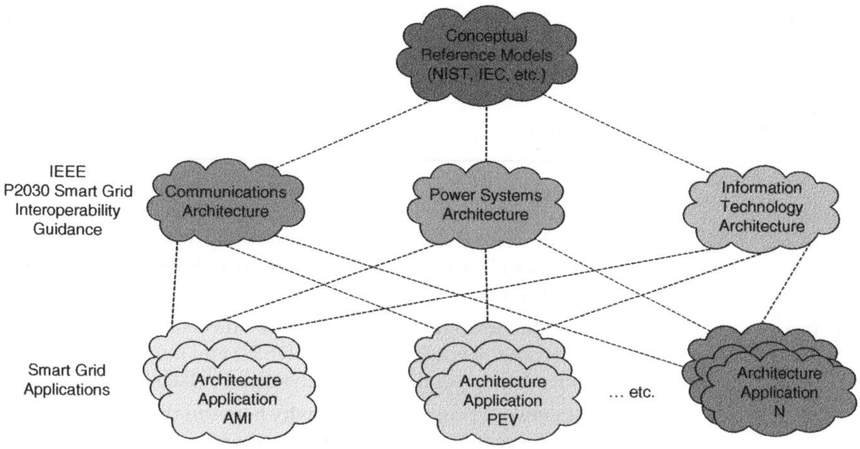

Figure 4.11 Scope of IEEE 2030 standardization process.

guidelines that allow the development and deployment of a robust and interoperable SG. As a result, in May 2011 the SGIP governing board established the Catalog of Standards (CoS). This CoS was updated several times and is available online through the National Institute of Standards and Technology (NIST) SG Collaboration (NIST, 2013). As of today, the CoS comprises 20 individual standards and five separate series containing 36 additional standards, which accounts for a total of 56 standards.

Besides the International Electrotechnical Commission (IEC), other technical professional organizations (see Table 4.7) have introduced their own standards and recommendations for development of a SG conceptual model. The prevalent contribution was done by the IEEE, which has more than 100 standards relevant to smart grids. Among them, over 20 IEEE standards were included in the NIST "Framework and Roadmap for Smart Grid Interoperability Standards," known collectively as Release 3.0, which was issued in 2014. The major standards from this document are listed in Table 4.7 (see appendix at the end of the chapter).

The scope of Institute of Electrical and Electronics Engineers (IEEE) 2030 standardization and overall reference architecture is illustrated in Figure 4.11 (IEEE, 2011). As described previously, the standardization process was initialized by conceptual reference models, such as NIST's Smart Grid conceptual model (cf. Figure 4.3) (NIST, 2014; NIST, 2012). To ensure the SG's interoperability, this process occurred in three architectural perspectives: communications, power systems, and information technology. As a result, various applications were created in a Smart Grid: advanced metering infrastructure, plug-in electric vehicle, etc. (Figure 4.11).

To define and elaborate upon such diverse functionality, a Smart Grid interoperability reference model (SGIRM) was created.

4.2.3 Smart Grid Interoperability Reference Model (SGIRM)

Once a conceptual model of the SG was defined (cf. Figure 4.3) (NIST, 2010; NIST, 2012), a reference architecture elaborating this model would be required. This architecture would define functional blocks and interfaces, thus bringing the developed conceptual model closer to implementation. IEEE's project 2030 pioneered in developing such

Table 4.2 Communications Networks Defined for the SG in the IEEE 2030 CT-IAP (IEEE, 2011; IEEE Smart Grid, 2015).

Communications Network	Description
xAN Energy services interfaces (ESIs)	Home area network (HAN), building area network (BAN), and industrial area network (IAN) encompassing all the intelligent electronic devices (IEDs) that allow monitoring and control of energy status and patterns within each context. ESIs represent logical gateways.
Neighborhood area network (NAN)	Last mile communications network that connects ESIs and smart meters, distributed energy resources (DERs) and microgrids to the utility control and operation center through the backhaul network
Backhaul	Backhaul network provides connectivity between the utility control and operation center and any communications network within the distribution, long-haul, and customer domains

reference architecture as displayed in Figure 4.11, leading to the SG interoperability reference model (SGIRM), well known in the community (IEEE, 2011). The SGIRM extends the NIST SG conceptual model defining three interoperability architectural perspectives (IAPs). IAPs represent the main areas of expertise involved in the SG: power systems (PS-IAP), information technology (IT-IAP), and communications technologies (CT-IAP) as depicted in Figure 4.11 (IEEE, 2011). Each IAP defines the main functional blocks required in each domain of the NIST SG conceptual model, the interfaces between functional blocks (intra-domain interfaces), and the interfaces between domains (inter-domain interfaces) (IEEE Smart Grid, 2015). The defined IAPs are further elaborated for the most important applications in the SG area, such as advanced metering infrastructures (AMIs) or plug-in electric vehicles (PEVs).

Being the main focus of this chapter, CT-IAP defines the communication networks that can be used in every domain. Communications networks domains and interfaces defined for the SG in IEEE's project 2030 are listed and described in Table 4.2 (IEEE, 2011; IEEE Smart Grid, 2015).

4.2.4 Smart Grid Architecture Model

The SG architecture model is the other representation of functional architecture of the SG that was developed by the Smart Metering-coordination Group (SMG) of the European Commission (EC) according to mandate M/490 (European Commission, 2011). According to SMG, the SG architecture model is defined as a three-dimensional architectural model comprising the domains, zones, and layers as displayed in Figure 4.12 (Smart Grid Coordination Group, 2012). The SG architecture model allows a technologically neutral representation of all the interoperability cases of the SG. The five defined layers of this model represent (top to bottom) the business objectives and processes, the functions, information exchange and data models, communication technologies and protocols, and its physical and logical components (Figure 4.12). The communication layer, the main focus and the core of the current chapter, is developed in a separate document (CEN, CENELEC, and ETSI, 2012). This document defines the

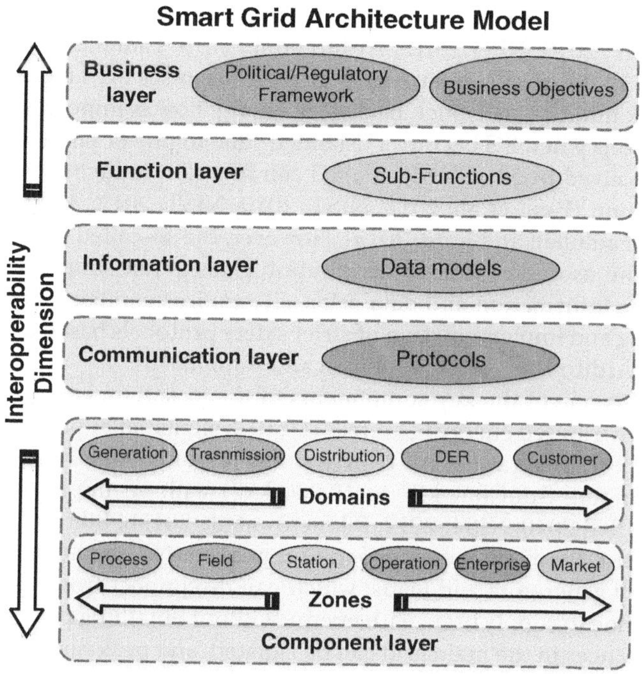

Figure 4.12 Smart Grid architecture model.

communications networks and their deployment at the component layer and maps the identified technologies and protocols onto these networks.

In spite of the fact that the SG architecture model was initially created to describe the functional architecture of the SG, it appears to be a well-formed conceptual model of the power system expanding the initial one developed by the NIST. Specifically, this SG architecture model includes domains from its central generation down to customers, zones of operation from individual processes up to the managing enterprise and the market, and with interoperability layers covering a whole SG system from the business layer down to smart grid components.

4.2.5 Energy Sources in the Smart Grid

In traditional power grids, electricity is generated by a few central energy sources and transmitted in a unidirectional fashion to a large number of users. In contrast, SGs use two-way flows of electricity supported by a bidirectional information transmission through an automated advanced network for distributed energy delivery (Fang et al., 2012; NIST, 2010; NIST, 2012; Bikmetov et al., 2015a; Bikmetov et al., 2015b; Bikmetov et al., 2016). Beside traditional energy generation stations, SGs have distributed renewable energy sources (RESs) and supporting battery energy storage (BES). The majority of RESs are solar and wind energy sources. BES can be designed and made up using a variety of technologies (Fang et al., 2012; 26th Annual Conference and Expo, 2015; Eurobat, 2013; Sebastián, 2016). The most popular types of BESs are lithium-based, nickel-based, and sodium-based batteries (26th Annual Conference and Expo, 2015; Eurobat, 2013).

As mentioned earlier, electricity in SGs can also be transmitted back into the grid by its users (Fang et al., 2012; NIST, 2010; NIST, 2012; Ardito et al., 2013; Fadlullah and Kato, 2015). As an example, SG's users with solar panels at their premises are able to generate energy and put it back into the grid. Such backward energy flow is important and extremely helpful in situations when SG becomes "islanded" due to power failures. Using such energy feedback received from SG's users, a SG can function reducing the level of its own energy generation (Fang et al., 2012; NIST, 2010; NIST, 2012; Ardito et al., 2013; Ardito et al., 2013; Fadlullah and Kato, 2015). However, the so-called backward flow of electricity from the user sites toward distribution grids is not simple or straightforward, contrary to the information and data grids—it requires compatibility of legacy hardware by upgrading and implementation of strict safety protocols based on information (Fang et al., 2012; Ardito et al., 2013; Fadlullah and Kato, 2015).

Distributed generation (DG) is one of the key components of the paradigm enabled by SG. DG takes advantage of distributed energy resources (DER) systems: solar panels, wind turbine farms, and gas/diesel generators. All DER systems are often small-scale power generators (typically ranging from few kW to 10,000 kW) with an improved power quality and reliability (NIST, 2010; NIST, 2012; Bikmetov et al., 2015b; Bikmetov et al., 2016; Bikmetov et al., 2014. From an energy generation perspective, the SG is a localized grouping of electricity generators and loads, which can disconnect from the main grid so that DG can continue to supply users with energy without obtaining power from outside. Thus, the disturbances in the main grid can be isolated, and peak-shaving techniques (Sebastián, R., 2016; Riffonneau et al., 2011; Dietrich et al., 2012) can be implemented. A study (International Energy Agency, 2002) from the International Energy Agency pointed out that a power system based on a large number of reliable small DGs can operate with the same reliability and a lower capacity margin than a system of equally reliable large generators.

A useful review of various distributed energy technologies such as microturbines (gas, diesel, etc.), photovoltaic panels, fuel cells, and wind power turbines can be found in Adinolfi et al. (2013). However, implementing DG in practice is not an easy scheme due to several reasons (Fang et al., 2012; Ardito et al., 2013; Smarter Grid Solutions UK, 2014; Bikmetov et al., 2016; 26th Annual Conference and Expo, 2015; Eurobat, 2013). First, DG involves large-scale deployments of RES, whose energy production is a subject to wide fluctuations due to weather and climate conditions. It was shown that the generation patterns from RES are far from being equal during a certain time span (Fang et al., 2012; Tuttle et al., 2016; Bikmetov et al., 2015b). Therefore, it is important to maintain an effective utilization of the DG in a way that is conscious of the variability of the yield from RES. The second challenge of DG's implementation is that the average operation costs of distributed generators for producing one unit of electricity are often higher than those of traditional large-scale central power plants (Pellicer et al., 2013; Fang et al., 2012; Tuttle et al., 2016).

Considering DG's potential benefits on power quality, it is essential to conduct a systematic research on how to balance the high capital costs and the reliable power supplies available through a DG paradigm. Although there is a limited penetration of DG in today's power system, the future SGs are expected to adopt a large number of distributed generators to increase the level of decentralization of existing power systems (Pellicer et al., 2013; Fang et al., 2012). As predicted in Pellicer et al., this process could include the following three stages:

1) accommodation of DG in the current power system;
2) introduction of a decentralized system of DG cooperating with the centralized generation system; and
3) delivery of most power by DG and a limited amount by central generation.

Since the localized DG enables the users to deploy their own generators, the large-scale deployment of DG will also change the traditional power grid design methodology, in which the generators are directly connected to the transmission grid (e.g., Figure 4.3). Due to such changes, a layer of energy generation and distribution control can be included between bulk generation and transmission.

The deployment of DG further leads to the concept of a virtual power plant (VPP), which manages a large group of distributed generators with a total capacity comparable to that of a traditional power plant (NIST, 2013). A VPP is a cluster of distributed generators that is collectively run by a central controller. The concentrated operational mode of VPPs allows reduction of a peak electricity load ("peak-shaving") and load-aware power generation at short notice (Fang et al., 2012; NIST Smart Grid Collaboration, 2013). The last benefit leads to real-time load scheduling and capacity planning within a SG (Bikmetov et al., 2015b; Bikmetov et al., 2016). A VPP cluster can replace a conventional power plant while providing higher efficiency and flexibility of energy distribution. Such advanced flexibility allows energy systems to react quicker to fluctuations in energy demand. Both benefits of VPPs (peak-shaving and load-aware generation) require a complex optimization, control, and secure communication methodologies.

Recently developed VPPs have been examined in numerous research studies (Andersen et al., 2008; Lombardi, Powalko, and Rudion, 2009; You et al., 2009a; You et al., 2009b). Anderson et al. (2008) focused on the investigation and description of a suitable software framework that can be utilized for implementation of the VPP concept in future power systems. The importance of service oriented architecture in implementing VPPs was emphasized in this work. Lombardi et al. (2009) focused on the optimization of VPP's architecture. Using an energy management system for this optimization, a VPP can be controlled to minimize the electricity production costs and to increase the utilization of renewable energy. S. You et al. (2009a) proposed a market-based VPP, which uses bidding and price signal as two optional operations. In this model, a VPP provides individual distributed energy resource units with the access to current electricity markets. S. You et al. (2009b), proposed a generic VPP model running under a liberalized electricity market environment and attempted to provide a summary of the main functions that are necessary for the efficient operation of the developed VPP model. A current integration of vehicle-to-grid (V2G) and VPP technologies was investigated in Zhang et al. (2017) as well. The architecture of V2G integrated in VPP was outlined providing a sketch of the distribution algorithm, and the associated optimization problem for the overall VPP system.

4.2.6 Energy Consumers in a Smart Grid

Conventionally, three types of users/customers' domains are considered within a SG paradigm: residential, commercial, and industrial (NIST, 2010; NIST, 2012; NIST, 2014; Bikmetov et al., 2014; Bikmetov et al., 2015a; Bikmetov et al., 2016). According to its definition, the residential domain consists of private dwellings (apartments, townhouses, etc.) of a smart city residents. The energy users within residential domains are single households, whose energy loads usually include the following set of appliances: washers, dryers, cooktops/ovens, dishwashers, water heaters, etc. The energy consumption level of a single residential user is typically less than 20 kW (NIST, 2010; NIST, 2012; Bikmetov et al., 2015a; Bikmetov et al., 2015b). It is common to combine residential users into clusters where energy management is performed by a single energy service provider (NIST, 2010; Bikmetov et al., 2015a; Bikmetov et al., 2015b). The commercial domain is formed by small- and medium-size businesses and enterprises: stores, restaurants, hotels, etc. The typical loads within commercial domains are commercial refrigerators, freezers and ovens, and HVAC systems of an entire building and a cluster. For a commercial user, the average level of energy consumption is about 20–200 kW (NIST, 2010; NIST, 2012; Bikmetov et al., 2015a; Bikmetov et al., 2015a; Bikmetov et al., 2015b). The industrial domain consists of plants, factories, and other manufacturing and engineering facilities with the following typical energy loads: heavy machinery, fabrication and manufacturing equipment, etc. Generally, the energy consumption level of a single industrial user is over 200 kW and less tolerant to fluctuations and load shedding (NIST, 2010; NIST, 2012; Bikmetov et al., 2015a; Bikmetov et al., 2015b).

The boundaries of these domains are generally considered at utility meters, such as smart meters (SMs), typically equipped the energy services interfaces (ESIs). For each user within its domain, energy management and control is performed through a single SM connected to energy loads by means of intelligent devices and sensors (NIST, 2010; NIST, 2012; Bikmetov et al., 2015a; Bikmetov et al., 2015b). These ESIs are secure interfaces for interactions between the energy service provider (ESP) and energy users (Figure 4.4). The ESIs were standardized during the development of the SG architecture model depicted in Figure 4.12. Besides, different domains can communicate through ESIs via the advanced metering infrastructure (AMI) or via another communication network implemented within a SG, such as an Internet infrastructure (NIST, 2012; Bikmetov et al., 2015; Bikmetov et al., 2014). In other words, the ESI can act as a bridge between the ESP and facility-based systems, such as a building automation system (BAS) or an energy management system (EMS) located at the users' domains of a SG (NIST, 2012; Bikmetov et al., 2015; Bikmetov et al., 2014). Therefore, each domain of users is equipped with a SM and an ESI that may reside in the SM, on the EMS, or in an independent gateway. In addition to that, the ESI allows each energy user to communicate with devices and systems located within the energy users' premises across a home area network or other LAN. In some cases, commercial and industrial users' domains encompass more than one energy management system (EMS) and, therefore, more than one communications path per energy consumer (NIST, 2012; Bikmetov et al., 2015; Bikmetov et al., 2014).

The EMS is the entry point for several applications running at the functional layer of the SG architecture model (cf. Figure 4.12): load control, monitoring and control of distributed generation, in-home display of customer usage, reading of non-energy meters, and integration with BASs and enterprises. The EMS provides users logging and auditing functions for cyber security purposes. Each user's domain is electrically connected to the distribution domain and to the ESP through a corresponding telecommunication network of an advanced metering infrastructure (AMI). Besides, each users domain communicates with the operations and market domains. Typical applications of an EMS within the customer domain are presented in Table 4.3.

Table 4.3 Typical Applications of EMS within the Users Domain.

Application	Description
Building/home automation	A system that is capable of controlling various functions within a building such as lighting and temperature control.
Industrial automation	A system that controls industrial processes such as manufacturing or warehousing.
Micro-generation	Includes all types of distributed generation including solar, wind, and hydro generators. Generation harnesses energy for electricity at a customer location. May be monitored, dispatched, or controlled via communications.

To ensure an efficient energy management and its reliable delivery, users domains are required to be in close interaction with their ESP. Based on the SG conceptual model, the service provider domain was introduced in Subsection 4.1.1. More details about this domain are given in the next subsection.

4.2.7 Energy Service Providers in the Smart Grid

Based on the developed SG conceptual model, an ESP or an energy service company (ESCO) can be defined as a commercial or nonprofit organization providing solutions for energy supply and generation, designing and implementing projects for energy efficiency improvement, energy retrofitting, and conservation (Ardito et al., 2013; Fadlullah and Kato, 2015; Nunna et al., 2017; Deng et al., 2015; Salinas et al., 2013). The main role of ESPs is to develop, design, build, and fund projects that produce or save energy, reduce energy costs, and decrease operations and maintenance costs at their customers' facilities (US Department of Energy, 2016). In general, ESPs act as project developers for a comprehensive range of energy conservation measures, assuming the technical and performance risks associated within their projects.

ESPs are acting at the business layer of the SG architecture model (in reference to Figure 4.12). Using the corresponding telecommunication connections and networks (Figure 4.3), they collect information about current, predicted, and required statuses of energy generation, transmission, and distribution from the operations domain of the SG. Besides, the ESPs inquire the data from markets about current energy prices for each source of its generation and the information about users' requirements and preferences from energy users' domains. All the data collected by the ESPs from various domains of a SG are utilized for continuous management of energy flow, which can be performed in real time (Bikmetov et al., 2015a; Bikmetov et al., 2015b). This process will be described in more detail in Section 4.3.

From a business point of view, ESPs are distinguished from other firms that offer energy-efficiency improvements in the fact that they use the performance-based contracting methodology: when an ESP implements a project, the ESP's compensation is directly linked to the actual energy cost savings. Many of the recent ESP's projects with substantial energy efficiency retrofits involve renewable energy technologies and advanced energy distribution techniques. Typically, such projects require large initial capital investments and have a relatively long payback period.

The substantial energy efficiency retrofits and renewable energy technologies inherent in energy savings performance contract (ESPC) projects typically require large initial capital investments and may have a relatively long payback period.

Each domain of the SG conceptual model (Figure 4.3) is characterized by a tight integration of flexible and secure communication networks (NIST, 2010; NIST, 2012; Bikmetov et al., 2015a; Bikmetov et al., 2015b). In addition to these networks, a large number of sensors and actuators are required to implement novel energy management techniques within a SG. Thus, a SG transforms into an intelligent electricity network, with building blocks represented by M2M communications. The entire monitoring and control infrastructure of a SG is deployed by means of M2M communications and the IoT paradigm.

4.3 Intelligent Machine-to-Machine Communications in Smart Grids

As described in the previous section, there are multiple functions that can be supported in a SG by intelligent M2M communications: smart metering, distribution and transmission control, etc. Smart metering in M2M interaction can facilitate flexible demand management where smart meters are two-way communicating devices that measure energy consumption and transmit that information via some ICT back to the local utility. With near real-time information available (e.g., the flow of energy in the grid), different levels of tariff can be calculated and made available for the users/consumers, who can make smarter and more responsible choices. Various large-scale WSANs are deployed in a SG to carry information about electric power system generation, transmission, distribution, and home applications for monitoring demand and response tasks.

4.3.1 Reference Architecture of Machine-to-Machine Interactions

M2M interactions are generally defined as a direct communication between devices using any available data transmission channel. M2M interactions enable various sensors or meters to communicate the sensed and recorded data to application software that can utilize this data to adjust certain industrial or manufacturing process (Dohler, Boswarthick, and Alonso-Zárate, 2012; Bush, 2013).

Figure 4.13 displays the central domains of the M2M reference architecture developed by ETSI (European Telecommunications Standards Institute, 2012). This is a resources-based, end-to-end architecture of M2M interactions identifying the functional entities and the related reference points. The M2M reference architecture can be used for the exchange of data and events between machines and entities involving communications across networks without requiring human intervention.

At the reference architecture (Figure 4.13), M2M interactions are described as a distributed system with service capabilities at both network and the M2M device domains level. These capabilities are defined in the specification and are used to put in communication applications, e.g., network, gateway, and device domains. The M2M device domain encompasses the so-called capillary networks (in ETSI terminology), i.e., the sensors and actuators networks (SANs) (López et al., 2014). The network domain represents the core of the M2M infrastructure and provides bidirectional bulk data exchange

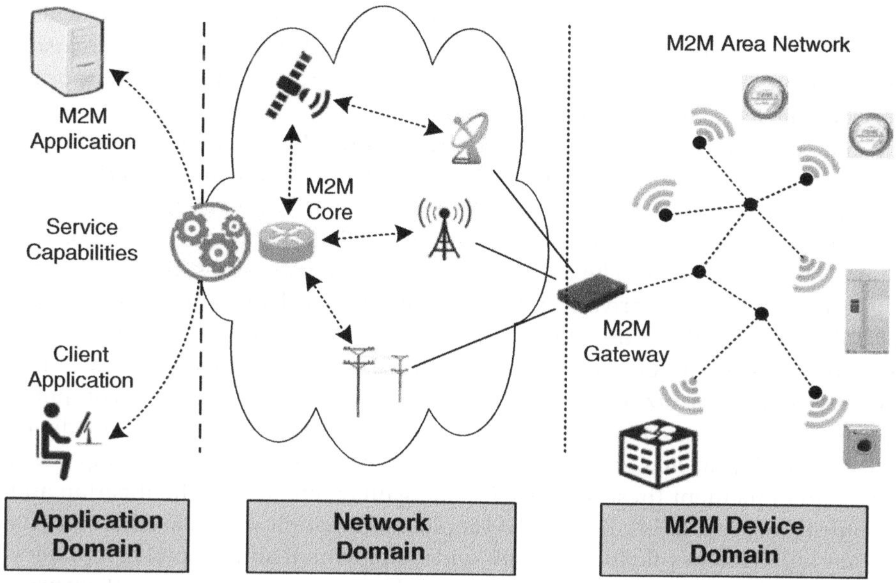

Figure 4.13 Main domains of the M2M reference architecture according to European Telecommunications Standards Institute.

over long distances. Finally, the application domain encompasses the services, which are delivered on the top of the M2M infrastructure.

4.3.2 Communication Media and Protocols

Although there are varieties of communications technologies available for use in M2M communication infrastructures for the SG (Gungor et al., 2011; Usman and Shami, 2013), wireless ones are currently of special interest. These avoid wired connectivity and enable mobility and identity. As a token of that, NIST set up a specific working group within the Priority Action Plan 2 (PAP2) to tackle the challenges and opportunities of wireless communications in the SG paradigms (NIST, 2011). Communications for the SG present specific requirements from both the technical and economic perspectives, such as described in (Güngör et al., 2013; Yan et al., 2013; Liu et al., 2012):

- Quality of Service (QoS) required for the target application. Notably, QoS policies are mainly oriented to traffic prioritization and resource allocation to face congestion situations. Some parameters that are widely used to quantify such QoS level are:
  - Latency defined as the end-to-end delay of the data.
  - Bandwidth providing an aggregated data rate high enough to carry the traffic associated to the target application. In general, this would depend on the volume of devices as well as on the size of the exchanged packets and the traffic pattern.
  - Reliability guaranteeing the correct performance during a given percentage of time: the more critical the application is, the higher such a percentage needs to be.
- Interoperability allowing equipment from different manufacturers to interact seamlessly. In order to achieve this goal, the main functional blocks comprising the communications infrastructure and the interfaces among them must be defined and

standardized. Standardization is crucial for effectively achieving this goal, which eventually fosters competition and thus yields more reliable products at lower cost.
- Scalability dictates that the communication architecture must be able to incorporate new services, devices, and infrastructure upgrades.
- Privacy and security is presented in the form of both physical and cybersecurity. In SGs, various aspects of users' privacy, such as identification, authorization, and access control, are considered (Liu et al., 2012; Dalipi and Yayilgan, 2016; Billewicz, 2016). Therefore, impersonation, data tampering, malicious software, denial of service, and cyberattacks need to be addressed as main security issues in SGs (Bekara, 2014). Along with security, privacy is a major concern for data transmission in SGs (Dalipi and Yayilgan, 2016). Initiatives for data privacy in SGs provide a multi-metrics approach to calculate system's privacy and dependability levels (Noll et al., 2015). Additionally, pseudo-nymizing (or differential privacy) and cryptographic computation approaches can be used for preserving privacy for IoT-enabled SGs (Chowdhery et al., 2015). An elliptic curve cryptography (ECC)–based session key technique was introduced to implement these approaches and gain an efficient authentication and access control (Ye et al., 2014). Because SG applications handle sensitive data, security and privacy represent key factors for their wide deployment and adoption. If privacy is not guaranteed, many users will not embrace the various new services. If security is not guaranteed, many service providers would not be implemented or rely on such new services. However, privacy and security are usually directly proportional to costs, so a trade-off would be required in order to obtain feasible solutions.

As a result, it is crucial to evaluate how different communication architectures and technologies meet such requirements before undertaking the important investments needed to deploy infrastructures on a large scale.

Recently, several divergent standards and protocols have been proposed for M2M communication and the IoT service platforms widely utilized in the SC concept implementation. Each protocol focuses on a specific aspect of M2M communication (Elmangoush, 2016). The lack of a protocol that can satisfy the heterogeneous requirements of M2M/IoT applications has resulted in a highly fragmented protocol stack in M2M/IoT systems as summarized in Table 4.4. Considering the variety in operating conditions and QoS requirements, it is impossible to depend on one protocol for all data streams or all applications. The core question that could be raised here is which transport protocol should be selected for a certain M2M application. In different studies (Smart Cities, 2007; Sun et al., 2014; Sun et al., 2015; Elmangoush, 2016), the

Table 4.4 Protocol Stack in M2M/IoT Communication System (Elmangoush, 2016).

Application layer	**REST/SOAP/SIP**							
	HTTP/HTTPS			**MQTT/STOMP/AMQP**				**CoAP**
Transport layer	TCP RCF 793						UDP RCF 768	
Network layer	6LoWPAN RCF 4944		IPsec RCF 2401	IPv4 RCF 791, IPv6 RCF 2460				DSCP RCF 2474
Physical and data link layer	IEEE 802.15.4	Bluetooth v4.0	IEEE 802.3	WLAN 802.11	WiMax IEEE 802.16	2.5G GPRS	3G UMTS	4G LTE 3GPP TF25.913

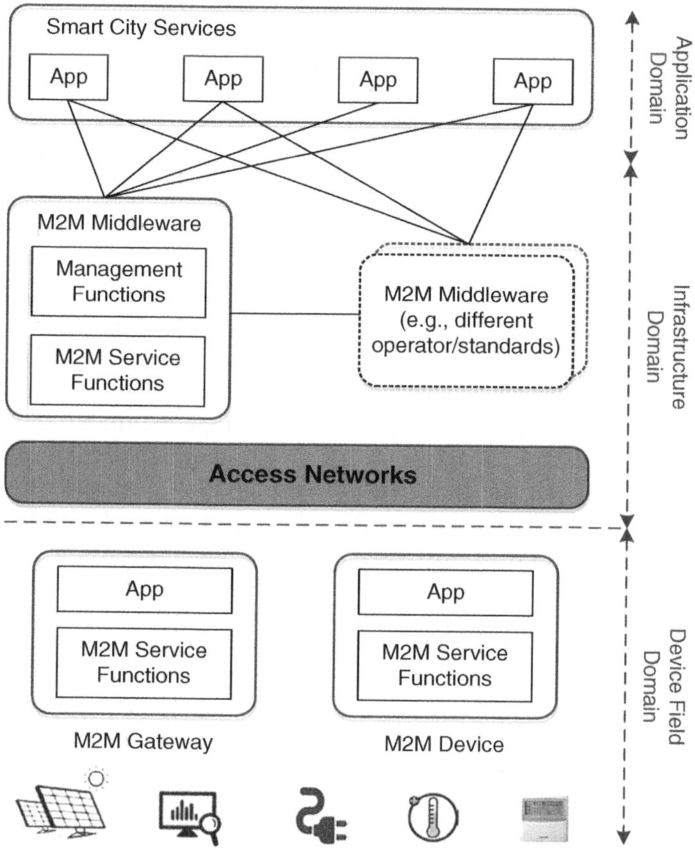

Figure 4.14 General architecture of M2M communication system in the smart city's infrastructure.

introduced platform enables the dynamic adaptation for transporting heterogeneous traffic of M2M applications and mediation with other M2M platforms.

To impose the protocol stack of an M2M/IoT communication system of SG at the M2M reference architecture, a general framework of M2M communication system in smart city's infrastructure was developed, e.g., Figure 4.14 (Elmangoush, 2016). The major objective of this framework is to increase the adaptability of M2M nodes in transporting sequences of requests from connected objects under different conditions, e.g., resource-constrained or resource-rich environments and QoS requirements. In Elmangoush (2016), these conditions were integrated to the open machine-type communication (OpenMTC) platform.

Currently, there is good momentum on M2M standardization efforts, which aim to achieve interoperability and compatibility in M2M systems independently of the vertical market solutions. Several standardization efforts related to M2M interactions in SGs have been carried out and have contributed to the current state of the art of this area (Elmangoush, 2016; IEEE. 2011; NIST, 2010; NIST, 2012; NIST, 2013; NIST, 2014).

From a structural point of view, the telecommunication network of a SG consists of three parts, namely, WAN, NAN, HAN (NIST, 2010; NIST, 2012; NIST, 2014, Bikmetov

Table 4.5 Telecommunication Media and Protocols for SGs.

<table>
<tr><th>WAN</th><th>NAN</th><th>HAN</th></tr>
<tr><td>Wi-MAX (IEEE 802.16)</td><td colspan="2">LoWPAN (IEEE 802.15.4) / PLC (IEEE 1901) / G.hn / Homeplug</td></tr>
<tr><td colspan="3">Cellular (GPRS/UMTS/LTE)</td></tr>
<tr><td colspan="3">WLAN (IEEE 802.11 a/b/g/n)</td></tr>
<tr><td>10GEPON (G.984) GPON</td><td>GPON</td><td>ADSL</td></tr>
</table>

et al., 2015a; Bikmetov et al., 2016). Each segment of such networks transmits different amounts of data, which increases from the smallest (within HAN) to the largest (within WAN), and has a corresponding bandwidth capacity.

As a mainstay, WANs support bidirectional communication that maintains distributed automation and power quality management. All power generation stations, substations, and transformers as well as data concentrators (DCs) communicate with the energy service provider (ESP) through the WAN. Such a network can be deployed/configured using several wired or wireless broadband technologies, listed in Table 4.5 (Bikmetov et al., 2015; Bikmetov et al., 2014; Budka et al., 2014; Khan et al., 2014). For all these technologies, the transmission bandwidths are determined by underlying protocols and often vary over a wide range, e.g., from 172 kb/s for GPRS to 10 Gb/s for 10G EPON (Bikmetov et al., 2014; Budka et al., 2014; Khan et al., 2014). A final choice of communication technology for the WAN would depend on the particular requirements of the SG's design. The physical connection is often presented by wireless, copper, and optical fiber media. The medium access control (MAC) is mostly used to form a communication link.

Deployed in a single user domain, HANs offer users a convenient capability for direct power demand and response management through monitoring and control of the intelligent devices described above. In most cases, such demand management can be performed through the usage of smart meters (SMs) as gateways to the utility (NIST, 2013; IEEE Smart Grid, 2015; Bikmetov et al., 2014; Budka et al., 2014; Khan et al., 2014). In other cases, the interfaces of data concentrators are used as gateways. For this demand management, a meter data management system (MDMS) application is used from users' end. Since the amount of data generated within a single user system is quite small, most of the technologies implemented at HAN have a limited bandwidth (Bikmetov et al., 2014; Budka et al., 2014; Khan et al., 2014), as summarized in Table 4.5.

NANs are central entities of an advanced metering infrastructure (AMI) to provide the information, obtained by SMs and collected by DCs, to the WAN (Fang et al., 2012; NIST, 2010; NIST, 2012; Bikmetov et al., 2014; Budka et al., 2014; Khan et al., 2014). Most common communication technologies implemented in NAN are listed in Table 4.5. All of these technologies use open standards to provide scalability and flexibility for SG communications and can be developed for small range coverage networks (Bikmetov et al., 2014; Budka et al., 2014). Similar to WAN, the implemented protocols define the neighborhood area network (NAN) transmission bandwidth that has currently an upper limit of 1 Gb/s.

Since hybrid passive optical networks (HPONs) (Bikmetov et al., 2015a; Bikmetov et al., 2015b) have broader transmission bandwidth than wireless networks and other

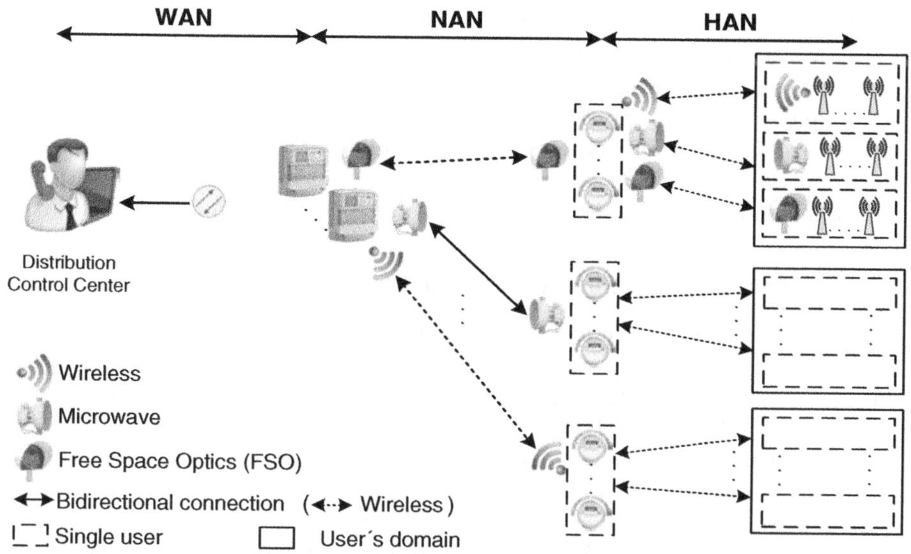

Figure 4.15 Schematic topology of a hybrid passive optical network (HPON) for an advanced metering infrastructure (AMI). HPON is implemented as a backbone network. Wireless, microwave, and free space optics technologies are considered for distribution and access networks.

protocols implemented in SGs, HPONs are able to provide a large number of clients per distribution line (Figure 4.15) remaining scalable for future upgrades (Bikmetov et al., 2014; Gungor et al., 2011; Chen, and Mahloo, 2014). Besides, taking into account low losses in fiber distribution systems (≤0.2 dB/km), HPONs can perform and maintain a broadband (over 100 Gb/s per client) connection to the remote devices located at distances of more than 40 km from the central office (Bikmetov et al., 2014). It is well established now that HPONs have high cost-efficiency (Chen, and Mahloo, 2014; Grobe et al., 2008; Dixit et al., 2011), which can be achieved using only passive components/elements in transmission paths and, on the other hand, by reducing the amount of components at the distribution control center (DCC) and optical network unit (ONU) modules in users' domains based on hybrid technology (Gungor et al., 2011; Chen et al., 2007a). The low cost of passive components (Chen, and Mahloo, 2014; Grobe et al., 2008; Dixit et al., 2011) and the significant reduction of spectral channels in the DCC and ONU modules with hybrid technology (Gungor et al., 2011; Chen et al., 2007a) in the transmission path provide the other key advantage of implementation of HPONs as telecommunication networks for SGs. The flexibility of protection schemes for HPONs is another advantage for their implementation (Bikmetov et al., 2014; Chen et al., 2007a; Chen et al., 2007b).

The scheme in Figure 4.15 shows a part of the general topology of the HPON for a SG, where the passive optical network (PON) is considered as a backbone. The distribution and access telecom networks are presented by hybrid (with PON) communication technologies: wireless, microwave, power line connections (PLC), and free space optics (FSO). Each SM gathers its data from the group of sensors with the corresponding interface (Bikmetov et al., 2015a; Bikmetov et al., 2015b). The final choice of communication technology for distribution and access networks will depend on a certain location and requirements of users and on the density of users' domain. For instance, PLC can be used for areas with low density of population and a well-developed cable infrastructure. On the other hand, wireless technologies can be implemented for densely populated territories with limited space. In the presence of a clear line of sight between the transmitting sensor and the access point, FSO communications will have an edge. The main advantage of this topology is the ability to maintain a continuous bidirectional information flow with a large bandwidth capacity (Bikmetov et al., 2015a; Bikmetov et al., 2015b).

4.3.3 Layered Structure of Machine-to-Machine Communications

Figure 4.16 shows the European Telecommunications Standards Institute (ETSI) board of directors architecture for a SG, which is formed by three main layers: the energy layer supporting production, distribution, transmission, and consumption; the control layer; and the service layer (European Telecommunications Standards Institute, 2012). The energy layer includes a large amount of sensors, electricity storage systems, and transmission and distribution systems and corresponds to the machine-to-machine (M2M) device domain in an M2M framework. The control layer connects the energy plane to the service plane and relates to the M2M network domain. Finally, the service layer provides all the SG-related services and corresponds to the M2M application domain in the M2M network architecture. All the architectural components located at the control layer can communicate with each other based on the developed standards, which are shown in Table 5 and Table 6, defining such M2M interactions. The corresponding interfaces of these components will follow these standards. In other words, the M2M infrastructure allows mapping the developed standards onto corresponding hardware implementation.

Smart buildings such as offices rely on a set of technology to enhance energy efficiency and user comfort factors as well as for monitoring and safety of the building. The M2M

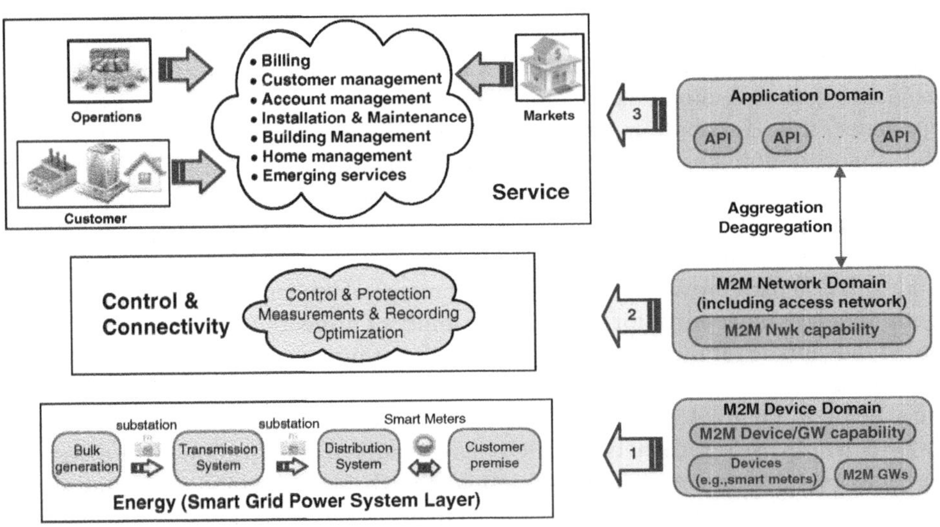

Figure 4.16 The structure of machine-to-machine (M2M) network for smart grids according to the European Telecommunications Standards Institute (ETSI). M2M domains are mapped onto the smart grid main layers.

Table 4.6 M2M Wireless Technologies and Standards Implemented in a SG (López et al., 2014; Elmangoush, 2016).

	802.15.4 (ZigBee/6LoWPAN)	Bluetooth/bluetooth low energy (LE)	802.11 (Wi-Fi)
Max data rate	250 kb/s	3 Mb/s (enhanced) 1 Mb/s (basic or LE)	22 Mb/s (802.11 g) 144 Mb/s (802.11 n)
Indoor range	10–20 m	1, 10, and 100 m classes, 5–15 m (LE)	45 m
Power	Low	medium low (LE)	high
Battery life	Years	days years (LE)	hours
Frequency band	2.4 GHz 868 MHz and 915 MHz	2.4 GHz	2.4 GHz, 3.6 GHz, and 5 GHz
Channel access	CSMA/CA (non-beacon based) or superframe structure (beacon based, non-contention)	frequency hopping or CSMA/CA	CSMA/CA
Applications	smart appliances smart meters lighting control home security office automation	voice smart meters data transfer game control health monitoring computer peripheral	networking between WAN and customer premises (M2M area networks) digital audio/voice

technology and wireless sensor and actuator networks (WSANs) are used in the building management system for lighting and heating, ventilation, and air conditioning (HVAC). They identify empty offices and then switch off devices such as monitors, lighting, and related IT peripherals and enable security and access systems.

Table 4.6 summarizes various wireless technologies for machine-to-machine (M2M) communications along with their attributes, advantages, and limitations. It is important to notice that the main requirement of the M2M devices in a home and office environment is their very low power consumption and advantage of mobility. That is because many devices can last years without requiring battery replacement. With the wide range of home/office devices that need to be networked, there is a need to support several different physical layer links.

Among different networking technologies, Ethernet, 802.15.4, Wi-Fi, Bluetooth, power line communications (PLC), and cellular all have a place in the home networking environment. The home M2M network will have to support all the different physical links and protocol stacks through the M2M gateways. The gateways also need to be equipped with corresponding interfaces for gathering information on what processing and energy resources are available in the M2M devices (usually with limited resources) and decide on how to disseminate data to optimize the resources. In general, the gateway capabilities include routing, network address translation (NAT), authentication, resource allocation, and so forth. Other capabilities of the M2M gateway are addressing remote entity management, security, history and data retention, transaction management, interworking proxy, and compensation brokerage. Smart building systems with wireless sensor and actuator networks (WSANs) are also expected to learn from the building environment and adapt the monitoring and control functions accordingly.

The right side of Figure 4.16 provides a graphical overview of how the three domains of the European Telecommunications Standards Institute (ETSI) M2M reference architecture are mapped onto a SG (López et al., 2014). The implementation of ETSI M2M communications architecture to SG scenarios is also described in Lu et al. (2012). Based on the layout of the SG's telecommunication network architecture, these domains can be imposed with corresponding parts of this architecture and presented as the overall system (Figure 4.17). Figure 4.17 shows the architecture of this system based on the information and communications technologies (ICT) platform developed under the scope of the European Union (EU) Framework Program 7 (FP7) project called Energy Saving Information Platform (ENERsip) (López et al., 2012). This project enables electricity consumption, energy costs savings, and proper integration of DERs (distributed energy resources) at the neighborhood level.

As is evident from Figure 4.17, the overall M2M system architecture for SGs is divided into four domains, which represent the main pillars from the ICT perspective (López et al., 2014). The building domain comprises the physical infrastructures owned by the customers of the power distribution grids, including consumption and generation equipment and the sensor and actuator networks (SANs) to monitor and control them.

The user domain (UD) encompasses the means through which the users and the system interact. Within the UD, energy efficiency can be achieved both through automated actions (e.g., demand response event) and by influencing users' behaviors. Thus, it is crucial to present information to the user in an easily understandable way. Also, the tools provided to the users for making their decisions should be user-friendly. The available applications may run in smartphones, tablets, or even in smart TVs.

The information system domain represents the "brain" of the system from the energy perspective, comprising the logic that allows the optimal use of the available resources at the neighborhood level at any time. Gathering the consumption and generation data of the same location or district at a given moment of time and processing them all together allows reaching global optimizations at the neighborhood level. It is more comprehensive than local optimizations at the household level, as it is the case in state-of-the-art

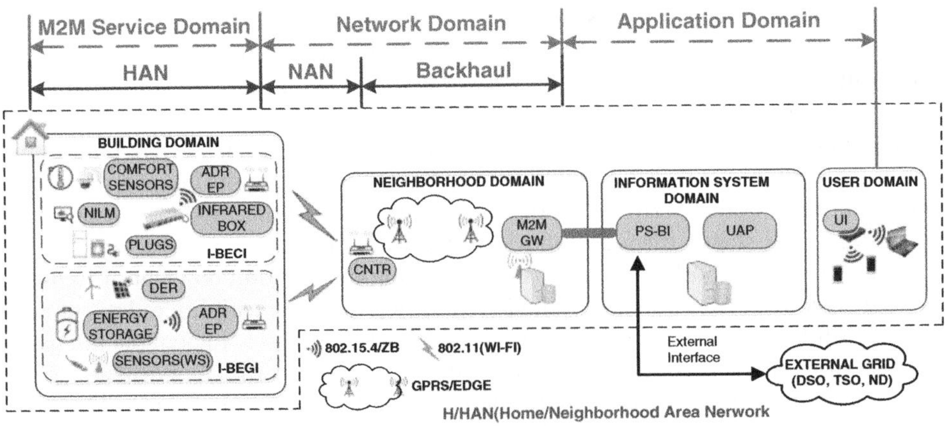

Figure 4.17 Overall system architecture, highlighting the relation with the standardization work.

home energy management systems (HEMSs) (Bikmetov et al., 2016; Perera, Halstensen, and Skeie, 2015; Hassoun and Dincer, 2015; Wu and Xu, 2013; Rafiqul, St-Hilaire, and Kunz, 2014). Additionally, since the users are still allowed to configure a set of parameters and thresholds and they are taken into account when running the optimization algorithms, local optimizations can also be reached.

The neighborhood domain represents the "workforce" of the system and encompasses the core communications infrastructure that carries data and commands back and forward, allowing that everything works correctly. Thus, the information system domain and the user domain are related with IT, whereas the building domain and the neighborhood domain are tightly related with mutual communications. As a matter of fact, the M2M communications architecture addressed in this chapter is spread across the latter domains, as illustrated in Figure 4.19.

The consumption and generation infrastructures are named as in-building energy consumption infrastructures (I-BECIs) (Carreiro et al., 2011) and in-building energy generation infrastructures (I-BEGIs) (López et al., 2013), respectively. I-BECIs and I-BEGIs may or may not be combined, giving rise to different profiles of customers:

- Consumers: users whose households or buildings are only composed of I-BECIs.
- Producers: users whose infrastructures comprise only I-BEGIs connected to the grid.
- Prosumers (Jablonska, 2014): those who own the so-called energy-positive households or buildings, which integrate both I-BECIs and I-BEGIs.

Every I-BECI and I-BEGI is equipped with the so-called automatic demand response end point (ADR-EP). The ADR-EPs work as communications gateways, aggregating and sending consumption or generation data and routing commands to the appropriate device(s). The ADR-EPs communicate directly with their associated concentrator. A given concentrator manages a group of ADR-EPs, forwarding the data coming from them and routing commands to the appropriate ADR-EP(s). Lastly, the M2M gateway has the global picture of the M2M communications infrastructure and works as operation support system (OSS), performing tasks such as network inventory, network components configuration, fault management, or service provisioning, as well as communications gateway to the information system (López et al., 2011).

As Figure 4.17 also illustrates, the communication within I-BECIs and I-BEGIs is based on IEEE standard 802.15.4/Zigbee. The communication between ADR-EPs and concentrators is based on user datagram protocol/Internet protocol (UDP/IP) on top of IEEE 802.11; and the communication between the concentrators and the M2M gateway is based on transmission control protocol (TCP/IP) on top of general packet radio service (GPRS). López et al. (2011) explain why these communications technologies are chosen.

The M2M communications architecture proposed in López et al. (2014) can be mapped onto the communications technologies of interoperability architectural perspectives (CT-IAP) of the overall IEEE 2030 smart grid interoperability reference model (SGIRM), as shown in Figure 4.17 in the continuous line of red arrows. The I-BECIs and the I-BEGIs represent the HANs, and the ADR-EP provides the functionality of the energy service interface (ESI); the communications segment comprising the ADR-EPs and the concentrators represents the NAN; and the communications segment composed by the concentrators and the M2M gateway represents the backhaul.

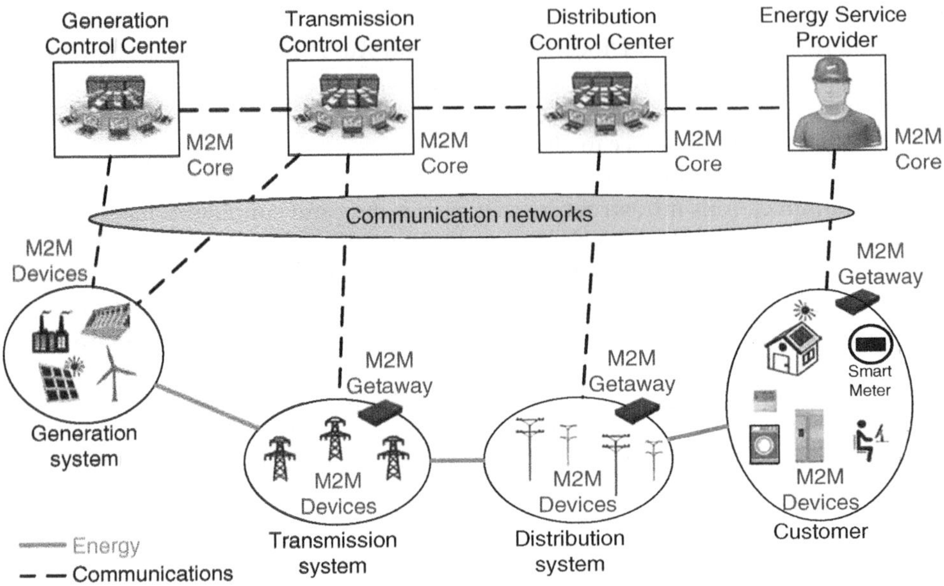

Figure 4.18 Mapping of the proposed M2M communications architecture onto the European Telecommunications Standards Institute (ETSI) M2M architecture applied to the Smart Grid.

In a dashed line, Figure 4.17 also shows the relationship between the M2M domains defined by ETSI (European Telecommunications Standards Institute, 2012). The SANs within the I-BECI and I-BEGI can be seen as capillary networks at the customer domain. The M2M communications architecture can be shown in two main sites (Figure 4.18) (Lu et al., 2012): M2M core and M2M devices are connected through corresponding communication networks. Each site consists of parts that belong to SG conceptual model (NIST, 2010; NIST, 2012).

Figure 4.19 shows a mapping of the proposed M2M communications architecture onto the power distribution infrastructure (López et al., 2014). In this figure, the automatic demand response end points are associated to the customers and the concentrators, which are also associated to the transformation points or feeders.

The M2M gateway is logically associated to the substation that manages the target neighborhood. However, using GPRS as backhaul technology allows the M2M gateway to be physically located at the substation or the data centers of the entity operating the platform: distributed system operator, retail electric provider, aggregator (López et al., 2014; European Commission, 2011; European Telecommunications Standards Institute, 2012; Lu et al., 2012).

The significance of M2M communication in SGs follows from the numerous benefits that it brings. First, real-time M2M communication establishes a close interaction between energy users and the ESPs, which in turn allows to reduce the retail energy price and improve the efficiency of energy generation, transmission, and distribution (Bikmetov et al., 2015b; Bikmetov et al., 2016; Perera, Halstensen, and Skeie, 2015; Federal Energy Regulatory Commission, 2017; Castro, Jara, and Skarmeta, 2013; Shelby, 2010; Bikmetov et al., 2017; Ramallo-Gonzáleza, 2015; Ji, et al., 2016; Harish and Kumar, 2016). Advanced data analytics enabled by M2M communication in combination with

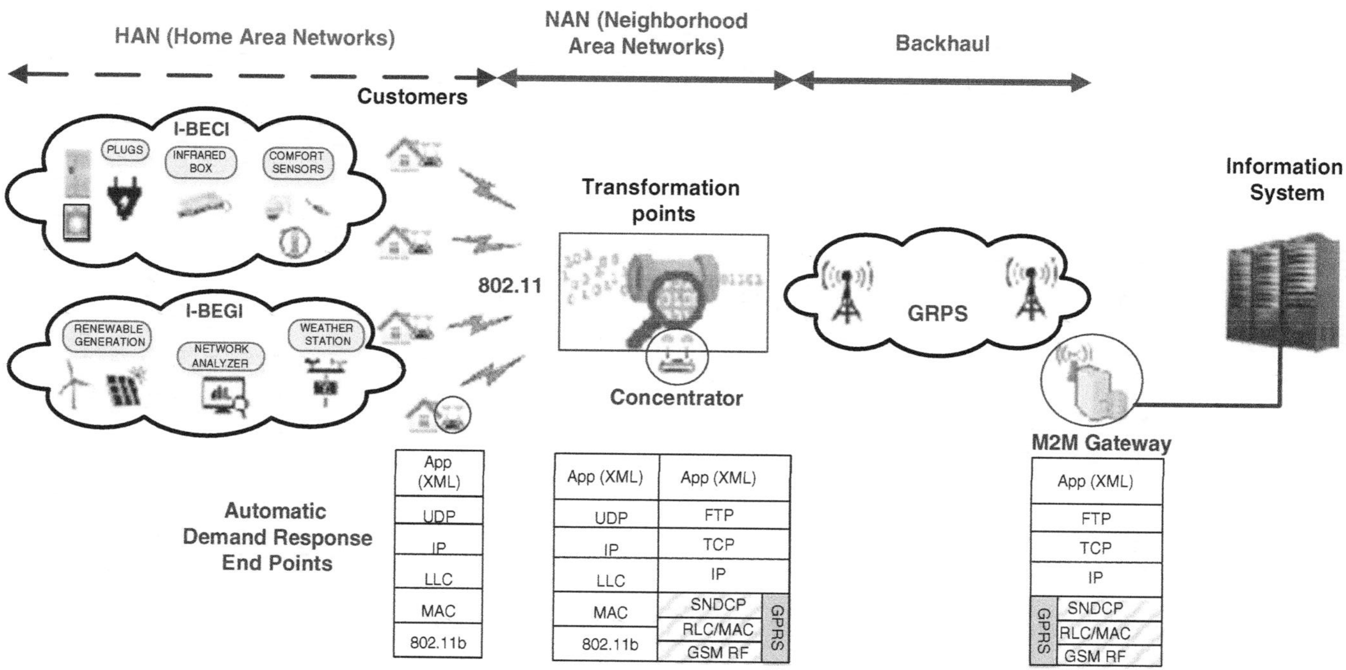

Figure 4.19 Mapping of the proposed M2M communications architecture onto the power distribution infrastructure.

IoT sensing is the other major benefit. The main applications of this benefit are proactive decision-making (e.g., demand response management by ESP) and future electricity price forecasting in SGs (Castro, Jara, and Skarmeta, 2013; Shelby, 2010; Donitzky et al., 2016; Wan et al., 2014; Rafiei et al., 2017; Liu et al., 2015; Vayá, and Andersson, 2015; Gungor et al., 2010). Additionally, the self-healing feature of SGs enabled by implementation of M2M communications increases the grid's reliability and improves its resiliency to failures (Ye et al. 2014).

4.4 Optimization Algorithms for Energy Production, Distribution, and Consumption

One of the latest achievements in the areas of distributed intelligence and M2M communication is a vast implementation of IoT sensors and emerging fusion in various aspects of the SC concept (Zanella et al., 2014; European Telecommunications Standards Institute, 2012; Jin et al., 2014; López et al., 2011; Lu et al., 2012). Utilization of SGs in the SC model (Figure 4.20) allows vast reduction of energy generated by traditional sources and high penetration of "green energy" (Xi et al., 2012; Igualada et al., 2014; Comodi et al., 2015; Yoon et al., 2014). These facts in turn lead to cost-efficient energy distribution and anticipated reduction of carbon dioxide emission. However, an overall demand of electric energy is still constantly growing in all sectors of the prototypical SC (Bikmetov et al., 2015b; Bikmetov et al., 2016; Igualada et al., 2014; Erol-Kantarci and Mouftah, 2015; Soroudi et al., 2016; Vardakas et al., 2015).

In order to improve the environment, a further reduction of carbon dioxide footprint in the atmosphere, efficient energy generation and distribution within SGs, the ESPs should be in a close and continuous interaction with the users. Such interaction can be established based on M2M communication and could reduce the requisite energy price and improve the efficiency of generation and distribution by renewables and traditional energy sources (see Figure 4.4) (Bikmetov et al., 2015a; Bikmetov et al., 2015b; Jin et al., 2014). Besides, IoT-friendly advanced metering infrastructure (AMI) supported by optimized telecommunication networks architecture is crucial in achieving these goals (Fang et al., 2012; CEN, CENELEC, and ETSI, 2012; Bikmetov, 2015a; Bikmetov, 2015b; Bush, 2013 European Telecommunications Standards Institute, 2012; Elmangoush, 2016; Rana and Li, 2015; Industrial Internet Consortium, 2018; Bikmetov et al., 2015.

There are several applications for optimization algorithms implementation within SGs. The major applications include energy consumption management for various types of users (residential, commercial, and industrial) that further developed into the "demand-response" (DR) concept, distributed multi-generation, real-time electricity pricing control, and home energy management based on comfort control.

In Liu et al. (2016), energy consumption management is performed based on dividing demands into essential and flexible, which are further categorized into delay sensitive and delay tolerant, classified by priorities based on historically given probabilities. In that work, an optimization problem is formulated to minimize the total electricity cost and the operation delay of flexible demands by obtaining the optimal energy management decisions. This problem was solved by a combination of centralized and distributed algorithms that were designed for practical implementation at neural networks (Liu et al., 2016). The optimal energy management method for residential buildings considering comfortable lifestyle was developed in Anvari-Moghaddam et al. (2015) using multi-objective mixed integer nonlinear programming model. A "thermal

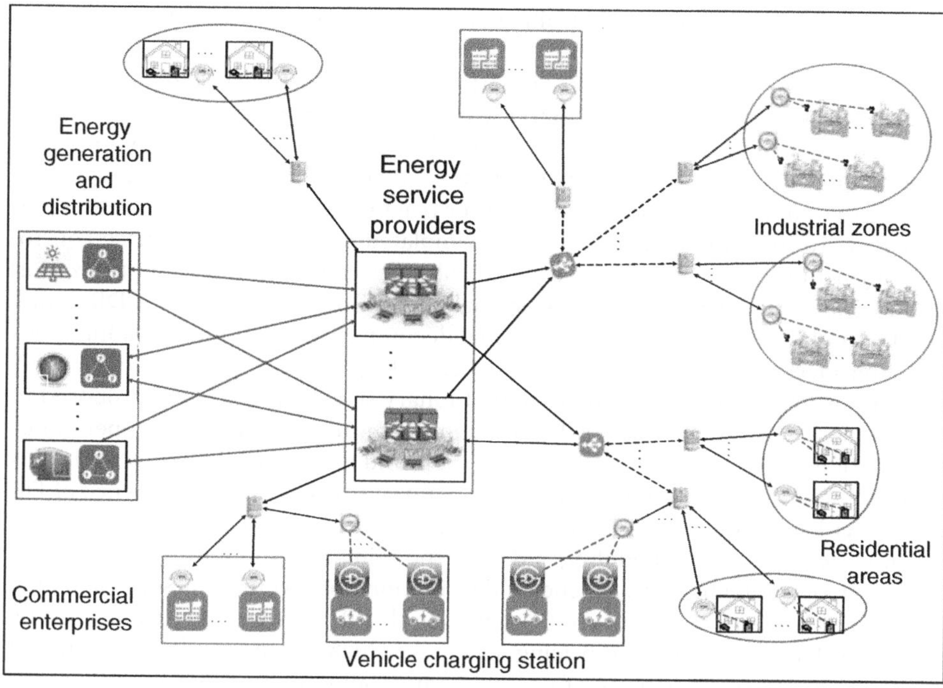

Figure 4.20 A typical architecture of a smart grid in the smart city paradigm.

comfort zone" algorithm was utilized to ensure an optimal task scheduling. The optimization problem was formulated to minimize the cost of the energy drawn from the external grid, while the usage of energy from internal grid (user's renewable sources and battery storage) for appliances is subject to individual delay constraints and a long-term average delay constraint. The developed algorithm requires some prediction of electricity prices assuming that energy can be sold from the battery to the grid.

A comprehensive framework was set up in Mancarella and Chicco (2013) to analyze distributed multi-generation (DMG) systems for the purpose of identifying and quantifying their potential to participate in real-time DR methodology and programs. In this work, the novel concept of electricity shifting potential is deployed within the DMG system to establish the upper limit for the possible reduction of the electricity flowing from the electrical grid to this system without affecting customers comfort level and experience (Mancarella, and Chicco, 2013). A third-party energy consumption control for a group of users was considered in Salinas et al. (2013) or formulation of the load-scheduling problem as a constrained multi-objective optimization problem (CMOP). The optimization objectives were to minimize energy consumption cost and to maximize the efficiency of a certain utility, which can be conflicting and non-commensurable. The modeling of an intelligent energy control center (ECC) for DGs using a multi-agent system has been presented in Manickavasagam (2015). In this work, a multi-agent system has been proposed to provide intelligent energy control and management in SGs based on the following benefits: extensibility, autonomy, and reduced maintenance. The DER model was created in a client and the ECC was created in the server. Communication between the server and clients is established using transmission control protocol/internet protocol (TCP/IP) (Manickavasagam, 2015).

The results show that the agent-controlling DER can be achieved from the server and clients.

In Qian et al. (2013), DR management is performed by a real-time pricing scheme that reduces the peak-to-average load ratio in SG systems. The proposed scheme addresses a two-stage optimization problem. On one hand, users react to prices announced by the retailer and maximize users' payoff. On the other hand, the retailer designs the real-time prices in response to the forecast user reactions to maximize its profit (Qian et al., 2013). To implement such scheme, a simulated-annealing-based price control (SAPC) algorithm is developed and used to solve the nonconvex price optimization problem (Qian et al., 2013). To address spatial and temporal constrains of DR, a distributed optimization algorithm based on dual decomposition without revealing user's private information was developed (Deng et al., 2015). A quick performance of this algorithm was obtained by implementation of binary search. The scheme proposed in Lampropoulos et al. (2013) captures the operational planning, the real-time operations, the verification of the energy and service provision, and the financial settlement. In this study, the emphasis is given to the real-time operations, where the convergence of DR system under load frequency control is investigated (Lampropoulos et al., 2013).

Dynamic electrical load scheduling and corresponding energy capacity planning are the other important techniques for energy distribution optimization in SGs (Bikmetov et al., 2015a; Bikmetov et al., 2015b; Bikmetov et al., 2016). These techniques can be implemented based on real-time pricing schemes, which can be modeled relying on users' preferences and availability of energy from various sources in a SG (Bikmetov et al., 2015a; Bikmetov et al., 2015b; Bikmetov et al., 2016).

4.5 Machine Learning Techniques in Efficient Energy Services and Management

As described previously in Sections 4.1 and 4.2, intelligent communication and networking must be designed to support SG's functionality (Figure 4.12). Utilization of these components of SG's architecture involves the development of artificial intelligence and machine learning techniques for the SGs highlight (Bush, 2013; Livani and Evrenosoglu, 2014; Li and Jayaweera, 2015) [119–120]. The main role of these techniques is to benefit from advances in machine intelligence via M2M and IoT interaction. During the past decade, M2M communication, the semantic web, cognitive radio, and cognitive networking served as attempts to incorporate intelligence into communication and, as an extension of it, into the SG concept (Figure. 4.3) (Bush, 2013; Li and Jayaweera, 2015) [119]. The fundamental nature of intelligence and communications implemented in the form of machine learning techniques plays a key role in power system information theory as the power grid itself becomes more active and "smart" (Bush, 2013; Livani and Evrenosoglu, 2014; Li and Jayaweera, 2015) [119].

One of the purposes of machine learning implementation in SGs is to create an effective algorithm selection between power system control algorithms depending on the state of a network. Such functionality can achieve better performance than the utilization of the same algorithm for every state (Bush, 2013) [119]. A novel method for creating algorithm selectors for power flow management on the IEEE 14- and 57-bus networks has been discussed in King et al. (2015). According to this method, the selectors were chosen from a diverse set of power flow management algorithms based on constraint satisfaction, optimal power flow, power flow sensitivity factors, and linear programming. The benefits of the developed method include minimization of overloads number and the curtailment applied to generators [119].

The other purpose of machine learning implementation in SGs is a real-time decision-making framework that can be effectively integrated with demand response schemes (see Section 4.3) that are prediction based and therefore inevitably lead to real-time power-load mismatches. To implement this functionality, the Baum-Welch algorithm based on Markov chains was adopted in Li and Jayaweera (2015) to learn the non-stationary dynamics of the environment. Instead of iterating the value function in each time step, the representational parsimonious sets were iterated by using the incremental pruning algorithm. The developed complex framework leads to optimal policies giving maximum rewards for the smart homes. A low-complexity real-time decision-making framework that allows adaptive incorporation of new observations, such as environment changes and weather conditions, was developed by the same research group.

Another area of machine learning implementation in SGs is automatic fault locations. In one of the most recent research works (Livani and Evrenosoglu, 2014), a single-ended traveling wave-based fault location method was developed for a hybrid transmission line, an overhead line combined with an underground cable. A discrete wavelet transformation technique was used for extraction of transient information from the measured voltages. The developed method is based on support vector machine (SVM), a machine learning technique (Livani and Evrenosoglu, 2014). As an extension of automatic fault location application, automatic early warning generation methods can be developed based on machine learning techniques. The development of these methods is very important in case of any grid disturbances, such as line contingency leading to cascade failure. A proactive blackout prediction model for a smart grid early warning system was proposed in Gupta et al. (2015) [120]. In this model, the probabilistic evaluation of system's performance is completed in steady state and under dynamical (line contingency) state that prepares a historical database for normal and cascade failure states. As in Livani and Evrenosoglu (2014), a SVM has been "educated" with this historical database and is used to predict blackout events and cascading failure in advance. A proactive prediction of cascading failure can significantly increase the resilience of SGs.

Machine learning techniques were not the focus of this chapter. Useful information about machine learning implementation in SGs can be extracted from the cited references and other specialized literature [121–126].

4.6 Future Perspectives

M2M communications in SGs is a fast-evolving field with numerous possibilities: smart metering, control of energy generation, transmission, and distribution, etc. After the description of this topic from various perspectives—reference architecture, layered structure, telecommunication media and protocols—there are several issues that outline the basis for future work.

Widely implemented in SGs, wireless communication systems including 3G and LTE cannot satisfy high demands of M2M services (Gungor et al., 2011; Usman and Shami, 2013) [127]. Being designed for SGs, 5G systems will accumulate multiple radio access technology networks (including LTE), with an increased flexibility in radio resource utilization. This combined architecture can enable the prioritization of immense M2M traffic, which allows meeting the ultra-low end-to-end latency requirements existing in SGs [128].

Within 5G systems, massive machine communications (MMC) can be considered as three different radio access types implemented for SGs [121]: MMC-D, MMC-A, and MMC-M. MMC-D is defined by a direct connection to the access network that is further used for communication between M2M devices. In MMC-A, M2M devices communicate through a connection to the access network via an accumulation point (dedicated gateway, smartphone connecting personal devices, or other dynamically selected device) collecting local traffic. MMC-M is a direct device-to-device (D2D) communication between M2M devices (Elmangoush, 2016). All these state-of-the-art architectures of MMC are optimized for specific applications [129]. The MMC-D type is the most suitable for devices with a high level of transmitting power, while the MMC-A solution is more common to be used for low-power devices. All the described types of MMC require a very high protocol efficiency (i.e., very low signaling overhead) and devices with long battery life. These aspects are open research issues that need to be addressed for the realization of MMC and should also bring significant improvements in development of emergency applications, such as vehicle-to-vehicle communication [130].

Security is another important characteristic of M2M communications implementation in SGs. Due to a quick integration of IoT in home appliances and commercial and industrial equipment and devices, new vulnerabilities arise almost daily [131]. Several cryptographic protocols designed for computer networks could be used to form authenticated and encrypted connections between M2M nodes (Dalipi and Yayilgan, 2016; Bekara, 2014) [132]. However, such protocols have been initially designed for the Web and entail further adaptation for their usage on embedded devices, which are very resource limited.

4.7 Appendix

Table 4.7 Standards identified by NIST "Framework and Roadmap for Smart Grid Interoperability" (NIST, 2014).

Standard		Application	Domains of SG
ANSI	C12.1	Establishes acceptable performance criteria for new types of smart meters, demand meters and registers, pulse and auxiliary devices. Describes acceptable in-service performance levels for meters and devices used in revenue metering.	Customer, service provider
	C12.18	Revenue metering end device tables.	Customer, service provider
	C12.19	Electricity meters—0.2 and 0.5 accuracy classes.	Customer, service provider
	C12.20	Transport of measurement device data over telephone networks.	Customer, service provider
	C12.21 /IEEE 1702	Protocol and optical interface for measurement devices.	Customer, service provider
	/ASHRAE 135/ISO 16484-5 BACnet	Defines an information model and messages for building communications at a customer's site. Incorporates a range of networking technologies, using IP protocols, to provide scalability from very small systems to multi-building operations that span wide geographic areas.	Customer

(continued)

Table 4.7 (Continued)

Standard		Application	Domains of SG
IEC	60870-6 -503	Performs Telecontrol Application Service defining the messages sent between control centers of different utilities.	Transmission, distribution
	60870-6-702	Defines a standard profile specifying which services and objects are mandatory and optional for compliance with the standard for implementing the application, presentation, and session layers. For a complete protocol implementation, this profile links to a connection-oriented transport profile specifying the transport, network, and data link layers.	Transmission
	60870-6-802	Formerly known as inter control center protocol (ICCP), the standard is used for communication of electric power system status and control messages between power control centers.	Transmission
	61850 Suite	Defines communications within transmission and distribution substations for automation and protection. It is being extended to cover communications beyond the substation to integration of distributed resources and between substations.	Transmission, distribution
	61968/61970 Suites	Define information exchanged among control center systems using common information models: application-level energy management system interfaces and messaging for distribution grid management in the utility space.	Operations
IEEE	1815 (DNP3)	Used for substation and feeder device automation, as well as for communications between control centers and substations.	Generation, transmission, distribution, operations, service provider
	C37.118.1/2	Defines phasor measurement unit (PMU) performance specifications and communications for it.	Transmission, distribution
	C37.238	Ethernet communications for power systems	Transmission, distribution
	C37.239	Defines a common format for interchange of power system event data	Transmission, distribution
	1547 Suite	Defines physical and electrical interconnections between the grid and distributed generation and storage.	Transmission, distribution, customer
	1588	Standard for time management and clock synchronization across the Smart Grid for equipment needing consistent time management.	Transmission, distribution
	1901/ ITU-T G.9972	Broadband over power line networks for home networking: medium access control and physical layer.	Customer

(continued)

Table 4.7 (Continued)

Standard		Application	Domains of SG
NAESB	REQ18, WEQ19	The standards specify two-way flows of energy usage information based on a standardized information model.	Customer, service provider
	REQ-21	Enables retail customers to share energy usage information with third parties who have acquired the right to act in this role.	Customer, service provider
OASIS	Energy interoperation	An information model and a communication model to enable demand response and energy transactions.	Markets
	Energy market information eXchange	An information model to enable the exchange of energy price, characteristics, time, and related information for wholesale energy markets, including market makers, market participants, quote streams, premises automation, and devices.	Markets
NEMA SG-AMI		Used by smart meter suppliers, utility customers, and regulators to guide both development and decision making in smart meter upgradeability.	Customer, distribution
OPC-UA industrial		A platform-independent specification for a secure, reliable, high-speed data exchange based on a publish/ subscribe mechanism	Customer
Open automated demand response (OpenADR)		Specification of messages exchanged between the demand response (DR) service providers and customers for price-responsive and reliability-based DR.	Operations, service providers
CEA-852.1		Provides a way to tunnel local operating network messages through an IP network using the user datagram protocol (UDP), thus providing a way to create larger internetworks	Customer, service provider
Smart energy profile 2.0		Home area network (HAN) device communications and information model.	Customer

References

26th Annual Conference and Expo, 2015, URL: http://26th-annual-conference.energystorage-events.org/agenda/26th

Adinolfi, G., Cigolotti, V., Graditi, G., and Ferruzzi G., 2013, 'Grid integration of distributed energy resources: Technologies, potentials contributions and future prospects'.

Al-Fuqaha, A., Guizani, M., Mohammadi, M., et al., 2015, 'Internet of Things: A Survey on Enabling Technologies, Protocols, and Applications', *IEEE Communications Surveys & Tutorials* 17, 2347–2376.

Ali, S., Wu, K., Weston, K., and Marinakis, D., 2016, 'A Machine Learning Approach to Meter Placement for Power Quality Estimation in Smart Grid', *IEEE Transactions on Smart Grid* 7, 1552–1561.

Andersen, P.B., Poulsen, B., Decker, M., Træholt, C., and Østergaard, J., 2008 , 'Evaluation of a generic virtual power plant framework using service oriented architecture', *In*: IEEE PECon'08, 1212–1217.

Anvari-Moghaddam, A., Monsef, H., and Rahimi-Kian, A., 2015, 'Optimal Smart Home Energy Management Considering Energy Saving and a Comfortable Lifestyle', *IEEE Transactions on Smart Grid* 6, 324–332.

Ardito, L., Procaccianti, G., Menga, G., and Morisio, M., 2013, 'Smart grid technologies in Europe: An overview', *Energies* 6, 251–281.

Bekara, C., 2014, 'Security Issues and Challenges for the IoT-based Smart Grid', *Procedia Computer Science* 34, 532–537.

Bikmetov, R., Raja, M.Y.A., and Kazi, K., 2017, 'Machine-to-Machine Communications in the Smart City — A Smart Grid Perspective', *In*: *Transportation and Power Grid in Smart Cities: Communication Networks and Services*, H. Mouftah, M. E. Kantarci, and M. H. Rehmani, Eds., UK: John Wiley.

Bikmetov, R., Raja, M.Y.A., et al., 2015b, 'Efficiency and robustness capabilities of MGs with AMI using dynamic algorithm of energy consumption prediction', *IEEE Explore.*

Bikmetov, R., Raja, M.Y.A., Kazi, K., et al., 2015, 'A perspective of RoF implementation for an advanced Smart Metering Infrastructure', *In*: North American Power Symposium (NAPS), Charlotte, US, 1–4.

Bikmetov, R., Raja, M.Y.A., Kazi, K., Sane, T.U., and Shevchenko, N., 2016, 'Dynamic energy capacity planning for distributed resources in Smart Microgrids', *In*: *2016 HONET-ICT*, 159–163.

Bikmetov, R., Raja, Y., et al., 2014, 'Role of PONs with the conventional communication systems used in today's smart- and MicroGrids', presented at HONET 2014, U.S.

Bikmetov, R., Raja, Y.M.A., Kazi, K., Chowdhury, B., and Enslin, J., 2015a, 'Dynamic Prediction Capabilities of Smart Metering Infrastructure', *In*: North American Power Symposium (NAPS), Charlotte, NC, USA.

Billewicz, K., 2016, 'Possibility of Internet of things technology implementation in smart power grids', *Energetyka,* 264–270.

Budka, K. et al., 2014, *Communication Networks for Smart Grids: Making Smart Grid Real.* Springer 2014, ISBN 978-1-4471-6302-2

Bush, S.F., 2013, 'Machine Intelligence in the Grid', *Smart Grid,* Wiley-IEEE Press (eBook).

Campbell, A.T. et al., 2008, 'The rise of people-centric sensing', *IEEE Internet Comput.* 12, 12–21.

Campbell, A.T., Eisenman, S.B., Lane, N.D., Miluzzo, E., and Peterson, R.A., 20116, 'People-centric urban sensing', *In*: 2nd Annual Int. Workshop Wireless Internet (WICON), New York, NY, USA.

Carreiro, A., López, G., Moura, P., et al., 2011, 'In-house monitoring and control network for the Smart Grid of the future', *In*: IEEE PES International Conference and Exhibition on Innovative Smart Grid Technologies, Manchester, UK, 2011, 1–7.

Castro, M., Jara, A.J., and Skarmeta, A.F., 2013, 'Smart Lighting Solutions for Smart Cities', presented at the 27th International Conference on Advanced Information Networking and Applications, 2013.

CEN, CENELEC, and ETSI, 2012, 'Smart Grid Reference Architecture—Annex F: Communication Architecture', *In*: Standardization mandate M/468, ed. Brussels, Belgium: EC.

CEN, CENELEC, and ETSI, 2012, 'Smart Grid Reference Architecture', *In*: Standardization mandate M/468, ed. Brussels, Belgium: EC.

Chen, J. and Mahloo, M., 2014, 'Toward Reliable Hybrid WDM/TDM Passive Optical Networks', *IEEE Communication Magazine* 52, 14–23.

Chen, J. et al., 2007a, 'Analysis of Protection Schemes in PON Compatible with Smooth Migration from TDM PON to Hybrid WDM/TDM PON', *Opt. Net.* 6.

Chen, J. et al., 2007b, 'A Novel Protection Scheme for a Hybrid WDM/TDM PON', *Proc. SPIE, Network Architectures, Management, and Applications* 6784.

Chowdhery, A., Bahl, V., Mahajan, R., et al., 2015, 'Privacy-preserving Data-aggregation for Internet-of-things in Smart Grid', *In*: IEEE Rising Stars Conference, 2015.

Chrysoulas C., and Fasli, M., 2016, 'A service oriented QoS architecture targeting the smart grid world & machine learning aspects', *In*: 2016 International Multidisciplinary Conference on Computer and Energy Science (SpliTech), 1–6.

Comodi, G., Giantomassi, A., Severini, M., et al., 2015, 'Multi-apartment residential microgrid with electrical and thermal storage devices: Experimental analysis and simulation of energy management strategies', *Applied Energy* 137, 854–866.

Corici, A., Ancuta, A., Shrestha, R., et al., 2015, 'A solution for provisioning reliable M2M infrastructures using SDN and device management', *In*: 3rd International Conference on Information and Communication Technology (ICoICT), 2015, 81–86.

Dalipi, F. and Yayilgan, S.Y., 2016, 'Security and Privacy Considerations for IoT Application on Smart Grids: Survey and Research Challenges', *IEEE Xplore*, 63 - 68.

Deng, R., Xiao, G., Lu, R., and Chen, J., 2015, 'Fast Distributed Demand Response With Spatially and Temporally Coupled Constraints in Smart Grid', *IEEE Transactions on Industrial Informatics* 11, 1597–1606.

Dietrich, K., Latorre, J.M., Olmos, L., and Ramos, A., 2012, 'Demand Response in an Isolated System With High Wind Integration', *IEEE Transactions on Power Systems* 27, 20–29.

Dixit A. et al., 2011, 'Flexibility Evaluation of Hybrid WDM/TDM PONs', presented at the IEEE Int'l. Conf. Advanced Networks and Telecommun. systems, 2011.

Dohler, M., Boswarthick, D., and Alonso-Zárate, J., 2012, 'Machine-to-machine in smart cities & smart grids', in *IEEE Global Communications Conference*, Anaheim, CA, USA, 2012.

Donitzky, C., Roos, O., and Sauty, S., 2016, 'A Digital Energy Network: The Internet of Things & the Smart Grid', URL: https://www.intel.com/content/dam/www/public/us/en/documents/white-papers/iot-smart-grid-paper.pdf.

Elmangoush, A.A., 2016, 'Application-derived Communication Protocol Selection in M2M Platforms for Smart Cities', Ph.D., Electrical Engineering and Computer science, Technical University of Berlin, Berlin, Germany.

Erol-Kantarci, M. and Mouftah, H.T., 2015, 'Energy-Efficient Information and Communication Infrastructures in the Smart Grid: A Survey on Interactions and Open Issues', *IEEE Communications Surveys & Tutorials* 17, 179–197.

Esmalifalak, M., Liu, L., Nguyen, N., et al., 2014, 'Detecting Stealthy False Data Injection Using Machine Learning in Smart Grid', *IEEE Systems Journal*. PP, 1–9.

Eurobat, 2013, 'Battery Energy Storage for Smart Grid Applications'.

European Commission (EC), 2011, 'Standardisation Mandate to CEN, CENELEC and ETSI Concerning the Charging of Electric Vehicle', *In*: Standardization mandate M/468, ed. Brussels, Belgium: EC.

European Telecommunications Standards Institute (ETSI), 2012, 'Applicability of M2M Architecture to Smart Grid Networks. Impact of Smart Grids on M2M Platform', Sophia-Antipolis, France.

Fadlullah, Z.Md. and Kato, N., 2015, 'Evolution of Smart Grids', Springer Briefs in Electrical and Computer Engineering.

Fang, X., Misra, S., Xue, G., et al., 2012, 'Smart Grid—The New and Improved Power Grid: A Survey', *IEEE Communications Surveys & Tutorials* 14, 944–980.

Federal Energy Regulatory Commission, 2017, 'Assessment of Demand Response & Advanced Metering'.

Ganti, R.K., Ye, F., and Lei, H., 2011, 'Mobile crowdsensing: Current state and future challenges', *IEEE Communications Magazine* 49, 32–39.

Grau, A., 2015, 'Can you trust your fridge?', *IEEE Spectrum* 52, 50–56.

Grobe, K. et al., 2008, 'PON in Adolescence: from TDMA to WDM PON', *IEEE Communication Magazine* 46, 26–34.

Gungor, V., Sahin, D., et al., 2011, 'Smart Grid Technologies: Communication Technologies and Standards', *IEEE Communication Magazine* 7, 529–539.

Gungor, V.C., Lu, B., and Hancke, G.P., 2010, 'Opportunities and challenges of wireless sensor networks in smart grid.', *IEEE Transactions on Industrial Electronics* 57, 3557–3564.

Güngör, V.C., Sahin, D., Kocak, T., et al., 2013, 'A survey on smart grid potential applications and communication requirements', *IEEE Trans. Ind. Inf*. 9, 28–42.

Gupta, S., Kambli, R., Wagh, S., and Kazi, F., 2015, 'Support-Vector-Machine-Based Proactive Cascade Prediction in Smart Grid Using Probabilistic Framework', *IEEE Transactions on Industrial Electronics* 62, 2478–2486.

Harish, V.S.K.V. and Kumar, A., 2016, 'Reduced order modeling and parameter identification of a building energy system model through an optimization routine', *Applied Energy* 162, 1010–1023.

Hassoun, A. and Dincer, I., 2015, 'Analysis and performance assessment of a multigenerational system powered by Organic Rankine Cycle for a net zero energy house', *Applied Thermal Engineering* 76, 25–36.

IEC, 2016, 'Smart Grid Standards Mapping Tool', URL: http://smartgridstandardsmap .com/.

IEEE Smart Grid, 2015, 'Approved & Proposed IEEE Smart Grid Standards.' URL: http:// smartgrid.ieee.org/component/finder/search#gsc.tab=0&gsc.q=Approved%20%26 %20Proposed%20IEEE%20Smart%20Grid%20Standards&gsc.page=1.

IEEE, 2011, 'Guide for Smart Grid Interoperability of Energy Technology and Information Technology Operation with the Electric Power System (EPS), and End-Use Applications and Loads', in *IEEE 2030*, ed. New York, NY, USA: IEEE.

Igualada, L., Corchero, C., et al., 2014, 'Optimal energy management for a residential microgrid including a vehicle-to-grid system', *IEEE Trans. on Smart Grid* 5, 2163–2172.

Industrial Internet Consortium, 2018, 'Communication & control testbed for microgrid applications', URL: http://www.iiconsortium.org/microgrid.htm.

International Energy Agency, 2002, 'Distributed generation in liberalised electricity markets'.

Jablonska, M.R., 2014, 'Microgrids technology review for the creation of urban prosumers societies', *Studies & Proceedings of Polish Association for Knowledge Management* 72, 16–27.

Ji, Y., Xu, P., et al., 2016, 'Estimating hourly cooling load in commercial buildings using a thermal network model and electricity submetering data', *Applied Energy* 169, 309–323.

Jin, J., Gubbi, J., et al., 2014, 'An Information Framework for Creating a Smart City Through Internet of Things', *IEEE Internet of Things* 1, 112–121.

Khan, M.F., Jain, A., Arunachalam, V., and Paventhan, A., 2014, 'Communication technologies for smart metering infrastructure', presented at the 2014 IEEE Students' Conference, Bhopal.

Khan, R. and Khan, J., 2013, 'A comprehensive review of the application characteristics and traffic requirements of a smart grid communications network', *Computer Networks* 57, 825–845.

King, J.E., Jupe, S.C.E., and Taylor, P.C., 2015, 'Network State-Based Algorithm Selection for Power Flow Management Using Machine Learning', *IEEE Transactions on Power Systems* 30, 2657–2664.

Lampropoulos, I., Baghină, N., Kling, W.L., and Ribeiro, P.F., 2013, 'A Predictive Control Scheme for Real-Time Demand Response Applications', *IEEE Transactions on Smart Grid* 4, 2049–2060.

Li, D. and Jayaweera, S.K., 2015, 'Machine-Learning Aided Optimal Customer Decisions for an Interactive Smart Grid', *IEEE Systems* 9, 1529–1540.

Li, D. and Jayaweera, S.K., 2015, 'Machine-Learning Aided Optimal Customer Decisions for an Interactive Smart Grid', *IEEE Systems Journal* 9, 1529–1540.

Liu, J., Xiao, Y., Li, S., et al., 2012, 'Cyber security and privacy issues in smart grids', *IEEE Commun. Surv. Tutor.* 14, 981–997.

Liu, M., Quilumba, F.L., and Lee, W.-J., 2015, 'Dispatch Scheduling for a Wind Farm With Hybrid Energy Storage Based on Wind and LMP Forecasting', *IEEE Transactions on Industry Applications* 51, 2375–2385.

Liu, Y., Yuen, C., Yu, R., Zhang, Y., and Xie, S., 2016 'Queuing-Based Energy Consumption Management for Heterogeneous Residential Demands in Smart Grid', *IEEE Transactions on Smart Grid* 7, 1650–1659.

Livani, H. and Evrenosoglu, C.Y., 2014, 'A machine learning and wavelet-based fault location method for hybrid transmission lines', *IEEE Transactions on Smart Grid* 5, 51–59.

Lombardi, P., Powalko, M., and Rudion, K., 2009, 'Optimal operation of a virtual power plant'.

López, G., Moura, P., Kantsepolsky, B., et al., 2012, 'European FP7 project ENERsip: Bringing ICT and energy together', *Glob. Commun. Newsl.* 15, 2–4.

López, G., Moura, P., Moreno, J.I., and Camacho, J.M., 2014, 'Multi-Faceted Assessment of a Wireless Communications Infrastructure for the Green Neighborhoods of the Smart Grid', *Energies* 7, 3453–3483.

López, G., Moura, P., Moreno, J.I., and de Almeida, A., 2011, 'M2M-based platform to enable energy efficiency within energy-positive neighborhoods', *In*: INFOCOM Workshop on M2M Communications and Networking, Shanghai, China, 2011, 217–222.

López, G., Moura, P., Moreno, J.I., de Almeida, A., et al., 2013, 'Monitoring system for the local distributed generation infrastructures of the smart grid', In: 22nd European Conference and Exhibition on Electricity Distribution, Stockholm, Sweden, 2013.

Lu, G., Seed, D., Starsinic, M., Chonggang, W., and Russell, P., 2012, 'Enabling smart grid with ETSI M2M standards', *In*: IEEE Wireless Communications and Networking, Paris, France, 2012, 148–153.

Mancarella, P. and Chicco, G., 2013, 'Real-Time Demand Response From Energy Shifting in Distributed Multi-Generation', *IEEE Transactions on Smart Grid* 4, 1928–1938.

Manickavasagam, K., 2015, 'Intelligent Energy Control Center for Distributed Generators Using Multi-Agent System', *IEEE Transactions on Power Systems* 30, 2442–2449.

National Institute of Standards and Technology (NIST), 2010, 'Framework and Roadmap for Smart Grid Interoperability Standards, Release 1.0', Gaithersburg, MD, USA.

National Institute of Standards and Technology (NIST), 2011, 'Guidelines for Assessing Wireless Standards for Smart Grid Applications: Priority Action Plan 2', Gaithersburg, MD, USA.

National Institute of Standards and Technology (NIST), 2012, 'Framework and Roadmap for Smart Grid Interoperability Standards, Release 2.0', Gaithersburg, MD, USA.

National Institute of Standards and Technology (NIST), 2014, 'Framework and Roadmap for Smart Grid Interoperability Standards, Release 3.0', Gaithersburg, MD, USA.

NIST Smart Grid Collaboration, 2013, 'SGIP Catalog of Standards', URL: http://collaborate.nist.gov/twiki-sggrid/bin/view/SmartGrid/SGIPCatalogOfStandards.

Noll, J., Garitano, I., Fayyad, S., et al., 2015, 'Measurable Security, Privacy and Dependability in Smart Grids', *Journal of Cyber Security* 3, 371–398.

Nunna, S.K., Battula, S., Doolla, S., and Srinivasan, D., 2017, 'Energy Management in Smart Distribution Systems with Vehicle-to-Grid Integrated Microgrids', *IEEE Transactions on Smart Grid* PP.

Pellicer, S., Santa, G., Bleda, A.L., and Maestre, R., 2013, 'A Global Perspective of Smart Cities: A Survey', presented at the Seventh International Conference on Innovative Mobile and Internet Services in Ubiquitous Computing, Spain, 2013.

Perera, D.W.U., Halstensen, M., and Skeie, N.-O. 2015, 'Prediction of Space Heating Energy Consumption in Cabins Based on Multivariate Regression Modelling', *International Journal of Modeling and Optimization* 6, 385–392.

Pouryazdan, M., Kantarci, B., Soyata, T., and Song, H., 2016, 'Anchor-assisted and vote-based trustworthiness assurance in smart city crowdsensing', *IEEE Access* 1, 1–3.

Qian, L.P., Zhang, Y.J.A., Huang, J., and Wu, Y., 2013, 'Demand Response Management via Real-Time Electricity Price Control in Smart Grids', *IEEE Journal on Selected Areas in Communications* 31, 1268–1280.

Rafiei, M., Niknam, T., and Khooban, M.-H., 2017, 'Probabilistic Forecasting of Hourly Electricity Price by Generalization of ELM for Usage in Improved Wavelet Neural Network', *IEEE Transactions on Industrial Informatics* 13, 71–79.

Rafiqul, A.M., St-Hilaire, M., and Kunz, M., 2014, 'A modular framework for cost optimization in smart grid', *In*: IEEE World Forum on Internet of Things (WF-IoT) 2014.

Ramallo-Gonzáleza, A.P., Blightb, T.S., and Coleyb, D.A., 2015, 'New optimisation methodology to uncover robust low energy designs that accounts for occupant behaviour or other unknowns', *Journal of Building Engineering* 2, 59–68.

Rana, M.M. and Li, L., 2015, 'Microgrid state estimation and control for smart grid and Internet of Things communication network', *Electronic letters* 51, 149–151.

Rapeepat, R., Prasad, A., Li, Z., et al., 2015, 'Recent Avdances in M2M Communications in 4G Networks and Evolution towards 5G', *In*: 18th International Conference on Intelligence in Next Generation Networks, 52–57.

Riffonneau, Y., Bacha, S., Barruel, F., and Ploix, S., 2011, 'Optimal Power Flow Management for Grid Connected PV Systems With Batteries', *IEEE Transactions on Sustainable Energy* 2, 309–320.

Salinas, S., Li, M., and Li, P., 2013, 'Multi-Objective Optimal Energy Consumption Scheduling in Smart Grids', *IEEE Transactions on Smart Grid* 4, 341–348.

Salinas, S., Li, M., et al., 2013, 'Dynamic Energy Management for the Smart Grid With Distributed Energy Resources', *IEEE Transactions on Smart Grid* 4, 2139–2151.

Sebastián, R., 2016, 'Application of a battery energy storage for frequency regulation and peak shaving in a wind diesel power system', *IET Generation, Transmission & Distribution* 10, 764–770.

Sheikhi, A., Rayati, M., and Ranjbar, A.M., 2015,'Energy Hub optimal sizing in the smart grid; machine learning approach', *In*: 2015 IEEE Power & Energy Society Innovative Smart Grid Technologies Conference (ISGT), 1–5.

Shelby, Z., 2010,'Embedded web services', *Wireless Communications, IEEE* 17, 52–7.

Smart Cities, 2007, 'Ranking of European Medium-Size Cities', Vienna University of Technology.

Smart Grid Coordination Group, 2012, 'Smart Grid Reference Architecture.'

Smarter Grid Solutions UK, 2014, 'Smart Grid Architecture Model', in *Power in Unity: A Joint Vision Approach*, Austin Court, UK, 1–5.

Soroudi, A., Siano, P., and Keane, A., 2016, 'Optimal DR and ESS Scheduling for Distribution Losses Payments Minimization Under Electricity Price Uncertainty', *IEEE Transactions on Smart Grid* 7, 261–272.

Sun Y. and Jara, A.J., 2014, 'An extensible and active semantic model of information organizing for the Internet of Things', *Pers. Ubiquitous Comput.* 18, 1821–1833.

Sun, Y. et al., 2014, 'Organizing and querying the big sensing data with event-linked network in the Internet of Things', *Int. J. Distrib. Sensor Netw.* 1.

Sun, Y., Lu, C., Bie, R., and Zhang, J., 2016, 'Semantic relation computing theory and its application', *J. Netw. Comput. Appl.* 59, 219–229.

Sun, Y., Song, H., Jara, A.J., and Bie, R., 2015, 'Internet of Things and Big Data Analytics for Smart and Connected Communities', *IEEE Access (Special Section on Smart Cities),* vol. 4, Dec 2015, 766–773.

Sun, Y., Yan, H., Lu, C., Bie, R., and Zhou, Z., 2014, 'Constructing the Web of events from raw data in the Web of things', *Mobile Inf. Syst.* 10, 105–125.

T. Maddox. (2016, Nov). *New forum seeks to unite 100 cities in standards to drive smart city innovation*. Available: http://www.techrepublic.com/article/new-forum-seeks-to-unite-100-cities-in-standards-to-drive-smart-city-innovation/

Tehrani, M.N., Nader, M., Uysal, M., et al., 2014, 'Device-to-device communication in 5G cellular networks: challenges, solutions, and future directions', *IEEE Communications Magazine* 52, 86–92.

Tuttle, D.P., Gülen, G., Hebner, R., King, C.W., et al., 2016, 'The History and Evolution of the U.S. Electricity Industry', The University of Texas at Austin, U.S..

US Department of Energy, 2016, 'Energy Service Companies', URL: https://energy.gov/eere/femp/energy-service-companies-0

Usman A., and Shami, S., 2013, 'Evolution of communication technologies for smart grid applications', *Renewable & Sustainable Energy Reviews* 19, 191–199.

Vardakas, J.S., Zorba, N., and Verikoukis, C.V., 2015, 'A Survey on Demand Response Programs in Smart Grids: Pricing Methods and Optimization Algorithms', *IEEE Communications Surveys & Tutorials* 17, 152–178.

Vayá, M.G. and Andersson, G., 2015, 'Optimal Bidding Strategy of a Plug-In Electric Vehicle Aggregator in Day-Ahead Electricity Markets Under Uncertainty', *IEEE Transactions on Power Systems* 30, 2375–2385.

Vienna University of Technology, 'Smart cities Ranking of European medium-sized cities', Vienna, 2007.

Wan, C., Xu, Z., Wang, Y., and et al., 2014, 'A Hybrid Approach for Probabilistic Forecasting of Electricity Price', *IEEE Transactions on Smart Grid* 5, 463–470.

Wu Z., and Xu, J., 2013, 'Predicting and optimization of energy consumption using system dynamics-fuzzy multiple objective programming in world heritage areas', *Energy* 49, 19–31.

Xi, F., Satyajayant, M., et al., 2012, 'Smart Grid—The New and Improved Power Grid: A Survey', *Communications Surveys & Tutorials* 14, 6–9.

Xu, H., Huang, H., Khalid, R.S., and Yu, H. 2016, 'Distributed machine learning based smart-grid energy management with occupant cognition', *In*: 2016 IEEE International Conference on Smart Grid Communications (SmartGridComm), 491–496.

Yan, Y., Quian, Y., Sharif, H., and Tipper, D., 2013, 'A survey on smart grid communication infrastructures: Motivations, requirements and challenges', *IEEE Communications Surveys & Tutorials* 15, 5–20.

Ye, N., Zhu, Y., et al., 2014, 'An Efficient Authentication and Access Control Scheme for Perception Layer of Internet of Things', *Applied Mathematics & Information Sciences* 8, 1–8.

Yoon, J.H., Baldick, R., and Novoselac, A., 2014, 'Dynamic Demand Response Controller Based on Real-Time Retail Price for Residential Buildings', *IEEE Transactions on Smart Grid* 5, 121–129.

You, S., Træholt, C., and Poulsen, B., 2009a, 'A market-based virtual power plant', *In*: International Conference on Clean Electrical Power, 460–465.

You, S., Træholt, C., and Poulsen, B., 2009b, 'Generic virtual power plants: Management of distributed energy resources under liberalized electricity market', 8th International Conference on Advances in Power System Control, Operation and Management, 1–6.

Zanella, A., Bui, N., Castellani, A., et al., 2014, 'Internet of Things for Smart Cities', *IEEE Internet of Things Journal*, vol. 1, 22–32.

Zhang, H., Hu, Z., Xu, Z., and Song, Y., 2017, 'Evaluation of Achievable Vehicle-to-Grid Capacity Using Aggregate PEV Model', *IEEE Transactions on Power Systems* 32, 784–794.

Zhuge H. and Sun, Y., 2010, 'The schema theory for semantic link network', *Future Generat. Comput. Syst.* 26, 408–420.

5

5G and D2D Communications at the Service of Smart Cities

Muhammad Usman[1], Muhammad Rizwan Asghar[2] and Fabrizio Granelli[1]

[1] Department of Information Engineering and Computer Science, University of Trento, Via Sommarive 9, 38123, Trento, Italy
[2] Department of Computer Science, The University of Auckland, Private Bag 92019, 1142 Auckland, New Zealand

5.1 Introduction

More than half of the world population lives in cities, and the trend of moving to bigger cities is increasing with the time (Chourabi et al., 2012). This increase in population of the cities is resulting in technical, physical, and material problems. Air pollution, traffic congestion, human health concerns, aging infrastructures, and difficulty in waste management are a few to mention (Washburn et al., 2009). The smart city is a recent concept, which aims at solving the aforementioned problems and improving the quality of life of citizens (Neirotti et al. 2014). A city is considered to be smart when information and communication technologies (ICT)–based solutions are employed in order to provide a high quality of life to its citizens, thus potentially achieving economic growth (Zanella et al., 2014). According to "European Smart Cities", an EU project (Neirotti et al., 2014), six areas of smartness are identified, including smart economy, smart people, smart governance, smart mobility, smart environment, and smart living.

ICT has become the nerve of smart cities enabling new ways of transport management, traffic control, environmental pollution monitoring, health care, public safety and security, surveillance and maintenance of public areas and preservation of cultural heritage Neirotti et al. (2014). The communication infrastructure in ICT represents the backbone of the smart city, connecting Internet of Things (IoT) over cellular networks. IoT in smart cities exhibits diverse requirements to communication infrastructure with heterogeneous data sizes, traffic demands and services Zanella et al. (2014). For example, applications like environmental monitoring and water management do not require very low latency communication. On the other hand, applications such as collision avoidance in smart cars impose stringent requirements on the communication requiring extremely low latency and very high reliability (Ziegler et al., 2016). The network must evolve to meet the diverse needs and preferences of the aforementioned services. For this evolution, a unified communication infrastructure is needed, which should not only support voice and data but also diverse users, services, and applications.

The conventional cellular networks such LTE-A are mainly designed for high data rates and large data sizes while IoT devices mostly exchange small data packets. It is

Transportation and Power Grid in Smart Cities: Communication Networks and Services, First Edition.
Edited by Hussein T. Mouftah, Melike Erol-Kantarci and Mubashir Husain Rehmani.

important to note that LTE-A is not optimized for IoT devices with respect to achieving energy efficiency. However, a new variant of LTE for machine-to-machine (M2M) communication, referred to as LTE-M, is developed to meet the needs of low power and low data rate of IoT. In addition, multiple competing technologies are proposed to fulfil the communication requirements of smart cities. These technologies include but are not limited to 5G, D2D communication, SigFox, IPv6 over low power wireless personal area networks (6LoWPAN), thread, bluetooth low energy (Bluetooth LE), low power wide area network (LPWAN or LoRaWAN), ZigBee and IEEE wireless technologies such as 802.11, 802.15.1, 802.15.3, 802.15.4, 802.15.6, and 802.16 (Yaqoob et al., 2017). A combination of these technologies can be employed, depending on the smart city scenario. However, this chapter focuses on 5G and more specifically D2D communication to act as a bridge between IoT and cellular communication (Steri et al., 2016). This is due to wide coverage of cellular networks, which are available in almost all urban environments. Moreover, the 3rd Generation Partnership Project (3GPP) has already started standardizing different technologies for various smart city scenarios from release 12 to 14 Nokia Networks white paper (2015), which will provide a path to integrate smart city applications in 5G networks (Lauridsen et al., 2017). These standardization efforts are summarized in Table 5.1.

D2D communication is a new paradigm in cellular networks, which enables direct interaction between nearby user equipments (UEs), minimizing data transmissions in radio access networks (Usman et al., 2015). In conventional cellular networks, the UEs communicate with each other through a common base station; whereas, in D2D, the UEs in close proximity can directly communicate with each other by establishing a peer-to-peer (P2P) link between them, as illustrated in Figure 5.1.

Exploiting D2D communication can potentially enhance the role of IoT in future smart cities. One such example is Internet of Vehicles (IoV), where vehicles communicate with each other in D2D mode, without traversing any data traffic to the base station. The nearby vehicles can be automatically alerted before any change of lane. This helps vehicles to better respond to emergency situations, thus avoiding potential

Table 5.1 An Overview of New Features of LTE-A Networks in 3GPP Releases 12 to 14 to Enable Different Smart City Applications.

Smart City Applications	LTE Variant
Internet of things(IoT) (M2M)	LTE-M
Proximity services (D2D)	LTE-D
Terrestrial TV (Broadcast)	LTE-B
Vehicle communication (Vehicle to X)	LTE for V2X
Connectivity for public transport WiFi access points	LTE for backhauling

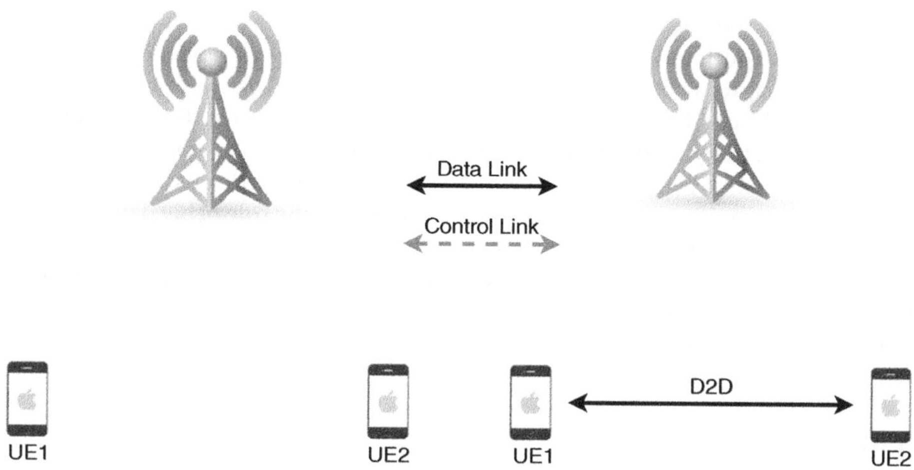

Figure 5.1 Conventional cellular communication (left side) versus direct D2D communication (right side): D2D communication minimizes data transmission in radio access networks, which improves spectrum efficiency.

accidents. Moreover, the traffic on the road can be prioritized. That is, school buses and ambulances can be assigned higher priorities over normal vehicles.

D2D communication has been addressed in release 12 (Astely et al., 2013) of 3GPP under the name of proximity services (ProSe). In particular, 3GPP radio access network (RAN) working group proposed two basic functions, ProSe discovery and ProSe communications, in TR 36.843, Rel. 12 "3GPP TR 36.843, Study on LTE device to device proximity services; Radio aspects, v12.0.1" (2014). However, 3GPP has initially targeted public safety applications in D2D communication. In this regard, Table 5.2 presents the supported ProSe functions (ProSe discovery and ProSe communications) for public safety and non-public safety applications in three different network scenarios. The in-coverage scenario represents a situation when all user devices lie in the coverage area of the cellular network. Similarly, in out-of-coverage scenario, all user devices are located outside the coverage area of the cellular network. The partial coverage scenario represents a situation when some user devices are located outside the coverage area of the cellular network. The devices at the edge of the coverage area relay the information of out-of-coverage devices to the base station or core network.

This chapter describes the smart city scenario, its requirements, and the potential impact on the life of citizens. By analyzing the types of services offered in a smart city environment, the chapter introduces the communication requirements to

Table 5.2 Supported ProSe Functions in 3GPP Release 12 to Enable D2D Communication in Public Safety and Non-Public Safety Applications.

Scenarios	Within network coverage	Outside network coverage	Partial network coverage
Supported applications	Supported ProSe functions		
Non-public safety	discovery	-	-
Public safety	discovery, communication	communication	communication

support smart city applications. Based on such analysis, the relevance of 5G and D2D communications are outlined, together with the related security aspects.

The rest of the chapter is organized as follows. In Section 5.2, we review the current literature in the field of 5G architecture and D2D communications. Then, we briefly present various smart city scenarios to identify the types of data and services required by a smart city infrastructure in Section 5.3. Next, we provide a discussion regarding communication ways to integrate smart city applications and the role of big data in smart cities as several services in smart cities require analysis and processing of great amount of heterogeneous data (Section 5.4). The chapter then highlights potential security and privacy problems, raised by the excessive use of ICT in smart cities (Section 5.4). Finally, Section 5.5 concludes the chapter.

5.2 Literature Review

Smart cities are often indicated as intelligent cities, virtual cities, digital cities, or information cities due to the pervasive usage of ICT and its services at their very core. ICT penetration in smart cities does not only include human-computer interaction (HCI) but also machine-type communication (MTC), where devices communicate massively with each other and with the information infrastructure (typically cellular) through some gateways (Albino et al., 2015; Schaffers et al., 2011). In this regard, various techniques and solutions have been proposed in the literature, which address the social and economic problems of cities using ICT.

In Batty et al. (2012), FutureICT solves the major problems in integrating smart city applications in the ICT sector. Piro et al. (2014) propose ICT for smart cities as information centric instead of user centric to fully exploit the available wireless communication systems. The authors propose an information centric network (ICN) able to (i) process real-time data packets according to information stored within them; (ii) provide content-aware caching and storage in network routers; (iii) efficiently transmit content-oriented data; and (iv) provide adaptive and flexible control mechanisms, which autonomously react to external events. Hernández-Muñoz et al. (2011) propose a user-driven approach, where IoT, Internet of Services (IoS), and Internet of People (IoP) act as major building blocks of the future Internet (FI) infrastructure for smart cities. The authors propose important design considerations of FI, which include capacity planning, scalability, inter-operability, and faster development of new and innovative applications.

In ICT, the communication infrastructure plays an important role in efficiently connecting smart city applications to the Internet. The cellular network, being the most ubiquitous deployed wireless infrastructure, is a potential candidate for this connection (Munoz et al., 2016). The need is to have a communication infrastructure that can successfully integrate diverse services and traffic demands of smart city applications, including applications requiring ultra low latency, such as the tactile Internet and remote surgery. A 5G network is supposed to satisfy the diverse requirements of smart city applications (Munoz et al., 2016). In this context, several architectures are proposed in the literature to cover the various aspects of 5G networks. Most of the proposals are based on software-defined networking (SDN), network functioning virtualization (NFV), and cloud computing concepts, which help to improve the network capacity, flexibility, scalability, energy efficiency, spectral efficiency, and management costs (Droste et al., 2015).

Zhang et al. (2015) propose SDN/NFV-based architecture for 5G networks. The architecture is implemented in LTE ultra-dense heterogeneous networks (HetNets). The virtualization and programmability make the network more flexible to adapt heterogeneous services and applications.

Droste et al. (2015) present a summary of the Mobile and wireless communications Enablers for Twenty-twenty (2020) Information Society (METIS) work on the 5G architecture. They consider three key aspects for the development of 5G architectures including flexibility, scalability, and service-oriented management. To accomplish these aspects of 5G networks, METIS proposes an information-centric SDN/NFV-based architecture for 5G, which can integrate IoT with a range of services and traffic demands.

Ziegler et al. (2016) propose an SDN/NFV-based architecture for 5G networks, called Cognitive and cloud Optimized Network Evolution (CONE), which provides an ability to integrate multiple radio access technologies (Multi-RAT) in licensed and unlicensed bands. The authors introduce the service enablement layer and the application layer in the architecture to fully support the heterogeneous requirements of smart city applications.

A similar kind of SDN/NFV-based architecture is proposed in Agyapong et al. (2014) with a tendency to integrate smart city applications. Most of the proposed architectures aim at improving performance of cellular networks, without focusing on integrating smart city applications as a primary goal. The smart city applications impose diverse requirements on cellular networks in terms of data size, Quality of Service (QoS), latency, and throughput. These requirements are different for what a cellular network is optimized for; thus a gap exists. D2D communication can be one of the options to bridge this gap by acting as a link between the cellular communication and smart city applications.

Steri et al. (2016) propose a D2D communication protocol that enables UEs as a hub of IoT traffic to upload data of smart city applications using mobile networks. The protocol enables multi-hop communication between D2D-enabled terminals with some security mechanism based on shared keys. The multi-hop communication extends the coverage of cellular networks to the devices/sensors that are not in the range of mobile networks.

Salpietro et al. (2015) propose a smart parking system using D2D communication, embedded sensors of a UE, and a remote server. The UE acts as both a sensor device and a gateway to disseminate parking information to nearby users using direct D2D links and to the remote server using a cellular link. The remote server stores information about the empty places in the parking, which can be later retrieved by a UE using a cellular connection.

Rigazzi et al. (2015) exploit D2D communication as an aggregation and trunking technique for the traffic generated by machine-type devices (MTDs). A UE aggregates the MTD, supplements it with its own data, and transmits it to the base station.

In the aforementioned works by Rigazzi et al. (2015), Salpietro et al. (2015), and Steri et al. (2016), D2D communication provides a communication path for MTDs to connect with the Internet. However, this creates a problem of resource allocation to D2D users (i.e., UEs) on their links to the cellular infrastructure and to MTDs or IoT. In the literature, different solutions have been proposed to solve the resource allocation problem in D2D communication focusing on IoT applications (Do-Yun and Choi, 2015; Huang et al., 2015; Shang et al., 2016; Zhou et al., 2016).

Some works in the literature propose D2D as an effective solution for disaster response networks (DRNs) and public safety applications. In this context, the most recent solutions include Al-Hourani et al. (2016), Ali et al. (2016), Apostolos et al. (2016), Kaleem et al. (2016), Usman et al. (2015), Muraoka et al. (2015), and Seo et al. (2016).

Despite all potential benefits of D2D communication, trust is a significant problem when communicating with unknown users. In this regard, social ties among users can potentially promote D2D communication. Considering this, social-aware D2D solutions are proposed in the literature, exploiting social ties and social trust among users. One such solution is presented in Datsika et al. (2016), whose authors propose an energy efficient and social-aware cooperative D2D medium access control (MAC) protocol, which exploits users' social connections in D2D cooperation. Moreover, the authors suggest that the green aspect of D2D networking should not be neglected and it should be incorporated in social-aware D2D cooperation in a way that the D2D QoS is improved.

Another work that stresses the need of social trust in D2D interactions is presented in Usman et al. (2017). The authors propose a combination of the pretty good privacy (PGP) and reputation-based model to bootstrap trust in D2D environments. In particular, the authors introduce a profiling server that keeps track of the reputation information of D2D users. Some other works that focus on social ties and social trust among D2D users are recently presented in Chen et al. (2013, 2015), Ometov et al. (2016), Ometov et al. (2016), Ometov et al. (2016), Wang, Cao, and Wu (2015), Wang and Wu (2015), Wang et al. (2015), and Wang et al. (2016).

More recently, Wang et al. (2016) consider the problem of reliability and secrecy enhancement for wireless content being shared between two communicating nodes. By exploiting social characteristics of multiple nodes in the presence of multiple independent eavesdroppers, they first investigate the impact of mobility for source node selection for transmission reliability, within a cooperative wireless network. Furthermore, the authors address social tie–based jammer node selections for cooperative jamming to provide secrecy, while allocating power appropriately to the source node and the cooperative jammer node to maximize the worst-case ergodic secrecy rate.

Ometov et al. (2016) propose a social awareness layer in D2D communication to build trust among D2D UEs. Before establishing D2D clusters, the UEs examine the social behavior patterns and interpersonal relationship of humans, thus forming trusted user groups. However, the authors claim that this trust and social-aware cooperation between UEs and with the network operators remains conditional to the incentives provided to participating UEs by the network operators. The authors identify three kinds of possible user incentives that apply to different D2D scenarios: pragmatic incentives (throughput gain, energy efficiency, or latency gain), indirect incentives (economic incentives), and social incentives (lend resources to friends and family).

Table 5.3 briefly summarizes the literature based on two main categories of networking strategies: ICN and user-centric networks.

Table 5.3 The proposed 5G Architectures in the Literature Can Be Divided into Broad Categories: Architectures Based on Information-Centric Networks (ICN) and Architectures Based on User-Centric Networks.

Current trends in 5G	Reference	IoT integration	Enabling technologies
Information-centric architectures	Hakiri et al. Hakiri and Berthou (2015), 2015	Yes	SDN/NFV
	Iwamura et al. Iwamura (2015), 2015	Yes	SDN/NFV
	Wang et al. Wang et al. (2014), 2014	-	NFV/D2D/eMBMS (Evolved Multimedia Broadcast Multicast Services)
	Su et al. Su and Xu (2015), 2015	Yes	HetNets (Heterogeneous Networks)
	Liang et al. Liang et al. (2015), 2015	Yes	SDN/NFV
	Droste et al. Droste et al. (2015), 2015	Yes	SDN/NFV/D2D /Cloud Computing
User-centric architectures	Bangerter et al. Bangerter et al. (2014), 2014	-	HetNets/D2D
	Peng et al. Peng et al. (2015), 2015	Yes	SDN/NFV/HetNets
	Pirinen et al. Pirinen (2014), 2014	Yes	SDN/NFV/Cloud Computing
	Liu et al. Liu (2015), 2015	Yes	CoMP (Coordinated Multi-Point)
	Jaber et al. Jaber et al. (2016), 2016	-	CoMP\SON (Self-Organizing Network)
	Datta et al. Datta et al. (2015), 2015	Yes	oneM2M (Machine-to-Machine Communication)

5.3 Smart City Scenarios

Smart cities employ multiple technologies to improve the living standard of citizens, providing citizens better services in health, transportation, energy, education, public safety, and security. These technologies include, but are not limited to, cloud computing, networking, SDN/NFV, big data, IoT, and cybersecurity architectures. Figure 5.2 (Khan et al., 2013) lists possible smart city application scenarios, which are connected to data acquisition and storage center using a communication infrastructure.

Figure 5.2 is divided into two parts. The upper part demonstrates different smart city scenarios. In each scenario, data generated by sensors is transmitted to a data acquisition and storage center, which can later be used to analyze and process the collected data. Cloud services and artificial intelligence can be utilized to analyze the big data collected

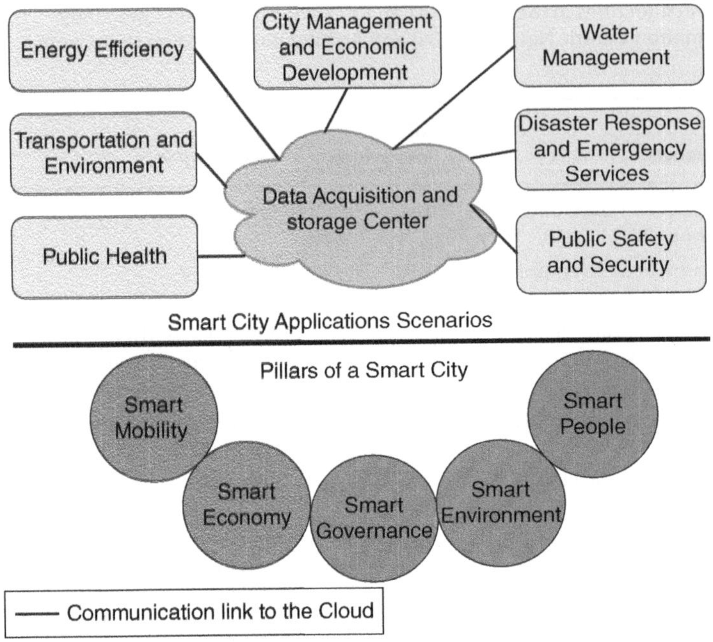

Figure 5.2 An abstract view of smart city applications and its pillars: the part above the horizontal line represents various smart city application scenarios, which are connected to data acquisition and storage center through the Internet. The lower part of the figure represents different pillars of a smart city (adapted from Khan et al., 2013).

from these sensors. Based on this analysis, various decisions can be made for the benefits of smart city residents. The lower part of the figure demonstrates pillars on which a smart city is built. These pillars play a vital role in building smart cities.

In the following, we will elaborate each application scenario and the communication infrastructure required.

5.3.1 Public Health

The technology has changed the way patients are treated in the modern world. According to some reports (Enbysk, 2015), the traditional ways of health care management are unable to handle the data of rapidly growing population in the world. There is a need to find smarter ways to handle this problem. The modern communication technologies and traditional health care processes can be integrated together in order to develop a better health care system. This includes wireless body area networks, communication networks, data analytics, and humans. The sensors acquire health information such as heartbeat, blood pressure, blood sugar, or any deterioration in health, which can be potentially transmitted using communication technologies to remote servers accessible by health care professionals for monitoring, diagnosis, or treatment purposes.

Many localities/councils in Europe, Asia, and America are working to provide innovative solutions for public health, making use of the aforementioned technologies (Createtomorrow, 2015). An overview of some of these solutions is provided below.

24/7 Social Care at Home Technologies for social care include solutions for older and disabled citizens, which automatically notify concerned staff about the problem such as a fall or deterioration in health conditions. Using these solutions, the residents requiring social care needs can independently stay at their home but could continuously be connected to 24/7 healthcare centers.

Wearables and Video Calling to Doctors This includes solutions to remotely consult a doctor using video conferencing. The patient can be remotely examined using wearable health patches (Zaragoza, 2015) that have the ability to continuously monitor and transmit vital signs such as heart rate, muscle movement, brain activity, hydration level, and temperature. Moreover, the prescription can also be automatically sent to a network of pharmacies for an automated delivery of medicine (Createtomorrow, 2015). This could dramatically reduce the physical visits to public health care centers.

Remote Surgery (Telesurgery) Remote surgery allows people to access world's leading expertise and health care services without travelling to the traditional health care facility. This can be imagined as a pool of experts, where any patient on earth can take advantage of being treated by the experts, provided that the Internet connection between the patient and the medical expert satisfies the latency requirements, necessary for carrying out certain surgery.

Note that all of the aforementioned applications impose different requirements on the communication infrastructure connecting them with the central services. For instance, social care and video calling do not demand stringent requirements of ultralow latencies. An uninterrupted Internet connection with sufficient throughput is enough for these applications. However, telesurgery imposes very stringent requirements on latency and reliability of the Internet connection. 5G networks can potentially fulfill the communication needs of a smart public health system as the network latency and reliability completely resonate with the requirements.

5.3.2 Transportation and Environment

The intelligent transportation system is an important part of smart cities, which not only affects the environment but also the quality of life of the citizens. For example, in a densely populated city, automobile sharing, bicycle sharing, electric vehicles, and robust local transport facilities can not only address the issue of CO_2 emission but also reduce traffic congestion in the city. All these facilities require an agile, robust, and secure communication network to allow vehicles to communicate with each other, with the infrastructure, and with the people.

The ubiquitous coverage of cellular systems makes it the most viable and trivial solution to solve the connectivity problem. Based on cellular technology, many initiatives have been taken to increase safety on the roads. For example, eCall is a European initiative that brings rapid assistance to the vehicles involved in accidents, by automatically informing E112 about the incident. The European Parliament has decided to make eCall mandatory to be included in all new cars sold in the European Union after April 2018 (Comission, 2015). The prototype of the solution has already been tested with general packet radio service (GPRS) and in-band signaling over cellular networks.

The cooperation between vehicles and intelligent mobility is one of the most powerful concepts of smart transport systems. It requires intelligent communication systems for wireless data exchange not only from vehicle to infrastructure (V2I) and vehicle to vehicle (V2V) but also between other road users such as motorcyclists, cyclists, and pedestrians. For this kind of communication, IEEE 802.11p has been proposed, which extends 802.11 in order to support the intelligent transport system. The amendments were approved and published under the name "Amendment 6: Wireless Access in Vehicular Environments" (Committee, 2015).

Vehicle-to-everything (V2X) communication is one the most important parts of an intelligent transportation system (ITS) that virtually connects vehicles on the road with almost everything in the proximity. V2X potentially changes the ecosystem of the current transportation system. For example, some researchers believe that connecting vehicles with everything will extinguish the need of traffic lights on the roads (Greengard, 2014). Instead, a traffic network will adjust the smart traffic lights, bus routes, light rail systems, and subways to optimize the traffic across the entire city. For this optimization, the status of virtual traffic lights and cues can be delivered to the drivers and pedestrians via smartphones (Greengard, 2014). The potential benefits of this optimized traffic network include, but are not limited to, reduced traffic congestion in the city, reduced CO_2 emission in the environment, reduced fuel wastage, and consequently the less economic loss. The economic loss solely was about \$121 billion in the US for the year 2011 (Greengard, 2014). In addition, by virtue of communication among sensors-empowered vehicles, cars can platoon themselves half a meter apart or less to stretch the road capacity and infrastructure.

Big data analysis will play a vital role in optimizing transportation of a smart city. For example, analyzing the macroscopic behavior of traffic across city is one of the biggest challenges for traffic engineers and computer scientists. This includes route modeling and demand modeling of a transport system subject to the behavior of vehicles and pedestrians in different weather conditions, sporting and entertainment events, office hours, etc. For this modeling, more sophisticated algorithms are needed based on big data approaches.

Other aspects of smart transport systems include the detection of toll evaders, driving on bus-only lane, red light crossing, and speeding using cameras, crime detection, and security (GSMA, 2015). All these require communications to a control center or among camera/sensors to correctly detect any misbehaving vehicle. Similarly, various services such as fleet management and PayAsYouDrive insurance currently use the GPRS system to transmit the driving style and vehicle health to insurance companies or car rentals. The use of 5G technology, in place of GPRS, will further improve the consumer experience with increased uplink speed and reduced latency (GSMA, 2015). This will help consumers to find an optimal route to the destination in real time.

Advertising traffic information through social media and mobile apps is another application of smart transportation systems. Mobile applications are cheaper and more real time than electronic displays, displaying possible congestions ahead. These kinds of applications are equally beneficial to transport authorities and mobile network operators (MNOs) for business and financial objectives, consumers for fuel-consumption and travel time reasons, and the climate by emitting less CO_2.

5.3.3 Energy Efficiency

Communication technology can play a major role in improving the energy efficiency of smart cities. The energy we use in our daily life mainly comes from the burning of fossil fuels, which emits CO_2 to the atmosphere. ICT can potentially improve the energy efficiency in the following ways.

Smart Buildings The intelligent building management system and wireless sensor networks can potentially help to control appliances of the building. The heating and cooling systems can be completely switched off when there is no one in the building. The entrance cameras can be used to detect any presence inside the building to turn on only specific appliances.

Smart Homes The communication infrastructure helps residents to stay connected with their homes even when they are away. The smart appliances and smart meters give the consumers awareness about their energy consumption in real time. They can change their behavior toward energy use and save both money and energy.

The communication infrastructure, which is an essential component of smart cities, itself consumes energy. For example, 3G, 4G, and 5G infrastructures, smartphones, tablets, sensor networks, cameras, and monitoring displays consume energy. In the smart city planning, the energy consumption of the ICT sector should also be appropriately considered to improve the overall energy efficiency.

5.3.4 Smart Grid

The so-called smart grid is a major source of energy efficiency in smart cities, which harmonizes the needs and capabilities of grid operators, electricity market stakeholders, and consumers to manage power networks as efficiently as possible (Al-Rubaye et al., 2016). A smart grid uses communication technologies to connect smart meters, smart appliances, and renewable energy resources with the control room in order to reduce the peak demand and better integrate energy from renewable resources into the power grid.

The renewable energy resources such as solar power and wind power are highly variable and so require more sophisticated control system and energy efficient solutions to better integrate them into power grids. For this integration, communication technologies remain at the heart of smart grids. D2D and 5G networks will play an important role in realizing the communication needs of smart grids. For example, smart meters and other sensors nodes in smart grids can be locally connected to a D2D access point to give it an access to cellular networks. After this, SDN techniques, such as network slicing, can be utilized to ensure the QoS required by the smart grids (Nakao et al., 2017). Concerning this, several solutions have been proposed in the literature focusing on 5G networks as a communication enabler of smart grids (Borkar and Pande, 2016; Khodashenas et al., 2016; Saxena et al., 2017).

The smart grid offers several benefits over traditional power grid networks, which include, but are not limited to, the following.

Reliability The smart grid makes the power grid more resistant against faults and engraves the process of fault-detection and self-healing in the power network. This consequently ensures reliable delivery of electricity and reduces vulnerability against attacks and natural disasters.

Energy Efficiency The smart grid makes the power network more energy efficient by including demand-side management, i.e., adjusting/balancing the load such that both consumers and electricity stakeholders get benefits. For example, at the consumer end, air conditioner can be turned off during short-term price spikes. At the stakeholder side, it reduces the truck rolls for meter reading and fault detection and recovery. The ultimate goal is to eliminate the redundancy in power grid networks and save energy and resources.

Flexibility The smart grid enables a bidirectional flow of energy, i.e., the electricity from local renewable energy resources can be added to the main power grid for the times when local demand is lower than production. This bidirectional flow of energy makes the network more flexible to save energy and also brings monetary benefits to the consumers that have their own local renewable energy resources.

5.3.5 Water Management

Water is one of the most important assets on our planet and thus should be managed properly to avoid possible wastage of this natural resource. The communication sector can be potentially utilized to track or forecast the water level in the rivers and to identify more resources of fresh water. More precisely, technologies such as remote sensing, semantic sensor web, geographical information systems (GISs), sensors, and cellular networks can be used innovatively. Communication technologies provide a significant way to obtain real-time information of various variables such as soil moisture level, rainfall, temperature, and forecast of floods and storms.

The smart metering and sensing system help water management authorities in providing real-time information about water usage and leaking in the water supply, thus providing better control over water management. This helps to manage the complete ecosystem of the water distribution network, from reservoirs to end users.

The ICT sector can help to manage water resources more smartly in the following ways (Report, 2015).

Water Resources and Weather Forecasting Communication technologies help map water resources to identify the current water supply and its characteristics to determine future water demands for sustainable economy growth. Technologies such as satellite-based remote sensing, GIS, Global Positioning System (GPS) are instrumental in carrying risk assessment on the hydrological cycle of water, identifying new water resources, and analyzing any environmental problems.

Irrigation in Agriculture and Landscaping The sensor semantic web is usually used to collect the sensor information and observations using web technologies. This data can be analyzed to save water in agriculture. The wireless sensors can be placed at different locations in crops and in the soil to monitor the moisture level. The water valve can be automatically activated based on the sensors information to water the plants at the right time with the right amount of water. The sensors data is transmitted to the farmer using mobile networks, which can monitor the daily consumption of water.

Water Distribution Network Sensors are installed throughout the water distribution network. These sensors are connected with a control room using communication networks to help in identifying any leakage in the distribution network. The leakage is monitored in real time, which assists companies to take immediate remedies on problems detected in the distribution network.

Early Warning System to Meet Water Demand Preservation of fresh water, flood management, and water recycling are the key parameters to be analyzed to meet the water demands of smart cities. For the cities located near the coast, there should be an early warning system to get information about the sea water level. The sensors network is deployed near the coast to check the weather conditions and water level. The extreme weather conditions and rise in the water level are communicated to control rooms using the cellular infrastructure.

5.3.6 Disaster Response and Emergency Services

Communication technologies can help greatly to save lives during natural disasters. According to a report (Report, 2015), over 7000 natural disasters occurred worldwide during 1980–2005, in which millions of lives were lost. Around 90% of those disasters were caused by weather and water-related incidents such as cyclones, floods, tsunamis, and droughts. Telecommunication can be used for weather forecasting, climate monitoring, and detecting and mitigating the effects of natural disasters. The world weather watch from the world meteorological organization divides the ICT into three core components:

Global Observing System (GOS) It provides measurements and observations of the atmosphere, at the ocean level, in the air, and in the space.

Global Telecommunication System (GTS) It consists of telecommunication networks and facilities to exchange real-time data of meteorological observations between national meteorological and hydrological services.

Global Data Processing and Forecasting System (GDPFS) It helps in processing meteorological data and provides analysis, warnings, and forecasts, which are generated by a network of meteorological centers. At the time of disaster, the network infrastructure also faces a communication breakdown. The disaster victims are no longer connected with the outside world. If the communication infrastructure is not affected by the disaster, even then a lot of people try to connect with their family members in the affected areas, causing congestion in the network. However, D2D communication in 5G networks provides support to get the people connected in the time of disaster (Usman et al., 2015).

5.3.7 Public Safety and Security

Reliable and robust communication networks are fundamental for public safety agencies to guarantee public protection and disaster relief (PPDR). The networks such as TETRA and P25 are widely used by public safety (PS) agencies to respond in critical situations (Kumbhar et al., 2016). Such networks provide mission-critical voice services to enable communication between emergency first responders. However, the mega incidents such as 9/11 revealed the failure of such networks mainly due to interoperability problems. The PS networks of different municipalities are incongruent with each other. On the other hand, their narrowband nature does not permit high data rate multimedia services such as videos that are instrumental for providing effective PPDR.

In response to the aforementioned problems, in 2012, the first responder network authority (FirstNet) was created in the US to build, operate, and maintain the first nationwide interoperable broadband wireless network for public safety (Farrill, 2012). The goal of FirstNet is to employ commercial cellular networks as a communication infrastructure for PS agencies nationwide. However, unlike narrowband PS networks, cellular networks provide non-mission-critical and often only one-to-one communication. The support for group communication remained very basic, meeting far less stringent constraints than those required for traditional PS networks.

In order to support public safety requirements in commercial cellular networks, the third-generation partnership project (3GPP) is working to enhance the LTE features. In this context, several initiatives are taken under different working groups. These initiatives include, proximity services (ProSe) and group communication in release 12 3GPP (2016c), mission-critical push-to-talk over LTE (MCPTT) in release 13/14 3GPP (2016d), and mission-critical video over LTE (MCVideo) and mission-critical data over LTE (MCDATA) in release 14 3GPP (2016a).

5.4 Discussion

All of the aforementioned smart city scenarios are divergent in terms of their requirements for latency, mobility, network reliability, network resiliency, and other properties. Integrating such heterogeneous applications in a communication infrastructure is not a trivial task. The communication infrastructure, fully supporting IoT and smart city applications, should include the following aspects, which are also a primary goal for realization of 5G networks (Ziegler et al., 2016).

5.4.1 Multiple Radio Access Technologies (Multi-RAT)

Supporting multi-RAT in licensed and unlicensed bands is one of the fundamental requirements of 5G networks. This allows 5G to fully utilize the entire frequency range from sub-6 GHz to 100 GHz (Ziegler et al., 2016). To enable multi-RAT, the concept of single radio controller (SRC) is introduced, which is an intelligent solution to automatically decide among available radio access technologies such as global system for mobile communications (GSM), universal mobile telecommunications system (UMTS), LTE, and Wi-Fi (Xing et al., 2013).

5.4.2 Virtualization

The communication requirements of different smart city applications are different from each other. For applications such as telemedicine and remote patient care, latency is very important, while applications such as waste management simply require a reliable Internet connection without having stringent requirements on latency. In order to support these varying requirements of smart city applications, the architecture of 5G networks should be service oriented (Kalyan, 2016). The service-oriented network architecture can be realized by "network slicing," where each service can be handled by its own network slice. To better manage the network slicing, virtualization is the best way, which uses SDN and NFV as enabling technologies. The SDN controller can configure and build network slices for each service in demand. This allows operators to rapidly set up services and scale them in response to changing demands.

5.4.3 Distributed/Edge Computing

IoT devices generate an unprecedented amount of data, which needs to be processed and analyzed in the cloud to act on it. For example, the connected vehicles in V2V communication not only communicate with each other but also interact with the sensors on stop lights and bus stops to get traffic updates and rerouting alerts. All these communications need lots of data to be processed, which is done in the cloud. After processing the data from different sources, the result is sent back to vehicles to avoid potential accidents. However, there might be a chance that by the time data makes its way to the cloud, the opportunity to act on it might be gone (i.e., it's too late). Moving the computing capabilities closer to IoT (at the network edge) is regarded as a potential solution to handle this problem of quick responsiveness for many smart city applications such as driverless cars, smart lighting, security cameras, and monitoring ill patients (Corcoran and Datta, 2016). This concept is known as edge computing or fog computing.

5.4.4 D2D Communication

Cellular networks are mainly designed for human-type communication (HTC) to support higher data rates and larger data sizes, while MTC in IoT typically exchanges smaller data packets. For example, the minimum size of a radio resource block that can be allocated to a device in LTE-A could be actually too big for the need of IoT applications. On the other hand, large energy consumption required by cellular communication is a major barrier in terms of its adoption as a connectivity platform for IoT applications in smart city scenarios (Orsino et al., 2016). D2D communication is considered as a viable solution to solve the aforementioned problems.

IoT devices can be clustered together based on their proximity. A smartphone can aggregate the traffic of the cluster to the cellular network to improve communication and energy efficiency (Orsino et al., 2016). As an example, Figure 5.3 presents a smart home scenario, where smart appliances are connected with a cellular network through an aggregator. Direct D2D communication is considered as a connectivity mechanism between smart appliances and the aggregator.

Enabling IoT with D2D communication encounters certain challenges, such as interference management, resource allocation, security, trust, and pricing. Along with these challenges, network operators require new business models to answer "pay for what" questions (Palattella et al., 2016). The device, which acts as an aggregator or a relay in D2D communication, will deplete its resources (such as battery, processing, storage, and communication). This requires innovative business models that can provide incentives

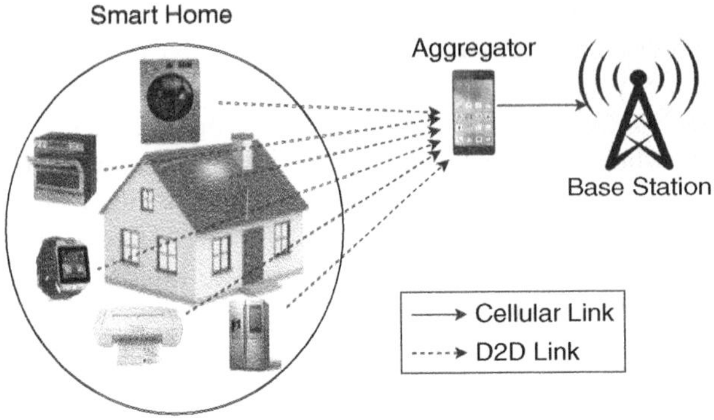

Figure 5.3 D2D communication as an aggregator for IoT traffic: home appliances are connected with a smartphone over a D2D link. The smartphone aggregates traffic from different sensing nodes and sends it to base stations when it has sufficient data to be transferred.

to the devices acting as aggregators or relays. An IoT operator, for the cases when a device is acting as an aggregator, can provide these incentives, as in this case an IoT operator is the main entity that is responsible for providing and managing the local connectivity. On the other hand, a mobile network operator can provide incentives to the devices acting as relays in licensed bands, as in this case, the network operator is the one mainly providing resources for connectivity.

However, from the point of view of IoT and mobile network operators, a business model is required to generate revenues. The case of an IoT operator is comparatively simpler as it provides connectivity to a constrained and local environment, e.g., a smart home or a smart building, while, in the case of network operators, the business model for IoT can be comparatively complicated as it is already relying on roaming and other agreements in the ecosystem. However, SDN can be adopted to reserve special data pipes for IoT traffic based on its requirements. In doing so, network operators can charge premium rates to IoT customers based on their QoS requirements and the resources they use. This flexibility of creating a service-oriented network for IoT applications is not easy to implement with other competing technologies (mentioned in Section 5. 1) with the same ubiquitous coverage as of mobile networks.

5.4.5 Big Data

IoT devices for different smart city applications generate a massive amount of data. The analysis and mining of this data is a significant step toward making a city smart (Smart Cities, 2013). Machine learning (ML) and artificial intelligence (AI) techniques are used to analyze the data. As an example, IoT devices and advanced ML and AI techniques could be used to observe the spread of life-threatening Zika virus in urban environments. Drones could be used to identify standing waters and observe Zika-infected Mosquitos. The mobility traces of smartphone reveal how humans are moving around the city. These traces along with the data collected from the drones can be processed and analyzed using ML and AI techniques to find clues on how Zika or other life-threatening diseases spread in the community (Steep, 2016). With the help of IoT and the big data ecosystem, municipalities can provide their citizens insights and information regarding these diseases.

5.4.6 Security and Privacy

Security is one of the most important elements to be considered in smart city applications prior to any implementation, such as web-based technologies for cities. For example, in a recent demonstration, Chinese students hacked a Tesla electric car and opened its doors while the car was moving on the road (GSMA, 2015). Moreover, some students demonstrated a hack for Toyota Prius and Ford Escape vehicles, which affected the steering and braking systems. Similarly, in a smart home, a malicious user can falsely control the communication and harm the appliances by forging possible instructions. Turning on the gas burner and a small lamp near it can potentially trigger fire in the home. To solve these security problems, end-to-end security should be inherently embedded in the network architecture of 5G cellular systems across all domains and layers.

Privacy is also a major concern in smart cities, where every sensor transmits some data, which can be private to citizens (Martínez-Ballesté et al., 2013). Privacy becomes a more critical issue when it comes to IoT as smart cities may deal with sensitive personal data such as health care information collected by wearable devices. This includes collection, transmission, processing, and sharing of sensitive data without appropriate privacy protection. Users may not be willing to expose their data to others, which hinders the processing and sharing of health data and users' experiences.

5.5 Conclusion

In this chapter, we provide a survey and useful considerations in the design and deployment of communications infrastructure for smart city applications. The chapter first reviews the current literature in the field of 5G and D2D communications with a reference to IoT integration. We identify the current trends in the designing of 5G networks and discuss possible integration of IoT and D2D communications. The chapter presents different application scenarios in smart cities with their communication needs. Finally, we highlight the enabling technologies of future communication infrastructure from the perspective of smart city integration.

References

3GPP (2016a), '3gpp TS 23.281, functional architecture and information flows to support mission critical video (MCVideo); stage 2', https://portal.3gpp.org/desktopmodules/Specifications/SpecificationDetails.aspx?specificationId=3087. Last accessed: 2016-11-25.

3GPP (2016b), '3gpp TS 23.282, functional architecture and information flows to support mission critical data (MCData); stage 2', https://portal.3gpp.org/desktopmodules/Specifications/SpecificationDetails.aspx?specificationId=3088. Last accessed: 2016-11-25.

3GPP (2016c), '3gpp TS 23.303 v12.2.0, proximity-based services (ProSe); stage 2 (release 12)', https://portal.3gpp.org/desktopmodules/Specifications/SpecificationDetails.aspx?specificationId=840. Last accessed: 2016-11-25.

3GPP (2016d), '3gpp TS 23.379, functional architecture and information flows to support mission critical push to talk (MCPTT); stage 2', https://portal.3gpp.org/desktopmodules/Specifications/SpecificationDetails.aspx?specificationId=3089. Last accessed: 2016-11-25.

3GPP TR 36.843, Study on LTE device to device proximity services; Radio aspects, v12.0.1 (2014). Last accessed: 2017-04-11.

Agyapong, P. K., Iwamura, M., Staehle, D., Kiess, W. and Benjebbour, A. (2014), 'Design considerations for a 5G network architecture', *IEEE Communications Magazine* 52(11), 65–75.

Al-Hourani, A., Kandeepan, S., and Jamalipour, A. (2016), 'Stochastic geometry study on device-to-device communication as a disaster relief solution', *IEEE Transactions on Vehicular Technology* 65(5), 3005–3017.

Al-Rubaye, S., Al-Dulaimi, A., and Cosmas, J. (2016), 'Spectrum allocation techniques for industrial smart grid infrastructure', *in* '2016 IEEE 14th International Conference on Industrial Informatics (INDIN)', pp. 1036–1039.

Albino, V., Berardi, U., and Dangelico, R. M. (2015), 'Smart cities: Definitions, dimensions, performance, and initiatives', *Journal of Urban Technology* 22(1), 3–21.

Ali, K., Nguyen, H. X., Shah, P., Vien, Q. T., and Bhuvanasundaram, N. (2016), 'Architecture for public safety network using D2D communication', *in* '2016 IEEE Wireless Communications and Networking Conference', pp. 1–6.

Apostolos, G., Konstantinos, K., Aikaterini, N., Foukalas, F., and Khattab, T. (2016), 'Energy efficient spectrum allocation and mode selection for mission-critical D2D communications', *in* 2016 IEEE Conference on Computer Communications Workshops (INFOCOM WKSHPS), pp. 435–440.

Astely, D., Dahlman, E., Fodor, G., Parkvall, S., and Sachs, J. (2013), 'LTE release 12 and beyond [accepted from open call]', *IEEE Communications Magazine* 51(7), 154–160.

Bangerter, B., Talwar, S., Arefi, R., and Stewart, K. (2014), 'Networks and devices for the 5G era', *IEEE Communications Magazine* 52(2), 90–96.

Batty, M., Axhausen, K. W., Giannotti, F., Pozdnoukhov, A., Bazzani, A., Wachowicz, M., Ouzounis, G., and Portugali, Y. (2012), 'Smart cities of the future', *The European Physical Journal Special Topics* 214(1), 481–518.

Borkar, S., and Pande, H. (2016), 'Application of 5G next generation network to internet of things', *in* 2016 International Conference on Internet of Things and Applications (IOTA), pp. 443–447.

Chen, X., Proulx, B., Gong, X., and Zhang, J. (2013), 'Social trust and social reciprocity based cooperative D2D communications', *in* Proceedings of the Fourteenth ACM International Symposium on Mobile Ad Hoc Networking and Computing, MobiHoc '13, ACM, New York, NY, USA, pp. 187–196.

Chen, X., Proulx, B., Gong, X., and Zhang, J. (2015), 'Exploiting social ties for cooperative D2D communications: A mobile social networking case', *IEEE/ACM Transactions on Networking* 23(5), 1471–1484.

Chourabi, H., Nam, T., Walker, S., Gil-Garcia, J. R., Mellouli, S., Nahon, K., Pardo, T. A., and Scholl, H. J. (2012), 'Understanding smart cities: An integrative framework', *in* System Science (HICSS), 2012 45th Hawaii International Conference on, IEEE, pp. 2289–2297.

Comission, E. (2015), 'eCall in all new cars from april 2018', https://ec.europa.eu/digital-single-market/en/news/ecall-all-new-cars-april-2018. Last accessed: 2017-04-17.

Committee, I. L. S. (2015), 'Part 11: Wireless LAN medium access control (MAC) and physical layer (PHY) specifications amendment 6: Wireless access in vehicular environments', URL: http://standards.ieee.org/getieee802/download/802.11p-2010.pdf. Last accessed: 2016-11-25.

Corcoran, P. and Datta, S. K. (2016), 'Mobile-edge computing and the internet of things for consumers: Extending cloud computing and services to the edge of the network', *IEEE Consumer Electronics Magazine* 5(4), 73–74.

Createtomorrow (2015), 'Where smart city tech and public health meet', URL: http://www.createtomorrow.co.uk/en/live-examples/technology-in-health. Last accessed: 2016-11-25.

Datsika, E., Antonopoulos, A., Zorba, N., and Verikoukis, C. (2016), 'Green cooperative device-to-device communication: A social-aware perspective', *IEEE Access* 4, 3697–3707.

Datta, S. K., Gyrard, A., Bonnet, C., and Boudaoud, K. (2015), 'oneM2M architecture based user centric IoT application development', *in* 2015 3rd International Conference on Future Internet of Things and Cloud, pp. 100–107.

Do-Yun, K. and Choi, Y. J. (2015), 'Spectrum coordination of multi-RAT D2D for IoT service', *in* 2015 International Conference on Information and Communication Technology Convergence (ICTC), pp. 671–673.

Droste, H., Zimmermann, G., Stamatelatos, M., Lindqvist, N., Bulakci, O., Eichinger, J., Venkatasubramanian, V., Dotsch, U., and Tullberg, H. (2015), 'The METIS 5G architecture: A summary of METIS work on 5G architectures', *in* 2015 IEEE 81st Vehicular Technology Conference (VTC Spring), pp. 1–5.

Enbysk, L. (2015), 'How smart technology is improving public health', URL: http://smartcitiescouncil.com/article/how-smart-technology-improving-public-health. Last accessed: 2016-11-25.

Farrill, F. C. (2012), 'Firstnet nationwide network (FNN) proposal', *First Responders Network Authority, Presentation to the Board* 25.

Greengard, S. (2014), 'Smart transportation networks drive gains', *Commun. ACM* 58(1), 25–27. **URL:** http://doi.acm.org/10.1145/2686742.

GSMA (2015), 'Mobilizing intelligent transportation systems (ITS)', URL: http://www.gsma.com/connectedliving/wp-content/uploads/2015/09/ITS-report-new.pdf. Last accessed: 2016-11-25.

Hakiri, A., and Berthou, P. (2015), 'Leveraging SDN for the 5G networks: Trends, prospects and challenges', *CoRR***abs/1506.02876**. **URL:** http://arxiv.org/abs/1506.02876.

Hernández-Muñoz, J. M., Vercher, J. B., Muñoz, L., Galache, J. A., Presser, M., Gómez, L. A. H., and Pettersson, J. (2011), 'Smart cities at the forefront of the future internet', *in The Future Internet Assembly*, Springer, pp. 447–462.

Huang, J., Yin, Y., Duan, Q., and Yan, H. (2015), 'A game-theoretic analysis on context-aware resource allocation for device-to-device communications in cloud-centric internet of things', *in* 2015 3rd International Conference on Future Internet of Things and Cloud, pp. 80–86.

Iwamura, M. (2015), 'NGMN view on 5G architecture', *in* 2015 IEEE 81st Vehicular Technology Conference (VTC Spring), pp. 1–5.

Jaber, M., Imran, M. A., Tafazolli, R., and Tukmanov, A. (2016), 'A distributed SON-based user-centric backhaul provisioning scheme', *IEEE Access* 4, 2314–2330.

Kaleem, Z., Li, Y., and Chang, K. (2016), 'Public safety users; priority-based energy and time-efficient device discovery scheme with contention resolution for prose in third generation partnership project long-term evolution-advanced systems', *IET Communications* 10(15), 1873–1883.

Kalyan (2016), 'Virtualization the real driver for 5G networks and IoT', URL: http://telecoms.com/opinion/virtualization-the-real-driver-for-5g-networks-and-iot/. Last accessed: 2016-11-25.

Khan, Z., Anjum, A., and Kiani, S. L. (2013), 'Cloud based big data analytics for smart future cities', *in* 2013 IEEE/ACM 6th International Conference on Utility and Cloud Computing, pp. 381–386.

Khodashenas, P. S., Aznar, J., Legarrea, A., Ruiz, C., Siddiqui, M. S., Escalona, E., and Figuerola, S. (2016), '5G network challenges and realization insights', *in* 2016 18th International Conference on Transparent Optical Networks (ICTON), pp. 1–4.

Kumbhar, A., Koohifar, F., Guvenc, I., and Mueller, B. (2016), 'A survey on legacy and emerging technologies for public safety communications', *IEEE Communications Surveys Tutorials* **PP**(99), 1–1.

Lauridsen, M., Gimenez, L. C., Rodriguez, I., Sorensen, T. B., and Mogensen, P. (2017), 'From LTE to 5G for connected mobility', *IEEE Communications Magazine* 55(3), 156–162.

Liang, C., Yu, F. R., and Zhang, X. (2015), 'Information-centric network function virtualization over 5G mobile wireless networks', *IEEE Network* 29(3), 68–74.

Liu, G. (2015), '5G-user centric network', *in* 2015 Asia-Pacific Microwave Conference (APMC), vol. 1, pp. 1–1.

Usman, M., Gebremariam, A. A., Raza U., and Granelli F. (2015), 'A Software-Defined Device-to-Device Communication Architecture for Public Safety Applications in 5G Networks', *IEEE Access* 3, 1649–1654.

Martínez-Ballesté, A., Pérez-Martínez, P. A. and Solanas, A. (2013), 'The pursuit of citizens' privacy: a privacy-aware smart city is possible', *IEEE Communications Magazine* 51(6), 136–141.

Mike Steep, M. N. (2016), 'Smart cities improve the health of their citizens', URL: http://www.forbes.com/sites/mikesteep/2016/06/27/can-smart-cities-improve-the-health-of-its-citizens/\#30564d507154. Last accessed: 2016-11-25.

Munoz, R., Mangues-Bafalluy, J., Vilalta, R., Verikoukis, C., Alonso-Zarate, J., Bartzoudis, N., Georgiadis, A., Payaro, M., Perez-Neira, A., Casellas, R., Martinez, R., Nunez-Martinez, J., Esteso, M. R., Pubill, D., Font-Bach, O., Henarejos, P., Serra, J., and Vazquez-Gallego, F. (2016), 'The CTTC 5G end-to-end experimental platform : Integrating heterogeneous wireless/optical networks, distributed cloud, and IoT devices', *IEEE Vehicular Technology Magazine* 11(1), 50–63.

Muraoka, K., Shikida, J., and Sugahara, H. (2015), 'Feasibility of capacity enhancement of public safety LTE using device-to-device communication', *in* 2015 International Conference on Information and Communication Technology Convergence (ICTC), pp. 350–355.

Nakao, A., Du, P., Kiriha, Y., Granelli, F., Gebremariam, A. A., Taleb, T., and Bagaa, M. (2017), 'End-to-end network slicing for 5G mobile networks', *Journal of Information Processing* 25, 153–163.

Neirotti, P., De Marco, A., Cagliano, A. C., Mangano, G., and Scorrano, F. (2014), 'Current trends in smart city initiatives: Some stylised facts', *Cities* 38, 25–36.

Nokia Networks white paper (2015), 'LTE-advanced evolution in releases 12–14: New services to pave the way to 5G', URL: http://resources.alcatel-lucent.com/asset/200175. Last accessed: 2017-04-11.

Ometov, A., Olshannikova, E., Masek, P., Olsson, T., Hosek, J., Andreev, S., and Koucheryavy, Y. (2016), 'Dynamic trust associations over socially-aware D2D technology: A practical implementation perspective', *IEEE Access* 4, 7692–7702.

Ometov, A., Orsino, A., Militano, L., Araniti, G., Moltchanov, D. and Andreev, S. (2016), 'A novel security-centric framework for D2D connectivity based on spatial and social proximity', *Computer Networks.*

Ometov, A., Orsino, A., Militano, L., Moltchanov, D., Araniti, G., Olshannikova, E., Fodor, G., Andreev, S., Olsson, T., Iera, A. et al. (2016), 'Towards trusted, social-aware D2D connectivity: Bridging across technology and sociality realms', *IEEE Wireless Communications.*

Orsino, A., Araniti, G., Militano, L., Alonso-Zarate, J., Molinaro, A., and Iera, A. (2016), 'Energy efficient IoT data collection in smart cities exploiting D2D communications', *Sensors* 16(6), 836.

Palattella, M. R., Dohler, M., Grieco, A., Rizzo, G., Torsner, J., Engel, T., and Ladid, L. (2016), 'Internet of things in the 5G era: Enablers, architecture, and business models', *IEEE Journal on Selected Areas in Communications* 34(3), 510–527.

Peng, M., Li, Y., Zhao, Z., and Wang, C. (2015), 'System architecture and key technologies for 5G heterogeneous cloud radio access networks', *IEEE Network* 29(2), 6–14.

Pirinen, P. (2014), 'A brief overview of 5G research activities', *in* 1st International Conference on 5G for Ubiquitous Connectivity, pp. 17–22.

Piro, G., Cianci, I., Grieco, L. A., Boggia, G., and Camarda, P. (2014), 'Information centric services in smart cities', *Journal of Systems and Software* 88, 169–188.

Report, I. T. W. (2015), 'ICT as an enabler for smart water management', URL: http://www.itu.int/dms_pub/itu-t/oth/23/01/T23010000100003PDFE.pdf. Last accessed: 2016-11-25.

Rigazzi, G., Pratas, N. K., Popovski, P., and Fantacci, R. (2015), 'Aggregation and trunking of M2M traffic via D2D connections', *in* 2015 IEEE International Conference on Communications (ICC), pp. 2973–2978.

Salpietro, R., Bedogni, L., Felice, M. D., and Bononi, L. (2015), 'Park here! a smart parking system based on smartphones' embedded sensors and short range communication technologies', *in* 2015 IEEE 2nd World Forum on Internet of Things (WF-IoT), pp. 18–23.

Saxena, N., Roy, A., and Kim, H. (2017), 'Efficient 5G small cell planning with eMBMS for optimal demand response in smart grids', *IEEE Transactions on Industrial Informatics* **PP**(99), 1–1.

Schaffers, H., Komninos, N., Pallot, M., Trousse, B., Nilsson, M., and Oliveira, A. (2011), 'Smart cities and the future internet: Towards cooperation frameworks for open innovation', *in The Future Internet Assembly*, Springer, pp. 431–446.

Seo, S. B., Kim, J. Y., and Jeon, W. S. (2016), 'Robust and fast device discovery in OFDMA-based cellular networks for disaster environment', *in* 2016 18th International Conference on Advanced Communication Technology (ICACT), pp. 498–502.

Shang, J., Gui, L., Zhou, H., Dong, M., and Chen, J. (2016), 'A group-based M2M multiple access scheme in massive MIMO MU-SCMA cellular networks', *in* 2016 8th International Conference on Wireless Communications Signal Processing (WCSP), pp. 1–6.

Smart Cities (2013), 'Trace analysis and mining for smart cities: issues, methods, and applications', *IEEE Communications Magazine* 121.

Steri, G., Baldini, G., Fovino, I. N., Neisse, R., and Goratti, L. (2016), 'A novel multi-hop secure LTE-D2D communication protocol for IoT scenarios', *in* 2016 23rd International Conference on Telecommunications (ICT), pp. 1–6.

Su, Z. and Xu, Q. (2015), 'Content distribution over content centric mobile social networks in 5G', *IEEE Communications Magazine* 53(6), 66–72.

Usman, M., Asghar, M. R., Ansari, I. S., and Granelli, F. (2017), 'Towards bootstrapping trust in D2D using PGP and reputation mechanism', *in* IEEE International Conference on Communications, ICC17, (Accepted).

Usman, M., Gebremariam, A. A., Granelli, F., and Kliazovich, D. (2015), 'Software-defined architecture for mobile cloud in device-to-device communication', *in* 2015 IEEE 20th International Workshop on Computer Aided Modelling and Design of Communication Links and Networks (CAMAD), pp. 75–79.

Wang, L., Cao, C., and Wu, H. (2015), 'Secure inter-cluster communications with cooperative jamming against social outcasts', *Computer Communications* 63, 1–10.

Wang, L., and Wu, H. (2015), 'Jamming partner selection for maximising the worst D2D secrecy rate based on social trust', *Transactions on Emerging Telecommunications Technologies*, pp. n/a–n/a.

Wang, L., Wu, H., Liu, L., Song, M., and Cheng, Y. (2015), 'Secrecy-oriented partner selection based on social trust in device-to-device communications', *in* 2015 IEEE International Conference on Communications (ICC), pp. 7275–7279.

Wang, L., Wu, H., and Stuber, G. L. (2016), 'Cooperative jamming aided secrecy enhancement in P2P communications with social interaction constraints', *IEEE Transactions on Vehicular Technology* **PP**(99), 1–1.

Wang, X., Chen, M., Taleb, T., Ksentini, A., and Leung, V. C. M. (2014), 'Cache in the air: exploiting content caching and delivery techniques for 5G systems', *IEEE Communications Magazine* 52(2), 131–139.

Washburn, D., Sindhu, U., Balaouras, S., Dines, R., Hayes, N., and Nelson, L. (2009), 'Helping CIOs understand smart city initiatives', *Growth* 17(2).

Xing, P., Yang, L., Li, C., Demestichas, P., and Georgakopoulos, A. (2013), 'Multi-RAT network architecture', *in* 'Wireless World Research Forum', White Paper, *Vol.* 2.

Yaqoob, I., Hashem, I. A. T., Mehmood, Y., Gani, A., Mokhtar, S., and Guizani, S. (2017), 'Enabling communication technologies for smart cities', *IEEE Communications Magazine* 55(1), 112–120.

Zanella, A., Bui, N., Castellani, A., Vangelista, L., and Zorzi, M. (2014), 'Internet of things for smart cities', *IEEE Internet of Things Journal* 1(1), 22–32.

Zaragoza, S. (2015), 'Wearable electronic health patches may now be cheaper and easier to make', URL: http://news.utexas.edu/2015/09/29/engineers-develop-new-method-for-making-wearable-electronics. Last accessed: 2016-11-25.

Zhang, J., Xie, W., and Yang, F. (2015), 'An architecture for 5G mobile network based on SDN and NFV', *in* 6th International Conference on Wireless, Mobile and Multi-Media (ICWMMN 2015), pp. 87–92.

Zhou, Z., Dong, M., Ota, K., Wang, G., and Yang, L. T. (2016), 'Energy-efficient resource allocation for D2D communications underlaying cloud-RAN-based LTE-A networks', *IEEE Internet of Things Journal* 3(3), 428–438.

Ziegler, V., Theimer, T., Sartori, C., Prade, J., Sprecher, N., Albal, K., and Bedekar, A. (2016), 'Architecture vision for the 5G era', *in* 2016 IEEE International Conference on Communications Workshops (ICC), pp. 51–56.

Section II

Emerging Communication Networks for Smart Cities

6

Software Defined Networking and Virtualization for Smart Grid

Hakki C. Cankaya

Fujitsu Network Communications, Richardson, TX

6.1 Introduction

The smart grids that connect smart homes and smart buildings have been built to collectively shape smart cities and will facilitate many new applications and services. The new and/or improved services and applications over smart-grid infrastructure will include but not be limited to Internet access, video surveillance, power distribution automation, security, and many more that we have not even think about yet. All these new services and applications will shape smart cities and improve lives. They require collecting ambient intelligence, situation awareness, and frequent and reliable communication between devices and distributed control centers (Rehmani, 2015). Machine-to-machine communication and IoT (Internet of Things) will be utilized regularly, and large volume of data (big data) that have been collected will have to be stored in the cloud and analyzed to make actionable decisions. Data centers and public and/or private clouds have been becoming parts of the utilities and energy sector IT infrastructure already. Security and privacy of the data and vulnerability of this complex and heterogeneous communication system are the main concerns for some time now, and there have been discussions on how to address them. Repeating attacks into networks of such systems are inevitable; however, correlating these individual attacks and understanding patterns are valuable means to provide intelligent prevention of planned attacks with known prediction algorithms.

Due to security reasons and specific application requirements (delay, jitter, QoS, etc.), utilities and power grids do not rely on the public Internet for their communication needs, and they mostly have their own telecommunications networks and/or receive dedicated private services from area service providers. Wireless access and point-to-point services are popular choices. One of the challenges for power grids is to constantly maintain and improve their dedicated telecommunications networks, which are administrated by multiple domains and technologies (Erol-Kantarci, 2015). Domains include the core network (WAN), the access networks (AN), the field area networks (FAN), and/or the neighborhood area networks (NAN) over multiple technologies. FAN is mostly a broadband wireless network that provides connectivity to ICT-enabled new-generation utilities field devices in a coverage area. NAN provides last-mile outdoor access to smart meters from WAN. These networks are owned by

Transportation and Power Grid in Smart Cities: Communication Networks and Services, First Edition.
Edited by Hussein T. Mouftah, Melike Erol-Kantarci and Mubashir Husain Rehmani.

utilities and/or distribution system operators or leased from network operators. It is also possible that they have a combination of both, which introduces additional administrative and management complexities. They use a variety of wireline and wireless technologies in these networks.

In this chapter, we describe the current status of the power grid and the motivation for modernization. We review software-defined networking and virtualization for both networking and functions. We then look at the challenges of the smart grid and how SDN/NFV will benefit the new applications. Last, we share some use cases and existing demo proposals of SDN/NFV for the smart grid from other groups.

6.2 Current Status of Power Grid and Smart Grid Modernization

Power and energy have been the main enablers of the growth and evaluation of human societies and civilization. We use energy from basic human needs such as lighting and heating to modern needs that include powering data centers for cloud applications and connecting people via internetworks. The power grid is basically an infrastructure that moves power from generation sites (power plants) to end consumers including homes, enterprises, production facilities, etc. In general, the power grid is made up of subsystems including (1) power generation, (2) transmission, (3) distribution, and (4) consumption (see Figure 6.1).

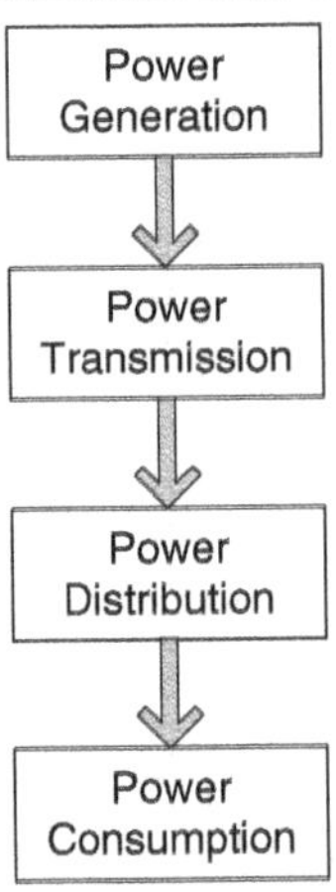

Figure 6.1 Traditional power grid.

The electricity is first generated at power stations by using different resources like hydro, coal, fuel-oil, wind, solar, or nuclear. The electricity is then transmitted over high-voltage transmission lines from generation sites to residential and industrial areas and locations where substations terminate the high-voltage feed and transfer it to low-voltage power. The distribution power network distributes this low-voltage power to the consumers in the designated area that is served by the substation. The residential or commercial users then use the electricity to power many of their power devices. The expectation is the reliability and availability of this power service.

The operation of this rather large and complex system requires large number of sensors, measurement devices, and controllers. All these devices need to communicate with each other through a series of communication networks and gather information to a control center where decisions to accommodate recent changes in the power grid are made and fed back to the devices in the grid.

6.2.1 Smart Grid

The introduction of the smart grid increases the amount of information exchange and the functionalities. The smart grid has the following information-generating and -consuming functions.

1) Consumers can now be producers of power. Therefore, power production information should be collected from consumers as well. This makes consumers a new source of information.

2) Accurate load balancing of the production requires dynamic load information exchange from all power generation stations.
3) Smart sensors collecting information from power transmission lines, transformers, distribution substations, etc.
4) Advanced smart meters collecting power usage information.
5) Information collected from electric vehicles (EV) and high-capacity energy storage devices, such as large batteries in EVs.

With the integration of the IoT to the smart grid, virtually all assets from anywhere can connect, and efficiencies and productivity can increase. This ubiquitous connectivity will transform our businesses; new services that improve our lives will be possible. IoT will offer new value to utilities enabling them to make better business decisions, use their resources more efficiently, and be the best service provider possible to their customers.

There are multiple types of data collected and moved in the smart grid with different requirements. These are: (1) time-critical data, and (2) volume data. Time-critical data has stringent time budget requirements for delay and is needed for decisions including protection switching. The volume data does not have such strict requirements for delay but needs high throughput; meter data is one of them.

The geographical footprint of the smart grid communication network is large and therefore employs heterogeneous networks in different sections of the entire coverage. Optical fiber networks are commonly used for connecting generation sites to substations and deployed along with transmission lines and are called core or wide area networks (WAN). In field area networks (FAN) and neighborhood area networks (NAN), the bandwidth is not high; however, flexibility and access are important. Power line communication and/or wireless access technologies are frequently used in distribution areas.

In the traditional power grid, the devices and protocols used in substations in the distribution areas were proprietary and closed systems. To increase the interoperability of many other new devices and functionalities that are coming in to the picture with the smart grid, a standardization of substations automation became a need. The TC 57 working group released a series of standards called IEC 61850 (IEC) family where inter- and intra-substation devices, communication, and functions have been reviewed and standardized.

In the process of modernizing the traditional power grid toward the smart grid, there is an increasing need for communication between machines as well as machine and human. This need puts information and communication technologies (ICT) in a vital role for this evaluation and modernization to happen smoothly.

In the smart grid, there are technology and business challenges. The integration of renewable distributed energy resources (DER) will require frequent and close monitoring of the entire system.

Some of the technology challenges of the smart grid include the following integration of systems and functionalities:

- demand side management (DSM);
- automated meter reading (AMR);
- customer energy management systems (CEMS) to manage new prosumers (consumer and producer) into the smart grid architecture;

- hybrid and plug-in electric vehicles with large batteries that require coordination for storage from the grid;
- intelligent electronic devices (IED) that are used for collecting, monitoring, and processing data;
- self-healing capability against failures and disasters (earthquakes, hurricanes, tornados, etc.);
- quick isolation of failures in microgrid to prevent large blackouts;
- providing predictive information to utilities as well as subscribers with recommendations about efficiencies in usage as well as production;
- interoperability among multiple heterogeneous networks and domains; the smart grid should be able to work multiple networks and platforms together with no inconvenience to users and other systems; and
- real-time analytic information processing systems to determine the current status and predictions for what is coming next.

Business challenges of smart grid include:

- increase energy trade, not only for traditional power plants, but from individual consumers/producers (prosumers); and
- developing business cases for new services and applications (predictive and/or historical analysis of usage and production for example).

The potential solutions for these challenges lead to high volume of data collection, transfer, and processing, which increase the complexity and the requirements of the ICT infrastructure. In addition to this complexity, the ICT infrastructure still requires even higher performance, availability, reliability (Abhishek, 2016), dependability, stability, and security. To combat this complexity, there have been proposals from researchers and practitioners in the industry, which include new networking concepts such as virtualization and the softwarization of functionalities. For a successful implementation of the smart grid system, one needs to seamlessly integrate these new concepts in to the existing grid system with ICT.

Moving to the smart grid requires a carrier-grade communication network, which enables scalability, self-healing, and organizing (Nakata, 2016), quality of service, management and monitoring, and security (confidentiality, integrity, and availability). Smart Grid is large scale, heterogeneous and distributed system. In Smart Grid, power distribution networks have large number of Distributed Energy Resources (DER) which are enabled by the new energy market where any consumer can become a producer of power. Information communication infrastructure should be enhanced matching these technological challenges of the Smart Grid.

In the smart grid, large data collection is required to make actionable smart decisions for the dynamic control of the intelligent devices. The new generation of devices such as phasor measurement units (PMUs) measure power quality with high accuracy. The measurements are frequent, resulting in up to hundreds of measurements per second, collecting large amounts of data.

6.3 Network Softwarerization in Smart Grids

6.3.1 Software Defined Networking (SDN) as Next-Generation Software-Centric Approach to Telecommunications Networks

The traditional communication networks are already quite complex and costly in operations and management. They are based on purpose-built hardware for each network function and use technology-specific interfaces. These issues and difficulties have motivated many researchers and industry toward new concepts and architectures. A relatively new paradigm change in communications networking seems to be taking the lead. Many startup companies, incumbent communication and networking equipment vendors, and medium and large network operators and service providers have invested both time and money in these new ways of designing and managing communications networks, called software-defined networking (SDN) and network function virtualization (NFV) as their next-generation software-centric approach to networking.

When it first started, there were multiple definitions of it offered by different organizations. Recently, these definitions have been converging based on some common points. Among all the organizations, the Open Networking Forum (ONF) has been one of the main organization behind SDN. They organize the education and implementation of this new paradigm and have working groups to make recommendations for different aspects of it. SDN is a programmable approach to designing, building, and managing reliable networks (Abhishek, 2016). It decouples network control from forwarding in network devices and offloads its functions to logically centralized SDN controller software, as seen in Figure 6.2.

The replacement of the control intelligence into logically centralized SDN controller(s) would capture the global view of the entire network and provide vendor-independent control to the day-to-day operators of the network. An immediate benefit is obvious simplification, which leads to cost reduction for network devices. Operating SDN-capable networks will be less complex. Network administrators can programmatically configure the entire network at the SDN control layer, instead of having to manually integrate all configurations in to devices scattered around the network. The SDN controller will provide the common network services such as routing, access control, dynamic bandwidth management, QoS, storage optimization, and policy management to different application-specific functions through open northbound APIs. SDN's unified control plane allows network abstraction and lets these applications be implemented easily and efficiently with required customization and optimization. The expectation is that these abstractions and simplifications will create a fruitful environment for innovation as turning up new services would be an easy task but needs to be invented. The same unified control plane facilitates the creation of a virtualized network and the slicing of pieces of the physical network or any other hardware source for specific functions and their virtualization, called virtualized functions. In this context combined with IT virtualization in data centers, network function virtualization (NFV) has been proposed and supported by major network operators and service providers as a platform to host and

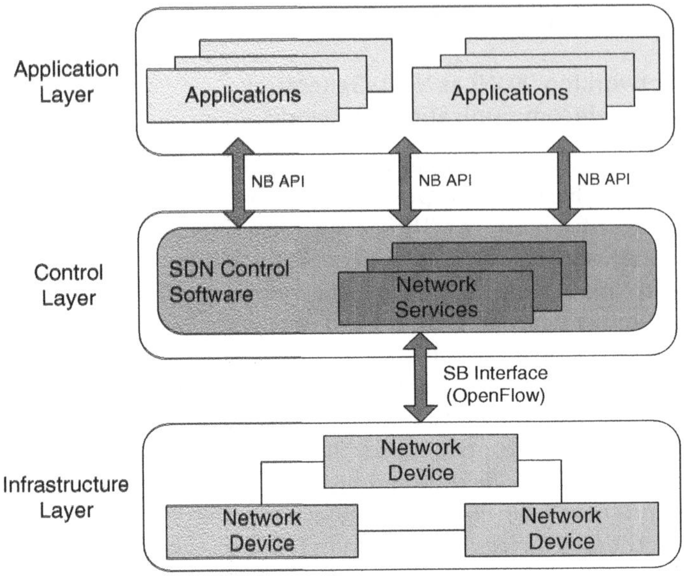

Figure 6.2 SDN framework.

Current Networks	SDN Networks	Benefit
Network provisioning complexity	Self provisioning	Additional revenue opportunity
Managed network functions	Automated network functions	Reduced OPEX
Technology-specific connections for the network	Technology-agnostic connections for the network	Reduced CAPEX and OPEX
Purpose-built hardware for each network function (routing, firewalling, etc.)	Software-based network functionality	Reduced CAPEX
Multi-vendor EMSs and their orchestration	Network OS with Southbound APIs	Reduced OPEX
Application-ware networks	Network-aware applications	Innovative applications and increased revenue opportunity

Figure 6.3 Generic SDN benefits.

operate network services as virtual network functions (VNFs). There have been studies suggesting that NFV results in considerable cost efficiencies in both Capex and Opex (Yu, 2015; Cerroni; 2014).

Flexibility is another benefit that SDN with NFV offers; mix-and-match solutions from different vendors would be accomplished easily with SDN. A summary of an incomplete list on how SDN can benefit current network is tabulated in Figure 6.3 (Fujitsu, 2015).

6.3.2 Adaptation of SDN for Smart Grid and City

There has been very focused research on alternative next-generation networking infrastructures and architectures such as SDN/NFV producing encouraging results; however, these are unlikely to be deployed by utility companies and distribution system operators without an implementation and demonstration of a successful large-scale prototype under real-world conditions to build smart cities.

Another reason for the slow pace of adopting these existing new communications concepts and prototypes into the real production network of any utility or energy company is the regulations and hefty fees enforced by NERC (North American Electric Reliability Corporation) for each day that the utility is in violation of standards. This is an understandable risk to be avoided because utilities networks are part of the critical infrastructure of smart cities.

The pathway to the adoption of these new concepts and architectures for smart grid and smart cities should involve the following steps:

1) idea generation on how SDN/NFV can be used to address the problems of utilities networks and smart city applications in general;
2) proof of concept in a lab environment;
3) testing with simulators/emulators;
4) testing under real conditions in a sliced isolated real network in field trials; and
5) Approval of results and migration to production network in incremental phases.

6.3.3 Opportunities for SDN in Smart Grid

Current challenges and push for modernization require open standards and non-proprietary solutions to mitigate the risk of rapid technology shifts and to reduce high OPEX cost. Following this trend, SDN and NFV not only provide solutions to some of the challenges of the utilities and energy sector for their networks but could also be a solid option for future-proof modernization including the smart grid.

SDN, as defined, provides a global end-to-end view of the network; therefore, with this new approach, the complexity of multiple domains and networks can be dealt with. SDN adopts mostly open standards and introduces technology abstraction, which provides a vendor-agnostic approach to configuring and maintaining various types of network elements. Hardware virtualization is one of the goals of SDN and is used for piecing up a physical network into virtual slices for ease of managing different networks while using resources efficiently. It will help connect multiple data centers in different physical locations and make their use efficient. Programmatic network control of SDNs serve for service abstraction and will simplify all configuration activities. Due to its holistic view of networks, an SDN network will be able to have a better control over delay and jitter in the network. This will help teleprotection of control traffic including SCADA. The bandwidth-on-demand feature of SDNs will solve the problem of elastic bandwidth need for new and emerging apps in the smart grid. This will also create an opportunity to increase revenue through accelerated service velocity in cases where the utility also serves as a service provider in the coverage area. It has been observed that more and more utility companies serve as Internet service providers in rural areas.

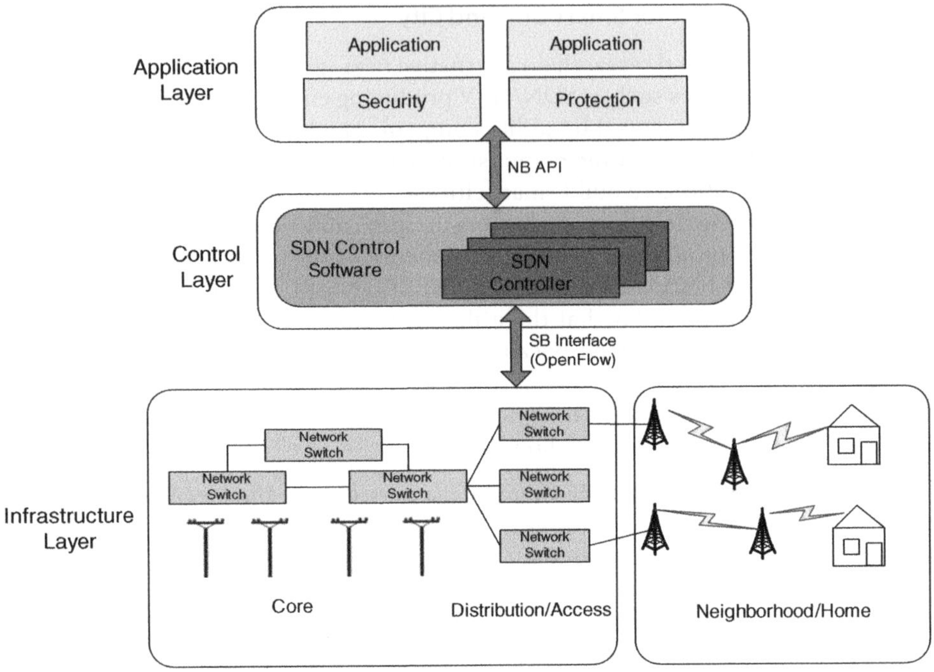

Figure 6.4 SDN-based smart grid communication infrastructure.

In this new networking infrastructure (see Figure 6.4), the utility-specific apps on the applications layer will have the opportunity to blossom. These apps could facilitate power/energy peak-shaving, analytics of system and customer data, M2M communication, fast restoration from a failure, better management of heterogeneous network devices, rapid diagnostics on large inventory of network devices, distributed automation enablement, AMI network monitoring, and many more one could think of.

SDN for resiliency and protection in smart grids: The smart grid maintains multiple large-scale heterogeneous communication networks to deliver services. One challenge is automatic fast recovery when failures occur by manmade or natural causes. There have been studies that demonstrate self-healing capability to the failures of links for inter-substation communication by using SDN architecture (Aydeger, 2016; Sydney2014; Ren, 2016; Ghosh 2016). All these failure recoveries are possible using the SDNs feature of having overall topology view of the entire network. To support multi-network fast recovery, there are proposed algorithms that are integrated into the SDN control layer (Dorsch, 2014). The algorithm proactively configures alternate routes for all possible failure scenarios and coordinates with SDN controller's monitoring function to be notified at a port status change (happens at a failure). Then, automatic switching takes place, and traffic start flowing on a re-routed path. There are some studies reporting improved recovery times compared to MPLS-based techniques (Aydeger, 2016). In addition to link failures, self-recoveries from node failures have been also reported by using the SDN architecture (Zhang, 2016).

Simple management: The simplicity of the management is a direct product of the separation of the control layer from the device infrastructure layer of the SDN architecture (see SDN architecture in Figure 6.2). An SDN controller at the control layer collects all topology information from devices and makes the collective management decision for the entire network. An example implementation of this management simplicity for the smart grid has been reported by Kim (2015) for an AMI application. AMI meters need forwarding information for the metered data they receive. In an SDN-based architecture, the SDN controller pushes the forwarding policy for all AMI devices in the data/infrastructure level, so that AMI meters do not carry the burden of making their own forwarding decisions; this reduces the computational burden on AMI devices as well as the topological updates AMI devices needed to receive to make those decisions.

Interoperability and programmatic interfaces: The SDN architecture offers standardized interfaces to be used seamlessly for different communications devices at the infrastructure layer. The northbound interfaces are used for the communication between the SDN controller and applications running for different network services. Due to the variety of such applications, there is no single interface for northbound. In fact, there are expected multiples of interfaces in addition to NETCONF and REST. The southbound interface is used for the communication between the SDN controller and the network switches and routers. The interface facilitates the dynamic control of switches and routers according to the network traffic demand by the SDN controller. The OpenFlow is the first and most used southbound interface in the field. Regardless of the vendors and models, these devices could be performing different functionalities as a physical device or implemented in software as a virtual device. All of the information provided by such devices can be transparently interpreted by the control layer and used by the service provider and/or network operator in an interoperable way; no special control stack should be maintained at the control level by the operator. This would give great flexibility to the smart grid operators for all different kinds of devices they need to have in their communications infrastructure. The SDN architecture offers the flexibility of adopting any change in the middleware control layer by the way it is designed based on software. Considering the rapidly changing protocols and new interfaces the smart grid needs in its growth cycle, this feature of SDN could be a good asset.

Network awareness: With SDN, all devices in the infrastructure layer may provide constant feedback to the controller in the control layer. The operator can define the network events based on the collective information, such as low performance, congestion, large delays, etc. These events could dynamically trigger functions such as load balancing, re-routing, and new arrangements of QoS parameters. The direct application of this in the smart grid could be the readjustments of AMI and IED devices in terms of reading frequencies and reconfigurations. These would not require any human interventions and eliminate any fat-finger problems.

Hardware as commodity: Currently, there are different types of equipment from multiple vendors in the traditional network architecture requiring frequent installs for updates and bug fixes. With the introduction of the new SDN architecture, hardware becomes a commodity, and types and vendors will not make a difference at the control layer. All changes and updates will be done at the control layer and pushed to all hardware at once by using the standardized interfaces.

Load balancing and quality of service: SDN offers load balancing based on flow information that is shared between the controller and application profiles. The traditional packet-based load balancing cannot relate the load balancing functionality to the applications through service profiles. This capability makes the new SDN-based network more application aware and able to prioritize flows for more efficient resource use. This feature of the SDN architecture could be quite useful to secure the minimum bandwidth and resource levels for high-priority information flows such as power-quality event warning messages so that these types of emergency event handlings can be prioritized. There are some specific studies that report such benefits of flow-based priority differentiation in smart grid applications (Sydney et al., 2014). Flow information that is used for load balancing in the SDN architecture can be easily leveraged for QoS and DiffServ applications. Once flow is recognized and prioritized by the SDN controller, depending on the arrival rate and delay requirements of the flow, appropriate queues can be modelled to accommodate any QoS mechanism in a dynamic manner. This flexibility and control in terms of QoS are quite useful in smart grid applications. There are proposals on heuristic methods for smart grid QoS applications where smart grid–specific AMI flows can be treated differently based on the application service profiles (Qin et al., 2014).

Security: A lot of attacks target critical infrastructures including utilities communications networks. More concerning is the fact that these types of activities are increasing, which justifies the ongoing awareness in and around the security of critical infrastructure networks and assets. These networks require large security coverage due to their footprint with multiple domains and numerous end points (Cankaya, 2015). Each end point counts for a potential point of attack and increases the risk and vulnerability of the system. There are many guidelines, best practices, and standards for security of utilities and energy sector administered by multiple organizations.

For a long time, we have been using the same technologies and practices for security, which have been strictly defense oriented. The "block all" vision that we had has been forcing us to drop a firewall appliance wherever the function is needed in the network. SDN has triggered a change for moving away from this trend. It can enable security developers to implement more innovative solutions for perceived threats as it provides better visibility of the entire network for both control and data. With logically centralized control, the confusion about where to place a security appliance will come to an end. SDN's security-enabled infrastructure allows network security administrators to route all traffic to a logically central firewall that can virtually be present wherever and whenever needed in the network (Ghosh et al., 2016). In addition to conventional brute-force techniques, the change will feed innovation for intelligent security applications such as smarter quarantine systems, faster emergency broadcasts, advanced Honeynets, and context-aware detection algorithms.

In general SDN can provide security based on software without a specialized hardware. There are proposals to use role-based access control systems in the SDN architecture (Pfeiffenberger and Du, 2014), where an SDN controller uses digital signature with shared keys for all control packets. Kloti et al. (2013) provide a good analysis for open-flow-based security for an SDN architected network and provides recommendations on security issues within OpenFlow deployments in SDN architectures.

Flexibility in network routing: An SDN centralized controller can continuously keep track of all network devices and maintain the topology of the network. Using this topology, it can also construct a routing information base that can compute any route with the network, instead of letting devices run their own copy of the distributed routing algorithms to build the routes. In this case, if a device needs a route, it receives the direction of the packet to move in terms of the forwarding information from the SDN controller. Dorsch et al. (2014) proposed a purpose-built SDN architecture for smart grids, called

SDN4SamrtGrid. In this new SDN architecture, the topology and the routes between the smart grid devices are constructed by the centralized SDN controller, and the forwarding information is disseminated to all required devices for next-hop decisions. The article also claims a superior performance of SDN4SMartGrid routing to the traditional open shortest path first (OSPF) routing.

Ease of deploying new services and applications: The SDN architecture makes the new service and application deployment relatively easy. Service providers and operators have the direct control of the SDN controller. The business case for a new service and/or application starts with simply a policy development for the new service. After the policy has been developed, the policy needs to be provisioned directly to the SDN controller without any hardware dependency in terms of installation and configuration on the infrastructure. The controller layer then monitors and manages the service continuously and decides to make changes only on the control layer to be pushed to the devices.

6.4 Virtualization for Networks and Functions

6.4.1 Network Virtualization

Network virtualization (NV) is a technique to create an independent isolated instance of a network service, called virtual network (VN), over a shared physical network infrastructure with customizable performance levels. A VN is composed of virtual nodes and links that form a virtual topology of the VN. A virtual node is mapped to a physical node; however, a virtual link can be mapped to a section of the physical network that may include multiple physical links and nodes. The owner of the VN can choose to use any protocol, mechanism, and/or control and management planes in this virtual network. The benefits of VNs are many. First, multiple VNs can share the same physical network infrastructure or multiple heterogeneous physical network infrastructures without functionally affecting each other. VNs are also isolated from each other meaning that any failure, misconfiguration, and security threats in one VN should not replicate themselves to the others. Another motivation for VNs is to ameliorate heterogeneity, which complicates the management and service delivery. One is the heterogeneity of the underlying network technologies. There may be WDM, Ethernet, and/or wireless technologies used at the physical infrastructure layer together. A VN is, by definition, transparent to these multiple technologies; only the topological connectivity is important to the administrator of the specific VN. Similarly, multiple VNs should be able to run different protocol stacks and deliver different services. In the smart grid, the amount of information being exchanged is quite large compared to the traditional grid. With the open platform that the smart grid provides, the sources of information have also been increasing. Power generation equipment feeds information for their status and load conditions from subscriber sites. All smart sensors monitoring transmission lines, substations, and transformers send information to control centers. Advanced meters are read frequently. Large capacity power storages at subscriber generation sites including electric vehicle batteries need to communicate their status to the controllers. Not only the count and size of these information sources but also where they communicate to matter. There are also different applications, mainly control and operations applications and data collection applications. The requirements for these applications are different. Control and operations applications are small in data volume but require real-time communication with limits on delay and availability; however, data collection may be in large bursts but do not have strict requirements on delay; however, they may benefit from high throughput.

Information exchange occurs on multiple physical network technologies, i.e., heterogeneous networks. Optical networking technology is used in the transmission part of the grid to connect substations and head-end power stations in a wide area. Power line communication (PLC) and wireless mesh (WM) are used in the distribution part of the grid due to the favorable cost point and flexibility of these technologies.

In comparison the requirements of the applications, optical transmission has reliability, low latency, and high data volume capabilities; however, it does not cover the distribution part of the neighborhood area and home networks due to high cost and ROI issues. On the other hand, PLC and WM are ubiquitous in home and neighborhood areas but don't have the robustness and reliability by themselves.

Because the smart grid delivers very diverse applications with a different set of requirements over heterogeneous networks and technologies, one can argue that NV is quite applicable to smart grid communication with considerable benefits. For example, a specific application can run on a tailored VN without affecting any other application running on different VNs. The main challenge in NV in the smart grid, like any other implementation of the NV, is the mapping of virtual networks to physical network. There have been some proposals for the mapping problem (Cai et al., 2010; Zhu et al., 2006; Chowdhury et al., 2009; Lischka and Karl, 2009).

In the smart grid, there are even multiple physical networks, i.e., PLS, WM, optical, for this mapping exercise. Houidi et al. (2011) report on mapping to multiple physical networks. Specifically for the smart grid, Pin et al. (2014) implement NV for smart grid applications by considering WM and PLC as underlying two physical networks and report benefits. The study provisions two VNs, one for real-time applications where latency and reliability are priority and the other is a data collection application where the throughput is more important. This multi-objective optimization lands itself into NP-hard problem set, and the authors propose a suboptimal heuristic solution to map these application-specific VNs to PLC and WM physical network infrastructures. They report considerable improvement in higher reliability by mapping logical links into diverse physical links from both the PLC and WM infrastructures. A similar diversification has been employed to increase throughput for data collection application VNs.

6.4.2 Network Function Virtualization

In traditional methods, network functions have been delivered by hardware products and devices; for example, various vendors make routers, firewalls, switches, load balancers, etc., for specific network functions and to be placed and installed at certain physical locations in networks. This traditional way of accomplishing network functions has been challenged by a new concept called network function virtualization (NFV) supported by the SDN framework. Other than being just a concept, the number of successful implementations in data center networks, wireless backhaul networks, and content delivery networks has been in the rise along with obvious benefits (Network Function Virtualisation, 2013; Chiosi et al., 2012).

The NFV framework has an extended coverage that includes NV and softwarization of functions (implementation of functions in software) to deliver network services. This will liberate offering services from strict dependency on purpose-built hardware devices in the traditional networking. The trend is to load the special functionalities to software and run the software on commodity compute, storage, and network hardware resources. The benefit is flexibility and agility in offering new and more complex services by using virtualized function chains. In this context, the coordination and life cycle management of VNFs are done by the VNF orchestrator/manager. The responsibility includes creating an instance of VNF and monitoring its performance and termination.

Niedermeier and Meer (2016) use NFV-based virtualization to implement virtualized advanced meter infrastructure (AMI) that reports increased reliability and availability with reduced cost, which were pain points for the smart grid AMI application without the implementation of virtualization.

6.5 Use Cases of SDN/NFV in the Smart Grid

In the smart grid, one of the features is to be able to dynamically respond to the varying levels of power/electricity demand. There are peak hours where the demand for power increases for a duration of time and then comes back to the regular level in a 24-hr. daily cycle. There are options to deal with these peaks. One option is to engage extra power generation temporarily, such as turning up the generators at multiple locations. The other option is to introduce intelligence to the control system by introducing power control agents, which dynamically collect measurements and make decisions about which device should use power and which device should be delayed in receiving power from the grid. For example, when people come back from work, their electric vehicles do not necessarily need an immediate power charge during the peak hours of 6:00–8:00 p.m., and it could be delayed till sometime after midnight to be ready by the next morning. Of course in these cases, the subscribers need to opt in to the application, which is a contractual issue. Collecting the measurements from smart devices for the application needs a reliable and real-time network and a control agent that can oversee the entire infrastructure. This application, therefore, is a good candidate for an SDN-based smart grid network infrastructure. First, an SDN-based network is able to provide protection and load balancing for dynamic traffic that is needed for this application. Sydney et al. (2014) propose an SDN-based platform for demand response that experimentally implements and reports benefits of automatic fail-over, load-balancing, and QoS, which are leveraged for the demand response use case of the smart grid.

Another challenge for the smart grid is to move the high data volume across domains and devices in the smart grid network. The smart grid has a large number of devices that include PMUs in the grid, intelligent electronic devices (IEDs) in substations, and smart meters at homes. These devices generate high frequency of data flow at hundreds of reads per second, which creates a scalability problem for the communication. To address this complexity there have been proposals for middleware solutions: GridDataBus (Kim et al., 2012), where producers and consumers of data are grouped based on certain features such as type and location; GridStat (Gjermundrod et al., 2009) is another study that manages the complexity by separating data and management planes. Management plane manages the communication network resources and failures; data plane provides the efficient data flow through data brokers. Software-based control of the SDN architecture can inherently become a good platform to facilitate these middleware-based solutions. Koldehofe et al. (2012) implements a similar middleware based on the SDN architecture by using the OpenFlow protocol. In this proposal the authors implemented the middleware in the SDN controller as a control agent (handler), which benefits from the holistic view of the network at the SDN controller and is able to cater for an efficient connection between data producers and consumers in the network by leveraging distributed systems tools such as group communication, while enjoying the other benefits of the SDN such as load balancing, fail-over, etc. It is reasonable to expect growth in these kinds of use cases to reduce the complexity of the smart grid communication (Hannon et al., 2016).

Rapid failure recovery and resiliency represent another use case for SDN in smart grids. Aydeger et al. (2015) claim a successful demonstration that integrates an emulated OpenDayLight controller-based SDN platform with smart grid applications such as manufacturing message specification data-flow on TCP. The demo setup uses Mininet for simulating the OpenFlow protocol and an ns-3 network simulator to generate the network topology and traffic. The aim was to increase the redundancy by using wireless connections as back-up for failures in primary wireline connections. The switch-over from wireline to wireless infrastructure is reported to be fast enough to sustain the connections for smart grid applications.

MicroGrids (Lasseter, 2010; Alparslan et al., 2017) and Virtual Power Plants (VPP) (Lukovic et al., 2010) have been proposed to alleviate the complexity of the distributed energy resources (DERs) in the smart grid, which is the main trend moving away from the centralized generation toward distributed generation. The expectation for this move is to capture efficiencies for the distribution of power and high reliability for trust. However, the trade-off is the complexity in the management of such distributed systems. In both MicroGrid and VPP the idea is to aggregate the DERs based on any criteria that makes sense for the specific application. There are proposals to aggregate based on geographical locations of DERs or the technology used at these locations. However, the control and management of such formations of clusters of DERs need high volume of information exchange where topology and data volume may be dynamic and require high availability of the networks. These features of this use case apparently could benefit from virtual networking where demand and topology changes can be accommodated by virtual networks and moving resources quite adaptively. Thus, the SDN infrastructure becomes an apparent fit for the solution. The parallel between VPP and virtual networks become quite obvious for future studies in smart grid communication.

Among the enablers of the smart grid are the intelligent electronic devices that have microprocessors and a control software. These devices perform all metering, monitoring, protection, and control functions for the smart grid. With the similar idea that stems from the compute virtualization, the software of all these devices could be removed from the individual devices and moved into grid data centers (grid cloud) on a virtual machine (VM)–based infrastructure and could be controlled and managed in a centralized way.

This would turn all these devices in to commodity hardware in substations and the smart grid; then SDN benefits that have been already realized for data centers and the cloud would become applicable to smart grid as well. Along with virtualized compute, the SDN-enabled VNs would create a multi-tenant environment for the smart grid, where a physical transmission system could serve multiple utilities that have virtual computer and network resources of their own, called slice of compute and slice of network, without any security concerns. This would create new business cases and revenue streams for smart grid admins as well as utilities that would use the virtual infrastructure as a service to create their services to end customers (subscribers) in an innovative and agile way.

6.6 Challenges and Issues with SDN/NFV-Based Smart Grid

Since SDN is creating a logically centralized control plane that keeps track of all devices in the infrastructure layer on a one-to-one communication basis, it creates an overhead on the communication infrastructure of the smart grid, which is not resource-rich. Therefore, an SDN-based smart grid could benefit from studies in reducing this overhead.

SDN can use centralized routing instead of using distributed OSPF-like routing algorithms. An SDN controller then distributes these routes in terms of flow-based forwarding information to the devices in the infrastructure layer. This works well except for some events, such as initiation of the network or after failure, where all or multiple devices need to be updated with the new forwarding information that creates a burst of data exchange. For such cases, traditional distributed routing may be kept as an alternative that may run in parallel to the centralized routing. Another solution is to employ multiple SDN controllers that manage and control some segments of the network. The coordination of multiple SDN controllers can use master-slave synchronization algorithms of distributed systems field.

IEC61850 provides a set of standards for substation automation and communication specifications for smart grid applications (Molina et al., 2015). IEC61850 also have recommendations for configuration language and information model for data representation. These recommendations should be integrated into the SDN-based smart grid communication architecture for vendor-agnostic implementations. Northbound interfaces between smart grid applications and the SDN controller layer need standardization to keep the vendor-agnostic nature of this trend.

The storage and processing of the collected high volume data and securing it over a large footprint are also challenges for the smart grid. Following this trend, there have been proposals that offer consolidated and comprehensive solutions for the smart grid that leverage the principles of SDNs. Jararweh et al. (2015) propose a new architecture for the mart grid that includes software-defined storage, software-defined security, and software-defined IoT along with SDN.

6.7 Conclusion

In conclusion, softwarerization and virtualization of networks and functions using SDNs and NFVs are inevitable developments for any communications networks. The smart grid operates on multiple large-scale heterogeneous communications networks to deliver new applications and services. To realize that, the operators of the smart grid need to work on how to best utilize the benefits of these new emerging technologies for their day-to-day needs in communication networking and share their expertise to shape the future. There are technological and business challenges to address, and there are promising recent academic and industrial proof of concepts and demonstrations to leverage. Moving forward, this is an exciting time in building smart cities and grids where innovation and opportunities will blossom in creating new applications that improve human lives and societies.

References

Abhishek, R. et al., 2016, 'SPArTaCuS: Service priority adaptiveness for emergency traffic in smart cities using software-defined networking', *Proceedings of IEEE International Smart Cities Conference (ISC2)*, Trento, 2016, 1–4.

Alparslan, M., et al., 2017, 'Impacts of microgrids with renewables on secondary distribution networks', *Applied Energy* (2017).

Aydeger, A., Akkaya, K., and Uluagac, A.S., 2015, 'SDN-based Resilience for Smart Grid Communications', *Proceedings of IEEE Conference on Network Function Virtualization and SDN*, 2015 Demo Track.

Aydeger, A., Akkaya, K., Cintuglu, M.H., Uluagac, A.S., and Mohammed, O., 2016, 'Software Defined Networking for Resilient Communications in Smart Grid Active Distribution Networks', *Proceedings of IEEE ICC SAC Communications for the Smart Grid*.

Cai, Z., Liu, F., Xiao, N., Liu,Q., and Wang, Z. 2010, 'Virtual network embedding for evolving networks', *Proceedings of IEEE GLOBECOM*, 2010, 1–5.

Cankaya, H.C., 2015, 'SDN as a Next-Generation Software-Centric Approach to Communications Networks', *OSP* Feb. 2015, vol. 33, issue 2.

Cerroni, W., and Callegati, F., 2014, 'Live migration of virtual network functions in cloud-based edge networks', *Proceedings of the IEEE International Conference on Communications (ICC 2014)*, 2963–2968.

Chiosi, M., Clarke, D., Willis, P., Feger, J., Bugenhagen, M., Khan, W., Fargano, M., Chen, C., Huang, J., Benitez, J., Michel, U., Damker, H., Ogaki, K., Fukui, M., Shimano, K., Delisle, D., Loudier, Q., Kolias, C., Guardini, I., Demaria, E., López, D., Salguero, Ramón, F.J., Ruhl, F., and Sen, P., 2012, 'Network functions virtualisation', introductory white paper. *Proceedings of SDN and OpenFlow World Congress.*

Chowdhury, N., Rahman, M., and Boutaba, R. 2009, 'Virtual network embedding with coordinated node and link mapping', *Proceedings of IEEE INFOCOM*, 2009, 783–791.

Dorsch, N., Kurtz, F., Georg, H., Hagerling, C., and Wietfeld, C., 2014, 'Software-defined networking for Smart Grid communications: Applications, challenges and advantages', *Proceedings of the 2014 IEEE International Conference on Smart Grid Communications (Smart GridComm)*, Venice, Italy, 3–6 November 2014, 422–427.

Erol-Kantarci, M., et al., 2015, 'Energy-Efficient Information and Communication Infrastructures in the Smart Grid: A Survey on Interactions and Open Issues', *IEEE Communications Surveys and Tutorials*, vol. 17, no. 1, 2015, 179–197.

Fujitsu 2015, 'Software Defined Networking for the Utilities and Energy Sector', URL: http://www.fujitsu.com/us/Images/SDN-for-Utilities.pdf.

Gjermundrod, H., Bakken, D.E., Hauser, C.H., and Bose, A., 2009, 'Gridstat: A flexible qos-managed data dissemination framework for the power grid', *IEEE Transactions on Power Delivery*, vol. 24, no. 1, 2009, 136–143.

Ghosh, U., et al. 2016, 'A Simulation Study on Smart Grid Resilience under Software-Defined Networking Controller Failures', *Proceedings of the 2nd ACM International Workshop on Cyber-Physical System Security (CPSS)*, ACM, 2016.

Hannon, C., et al., 2016, 'DSSnet: A smart grid modeling platform combining electrical power distribution system simulation and software defined networking emulation', *Proceedings of the 2016 annual ACM Conference on SIGSIM Principles of Advanced Discrete Simulation.*

Houidi, I., Louati, W., Ben Ameur, W., and Zeghlache, D., 2011, 'Virtual network provisioning across multiple substrate networks', *Computer Networks*, vol. 55, no. 4, Mar. 2011, 1011–1023.

IEC, *Communication networks and systems in substation—Specific communication service mapping, IEC 61850.*

Jararweh, Y., Darabseh, A., Al-Ayyoub, M., Bousselham, A., Benkhelifa, B. 2015, 'Software Defined Based Smart Grid Architecture', *Proceedings of 12th IEEE International Conference of Computer Systems and Applications (AICCSA).*

Kim, Y.J., Lee, J., Atkinson, G., and Thottan, M., 2012, 'Griddatabus: Information-centric platform for scalable secure resilient phasor-data sharing', *Proceedings of IEEE Computer Communications Workshops (INFOCOM WKSHPS)*, 115–120.

Kim, J., Filali, F., and Ko, Y., 2015, 'Trends and Potentials of the Smart Grid Infrastructure: From ICT Sub-System to SDN-Enabled Smart Grid Architecture', *Applied Sciences*, 5, 2015, pp 706–727.

Kloti, R., Kotronis, V., and Smith, P., 2013, 'Openflow: A security analysis', *Proceedings of the 2013 21st IEEE International Conference on Network Protocols (ICNP)*, Göttingen, Germany, 7–10 October 2013, 1–6.

Koldehofe, B., et al. 2012, 'The power of software-defined networking: line-rate content-based routing using OpenFlow', *Proceedings of 7th Workshop on Middleware for Next Generation Internet Computing*, Montreal, Canada, December 2012.

Lasseter, R.H., 2010, 'Microgrids and distributed generation', *Proceedings of Intelligent Automation & Soft Computing*, 2010, 16(2): 225–234.

Lischka, J., and Karl, H., 2009, 'A virtual network mapping algorithm based on subgraph isomorphism detection', *Proceedings of the 1st ACM Workshop on Virtualized Infrastructure System Architecture*, 2009, 81–88.

Lukovic, S., et al. 2010, 'Virtual power plant as a bridge between distributed energy resources and smart grid', *Proceedings of 43rd Hawaii International Conference on System Sciences (HICSS)*, Kauai, HI, USA, January 2010.

Molina, E., Jacob, E., Matias, J., Moreira, N., and Astarloa, A., 2015, 'Using software defined networking to manage and control IEC 61850-based systems', *Computer Electronics Engineering*, 43, 142–154.

Nakata S., et al., 2016, 'A smart grid technology for electrical power transmission lines by a self-organized optical network using LED', *Proceedings of SPIE 9948, Novel Optical Systems Design and Optimization XIX*, 99481E, September 2016.

Network Functions Virtualisation (NFV) 2013, 'Use Cases. Technical report', ETSI GS NFV 001 V1.1.1 (2013-10), 2013.

Niedermeier, M., and Meer, H., 2016, 'Construction Dependable Smart Grid Networks sing Network Functions Virtualization', *Journal of Network and Systems Management*, 24, 2016, 449–469.

Pfeiffenberger, T., and Du, J.L., 2014, 'Evaluation of software-defined networking for power systems', *Proceedings of the 2014 IEEE International Conference on Intelligent Energy and Power Systems (IEPS)*, Kyiv, Ukraine, 2–6 June 2014, 181–185.

Pin L., Wang, X., Yang, Y., and Xu, M., 2014, 'Network Virtualization for Smart Grid Communication', *IEEE Systems Journal*, Vol. 8, No.2, June 2014.

Qin, Z., Denker, G., Giannelli, C., Bellavista, P., and Venkatasubramanian, N. A., 2014, 'Software defined networking architecture for the internet-of-things', *Proceedings of the 2014 IEEE Network Operations and Management Symposium (NOMS)*, Krakow, Poland, 5–9 May 2014; 1–9.

Rehmani, M.H., et al. 2015, 'Smart Grids: a Hub of Interdisciplinary Research', *IEEE Access*, vol. 3, 2015, 3114–3118.

Ren, L., et al., 2016, 'Enabling Resilient Microgrid through Programmable Network', *IEEE Transactions on Smart Grid*, Issue 99, 2016.

Sydney, A., Ochs, D.S., Scoglio, C., Gruenbacher, D., and Miller, R., 2014, 'Using GENI for experimental evaluation of Software Defined Networking in Smart Grids', *Computer Networks, n.* 63, 2014, 5–16.

Yu, R., Xue, G.,Kilari, V.T., and Zhang, X., 2015, 'Function virtualization in the multi-tenant cloud', *IEEE Network*, vol. 29(3): 42–47.

Zhang, X., Wei, K., Guo, L., Hou, W., and Wu, J., 2016, 'SDB-based Resilience Solutions for Smart Grids', Software Networking (ICSN), 2016 International Conference on, May 2016.

Zhu Y., and Ammar, M., 2006, 'Algorithms for assigning substrate network resources to virtual network components', *Proceedings of IEEE INFOCOM*, 2006, vol. 2, 1–12.

7

GHetNet: A Framework Validating Green Mobile Femtocells in Smart-Grids

Fadi Al-Turjman

Antalya Bilim University, Antalya, Turkey

7.1 Introduction

Mobile networks have been experiencing an explosive revolution in the recent decade over numerous radio access technologies such as GSM, LTE-Advanced, Wi-Fi, LiFi, etc., which provide a resilient tie to the Internet. This in turn has led to a massive amount of mobile applications among which the smart grid starts to be the foremost one [1, 2]. The smart grid has been evolved recently in managing our vast electricity demands in a sustainable, smart, and economic manner, while utilizing already existing heterogeneous networks' (HetNets) infrastructures. The smart grid is simply an energy network that can automatically monitor the flow of electricity in a city and adjust to changes in users' demands accordingly. It comes with smart meters, which are connected to the Internet to provide consumers/suppliers with smart decisions on their ongoing energy usage/production. For example, a number of smart home appliances such as the dishwasher and air conditioners can communicate with the grid using these smart meters and automatically manage their electricity usage to avoid peak times and make more profit. Within a smart and green grid, mobility has to be intended as the way in which customers can access and explore the grid resources using advanced and eco-friendly electricity modes. This implies being aware of the available energy resources and their real values (in term of cost, time, and carbon emissions), as well as a simple and unified access to mobile payments. Supporting such an interconnected, heterogeneous, and dynamic grid system with lower environmental impact requires the adoption of innovative and sophisticated solutions from the utility management centers and city administration to control the energy resources and policies in order to proactively enhance/protect them. In order to manage this energy, there is a need for a wireless HetNet that can connect the different system entities, including mobile users' power lines and appliances at home. These wireless HetNets can be home area network (HAN) and/or neighborhood area network (NAN). HANs consists of three components, which measure, accumulate, and examine energy usage from smart devices [3, 4]. NAN connects multiple HANs to local/regional access points (i.e., access points within the range of instantaneous transmitters), where transmission lines carry the data to the utility [3, 7]. HANs need communication infrastructure in order to connect these elements.

Transportation and Power Grid in Smart Cities: Communication Networks and Services, First Edition.
Edited by Hussein T. Mouftah, Melike Erol-Kantarci and Mubashir Husain Rehmani.

One potential solution for such rapidly growing HetNets is the femtocell [1, 5]. A femtocell is a small cell deployed most often for enhanced coverage and energy-efficient communications. It uses between 8 and 120 mW, which is even lower than the Wi-Fi access point. Since it can be used as a communication mechanism in HANs and NANs, users can save energy and money using these femtocells in smart grid systems, where the connected home appliances can be programmed to reduce the home electricity demands over the grid as well as the consumers' electricity bill. This can be generalized for homes spanning a country scale, for example, where mobile outdoor femtocells can be utilized to achieve ULS coverages. However, this can overwhelm the existing HetNets' infrastructure and makes our expectations unrealistic in proximity of the green planet vision [1, 39]. Since the number of these femtocells is rapidly growing in several countries nowadays, lots of energy consumption demands and carbon emissions shall be carefully considered over the grid. Moreover, the continuously radiated wireless signals from these dramatically growing HetNets necessitate massive energy supply. This means energy is used for just-in-case scenarios where a user might or might not exist. Therefore, femtocells are recommended as powerful candidates vanishing the unnecessary energy loss. Unlike the macrocell, a femtocell can provide local coverage based on users' demands. Meanwhile, the user equipment (UE) can save more energy while connected to a closer femtocell. Thus, femtocell is more energy efficient in addition to providing more coverage, capacity, and better QoS. Nevertheless, significant energy amounts can be wasted in data (re)transmission unless a reasonable load balance is applied between the deployed static/mobile FBSs, which are typically planned to support vast counts of static/mobile users in the grid. This load balance is necessary to avoid overwhelming a femtocell while other feasible femtocells in the same region can still admit the incoming call/data requests. Accordingly, mobile calls/data requests may move to neighboring cells while being served by the FBS, which is a typical departure scenario in the dynamic smart grid. This departure and unavailability of an FBS can degrade the performance of the grid system unless it has been carefully considered at the early planning and validation stages of the femtocellular system design. This would not be achieved without a realistic case study analysis and an accurate analytical model that can predict the system performance under such setups.

In this chapter, a HetNet composed of multiple macrocells and numerous femtocells is deliberated as a use case toward greener future HetNets. Detailed analysis of the system is given for the mobile FBS based on queuing theory concepts. Results are analyzed and elaborated in terms of key performance characteristics of the system such as response time, throughput, waiting time, and energy consumption. The analytical model and solution approach is also validated through extensive measurement-based simulation results.

7.2 Related Work

There are several attempts in the literature toward validating the effectiveness of HetNets telecommunication systems in general and specifically femtocells. All these attempts can be classified into one of the following performance evaluation techniques: (1) analytical techniques, (2) simulation techniques, and (3) benchmarking (experimental) techniques.

Analytical techniques validate the behavior of a femtocell system using mathematical concepts and languages. For example, [8] proposed a closed-form expression that describes the derived spectral and energy efficiency in mobile femtocell networks. They investigate the spectral efficiency for multi-user system-level using opportunistic scheduling schemes/algorithms. In [9], exponential path loss and fast-fading models have been proposed to predict the energy consumption metric in static femtocells. In [10], the authors used an M/M/1 queuing model for their analytical study and the Matrix Geometric Method for analyzing the femtocell performance in terms of delay and energy consumption. Similarly, [11] analyze the performance of a limited capacity femtocellular network in terms of blocking probability, delay, and buffer size utilization. They used an M/M/1/K queue and their results assure that the aforementioned metrics are highly dependent on the experienced traffic volumes as well as the buffer size.

On the other hand, simulation techniques provide programming-based environments to emulate specific parameters in practice while studying the femtocell performance. For example, in [33] simulated a network of varying cells' dimensions. Based on coverage and traffic load parameters, energy consumption was minimized without conceding the quality of service (QoS). They focused on power consumption in five different setups. In the first setup, they assume a femtocellular network instead of the macro one in a region that is fully covered. Under this setup they achieved up to 88% reduction in power consumption. In the second setup, they divided the deployment area into three regions: urban, suburban, and rural. They considered varying mobile users' density and demanded coverage using static femtocells in the urban regions, macrocells in suburban regions, and mobile femtocells in rural regions. Consequently, they achieved up to 80% reduction in power consumption in this setup. In a third setup, femtocells have been allocated to densely populated urban areas, picocells to lightly populated urban regions, microcells to suburban regions, and mobile femtocells to rural regions. Accordingly, the reduction in power consumption rate was about 9%. In the fourth setup, they allocated microcells and femtocells at the borders and macrocells at the remaining parts of the targeted region. The reduction in power consumption was about 5%. In the fifth one, femtocells were allocated at the boundaries of the macrocells, where the radio signal is not enough for making a call. Consequently, a reduction of almost 2% was achieved. Simulation provides fairly precise results, but this approach requires high computation times. Compared to simulation, an analytical modelling technique is computationally more efficient.

Meanwhile, experimental studies (benchmarks) form another technique that aims at performing actual measurements using testbeds in the laboratory or the targeted field. For instance, in the analysis conducted by Bell Labs, the efficiency of a hybrid network of both femtocell and macrocell was practically analyzed [12]. In this validation technique, macrocells were used to assure coverage and femtocells to offload the macrocell traffic. The implementation field was spanning a 10×10 km^2 urban area, and the population of users was about 200,000. They deployed randomly a varying number of femtocells that can serve up to 8 users in a 10,000 m^2 area with 15 W energy consumption. The reason of the higher consumption rate compared to other analyses is that it is an open-access area. They used continuously the operated macrocells with 2.7 kW energy consumption. The result of this study was dependent on the network usage. When femtocells were used mainly for voice calls, there was no big saving in terms of energy. However, when femtocells were used for data connections, the total energy consumption was reduced by 60%. Another one was performed by Ofcom, a UK telecommunication company [13]. They deployed femtocells to 8 million households, which is nearly 25% of the area of UK. Each femtocell consumed 7 W per day and a total annual energy consumption equal to 490 GWh. They used 30,000 base stations in order to provide the full coverage. Their results conclude that it takes 40 times more energy to deliver a signal to an indoor user from the macrocell compared to the femtocell. As a result, in order to provide the same coverage provided by femtocells, the total annual energy consumption of a macrocell was about 700 GWh per operator. Thus the ratio of energy consumption in order to provide the same indoor coverage in the UK was 7:1 over using macrocells. Moreover, [34] have used benchmarking for the performance evaluation of the femtocell networks. They propose a resource management algorithm to address this problem and study the system performance by experimental work/test-beds. Results from this experimental work show that the proposed algorithm can significantly improve performance compared to the randomized placement algorithms. Furthermore, the use of femtocells for medical issues in moving vehicles such as the ambulance has been practically studied in [35]. A femtocellular BS is placed inside the ambulance. Data are transmitted/received to/from the backhaul macrocellular network through a transceiver installed on the roof of the ambulance in which it is connected to FBS by the wired network. The obtained results presented reveal that significant performance advantages in terms of packet loss, throughput, and delay are achieved by the use of femtocells in comparison to the macrocellular-based networks. The main problem associated with use of testbeds in practice is the difficulty in extrapolation of obtained results to various scenarios in terms of cost, time, and efforts [15].

All these validation techniques—analytical, simulation, and experimental ones—vary in the targeted performance metrics, including the average resource utilization, the average number of user equipment in the system, the average power consumption, the average waiting time, throughput, etc. However, all of them can be categorized into either static or dynamic techniques. By static and dynamic techniques, we refer to systems with/without mobile femtocells. In static models such as the ones presented in [18], performance characteristics of cellular networks have been investigated without considering mobility, and thus we call them static systems. Unlike static models, in the dynamic models, mobility is considered as one of the utmost important issues in assessing the grid performance [22].

7.2.1 Static Validation Techniques

In static techniques, FBSs are mostly deployed in indoor environments such as an office, shopping mall, or a smart home to enhance the indoor coverage and improve users' experience. In the literature, several studies have been proposed to analyze and evaluate the performance of typical femtocells in such static setups. For example, in [17], proposed a simple analytical model to predict a static FBS power consumption based on specific traffic load. They assumed only one femtocell that supports up to four simultaneous channels to predict energy consumption of voice and FTP messages. However, radio energy consumption was neglected while performing specific communication setups such as the downlink transmissions. In [8], authors investigated the energy consumption of a Wi-Fi network on a university campus. A simple queuing model is used to save energy in Wi-Fi by considering sleep modes for the femtocell according to user demands. Presented results show that by using sleep modes for the femtocell, a considerable amount of power consumption was saved with small number of connected users. The authors in [19] analyze the behavior of static femtocells while applying adaptive modulation and coding techniques with sleep modes using queuing theory. They are interested in evaluating energy consumption rates per packet, average delays, and packet loss. Furthermore, in [20], an admission control problem in an LTE radio network is addressed. The authors proposed a model for two video services: video conferencing and video on demand. Teletraffic and queuing theories are applied to obtain a recursive algorithm in order to assess the system performance using different metrics such as blocking probabilities and the average system throughput. However, and with the rapid increase in demanding coverage and bandwidth everywhere, mobile femtocells deployments and validation techniques are of utmost importance in providing green systems.

7.2.2 Dynamic Validation Techniques

Several attempts have been made toward realizing femtocell-based HetNets in the literature. However, among the most significant issues in performance evaluation of such HetNets is the mobility factor. The authors [22] propose the idea of deploying FBSs in moving vehicles in order to improve the uplink throughput for mobile users. Results show that mobile FBSs can enhance the QoS and maintain an acceptable level of signal strength. In [23], proposed seamless multimedia service for mobile user equipment (UE) travelling in high-speed trains by deploying FBSs onboard. The onboard FBSs communicate with macrocells to facilitate the seamless handover. In [24], an integrated cellular/Wi-Fi system is modelled for high mobility using a two-phase open queuing system with guard channel and buffering to obtain acceptable levels of QoS in heterogeneous environments. An exact analytical approach of the targeted system is given using the spectral expansion method that can be useful for vertical handover decision management. Similarly, authors in [25] model an integrated cellular/Wi-Fi HetNet in order to study specific performance characteristics such as the average queue length, throughput, and blocking probability. The FBS system is demonstrated as a two-phase open queuing network, and the exact solution is presented using the spectral expansion method. Authors in [26] propose a load-balancing scheme based on the experienced handover performance and dropped call rate while varying the number of user equipment (UE) and their corresponding mobility speeds. This approach demands FBSs, which are typically positioned in multiple macrocells based on the velocity of the UE while considering the number of UE per femtocell. The handover process is determined by using a simulation-based technique. In [27], authors propose an effective mobile femtocell application for power-saving purposes in cellular networks. In this application, the signal-to-noise-plus-interference ratio is enhanced in comparison to traditional schemes. The base station failure probability and energy consumption in the overall network have been analyzed. However, other QoS metrics such as the average delay and packet loss/retransmission have been ignored. In [28], heterogeneous networks and the smart grid are jointly studied to improve the power consumption. The authors formulate the decision problems of the power allocation, electricity, and interference price as a three-level Stackelberg game. Then, they use a homogeneous Bertrand game to model the price decisions. Simulation results show that their proposed grid-based approach can considerably reduce operational costs and CO_2 emissions. In [29], a mobility extrapolation approach based on users' trajectories is proposed. It considers the received signal quality with mixed real-time (RT) and non-RT traffic model. Via simulation results, the authors reveal that the proposed scheme has clearly reduced delay by 29%, and maintained the desired QoS. However, it doesn't provide a closed analytical form for performality analysis while considering other metrics such as system throughput and number of users per cell. In [30], handover approaches considering the signal strength and users' velocities are proposed. Different handover approaches have been simulated and deliberated. Simulation results show that the proposed approaches can obviously eliminate unnecessary handovers compared with conventional soft handover at the same signal level. However, delay and cost metrics have been neglected. In [31], a modified signaling procedure for the mobile user in a gateway-based femtocell architecture is proposed. A new handover approach based on the user's speed and required QoS is proposed. The comparison between the proposed approach and traditional handover approaches shows that the approach proposed in [31] performs better in reducing the unnecessary handovers. In [40], mobile wireless cellular networks are modelled using a Markov reward model, where multiple channels per cell in homogeneous systems are considered. Performance characteristics of the system such as mean queue length and blocking probability are presented using an analytical model. Moreover, two different handoff schemes with/without preemptive priority procedures for heterogeneous mobile networks are proposed and analyzed in [36]. A 3-D Markov chain is used to model the system and analyze the HetNet performance in terms of average delay, blocking probability, and the probability of forced termination [42]. The following table further summarizes the comparisons between the aforementioned studies.

Nevertheless, existing research efforts, as depicted in Table 7.1, do not assess the performance gains of green small cells in a systematic closed framework (analytical, test-bed, and simulation based) that can adapt to varying design factors in practice (e.g., energy, mobility, throughput, delay, etc.) [21]. A measureable performance analysis of such gain is definitely needed nowadays for better energy utilization and more green applications. And hence, the work in this chapter is proposed. Particularly, we consider the different velocity effect of mobile femtocells while applied in typical applications such as the smart grid. For more practicality, we specifically considered the e-mobility project led by Siemens as a use-case study [37]. Femtocells' base-stations (FBSs) [37] are designed to be static/mobile stations on buses, taxis, trains, etc., over the electric grid of a city to provide better coverage and capacity to dramatically increasing mobile users while saving their energy. In fact, e-mobility represents the concept of using electric powertrain technologies, in-vehicle information, and communication technologies and connected infrastructures to enable the electric propulsion of vehicles and fleets. It is motivated by the need to address corporate fuel efficiency and emission requirements, as well as market demands for lower operational costs [37]. It aims at reducing CO_2 emissions by promoting public and green transport choices. This can ultimately lead to green transportation, which means any eco-friendly transport practice or vehicle that does not have any negative impact on the environment. Accordingly, electrical vehicles/trains have been used in this case study, where they rely heavily on the mobile FBSs in exchanging their energy status during the day in addition to tracking massive

Table 7.1 A Summary of the Aforementioned Related Works in the Literature.

Ref.	MQL	Through-put	Energy	Mobility	Delay	Analy-tical	Simu-lation	Test-bed	Application
[8]	-	-	√	-	√	√	-	√	Univ. campus (outdoor)
[20]	√	√	√	-	-	√	-	-	Video streaming (outdoor)
[27, 30]	-	-	√	√	√	√	√	-	Transportations (outdoor)
[31, 40]	√	√	-	√	-	-	√	-	Transportations (outdoor)
[6, 24, 25]	√	√	-	√	√	√	-	-	Transportations (outdoor)
[22, 23]	-	√	-	√	-	√	√	-	Transportation/ multimedia (outdoor)
[34]	-	-	√	-	√	-	-	√	Office/home (indoor)
[35, 37]	-	-	√	-	√	-	-	√	Transportations (outdoor)
[26, 29]	-	-	√	√	-	-	√	-	Real-time/ transportations (outdoor)
[28]	-	-	√	√	-	-	√	-	Electricity pricing (indoor)

counts of smart home appliances. And thus, a careful FBS traffic load study is indeed required. In this study, we propose a green HetNet framework that can analyze the above mentioned performance metrics in presence of service interrupt due to mobile FBSs departures from the neighborhood of the UEs.

7.3 System Models

This section presents the proposed model for the performance assessment of a green HetNet composed of multiple macrocells and mobile FBSs. As shown in Figure 7.1, a set of mobile femtocells are deployed inside the coverage area of the macrocells. The considered network comes with a wireless transmission link via Wi-Fi and other interfaces to the Internet backhaul in an urban area with multiple macro base stations (macroBSs). Each macroBS serves mobile users in the circular macrocell of the figure. We also consider a set of N mobile FBSs deployed in public transportation vehicles (e.g., buses) to provide onboard coverage, and offload data traffic created by UEs in the

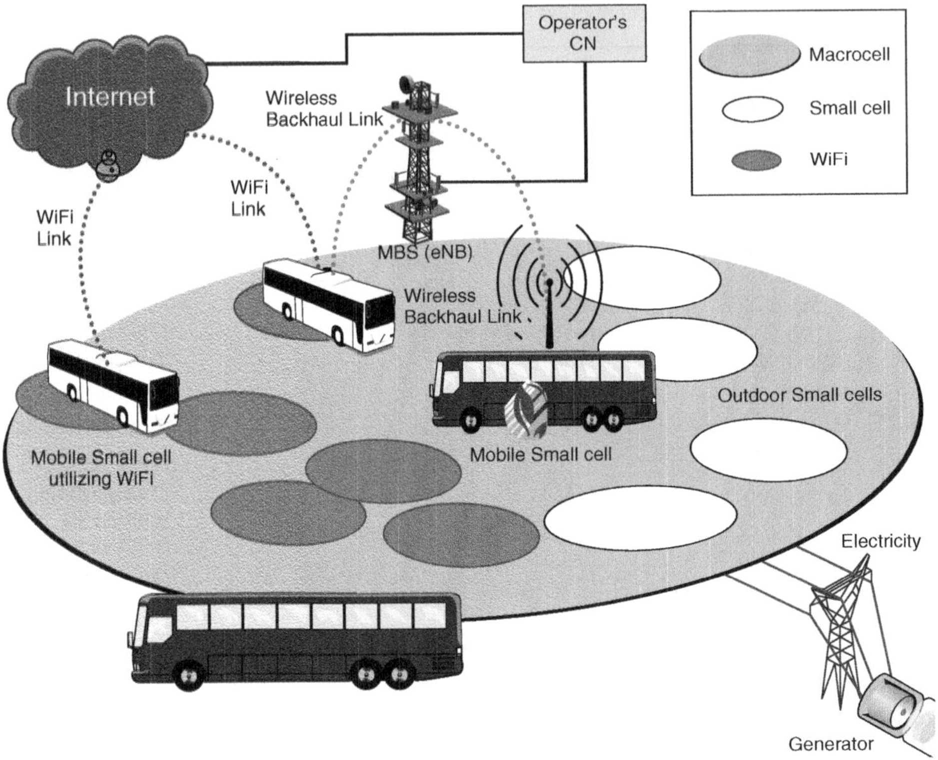

Figure 7.1 A typical grid-based HetNet and a set of mobile FBSs in e-mobility.

vicinity of a FBS. The buses move in an urban city, with varying velocities. Mobile FBSs' received signal strength indicators (RSSI) is the triggering condition for an offload operation from the macrocell BS. As in any cellular network, UEs should communicate with the base station that has the highest RSSI. Therefore, when a UE_i enters the bus, it will sense the mobile FBS's high RSSI, due to the short distance between the UE and the mobile FBS. Then, the UE_i will report the mobile FBS's RSSI to the serving macro BS and send a request to be offloaded.

The system is modeled as an M/M/N/L queuing system in Figure 7.22. There are N identical channels available in a multi-interface FBS, where each FBS comes nowadays with different communication standards, such as Bluetooth, Wi-Fi, Ethernet, etc. Requests are assumed to arrive independently following a Poisson distribution process as in [38]. When the femtocell channels are all busy, the incoming user requests are buffered and form a queue of size W. The symbols used in this work are briefed in Table 7.2.

In this work, not only the FBS is mobile, but also users can move to neighboring cells while they are being served in the grid system. It is typical in HetNet systems to experience some FBSs' departures due to many different reasons including load balance and/or energy saving. These departures and unavailability of an FBS may degrade the performance of the grid. It is assumed that a single FBS arrival is available for every FBS departure, to replace the departed FBS. Similar to the previous attempts [6, 22], each

Table 7.2 Summary of Symbols.

Symbol	Definition
r	radius of the femtocell
V	velocity of the mobile users
P	perimeter of the femtocell
A	area of the femtocell
N	total number of channels in the FBS
W	queue capacity of the cell
L	maximum number of requests in the cell
λ_p	total arrival rate of requests in the cell
μ	total service rate of completed requests in the cell
μ_p	average service rate of UE requests in the cell
λ_v	departure rate of a mobile FBS deployed on a vehicle.
μ_v	mean arrival rate of a mobile FBS deployed on a vehicle.
M	number of mobile FBS in the system.

macrocell is a circle with radius R and is served by a base station placed at the center. The femtocells, which are deployed within the coverage area of a macrocell, are assumed to be circular with radius r and are served by femtocell base stations (FBSs).

7.3.1 Markov Model

In order to satisfy energy requirements for HetNets applications, a Markov discrete-time stochastic process M/M/N/L queuing model is assumed. The queuing capacity of the system is denoted by W as we mentioned, and L represents the maximum capacity, which includes the number of FBSs in the system (M), and thus, $L = M \times (W + N)$. Similar to [38–41], arrivals to the system are assumed to follow a Poisson distribution with rate λ_p, and the service time of the FBS is exponentially distributed with rate μ. Our system is proposed under a realistic assumption of a mobile FBS queue that can hold waiting requests as long as they are within the required communication range. Mobile FBS may move and thus user requests will handover to neighboring cells with rate μ_p while they are either in the queue or being served in the system. The departure rate of the FBS is assumed to be exponentially distributed and is denoted by λ_v [38]. Following the departure, the departed FBS is replaced by an arriving one with arrival rate μ_v with exponential distribution. Figure 7.2 represents our FBS queuing system under this study.

7.3.2 Service-Rate Model

According to [14], the dwell time of a mobile FBS is the time that the mobile node spends in a given system. The dwell time is assumed to have an exponential distribution with a mean rate μ_p. And thus, the service rate due to mobility can be calculated as $\mu_p = \frac{P.E[v]}{\pi A}$,

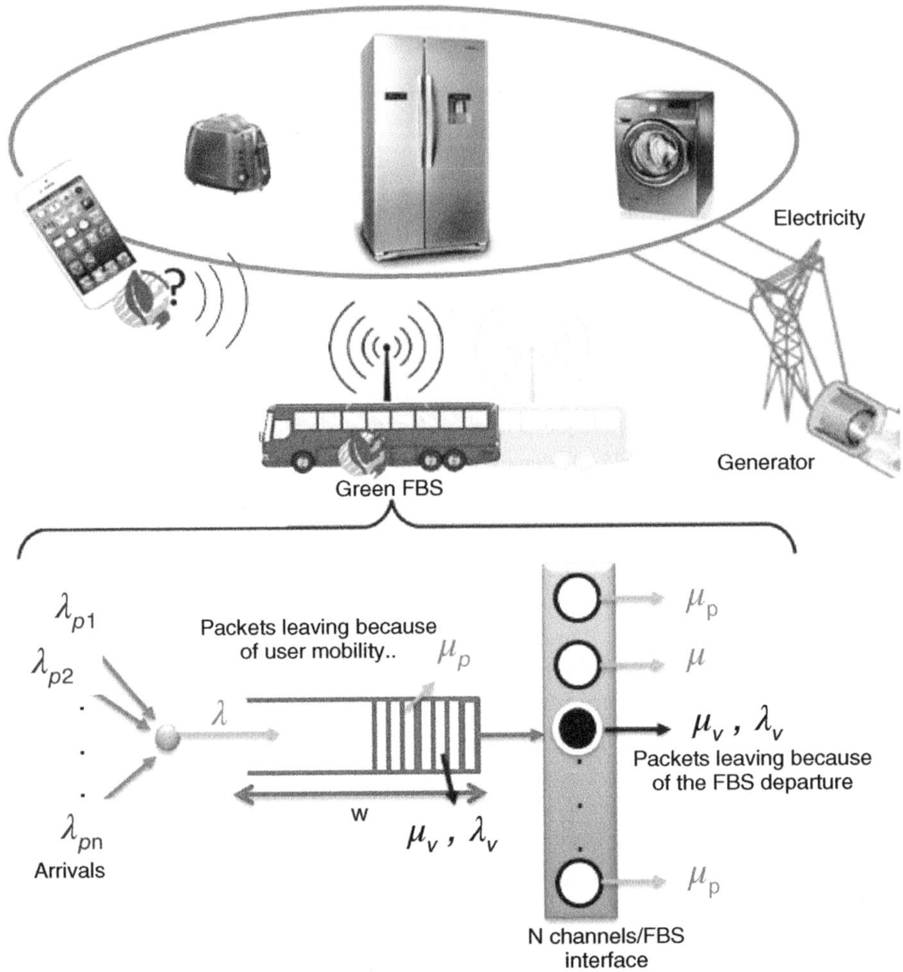

Figure 7.2 The queuing system considered with FBS departures and arrivals.

where $E[v]$ is the average expected velocity of the mobile femtocell, and P and A are the length of the perimeter of the macrocell and area of the macrocell, respectively.

7.3.3 Communication Model

Typically, the signal level at a distance d from the transmitter depends on the surrounding environment and obstacles. This signal strength can be captured through what we call a log-normal path-loss model. According to this model, the received signal power (P_r) at the distance d follows a log-normal distribution centered on the average power value [18, 44]. This can be formulated as follows:

$$P_r = K_0 - 10\rho \log(d) - \chi, \tag{7.1}$$

where d is the distance in km between the center of a candidate UE and its associate FBS, χ is a Gaussian distributed random variable with zero mean and variance σ^2 , i.e. $\chi \sim N(0,\sigma^2)$, ρ is the path loss exponent calculated based on experimental data, and K_0 is a constant calculated according to the average heights of the transmitter and receiver. Hence, the achievable throughput at each FBS site can be approximated using Shannon's capacity equation with an SNR clipping at 20dB for practical modulation as follows:

$$R_i = B\log_2\left(1+\frac{P^i{}_{RX}}{N_0B}\right), \tag{7.2}$$

where R_i is the data rate at UE_i, B is the FBS bandwidth, $P^i{}_{RX}$, which is computed based on Equation (7.1) is the received power at candidate femtocell i, and N_0 is the background noise power spectral density.

7.4 The Green HetNet (GHetNet) Framework

The green HetNet (GHetNet) framework not only assesses the performance of the grid but also can enable the smart grid operators to offload a portion of the data traffic from the overburdened macrocells to the mobile FBSs installed on top of public transportation buses and trains. The mobile FBS is usually connected to an energy-limited source and multiple transmitters' interfaces including Wi-Fi, Bluetooth, 4G, etc. This multi-interface transmitter provides the backhaul for the mobile FBS. Accordingly, our GHetNet framework is composed of four main entities: UEs, mobile FBSs, macro-BSs, and the multi-interface transmitters.

- UE: a handheld device and/or home appliance that has a cellular/Wi-Fi interface in order to connect to the grid backhaul network.
- Mobile FBS: an enterprise FBS that can serve up to N users, simultaneously. It is registered and preconfigured in all ubiquitously accessible access points to the backhaul.
- Macro-BS: a regular existing macro-BS in the city; there are usually more than one.
- Multi-interface transmitter: a multiple I/O module that is integrated with the FBS circuit and installed together on the roof of the public transportation vehicle.

These components are utilized by the GHetNet framework in four key phases: (1) trigger phase, (2) classify phase, (3) decide phase, and (4) offload phase. In the trigger phase, once a UE enters the coverage of a mobile FBS, a trigger to offload its traffic from the macro-BS is initiated based on the mobile FBS's signal strength (i.e., the RSSI). When the trigger condition is fulfilled, the UE sends a request to be accepted by the mobile FBS. In the classify phase, the serving macro-BS receives the request of the UE and checks its feasibility, which can be determined by the current MQL, throughput, and experienced energy consumption of the system. In our framework, we define a number of user classes $C = \{C_1, C_2, C_3, \ldots, C_m\}$, where C_m has the highest importance in C. For example, a user who can experience the best in terms of throughput and delay if it has been offloaded from the macro-BS to the mobile FBS will be given the highest-class, C_m. In the decide phase, after the mobile FBS receives the candidate UE, it compares against other received candidates to be offloaded from the macro-BS based on three main metrics: the present MQL, the instantaneous throughput of the FBS, and the average energy consumed per packet. Once these metrics are checked, the mobile FBS can

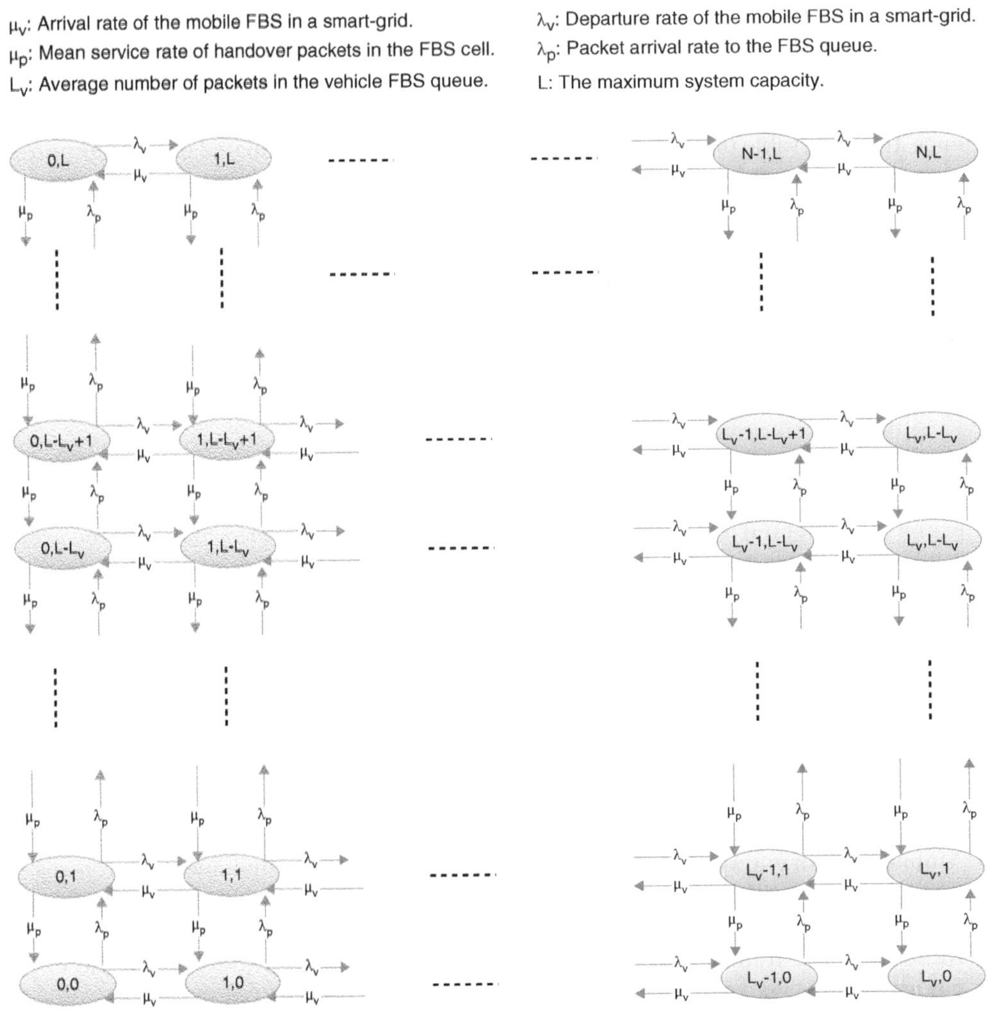

Figure 7.3 A finite state diagram for the HetNet system. It describes all the FBSs' states and their transitions in a smart grid setup where not only users are mobile but also the FBS itself.

accept the most appropriate candidate and inform the macro-BS. Finally, in the offload phase, the macro-BS will transfer the accepted candidate to the mobile FBS and update the grid system. There are some scenarios where the Wi-Fi coverage is not sufficient or the signal strength degrades below a certain threshold. In such scenarios, the mobile FBS can use any other communication interface it has to continue. This system can be considered in Q-theory as a two-dimensional Markov process on a finite lattice strip, as shown in Figure 7.3. The Markov chain is denoted by $Z = \{[P(t), V(t)]; t \geq 0\}$ and is used for performance assessment of the considered HetNet system in this chapter. We assume Z is irreducible [1] with a state space of $\{0, 1, \ldots, N\} \times \{0, 1, \ldots, L\}$. Furthermore,

1 A Markov chain for which there is only one communication class is called an irreducible Markov chain. And since we have finite states for the smart grid system, each state will be visited an infinite number of

we assume that the number of channels in an FBS, P(t),is represented in the horizontal direction, and the total number of UE candidates, V(t), is represented in the vertical direction of a finite lattice strip. In this study, an spectral expansion approach can be employed, where A is the matrix of purely horizontal transitions with zeros on the main diagonal, and one-step upward and one-step downward transitions are represented in matrices B and C, respectively.

Lemma 7.4.1 *Elements of matrix A depend only on the departure and arrival rates of the FBSs, λ_v and μ_v, respectively. The transition rate matrices A, B, and C are square matrices each of size $(N+1)\times(N+1)$, and given in Equations (7.3)–(7.4). The matrix C depends on the number of UE candidates in the system for $v = 0, 1, \ldots, L$. And thus, the threshold M is taken as $M = L$.*

$$A = \begin{pmatrix} 0 & \min(\mu_v,\mu_p) & 0 & 0 & 0 & 0 & 0 & 0 \\ \lambda_v & 0 & \min(\mu_v,\mu_p) & 0 & 0 & 0 & 0 & 0 \\ 0 & 2\lambda_v & 0 & \min(\mu_v,\mu_p) & 0 & 0 & 0 & 0 \\ 0 & 0 & 3\lambda_v & 0 & \ddots & 0 & 0 & 0 \\ 0 & 0 & 0 & \ddots & 0 & \ddots & 0 & 0 \\ 0 & 0 & 0 & 0 & \ddots & 0 & \min(\mu_v,\mu_p) & 0 \\ 0 & 0 & 0 & 0 & 0 & (N-1)\lambda_v & 0 & \min(\mu_v,\mu_p) \\ 0 & 0 & 0 & 0 & 0 & 0 & N\lambda_v & 0 \end{pmatrix}$$

$$B = \begin{pmatrix} \lambda_p & 0 & 0 & 0 & 0 & 0 & 0 \\ 0 & \lambda_p & 0 & 0 & 0 & 0 & 0 \\ 0 & 0 & \lambda_p & 0 & 0 & 0 & 0 \\ 0 & 0 & 0 & \ddots & 0 & 0 & 0 \\ 0 & 0 & 0 & 0 & \ddots & 0 & 0 \\ 0 & 0 & 0 & 0 & 0 & \lambda_p & 0 \\ 0 & 0 & 0 & 0 & 0 & 0 & \lambda_p \end{pmatrix} \tag{7.3}$$

$$C = \begin{pmatrix} \min(0,v)\mu + v\mu_p & 0 & 0 & 0 & 0 \\ 0 & \min(1,v)\mu + v\mu_p & 0 & 0 & 0 \\ 0 & 0 & \ddots & 0 & 0 \\ 0 & 0 & 0 & \ddots & 0 \\ 0 & 0 & 0 & 0 & \min(N,v)\mu + v\mu_p \end{pmatrix} \quad 0 \le v \le L, \tag{7.4}$$

Lemma 7.4.2 *The steady-state probabilities of Markov chain states presented in Figure 7.3 can be expressed as:*

$$Pp, v = \lim_{t\to\infty} (\Pr ob(P(t) = p, \quad V(t) = v)), \quad 0 \le p \le N and 0 \le v \le L \tag{7.5}$$

Based on what we stated in Lemma 7.1 and Lemma 7.2, the following theorem can be derived.

---

times regardless of the initial state of the grid. This property is of utmost importance to achieve the complete solution that covers all UEs scenarios in the grid.

Theorem 7.4.3 *Assume that β and ψ, are the independent probability of a UE candidate being admitted by an FBS and the inter-arrival time of the incoming UE packets at the same FBS, respectively. The probability of being in a state* i, j *of Figure 7.3 is given by*

$$Pi,j = \sum_{i=0}^{N} \left(\psi_{i-i+1}\lambda^{j-i+1} + \psi_{j-i}\beta^{j-i}\right), \quad 0 \leq j \leq L \tag{7.6}$$

Proof: This can be proved based on the above described Markov queuing model. Assuming the Poisson process packet arrival λ, all transitions from an empty-queue status to a non-empty status accordingly can be represented by:

$$P_{0,i} = \lambda_i * \psi, \quad \forall i = 0, \ldots, L \tag{7.7}$$

$$P_{0,L} = \lambda_{\geq L}, \tag{7.8}$$

Since the transition probability of the arrived packet to a FBS can be represented by:

$$P = \beta * \lambda_0 * \psi, \quad \forall i = 0, \ldots, L \tag{7.9}$$

The probability of having a non-decreasing queue (i.e., non-transition probability), which can be obtained from two terms depending on the oldest packet in the queue and winning the contention to access the media (first term) or otherwise (second term) can be achieved by Equations (7.10) and (7.11).

$$P_{i,i-1} = \beta \times \psi \times \lambda_{j-i+1} \times \psi + (1 - \beta \times \psi) \times \lambda_{j-i} \times \psi, \qquad \forall i = 1, \ldots, L-1 \tag{7.10}$$

$$P_{i,L} = \beta \times \psi \times \lambda_{\geq L-i+1} \times \psi + (1 - \beta \times \psi) \times \lambda_{\geq L-i} \times \psi, \quad \forall i = 1, \ldots, L \tag{7.11}$$

Now, let us assume a Markov model with the aforementioned finite set of transitions and transitional probabilities $\pi_w(t)$ in matrix P of having m packets, as illustrated in Figure 7.3. The steady-state equations for each packet can be described as follows:

$$\pi_0(t) = \pi_0 \times (1 - \lambda \times \psi) + \pi_1 \times \beta \times + o(\psi) \tag{7.12}$$

$$\pi_L(t) = \pi_{L-1} \times (t) \times \beta \times \psi + \pi_L(t) \times (1 - \beta \times \psi) + o(\psi) \tag{7.13}$$

$$\begin{aligned}\pi_w(t) = \pi_{w-1} &\times (t) \times \beta \times \psi \\ &+ \pi_w(t) \times (1 - \lambda \times \psi - \beta \times \psi) + \pi_{w+1}(t) \times \beta \times \psi + o(\psi), \forall w \neq 0\end{aligned} \tag{7.14}$$

where $\pi_i(t)$ denotes the steady state of the packet at time t, and $o(\psi)$ is defined as a function of ψ such that $\lim_{\psi \to 0} \frac{o(\psi)}{\psi} = 1$. Since the proposed model is considered to be an irreducible, periodic, and recurrent non-null Markov chain, the model possesses the unique stationary probability $\Pi_w(t) = \{\Pi_0(t), \ldots, \Pi_L(t)\}$, where $\sum_{w=0}^{L} \Pi_w(t) = 1$, which strictly provides the probability that the mean rate of arrivals per state is less than the mean rate at which packets are obtained by the FBS per state. Moreover, the average queue length will become stable, and the number of packets in the queue will be finite. Thus, both the packet arrival rate λ and the probability of successful transmission β for a specified packet processing in the smart grid system become variables in the transition matrix P. □

Consequently, the state probabilities in Theorem 7.1 can be used in computing important performance measures such as delay and throughput (γ). According to the probability distribution of the proposed Markov chain, the mean queue length (or MQL) of the admitted packets can be calculated directly from Little's law by

$$\text{MQL} = \frac{\sum_{w=0}^{\infty} w \times \Pi_w(\text{t})}{\lambda \times \beta} \tag{7.15}$$

and the FBS throughput to be

$$\gamma = \sum_{w=1}^{\infty} \Pi_w(\text{t}). \tag{7.16}$$

Furthermore, the energy consumed by the proposed green HetNet system can be found based on the assumed Markov chain while considering the aforementioned path-loss model in Equation (7.1). Let us assume that every FBS node in the grid has the following four key power transition states: idle, transmit, receive, and sleep, which can be represented by P_{TX}, P_{RX}, P_{idle}, and P_{sleep}, respectively. Furthermore, P_{total} is defined as the total power spent for all these power states during the FBS operational lifetime. Accordingly,

$$P_{total} = \frac{PktLength}{R} \times P_{RX} + N \times P_{TX} + P_{idle} + P_{amp} + P_{sleep}, \tag{7.17}$$

where Pkt_{Length} represents the received packet length, and P_{amp} is the amplifier power denoted by $P_{amp} = \frac{c*\gamma*\left(\frac{d_{max}}{N}\right)^{\alpha}}{\eta}$, where η is a performance parameter in the RF amplifier, γ is the sufficient RSSI level, R indicates the data packet transmission rate, d_{max} is the total distance between the UE and the FBS, and c is a constant value proportional to the packet length, channel attenuation, and nonlinear effects of the power amplifier. Accordingly, the total energy consumption during the FBS operational period varies according to which status it will be at. Please note that the FBS might be transmitting a packet with a successful transmission probability according to the steady state of the Markovian model. Also, the FBS might be in a receive mode, receiving a packet successfully with a varying probability based on the embedded path-loss model instead of a random selection. Because of the queue overflow, the FBS may fail to receive the packet; and thus, it should go into an idle mode. Consequently, the expected energy consumption for a multi-interface FBS can be expressed as follows:

$$\begin{aligned} E(\Pi_m(t)) = {} & (N-1) \times \Pi_m(t) \times P_{RX} + N \times \Pi_m(t) \times P_{TX} + \Pi_m(t) \\ & \times (P_{idle} + P_{sleep} + P_{amp}), \end{aligned} \tag{7.18}$$

and thus, the energy consumption of a successful communication between the UE and the FBS is given by

$$Energy = \sum_{(i=1)}^{n} \frac{1}{(1 - PER_i)} \times E(\Pi_m(t)), \tag{7.19}$$

where PER_i is the packet error rate at the i-th FBS in the grid.

7.5 A Case Study: E-Mobility for Smart Grids

E-mobility is a real-case scenario for applied mobile FBS in a smart grid to optimize the electricity usage in the grid while experiencing peak demands 37. Electrical cars in this project are relying on mobile FBS in exchanging their energy status during the day in addition to other smart home appliances, and thus, heavy data traffic is expected to be generated. In such a comprehensive mobile HetNet model, an energy-aware design and validation framework is a must for more optimized femtocells' usage in serving the vast incoming requests per time unit. This framework shall be able to overcome defies related to users' mobility as well as FBSs' departures and/or arrivals. Hence, we testify our proposed GNetNet framework in analyzing and improving QoS-related metrics under real-world conditions and setups in the grid application. In this use case study, we assume mobile energy-hungry UEs, which can be static/mobile commuting in outdoor public transportation systems. FBSs can be static/fixed on the road infrastructure or mobile on top of the roof of a public bus as shown in Figure 7.4. Typically, UEs are presumed to be equally distributed in the city under their serving cell coverage. For more realistic studies, we embrace typical LTE-values that have been used in [43], as depicted in Tables 7.3 and 7.4.

Table 7.3 Specifications of a Femtocell Base Station parameters [43].

Parameter	*Value*
P (mW)	20
BW (MHz)	5
N_0 (W/Hz)	4×10^{-21}
P_{mp} (W)	3.2
P_{FPGA} (W)	4.7
P_{trans} (W)	1.7
P_{amp} (W)	2.4

Table 7.4 Components of the considered use-case scenario [43].

HetNet component	*Value*
FBS radius (m)	30
FBS velocity (km/hr)	low, medium, high
FBS channels	8
Expected departure rate per hour (λ_v)	0.001
Expected service rate per hour (μ)	200
Expected arrival rate per hour (μ_v)	0.5

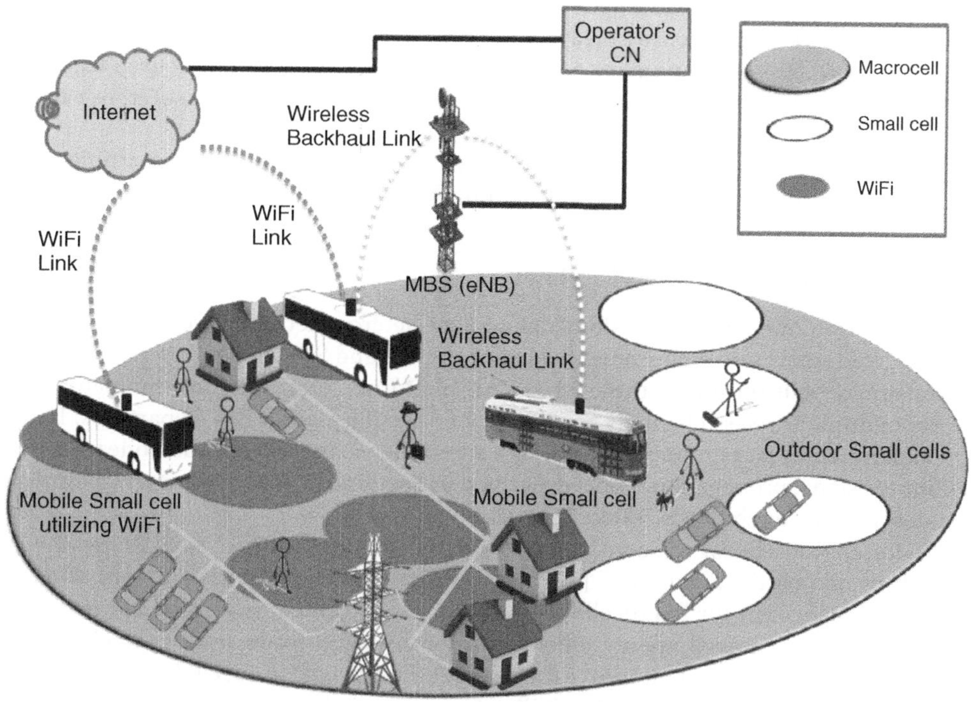

Figure 7.4 Mobile FBS serving mobile/static users in a smart grid setup.

7.5.1 Performance metrics and parameters

In order to assess our proposed GHetNet framework, we consider the following performance metrics:

- Energy (E): is the average power consumption rate per packet as a meter for the grid energy saving. This metric is chosen to evaluate the core infrastructure of the grid in terms of the amount of energy consumed by FBSs. This metric is measured in joules.
- Throughput (γ): is the average percentage of successfully received packets from the FBS. This metric has been chosen to reflect the effectiveness of the proposed GHetNet framework in the grid and is set here as a quality metric. It is measured in packets per hour "pkts/h."
- Response time (R): is the average time a UE request is experiencing from arrival until the departure from the FBS and is measured in seconds. It plays a significant role in performance evaluation since it incorporates all the delays involved per UE request.

These metrics are assessed while varying the following system parameters:

- Mobile FBS velocity: is the average experienced speed per FBS while it moves in the grid system. It can be either: (i) low speed, between 0 and 15 km/h, (ii) medium speed, between 15 and 40 km/h, and (iii) high speed, above 40 km/h.

- MQL: is the average queue length per FBS in the grid. It is measured in "packets," and it represents the experienced delay in an FBS.
- Arrival rate (λ_p): is the number of packet arrivals per time unit. It is used to reflect the effect of different traffic loads on the grid performance. It is measured in packets per hour "pkts/h."
- Service rate (μ): is the rate of packets served per FBS equipped with multiple communication interfaces. It is measured in number of served users per hour "user/h".

7.5.2 Simulation Setups and Baselines

In this section, we simulate randomly generated heterogeneous networks to represent the targeted smart grid environment using MATLAB and Simulink 8.7. A discrete event simulator is built on top of these MATLAB platforms. Our Simulink simulator supports multiple wireless channels per FBS, users' mobility, and FBSs' departures. Based on experimental measurements taken in a site of dense heterogeneous nodes [43], we adopt the signal propagation model described in Section 7.3, where we set the path-loss model variables as shown in Table 7.5.

As for the baselines, we consider two related femtocell-based approaches in the literature, namely the FemtoWoC [32] and the GPowCon approaches [33]. The wireless over cable approach for femtocells (FemtoWoC) is projected as a green approach for improving the femtocell energy efficiency. FemtoWoC introduce a novel idle mode energy-saving procedure to improve the energy efficiency of femtocellular base stations. In [33], a green power consumption (GPowCon) approach is proposed to minimize the average energy consumption based on traffic and coverage parameters. Both of them aim at assessing and improving the femtocell-based network performance in terms of energy, and hence, they have been chosen in order to evaluate our proposed GHetNet framework. The simulation results obtained from our MATLAB code are presented comparatively with the analytical results from our queue model and validated to reflect the performance of the actual system.

7.5.3 Results and Discussion

Obtained results for the proposed GHetNet framework are divided into two parts in this section. First, we discuss the impact of mobility-speed factors on an FBS performance in terms of the response time, throughput, and queue length. Second, the impact of traffic load on the FBS energy consumption is considered. Please note that our achieved simulation results are within the confidence interval of 5% with a confidence level of 95%.

Table 7.5 Parameters of the Simulated Networks.

Parameter	*Value*
ρ	4.8
δ^2	10
K_0	42.152
r	30 m

7.5.3.1 The Impact of Velocity on FBS Performance

In this subsection, performance metrics of the femtocell-based grid system are compared under three different velocities, namely the low, medium, and high velocity, while varying the arrival rate (λ_p) from 3500 to 6000 users/h. Consequently, Figure 7.5 shows the effect of the mobile FBS velocity on the mean queue length for various arrival rates. It is clear from the figure that when the system is congested, the mean queue length will also grow. This is due to the increasing service demands from the FBS, simultaneously. As the mobile FBS moves quicker, the mean queue length decreases. This is due to the fact that the service rate, μ_p, is directly proportional to the expected velocity of mobile FBS. Therefore, as the velocity increases, users' requests will leave the FBS queue sooner, and the mean queue length will decrease. For instance, when the arrival rate λ_p is equal to 6000, the mean queue length is very close to the assumed queue size at velocity of 3 km/h. However, when the mobile FBS starts moving faster at a speed of 60 km/h, the average queue length is equal to 7 packets.

In Figure 7.6, we assume the same setups as the one used in Figure 7.5. It compares throughput of the system versus the average FBS velocity while varying the arrival rate parameter (λ_p). As expected, when the arrival rate increases, more packets are served, and thus, throughput increases accordingly. It has been also observed that as the mobile FBS moves faster, the system throughput decreases. This can be due to the increasing number of removed UE requests from the mobile FBS queue when the velocity increases and the FBS moves faster. And hence, the number of served packets will decrease, and throughput will decrease as well. In order to validate our GHetNet framework, analytical results as well as simulation results have been presented comparatively for both figures: Figure 7.5 and Figure 7.6. The maximum discrepancies between them are 1.96% and 0.07% for Figure 7.5 and Figure 7.6, respectively. Both are less than the aforementioned confidence interval of 5%.

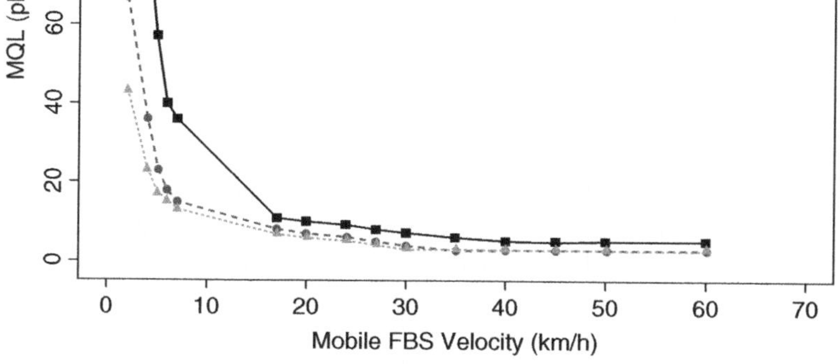

Figure 7.5 The effect of FBS velocity on MQL.

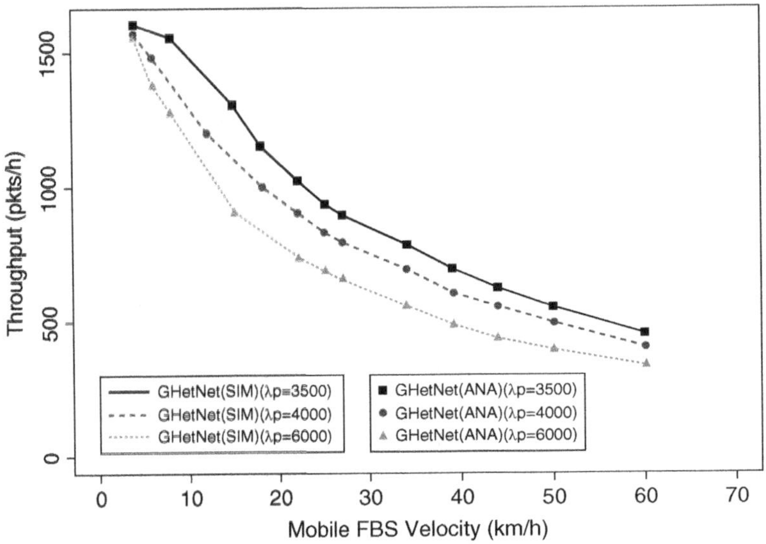

Figure 7.6 The effect of velocity of mobile users on throughput.

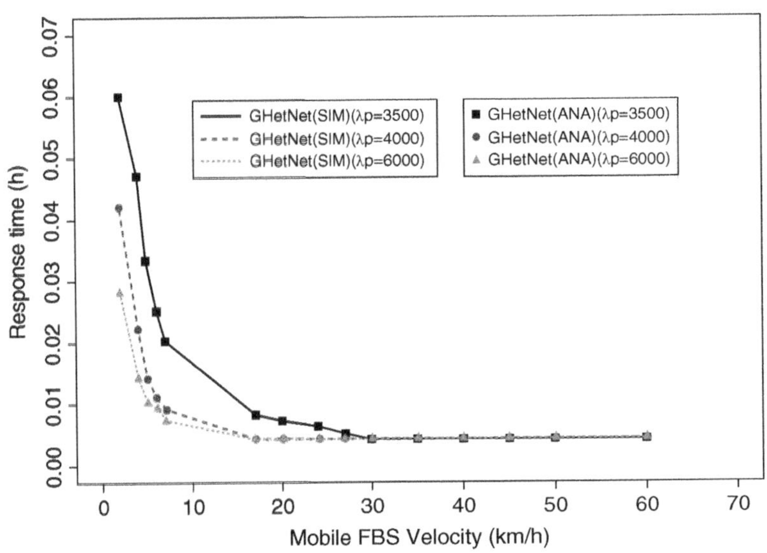

Figure 7.7 The effect of the FBS velocity on response time.

In Figure 7.7, the effects of mobile FBS velocity on the response time are shown. Obviously, the arrival rate is the most effective factor that influences the response time while experiencing low velocities. As the velocity increases, the main factor affecting response time will be the speed value since the departure of incoming user packets becomes considerably higher than the arrival rate. Consequently, for medium and high speeds, arrival rate does not affect significantly the response time, as depicted in Figure 7.7.

7.5.3.2 The Impact of the Grid Load on Energy Consumption

Generally, when $d \leq R$ a single communication interface is considered to be more energy efficient. Especially while experiencing low path-loss exponents, the communication channel condition is close to the perfect value of required SNR. The assumed reduction in energy consumption is considered a vital issue and should be utilized in terms of reducing the power overhead via the GHetNet framework, where FBSs can be placed everywhere to provide a multi-interface communication in connecting heterogeneous nodes, e.g., sensors, PDAs, and tablets. And thus, an FBS with multiple communication channels can dramatically decrease the power consumption of the grid. For example, in Figure 7.8, we observe the amount of energy consumed while the queue length of the mobile FBS is increasing. Obviously, the increment in the queue length causes undesired exponential energy increase.

And thus, more attention shall be given for the queue length in HetNets. Consequently, we consider the MQL factor in our proposed GHetNet, and thus, unnecessary energy waste has been avoided in comparison to other alternatives in the literature such as the FemtoWoC. Meanwhile, the GPowCon approach consumes less amount of energy per hour due to ignoring the QoS requirements (e.g., delay and throughput) that have been set as mandatory constraints by GHetNet and FemtoWoC. In general, there was a steady increase in the energy spend per hour in all categories, which supports the validity of the proposed GHetNet framework. Moreover, it verifies the GHetNet accuracy in predicting/measuring the energy consumption amounts.

In Figure 7.9, we further position the proposed GHetNet framework against the other two alternatives in terms of energy while varying the average service rate per user. This figure studies the effect of the FBS users' density against energy consumption as well. The service rate μ is used to reflect the traffic load in the vicinity of an FBS. In general, we observe a linear increase of energy spend per hour in all the categories, where the highest energy consumption trend is recorded by the FemtoWoC approach again, followed by

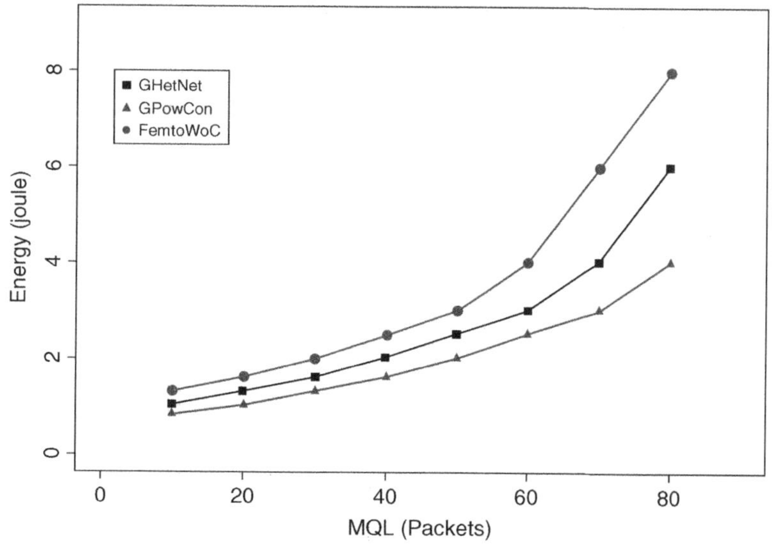

Figure 7.8 Energy spent per hour vs. the average MQL.

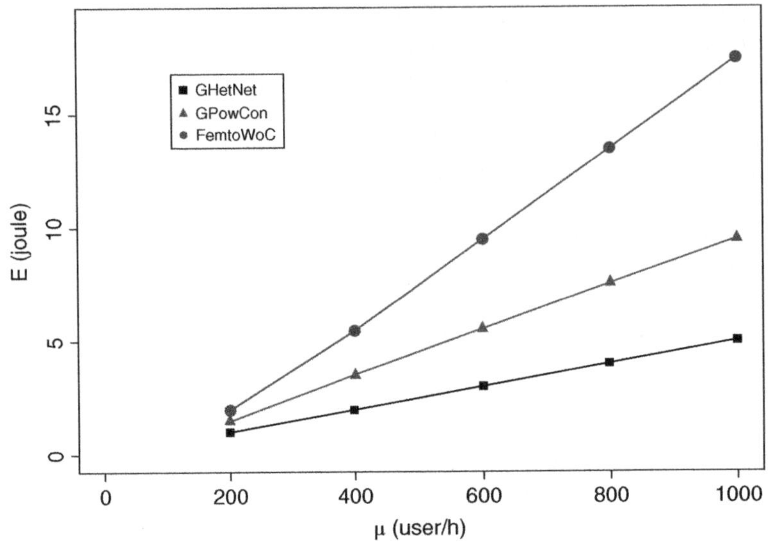

Figure 7.9 Energy spent per hour vs. the average service rate μ.

GPowCon, and the lowest by GHetNet. FBSs in FemtoWoC completely switch off their radio transmitters and related processing when they are not considered in an active call. This necessitates further processing and energy consumption when the FBSs wakeup. We generally notice that even with relatively large increments in user/arrivals counts, the FBS energy consumption is not that much affected. This can be returned to the careful consideration of the average service time factor in our Markov-based framework while considering the mobile FBS departures. Unlike the effect of MQL, number

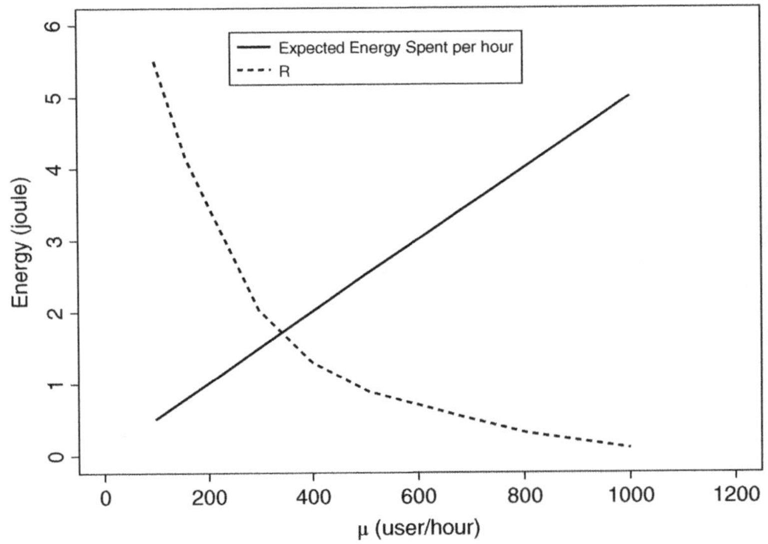

Figure 7.10 Response time and energy consumed as a function of the service rate.

of served users can barely affect the consumed FBS energy. This can be returned to the mobility factor, where a great portion of these arrivals are leaving the queue due to communication range conditions.

Moreover, we show the effect of the service rate (μ) on the response time (R) and the expected energy consumption of the incoming packets per hour in Figure 7.10. At μ equal to 100, the expected energy consumption is ~0.53 joules per hour. As the service rate increases, the response time starts to decrease, and at the same time more energy is consumed. For instance, when μ is equal to 1000, the response time is equal to 5.4 seconds, which is ~60 times less than the energy consumed once μ is equal to 100. Using the same figure, we can control as well the optimum service rate at which we can achieve the best response time and energy consumption (e.g., when $\mu = 300$ in Figure 7.10). Figure 7.10 also depicts that there is a trade-off between the average power consumption and the efficiency of the FBS. Furthermore, the proposed Markov chain can play a key role in specifying the operative space and performance level, as well as the average energy consumption.

7.6 Conclusion

The emerging paradigm of the smart grid is rapidly growing nowadays, intending to improve the quality of our life by connecting vast smart devices, via multiple technologies. Femtocells and heterogeneous networks (HetNets) play a significant role in realizing this vision. In this chapter, a green HetNet consisting of several mobile customers using the smart grid is considered in presence of multiple macrocells and FBSs' departures and arrivals. Accordingly, a new smart grid model is presented in order to validate the usage of mobile FBSs while supporting extreme data traffic in green applications. We believe that the aid of highly available FBSs in such setups in the smart grid will become very important in the coming decades. Furthermore, the obtained results show that traffic load (or UE arrival rate per time unit), the velocity of mobile FBSs, and the multi-interface FBS are crucial parameters. They play a key role in predicting the best average energy consumed by the grid system. For instance, increasing the users' arrival rate in the femtocell can lead to an increment in the average waiting queue length, throughput, and FBS response time. Moreover, achieved results show that for high and moderate velocity mobile FBSs, traffic load does not significantly affect the response time. Our proposed GHetNet framework can actually provide a threshold on the average energy consumption while maintaining a specific response time. Such a framework can be quite beneficial in specifying the optimum operational conditions in a femtocell-based grid system.

References

1 VNI Mobile Forecast Highlights, 2015–2020, URL: http://www.cisco.com/assets/sol/sp/vni/forecast_highlights_mobile/index.html, 2016.

2 Hasan, M.Z. et al., "Optimized Multi-Constrained Quality-of-Service Multipath Routing Approach for Multimedia Sensor Networks", *IEEE Sensors Journal*, vol. 17, no. 7, 2017, 2298–2309.

3 Bou-Harb, E., Fachkha, C., Pourzandi, M., Debbabi, M., and Assi, C., 'Communication security for smart grid distribution networks', *IEEE Communications Magazine*, 51(1), 2013, 42–49.
4 Al-Turjman, F., 'Identification Framework for Smart Environments in the Era of Cloud-IoT', *in Proc. Of the Inter. Conf. on Internet of Things, Data and Cloud Computing (ICC)*, University of Cambridge, United Kingdom, 2017.
5 Qutqut, M. et al., 'Dynamic small cell placement strategies for LTE Heterogeneous Networks', *in Proc. Of the IEEE ISCC*, 2014, 1–6.
6 Qutqut, M.H. et al., 'MFW: Mobile femto-cells utilizing WiFi: A data offloading framework for cellular networks using mobile femtocells', *in Proc. of the IEEE International Conference on Communications (ICC)*, 2013, 6427–6431.
7 Al-Turjman, F., 'Hybrid Approach for Mobile Couriers Election in Smart-cities', *in Proc. of the IEEE Local Computer Networks (LCN)*, Dubai, UAE, 2016, 507–510.
8 Haider, F., Wang, C.X., Ai, B., Haas, H., and Hepsaydir, E., 'Spectral/energy efficiency tradeoff of cellular systems with mobile femtocell deployment', *IEEE Transactions on Vehicular Technology*, vol. 65, no.5, 2016, 3389–3400.
9 Cao, F. and Fan, Z., 'The tradeoff between energy efficiency and system performance of femtocell deployment', *in Proc. the IEEE Int. Symposium on Wireless Communication Systems (ISWCS)*, 2010, 315–319.
10 Kumar, W., Kumar, P., and Halepoto, I., 'Performance Analysis of an Energy Efficient Femtocell Network Using Queuing Theory', *Mehran University Research Journal of Engineering & Technology*, vol. 32, no. 3, 2013, 535–542.
11 Kumar, W., Aamir, S., and Qadeer, S., 'Performance Analysis of a Finite Capacity Femtocell Network', *Mehran University Research Journal of Engineering & Technology*, vol. 33, no. 1, 2014, 129–136.
12 Feng, Z., Yuexia, Z., 'Study on smart grid communications system based on new generation wireless technology', *in Proc. of the Int. Conf. on Electronics, Communications and Control (ICECC)*, 2011.
13 Baines, R. (n.d.). 'Femtocells - Reducing Power Consumption in Mobile Networks', Retrieved Dec., 2016, URL: http://www.low-powerdesign.com/article_baines_092811.html.
14 Bolch, G., Greiner, S., de Meer, H., Trivedi, K. S., *Queueing Networks and Markov Chains: Modeling and Performance Evaluation with Computer Science Applications*, John Wiley & Sons, 2006.
15 Hasan, M.Z. et al., 'A Survey on Multipath Routing Protocols for QoS Assurances in Real-Time Multimedia Wireless Sensor Networks', *IEEE Communications Surveys and Tutorials*, 2017, doi: 10.1109/COMST.2017.2661201.
16 Next Generation Mobile Networks (NGMN), 'NGMN radio access performance evaluation methodology,' Jan. 2008, URL: http://www.ngmn.org/uploads/media/NGMN Radio Access Performance Evaluation Methodology.pdf.
17 Riggio, R. and Leith, D. J., "A measurement-based model of energy consumption in femtocells", *IFIP Wireless Days*, 2012.
18 Al-Turjman, F., 'Information-centric sensor networks for cognitive IoT: an overview', *Annals of Telecommunications*, 2016 1–16.
19 Gong, J., Zhou, S., Niu, Z., 'Queuing on energy-efficient wireless transmissions with adaptive modulation and coding', *in Proc. Of the IEEE Int. Conf. on Communications (ICC)*, 2011, 1–5.

20 Borodakiy, V.Y. et al., 'Modelling and performance analysis of pre-emption based radio admission control scheme for video conferencing over LTE', *in Proc. of the ITU Kaleidoscope Academic Conference on Living in a converged world-Impossible without standards*, 2014, 53–59.

21 Al-Turjman, F., 'Cognitive Routing Protocol for Disaster-inspired Internet of Things', *Elsevier Future Generation Computer Systems*, 2017, doi: 10.1016/j.future.2017.03.014.

22 Chowdhury, M., Lee, S., Ru, B., Park, N., and Jang, Y., 'Service quality improvement of mobile users in vehicular environment by mobile femtocell network deployment,' *in Proc. Int. Conf. on ICT Convergence (ICTC)*, 2011, 194–198.

23 Karimi, O., Liu, J., and Wang, C., 'Seamless wireless connectivity for multimedia services in high speed trains,' *IEEE J. Sel. Areas Commun.*, vol. 30, no. 4, May 2012, 729–739.

24 Kirsal, Y., Ever, E., Kocyigit, A., Gemikonakli, O., and Mapp, G., 'Modelling and analysis of vertical handover in highly mobile environments', *The Journal of Supercomputing* 71 (2015) 4352–4380.

25 Baloch, R., Awan, I., and Min G., 'A mathematical model for wireless channel allocation and handoff schemes', *Telecommunication Systems*, vol. 45, no. 4, 2010, 275–287.

26 Yusof, A.L. et al., 'Handover adaptation for load balancing scheme in femtocell Long Term Evolution (LTE) network', *in Proc. of the IEEE Control and System Graduate Research Colloquium*, Shah Alam, 2014, 242–246.

27 Nasim, M., Chowdhury M., and Sefat, S., 'Energy-efficient cellular communications powered by smart grid technology', *in Proc. of the Int. Conf. on Electrical Information and Communication Technologies (EICT)*, Khulna, 2015, 298–303.

28 Bu, S. and Yu, F. R., 'Dynamic energy-efficient resource allocation in cognitive heterogeneous wireless networks with the smart grid', *in Proc. of the Int. IEEE Global Communications Conference (GLOBECOM)*, Anaheim, CA, 2012, 3032–3036.

29 Tabany M.R. and Guy, C.G., 'A mobility prediction scheme of LTE/LTE-A femtocells under different velocity scenarios', *in Proc. of the Int. Conf. on Computer Aided Modelling and Design of Communication Links and Networks (CAMAD)*, Guildford, 2015, 318–323.

30 Wu, S., Zhang, X., Zheng, R., Yin, Z., Fang Y., and Yang, D., 'Handover Study Concerning Mobility in the Two-Hierarchy Network', *in Proc. of the Int. IEEE Conf. on Vehicular Technology*, Barcelona, 2009, 1–5.

31 Zhang, H., Wen, X., Wang, B., Zheng, W., and Sun, Y., 'A Novel Handover Mechanism Between Femtocell and Macrocell for LTE Based Networks', *in Proc. of the Int. Conf. on Communication Software and Networks*, Singapore, 2010, 228–231.

32 Gambini, J., Spagnolini, U., 'Wireless over cable for femtocell systems', *IEEE Communications Magazine*, 2013, 178–185.

33 Mukherjee, A., Bhattacherjee, S., Pal, S., and De, D., 'Femtocell based green power consumption methods for mobile network', *Computer Networks*, vol. 57, no. 1, 2013, 162–178.

34 Kulkarni, P., Gormus, S., Chin, W.H., and Haines, R.J., 'Distributed resource allocation in small cellular Networks-Test-bed experiments and results', *in Proc. of the Int. Conf. on Wireless Communications and Mobile Computing*, July 2011, 1262–1267.

35 Rehman, I.U., Philip, N.Y., and Istepanian, R.S., 'Performance analysis of medical video streaming over 4G and beyond small cells for indoor and moving vehicle (ambulance) scenarios', *In Proc. of the Int. Conf. on Wireless Mobile Communication and Healthcare (Mobihealth)*, 2014, 211–216.
36 Zeng, Q.-A. and Agrawal D.P., 'Modeling and efficient handling of handoffs in integrated wireless mobile networks', *IEEE Transactions on Vehicular Technology*, vol. 51, 2002, 1469–1478.
37 [online] http://w3.siemens.com/smartgrid/global/en/products-systems-solutions/software-solutions/solutions_for_emobility/Pages/overview.aspx.
38 Trivedi, K. S., Dharmaraja, S., Ma, X., 'Analytic modeling of handoffs in wireless cellular networks', *Information Sciences*, vol. 148, no. 1, pp. 155–166, 2002.
39 Al-Turjman, F., 'Impact of User's Habits on Smartphones' Sensors: An Overview', *HONET-ICT International IEEE Symposium*, Kyrenia, Cyprus, Oct. 2016, 70–74.
40 El Bouabidi, I., Zarai, F., Obaidat, M. S., Kamoun, L., "Design and analysis of secure host-based mobility protocol for wireless heterogeneous networks", *Journal of Supercomputing*, vol. 70, pp. 1036–1050, 2014.
41 H. Beigy, M. R. Meybodi, "A learning automata-based adaptive uniform fractional guard channel algorithm", *Journal of Supercomputing*, vol. 71, pp. 871–893, 2015.
42 F. Al-Turjman and H. Hassanein, "Towards augmented connectivity with delay constraints in WSN federation", *International Journal of Ad Hoc and Ubiquitous Computing*, vol. 11, no. 2, 2012, 97–108.
43 Zhang, J., De la Roche, G., et al., 'Femtocells: technologies and deployment', *Wiley Online Library*, 2010.
44 Singh, G. and Al-Turjman, F., 'Learning Data Delivery Paths in QoI-Aware Information-Centric Sensor Networks', *IEEE Internet of Things Journal*, vol. 3, no. 4, 2016 572–580.
45 Al-Fagih, A.E. et al., 'A priced public sensing framework for heterogeneous IoT architectures', *IEEE Transactions on Emerging Topics in Computing*, vol. 1, 2013, 133–147.
46 Singh, G.T. and Al-Turjman, F. M., 'A data delivery framework for cognitive information-centric sensor networks in smart outdoor monitoring', *Computer Communications*, vol. 74, 2016, 38–51.

8

Communication Architectures and Technologies for Advanced Smart Grid Services

Francois Lemercier[1], Guillaume Habault[2], Georgios Z. Papadopoulos[2], Patrick Maille[2], Nicolas Montavont[2] and Periklis Chatzimisios[3]

[1] *Itron and IMT Atlantique - IRISA, 2 rue de la Chataigneraie, 35510 Cesson-Sévigné, France*
[2] *IMT Atlantique - IRISA, 2 rue de la Chataigneraie, 35510 Cesson-Sévigné, France*
[3] *Department of Informatics, Alexander T.E.I. of Thessaloniki PC 57400, Sindos, Thessaloniki, Greece*

8.1 Introduction

Over the past years, the demand for electricity has faced a drastic growth, as the number and heterogeneity of electrical devices is continuously increasing. In parallel, the power sector is undergoing major changes, mostly by the switch from fossil to renewable energies, the evolving energy policies, and the emergence of less reliable renewable micro-generation. As stated in a 2015 Eurelectric survey [1], the grid requires taking into consideration these modifications while ensuring secure, sustainable, competitive, and affordable energy for any individual and business.

Proper operation of the electrical network is based on the balance between production and consumption, a great challenge for the network management. Actually, the grid structure is evolving from a rigid and centralized architecture with large production units at the top satisfying demand at the bottom to a more distributed one with individual premises equipped with local renewable production units. Electrical production is therefore getting more decentralized but at the same time less predictable as renewable sources are sporadic. In order to efficiently balance production and consumption, real-time measurements, predictions, and control capabilities are needed in a widespread management system.

Furthermore, electrical devices have also evolved in recent years. Some are now mobile such as electric vehicles (EVs) making demand prediction more difficult; others, such as connected heating/cooling devices, offer remote management capabilities to their owners. Together, their growing numbers, and in particular the increasing penetration of EVs, make the management of the system even more complex. Today, private EVs are charged as soon as they are plugged into the grid, without any management system. And most of these charging processes occur at peak hours, in the evening, which is very challenging for the grid. While there are only few EVs, the electrical grid can afford to provide this power. But if EVs are generalized, we will need to avoid charging all of them on peak hours and so, shift these demands on a time window that will contain the needed power below a given threshold and/or align the consumption

Transportation and Power Grid in Smart Cities: Communication Networks and Services, First Edition.
Edited by Hussein T. Mouftah, Melike Erol-Kantarci and Mubashir Husain Rehmani.

to the production periods. As a consequence, it is vital for the smart grid to benefit from a control system of the charging periods to balance the various energy demands over different periods of time. As presented by Lefrancois et al. [2], several mechanisms are required to provide smart-charging services in an automated way. Moreover, EVs can be seen as mobile batteries from the electrical network. In order to integrate such devices into the system, mobility mechanisms are needed. Therefore, with appropriate ICT systems and services – such as smart charging ones – the smart grid could be able to efficiently use these batteries. For instance, the electrical grid could encourage them to charge at given periods – and compensate for overproduction – or it could ask them to re-inject their electricity when and where it is needed [3].

Nevertheless, electrical devices, such as EVs or home appliances, are not components of the smart grid yet. Their presences and types depend on users' will. Even though utilities could benefit from managing these devices, the complexity arising from them is out of their scopes [4]. As a consequence, a smarter and extended grid is required more than ever to support utilities efficiently, monitor, and manage this system [5], as well as "connecting" the grid with devices beyond.

The ongoing transition in the energy sector has propelled innovation forward. The electrical grid is already getting smarter: it enables both automated meter reading and indirect controlling demand with real-time pricing signals. Under a smarter system, new business opportunities can be envisioned. Indeed, a system offering optimized and automated energy production, distribution, and consumption will enable the appearance of flexible services (selling flexibility in consumption or production). Together with appropriate incentives, this would lead to the creation of new energy markets that would take place at various scales and levels.

These objectives can only be accomplished with a proper ICT architecture to first interconnect all energy actors. As long as this interconnection is ensured, they will be able to exchange energy information (consumption, production, or storage) and use available energy services. Such an architecture must address several requirements. First, it has to take into consideration the diversity of electrical devices that are plugged into the grid. The architecture must be flexible enough to adapt to different hardware constraints and software choices and absorb their evolution over time. Scalability is another must for the architecture, since the future smart grid involves a huge number of devices, including consumption devices and production sites. Finally, real-time measurements, prediction, and control capabilities are required to better plan for the production and consumption alignment. Consequently, the architecture must provide efficient mechanisms to 1) collect and process all the produced data; and 2) make decisions and control end points accordingly. Note that some actors might not require data in real time, nor detailed data (ensuring data privacy). Lastly, it must be secured in order to avoid any entity both taking control of a device without being authorized and collecting data without being granted access to it.

There are also several issues to solve before implementing such an innovative system. While the remote control and coordination for electrical loads of homes, office buildings, and industrial premises have been possible for decades, such control is not yet widely enough adopted as it involves handling the issue of control of a large volume of distributed nodes. Furthermore, such a system should deal with the challenges of new electrical networks in which any party can act as an energy producer and/or consumer; hence the term "prosumer." These challenges require to deal with the following issues:

- finding a given party in a large structure;
- accessing a given resource (data or control of nodes);
- incentivizing for this access; and
- implementing technical compatibility with any system.

To properly cope with these problems, ICT is needed for energy actors to interconnect and better manage energy usage. Future smart grid systems should provide the tools for an efficient integration of the following:

1) local renewable energy production;
2) management systems to retrieve electrical production, consumption, and storage information; and
3) other energy actors to provide systems, mechanisms, or services that use such information for electricity management and control.

In order to achieve these goals, technology is required to:

- offer automated access to any measurement point and load, in order to lower the cost of control; and
- exchange data in such a way that it will allow data availability for large-scale scenarios as well as to make possible application-specific extensions in order to realize the management of heterogeneous infrastructures.

In this chapter, we focus on presenting communication architectures, technologies, and protocols employed in the smart grid environment to solve the problems that a balancing system must face. In Section 8.2 we provide an overview of the existing communication systems currently utilized in the smart grid, and then we present what is required in order to meet a next-generation smart grid system. Before concluding, Section 8.3 develops the routing issues and certain existing solutions to efficiently transmit information while using constrained technologies.

8.2 The Smart Grid Communication Architecture and Infrastructure

In smart grids, the legacy communication architecture enabling data collection and device management is called the *advanced metering infrastructure (AMI)* [6, Chap. 7]. This architecture is an evolution of the automated meter reading (AMR) by adding a bidirectional communication framework, which was deployed to facilitate meter reading, billing, and consumption planning. This two-way communication feature of AMI also offers additional operations on a network. It sure helps the utilities better control their network, but several other opportunities are foreseen with such architecture especially if using high-speed Internet protocol (IP)-based technologies.

In this section, we describe the AMI architecture that involves the main elements of the electric supply chain (smart meters, distribution units, production units). While AMI communications are operated by utilities and allow basic services such as monitoring and some control, we will show that they are not enough to provide the grid with fine-grained and optimized capabilities such as appliance-level demand-response. For such advanced possibilities, we claim that other communication systems are needed and that the currently developing paradigm [7] of the Internet of Things (IoT) is an excellent candidate. Hence, we develop that view later in this section.

The IoT consists of a set of smart "things" that are able to connect to the Internet – directly or via a gateway – and feed other devices with their collected information (often referred to as big data due to their high volume). As long as these things can be uniquely identified and provide some empirical data on our environment –in this case, electric information (consumption, production, or even storage status)– anything can be a "thing." But they may have different capabilities in terms of hardware and control (e.g., a sensor compared to an EV).

In smart grid communication systems, like other communication systems, we can distinguish a core part and a last-mile[1] part. The former involves large storage and computational capabilities used to collect, organize, and process data in order to coordinate devices through remote management commands. The latter part consists of uniquely identified and connected objects (e.g., sensors, actuators, and smart devices) as well as communication links between them, possibly by employing various wired or wireless technologies.

This section aims at detailing the characteristics and technologies of the different communication networks that comply with smart grid requirements. We here present the current state of advancement in this domain, as well as certain ongoing efforts to make the grid smarter.

8.2.1 DSO-Based Communications

The legacy communication network used in smart grids is the aforementioned AMI. The main elements for such infrastructure are the so-called *smart meters*, i.e., power meters embedded with communication capacities that are able to activate some generic functions from remote control. These meters are deployed by utilities and enable distribution system operators (DSOs) to both meter consumption and command end-points remotely. This bidirectional communication helps DSOs efficiently manage their networks.

Hereafter, we describe the communication organization in such DSO-based smart grids, as well as the limits that this structure imposes on the services that can be built.

8.2.1.1 The Existing AMI Organization

Energy distribution networks have a tree topology with large production units on top producing most of the required energy, which is then transported via a widespread distribution network toward consumers (called the end points). Figure 8.1 illustrates this top-down configuration in which consumers can be reached after passing through different aggregating nodes, i.e., transformers. Utilities forecast the consumption of end points based on their historical consumption and adjust these forecasts based on automated metering. These forecasts are therefore very sensitive to any modification of end-points behavior.

Currently there are several notable trends taking place in the energy market. Some examples are the shift away from fossil to renewable energy, the steadily increasing number and heterogeneity of electrical appliances and devices, and the decreasing price of various distributed energy production technologies.

DSOs require a bidirectional communication with smart meters in order to efficiently manage their networks and face these trends. In fact, one consequence of these upcoming changes is the growing demand at the consumer side. Therefore, utilities

1 specifically, last-hop since some radio technologies allow communications over more than ten miles

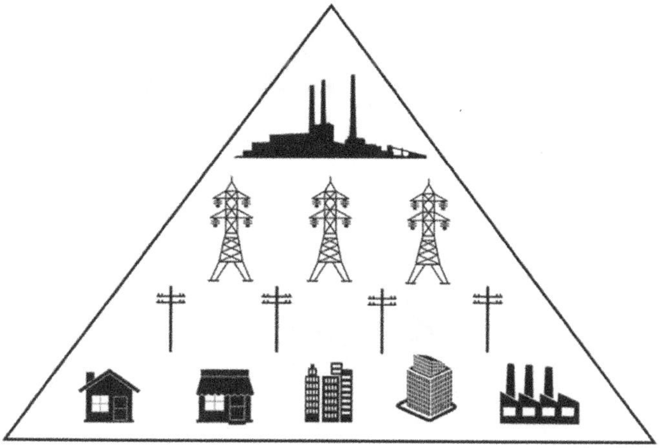

Figure 8.1 The current distribution network topology.

have to plan such growth, increase the amount of electricity produced, and transport this electricity toward the end points. However, as the distribution network capacity is limited, a question that arises is what will happen when reaching this limit. Utilities should encourage end points to balance their consumption during a 24 h period; hence the bidirectional communication with smart meters. In addition, consumers are getting more and more equipped with micro-generation systems. As a consequence, the grid evolves toward a system in which large energy production facilities – dams, nuclear power plants, etc. – must coexist with a myriad of smaller, less reliable systems in the same network. Therefore, and without more information from these "prosumers" – at the same time consumer and producer – it will become more and more difficult for DSOs to estimate consumption of such end points. The ideal solution in order to take into consideration these trends would be to modify the whole distribution network. First, by increasing its capacity and secondly, by allowing *prosumers* to re-inject their production excess into the network – and have smart meters metering it. However, these modifications are very costly and so inconceivable for now or should be planned and spread over several years. For these reasons, DSOs prefer to first set up means to efficiently meter as well as send signals to individuals to balance their consumption.

Therefore, initial smart grid efforts, mainly directed by DSOs, use the AMI in order to offer better management of a distributed energy production. And the smart meter is thus a central piece of the needed communication framework. On the one hand, it allows real-time monitoring in order to collect accurate data about the consumption (and possibly the production). On the other hand, it allows receiving real-time electricity pricing signals, which trigger a certain period of time where the energy is cheaper. Relays can be used in user premises to activate or deactivate devices that are plugged into a specific circuit (e.g., hot water tank, which could be switched on during the night).

As illustrated in Figure 8.2, the AMI consists of a tree of smart meters, placed at different positions of the network. Data from all these meters are collected, stored, processed, and analyzed by the utility center. Routing protocols are thus required from the smart meters (i.e., nodes of the tree) to the utility center (i.e., root of the tree) in order to forward the data hop by hop. Basic commands issued by the utility center to announce the price in real time are actually broadcast to the whole network and do not require unicast transmission.

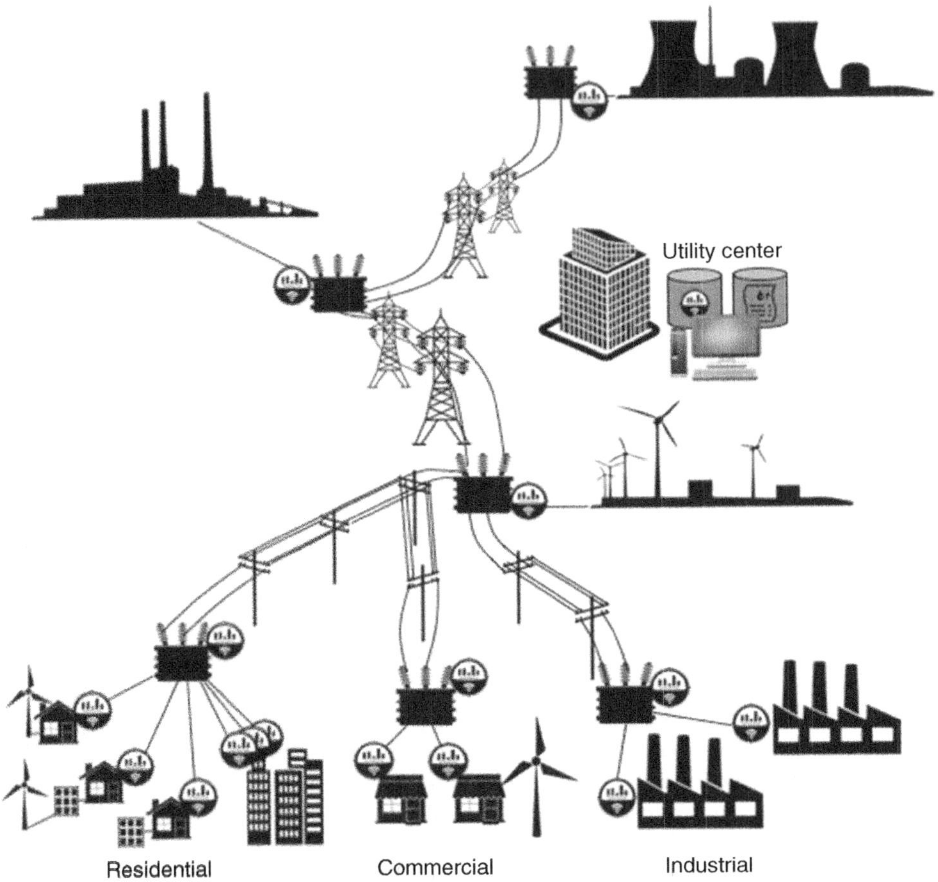

Figure 8.2 Illustration of an AMI.

8.2.1.2 Communication Technologies used in the AMI

Currently, the communication taking place within the core system is based on Internet technologies with IP above a variety of underlying technologies such as optical fiber, cellular network, or power line communication (PLC). With some of them, one aggregating node or a relay node can transfer data on behalf of hundreds of smart meters behind it. To reach the core of the AMI, data issued by (or intended to) the smart meters typically have different "hops" distance to perform. The distance to cover in one hop will vary depending on the network configuration. The technology used to communicate depends on this distance and the environment characteristics. In particular, the technologies used for the last-hop heavily depend on a) the meter capabilities in terms of communication and processing power, b) the energy requirements for this communication, and c) the relative performance of the available options – such as the distance with a wireless base station, the quality of PLC transmission, etc. A non-exhaustive list of commonly used technologies is given in Table 8.1, along with their advantages and drawbacks [8–10]. For wireless technologies, the "energy cost" column refers to the energy consumption associated with the technologies: it includes the transmission power, as well as the processing cost of the corresponding protocol stack and the specific medium

Table 8.1 Communication technologies for the AMI last-hop.

Technology	Range	Throughput	Energy cost	Monetary cost
802.15.4 [11, 12]	up to 100 m	up to 250 $kbit/s$	cheap	cheap
802.11ah [8]	up to 1 km	Tens of $Mbit/s$	expensive	cheap
LoRaWAN [9]	5 km (urban) to 15 km (rural)	Tens of $kbit/s$	cheap	cheap
LTE-Cat M [10]	2 km (urban) to 5 km (rural)	$\approx$200 $kbit/s$	medium	expensive
Narrowband PLC [13]	several km	up to 500 $kbit/s$	cheap	cheap
Wired Internet	–	up to 1 $Gbit/s$	cheap	expensive

access control procedures. The "monetary cost" reflects the cost of the hardware components needed to implement a technology, including SIM cards, as well as potential spectrum license fees.

We need to bear in mind that all technologies listed in Table 8.1 are not always available: due to hardware limitations and/or environmental constraints, the choice can be limited.

8.2.1.3 AMI Limitations

Up to now, on the demand side, the smart grid is mostly limited to the smart meters. Their deployment within the AMI is mapped onto the distribution structure. Furthermore, the intelligence in meters or in intermediate nodes is very limited, leaving all the intelligence in the network core, i.e., the utility center. As a consequence, the AMI lies into a centralized system as previously illustrated: all the data transit through data centers controlled by a DSOs, where all storage and analysis actions are performed; in other words, the AMI is fully operated and controlled by utilities.

With respect to this status, we see three kinds of limitations, all in favor of the use of alternative communication systems.

First, we see issues in terms of **incentives**. Indeed, DSOs will not expect significant gains from improving the communication capabilities, since those gains will go to the entities exploiting them, such as flexibility operators. Conveying the data is not the core mission of DSOs, and they will probably be reluctant to invest in communication improvements if they do not get a return on investment.

Second, the current situation raises some **privacy** concerns: indeed, DSOs have now access to a precise electricity consumption over time, which may raise some privacy concerns. And, if the infrastructure evolves in a way that even more information is transferred to the DSO, e.g., with a per-device monitoring, such a situation is unlikely to be accepted by users. On the contrary, they should decide which entity (if any) to trust with those data and at which granularity level.

Third, we envision some **interoperability** issues:

all the gathered data, which we recall are mainly used for billing and planning purposes, are very sensitive. Therefore, DSOs will not likely let users, other organizations, or entities manipulate them. Consequently, the AMI is currently a closed network where there is no interconnection with other systems and architectures – resulting in a situation where users can only visualize their data. This missing interconnection limits the possibilities to cross-reference or use information coming from other systems. For instance, senior monitoring systems would be more accurate if health information could be crossed with electric information, and DSOs could benefit from having external information to better plan consumption or production.

Nevertheless, smart meters will probably embed more functionalities in the future, such as data storage and decision / management mechanisms. However, adding such capabilities will increase the cost to produce and deploy such meters. Therefore, DSOs will have to determine a business model that will push them into deploying such a costly new system. Although, this suggestion appears to be innovative, if there is not a significant gain for DSOs, these "smarter" smart meters will never be deployed.

8.2.2 Internet-Based Architectures

As stated before, other communication networks used in the smart grid are centered on the Internet and, thus, employ various IoT standards. The main elements for such architecture are the *smart appliances*, i.e., devices and/or appliances embedded with communication capabilities, as well as actuators or sensors. These new connected devices are "deployed" – or should we say, sold – by manufacturers and enable a user to have controllable devices. This actually gives the users a better control and knowledge of their appliances behavior – i.e., consume, produce, or store. In addition, it offers additional flexibility for users to change the way that their devices behave.

In the future, a large number of these smart appliances [14] is expected to appear – consuming, storing, and/or producing energy. These smart appliances provide their energy information (consumption or production) along with controlling capabilities. These smart devices are usually connected to the Internet, and if not – probably due to specific constraints – they are still accessible via gateways. This enables users to remotely access, control, and/or monitor smart devices. For instance, almost all of them come with a monitoring/controlling mobile application.

Furthermore, as the main purpose of the Internet is to interconnect systems, connected devices have access to a plethora of services and information. As a result, these smart appliances can use the Internet to have access to external information such as weather information. Moreover, manufacturers can develop management services on top of the Internet for users that prefer to leave the management of their smart appliances to third parties. Such Internet-based communication networks offer many possibilities for smart grid applications and very likely some we have not even thought of yet.

An example of smart grid services that can be deployed over such architecture is the automatic shedding of certain industrial loads during peak hours. In fact, with such system, an operator could sell the flexibility of a pool of non-critical industrial load as adjustment energy reserve for the grid. As a result, on grid demand – for instance during peak hours–, this flexibility operator (FO) can switch off a set of its managed load in order to reduce the peak. This peak reduction enables utilities to save money – e.g., avoiding to buy external electricity – which in return pay the FO for its assistance. Internet-based architecture enables a FO to aggregate industrial load from different places and thus, to manage a large portfolio of flexibility. It can be considered as a full win-win scenario as industry will also get a share of the FO's remuneration while decreasing their electricity bill.

In an Internet-based architecture, the core domain is the Internet with all the protocols and standards it provides. Communication within the core domain is therefore

IP-based just like in the AMI. Moreover, last-mile communications also rely on IP within the local area network (LAN) – or home area network (HAN), premise area network (PAN), field area network (FAN), or even business or industrial area network (BAN) – or they might use gateways to connect to IP.

The universal nature of the Internet and the flexibility of the IP stack provide lots of opportunities. Such architecture enables any provider to offer various services toward smart appliances – that might not be directly related to energy – as well as enhancing smart grid possibilities.

8.2.2.1 IP-Based Architecture Limitations

Energy and smart grids are potential application domains for IoT among others such as transportation, health care, or environmental monitoring. One issue of such an Internet-based solution is that there are several ways to implement an architecture for an enhanced smart grid. In addition, there is not only one standard architecture, leading the standardization efforts to be scattered [15, 16]. Furthermore, energy-related information is fragmented toward several standards and metrics, leading almost all smart appliances to be manufacturer specific. This diversity leads to a fragmented landscape of architecture models and isolated "silos" implementation . Interconnection of these silos is very difficult, if not impossible, as they often use their own architecture, models, and mechanisms – sometimes even between silos belonging to the same application domain.

Another concern with this type of solution is the lack of security around the deployed devices. Everything and everybody is now connected on the Internet, but not all of them are benevolent. In fact, as most IoT architectures are not mature enough, they do not include strong security mechanisms to protect underlying networks. As a result, connected devices are more and more targeted in order to break in private networks and user privacy[2].

Last but not least, there is no actual interconnections with utilities networks apart from a potential interaction with the centralized management system in order to realize demand response (DR) mechanisms. Nevertheless, enabling crossed information from both smart meters and smart appliances could greatly enhance smart grid systems.

8.2.3 Next-Generation Smart Grid Architecture

As mentioned in the previous sections, there are currently two types of smart grid systems, each using different types of communication technologies: 1) One relying on smart meters and mostly used by utilities; 2) The other one relying on smart appliances and mostly used by end users. However, each system has some drawbacks and would benefit from the other. The next-generation grid communication network has a future that is all planned out. It is expected to interconnect both systems in order to provide an enhanced one as depicted in Figure 8.3. In this figure, each site has both a smart meter, an energy management system (EMS) (depicted as a gray server), along with Internet access. It receives information from the grid (solicitation or demand) via the smart meter (which is also used for billing). Each EMS aggregates data collected from both smart appliances

2 http://arstechnica.com/security/2016/10/double-dip-internet-of-things-botnet-attack-felt-across-the-internet/

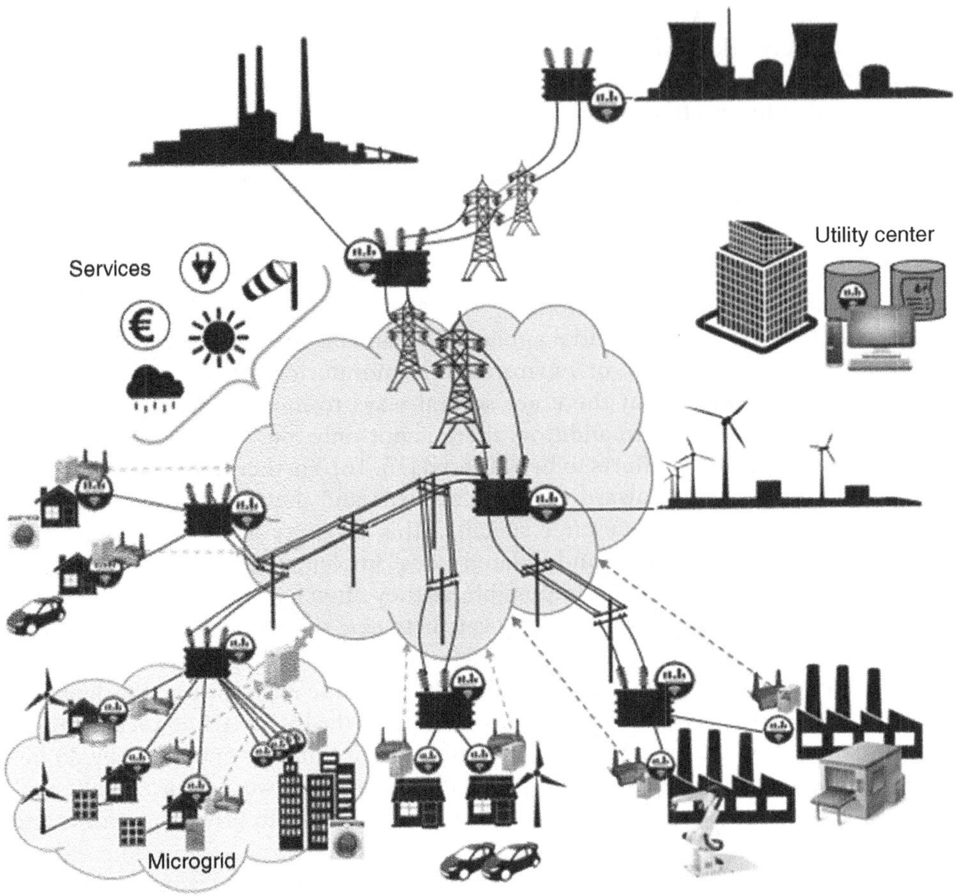

Figure 8.3 Illustration of the next-generation smart grid system.

and the smart meter in order for the end user to have an almost real-time visualization of its consumption, production, and storage (if any). An EMS can also inform the grid of its capabilities (production, load shedding, etc.) via the smart meter. EMSs can operate collectively and depend on a "higher" EMS to form a microgrid – a smaller managed grid, as depicted by the gray cloud. This microgrid will also have its own EMS that will be able to coordinate "lower" ones in order to respond to the grid more efficiently and on a larger scale.

This next-generation grid should involve more consumers and prosumers in the management decisions of their own sites. In order to optimize consumption and production both at the premises and on a global level, it is crucial to simultaneously utilize inputs from a) the grid, such as pricing and demand; b) the user, providing its needs and requirements; c) smart appliances, informing on their specificity and capabilities – e.g., shedding potential. Consequently, all these data are combined in order to determine the best usage for given smart appliances. Moreover, being connected to the Internet, this system will also have access to several services and information coming from non-energy related connected objects (such as outside temperature, weather forecasts, etc.) that might also help in the decision making. Nevertheless, in order to reach such a complex system, several challenges need to be addressed.

8.2.3.1 Technical Issues for Next-Generation Smart Grids

The number of smart objects is expected to explode in the near future [17]. Many of them will be electrical appliances, leveraging connectivity to provide daily-life services to their owners and possibly to the grid (through load shedding or storage for instance). This plethora of smart devices will use different types of access technologies, protocols, and information formats. As a consequence, this next-generation system should be flexible, adaptable, and dynamic and will enable automation.

In order to support scalability, this diversity of devices has to be handled locally. It will first lower the complexity of such system and then enable local management. However, smart meters currently deployed in the AMI have not been designed for managing and aggregating data on behalf of several appliances, nor for enabling such elaborated services – i.e., collecting input from users and from other devices/services in order to analyze data and control appliances.

8.2.3.2 Handing Back the Keys to the User: Energy Management Should Be Separated from the Smart Meter

We insist here on the distinction between the devices interacting with the user (to make decisions within its area) and the devices used by the utility to monitor and bill the area consumption, as well as inform about the grid demands.

In order to enable a more effective energy management, end users have to be more involved with respect to their data as well as to make decisions about energy usage. Their involvement should not anymore only be on a monitoring basis. They have to understand how they consume (and produce) and at the same time how they can act for both their own and the grid benefits. Therefore, they will be part of local decision making regarding the balance of their own consumption and production, and/or at the same time they could participate in the balance optimization of the whole grid by shifting, postponing, or switching off their appliances upon request. Therefore, the next-generation smart grid architecture should provide users with the proper tools to control both their data and energy usage.

In order to realize all these actions, a dedicated management equipment is required. The smart meter could play this role if it can be equipped with new functionalities (storage, analysis, and user interactions). However, as previously mentioned, this will initially be more costly for the DSO and also implies that the user could possibly alter the functioning of the smart meter through interactions. Let us recall that the main purpose of smart meters is to ensure billing and planning for utilities; it is therefore not acceptable to affect the communications with the utility management center. For these reasons, it is very unlikely that the smart meter would act as a local EMS in the future.

On the contrary, it is more probable to see users acquiring an independent EMS in order to have access to energy services. Smart meters would still be in use, first by the DSO to monitor, bill, and send grid-related requests (consumption reduction or increase, load shifting) and furthermore, by the EMS in order to be aware of the site global consumption and possibly to share certain data with the DSO (e.g., consumption forecast). In any case, the per-device decisions should be left to the EMS, either directly operated by the user or by a specific service provider.

This may enable new open markets for energy management in which the users should be able to choose who will get access to their detailed energy data as well as control of their smart appliances. Similarly, the users should be free to decide whether to let a third party manage their consumption/production and if positive, to select one. Switching among "energy management service providers" should be simple enough so that market competition would lead to a desirable outcome.

8.2.3.3 To Build an Open Market, Use an Open Network

As previously stated, the communication network used for energy management should be adaptive, dynamic, flexible, and easily accessible to allow fair competition. The Internet naturally appears as a good candidate as it is built on open standards that provide almost all these features while adapting to smart grid applications through various IoT standardization efforts. The openness of the Internet and its standards seem preferable than letting energy operators define proprietary networks and protocols. The latter approach might potentially lead to several isolated platforms with few potential interconnections. With the use of the right tools, the Internet will provide all the required flexibility to handle a large volume of distributed consumption and production points.

As a result, the next-generation smart grid core network should be decentralized and mainly focus on ensuring the interconnection of these end points, while the intelligence should be distributed toward appliances, equipment, systems, and aggregating units, as we will detail in the following subsection.

8.2.3.4 Multi-Level Aggregation

In this next-generation model, an EMS can aggregate data from different devices of a given site. Communications between EMS and appliances are based on the IoT concept, which specified that all devices must be uniquely identified. An EMS can therefore analyze and process collected data, which results in making decisions and controlling corresponding devices . The same principle could be performed at different levels (a set of EMSs could be managed by another one). Actually, using Internet-based technologies (such as peer-to-peer tools) simplifies the communication between different EMSs. As a result, it allows to define several types of aggregation levels, corresponding to different scales for the appliance management grain as well as for global energy balance and economic relationships.

For instance, a hierarchical and geographical aggregation can be defined, where several individual users from the same area gather to form a microgrid. At the individual site level, a fine-grained management can be parameterized, along with an EMS, and accordingly to each individual's preferences. These different sites – with potential storage and/or production capabilities – coordinate themselves and simultaneously provide their general energy information to an EMS taking care of this microgrid. At the end, this smaller grid is seen as a single entity to the rest of the grid. The microgrid EMS manages lower-level EMSs. It might receive demands from the grid and determine if and how it can satisfy them, using available storages within the microgrid or in turn sending demands to lower EMSs. Nevertheless, the shedding potential of this sites' gathering is more important than each individual site.

Another example is to define an aggregation of appliances belonging to different sites. In fact, controlling one appliance may not be significant at the grid level. However, remotely aggregating the flexibility supply from many appliances offers a great potential – e.g., with a fleet of EVs.

The IoT paradigm is therefore allowing to gather all devices, from different areas but with similar behavior. These aggregation mechanisms offer more significant potential than what each individual could offer. Such next-generation grid communication systems are encouraging the emergence of FOs that sell negative energy (also called "negawatts") – i.e., selling flexibility in consumption, load shedding capacity, or vehicle-to-grid transfer system [5]. Additionally, an open network simplifies the operation of new markets, since the layered structure of the Internet allows to create new services upon existing ones. Actually, the proper incentives can be computed based on the available data and transmitted to actors at each envisioned scale.

Therefore, the main elements for such model are the EMSs that have data storage, analysis, and management capabilities as well as decision and control functionalities. They rely on available smart appliances and smart meter data in order to efficiently optimize the energy balance of their managed group. The same communication technologies listed in Table 8.1 can still be implemented. However, peer-to-peer communication as well as shared directories will be required to ensure an efficient coordination of these elements while reducing the risks of a single point of failure affecting the system.

8.2.3.5 Security Concerns

We also stress the importance of security required in such smart grid systems. While we still regularly observe new attacks occurring on the Internet, and recently targeting connected objects, the redundancy and resilience of the Internet, together with various improvements in Internet security, still make it the best candidate for smart grid applications. In fact, the Internet, and in particular the IoT, benefits from the efforts of many specialists as regards security issues [18].

In any case, this next-generation system will be required to use the following building blocks for providing security:

- Authentication: identifying a node before letting it access the system;
- Authorization: process to determine if a node has the authorization to perform certain tasks;
- Access control: ensuring that a node can access certain resources (data and/or devices).

For instance, these mechanisms should help the system ensure that a given third party is really who it claims to be, that it can interconnect with other parties of the system, and perhaps, that it can access data from a given EMS.

8.2.3.6 Ongoing Research Efforts

Research efforts are required in order to study, standardize and test such next-generation smart grid communication systems. Before being adopted, they have to demonstrate their benefits and above all that they are considered secure for both the end users and utilities.

In the IoT paradigm, several groups are trying to standardize a common IoT architecture. Standardization and uniformity are required in order to be able to collect information from different application domains and cross them in order to enhance the system.

For instance, machine learning can be used to determine predictive information based on gathered data. But, as aforementioned, the IoT landscape is currently fragmented, and standardization efforts are scattered. As a matter of fact, most of the IoT architectures currently used are deployment-specific and do not enable simple interconnection. In the following, we cite certain IoT architectures:

- *oneM2M*: A global organization composed of standardization bodies as well as industries. It was created to provide technical specifications addressing the need for a common architecture to connect the myriad of machine-to-machine (M2M) devices. oneM2M *Functional Architecture* [19] mainly focuses on the collection of data from field nodes and interconnection of M2M systems. This group is still active and have ongoing research on semantics, which should provide automation to their Functional Architecture. However, it does not plan to have end users interacting with end nodes.
- *Internet of Things - Architecture* (IoT-A): A European research project that aims at providing a reference architecture model for IoT applications and business. This project develops guidelines (common understanding, common grounding, standardized interfaces, and best practices) and a reference model for building compliant IoT solutions. The resulting *Architecture Reference Model* (ARM) [20] helps to build and interconnect systems. However, it carries on the silos model.
- *Industrial Internet Consortium* (IIC): An organization composed of several companies created to promote open standards and interoperability for technologies used in industrial and M2M environments. They designed an architecture, *Industrial Internet Reference Architecture* (IIRA) [21], enabling the set up of industrial IoT systems, which will be compatible and interoperable with other industrial systems. It is based on specific publish-subscribe mechanisms that help the interconnection ensure reliability, performance, and scalability. However, IIRA supposes that all deployments have a centralized management domain.
- *Smart Energy Aware Systems* (SEAS): A European project aiming to provide the ICT tools to interconnect energy actors in order to better manage, coordinate, and optimize energy consumption, production, and storage. The proposed *SEAS reference architecture model (S-RAM)* [22] derives from oneM2M and is based on distributed core services to interconnect both energy actors and management systems. These core services should take care of the following:
  - finding other parties in a simple way;
  - automatically learn from them using semantics;
  - ensuring security of the system;
  - ensuring monetary compensation for compliance to commitments.

  This model defines a way to divide the system in various groups, which therefore provides different management levels.
- *Alliance for Internet of Things Innovation* (AIOTI): An alliance initiated by the European Commission based on the observation that there was no common European IoT market. The aim of this alliance is to strengthen interaction among IoT players in Europe and to contribute to the creation of a dynamic European IoT ecosystem. In order to meet this target, they suggest to define a *High Level Architecture*[3], providing minimal requirements, using semantics, and compatible with previous architectures.

3 No document published at the time this chapter was written

All the above approaches attempt to provide a common reference model for IoT-related solutions, apart from SEAS, which is dedicated to the energy domain – but S-RAM could be used for other application domains.

The next-generation smart grid systems that would interconnect AMI with an Internet-based communication network might probably rely on one of these architectures. The most promising solutions from the ones cited previously would be a) IoT-A, but it will not provide the openness required to support open energy market; and b) SEAS, as it appears to provide all the required building blocks.

No matter which solution is chosen to support such next-generation smart grid systems, it would have to demonstrate the advantages of such a complex system through research and test-bed implementations. In the next part, we give more details on the communication framework and especially the routing part that nowadays allows smart meters to communicate with the utility center in an efficient and reliable way.

8.3 Routing Information in the Smart Grid

In smart grid networks, the majority of smart meters are located in the neighborhood area network (NAN), without having direct communication with the data concentrator. As a result, routing paths need to be established for smart meters to reach the data concentrator (sink) of the network.

An AMI connects the wide area network (WAN) and the NANs, acting as a gateway between the smart meters and the utilities. As it was presented in Section 8.2.1.2, those networks mostly utilize technologies that are sensitive to perturbations and could be considered as low power and lossy networks (LLNs). Considering the fact that most of the smart meters that are used in AMI networks come with a single communication interface, the path to the the sink needs to be carefully designed, as no redundancy exists.

In narrowband PLC and wireless networks, most of the nodes that exist in the infrastructure cannot communicate directly with the sink or with other nodes due to the limited transmission capacity (long distance, external interference, and noise). Therefore, the nodes need to collaborate together to forward the data packets to the final destination. Similarly, in a smart grid network, the nodes are the smart meters that route metering information to the data concentrator.

Typically, a routing protocol constructs and maintains the best paths in the network for the packets to be routed toward the destination. To do so, routing protocols propagate a routing information message using either proactive or reactive models.

A high number of hops degrades the network performance as it introduces additional delay in reactive routing or additional overhead in proactive approaches. To minimize the impact of routing, it is essential to minimize the number of hops in the network. However, it is also important to carefully select the optimal path to the destination according to an objective function and appropriate metrics. Note that the shortest path is not always the optimal solution, i.e., the expected transmission count (ETX) is a popular metric in IoT networks.

8.3.1 Routing Family of Protocols

Before the emergence of LLNs, several routing protocols have been presented and studied. However, none of them was meeting the requirements of such wireless and lossy networks. Levis et al. (2008) show in[23] that most of the routing protocols do not fulfill the requirements of LLN such as footprint or maximum transmission unit (MTU) limitation [24]. In the smart grid network community, mainly two routing protocols have emerged and are widely studied and deployed. J.Yi et al. (2013) [25] present how critical the routing protocol is for smart grid networks and how routing protocol for LLNs (RPL) and lightweight on-demand ad hoc distance-vector routing protocol next generation (LOADng) could tackle the specifics of smart grid applications. These protocols have routing metrics that deal with the characteristics of LLN that make them specifically suitable for smart grid networks. For instance, the link reliability metric is not used in the Internet routing protocols because technologies employed in such networks are extremely reliable, and fast recovery mechanisms exists for failure. But in LLN, taking into account the reliability of the links to build the path is significant because link quality quickly changes over time.

In addition, using the node's energy consumption as a metric allows to consider how the node is powered and what is its remaining lifetime. Such metric is a key enabler to enhance the lifetime of a wireless network where devices are mostly battery operated. Hereafter, we provide a detailed description of two leading families of routing protocols, based on the propagation of the routing information in the network, namely the proactive and the reactive routing protocols, respectively. Moreover, we present a performance comparison of the most popular routing protocols such as RPL, ad hoc on-Demand distance vector protocol (AODV), and LOADng for LLNs.

8.3.1.1 Proactive Routing Protocol

In proactive routing protocols, routes are built *a priori*, and as a result, all nodes in a network are aware of the routes to any destination at any time. Thus, a node may transmit a data packet to any destination with no delay, since all routes are stored in the routing tables. However, periodic routing-related control packets need to be transmitted to maintain the routing table updated. Furthermore, to control the network overload, the periodicity of these control packets must be accurately defined.

RPL[26] is today the main protocol in the proactive family of routing protocols chosen in LLN. It is actually a distance vector routing protocol specified by the Internet Engineering Task Force (IETF) ROLL working group [27]. RPL is defined as link-layer agnostic, so it can operate over wireless or PLC networks for example.

8.3.1.2 Topology Management under RPL

In a LLN, the topology is not predefined and, thus, RPL is in charge of discovering and carefully selecting nodes in order to construct optimal routes. The topology is organized based on a directed acyclic graphs (DAG), a graph where the connections between nodes have a direction and a non-circular property. Based on the "acyclic" nature of the DAG, the graph comprises at least one root, a node with no outgoing edge. In Figure 8.4 (a), a DAG composed of ten nodes and three DAG roots is illustrated. To construct a routing topology, RPL employs an extension of DAG: the destination oriented DAG (DODAG), which is similar to DAG with a single DAG root. In a smart grid scenario, the root of a RPL network could be the data concentrator that gathers the metering information. Figure 8.4 (b) depicts a DODAG topology that consists of eight nodes with one root.

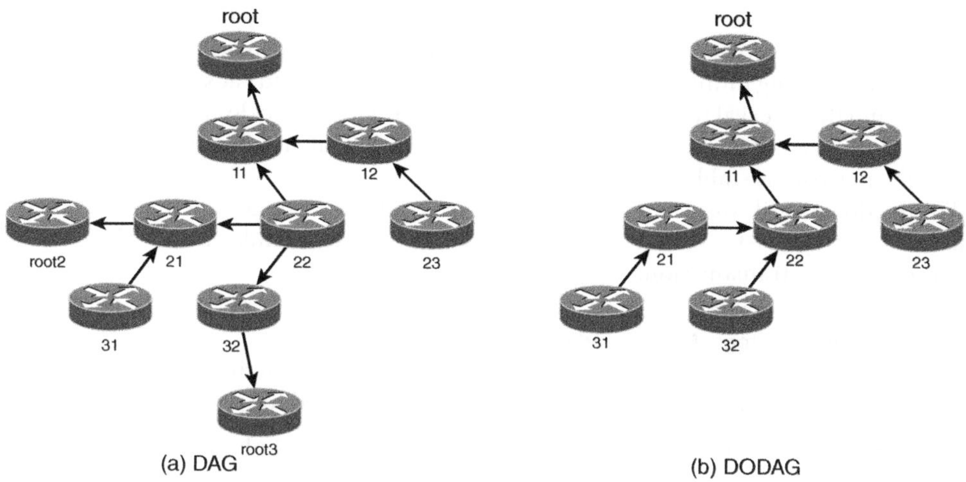

Figure 8.4 Example of a DAG and a DODAG.

To establish and maintain routes, RPL uses three different types of ICMPv6 control packets:

- DAG information object (DIO)
- DAG information solicitation (DIS)
- destination advertisement object (DAO)

The upward route construction, the one used between smart meters and the core network, is managed by transmitting DIO messages in multicast. DIO messages contain information that allows discovering a RPL instance, calculating its own rank and choosing parents in the DODAG. The rank contained in the DIO message is the rank of the node sending the DIO message and determines the relative position of a node in the DODAG. The rank is computed by the objective function using routing metrics, and its purpose is to avoid loops. The downward route construction, which is optional in RPL, is managed by the DAO messages to propagate information about the destination in the upward direction. To construct the downward routes, there are storing and non-storing modes. Finally, DIS control packets are utilized to solicit a DIO message from a RPL node.

8.3.1.3 Routing Table Maintenance under RPL

As previously stated, DIO messages are periodically transmitted to build and maintain the RPL DODAG. However, if the network is stable, the DIO message frequency is decreased to reduce the overhead of signaling messages. On the contrary, if the condition of the network is not stable, more DIO messages have to be transmitted. This timing function is called trickle timer [28]. If a received DIO message does not imply any change on the receiver in terms of rank, parent set, or preferred parent, the DIO is considered consistent. As long as consistent messages are received, the interval between DIO messages is exponentially doubled to reduce the overhead of periodic messages. Conversely, when the network is not stable and DIO messages are inconsistent with the known topology, more DIS and DIO messages are needed to update the node routing tables. Messages such as multicast DIS without a solicited information option or DIO messages containing infinite rank are considered inconsistent and cause the trickle timer to reset, and the interval time is set to its minimum value. The trickle algorithm allows to be reactive in case of a change or failure in the network while minimizing the overhead when the network is stable.

For the downward route construction, a DelayDAO is sent to govern the emission of the DAO messages. At each transmission of a DAO message, a random interval is chosen before the actual transmission.

8.3.1.4 Routing Strategy: Metrics and Constraints

A metric in RPL is a quantitative value, and it is used to evaluate the path cost. Vasseur et al. (2012) [29] define two kinds of metrics that can be used for path calculation:

- the **link metric**, which concerns the link's attributes, e.g., link quality level (LQL), ETX, latency, and throughput; and
- the **node metric**, which takes into account the node state and attribute (NSA) such as energy (remaining energy, power source) or min-hop (number of hops to the root).

RPL supports also a constraint-based routing where the constraint may be applied on both link and nodes. If a link or a node does not satisfy a constraint, it is discarded from the parent set.

This constraint is used to include or eliminate a link or a node that does not meet a specific criterion. For instance, the objective function will not choose a path that traverses a node that is battery powered or a link with low ETX. The RPL objective function could combine metrics and constraints to compute the best path.

8.3.1.5 Path Computation under RPL

To compute the optimal path, the objective function plays a major role in the RPL protocol. To this aim, the two following algorithms have to be defined:

- the computation of the node's rank according to one or several metrics; and
- the parent selection operation according to metrics and constraints.

Two objective functions have been defined by the ROLL working group: objective function zero (OF0) and minimum rank with hysteresis objective function (MRHOF) and are presented next.

The Objective Function Zero The OF0 [30] works by computing the rank based on the addition of a scalar, representing the link properties to the rank of the preferred parent. The scalar value is normalized between 1 and 9 for expressing the link properties, with 1 for excellent, and 9 for very poor. Note that any kind of metric could be used for the scalar value. This objective function allows to find the closest grounded root (a root that offers connectivity to the application goal) by selecting a preferred parent and a backup successor if available.

The rank computation is given by the algorithm below:

$$R(N) = R(P) + rank\_increase \tag{8.1}$$

$$rank\_increase = ((Rf * Sp + Sr) * MinHopRankIncrease) \tag{8.2}$$

where:

- $R(P)$ is the preferred parent's rank
- Sp (the step_of_rank), Sr (stretch_of_rank), and Rf (rank_factor) are respectively the expression of the link properties normalized between 1 and 9, the maximum augmentation to the step_of_rank of a preferred parent to allow the selection of an additional feasible successor and a value used to increase the importance of the link properties.
- MinHopRankIncrease is a multiplying factor that plays a major role in the rank computation by reflecting the impact of the metric on the rank increase. The default value is **256**, as it is described in[26].

OF0 parent selection is governed by several rules (see Section 4.2.1 of [30]), but the most important is that the selected parent must be the one that causes the lesser resulting rank for the node. This selected parent becomes the "preferred" parent.

The Minimum Rank Hysteresis Objective Function MRHOF [31] optimizes the path to the root that minimizes a defined metric. However, it avoids changing this path frequently. Light metrics variations cause changes in the network that are decreased by introducing a hysteresis. MRHOF works with additive metrics and introduces the path cost for the rank computation, which specifies the property of the path to the root regarding the employed metric. The path cost is calculated by the sum of the path cost advertised by the parent and the link metric cost to the parent.

The rank computation for MRHOF is given by the algorithm below:

$$path_{cost} = parent_{path\_cost} + link\_cost \tag{8.3}$$

$$rank = func(path_{cost}) \tag{8.4}$$

where:

- $parent_{path\_cost}$ is advertised by the parent and represents the pathcost of the parent; and
- *link_cost* is the cost associated with the parent's link regarding the selected metric.

MRHOF parent selection is governed by a hysteresis function given by the equation below where $P1_{path\_cost}$ and $P2_{path\_cost}$ are respectively the path cost to parent 1 and parent 2. PP is the selected parent designated as Preferred Parent, P1 is the current best parent, and P2 is a candidate parent.

$$PP = \begin{cases} \mathrm{P2} & if \quad P1_{\mathrm{path\_cost}} + Threshold > P2_{\mathrm{path\_cost}} \\ \mathrm{P1} & else \end{cases} \tag{8.5}$$

where *Threshold* is the hysteresis function, i.e., the minimum difference between the cost of the path through the preferred parent and the cost path of a candidate parent to trigger the selection of a new preferred parent. This objective function allows for selection of the route toward the root with the lowest path cost, e.g., minimum hop counts if the hop-count metric is used.

8.3.1.6 Summary of the RPL DODAG construction

Figure 8.5 shows an example of the upward route construction using the hop-count metric. Once the trickle timer is expired, RPL root will broadcast a DIO message containing

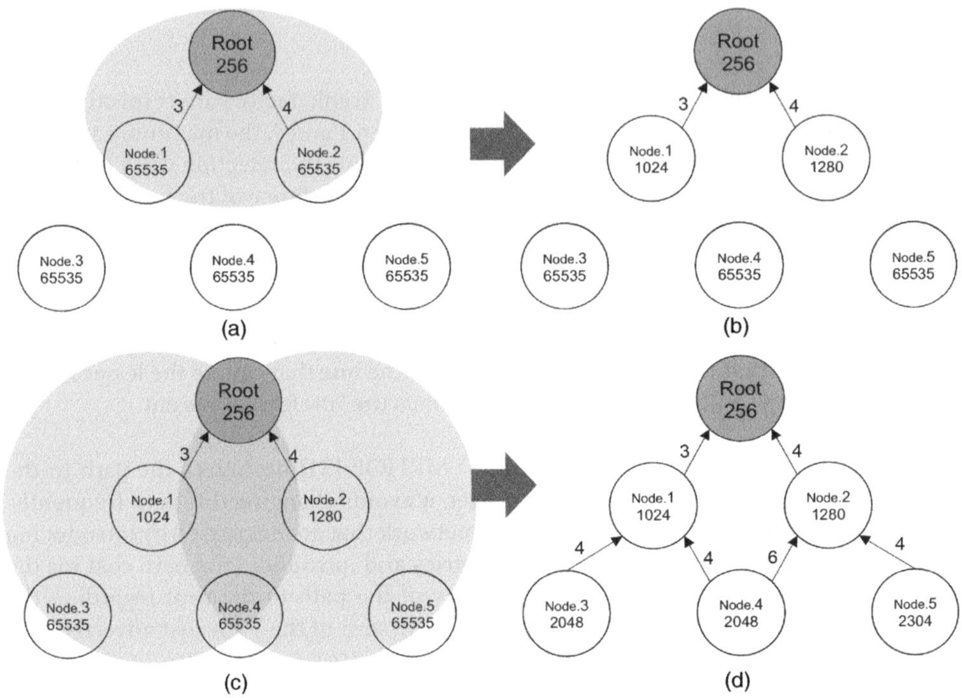

Figure 8.5 Example of an upward route construction with RPL.

its rank. Nodes in the coverage area of the root (i.e., yellow circles) will receive the DIO message and process it. If the DIO message had been corrupted, it would have been discarded. Since the root is the sink of the network, nodes 1 and 2 can not be closer to the root, so they will add the root as their preferred parent and compute their rank. To test if a candidate neighbor is eligible to be a preferred parent, a node will verify if the rank contained in the received DIO message added to a RPL parametric value (min_hop_rank_increase) is less than its rank. Then node 1 and 2 will broadcast their own DIO message with their new computed rank. Note that since the root has a smaller rank than the one advertised in nodes 1 and 2 DIO messages, nodes 1 and 2 will not be considered as potential parents for the root. It is worth mentioning that ranks shown under node names in this example depend on the objective function, and values shown beside edges represent the link quality (i.e., ETX). The arrows between nodes represent the upward route, and when a node installed at least one of them, it is considered to have joined the DODAG. It has to be noted that a node may either stay silent and wait for a DIO message or it may send a DIS message during the initialization process.

8.3.1.7 Reactive Routing Protocol

In a reactive-based routing protocol, routes are built and maintained only when they are requested, which means that there is no need to maintain a route if there is no traffic. Thus, a delay is added before transmitting a data packet due to the route construction. Contrary to proactive protocols, reactive protocols do not need to send routing information periodically and, thus, will require less energy or CPU resources. However, the quantity of routing messages will greatly depend on the frequency of the traffic in the network.

8.3.1.8 Topology Management under AODV

AODV [32] is a well-known reactive routing protocol designed for use in mobile ad hoc networks (MANET). It floods the network with broadcast route-request messages when a needed route that does not exist.

To establish and maintain routes, AODV uses five types of messages:

- RREQ: route request
- RREP: route reply
- RERR: route error
- RREP-ACK: route reply acknowledgment
- HELLO: link status monitoring

When a source node expects to establish a route to a destination, it broadcasts a RREQ packet. Once the destination is reached (or an intermediate node that knows the route to the destination is reached), a RREP message is sent back to the RREQ sender, which ends the route discovery process. If a RREP message is received, the route discovery operation is over. Otherwise, after certain period, it repeats the RREQ message and increases the waiting period. If there is no RREP message, this process can be repeated several times (by default, RREQ_RETRIES = 2). If there is still no response after three attempts, the route search process is aborted. Consequently, a new route request will be initiated after ten seconds. A node receiving a RREQ packet will send a RREP (route reply) packet if it is the destination or if it has a route to the destination with a sequence number greater or equal to the RREQ packet; otherwise, it rebroadcasts the RREQ packet. Each node keeps a trace of the source IPs and the identifiers of the RREQ packets. In case of receiving a RREQ packet that they have already processed, they delete it.

Once the source has received the RREP packets, it can start sending data packets to the destination. If the source subsequently receives a RREP containing a higher or equal sequence number but with a smaller number of hops, it will update its routing information to that destination and start using the best route. A route is maintained as long as it continues to be active, in other words, as long as data traverse between the source and the destination. The link expires when there is no more data in transit on the link and after a predefined delay. If the link is cut, the end node sends a RERR (route error) packet to the source node to warn that the destination is currently unreachable. If the source node still wants to get a route to that destination, it must start the route discovery process again.

Concerning the routing table, each entry contains nine fields. In addition to IP address of the destination node, the fields contain routing information and information related to the qualitative state of the route for maintenance purposes. Unlike other protocols, AODV only maintains information about the next hop in the route, not the entire routing list. This saves memory and decreases overhead for route maintenance. The routing table also contains information enabling the host to share information with other nodes when link states change. To ensure the information is the latest one available in the route table entry, a sequence number for the IP address is included in the message. This sequence number is called the "destination sequence number." It is updated each time a node receives a RRER, RREP, or RREQ message.

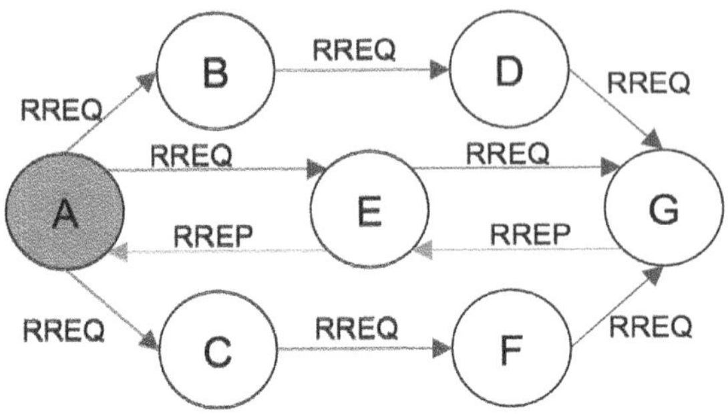

Figure 8.6 Example of an AODV route detection between node A and G.

To offer connectivity information, nodes that are part of an active route can broadcast local Hello messages. Every HELLO_INTERVAL, the node will check if it has sent a broadcast message during the last interval, and if it has not, it will broadcast a RREP message with a TTL set to 1. Within a dedicated period, if a node that has received a Hello message from a neighbor does not receive any packet from that neighbor, the node will assume the link is lost and will send a RRER route error message. Figure 8.6 shows a route search on the initiative of the node A in the direction of J. The RREQ message is broadcast from node A to all its neighbors. When node G receives the message, it returns a RREP message to node A through node E.

- The RREQ route request message is sent to search for available routes; it is made of frame size 24 *bytes* in length.
- The RREP route response to demand message is sent to indicate available routes to the originator of the demand; the frame consists of 20 *bytes*.
- The RRER is sent to report routes with potential errors to the originator of the demand; it consists of 20 *bytes*.

A route reply acknowledgment (RREP-ACK) message is sent in response to a RREP message with the 'A' bit set to 1 when there is danger of unidirectional links preventing the completion of a route discovery cycle. It indicates that another available route is already used.

8.3.2 Reactive Routing Protocol in a Constrained Network

Several proposals emerged to simplify and adapt AODV for LLNs. In 2011 and 2012, with the use of an adaptation of AODV in G3-PLC standard in smart grids networks, a single LOADng specification emerged, as the next version of AODV.

6LoWPAN Ad Hoc On-Demand Distance Vector Routing (LOAD) and Lightweight On-demand Ad Hoc Distance-vector Routing Protocol Next Generation (LOADng) are both routing protocols based on the AODV reactive routing protocol. LOADng is the latest version of LOAD, where many features have been reviewed to make LOADng more efficient and extensible. In LOADng several extensions have been included to improve the performance under specific scenarios such as LLN, by reducing the network overhead. Thus, LOADng, LOAD, and AODV share many common points:

1) A node that has data to transmit to a destination but has not any information related to this destination in its routing table: it sends a RREQ message. If the message is received by a node that has been already transmitted by itself, it will discard the

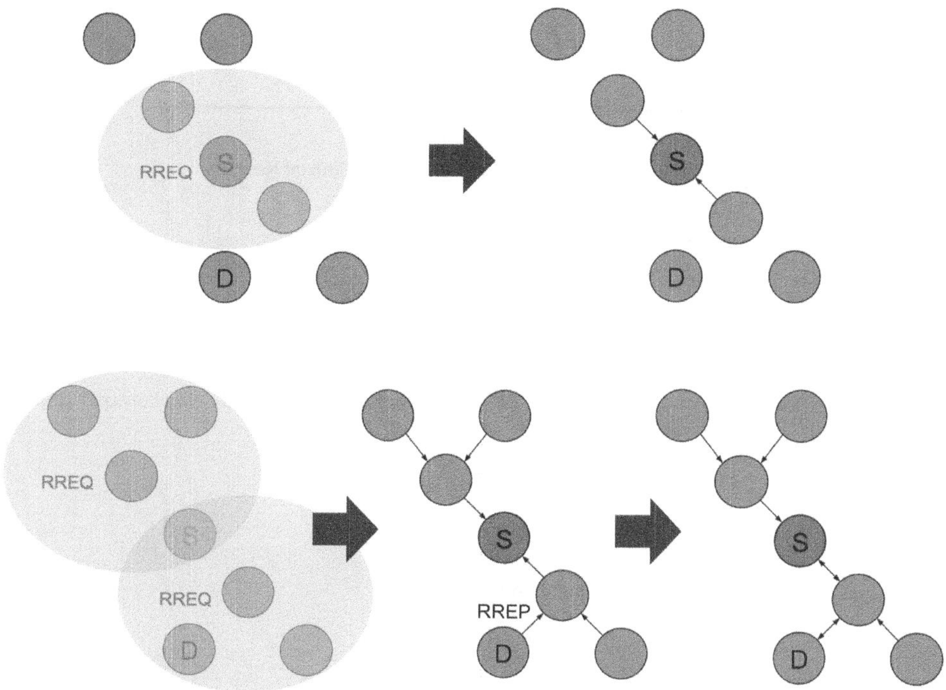

Figure 8.7 Example of a route construction with LOADng.

message to avoid breaking existing routes. Intermediate nodes construct the reverse route upon the reception of the RREQ message.

2) When the destination node receives the RREQ message, it can generate a RREP message immediately and, thus, minimize the time to establish the path. It could also wait to receive several RREQ messages with better metric to optimize the path at the cost of a longer path establishment delay.
3) To detect broken or asymmetric links, intermediate nodes can request an acknowledgment during the forward route to the destination construction.
4) Except for LOADng, when a node is no longer able to forward packets to the next hop, a local repair mechanism is triggered to solve the problem. In case the local repair mechanism fails, a RRER message is sent to the originator of the message.

In Figure 8.7 an example of route construction using LOADng routing protocol is illustrated.

8.3.2.1 Performance Evaluation

As LOAD and RPL are both specifically designed for LLN, hereafter, we will present a performance evaluation comparison of these two protocols, Table 8.2 summarize their specificities.

Comparing RPL and LOAD will mainly depend on the topology (e.g., density). In a stable network, the round-trip time for a data request (i.e., end-to-end delay) will tend to be better with RPL, due to the time needed to build the path using the LOAD RREQ message. Thus, thanks to the trickle timer, RPL will decrease significantly the

Table 8.2 LOADng / RPL comparison.

Protocol	LOADng	RPL
Type	On-Demand	Proactive
Algorithm	Distance Vector	Distance Vector Source Routing
Local repair	Yes	Yes
Mobility	Static, Mobile	Static, Mobile
Scalability	High	High
Supported traffic	P2P	P2P, P2MP, MP2P

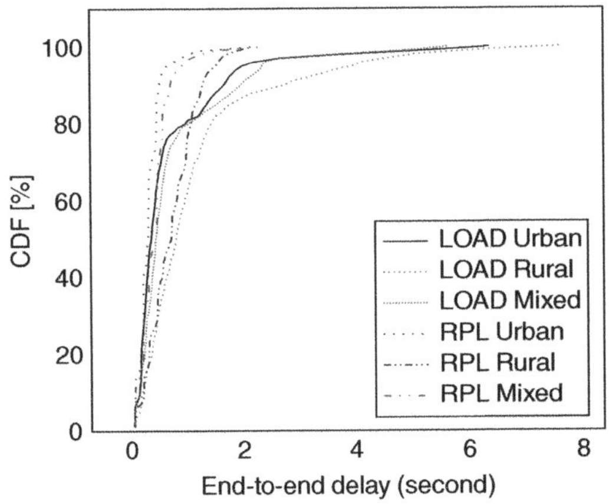

Figure 8.8 End-to-end delay comparison.

control traffic as long as the network stays stable. When the condition of the network evolves unexpectedly (i.e., a node loses its parent, a path cost changes), RPL will reset the trickle timer and send more DIO messages to recompute the DODAG. This explains the additional control traffic in RPL, and may result in a broadcast storm, caused by the issue of DIO messages with increased DODAG version number (global repair). A global repair of the DODAG is triggered by the root. It re-computes the DODAG and increases the end-to-end delay where LOAD will be less affected by the network variation. In smart grid applications, end-to-end delay tolerance could vary from below 10*ms* to more than two seconds (e.g., smart meter reading). On the other hand, a teleprotection, for instance, which ensures the protection of network equipment from severe damage by managing the grid load, requires fast signals to pilot protective relays: no more than 10*ms* [33]. Figure 8.8 shows the end-to-end delay comparison between LOAD and RPL under different smart grid scenarios.

Since in LLNs, nodes have constrained memory, smart meters will only store a dozen entries, whereas the routing table usually contains hundred of thousands of routes in IP core routers. As a consequence of flooding, each smart meter in a LOAD network receiving a RREQ message will install a route toward the sender, resulting in a large number of unnecessary routing entries. The same issue occurs when a node is situated on a route of a RREP message. On the contrary, most routers in RPL network have the default entry toward the preferred parent. However, when RPL operates in storing-mode, nodes that are chosen as preferred parent have to store the downward route and may cause critical issues such as loops, in case a node runs out of energy.

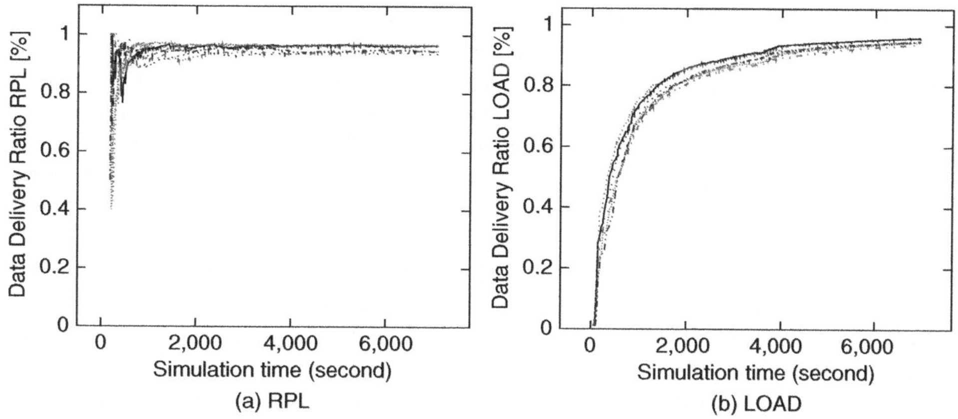

Figure 8.9 Data delivery ratio comparison.

Concerning the path efficiency, since RPL computes a DODAG from a sub-topology of the physical network, the traffic has to follow paths along the DODAG even if a more optimal path exists in the physical world. Those protocols produce a sub-optimal solution, which can be improved by carefully selecting parameters for the metrics used to arbitrate the chosen links. For instance, LOAD uses the LQI (link quality indicator) of the 6LoWPAN physical layer in addition to the Hop distance.

Figure 8.9 shows the root data delivery ratio of a 100-node topology for RPL and LOAD, respectively, after two hours of simulation. As the traffic is set to start when the simulation initiates, RPL demonstrates additional delay before the actual data packet reception. However, RPL attains high performance once the DODAG is established. Concerning LOAD, results indicate that data is received quickly once the network is initiated; however, it takes time for LOAD to reach the same DDR as RPL.

In RPL, packets are sent only after the DODAG is constructed. Thus, if the metric chosen for constructing the DODAG is hop count, then RPL will compute a DODAG with minimum hops. Due to the flooding mechanism of LOAD, nodes construct the path using the first RREP message arrived, which is not necessarily the optimal one in terms of hops. The packet will follow a non-optimal route until subsequent RREP message reception to update the path.

In terms of overhead, differences between the protocols will mainly depend on the implementation and employed parameters. The stability of the network has a significant impact on RPL protocol, since its parameters should be carefully selected to handle specific network circumstances. In LOAD, route hold time (RHT) will greatly impact the frequency of the flooding and, consequently, will increase the overhead. The number of nodes in the network is also critical in LOAD since high density in the network will increase the overall overhead. In RPL, we expect the maximum overhead at the beginning of the DODAG construction and then a reduction as the network becomes stable, due to the behavior of the trickle algorithm.

8.3.2.2 Summary on Routing Protocols

Choosing between reactive and proactive routing protocols in a smart grid network depends on multiple factors. The application, which identifies the type of traffic, has a major role in choosing the routing protocol and its corresponding parameters. Several parameters will also depend on the density of smart meters and the type of topology, i.e., number of maximum hops to the root. Furthermore, the priority of the traffic has an impact on the routing strategy as well, i.e., if the application is tolerant to high end-to-end delay. However, each protocol has different implementation issues attaining, thus, different performances. As a result, several parameters need to be properly configured in order to satisfy the requirements of the considered network and application.

RPL, for instance, is known to work well in multi-point-to-point applications, a typical scenario in smart grid network, where data concentrators will receive the data from a large amount of smart meters in the NANs. LOADng address 6LoWPAN ad hoc on-demand distance vector routing (LOAD) multi-point-to-point issue and offer a similar performance to RPL at the cost of delay for the route discovery process. In a smart grid scenario, for typical monthly readings, the delay could not be a critical issue, so both protocols could be chosen depending on the acceptable overhead. The frequency of the traffic is also a major issue in smart grid networks, where a stable routing graph, such as RPL constantly maintained DODAG will greatly impact the delay at the cost of energy consumption. On the other hand, if the smart grid traffic is sparse, the need of maintaining a routing graph at the cost of high control traffic is not essential. For example, millions of smart meters using RPL have been installed in California mandated by the California Public Utilities Commission (CPUC) while French Enedis has chosen LOADng for the widely deployed Linky smart meters in France.

Furthermore, today smart grid devices have multiple heterogeneous communication interfaces, which leads to hybrid networks. Such a feature allows to enhance reliability and robustness by taking advantages of all available technologies (i.e., PLC and 802.15.4) [34].

8.4 Conclusion

This chapter was devoted to present and discuss both infrastructures and communication architectures as well as technologies and protocols employed in smart grid systems. We especially detailed the necessary requirements next-generation smart grid systems should address in order to enable dynamic and evolving architectures. In addition, we compared certain existing routing families that efficiently fulfill the requirements of a constrained-based smart grid environment.

Current smart grid systems, which rely on advanced metering infrastructure (AMI), are facing several issues as prosumers demands are constantly growing while simultaneously production capacity is less predictable due to the increasing popularity of local renewable production units (i.e., a decentralized energy production system). In order to have a better management of the network, utilities require tools to forecast both consumption and local production of end points, as well as mechanisms to remotely control them. As a consequence, current smart grid systems need to open up to Internet-based architectures. They will therefore move from a centralized to a distributed configuration in which end users could be more involved and that might have different management levels.

Such configuration requires dedicated energy management system (EMS), which will collaborate with both end users and smart meters deployed by a DSO, in order to locally handle the management of a given set of nodes. Using such EMSs will help any provider deploy new energy services while ensuring users' data privacy. However, security concerns arise from this configuration and must be considered along with its deployment.

Regardless of the configuration chosen, smart grid networks mostly remain on constrained devices used in large topologies As a result, we also insist in this chapter in presenting two families of routing protocols that efficiently manage the routing issues in such topologies. Proactive routing protocols, such as RPL, construct routes *a priori*, i.e., before they are required, and all nodes are perfectly aware of the path to any destination. Applications that need low end-to-end delay will benefit from such a proactive protocol, as once the routing graph is built, traffic could be sent without any additional time. On the other hand, reactive routing protocols, such as LOAD, construct the routes only when required. Therefore, these protocols allow applications to overcome any failure in the network without waiting for a complete reconstruction of the routing graph. However, they add delay in the data traffic.

References

1 Eurelectric (2015), Power Statistics and Trends: The Five Dimensions of the Energy Union.

2 Lefrancois, M., Habault, G., Ramondou, C., and Francon, E. (2016) Outsourcing Electric Vehicle Smart Charging on the Web of Data, in *GREEN 2016 IARA International Conference.*

3 Ouya, A., Martinez De Aragon, B., Bouette, C., Habault, G., Montavont, N., and Papadopoulos, G.Z. (2017) An Efficient Electric Vehicle Charging Architecture based on LoRa Communication, in *Proc. of the IEEE International Conference on Smart Grid Communications (SmartGridComm).*

4 Habault, G., Lefrancois, M., Lemercier, F., Montavont, N., Chatzimisios, P., and Papadopoulos, G.Z. (2017) Monitoring Traffic Optimization in Smart Grid Architectures. *IEEE Trans. Industrial Informatics.*

5 Tsoleridis, C., Chatzimisios, P., and Fouliras, P. (2016) *Vehicle-to-Grid Networks: Issues and Challenges*, CRC Press, pp. 347–369.

6 Bush, S.F. (2014) *Smart Grid: Communication-Enabled Intelligence for the Electric Power Grid*, Wiley-IEEE Press.

7 Ashton, K. (2009), "That 'Internet of Things' Thing", http://www.rfidjournal.com/articles/view?4986.

8 IEEE Computer Society (2016), IEEE Draft Standard for Information Technology-Telecommunications and Information Exchange Between Systems-Local and Metropolitan Area Networks-Specific Requirements-Part 11: Wireless LAN Medium Access Control (MAC) and Physical Layer (PHY) Specifications: Amendment 2: Sub 1 GHz License Exempt Operation.

9 LoRa Alliance (2015), LoRaWAN Specification v1.0, https://www.lora-alliance.org/portals/0/specs/LoRaWAN%20Specification%201R0.pdf.

10 3GPP (2016), LTE Category M (Cat M). Release 13.

11 IEEE Computer Society (2016), IEEE Standard for Low-Rate Wireless Personal Area Networks (LR-WPANs), IEEE Std 802.15.4-2015 (Revision of IEEE Std 802.15.4-2011).

12 IEEE Computer Society (2012) IEEE Standard for Local and metropolitan area networks–Part 15.4: Low-Rate Wireless Personal Area Networks (LR-WPANs) Amendment 3: Physical Layer (PHY) Specifications for Low-Data-Rate, Wireless, Smart Metering Utility Networks.

13 Galli, S., Scaglione, A., and Wang, Z. (2011) For the Grid and Through the Grid: The Role of Power Line Communications in the Smart Grid. *Proceedings of the IEEE,* **99** (6), 998–1027.

14 Lawton, G. (2004) "Machine-to-machine Technology Gears up for Growth". *Computer,* **37** (9), 12–15.

15 Fan, Z., Kulkarni, P., Gormus, S., Efthymiou, C., Kalogridis, G., Sooriyabandara, M., Zhu, Z., Lambotharan, S., and Chin, W. (2013) Smart grid communications: Overview of research challenges, solutions, and standardization activities. *IEEE Communications Surveys & Tutorials,* **15** (1), 21–38.

16 Chatzimisios, P., Stratogiannis, D., Tsiropoulos, G., and Stavrou, G. (2013) A survey on smart grid communications: from an architecture overview to standardization activities. *Handbook of Green Information and Communication Systems.*

17 Camhi, J. (2015), BI Intelligence projects 34 billion devices will be connected by 2020, http://www.businessinsider.com/bi-intelligence-34-billion-connected-devices-2020-2015-11?IR=T.

18 Russell, B. and van Duren, D. (2016) *Practical Internet of Things Security,* Packt Publishing.

19 oneM2M (2015), Functional Architecture, http://www.onem2m.org/images/files/deliverables/TS-0001- Functional_Architecture-V1_6_1.pdf.

20 IoT-A (2013), Final Architecture Reference Model, http://www.iot-a.eu/public.

21 Industrial Internet Consortium (2015), The Industrial Internet Reference Architecture, https://www.iiconsortium.org/IIRA-1-7-ajs.pdf.

22 Habault, G., Bonnin, J.M., and Hursti, J. (2016) Defining a Distributed Architecture for Smart Energy Aware Systems, in *CSD&M 2016 : 7th International Conference on Complex Systems Design & Management.*

23 Vasseur, J.P. (2008) Overview of Existing Routing Protocols for Low Power and Lossy Networks, *Internet-Draft draft-levis-roll-overview-protocols-00,* Internet Engineering Task Force. URL https://datatracker.ietf.org/doc/html/draft-levis-roll-overview-protocols-00, work in Progress.

24 Vasseur, J.P. and Cullerot, D.L. (2007) Routing Requirements for Low Power And Lossy Networks, *Internet-Draft draft-culler-rl2n-routing-reqs-01,* Internet Engineering Task Force. URL https://datatracker.ietf.org/doc/html/draft-culler-rl2n-routing-reqs-01, work in Progress.

25 Yi, J., Clausen, T., and Igarashi, Y. (2013) Evaluation of routing protocol for low power and lossy networks: Loadng and rpl, in *IEEE Conf. Wireless Sensor (ICWISE),* IEEE, pp. 19–24.

26 Winter, T., Thubert, P., Brandt, A., Hui, J., Kelsey, R., Levis, P., Pister, K., Struik, R., Vasseur, J., and R., A. (2012), RPL: IPv6 Routing Protocol for Low-Power and Lossy Networks, RFC 6550.

27 Ancillotti, E., Bruno, R., and Conti, M. (2014) Reliable data delivery with the IETF routing protocol for low-power and lossy networks. *IEEE Trans. Industrial Informatics*, **10** (3), 1864–1877.

28 Levis, P., Clausen, T., Hui, J., Gnawali, O., and Ko, J. (2011), The Trickle Algorithm, RFC 6206.

29 Vasseur, J., Kim, M., Pister, K., Dejean, N., and Barthel, D. (2012), Routing metrics used for path calculation in low-power and lossy networks, RFC 6551.

30 Thubert, P. (2012), Objective Function Zero for the Routing Protocol for Low-Power and Lossy Networks (RPL), RFC 6552 (Proposed Standard).

31 Gnawali, O. and Levis, P. (2012), The Minimum Rank with Hysteresis Objective Function, RFC 6719 (Proposed Standard). URL http://www.ietf.org/rfc/rfc6719.txt.

32 Perkins, C., Belding-Royer, E., and Das, S. (2003), Ad hoc On-Demand Distance Vector (AODV) Routing, RFC 3561 (Experimental). URL http://www.ietf.org/rfc/rfc3561.txt.

33 (2009) Wireless Systems for Industrial Automation: Process Control and Related Applications (ISA100.11a), International Society of Automation (ISA), Sep. 2009. *ANSI/ISA-100.11a-2011*, pp. 1–792.

34 Lemercier, F., Montavont, N., Toutain, L., Vijayasankar, K., Vedantham, R., and Itron, P.C. (2016) Support for hybrid network in RPL, in *SmartGridComm*, IEEE, pp. 527–532.

10

A Prediction Module for Smart City IoT Platforms

Sema F. Oktug, Yusuf Yaslan and Halil Gulacar

Computer Engineering Department, Istanbul Technical University, Istanbul, Turkey

10.1 Introduction

Today, nearly all types of sensors have a wireless network connection with data sending and receiving capacity, which brings the device connectivity, namely Internet of things (IoT) technologies, to our daily life. IoT is a paradigm that enables communication between objects (anything and anyone) at any time and any place. IoT is widely used in different application areas such as home automation, health, traffic, manufacturing, etc. With the advance of the IoT technologies, increasing amount of data became available from sensors distributed around the cities. Big data technologies and machine learning algorithms enable services and solutions to citizens and decision makers by using these urban data.

The rapid urbanization in the last century required the decision makers to increase the citizens' quality of life and make appropriate investments in infrastructure. Nowadays, utilization of energy resources, smart traffic management, air pollution, waste management, and other urban-related issues are critical. The prediction systems are required in these fields in order to improve the quality of services and solutions. For instance, a smart traffic application that has prediction capability can help dwellers to forecast the traffic jams and arrange their routes before going out. Similarly, prediction of the solid waste generation also helps to plan and design waste management systems. Prediction of air pollutants such as carbon monoxide (CO), ozone (O_3), sulphur dioxide (SO_2) and nitrogen dioxide (NO_2) in the air is crucial for public health, and recently urban air pollution is becoming a more serious problem (Mehta et. al., 2016). Therefore, air quality systems should include quality predictions for government authorities to take cautions against pollution. Another important aspect of smart cities is an efficient energy management system and smart grid concept where more renewable energy resources are utilized. In order to have a better energy management, it is crucial to effectively predict the renewable power generation from solar and wind resources (Wang et. al., 2015). Besides, it is also important to forecast the electric load demands for energy management. Like in power grid systems, in all of the demand-and-supply-based management systems and applications having similar structure, predicting demand is very important. If you predict the demand of the next day and see that it is more than the supply available and if you could not increase the supply, you should find ways to reduce

Transportation and Power Grid in Smart Cities: Communication Networks and Services, First Edition.
Edited by Hussein T. Mouftah, Melike Erol-Kantarci and Mubashir Husain Rehmani.

demand in order not to cause shortages. In addition to the aforementioned applications, another problem for drivers in crowded cities is to find parking places in the city. Hence, parking availability prediction becomes crucial (Vlahogianni et. al., 2015). Most of these examples use predictions in practice, and the number of such examples that are showing the importance of prediction could be increased. Hence, we could say that the prediction-based services/solutions, especially the ones related to the cities, are very valuable for the citizens.

Nowadays, most of these services are provided by IoT platforms that are developed by using different technologies for different purposes. Many IoT platforms have been previously developed, and a comprehensive comparison of most of the IoT platforms are given in Köhler, Wörner, and Wortmann (2014). However, that work is not dedicated to smart city IoT platforms, and as the technology evolves rapidly, the number of platforms and their capabilities are changing day by day. Hence, the need for systematic evaluation of the capabilities of smart city IoT platforms is increasing. On the other hand, cloud computing environments enable scalable smart city platforms that combine multiple stakeholders' services. Cloud platforms also support rapid application development by serving programming tools, environments, and data storage area. Hence, cloud support to any IoT platform recently became inevitable.

In this work, we first introduce recently developed IoT platforms for smart city applications. These platforms could be listed as ARMmbed (Mbed, 2016), Cumulocity (Cumulocity, 2016), DeviceHive (DeviceHive, 2015), Digi (Digi, 2016), Digital Service Cloud (Digital Service Cloud, 2015), FiWare (FIWARE, 2011), GSN (GSN, 2016), IoTgo (IoTgo, 2014), Kaa (Kaa, 2014), Nimbits (Nimbits, n.d.), RealTime.io (RealTime .io, 2016), SensorCloud (SensorCloud, 2016), SiteWhere (SiteWhere, 2009), TempoIQ (TempoIQ, 2016), Thinger.io (Thinger.io, 2016), Thingsquare (Thingsquare, n.d.), ThingWorx (ThingWorx, 2016), Xievly (Xively, 2016), Vital (VITAL, 2016), chronologically. We compare them with respect to the cloud systems employed, communication protocols, authentication methods, database types, and prediction capabilities.

Next, we focus to the prediction module developed for the Vital platform by

- providing information about the prediction methods employed;
- showing the results obtained for the traffic speed sensors of the Istanbul Metropolitan Municipality, and evaluating the results by using the root mean squared error measure; and
- explaining the integration of the prediction module to the Vital platform.

The acronyms used throughout the chapter are listed below:

- IoT: Internet of things
- ARM: Acorn-advanced RISC machine
- OS: operating system
- Bluetooth LE: Bluetooth low energy
- Wi-Fi: wireless fidelity
- CoAP: constrained application protocol
- MQTT: message queue telemetry transport
- HTTP: yyper-text transfer protocol
- REST: representational state transfer
- OWASP: open web application security project

- HDFS: hadoop-distributed file system
- CoAP: constrained application protocol
- XMPP: extensible messaging and presence protocol
- API: application programming interface
- PaaS: platform as a service

The chapter is organized as follows. In Section 10.2, we give details about the IoT platforms for smart cities and compare them in a table. Next, Section 10.3 introduces the developed prediction module and the VITAL IoT platform. Section 10.4 details the use case application developed for speed prediction and presents the experimental results obtained. The chapter is concluded by summarizing the results obtained and giving future directions.

10.2 IoT Platforms for Smart Cities

In this section, we will cover 18 IoT platforms introduced for smart cities in terms of their cloud support, communication protocols, authentication mechanisms, databases, and prediction capabilities. First, we introduce the IoT platforms by giving brief information about their capabilities. Detailed comparison of these platforms is given in Table 10.1.

10.2.1 ARM Mbed

Mbed (Mbed, 2016) IoT platform and operating system (OS) was released by ARM and licensed under Apache 2.0 in 2009. ARM is the company that produces microprocessors for mobile, IoT, wearable systems, automotive, health care, and other markets. Mbed OS is designed for IoT devices and includes the features necessary to develop a connected product based on ARM Cortex-M microcontroller. Mbed OS is the core of the system and supports sensors and actuators and different communication technologies such as Wi-Fi and Bluetooth LE. It also has a web-based development environment. Mbed has its own cloud, named ARM Mbed Cloud. The source code written through web browsers is compiled in the cloud environment. Communication is provided by using CoAP, MQTT, and HTTP/RESTful protocols.

10.2.2 Cumulocity

The term "Cumulocity" is a compound of "Cumulus" (a type of cloud) and "velocity." It is Nokia's IoT spin-off, founded in 2010 in Silicon Valley, California (Cumulocity 2016). Cumulocity GmbH continues development in Düsseldorf, Germany, with the support of Nokia Siemens Networks.

Cumulocity is an application-centric IoT platform whose functionality is available for public via open APIs. Cumulocity has its own cloud system, and sensor nodes are clients that connect to this system through RESTful, HTTPS, and API. Data collection from devices and their remote management is done through the Cloud Fieldbus application. HTTP(S)/RESTful is the only supported communication protocol in Cumulocity (Derhamy et. al., 2001). It implements OAuth2 protocol for authentication. There is no information about the database system used. On the other hand, it has a system for real-time data analytics.

Table 10.1 Comparisons of IoT-Based Smart City Platforms.

Platforms	Cloud	Communication protocols	Authentication	Database	Prediction opportunity
ARMmbed	ARMmbed cloud	CoAP; MQTT; HTTP/REST	-	-	-
Cumulocity	Cumulocity cloud	HTTP/REST	OAuth2	-	Cumulocity analytics
DeviceHive	On-premises; cloud; hosted; hybrid	MQTT; HTTP/REST	OAuth2; HTTP basic authentication	MongoDB	ElasticSearch, Apache Spark, Cassandra, and Kafka
Digi	Digi Cloud	HTTP/REST	HTTP basic authentication	-	-
Digital Service Cloud	Microsoft Azure	HTTP/REST	OWASP	MongoDB; SQL Server	Azure machine learning
FiWare	FiWare cloud	MQTT; HTTP/REST	OAuth2	Hadoop; MongoDB	TrafficHeat (In FiWare Hackathon big data)
GSN	On-premises	HTTP/REST, SOAP; MQTT; CoAP	HTTP basic authentication	MySQL	Weka
IoTgo	On-premises; cloud; Hosted; hybrid	HTTP/REST; WebSocket	HTTP basic authentication	MongoDB	-
Kaa	On-premises; cloud; hosted; hybrid	MQTT; HTTP/REST; CoAP; XMPP	OAuth	Cassandra; MongoDB	Apache Spark
Nimbits	On-premises; cloud; hosted	MQTT; XMPP; HTTP/REST	HTTP basic authentication	MySQL	-
RealTime.io	ioBridge cloud	UDP-based ioDP; HTTP/REST	HTTP basic authentication	-	-
SensorCloud	SensorCloud; Amazon Web Services	HTTP/REST	API key; HTTP basic authentication	SQL Azure	MathEngine; IPython and SciPy
SiteWhere	On-premises; cloud; hosted; hybrid	MQTT; AMQP; Stomp; HTTP/REST	HTTP basic authentication	MongoDB; Apache HBase; InfluxDB	-
TempoIQ	CODA Cloud	MQTT; HTTP/REST	-	DataIQ	Analytic Composer
Thinger.io	Thinger Cloud	MQTT; HTTP/REST	OAuht2	MongoDB	-

(Continued)

Table 10.1 (Continued)

Platforms	Cloud	Communication protocols	Authentication	Database	Prediction opportunity
ThingSquare	Thingsquare Cloud	HTTP/REST	HTTP basic authentication	-	-
ThingWorx	On-premises; hosted; hybrid; ThingWorx cloud; Amazon Web Services; Microsoft Azure	CoAP; WebSockets; MQTT; HTTP/REST	HTTP basic authentication	Neo4j; Apache Casandra	ThingWorx Analytics
VITAL	On-premises; cloud; hosted	HTTP /REST	HTTP basic authentication	MongoDB	Weka
Xively	Xively cloud	MQTT; HTTP/REST; WebSockets	OAuth2; HTTP basic authentication	Blueprint	-

10.2.3 DeviceHive

DeviceHive (DeviceHive, 2015) is an open-source IoT data platform, which was introduced by an IoT research and development company, DataArt, in 2012. It supports the AllJoyn system, which allows devices to communicate with other devices around them. AllJoyn provides cloud connectivity for AllJoyn-supported devices, and it can be used for as a bridge to connect with third-party protocols.

DeviceHive is compatible with on-premise, cloud, hosted, and hybrid systems. It works in public and private clouds: Microsoft Azure, Amazon Web Services, Apache Mesos, and OpenStack. It uses MQTT and HTTP/RESTful protocols for communication. DeviceHive has two authentication protocols: OAuth2 and HTTP basic authentication. It stores data in MongoDB and has support to many data analytics and big data tools such as ElasticSearch, Apache Spark, Cassandra, and Kafka. It supports real-time and batch processing data analytic solutions.

10.2.4 Digi

Digi Inc. (Digi, 2016), which was founded in 1985, focused on PC boards in the first years. Then the company shifted toward M2M solutions such as satellite communication devices, integrated circuits, and gateways. Digi has started to proceed on its business with Etherios Inc., founded in 2008, and focused on the cloud sector since 2012. Respectively, the iDigi Energy and iDigi Tank platforms were developed for creating smart energy networks and monitoring tank storages. After adding the visualization tools of Thingworx and the cloud capabilities of Etherios, Digi became an advanced IoT platform (Köhler, Wörner, and Wortmann, 2014).

Digi has its own cloud, named Digi Cloud, for device integration and monitoring. It uses the HTTP/RESTful communication protocol and implements HTTP basic authentication protocol. On the other hand, there isn't any information about Digi's database system and data analytic services.

10.2.5 Digital Service Cloud

Digital Service Cloud Open IoT Platform (Digital Service Cloud, 2015) was launched by iYogi, New Delhi, in May 2015. It is built on Microsoft Azure and uses various Azure modules such as Service Bus, Event Hub, DocumentDB, and Machine Learning, for data analytics operations. Digital Service Cloud has more than 12 million connected devices, and it has the highest number of devices among digital service management platforms.

The infrastructure has the HTTP/RESTful communication protocol and OWASP for authentication service. Digital Service Cloud stores collected data in MongoDB and user data in Microsoft SQL Server.

10.2.6 FiWare

FiWare (FIWARE, 2011) is a middleware IoT platform developed by the FiWare community, which is an independent open community that aims to develop smart applications for various areas by constructing sustainable, open, and application-driven middleware platform standards. It was supported by the European Union Future Internet Public-Private Partnership (FI-PPP) project's Framework Programme 7 (FP7) between 2011 and 2014 (COMM 2016). FiWare has its own cloud infrastructure, named FiWare Cloud, which is based on OpenStack. It uses MQTT and HTTP(S)/RESTful protocols for communication and the OAuth2 protocol for authentication. FiWare uses an HDFS-based system for data storage and has a module named Cosmos for big data analysis. The TrafficHeat project used this module for traffic predictions in FiWare Big Data Hackathon (Santander-Fiware, 2016).

10.2.7 Global Sensor Networks (GSN)

The Global Sensor Network (GSN, 2016) project is one of the initial projects that have been started in 2004 by Distributed Information Systems Laboratory LSIR at École Polytechnique Fédérale de Lausanne. It's an open-source platform and is written in the Java programming language (Petrolo, Loscri, and Mitton, 2015). The main aim was to develop a middleware platform for sensor data integration and query processing. Later GSN is supported by other projects and institutions such as the OpenIoT project, the HYDROSYS project, and others. As it is one of the premise projects, it doesn't have any special cloud support and runs the on-premise principle. While its first version uses HTTP/SOAP and RESTful communication protocols, MQTT and CoAP communication protocols were added in the second version. GSN implements HTTP basic authentication for authentication of users and uses a MySQL database for data storage. Although GSN has a Weka library that enables data analysis and mining operations, no application that predicts sensor observations by using this library was identified.

10.2.8 IoTgo

IoTgo (IoTgo, 2014) is an open-source IoT platform founded by ITEAD Intelligent Systems Co. Ltd, which is headquartered in Shenzhen, China.

IoTgo is compatible with on-premise, cloud, hosted, and hybrid systems. WebSocket and HTTP/RESTful are two communication protocols supported by IoTgo. IoTgo implements HTTP basic authentication method for authentication, and MongoDB is used for data storage. No information about data analytics has been found.

10.2.9 Kaa

Kaa (Kaa, 2014) is a middleware open-source IoT platform that aims to collect information from various sources, and it is licensed under Apache 2.0. Since 2014, the Kaa project is an enterprise of CyberVision, a software company. Kaa has on-premise, hosted, hybrid, and cloud deployments. It supports MQTT, HTTP(S)/RESTful, CoAP, and XMPP communication protocols and implements the OAuth authentication protocol. MongoDB and Cassandra databases are used for data storage. Collected data could be analyzed employing the Apache Spark extension of the platform. The Kaa project has endpoint SDKs for Java, C, C++, and Objective-C.

10.2.10 Nimbits

Nimbits (Nimbits, n.d.) is an open-source PaaS for IoT licensed under Apache 2.0. The Nimbits platform is compatible with Raspberry Pi, J2EE Web Application Servers, Amazon EC2, and the Google App Engine. It supports MQTT, XMPP, and HTTP/RESTful communication protocols and provides authentication by implementing HTTP basic authentication protocol. MySQL database is used to store the data. It provides Java libraries and SDK for development (Mazhelis and Tyrvainen, 2014). There is no information about data analytics and prediction systems in Nimbits.

10.2.11 RealTime.io

RealTime.io (RealTime.io, 2016) is a PaaS of ioBridge for IoT systems. The purpose of ioBridge RealTime.io is easy and cheap connection of products to the Internet and building a bridge between embedded systems and user applications.

RealTime.io has its own cloud system for storage and management. It uses UDP-based ioDP (Mazhelis and Tyrvainen, 2014) and HTTP/RESTful protocol for communication and HTTP basic authentication for user and device authentication. There is no information about the database system employed and the data analytic tools of RealTime.io.

10.2.12 SensorCloud

SensorCloud (SensorCloud, 2016) is an IoT platform of US-based LORD MicroStrain corporation. It was introduced by MicroStrain, founded in 1987. In the first years, the company focused on sensors used in biomedical research. MicroStrain released their first inertial product, the 3DM, in 1999. While it was initially thought of as a biomedical product, it was later made compatible with navigation systems, civil engineering projects, and industrial equipment monitoring. In 2001, MicroStrain released the first wireless sensor of the company. Owing to such a background, the company launched SensorCloud that enables to store, fast monitor, and analyze data in 2011. One year later, MicroStrain was acquired by LORD Corporation.

Sensor Cloud provides a cloud platform hosted by Amazon Web Services (AWS). It communicates via the HTTP/RESTful protocol and authenticates users via custom API key and HTTP basic authentication. It stores data in SQL Azure database. SensorCloud provides support to Python, Java, C#, and C++ (Mazhelis and Tyrvainen, 2014). It has also MathEngine, IPython, and Scipy support for data analysis and machine learning operations.

10.2.13 SiteWhere

SiteWhere (SiteWhere, 2009) is an IoT platform launched by SiteWhere LLC in 2009. The platform was initially aimed to be a telecommunications-processing platform. Then real-time location data has been integrated. And finally it became a more generalized IoT platform, and it is open-sourced since June 2013.

SiteWhere is compatible with on-premise, cloud, hosted, and hybrid systems. It communicates via the MQTT, AMQP, Stomp, and HTTP/RESTful protocols and implements HTTP basic authentication protocol. MongoDB, Apache HBase, and InfluxDB technologies are used for data storage. There isn't any information about predictive analytics usage by the SiteWhere platform.

10.2.14 TempoIQ

TempoIQ (TempoIQ, 2016), is an IoT platform developed with the purpose of providing fast, flexible, and agile device data collection, storage, and monitoring. TempoIQ uses the CODA cloud system for monitoring and controlling devices and data. Its communication protocols are MQTT and HTTP/RESTful. It provides support for C, Java, Python, Ruby and .NET (Mazhelis and Tyrvainen, 2014). TempoIQ stores data in the DataIQ module and has the Analytic Composer for data analysis in the AnalyticIQ module.

10.2.15 Thinger.io

Thinger.io (Thinger.io, 2016) is an IoT platform that has its own cloud infrastructure to connect and manage devices. It uses MQTT and HTTP/RESTful communication protocols. Users/customers can integrate it to their business logic by using REST API. Thinger .io authenticates users and devices by implementing the OAuth2 protocol, and collected data are stored in MongoDB. No information about prediction of device data has been found.

10.2.16 Thingsquare

The development of the Contiki Operating System, which is compatible with low-power and less memory and suitable for connected devices, led to the foundation of ThingSquare (Thingsquare n.d.). Contiki and its commercial type ThingSquare Mist are released under the BSD license and open source (Köhler, Wörner, and Wortmann, 2014). ThingSquare focuses on mostly the automation, controlling, and monitoring of things via the Internet than integrating applications and analyzing collected data (Derhamy et. al., 2001). ThingSquare has its own cloud for deployment. It uses the HTTP/RESTful communication protocol and HTTP basic authentication method for authentication. ThingSquare has an IDE named "ThingSquare Code," which enables development of applications for ThingSquare Mist. There isn't any information about its database and data analytics capabilities.

10.2.17 ThingWorx

ThingWorx (ThingWorx, 2016) is an IoT platform that reduces complexity for non-technical people (Köhler, Wörner, and Wortmann, 2014). It focuses on integration, transformation, and presentation of collected data instead of communication between nodes and collection of data. Four main blocks of ThingWorx were built in 2011. The first block, SQUEAL, enables to search data and devices. The second one, Mashup Builder block, enables application development by "drag and drop." The third block, Composer, was developed for composition of different data visualizations, storages. and business logics. Finally, the social networks block is used for crowdsourcing (Köhler, Wörner, and Wortmann, 2014). ThingWorx enables on-premise, hosted, and hybrid deployments (Mazhelis and Tyrvainen, 2014). Besides, it is also compatible with third-party clouds such as Amazon Web Services and Microsoft Azure. Thingworx uses CoAP, WebSockets, MQTT, and HTTP/RESTful communication protocols (Derhamy et. al., 2001) and implements HTTP basic authentication as authentication protocol. It stores model data in Apache Casandra, supports Neo4j graph database, and has a module named ThingWorx Analytics for data analytics.

10.2.18 VITAL

The VITAL (Virtualized programmable InTerfAces for smart, secure, and cost-effective IoT depLoyments in smart cities; VITAL, 2016) platform aims to integrate inter-connected objects (ICOs) among multiple IoT platforms. The detailed description of the platform will be given in the following sections.

10.2.19 Xively

Xively (Xively, 2016) is another a PaaS platform that enables real-time communication, device monitoring, and data storage and distribution (Mazhelis and Tyrvainen, 2014). It was initially developed in 2007 to connect things to things and was named Pachube. In 2011 it was widely used in Japan to monitor radioactive fallout across the country after the nuclear accident in Fukushima. Then in the same year it was acquired by Log-Meln and renamed Cosm. Later on in 2013, it was renamed Xively (Köhler, Wörner and Wortmann, 2014). Xively supports MQTT, HTTP(S)/RESTful, and WebSocket communication protocols, and it has its own public cloud. HTTP basic authentication protocol and OAuth2 are the two supported protocols to authenticate users. Xively uses Blueprint database and has libraries for Python, Ruby, and Java. It doesn't have any sensor data prediction support.

It could be summarized that most of the IoT middleware platforms only provide monitoring, controlling and planning of sources/sensors for the moment. However, it should be kept in mind that the services/applications supporting prediction are becoming very important in order to manage, plan, and prepare cities to the future.

10.3 Prediction Module Developed

By using IoT and smart city technologies, it becomes easier to manage and organize cities. Nowadays, huge amounts of data can be collected using sensors deployed in cities.

Using prediction modules these data can be used for reasoning and better governance of the cities (Pradhan et al., 2016; Strohbach et al., 2015). In this work, we will mainly focus on the prediction module of the VITAL platform, which was developed with the support of EU 7th Framework Program. First, this subsection provides brief information about the architecture of the VITAL platform. Then, the details of the prediction module will be presented.

10.3.1 The VITAL IoT Platform

The VITAL (Virtualized programmable InTerfAces for smart, secure, and cost-effective IoT depLoyments in smart cities) platform aims to integrate inter-connected objects (ICOs) among multiple IoT platforms (Petrolo, Loscri and Mitton, 2015), as previously explained. The management and access of heterogeneous objects are achieved with virtualization of interfaces. The VITAL architecture is given in Figure 10.1. As shown in the figure, the data and service access to VITAL are implemented using VUIAs (virtualized universal access interfaces), which makes VITAL platform-agnostic. The VUIA layer has several connectors enabled to communicate different IoT platforms and clouds. VITAL integrates heterogeneous data and functionalities of different sensors in a platform-agnostic way with the help of linked data standards. Different sensor clouds are deployed to the VITAL platform via the platform provider interface (PPI). In order to model different IoT systems, the VITAL consortium selected four platforms, namely, X-GSN, Xively, Hi Reply, andFIT IoT Lab. On the top of the VITAL core platform there is a security module to authenticate all users. Sensor observations are stored and managed via the data management services (DMS) module, and MongoDB is used as database. Detailed description about the VITAL platform can be found in Petrolo, Loscri and Mitton (2015) and VITAL (2016).

10.3.2 VITAL Prediction Module

Generally, most of the collected data in city data repositories can be considered as time series; hence, time series analysis (Gulacar, Yaslan and Oktug, 2016) or interpolation techniques are used (Dobre and Xhafa, 2014) for analysing them. On the other hand, one can also obtain feature vectors from sensor observations, and depending to the application domain, successful machine learning algorithms can be applied for classification, clustering, and regression applications. In order to predict sensor observations, regression algorithms can be used, where the aim is to find a continuous function between a dependent variable and one or more independent variables (Pan et al., 2013). Besides, regression relies on a loss function, which is generally the difference between the predicted value and the actual value to learn the best function, the one with least error. In this subsection, we will use the most common and successful regression models, such as support vector regression, regression trees, etc., for traffic sensor data prediction in Istanbul (Gulacar, Yaslan, and Oktug, 2016). We consider various representations of the feature vectors and their effects on performance as well.

The prediction module explained is an extension to the VITAL platform architecture given in Figure 10.1. The VITAL platform easily enables to extend services when the security issues are implemented. The extended prediction module and its VITAL integration are shown in Figure 10.2. As shown in the figure, the prediction module

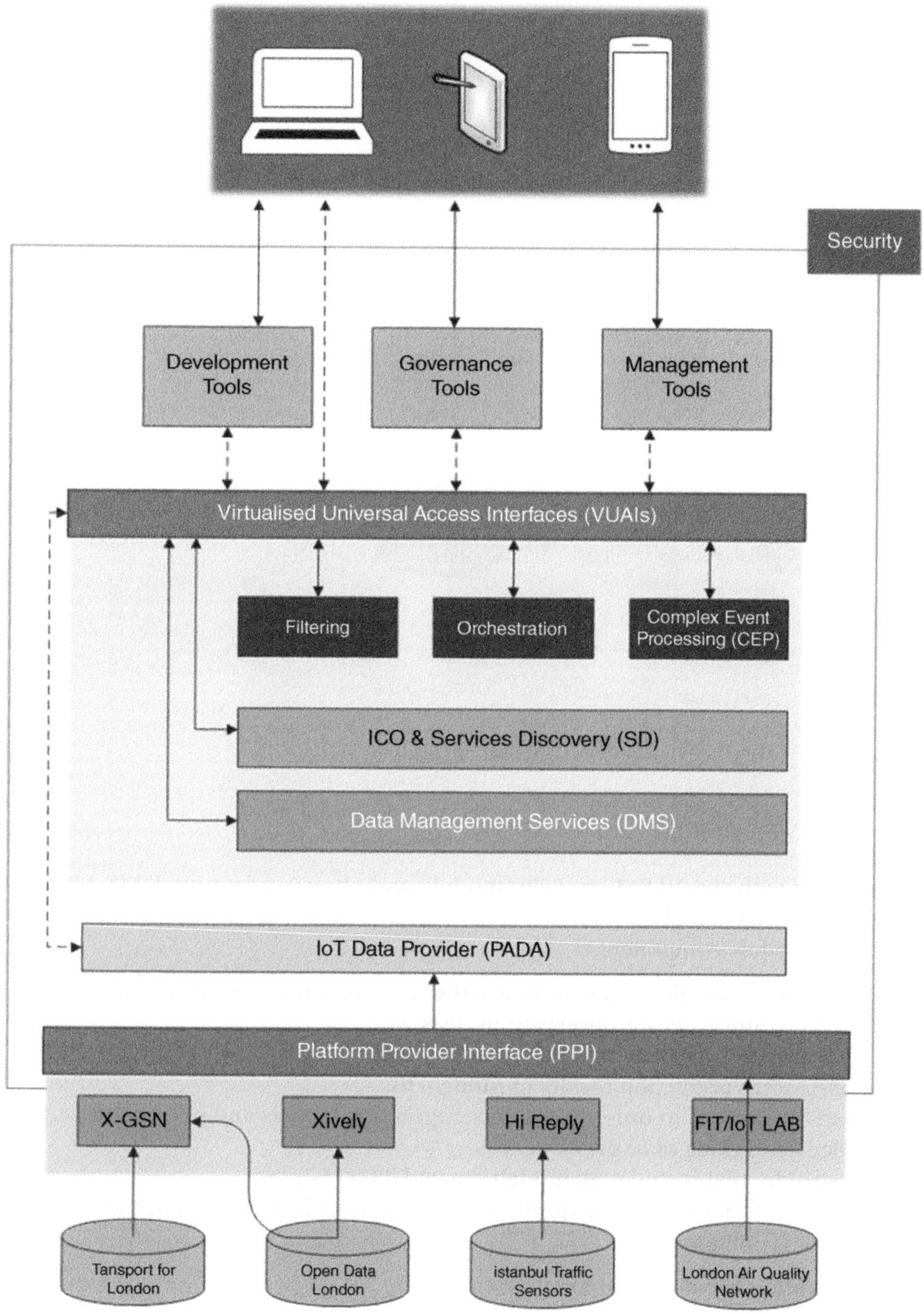

Figure 10.1 The VITAL platform architecture.

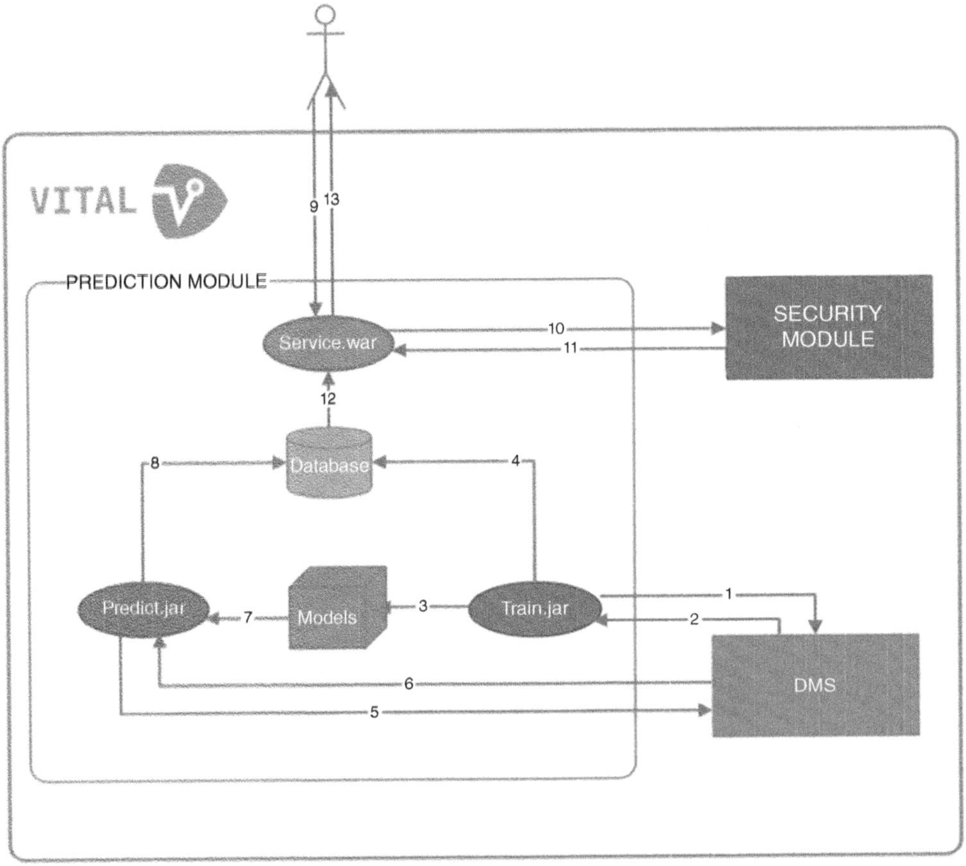

Figure 10.2 The VITAL prediction module extension.

communicates with the VITAL security module and observes data via DMS and has core components that are listed below:

Prediction module components:

1) VITAL security module: Determines whether an access token is valid or not.
2) VITAL DMS: Stores sensor observations, lists, etc.
3) Service.war: Gets, controls, and responds to client requests.
4) Database: Stores prediction results in MongoDB.
5) Train.jar: Trains data in order to produce regression models and saves models. It is scheduled as working once every week.
6) Models: Folder that includes all models created after training.
7) Predict.jar: Gets data and makes prediction by using pickle. It is scheduled as working once every 15 minutes.

Communication between components (edges):

1) Training data request: Send request DMS in order to get sensor observations for training.
2) Training data response: Get sensor observations from DMS.

3) Save model: Regression models created in training phase are saved into Models folder in order to reuse by Predict.jar.
4) Weekly prediction results: Train.jar creates regression models. Additionally, it predicts weekly (long-term) data and inserts into database.
5) Observation request: Send request to DMS in order to get sensor observations.
6) Observation response: DMS response that includes sensor observations.
7) Get Model: Gets regression models for prediction.
8) Insert prediction results: Insert prediction results to database.
9) Client Request: Client must send request whose header includes vitalAccessToken and body includes the following item(s):
   a. city: Shows which PPI's prediction results are requested.
   b. sensor*: List of sensors' id.
   c. property*: List of observation property or properties.
   d. gt*: Lower bound of time interval.
   e. lt*: Upper bound of time interval.
10) Security request: Server.war sends request to VITAL/securitywrapper in order to determine whether vitalAccessToken is valid or not.
11) Security response: VITAL/securitywrapper response whether vitalAccessToken is valid or not.
12) Related prediction results: Asks prediction results from database according to user query.
13) Prediction response: Prediction response is a list of nodes including the following items:
   a. city: PPI name
   b. sensor: sensor ID
   c. property: observation property
   d. value: predicted observation value
   e. prediction_time: shows when the module predicted that value
   f. minute_later: shows which prediction time mode (15 min., 30 min. ahead, or long-term(999))
   g. predicted_time: shows for which time the module predicted that value.

In order to observe prediction results from the VITAL platform, one needs to implement the post method given in Table 10.2.

10.4 A Use Case Employing the Traffic Sensors in Istanbul

Istanbul, with population of 14 million, is one of the most crowded metropolises in the world, and its 856-kilometer-long road network contains approximately 3 million cars (Gulacar, Yaslan, and Oktug, 2016). This huge amount of cars cause traffic jams on the complex road network of the city. Therefore, monitoring the traffic is crucial for management, and for this purpose, Istanbul Metropolitan Municipality (IMM) deployed many traffic speed sensors around the roads.

In this work, we used sensor observations from the D100 road, which connects the Asian part and the European part of Istanbul and has an airport at either end. The D100

* It is not necessary. If it does not exist in request, then results for all alternatives are returned.

Table 10.2 VITAL Prediction Module Access Method.

	Query Predictions from Prediction Module	
Description	This interface is used to query prediction results.	
Method	POST	
URL	PREDICTION_BASE_URL/prediction	
Request headers	Content-Type	application/json
	Cookie	vitalAccessToken

Request body	Example
	{
	"sensors": [
	"http://vital-integration.atosresearch.eu:8280/hireplyppi/sensor/vital2-I_TrS_32Dir0"
	],
	"city": "hireplyppi",
	"lt": "2016-11-23T15:34:00+00:00",
	"property": [
	"Traffic"
	],
	"gt": "2016-11-23T14:34:00+00:00"
	}	

Response header	Content-Type	application/json

Response body	Example
	[
	{
	"predicted_time": "2016-11-23T15:30:00Z",
	"city": "hireplyppi",
	"property": "Traffic",
	"prediction_time": "2016-11-23T14:30:00Z",
	"sensor": "http://vital-integration.atosresearch.eu:8280/hireplyppi/sensor/vital2-I_TrS_32Dir0",
	"value": 85,
	"minute_later": 60
	}
	]	

Notes	• Request must include vitalAccessToken Cookie. • Request does not have to include "sensors," "property," "gt," and "lt" items.	

road segment is shown in Figure 10.3. This data set contains approximately 30 million speed observations collected by 122 sensors located on the D100 road from Jan 1, 2014, to Dec 31, 2014. These observations are obtained from Autoscope Terra (image processing sensor), RTMS (Radar Sensor), Smart Sensor HD (Radar Sensor) type sensors which send observations every 1–2 minutes (Gulacar, Yaslan, and Oktug, 2016).

10.4.1 Prediction Techniques Employed

In our previous work (Gulacar, Yaslan, and Oktug, 2016), we studied in detail the effects of using different feature vectors and clustering and regression methods on traffic speed prediction results. We determined that while the weather feature does not increase the prediction performance beyond our expectations, binary record time vectors increased

Figure 10.3 D100 road segment and deployed speed sensors.

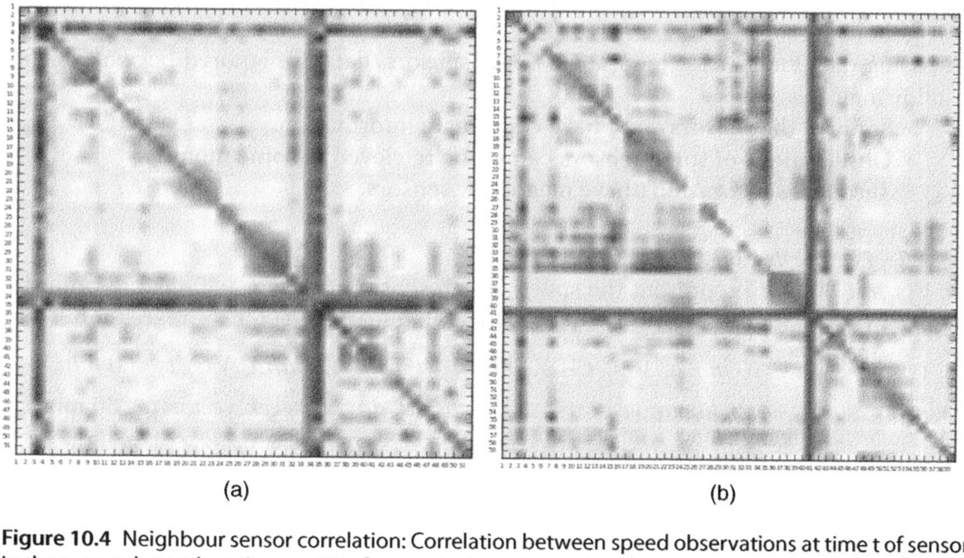

Figure 10.4 Neighbour sensor correlation: Correlation between speed observations at time t of sensor in the row and speeds at time t − 30 of sensor in the column. Sensors are sequential in direction from (a) Europe to Asia and (b) from Asia to Europe.

the accuracy. Therefore, in this work we applied prediction methods on feature space that include previous speed observations and binary record time vectors. We also did extensive analysis on the data set employed. We showed that the observations of sensors at time t are strongly correlated with the observations of their neighbor sensors at time t − 30, as shown in Figure 10.4. In this experiment we aimed to find whether there is a relationship between the average values of sensor observations. Mostly, neighbouring sensors have relationship with the next sensor after 30 min. Thus this speed information is added as an extra feature in our predictions. Additionally, we analyzed the effect of adding neighbor sensor observation(s) to the feature vectors.

In this work, the clustering step is neglected since the module is compatible with a real IoT platform VITAL. While new sensors and/or PPIs can be added to the platform, determining a cluster is an extra burden for the platform. Besides, there is no significant differences between success of using clustering methods.

In the rest of this section, firstly, we describe data preprocessing operations. Then, constructions of feature vectors and normalization steps are described. Later, we describe training and test partitioning, and evaluation metrics are given. Finally, we give the applied regression methods and experimental results.

10.4.1.1 Data Preprocessing

In order to prepare data to create feature vectors, the following preprocessing operations are employed:

- 5-min average data: We obtained the speed average data within 5 minutes in order to avoid high fluctuations of speed data.
- Neighbour sensor detection: The raw data did not contain any information about neighbouring sensors. Only sensor directions (Asia to Europe or vice versa) and sensor coordinates were available. Therefore, by using this information we detected neighbouring sensors.
- Missing data and sensor ignorance: Some sensors' data are ignored because of the following causes:
  - Some of the sensors don't have coordinate information.
  - Observations of some sensors cannot be retrieved in some months.
  - Terminal sensors don't have neighbour sensors.

10.4.1.2 Feature Vectors

Let $V(s_i, t + 30\ min.)$ represent the speed of i-th sensor at time $t + 30min.$; in order to predict $V(s_i, t + 30\ min.)$ we construct the following feature vector for regression:

$$\begin{aligned}[V(s_i, t), V(s_i, t-15\ min.),\ V(s_i, t-30\ min.),\ V(s_i, t-45\ min.),\ V(s_i, t-60\ min.)]\\ + neighbour\_vector(s_i,\ t, j\_vector) + day\_of\_week\_vector(t+30\text{ min.})\\ + hour\_of\_day\_vector(t+30\text{ min.}) + time\_slice\_vector(t+30\text{ min.})\end{aligned}$$

where:

- $+$: vector concatenation operator,
- $V(s_i,\ x)$: speed of i-th sensor at time x,
- $neighbour\_vector(s_i, t, j_{vector})$: returns vector that includes speeds of $(i+j)^{th}$ sensor(s) that is/are neighbour(s) of s_i at time t. For example,
- $neighbour\_vector(s_3, 01.01.2014T14:24:00+00:00, [-2,\ +2])$ returns
- $[V(s_1,\ 01.01.2014T14:24:00+00:00),\ V(s_5,\ 01.01.2014T14:24:00+00:00)]$,
- $day\_of\_week\_vector(x)$: returns 7-bit binary vector that represents day-of-week value of time x. For example, $day\_of\_week\_vector(01.01.2014T14:24:00+00:00)$ returns [0, 0, 1, 0, 0, 0, 0],
- $hour\_of\_day\_vector(x)$: returns 24-bit binary vector that represents hour-of-day value of time x. For example, $hour\_of\_day\_vector(01.01.2014T14:24:00+00:00)$ returns a 24-bit binary vector containing zeros except the leftmost 14-th bit,

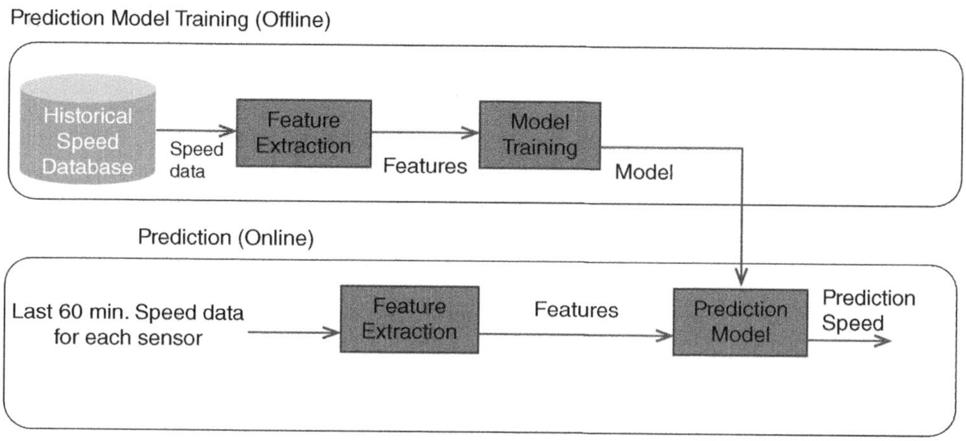

Figure 10.5 Prediction model training and online speed prediction for each sensor.

- *time_slice_vector*(x) : returns 4-bit binary vector that represents 15 min. time slice of hour value of time x. For example, *time_slice_vector*$(01.01.2014T14 : 24 : 00 + 00 : 00)$ returns [0, 1, 0, 0].

After constructing the feature vectors, non-binary features are normalized by using z-score normalization. These feature extraction and model training and testing phases are given in Figure 10.5.

10.4.2 Results

Our experimental prediction results are obtained using various regression algorithms, namely: AdaBoost.R2 regressor (ADA; Solomatine and Shrestha, 2004), decision tree regressor (DTR; Breiman, Friedman, and Stone, 1984), Gradient boosting regressor (GBR; Friedman, 2002), K-nearest neighbors regressor (KNN; Altman, 1992), kernel ridge regressor (KRR; Schölkopf, Luo, and Vovk, 2013), random forest regressor (RFR; Breiman, 2001) and support vector regressor with RBF kernel (SVR; Vapnik, 1995).

Adaboost regressor is an ensemble model and starts with fitting a model on the initial data set. Then, based on the error on the data set, it updates the selection probability of the samples and trains new prediction models on difficult examples.

Decision tree regressor is a hierarchical model that partitions the features based on a split measure such as mean square error. Decision trees are simple to interpret and handle both numerical and categorical variables.

Gradient boosting regressor is an ensemble model where several weak learners are learned on the residual of the previous models.

K-nearest neighbors regressor estimates the target value of a data sample by calculating the average target values of the K nearest neighbors.

Kernel ridge regressor uses the least squares regression algorithm with regularization and applies Kernels functions.

Random forest regressor trains a number of decision tree regressors on various subsamples of the data set and combines their outputs to find the final result.

Support vector regressor is the generalization of support vector machines for regression problems. It aims to minimize the ε-sensitive errors on the training

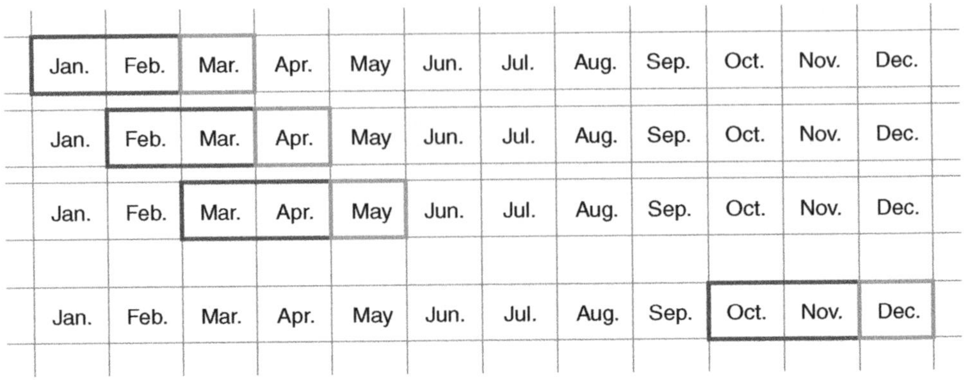

Figure 10.6 Training and test data set splitting. Two-month training months (blue) and one-month test (green) sets.

set. Generally, a nonlinear kernel function is used to project initial data into a higher-dimensional space.

Each regression algorithm is trained on a two-month training set and evaluated on the upcoming test month using the aforementioned feature vectors. The training sets and test sets that are obtained using a sliding window is given in Figure 10.6.

Experimental results are evaluated using root mean square error (RMSE):

$$RMSE = \sqrt{\frac{\sum_{i=1}^{N} (predicted\_speed[i] - real\_speed[i])^2}{N}} \tag{10.1}$$

10.4.2.1 Regression Results

The first experimental results are obtained without considering the neighbor sensors ($j_{vector} = [\,]$). These results for different regression methods are given in Table 10.3. The best accuracy is obtained with the kernel ridge regression method. On the other hand, the worst performance is obtained with the Adaboost.R2 regression algorithm.

Table 10.3 RMSE for Different Methods without Neighbor Speed Information.

	ADA	DTR	GBR	KNN	KRR	RFR	SVR
(10, 11) -> 12	16.85	15.62	14.03	15.21	13.91	14.38	14.77
(9, 10) -> 11	17.89	16.82	15.33	16.41	15.21	15.65	16.02
(8, 9) -> 10	17.95	17.10	15.38	16.63	15.26	15.75	15.71
(7, 8) -> 9	17.13	17.59	15.84	16.48	15.64	16.28	16.18
(6, 7) -> 8	17.02	15.94	14.07	15.20	14.01	14.47	14.59
(5, 6) -> 7	18.54	16.39	14.59	15.67	14.71	15.06	14.70
(4, 5) -> 6	16.57	15.36	13.54	14.60	13.49	14.03	13.92
(3, 4) -> 5	15.20	13.78	12.18	13.39	12.12	12.55	12.74
(2, 3) -> 4	15.37	13.84	12.20	13.42	12.21	12.59	12.85
(1, 2) -> 3	15.19	13.70	12.06	13.02	12.06	12.39	12.64
Regressor averages	16.77	15.61	13.92	15.00	**13.86**	14.32	14.41
Total average	14.84						

Table 10.4 RMSE for Different Methods with Observations of the Previous Sensor.

	ADA	DTR	GBR	KNN	KRR	RFR	SVR
(10, 11) -> 12	16.78	15.61	13.90	15.02	13.75	14.25	14.57
(9, 10) -> 11	17.49	16.79	15.12	16.18	15.02	15.42	15.73
(8, 9) -> 10	17.87	17.36	15.45	16.48	15.17	15.77	15.57
(7, 8) -> 9	17.11	17.56	15.63	16.26	15.44	16.07	15.98
(6, 7) -> 8	16.86	16.02	13.97	15.01	13.88	14.35	14.36
(5, 6) -> 7	18.91	16.89	14.87	15.71	14.86	15.33	14.72
(4, 5) -> 6	16.69	15.78	13.70	14.63	13.58	14.11	13.98
(3, 4) -> 5	15.11	13.95	12.21	13.39	12.14	12.56	12.74
(2, 3) -> 4	15.63	14.29	12.39	13.51	12.36	12.77	12.94
(1, 2) -> 3	15.40	14.00	12.13	13.02	12.09	12.50	12.63
Regressor Averages	16.79	15.82	13.94	14.92	**13.83**	14.31	14.32
Total average	14.85						

Table 10.5 RMSE for Different Methods with Observations of the Previous Two Sensors.

	ADA	DTR	GBR	KNN	KRR	RFR	SVR
(10, 11) -> 12	16.47	15.61	13.85	15.01	13.72	14.20	14.56
(9, 10) -> 11	17.64	16.83	15.15	16.23	15.05	15.47	15.80
(8, 9) -> 10	17.88	17.29	15.40	16.45	15.20	15.75	15.61
(7, 8) -> 9	17.01	17.65	15.72	16.35	15.49	16.16	16.08
(6, 7) -> 8	16.64	16.10	14.06	15.08	13.91	14.48	14.43
(5, 6) -> 7	18.54	16.65	14.61	15.52	14.67	15.06	14.61
(4, 5) -> 6	16.54	15.63	13.49	14.39	13.39	13.95	13.78
(3, 4) -> 5	15.21	14.07	12.26	13.41	12.17	12.65	12.75
(2, 3) -> 4	15.63	14.17	12.33	13.46	12.30	12.70	12.91
(1, 2) -> 3	15.45	13.97	12.09	12.96	12.01	12.47	12.55
Regressor Averages	16.70	15.80	13.90	14.89	**13.79**	14.29	14.31
Total average	14.81						

Next, we analyzed the effect of incorporating the previous sensor observations to the feature vector. The RMSE values obtained with the feature vector extended with the previous 1 sensor ($j_{vector} = [-1]$) observations are given in Table 10.4. It can be seen from the table that the kernel ridge regression performs best, and incorporating the previous sensor observation slightly improved the performance.

The next experiment is obtained by incorporating the two previous sensor observations to the feature vector. The RMSE values obtained with the feature vector extended with the previous 2 sensor ($j_{vector} = [-2]$) observations are given in Table 10.5. Note that incorporating the previous two sensors observations improves the prediction performance of all algorithms.

The last experimental results are obtained by extending the feature vector the with the observations of the previous two sensors. The RMSE values obtained with the feature vector extended with the observations of the previous two sensors ($j_{vector} = [-1, -2]$) are given in Table 10.6. Although the result for KRR is better than the results given in

Table 10.6 RMSE for Different Methods with Observations of the Previous Two Sensor.

	ADA	DTR	GBR	KNN	KRR	RFR	SVR
(10, 11) -> 12	16.37	15.64	13.81	14.91	13.65	14.13	14.44
(9, 10) -> 11	17.59	16.78	15.04	16.12	14.96	15.35	15.62
(8, 9) -> 10	17.87	17.73	15.52	16.42	15.19	15.94	15.54
(7, 8) -> 9	16.95	17.62	15.61	16.23	15.41	16.09	15.96
(6, 7) -> 8	16.36	16.26	13.93	14.94	13.80	14.34	14.23
(5, 6) -> 7	18.91	17.11	14.82	15.60	14.90	15.25	14.68
(4, 5) -> 6	16.66	15.79	13.65	14.52	13.54	14.10	13.91
(3, 4) -> 5	15.31	14.29	12.35	13.50	12.27	12.72	12.83
(2, 3) -> 4	15.58	14.58	12.58	13.65	12.54	12.95	13.12
(1, 2) -> 3	15.41	14.29	12.25	13.10	12.18	12.65	12.69
Regressor Averages	16.70	16.01	13.96	14.90	**13.84**	14.35	14.30
Total average	14.87						

Table 10.3, it is less than the results obtained with the feature set extended with the observations of the previous 2 sensor.

Note that we also obtained experimental prediction results by extending the feature set using the following sensor observations ($j_{vector} = [1]$, $j_{vector} = [2]$, $j_{vector} = [1, 2]$, $j_{vector} = [-1, 1]$, $j_{vector} = [-1, -2, 1, 2]$); however, none of these extensions performed better than the results obtained with the feature vector extended with the observations of the previous 2 sensor ($j_{vector} = [-2]$). Therefore, we don't report them. The best performance for prediction is obtained with kernel ridge regression employing the feature vector extended with the observations of the previous 2 sensor ($j_{vector} = [-2]$).

10.5 Conclusion

In this work, we summarized and compared most of the common smart city platforms with respect to the employed cloud support, communication protocols, authentication mechanisms, database systems, and prediction capabilities. Although, prediction has an important role in our daily lives, we see that only half of the platforms have prediction capabilities. We also introduce the VITAL smart city platform and its prediction module extension, which is implemented by employing the security mechanism in VITAL. The proposed module is evaluated for many traffic speed sensors deployed in Istanbul by Istanbul Metropolitan Municipality. We have compared the experimental prediction results obtained from various regression algorithms employing different feature vectors. It is observed that the best performance is obtained with kernel ridge regression employing the feature vector extended with the observations of the previous 2 sensor.

Acknowledgment

The tool and the use case presented in this chapter were developed with support from the EU Project with contract number CNECT-ICT-608682 and title "Virtualized programmable InTerfAces for smart, secure and cost-effective IoT depLoyments in smart cities (VITAL)."

References

Altman, N.S. 1992, 'An introduction to kernel and nearest-neighbor nonparametric regression', *The American Statistician*, 46(3), pp. 175–185. doi: 10.1080/00031305.1992.10475879.

Breiman, L., Friedman, J. and Stone, C.J. 1984, *Classification and regression trees*. New York, NY: Chapman and Hall/CRC.

Breiman, L. 2001, *Machine Learning*, 45(1), pp. 5–32. doi: 10.1023/a:1010933404324.

COMM 2016, *European commission: CORDIS: Projects & results service: FI-WARE: Future Internet core platform*, viewed 5 December 2016, http://cordis.europa.eu/project/rcn/99929_en.html

Cumulocity 2016, Cumulocity GmbH, viewed 5 December 2016, https://www.cumulocity.com

Derhamy, H., Eliasson, J., Delsing, J. and Priller, P. 2015, 'A survey of commercial frameworks for the Internet of things', *2015 IEEE 20th Conference on Emerging Technologies & Factory Automation (ETFA)*. doi: 10.1109/etfa.2015.7301661.

DeviceHive 2015, DataArt Solutions, viewed 5 December 2016, http://devicehive.com

Digi 2016, Digi International Inc., viewed 5 December 2016, https://www.digi.com

Digital Service Cloud 2015, viewed 2 November 2016, http://www.digitalservicecloud.com

Dobre, C. and Xhafa, F. 2014, 'Intelligent services for big data science', *Future Generation Computer Systems*, 37, pp. 267–281. doi: 10.1016/j.future.2013.07.014.

FIWARE 2011, viewed 5 December 2016, https://www.fiware.org

Friedman, J.H. 2002, 'Stochastic gradient boosting', *Computational Statistics & Data Analysis*, 38(4), pp. 367–378. doi: 10.1016/s0167-9473(01)00065-2.

GSN 2016, viewed 5 December 2016, https://github.com/LSIR/gsn/wiki

Gulacar, H., Yaslan, Y. and Oktug, S.F. 2016, 'Short term traffic speed prediction using different feature sets and sensor clusters', *NOMS 2016 - 2016 IEEE/IFIP Network Operations and Management Symposium*. doi: 10.1109/noms.2016.7503000.

IoTgo 2014, ITEAD Intelligent Systems Co. Ltd., viewed 5 December 2016, http://iotgo.iteadstudio.com

Kaa 2014, CyberVision Inc., viewed 5 December 2016, http://www.kaaproject.org

Köhler, M., Wörner, D., and Wortmann, F. 2014, 'Platforms for the internet of things – an analysis of existing solutions', *5th Bosch Conference on Systems and Software Engineering (BoCSE)*.

Mazhelis, O. and Tyrvainen, P. 2014, 'A framework for evaluating Internet-of-Things platforms: Application provider viewpoint', *2014 IEEE World Forum on Internet of Things (WF-IoT)*. doi: 10.1109/wf-iot.2014.6803137.

Mbed 2016, ARM Ltd., viewed 5 December 2016, https://www.mbed.com/en

Mehta, Y., Pai, M.M.M., Mallissery, S. and Singh, S. 2016, 'Cloud enabled air quality detection, analysis and prediction - A smart city application for smart health', *2016 3rd MEC International Conference on Big Data and Smart City (ICBDSC)*. doi: 10.1109/icbdsc.2016.7460380.

Nimbits (n.d.), viewed 5 December 2016, http://bsautner.github.io/com.nimbits

Pan, G., Qi, G., Zhang, W., Li, S., Wu, Z. and Yang, L. 2013, 'Trace analysis and mining for smart cities: Issues, methods, and applications', *IEEE Communications Magazine*, 51(6), pp. 120–126. doi: 10.1109/mcom.2013.6525604.

Petrolo, R., Loscrì, V. and Mitton, N. 2015, 'Towards a smart city based on cloud of things, a survey on the smart city vision and paradigms', *Transactions on Emerging Telecommunications Technologies*, , p. n/a–n/a. doi: 10.1002/ett.2931.

Pradhan, S., Dubey, A., Neema, S. and Gokhale, A. 2016, 'Towards a generic computation model for smart city platforms', *2016 1st International Workshop on Science of Smart City Operations and Platforms Engineering (SCOPE) in partnership with Global City Teams Challenge (GCTC) (SCOPE - GCTC)*, . doi: 10.1109/scope.2016.7515059.

RealTime.io 2016, viewed 5 December 2016, http://realtime.io

Santander-fiware 2016, *Santander-fiware/hackathon*, viewed 5 December 2016, https://github.com/santander-fiware/hackathon

Schölkopf, B., Luo, Z. and Vovk, V. 2013, *Empirical inference Festschrift in honor of Vladimir N. Vapnik*. Berlin: Springer Berlin.

SensorCloud 2016, LORD MicroStrain, viewed 5 December 2016, http://www.sensorcloud.com

Solomatine, D.P. and Shrestha, D.L. 2004, 'AdaBoost.RT: A boosting algorithm for regression problems', *2004 IEEE International Joint Conference on Neural Networks (IEEE Cat. No.04CH37541)*. doi: 10.1109/ijcnn.2004.1380102.

SiteWhere 2009, SiteWhere LLC, viewed 5 December 2016, http://www.sitewhere.org

Strohbach, M., Ziekow, H., Gazis, V. and Akiva, N. 2015, 'Towards a big data Analytics framework for IoT and smart city applications', in *Modeling and Processing for Next-Generation Big-Data Technologies*. Springer Science + Business Media, pp. 257–282.

TempoIQ 2016, viewed 5 December 2016, https://www.tempoiq.com

Thinger.io 2016, viewed 5 *December* 2016, https://thinger.io

Thingsquare (xxxxn.d.), viewed 5 December 2016, http://www.thingsquare.com

ThingWorx 2016, PTC, viewed 5 December 2016, https://www.thingworx.com

Vapnik, V.N. 1995, *The Nature of Statistical Learning Theory*.

VITAL 2016, The VITAL Consortium, viewed 5 December 2016, http://vital-iot.eu

Vlahogianni, E.I., Kepaptsoglou, K., Tsetsos, V. and Karlaftis, M.G. 2015, 'A real-time parking prediction system for smart cities', *Journal of Intelligent Transportation Systems*, 20(2), pp. 192–204. doi: 10.1080/15472450.2015.1037955.

Wang, Y., Cao, G., Mao, S. and Nelms, R.M. 2015, 'Analysis of solar generation and weather data in smart grid with simultaneous inference of nonlinear time series', *2015 IEEE Conference on Computer Communications Workshops (INFOCOM WKSHPS)*. doi: 10.1109/infcomw.2015.7179451.

Xively 2016, LogMeIn Inc., viewed 5 December 2016, https://www.xively.com

Section III

Renewable Energy Resources and Microgrid in Smart Cities

11

Integration of Renewable Energy Resources in the Smart Grid: Opportunities and Challenges

Mohammad Upal Mahfuz[1], Ahmed O. Nasif[2], Md Maruf Hossain[1] and Md. Abdur Rahman[3]

[1] *Department of Natural and Applied Sciences (Engineering Technology Program), University of Wisconsin-Green Bay, 2420 Nicolet Drive, Green Bay, Wisconsin, 54311, USA*
[2] *Department of Engineering Technology, University of Wisconsin-Oshkosh, 800 Algoma Blvd. Oshkosh, Wisconsin, 54901, USA*
[3] *Department of Electrical and Electronic Engineering, American International University-Bangladesh, House 83/B, Road 4, Kemal Ataturk Avenue, Banani, Dhaka 1213, Bangladesh*

11.1 Introduction

Electrical power grids are considered as an enormously important subject in the path of human civilization. An electrical power grid has mainly four components, namely, the generation, transmission, and distribution systems, and the customers. Apart from the customers, who are the users of electricity, the generation, transmission, and distribution systems are responsible for bringing electrical power from the generation sites through a complex network of transmission lines and distribution systems to the customer premises (Kirtley, 2010). It goes without saying that electrical power grids have contributed a lot to our individual lives, society as a whole, and its continuous development through many industrial and manufacturing processes. However, the existing power grids have been facing several issues at the present time for which it is now necessary to modernize them through the use of new power system and information and communication technologies. For instance, it has been known that the power grids faced more voltage sags, blackouts, and overloads in the past decade than in the past 40 years (Gao et al., 2012), where in most of the cases these blackout events are thought to be caused by slowness of the response time when the devices communicate over the grid. As the population is increasing, the electrical power consumption is also increasing. This is further true when, due to the rise in the number of personal devices needing electrical power, there is a high demand of electricity to the existing power grid, which makes the existing power grid more unstable in terms of its performance. Finally, old power grids are a significant source of carbon emissions in the environment, which is not encouraging in terms of the global movement toward sustainability and green electricity. For example, in the United States, the power system alone produces 40% of all nationwide carbon emissions. These factors, when combined, are sufficient to consider a suggested change from the old existing power grid to a modernized power grid (Gao et al., 2012).

Transportation and Power Grid in Smart Cities: Communication Networks and Services, First Edition.
Edited by Hussein T. Mouftah, Melike Erol-Kantarci and Mubashir Husain Rehmani.

In recent years, the concept of the smart grid (SG) has been very popular in order to envision the future of energy-efficient power grid better equipped with efficient support systems including sensor and actuator networks to sense and actuate individual "smart" devices respectively, demand management systems, and renewable energy options for environment-friendly power grid operation (Fang et al., 2012; Guizani and Anan, 2014; Shafiullah et al.; 2010, Aroge, 2014; Rehmani et al., 2016). Unlike in a conventional power grid, in the SG, the power generation, transmission, and distribution components as well as a huge number of sensors and actuators can both transfer electrical power between each other and communicate with each other through information flow taking place between them. The SG has faced several challenges from multiple viewpoints ranging from electrical power generation through transmission and distribution networks to consumers and utility management sides (Aroge, 2014). The era of traditional fossil fuel–based electrical power generation systems would perhaps come to an end soon as the depletion of natural resources and increased level of environmental pollution have established a growing threat for the present-day world (Aroge, 2014). The field of SG is an important hub where researchers from a diverse background can combine their research, expertise, and experience to improve the traditional electrical power grid toward the SG paradigm (Rehmani et al., 2016).

Renewable energy (RE) resources, namely, wind, solar, and hydro, provide alternative sources of electrical power generation and sustainable solutions and controls to our increasing environmental pollution. Technological advancement of RE-harnessing systems and growing awareness and an environment-friendly attitude of the human society toward a sustainable future have realized an increased level of RE systems used today. Therefore, in today's world, RE systems have become an inevitable option for electrical power generation support to the grid, which demands that RE systems should be integrated in future power grids and hence the SG.

Since the electrical power generation potential of RE-based systems entirely depends on variation of RE resources involved, the intermittent nature of the RE systems poses a major challenge in the integration of RE systems in the SG. For the same reason, unlike traditional fossil fuel–based power generation systems, RE systems in general cannot be dispatched easily and are not easily controllable, and so enhanced technical, financial (cost-benefit), and regulatory research are more likely to take place when integration of RE in the SG is concerned (Aroge, 2014).

In this chapter, the prospects and challenges of integrating RE resources in the SG have been addressed. The chapter is organized as follows: Section 11.2 describes the SG paradigm briefly, followed by Section 11.3 discussing the issue of integrating RE resources in the SG. Prospects and challenges of integrating RE resources have been discussed in Section 11.4. Section 11.5 briefly presents some case studies within the subject matter. Finally, Section 11.6 concludes the chapter.

11.2 The Smart Grid Paradigm

11.2.1 The Smart Grid Concept

The SG is a new power grid paradigm. The SG makes use of the digital technologies in all the system components as well as the latest information and communication technologies in order to save energy, reduce cost, and provide enhanced reliability and ensures quality electrical power to all, anytime, anywhere within the power grid (Thomas and McDonald, 2015). As the SG has several stakeholders, some of which are from energy economics and business, different perspectives of the SG among different stakeholders may be possible. However, the main objectives and the general concept of the SG are to motivate the customers to take a role in electrical power systems and to include them in the consumption-decision process, resist attack and ensure security, provide quality power to numerous devices connected in home area networks (HAN), accommodate energy storage and generation options, e.g., renewable energy resources, optimize assets, operations, and functionality, ensure reliability, create business, and be able to heal automatically (Thomas and McDonald, 2015). The basic concept of the SG is to incorporate enhanced digital information capabilities, automation, and communication technologies in the existing electrical power grid so that it can work smartly, provide useful solutions, and contribute to sustainability reducing the carbon footprint. A SG would be able to optimize the entire energy management system, which would ensure that customers be benefitted financially by reducing the utility bills without changing their lifestyle (Thomas and McDonald, 2015). As the SG is by default equipped with intelligence through digital technologies, communication systems, and automation, it is anticipated that one of the major benefits that the SG would bring about is the high penetration of renewable energy resources in the power grid ensuring less carbon emissions. In the SG framework, all the units involved can communicate with one another by means of advanced communication systems and networks. In addition, the level of automation and the use of digital technologies are anticipated to be higher in the SG framework in comparison with the same in a traditional grid framework. This could bring a huge potential to better handle the intermittent nature of RE resources in the SG than in a traditional grid.

Figures 11.1a and 11.1b show the conceptual diagrams of a traditional power grid and the SG respectively. As seen from Figure 11.1, both traditional power grid and the SG have four basic units, namely, power generation, transmission, and distribution systems, and customers. In Fig. 11.1a, the power generation unit includes not only fossil fuel–based but also all forms of RE electrical power generation units. While in a traditional power grid, the electrical power flow is unidirectional from the generation side to the customers, in the SG, electrical the power flow is bidirectional, allowing the power consumers to play a role to send electrical power back to the grid (Fang et al., 2012). Another important aspect of the SG concept that makes the SG a unique system is that there is a bidirectional flow of communication information among the components of the SG. This means, the SG units can communicate with each other through a communication network of sensors and actuators so that power flow can be controlled and monitored by information flow among the components. As shown in Figure 11.1b, the SG includes three additional components, namely, the market, operations and management, and service providers, where the four basic units continuously communicate to each order to ensure proper functionality of the system. In Figure 11.1a, the unidirectional arrows denote the unidirectional power flow of the grid. In Figure 11.1b, the bidirectional red and blue arrows denote the bidirectional power flow and communication information flow in the grid, respectively. In addition, in Fig. 11.1b) the black and the red solid lines respectively denote the interaction among RE units and that between an RE unit and the SG, where the RE unit is integrated in the SG.

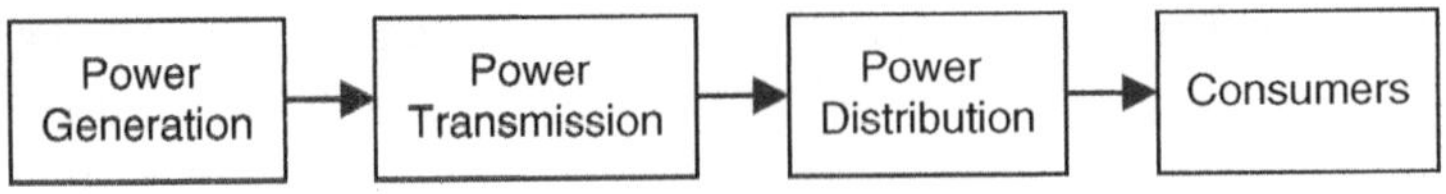

(a) Regular power grid

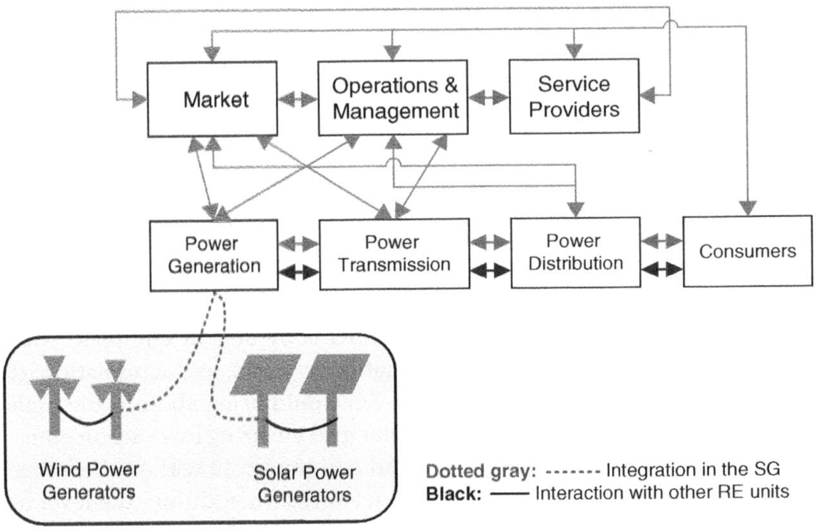

(b) The smart grid: gray-shaded area represents RE integration

Figure 11.1 Conceptual diagrams of a regular power grid and the smart grid.

11.2.2 System Components of the SG

The SG principally consists of three major systems, namely, the smart infrastructure system, the smart management system, and the smart protection system (Fang et al., 2012). In the following, these systems have been described briefly alongside their importance in the possible integration of RE resources in the SG.

(1) The smart infrastructure system: The smart infrastructure system provides the energy, information, and communication infrastructure of the SG, which is responsible for ensuring reliable two-way flow of both information and electrical energy. In a traditional power grid, there is no information flow between system components and the power flow is entirely unidirectional, meaning that the power generated by the generation system passes through the transmission and distribution systems and is finally delivered to consumers. Since the power passes through the distribution transformers, if any system fault has taken place, the traditional power grid has no measures to escape from the fault but to suffer power outages in the corresponding units. On the other hand, in the SG, the system components are capable of sending the power back to the grid system and hence the two-way power flow. Apart from that, the SG system components communicate information through sensor and actuator networks, bringing intelligence to them, which helps the power grid to escape from any faulty part of the grid system, thereby ensuring the uninterrupted power supply to the consumers. Based on the functionality, the smart infrastructure system can be divided into three subsystems, namely, the smart energy subsystem, the smart information subsystem, and the smart

communication subsystem, which develop the SG's underlying infrastructure related to energy flow, information flow, and communication flow, respectively (Fang et al., 2012; Guizani and Anan, 2014).

The smart energy subsystem is mainly concerned about the energy aspects of the SG, e.g., advanced techniques for electrical energy generation, transmission, and distribution, as well as efficient ways to consume electricity at the customers' ends. With the integration of RE electrical power units in the SG, this subsystem now needs to incorporate the additional RE supplies in the SG system, handle the intermittent nature of the RE electrical power generations and the associated variability characteristics over daily, weekly, and monthly time frames as well as the transmission, distribution, and consumption of these RE resources at various stages of the SG system.

The advanced information subsystem mainly relies on advanced techniques that are needed to handle information efficiently in the aspects of the SG, which include the advanced metering infrastructure (AMI), information monitoring and management in the customer premises with HANs, and similar technologies (Thomas and McDonald, 2015). With RE resources that are located at the customer premises or a nearby location from the customer premises, by virtue of the bidirectional energy flow possible in the SG, the AMI and HAN systems would be required to keep track of the excess power that would be delivered to the SG system from the customer on a timely manner. In such a scenario, a part of the RE power, which would be available in excess to the power required and stored by the customer, would not be lost and thus be delivered to the power system for use by other customers who need extra power.

The smart communication subsystem mainly focuses on establishing and maintaining reliable communication among all system components in the power system spanning from the generation units through transmission and distribution networks to the customer premises. At different power system components, the communication systems may not be exactly the same but should depend on the type of the devices to communicate. For example, the communication medium from the generator to the transmission and distribution networks may be different from the same at the customer premises where a smart home (Komninos et al., 2014; Pătru et al., 2016) enables all smart appliances to be connected through a reliable and secured HAN. With the inclusion of the RE resources in the SG system, due to the variability of RE resources to be handled, it is anticipated that the smart communication subsystem would need to handle more communication data than before, as the SG would now need to communicate with various subsystems more frequently for these additional RE generation systems.

(2) Smart management system: This system component provides the SG with overall advanced management and control services as well as functionalities that are at the heart of ensuring a reliable experience with the SG. Taking the benefits of modernized capabilities of the existing system and the capabilities of newly deployed systems, it has now become possible to provide advanced management and control services and improved functionalities. The potential services and functionalities that are most commonly achieved with a smart management system are better energy efficiency, reduced operational costs, balance between supply and demand, and maximizing utility usage (Fang et al., 2012). In order to handle the variable nature of the RE resources efficiently, it is anticipated that the smart management system would be required to perform comparatively more management and control services for ensuring a consistently reliable experience with the SG.

(3) Smart protection system: Being a smart system, the SG ensures protection against certain system or grid failures, addresses cyber security issues, and preserves security and privacy of consumer information very strictly. The SG achieves these by making a good use of the smarter infrastructure behind the SG concept (Fang et al., 2012; Guizani and Anan, 2014). The smart protection system needs to be materialized from two points of views, namely, from the SG itself and from the customers. For instance, the subsystem is based upon intelligence that would protect the system from unreliable events and faults and unstable operation at every system component. In the case of failures and faults, the SG has the opportunity to do a post-analysis of the events and thus make the grid even more reliable for the future. From customers' points of view, since the SG handles customers' data including individual usage patterns through AMI and HAN, security and privacy of customer data is of utmost importance in the SG. With the inclusion of RE resources in the SG, due to the variability of the RE resources, each system component becomes more affected in terms of risks of unstable operation and a higher number of events for post-analysis, and thus the workload of the smart protection system increases in order to provide an overall reliable system operation in the SG. A smart protection system should take all these into account.

11.3 Renewable Energy Integration in the Smart Grid

11.3.1 Resource Characteristics and Distributed Generation

The demand for electrical energy in the power grid varies over time during the day and also the period of the month and/or year. Daily demand patterns generally peak in the morning or evening and gradually goes down at night. If the electrical energy was to be generated from fossil fuel only, it would have been possible to control the fuel input according to the electrical power demand. Renewable energy resources depend mostly on the environment and weather conditions. Due to the intermittent behavior of RE-based electrical power generation systems, both the customers' demand and the natural resources of RE vary over time, making management of both sides very challenging when RE-based systems are integrated in the power grid. Intermittency in RE resources can be broadly categorized into the characteristics that are somewhat periodic and those with random irregularities. It also depends on the particular type of RE resource under consideration. For instance, wind energy can be very variable based on the time of the day, month, terrain type, and any particular weather condition, whereas tidal energy can be quite accurately predictable (Twidell and Weir, 2006). Electrical power output varies with the change in wind speed. Also, solar energy can be very predictable in some regions of the world but somewhat unpredictable in other regions. Solar irradiation profiles also impact the output of solar power generators at any location. For hydropower, reservoir height and flow are the two variables that make the electrical output power intermittent. For tidal power systems, tidal range, contained area, and tidal current are the major variables (Twidell and Weir, 2006). Above all, the intermittent nature of RE resources is evident from resource assessment data in almost every location. Electrical power generation from RE resources does not cost anything from the resource point of view; however, their intermittent nature makes integration to the grid challenging. The abundance of RE-based electrical power generation opportunities along with their limited power output makes them an ideal candidate for distributed generation systems (Thomas and McDonald, 2015). However, interconnected distributed RE power generation systems could also be effective producing comparable power outputs as in in conventional power generation systems (Guizani and Anan, 2014; Molderink et al., 2010; Thomas and McDonald, 2015).

11.3.2 Why Is Integration Necessary?

In today's world, sustainable engineering design is highly regarded in every aspect of social and economic development. Consumption of electrical energy in an efficient manner saves the fossil fuels that are used in producing electrical power in traditional fossil fuel–based power plants. Technology has progressed much in the field of energy-efficient systems such that energy-efficient and smart appliances are more available today than they were in the past. RE-based electrical power generation is a sustainable way of electrical power generation for the society. Renewable resources are so abundant in nature that, if harnessed properly, they could provide the necessary electrical power for a part of the current world population. Therefore, sustainable use of resources is considered as one of the driving forces for realizing the integration of RE resources in the SG. In fact, it goes without saying that one of the objectives of modernizing the old power grid and developing the new SG is to reduce loss of electrical power consumption by using energy-efficient and smart appliances. That is how RE and its integration to the SG are tied together through the sense of sustainability, there being the fact that cleaner energy utilization is one of the objectives of RE and its integration to the SG.

Awareness of environmental pollution control is another significant factor for integrating RE-based electrical power systems into the SG. Fossil fuel–based traditional power generation systems are a significant source of environmental pollution at the present time. As mentioned earlier, traditional power systems are responsible for a large amount of carbon emissions in the present world and hence the corresponding aftereffects, e.g., global warming and subsequent degradation of the natural environment. An increasing level of environmental awareness among the population has encouraged them to use more RE-based electrical power generating systems, which, as a result, have facilitated the integration process of many RE-based power generation systems into the power grid and thus the SG. Unlike the SG, the currently existing power grid is in most cases not capable of handling all these different technologies integrated to it (Guizani and Anan, 2014) while at the same time increasing its energy efficiency and being more environmentally friendly. Therefore, while it is evident that the existing power grid must be modernized and made more environmentally friendly, at the same time it is also necessary to integrate all the RE-based power systems in the existing power grid and through that to the SG paradigm. Integrating RE resources in the SG is a necessary step toward sustainable development of the human society.

11.4 Opportunities and Challenges

The intermittent nature and the variability of RE resources pose a significant level of challenge in their integration in the power grid, which could possibly be solved with the

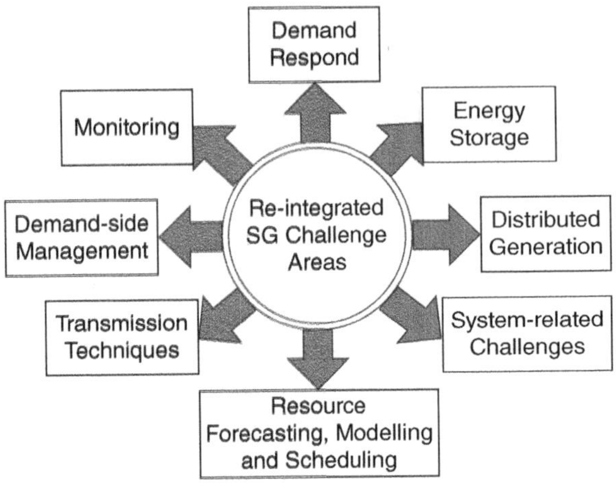

Figure 11.2 Challenge areas of the RE-integrated SG system.

technologies and concepts of the SG. In this section, the opportunities of integrating RE resources have been identified and the corresponding challenges have been explained. Figure 11.2 presents the key prospects of integrating RE resources in the SG.

11.4.1 Energy Storage (ES)

The intermittent nature (due to meteorological fluctuations) of RE resources causing voltage and frequency variations in electrical power generation makes an accompanying energy storage device a necessity for the RE system (Maharjan et al., 2012; Roberts and Sandberg, 2011). Since the SG allows for intelligence and computing resources including bidirectional information and power flows, there exists an enormous opportunity for energy storage in integrating RE resources in the SG. The intermittency of the RE power generation may be absorbed in the grid itself through proper energy management systems. The SG should have a highly efficient energy management algorithm in place that would not only take care of the variation in the power output of RE systems but also manage it properly so that this variation can be handled efficiently among the system units. The main use of distributed storage systems include peak-load deduction (peak shaving) at substations, storage of off-peak wind energy, power smoothing (for large solar arrays, ancillary services), frequency regulation, and transmission, distribution feeder reliability improvement, and customer feeder load management.

11.4.1.1 Key Energy Storage Technologies

Examples of some of the main energy storage options for the RE-integrated SG system would be battery storage (Sparacino et al., 2012; Maharjan et al., 2012), pumped hydropower (Sparacino et al., 2012; Whittingham, 2012), and flywheel technology (Whittingham, 2012; Daoud et al., 2012), while some additional energy storage options would be supercapacitors, superconducting magnetic energy, and compressed air storage technology (Aroge, 2014). For an ES system to be widely deployable in the SG, it must have high capacity, long discharge times, be cost effective, and not have special site restrictions.

Battery storage (BS): BS systems can be used for renewable integrations, peak shaving, load shifting and levelling, distributed energy storage, power quality management, and temporary/emergency backup. Moreover, a typical BS can be installed at any traditional substation, which consists of three main parts: battery module (stores and dispenses energy), DC-to-AC converter (power electronics), and power management unit (to optimize charging and discharging; Sevilla et al., 2016; Pourbeik et al., 2015). Traditional battery storage systems use a multipulse converter with a complicated zig-zag transformer, whereas modern BS systems use a multilevel converter with an advanced battery technology, such as lithium (Li)-ion, sodium sulphur (NaS), nickel metal hydride (NiMH), and so on (Maharjan et al., 2012). The multilevel cascade converter is viewed as one of the most promising topologies for BS systems (Rodriguez et al., 2007). For a detailed account on state-of-the-art and future research efforts on battery systems, see Whittingham (2012).

Pumped Hydropower (PHP): PHP operates by storing potential energy during low-demand hours by pumping water at a higher elevation, which can subsequently be used to increase the generation of power using turbines in peak times. This is, by far, the most common, cost effective, and mature ES system existing at present. However, the problem with PHP is that the energy consumers may be located in zones where the PHP layout is not suitable, and therefore, the SE from PHP may need to be brought to the consumers from far away over long transmission lines (Sparacino et al., 2012; Whittingham, 2012).

Flywheel technology (FW): FW stores energy in kinetic energy form, the amount of stored energy being proportional to the FW rotation speed. Although this type of technology is finding applications in frequency smoothing, it seems to be not very suitable for large scale, long-term ES (Whittingham, 2012; Daoud et al., 2012).

Compressed air storage technology (CAS): In this type of technology, off-peak power is used to compress air that is stored in tanks or even the natural gas pipe network, which makes it a difficult choice for distributed ES option. Also, the transformation of energy from electricity to another physical form, namely, the gaseous form, incurs unavoidable conversion losses (Sevilla et al., 2016; Jilek et al., 2015).

Supercapacitors (SC): Supercapacitors (SC) greatly improve energy storage capabilities of traditional capacitors by creating charge separation in two electrodes at small distances apart immersed in electrolytes, as it is in batteries. The advantages of SCs are that they have small inner resistance, large capacity, long life compared to batteries, and are pollution free (ElKady and Kaner, 2015; Whittingham, 2012). Table 11.1 shows a quick overview of ES systems and their performance parameters, namely, power, specific energy, discharge time, efficiency, and capital costs in the aspects of their potential integration in the SG.

11.4.1.2 Key Energy Storage Challenges in SG

Energy storage devices built in the SG system need to provide a reliable and uninterrupted power supply, better power quality, and energy management throughout a period of electrical fault, break during switching, and power outage (Aroge, 2014). Below we mention the main challenges of ES in the SG.

1) Although energy storage apparently encourages RE systems to be integrated in the SG, as more and more RE-based distributed generations are integrated in the SG, it could appear as a technical challenge to increase the efficiency and the capacity of the energy storage devices and to invent and develop new technologies in this regard (Guizani and Anan, 2014; Boksha et al., 2014; Perez et al., 2015).

Table 11.1 Overview of Storage Systems and Their Parameters (Jilek et al., 2015; Breeze, 2005).

Storage system	Power of system	Specific energy	Discharge time	Efficiency	Capital costs
Accumulators (lead, NiCd, NiMH, Li-on)	kW–500 MW	MWh–100 MWh	1–8 h	88–92%	500–1,000 $/kW
Flywheels	500 kW–1 MW	100 kWh–100 MWh	<5 min	90–95%	2,000 $/kW
Pumped storage hydropower plant	100–4,000 MW	500 MWh–15 GWh	4–12 h	75–80%	800–3,500 $/kW
Compressed air energy storage	25–3,000 MW	200 MWh–10 GWh	1–20 h	>60%	400 $/kW
NaS	1 MW	1 MWh	1 h	about 89%	200 $/kW
Supercapacitors	<250 kW	10 kWh	<1 min	>95%	–
Flow batteries	100 kW–10 MW	1–100 MWh	10 h	75–85%	800–1,200 $/kW
Accumulation to hydrogen	10 MW	unlimited	>5 h	<60%	6,000 $/kW
Superconducting induction accumulators	10 kW–10 MW	10 kWh–1 MWh	1–30 min	90%	2,000–3,000 $/kW

2) Planning for optimal placement, capacity, and type of ES systems remains an open problem (Sardi and Mithulananthan, 2014). Also, the scaling up of more and more distributed energy sources and the associated ES systems to the grid need to be explored (Sardi and Mithulananthan, 2014).
3) The economic challenges of energy storage devices include deciding on whether the energy storage device would be located profitably at the power generation, transmission, and distribution sides or the consumer side (Guizani and Anan, 2014; Tushar et al., 2016; Perez et al., 2015; Zhu et al., 2014). Community ES has become more common because of the benefit and increased number of RE sources at the customers' end (Sardi and Mithulananthan, 2014). However, it is rather costly to install ES units at residential areas where the customer-owned RE sources are located. Models of energy management and pricing schemes profitable to both residential users and power utilities need to be proposed. Solutions based on game theory, prospect theory, and auction models, which employ two-way communications between residential customers and power utilities have been proposed (Tushar et al., 2016; Nguyen et al., 2015; Wang and Saad, 2015; Wang et al., 2014).
4) There exist technical challenges in developing methods of integrating ES-coupled RE sources so that overall stability and robustness of the grid can be improved (Hill et al., 2012; Mohd et al., 2008; Rahbar et al., 2015). Some of the key factors that the new integration methods must provide are rapid response, voltage stability, rate variation control, dynamic VAR support and frequency regulation.

11.4.2 Distributed Generation (DG)

One of the main goals of the SG is to have robust contingencies against loss of transmission and generation, as well as load variations over the short and long terms. In fact,

Table 11.2 Types of Generators (Su and Gamal, 2011).

Generator type	Response time scale	Example
Ramping generators	Slow/moderate	Base load generators (coal-fired, hydro, and nuclear power plants), intermediate generators (combined-cycle combustion turbine), and peaking generators (gas turbines)
Renewable generators	Output is intermittent	Wind, solar
Fast-ramping generators	Short	Gas turbines

according to the Energy Independence and Security Act (EISA07), a major attribute of a SG is the development and integration of distributed energy resources, especially renewable resources, to increase efficiency in generation and utilization (Vaziri et al., 2011). Since RE sources are highly variable and usually not forecastable, it is crucial to have flexible, dispatchable, quick, and distributed energy generation options available. The idea is to have users, whether industrial, commercial, or residential, be supplied with energy on demand from a group of sources varying widely in terms of their location, service times, and energy generation and storage methods.

11.4.2.1 Key DG Sources and Generators

There are four main types of energy sources used in DG, namely, chemical, wind, hydraulic, and solar (Hernández et al., 2012). Chemical energy harvested from biomass, animal waste, natural gas, coal, and diesel, which are transformed into electric power by AC generators using microturbines and internal combustion engines. Windmills harness energy from wind using aerogenerators (wind turbines). In hydraulic turbines, hydraulic potential is exploited to produce energy from water. In solar cells, solar radiation is harvested by means of the photovoltaic effect.

Table 11.2 shows the main types of generators used in harvesting DG sources.

11.4.2.2 Key Parts and Functions of a DG System and Its Distribution

The RE sources in SG are distributed by a network of feeders, which will connect customers to the distribution system via smart meters. A smart meter is envisioned to have one- or two-way communication with the electric control, metering, and billing network. In addition, smart meters can communicate with each other to share information for automatic demand response and self-healing switching (Clarizia, 2016; Hurwitz and Ki, 2011).

Microgrids and feeder networks are important parts to realize distribution in the SG. A microgrid is a company or a community powered, in addition to the main utility grid, by its own collection of RE sources, such as solar panels, fuel cells, stand-alone generators, wind generators, biofuel generators, etc. Feeder networks connected to multiple substations can automatically detect and isolate fault areas, reroute powers from other sources and microgrids to minimize outages and increase reliability and efficiency.

A key characteristic to achieve this is via distributed automation, which accomplishes monitoring, control, and data communications. Distributed automation involves recognition and integration of distributed resources, such as DG sources, storage systems, and loads, with active and passive interfaces for energy dispatch controlled by distributed algorithms (Huang et al., 2011).

Intelligent fault management schemes also have to be in place, which may employ fault isolation devices for fast response to fault events. To have optimal power flow through the SG, safe and reliable interconnection of DG should be established for load forecasting and control. Data networks will have to be used to implement algorithms to control power flows, distributed automation, DG output levels, demand response, protection systems, and customer feedback and ensure cyber security (Vaziri et al., 2011; Olivares et al., 2014).

11.4.2.3 DG and Dispatch Challenges

Although there exists well-documented literature on stand-alone large-capacity RE systems, DG fits well for RE systems especially because RE resources are very much location-specific that could contribute to DG of RE. For instance, community PV installation, solar home systems, stand-alone wind energy systems, community wind farms, and small hydropower generations could be set up to serve the community electricity needs, and the additional power could be contributed to the SG through appropriate SG infrastructure and ES facilities with enhanced energy management algorithms such that, in addition to serving locally, the DGs could also contribute to the SG infrastructure for communities and people in need of electrical power (Zhang et al., 2016; Wang et al., 2016; Tushar et al., 2015).

DG can provide efficient energy management and reliable grid operation. The instability arisen from the distributed RE generations could be handled by means of the SG infrastructure through intelligent algorithms and computing systems existing in the SG. While more and more DGs would be integrated in the SG, providing an intelligent algorithm to manage all these would pose a challenge for the SG (Iqbal et al., 2016; Becerra et al., 2015; Suslov et al., 2015; Biswas et al., 2013; Hubert and Grijalva, 2011; Qiu et al., 2011; Kim et al., 2011; Caron and Kesidis, 2010).

To incorporate renewable DG into SG, several new dispatch challenges will need to be addressed (Cheung et al., 2010; Vaziri et al., 2011). For instance, distributed RE resources that are basically less predictable need to be incorporated in the SG according to a price-based scheme; the RE resources need to be fulfilling an active and dynamic energy demand; the integration of RE resources in the SG would require the modeling parameters to be adapted; a complicated optimization problem need to be solved in terms of security and meeting the energy demand properly; an efficient and real-time dispatch mechanism needs to be developed; proper mechanisms need to exist to handle emergency situations; and a post-event analysis technique needs to be available.

If DG is not implemented properly, it may lead to instability of the grid (Chen et al., 2010; Kroposki et al., 2008; Driesen and Katiraei, 2008). Innovative architectures, e.g., microgrid (Lasseter and Paigi, 2004) and LoCal grid (He et al., 2008), have been proposed to virtualize a local generator as a constant load, source, or zero load to the grid, greatly simplifying its impact on the grid. In addition, realistic modeling and analysis of DGs (Andrew et al., 2011), quality of supply, reliability (Blajszczak et al., 2011), synchronization technology (Abbasi et al., 2012), energy management (Rugthaicharoencheep and Boonthienthong, 2012), information aggregation (Alam et al., 2013), monitoring (Rana and Li, 2015; López, 2013) and opportunity identification (Mutule et al., 2013) are some key factors that have to be leveraged to maximize the benefits of DGs.

To incorporate DG in the SG, optimization plays an important role. Detailed optimization schemes need to be developed to characterize DG sources in terms of their configuration, compatibility, and capacity (Cui and Dai, 2011; Sharma and Kaur, 2016). The objective of the optimization could be to minimize DG operation cost, minimize network loss, and maximize environmental benefits. Some constraints for the optimization may include characteristics of active and reactive power, voltage upper and lower limits, range of power flow changes, impact of uncertainties of RE sources, and effect on surrounding users of generation startup or shutdown (Cui and Dai, 2011). The key challenge is to formulate objective functions and constraints that capture the benefits, risks, and uncertainties of a highly complex SG for both centralized and distributed optimization.

Dynamic pricing can be used as a mechanism to create cooperation, competition, and coalition among different DG sources, which can considerably increase the effectiveness of the SG. Although pricing in the SG has been well studied, most works are done with a consumer-centric perspective (Chen et al., 2010b; Sotkiewicz and Vignolo, 2006; Wang et al., 2010; Kasbekar and Sarkar, 2011; Wang et al., 2010). Research in this area needs to go beyond the interaction between a single energy source and the customers, to multiple DG sources that are in competition and cooperation with each other in a deregulated energy market. For example, it is possible for microgrids to achieve higher payoffs by rational cooperation among each other, compared to selfishly competing with each other (Huang and Sarkar, 2013).

Novel power electronics devices need to be developed for voltage and frequency control to improve grid stability and maximize power transfer (Yu et al., 2011). DG based on rotating machines need fault current limiters and other advanced protection devices (Hidalgo et al., 2011; Divan et al., 2014; Sirisukprasert, 2014; Kobayashi, 2014; Alain et al., 2015; Cataliotti et al., 2015; Kisacikoglu et al., 2015; Isa et al., 2015; Kinsky et al., 2013).

11.4.3 Resource Forecasting, Modeling, and Scheduling

RE resources forecasting and scheduling are important in order to ensure grid stability in the SG. Utilities need to apply resources forecasting and scheduling in order to take necessary actions or preventive measures to better prepare for any event that take place in the grid system keeping a good time frame. For instance, wind, solar, and sometimes, hydro energy resources are highly dependent on location and weather. Resource forecasting could include collecting resource characteristic data, local geographical information, and weather data. Having performed RE resources forecasting through an effective modeling method and scheduling the resources accordingly, the SG can provide an efficient way for the RE systems to be integrated in the SG, and the consequences that the RE systems could bring during these integrations are taken care of.

11.4.3.1 Resource Modeling and Scheduling

Below we discuss the main issues of resource modeling and scheduling, which include modeling and power flow control of ES systems, information management, scheduling of intermittent RE sources, and the use of computational intelligence in the SG.

1) Modeling and analysis are critical for gainfully deploying ES systems to reduce RE variability in an environmentally friendly manner. Power flow control algorithms

based on proper modeling techniques for both long time scale (days, hours) bulk ES, such as pumped hydroelectric and compressed air energy storage, and short time scale (minutes, seconds) fast-response ES, such as electric vehicle batteries and flywheels, need to be developed. In Su and Gamal, (2011), the optimal power flow problem is formulated as an infinite horizon average-cost dynamic program with the cost function taken as a weighted sum of the average fast-ramping generation and the loss of load probability.

2) Information management is another important enabler of the SG. It is one of the three main parts of the SG architecture, which comprises an electric power system (includes generation, transmission, distribution, and consumers), a communication system (establishes connectivity among the different systems and devices), and an information system (stores and processes data and information). (IEEE Std., 2030, 2011.) The information management system in the SG is vastly more complex than traditional bulk generation and transmissions systems and uses a supervisory control and data acquisition (SCADA) system.
   Some key challenges for the information system in the SG include the stochastic nature of RE sources, buffering effect of ES systems, behavior patterns of customers in the demand side management, and high mobility of electric vehicle increasing uncertainty in demand estimation. The first step toward an information management system is to develop appropriate stochastic models for the different components in the SG, such as wind power generation, solar power generation, energy demand, vehicle mobility, component outage, and energy storage. Based on these component-level stochastic models, system-level stochastic information management can be formed, whose key functions include system planning, system maintenance, unit commitment, economic dispatch, and regulation control and protection, all of which can have widely varying time frames. The system-level information management consists of mainly four parts: (i) bulk generation and transmission, (ii) distribution system and microgrid, (iii) demand-side management, and (iv) electric vehicle integration (Liang et al., 2014).
3) Novel energy scheduling is the hallmark of the SG, which can usually be posed as a constrained optimization problem. Typically, increasing energy production using traditional energy sources leads to increased price. However, in the SG, with RE sources equipped with DG and ES, it is possible to develop device scheduling algorithms that not only take advantage of RE sources during peak hours but also modify consumer demand profile by implementing real-time pricing. The main challenge becomes the intermittent nature of RE sources, which makes it difficult to create accurate scheduling schemes. Most studies assume that the user demand is known ahead of time and with certainty.
   Uncertainty in energy production and user demand can be modelled as random processes. In this context, the user demands can be categorized as delay tolerant and delay intolerant. The goal is to schedule all energy sources so that the long-term time-averaged total cost of the utility company is minimized given that all consumers' load demand is satisfied. In the literature, several numerical optimization algorithms have been proposed, including genetic algorithms (Chen et al., 2011; Logenthiran and Srinivasan, 2009), particle swarm optimization (Alrashidi and El-Hawary, 2008), simulated annealing (Brown et al., 1996; Jonnavithula and Billinton, 1996), and network-flow programming (Tang, 1996).

In Salinas et al. (2013), such a dynamic energy management problem is formulated as a Lyapunov optimization problem, where it is shown that all the delay-tolerant demand can be met within the user-specified deadlines, and bounds on the optimal strategy have been found. In Zhao et al. (2014), in the context of microgrids, consisting of a set of DGs, the least-cost unit commitment is found given that load, environmental, and various system restrictions are met. Unit commitment is the problem of finding the optimal decision on which power generator should be on-line (or active) over time, which minimizes the cost of operation (Liang et al., 2014).

4) Holistically speaking, the SG represents a very complex adaptive system under semi-autonomous distributed control. It is characterized by complexity, non-convexity, nonlinearity, non-stationarity, variability, and uncertainties far greater than the existing traditional power grid. Therefore, there is a need to transform data into information, information into knowledge, and knowledge into understanding, which leads to intelligent decision making and action in real time. It is envisioned that computational intelligence, which is a framework that leverages evolving, uncertain, and complex interactions of distributed parts and systems in the SG for learning, sense-making, decision making, and adaptation, will play an important role (Venayagamoorthy, 2011; Reddy et al., 2011; Chandler and Hughes, 2013; Werbos, 2011; He, 2010; Ma et al., 2009). Some of the main attributes of computational intelligence are identification/prediction of nonlinear dynamics, robust/adaptive/optimal/coordinated nonlinear control, complex, and large-scale dynamic optimization, scalable, fast, and accurate decision making, self-healing, immunity, and behavioral modeling (Venayagamoorthy, 2011).

11.4.3.2 Resource Forecasting (RF)

Energy management in the SG is mainly concerned with optimal power scheduling, which requires information about resource forecasting (RF). In a sense, any power dispatch algorithm can only be satisfactory if the forecasting information that is fed to it is accurate. The issue of RF becomes vitally more sensitive for stand-alone microgrids that are characterized by a limited number of generators with highly variable energy outputs. As a result, RF suitable for SG needs to be able to capture this variability of RE sources. Prediction of solar or wind power is a complex problem because of high spatial and temporal dependencies.

There are two types of forecasting approaches: (i) forecast the power generation directly (Kelouwani and Agbossou, 2004; Chatterjee and Keyhani, 2012), (ii) forecast the key variables affecting the power generation and then forecast the power generation using an empirical formula (Kavasseri and Seetharaman, 2009; Deng et al., 2010). For solar power, solar radiation and temperature are predicted, whereas for wind power, wind speed and direction are predicted. However, both of these approaches do not directly model the variability of the prediction.

Some traditional but less accurate forecasting methods include multiple linear regressions and stochastic time series methods, e.g., autoregressive (AR), moving average (MA), autoregressive moving average (ARMA), autoregressive integrated moving average (ARIMA), and exponential smoothing; see Zhao and Tang (2016) and the references therein.

Due to the highly nonlinear characteristics of weather prediction, tools of computational intelligence have been applied effectively for developing forecasting algorithms

(Xu et al., 2010; Chan et al., 2011; Palma-Behnke et al., 2011; Bashir and El-Hawary, 2009; Hinojosa and Hoese, 2010; Amjady et al., 2010). Among these, artificial neural networks have been most widely used (Xu et al., 2010; Chan et al., 2011; Palma-Behnke et al., 2011). Resource and load forecasting with prediction intervals based on fuzzy modeling to characterize the inherent variability of RE source forecasting have been presented in Sáez et al. (2015) and Hong and Wang (2014). These techniques use weather and performance data to learn the relationship between weather and system performance.

11.4.4 Demand Response

According to the Federal Energy Regulatory Commission, demand response (DR) is defined as "Changes in electric use by demand-side resources from their normal consumption patterns in response to changes in the price of electricity, or to incentive payments designed to induce lower electricity use at times of high wholesale market prices or when system reliability is jeopardized." (F. E. R. Commission, 2011). Processed information received as output from the resources forecasting, modeling, and scheduling of the RE resources could be passed to the consumers so that they could initiate efficient demand response. As an example, based on weather forecast, high solar energy can be predicted on a specific day or time of the month, which would be understood from the consumer side as to when the energy storage would be at maximum for planning a high energy usage for that time of the month. In a similar manner, the consumer could also plan on when to depend more on storage device, especially during the cloudy periods of the month (Chen, 2016).

DR is essentially a technology that enables loads to respond to generation, transmission, and distribution. It can be realized by direct load control like reduction of load by curtailing it, as well as complete load interruptions, e.g., residential air conditioners. Some benefits of DR are (Hamidi et al., 2010):

1) stabilization of pricing;
2) reduction of the chance of blackouts;
3) improvement of system security;
4) reduction of system congestion;
5) reduction of greenhouse gas emission;
6) improvement of market efficiency by response from consumers; and
7) allowing more time for upgrading the SG and network.

DR is implemented by installing DR modules in consumer appliances, which can communicate with the SG via programmable logic controllers (PLCs) or via wireless communication. The module may be smart enough to turn on an appliance after a given period of time or notify the consumer to do so. The charging time of EVs, temperature control on houses and buildings, and washing machine time span are just a few examples. Based on the forecasting of energy cost, the consumer may be given different pricing options to adjust their consumption of energy. For example, in Europe some of the domestic customers are given three color-coded options, namely, blue, white, and green, representing low, medium, and high prices of energy cost respectively. Such DR has been shown to reduce price and remove fluctuations from the SG. Data centers are seen as good candidates for participating in DR and have the potential to significantly contribute to peak-load shaving and facilitate the incorporation of RE sources (Wierman, 2014).

In North America, there are 10 independent system operators (ISO) and regional transmission operators (RTO) that control the management of the regional power grids by implementing DR. Rahimi and Ipakchi (2010), give a summary of the policies, programs, and rules of the different ISO/RTO authorities.

Two types of DR models can be considered: price based and incentive based. In price-based DR, the consumer can reduce the electricity bill by choosing to use appliances during off-peak hours, instead of peak hours. In incentive-based DR, the utility company provides some sort of monetary incentive to the consumer to turn off some of its loads during peak demand hours. In both cases, the role of the consumer is passive, that is, they only respond to the utility company by turning off appliances or shifting the time of usage.

As electric vehicles (EVs) become more common, the role of consumers will change from being passive to active. EV users can become active players in the SG, supplying energy to the grid during periods of high demand and make monetary gains or reducing their overall energy cost. Recently, there has been interest in using multi-agent systems (MASs) to actively take part in DR. MASs are defined as a collection of computational systems that are loosely coupled and can interact with each other for solving a specific problem (Khosla et al., 2004). The MAS is seen as a promising platform to create automatic load management in the smart home/building/community of the future (Li and Nair, 2015). Some key challenges facing DR are:

1) Robust control structures and advanced forecasting are needed to effectively leverage the highly fluctuating RE sources.
2) The injection of power from EVs to the grid needs to be facilitated by changing the existing electricity market policies and rules.
3) New MAS structures need to be developed to realize their full potential in DR. The MAS existing so far simply performs a fixed set of communication and control operations. Therefore, there is a need to devise more sophisticated MAS.

To gain significant benefits from load management, a large number of consumers need to participate in DR. This requires an infrastructure that is capable of handling large volumes of control information throughout the SG.

11.4.5 Demand-Side Management (DSM)

Demand-side management (DSM) is the planning and implementation of a set of policies imposed by the utility companies to regulate the energy consumption by the consumers, typically by shifting the electricity consumption from peak to off-peak demand times. Though appropriate DSM helps to achieve least cost power and maintain a balance between the power generation and power consumption in order to preserve the stability of the power system networks, adopting an efficient DSM is a prevailing challenge in the SG systems.

The main challenge to develop an effective DSM is the uncertain pattern of power generated by the RE sources unlike the conventional power plants. According to the International Energy Agency (IEA), 25% of the world's electricity will be supplied by renewable energy sources by 2035 (Hossain and Ali, 2016). As the penetration of RE sources are gradually growing, to overcome the problem of power generation uncertainty from the RE sources, it is possible to have the RE forecast results available to achieve a better DSM in the grid from the utilities point of view. Through RE resource assessment and modeling, the utilities can estimate the time of day, month, or even year when the RE resources are at a higher or lower profile. Based on the resources data, utilities can adjust their generation system to optimize the overall natural resources, e.g., the consumption of fossil fuel. In addition, it is possible for the utility companies to translate these data to the pricing information per kWh of the energy that consumers pay for during a specific period, which would influence the decision of the consumers on consuming energy during the peak loads and vice versa (Aroge, 2014).

To influence the consumers to take part in the DSM program, the utility companies can implement proper pricing strategies such as TOU (time-of-use) pricing, real-time pricing with lower rates during off-peak periods, AMI (advanced metering infrastructure), and smart meters (Jalali and Kazemi, 2015). Additionally, there are many forecast methods that can provide acceptable results in price (or load) forecasting such as autoregressive integrated moving average (ARIMA), generalized autoregressive conditional heteroscedasticity (GARCH) model, empirical mode decomposition (EMD) + extended Kalman filter (EKF) + extreme learning machine with kernel (KELM) + particle swarm optimization (PSO), mutual information (MI) + wavelet transform (WT) + least square support vector machine (LSSVM), MI +WT+ fuzzy neural network (FNN), recursive dynamic factor analysis (RDFA), hybrid method based on modified neural network (NN), WT + extreme learning machine (ELM) + partial least squares regression, WT + grey model, hybrid evolutionary fuzzy, multiple seasonal patterns and modified firefly algorithm, cuckoo search (CS) + optimally pruned extreme learning machine (OP-ELM), artificial neural networks (ANN), and wavelet packet transform + LSSVM + artificial bee colony (ABC) (Ghasemi et al., 2016). Besides an effective communication infrastructure with the capabilities of handling big data, appropriate intelligent devices are the key components to set up an effective DSM for the smart grid system. Overall, it is possible for the utilities to influence the demand of electricity at some point of time by applying an efficient DSM policy and so achieve a seamless integration of RE resources in the SG system.

11.4.6 Monitoring

Besides the appropriate forecasting tools of the RE sources and customer energy demand, real-time monitoring is one of the key contributions that the SG makes in developing an environment-friendly power grid concept. Integrating RE resources in the SG would require a high level of real-time monitoring. This is because real-time monitoring would have to be focusing on both the SG for its regular monitoring and the RE system infrastructures for an extended checkup on their operational systems especially due to their variable characteristics. One of the system components of the SG, namely, the smart communication subsystem, provides the appropriate communication infrastructure based on a large number of sensors and actuator networks working on bidirectional information communication. The benefits offered by these technologies hold more accurate and timely broadcasting of the information about the grid, which help the grid operation programs carry out exact and efficient real-time scheduling to minimize the problems created by the unpredictability of renewable generation and variations in customer demand (Wen and Li, 2014). Additionally, real-time monitoring of the grid condition provides better reliability and protection in the grid. For example, in the occurrence of a power line fault or failure of the equipment, the grid can sense the problem with the large number of sensors in the system. Next, by applying SG's computational intelligence, the SG can divert the electrical power to a desired destination through an alternative path, providing uninterrupted power to the destination; thus, reliability of the power system is ensured. As an additional challenge, in the SG systems, real-time monitoring would also cover the RE generation plants for maintenance, planning, and security issues (Aroge, 2014). During system monitoring, as the SG involves bidirectional communication between consumers and other components of the SG system, ensuring information security for the entire network is an important issue for the SG system. In home-monitoring applications, wireless sensor networks could have huge potential, allowing them to be widely used to enhance cost efficiency through the use of in-home energy management systems in the SG scenario (Erol-Kantarci and Mouftah, 2011).

11.4.7 Transmission Techniques

The SG ensures efficient power transmission from generation sites to the consumers. Being more sustainable and environment-friendly requires the SG to transmit power from the generator to consumers with minimum loss by means of flexible AC transmission systems (FACTS) and high-voltage DC (HVDC) transmission techniques (Fangxing et al., 2010; Liserre et al., 2010). As the RE systems are very much location specific, the RE resources, e.g., on-shore and off-shore wind and solar power generators, are sometimes located in a remote location considerably far away from the load centers. An HVDC transmission scheme can be considered for power transfer for the long distance RE sources due to the following advantages over high-voltage AC (HVAC) power line (Bresesti et al., 2007): low cable cost, lower voltage drop in the power line with lowest power losses, presence of no resonance between the cables and other AC equipment, faster power control, etc. However, the main drawback of an HVDC system is the high cost substation due to the presence of costly power electronic devices and cooling systems.

On the other hand, during power transfer through the HVAC power line, different types of FACTS devices are used to improve the power systems stability, power quality, and power systems' reliability. The commonly used FACTS devices considered in the power systems are static synchronous compensator (STATCOM), static VAR compensator (SVC), thyristor-controlled reactor (TCR), thyristor-switched reactor (TSR), thyristor-switched capacitor (TSC), superconducting fault current limiter (SFCL), bridge-type fault current limiter (BFCL), resonant type fault current limiter, series dynamic braking resistor (SDBR), etc. (Hossain and Ali, 2015). Compared to an HVDC transmission system, the major advantage of HVAC is low-cost substations, as no power electronic devices are required during the power conversion (Bresesti et al., 2007).

11.4.8 System-Related Challenges

In addition to the specific challenges as outlined above, some system-related challenges include the following (Shafiullah et al., 2010). First, design of an RE system's generator could be an important factor in its integration in the SG. Second, the interaction between one RE system with another RE system of the same or different type could also affect its integration to the SG. The position of the RE plant with respect to the SG is another important factor in the integration process. Admittedly, the nearer the location of the RE system to the grid, the better would be the electrical power strength due to reduced loss. However, for distributed generation of RE, the location of the RE system should be investigated and chosen carefully. Finally, the energy cost is another important factor that should also be investigated while an RE system is integrated in the SG (Shafiullah et al., 2010).

Furthermore, in the smart grid technology, the growing number of random parameters will complicate the mathematical computations of the control algorithms and reliability indexes of the whole power system (Hossain and Ali, 2015). As the smart grid integrates various generation sources, working with the appropriate performance indexes to evaluate the reliability, stability, and sustainability of the whole system is one of the important issues (Hossain and Ali, 2015) that needs to be considered for the SG system.

In another aspect of the SG, as information and communication networks interact with power system networks, there would be several challenges for information and communication systems and networks as well as data centers, which are out of the traditional power system–related operations. For example, energy efficiency in communication networks in the aspects of the SG could be considered as a potential challenge area in the SG framework (Erol-Kantarci and Mouftah, 2015), which would definitely need more research focus when RE resources are to be incorporated in the SG.

11.4.9 V2G Challenges

1) What is V2G?
   V2G refers to the capability of plug-in EVs (PEVs) to provide energy to the grid. The V2G option can be viewed as another form of distributed ES system in the SG. In V2G, the PEVs are equipped with bidirectional energy delivery capability so that, based on demand, the EV can be drawing energy (charging) from the SG during off-peak hours, as well as providing energy (discharging) to the SG during peak hours. Although the conventional thinking is that PEVs can be charging at home from evening till morning, we need to consider that most vehicles stay parked 90% of the time. So there is ample opportunity to allow PEVs to interact with the SG for cost minimization and load shaving. However, this process needs to be carefully controlled by using sophisticated communication mechanisms and infrastructure (Kempton and Tomić, 2005; Peterson et al., 2010; Lund and Kempton, 2008; Turton and Moura, 2008; Zhang et al., 2016).
2) V2G Opportunities and Applications
   Some important opportunities and applications of V2G are outlined below (Hosseini et al., 2012).
   a) Virtual power plant (VPP): A fleet of PEVs can be viewed as a VPP, which can serve as a substitute for intermittent RE sources when needed.
   b) Optimal placement of EVs can be made in the distribution part of the SG to make it more secure and resilient against outages.
   c) The PEVs can reduce the uncertainties associated with RE resources by charging from the SG whenever such resources are available.

d) V2G can also be used for vehicle-to-building (V2B) and vehicle-to-house (V2H) applications.
e) Smart parking lots: Smart parking lots are envisioned to house a fleet of PEVs in a secure environment, which can serve as an interface that provides both charging and discharging services. The PEVs can simply connect to the SG infrastructure to start charging for the duration of the parking, or, get a quick battery replacement by switching uncharged batteries with fully charged batteries available at the lot. The parking lot could also serve as a VPP and aggregate energy from the parked cars to supply energy to the SG.

3) V2G Challenges
Below we discuss some of the key challenges in realizing V2G technology.
a) V2G systems face two major challenges, namely, load shaving and frequency regulation. Load shaving refers to meeting the peak time load by providing energy to the grid during peak demand. For the consumer with PEV, it is advantageous to discharge during peak hours and charge during off-peak hours so that the overall operation cost of the PEV user is minimized. Frequency regulation refers to fine-tuning of the frequency of the grid at a smaller time scale. There is also environmental gain in using V2G, since it reduces the need for fossil fuel resulting in less carbon emissions. However, there are several factors that limit the amount of power that a PEV can provide to the grid, such as battery size, amount of charge in the PEV, capacity of the plug circuit, battery life span, and battery charging/discharging capacity (Hosseini et al., 2012).
b) Energy cost minimization in V2G is a key problem. In V2G, factors like bidirectional energy flow between the vehicle and the grid, vehicle's mobility pattern, battery model and its feasibility of use, and time-of-use-based pricing of consumed electrical energy can be considered as influencing factors in the overall energy cost minimization and therefore considered as open issues in this field (Liang et al., 2012).
c) Fast PEV chargers: Depending on its size and structure, a PEV can consume 4–16 kWh per day. So there is a need for developing fast chargers that are comparable to refueling time (Kisacikoglu et al., 2015). Also, induction and conduction-based methods are available for charging. Induction-based chargers do not require physical contact and therefore are more popular.
d) The energy market policies and regulations are likely to be transformed by the V2G technology. However, to make V2G a significant player in practice, there is a need for an aggregator entity (Xing et al., 2016). This entity will satisfy the demand of the SG by monitoring, controlling, and managing fleets of PEVs.
e) The PEV impacts the voltage regulation and phase balance of the grid, which determines the quality of power. Therefore, careful requirements need to be developed for determining how PEVs are integrated into the SG (Dharmakeerthi et al., 2011).
f) PEV also introduces harmonics into the grid for both low and high penetration levels. Proper charger technology to network filtering needs to be employed to mitigate this (Dharmakeerthi et al., 2011).
g) PEV leads to more power flow in the SG, which leads to more power losses. Charging during off-peaks, proper phase balancing, and using DGs in nearby locations can be used to counter this issue (Dharmakeerthi et al., 2011).

h) Cyber security is an important concern. Confidentiality of transmission information needs to be assured, and tampering of data needs to be prevented at all costs. Security mechanisms to counter suspicious traffic patterns and distributed denial-of-service attacks must be ensured as well. In Zhang et al. (2016), a framework based on software defined radio architecture to implement secured communication and control for V2G has been proposed.

11.4.10 Security Challenges in the High Penetration of RE Resources

Increased power generation from the highly penetrated RE sources offers a greener and more sustainable future. At the same time this also introduces new risks and challenges for the transmission system, e.g., limited control on power generation from the high penetration of RE resources with additional challenges in scheduling and dispatching controllable resources to follow the net load and control the power balance in the system (Navid and Rosenwald, 2012), which needs to be mitigated to maintain the power system stability, power quality, and system reliability. In addition, it has been noticed that RE sources deteriorate the operations of the protective relays during the network faults. RE sources can introduce additional faults on adjacent feeders even after opening the utility protection devices such as protective relay, circuit breaker, and fuse (Baran et al., 2012). This scenario can be more vulnerable to the power system network due to the presence of highly penetrated RE sources. However, this problem can be minimized with the help of proper coordination and optimal operation of the power system protective devices and with the presence of other compensating devices such as FACTS devices, energy storage systems, etc. In addition, with the centralized energy management system of the SG, decentralized control of the high penetration of RE resources and secured communication without delay will facilitate the SG system to ensure the power system protection and stability, power quality, and overall power system reliability.

11.5 Case Studies

In this section, some case study examples have been shown where RE-based power generation systems have been integrated into the SG. Singapore is a city-state with 712.4 square kilometers and 5.183 million people that has undertaken a project to build a microgrid in one of its islands named Pulau Ubin with an area of 10 square kilometers located north-east of the mainland where 100 villagers live (Koh et al., 2012). The electrical consumption of Pulau Ubin island is estimated to be 1.7 MW with approximately 2,500 MWh annual energy consumption in 2009 (Koh et al., 2012). The Energy Market Authority (EMA) of Singapore, a statutory board whose main goal is to provide reliable electrical power to residents, has planned for a RE-based power system. To that plan, they carried out a survey to collect data on load point, electricity consumption, and daily load pattern and applied industry methods to forecast the growth of electrical power consumption till 2020. In the absence of any electrical power grid in the island, the residents use diesel generators to fulfil their electricity demand. Even though diesel generators are expensive, unreliable, somewhat inconvenient, building transmission network from Singapore to Pulau Ubin proved to be economically infeasible due to modest load demand in the island. Thus the authority anticipates that the project would be a "living laboratory" relying on RE in a sustainable manner (Koh et al., 2012). In that context, the main intention is to invite companies to build a microgrid infrastructure where the interested companies would "design, build, own, and operate" their microgrid in a competitive and cost-effective manner.

Another example of integrated RE resources in the SG is the power grid network of India. As of August 2011, India has an installed capacity of 21,000 MW of RE generation with a total estimated potential of 188,700 MW (Mukhopadhyay et al., 2012). In India, wind energy is the top of all RE resources with approximately 15,000 MW out of the 21,000 MW installed capacity. Near future plans of RE projects in India also include creating 20,000 MW solar power grid including 20 million solar lights and 20 million square meters of area for solar thermal energy purposes by 2022 (Mukhopadhyay et al., 2012). The project hopes to be complete in three phases, with phase 1 aiming at setting 1,100 MW of capacity of grid-connected solar power systems, 100 MW of capacity of small, rooftop solar plants, and 200 MW equivalent of off-grid solar power systems (Mukhopadhyay et al., 2012). Energy storage systems, namely, pumped hydro and battery storage, were used in conjunction with RE systems to properly take care of their intermittent nature along with the customer participation, which would necessitate a smart power grid with advanced ICT to optimize RE generation and storage systems as well as their interconnectivity and integration to the SG. In India, initiatives have been taken to meet these goals regarding the integration of RE resources in the SG (Mukhopadhyay et al., 2012).

11.6 Conclusion

Integrating RE resources in the SG is a promising yet challenging task. In this chapter, prospects and the corresponding challenges of integrating RE resources in the SG have been presented. Having provided a quick overview of the SG framework, integration of RE resources has been presented. The opportunities and the corresponding challenges of integrating RE resources in the SG have been categorized and presented separately with supporting references for interested readers in the crossroads of RE and the SG fields. On the search for sustainable energy solutions of emerging societies, it is inevitable that RE resources be integrated with the SG system. To that end, this chapter could certainly serve as a starting point for the students and practicing engineers to start research and development in the interdisciplinary field of RE and the SG. The chapter also provides some case studies on this topic. The RE-integrated SG infrastructure can contribute to the progress toward a sustainable power grid and its efficient management. In particular, a huge amount of opportunity exists for the RE-integrated SG infrastructure for developing countries where system loss in the power grid can be a concern. Although a RE-integrated SG system may seem costly in its installation, in the long run it should become a feasible and environment-friendly solution.

A detailed analysis on modeling RE integration in the SG and its performance evaluation are among the future directions of this field of research. To this end, modeling the effects of variability of RE resources on the performance and stability of the SG system would be worth investigating. It should be noted that the inclusion of RE resources would bring along the most important challenge, namely, the intermittent nature of RE resources themselves, which would require the support systems of the SG to be capable of handling the variability of the RE resources efficiently. Thus, it would be worth investigating the performance of the electrical machines and power electronic circuits necessary in the SG system when RE resources would be integrated. Another possible direction could be to investigate how the business model of the SG system would be affected with the inclusion of RE resources, especially from both utilities' and consumers' points of view. Certainly, the research and development of the RE integration in the SG will be continued to establish this technology intensely to meet up the world electricity demand efficiently.

References

Abbasi, M.A., Ali, E., Islam, M., and Shahid, M.B., 2012, 'Strategies for synchronization and optimization of distributed generation system in a smart grid', Paper presented at: Green and Ubiquitous Technology (GUT), 2012 International Conference on IEEE.

Ahmad, A., Javaid, N., Guizani, M., Alrajeh, N., and Khan, A., 2016, 'An Accurate and Fast Converging Short-Term Load Forecasting Model for Industrial Applications in a Smart Grid', *Industrial Informatics, IEEE Transactions*, 99.

Alain, G., Dominique, R., and Hans-Peter, B., 2015, 'AC line voltage controller for grid integration of renewable energy sources', Paper presented at: Power Electronics and Applications (EPE'15 ECCE-Europe), 2015 17th European Conference on IEEE.

Alam, S.S., Natarajan, B., and Pahwa, A., 2013, 'Impact of correlated distributed generation on information aggregation in smart grid', Paper presented at: Green Technologies Conference, 2013 IEEE.

Alrashidi, M., and El-Hawary, M., 2008, 'Impact of loading conditions on the emission-economic dispatch', *World Academy of Science, Engineering and Technology* 29, 148–151.

Amjady, N., Keynia, F., and Zareipour, H., 2010, 'Short-term load forecast of microgrids by a new bilevel prediction strategy', *IEEE Transactions on Smart Grid* 1, 286–294.

Andrew, X.R.I., Anand, M., and Sivakumar, P., 2011, 'Modelling and analysis of emulator for distributed generation sourced smart grid using digital signal controller (TMS320F28335)', Paper presented at: Innovative Smart Grid Technologies-India (ISGT India), 2011 IEEE PES.

Aroge, A., 2014, 'The Smart Grid and Renewable Energy Integration in Nigeria', Global Energy Network Institute (GENI).

Atasoy, T., Akinc, H.E., and Ercin, O., 2015, 'An analysis on smart grid applications and grid integration of renewable energy systems in smart cities', *In* 2015 International Conference on Renewable Energy Research and Applications (ICRERA), 22–25 Nov. 2015, 547–550.

Atzeni, I., Ordóñez, L.G., Scutari, G., Palomar, D.P., and Fonollosa, J.R., 2013, 'Demand-side management via distributed energy generation and storage optimization', *IEEE Transactions on Smart Grid* 4, 866–876.

Baran, M.E., Hooshyar, H., Shen, Z., and Huang, A., 2012, 'Accommodating high PV penetration on distribution feeders', *IEEE Transactions on Smart Grid*, 3(2), 1039–1046.

Bashir, Z., and El-Hawary, M., 2009, 'Applying wavelets to short-term load forecasting using PSO-based neural networks', *IEEE Transactions on Power Systems* 24, 20–27.

Becerra, A.C., Zeng, W., Chow, M.-Y., and Rodríguez-Andina, J.J., 2015, 'Green city: A low-cost testbed for distributed control algorithms in Smart Grid', Paper presented at: Industrial Electronics Society, IECON 2015-41st Annual Conference of the IEEE.

Bera, S., Ojha, T., Misra, S., and Obaidat, M.S., 2015, 'Cloud-Based Optimal Energy Forecasting for Enabling Green Smart Grid Communication', Paper presented at: 2015 IEEE Global Communications Conference (GLOBECOM, IEEE).

Biswas, S.S., Shariatzadeh, F., Beckstrom, R., and Srivastava, A.K., 2013, 'Real time testing and validation of Smart Grid devices and algorithms', Paper presented at: IEEE Power and Energy Society General Meeting (PES).

Blajszczak, G., Antos, P., and Wasiluk-Hassa, M., 2011, 'Smart grid accommodation of distributed generation for more reliable quality of supply', Paper presented at: Electrical Power Quality and Utilisation (EPQU), 2011 11th International Conference (IEEE).

Boksha, V.V., Foster, R.O., Ignatiev, A.N., Chow, C., Ryjov, A.P., Bulai, P.M., Progler, C., Hague, E., and Wong, A.C., 2014, 'From energy storage to EnerNet: Smart grid for abundant energy', Paper presented at: Smart Grid Conference (SASG), 2014 Saudi Arabia (IEEE).

Boulanger, A.G., Chu, A.C., Maxx, S., and Waltz, D.L., 2011, 'Vehicle electrification: status and issues,' *Proceedings of the IEEE* 99, 1116–1138.

Boutsika, T. N., and Papathanassiou, S. A., (2008), 'Short-circuit calculations in networks with distributed generation', *Electric Power Systems Research* 78, 1181–1191.

Breeze, P., 2014, *Power Generation Technologies*, Newnes.

Bresesti, P., and Cerretti, A., 2007, 'SDNO: Smart Distribution Network Operation Project', Paper presented at: Power Engineering Society General Meeting, 2007 (IEEE).

Bresesti, P. et al., 2007, 'HVDC connection of offshore wind farms to the transmission system', *IEEE Transactions on Energy Conversion*, 22(1), 37–43.

Brown, R., Gupta, S., Christie, R., Venkata, S., and Fletcher, R., (1996), 'Automated primary distribution system design: reliability and cost optimization', Paper presented at: Transmission and Distribution Conference, 1996, Proceedings, 1996 (IEEE).

Cao, J.W., Yu-Xin W., Guo-Yu T., Shu-Qing Z., Ai-Xuan X., Ai-Xuan X., Zhen C., and Chao L., 2013, 'Information system architecture for smart grids', *Chinese Journal of Computers* 36, 143G167.

Caron, S. and Kesidis, G., 2010, 'Incentive-based energy consumption scheduling algorithms for the smart grid', Paper presented at: Smart grid communications (SmartGridComm), 2010 First IEEE International Conference on IEEE.

Cataliotti, A., Di Cara, D., Marsala, G., Pecoraro, A., Ragusa, A., and Tin, G., 2015, 'High frequency modeling technique for three phase power electronics module', Paper presented at: 2015 5th International Conference on Electric Utility Deregulation and Restructuring and Power Technologies (DRPT, IEEE).

Chan, P.P., Chen, W.-C., Ng, W.W., and Yeung, D.S., 2011, 'Multiple classifier system for short term load forecast of Microgrid', Paper presented at: Machine Learning and Cybernetics (ICMLC), 2011 International Conference on IEEE.

Chandler, S.A. and Hughes, J.G., 2013, 'Smart grid distribution prediction and control using computational intelligence', Paper presented at: Technologies for Sustainability (SusTech), 2013 1st IEEE Conference on IEEE.

Chatterjee, A. and Keyhani, A., 2012, 'Neural network estimation of microgrid maximum solar power', *IEEE Transactions on Smart Grid* 3, 1860–1866.

Chen, C., Duan, S., Cai, T., Liu, B., and Hu, G., 2011, 'Smart energy management system for optimal microgrid economic operation', *IET Renewable Power Generation* 5, 258–267.

Chen, H., 2016, *Power Grid Operation in a Market Environment: Economic Efficiency and Risk Mitigation*, John Wiley & Sons.

Chen, X., Dinh, H., and Wang, B., 2010, 'Cascading failures in smart grid-benefits of distributed generation', Paper presented at: Smart Grid Communications (SmartGridComm), 2010 First IEEE International Conference on IEEE.

Cheung, K., Wang, X., Chiu, B.-C., Xiao, Y., and Rios-Zalapa, R., 2010, 'Generation dispatch in a smart grid environment', Paper presented at: Innovative Smart Grid Technologies (ISGT), 2010 (IEEE).

Clarizia, F., Gallo, D., Landi, C., Luiso, M., and Rinaldi, R., 2016, 'Smart meter systems for smart grid management', Paper presented at: Instrumentation and Measurement Technology Conference Proceedings (I2MTC), 2016 IEEE International.

Cui, H., and Dai, W., 2011, 'Multi-objective optimal allocation of distributed generation in smart grid', Paper presented at: Electrical and Control Engineering (ICECE), 2011 International Conference on IEEE.

Daoud, M. I., Abdel-Khalik, A., Massoud, A., Ahmed, S., and Abbasy, N. H., 2012, 'On the development of flywheel storage systems for power system applications: A survey', Paper presented at: Electrical Machines (ICEM), 2012 XXth International Conference on IEEE.

Deng, F., Su, G., Liu, C., and Wang, Z., 2010, 'Global solar radiation modeling using the artificial neural network technique, Paper presented at: 2010 Asia-Pacific Power and Energy Engineering Conference (IEEE)'.

Dharmakeerthi, C., Mithulananthan, N., and Saha, T., 2011, 'Overview of the impacts of plug-in electric vehicles on the power grid', Paper presented at: Innovative Smart Grid Technologies Asia (ISGT), 2011 IEEE PES.

Divan, D., Moghe, R., and Prasai, A., 2014, 'Power Electronics at the Grid Edge: The key to unlocking value from the smart grid', *IEEE Power Electronics Magazine* 1, 16–22.

Driesen, J., and Katiraei, F., 2008, 'Design for distributed energy resources', *IEEE Power and Energy Magazine* 6, 30–40.

Erol-Kantarci, M. and Mouftah, H.T., 2011, 'Wireless Sensor Networks for Cost-Efficient Residential Energy Management in the Smart Grid', *IEEE Transactions on Smart Grid*, 2(2), 314–325.

Erol-Kantarci, M. and Mouftah, H.T., 2015, 'Energy-Efficient Information and Communication Infrastructures in the Smart Grid: A Survey on Interactions and Open Issues', *IEEE Communications Surveys & Tutorials*, 17(1), 179–197.

Fang, X., Misra, S., Xue, G., and Yang, D., 2012, 'Smart Grid—The New and Improved Power Grid: A Survey', *IEEE Communications Surveys & Tutorials* 14(4), 944–980.

Fangxing, L., Wei, Q., Hongbin, S., Hui, W., Jianhui, W., Yan, X., Zhao, X. and Pei, Z., (2010), 'Smart Transmission Grid: Vision and Framework', *IEEE Transactions on Smart Grid* 1(2), 168–177.

Gao, J., Xiao, Y., Liu, J., Liang, W., and Chen, C.P., 2012, 'A survey of communication/networking in Smart Grids', *Future Generation Computer Systems* 28(2), 391–404.

Ghasemi, A. et al., 2016, 'A novel hybrid algorithm for electricity price and load forecasting in smart grids with demand-side management', *Applied Energy* 177, 40–59, URL: http://dx.doi.org/10.1016/j.apenergy.2016.05.083.

Guizani, M. and Anan, M., 2014, 'Smart grid opportunities and challenges of integrating renewable sources: a survey', 2014 International Wireless Communications and Mobile Computing Conference (IWCMC), 4–8 *Aug.* 2014, 1098–1105.

Grond, L., Schulze, P., and Holstein, J., 2013, 'Systems analyses power to gas: a technology review', DNV KEMA Energy & Sustainability, Groningen.

F.E.R. Commission, 2011, 'Assessment of demand response and advanced metering'.

Fangxing, L., Wei, Q., Hongbin, S., Hui, W., Jianhui, W., Yan, X., Zhao, X., and Pei, Z., 2010, 'Smart Transmission Grid: Vision and Framework', *IEEE Transactions on Smart Grid* 1(2), 168–177.

Hamidi, V., Smith, K.S., and Wilson, R.C., 2010, 'Smart grid technology review within the transmission and distribution sector', Paper presented at: 2010 IEEE PES Innovative Smart Grid Technologies Conference Europe (ISGT Europe, IEEE).

He, H., 2010, 'Toward a smart grid: integration of computational intelligence into power grid', Paper presented at: The 2010 International Joint Conference on Neural Networks (IJCNN, IEEE).

He, M.M., Reutzel, E.M., Jiang, X., Katz, R.H., Sanders, S.R., Culler, D.E., and Lutz, K., 2008, 'An architecture for local energy generation, distribution, and sharing', Paper presented at: Energy 2030 Conference, 2008. ENERGY 2008, IEEE.

Hernández, J., Blanco, A.M., and Luna, L.E., 2012, 'Design and installation of a smart grid with distributed generation. A pilot case in the Colombian networks', Paper presented at: Photovoltaic Specialists Conference (PVSC), 2012 38th IEEE.

Hidalgo, R., Abbey, C., and Joós, G., 2011, 'Integrating distributed generation with Smart Grid enabling technologies', Paper presented at: Innovative Smart Grid Technologies (ISGT Latin America), 2011 IEEE PES Conference on IEEE.

Hill, C.A., Such, M.C., Chen, D., Gonzalez, J., and Grady, W.M., 2012, 'Battery energy storage for enabling integration of distributed solar power generation', *IEEE Transactions on Smart Grid* 3, 850–857.

Hinojosa, V., and Hoese, A., (2010), 'Short-term load forecasting using fuzzy inductive reasoning and evolutionary algorithms', *IEEE Transactions on Power Systems* 25, 565–574.

Hong, T., and Wang, P., (2014), 'Fuzzy interaction regression for short term load forecasting', *Fuzzy Optimization and Decision Making* 13, 91–103.

Hossain, M. M., and Ali, M. H., (2015), 'Future research directions for the wind turbine generator system', *Renewable and Sustainable Energy Reviews* 49, 481–489. URL: http://www.sciencedirect.com/science/article/pii/S1364032115003962.

Hossain, M.M. and Ali, M.H., 2016, 'Transient Stability Improvement of Doubly Fed Induction Generator Based Variable Speed Wind Generator Using DC Resistive Fault Current Limiter', *IET Renewable Power Generation* 10(2), 150–157.

Hosseini, S.S., Badri, A., and Parvania, M., 2012, 'The plug-in electric vehicles for power system applications: The vehicle to grid (V2G) concept', Paper presented at: Energy Conference and Exhibition (ENERGYCON), 2012 IEEE International.

Huang, A.Q., Crow, M.L., Heydt, G.T., Zheng, J.P., and Dale, S.J., 2011, 'The future renewable electric energy delivery and management (FREEDM) system: the energy internet', *Proceedings of the IEEE* 99, 133–148.

Huang, C. and Sarkar, S., 2013, 'Dynamic pricing for distributed generation in smart grid', Paper presented at: Green Technologies Conference, 2013 IEEE.

Hubert, T., and Grijalva, S., (2011), 'Realizing smart grid benefits requires energy optimization algorithms at residential level', Paper presented at: Innovative Smart Grid Technologies (ISGT), 2011 IEEE PES.

Hurwitz, J., and Ki, W.-H., 2011, 'ES6: Technologies for smart grid and smart meter', Paper presented at: 2011 IEEE International Solid-State Circuits Conference (IEEE).

Iqbal, Z., Javaid, N., Khan, M.R., Ahmed, I., Khan, Z.A., and Qasim, U., 2016, 'Cost and Load Reduction Using Heuristic Algorithms in Smart Grid', Paper presented at: 2016 30th International Conference on Advanced Information Networking and Applications Workshops (WAINA, IEEE).

Isa, N.B.M., Wei, T.C., and Yatim, A.H.M., 2015, 'Smart grid technology: Communications, power electronics and control system', Paper presented at: Sustainable Energy Engineering and Application (ICSEEA), 2015 International Conference on IEEE.

Jalali, M.M., and Kazemi, A., 2015, 'Demand side management in a smart grid with multiple electricity suppliers', *Energy* 81, 766–776. URL: http://dx.doi.org/10.1016/j.energy.2015.01.027.

Jílek, R., Gurecký, J., Rusnok, S., and Sobota, P., 2015, 'Electricity storage systems using compressed air', Paper presented at: Electric Power Engineering (EPE), 2015 16th International Scientific Conference on IEEE.

Jonnavithula, S. and Billinton, R., 1996, 'Minimum cost analysis of feeder routing in distribution system planning', *IEEE Transactions on Power Delivery* 11, 1935–1940.

Kasbekar, G. S., and Sarkar, S., (2012), 'Pricing games among interconnected microgrids', Paper presented at: 2012 IEEE Power and Energy Society General Meeting (IEEE).

Kavasseri, R.G. and Seetharaman, K., 2009, 'Day-ahead wind speed forecasting using f-ARIMA models', *Renewable Energy* 34, 1388–1393.

Kelouwani, S., and Agbossou, K., 2004, 'Nonlinear model identification of wind turbine with a neural network', *IEEE Transactions on Energy Conversion* 19, 607–612.

Kempton, W., and Tomić, J., 2005, 'Vehicle-to-grid power implementation: From stabilizing the grid to supporting large-scale renewable energy', *Journal of Power Sources* 144, 280–294.

Khosla, R., and Ichalkaranje, N., 2013, *Design of intelligent multi-agent systems: human-centredness, architectures, learning and adaptation*, Springer.

Kiani, A. and Annaswamy, A., 2012, 'Distributed hierarchical control for renewable energy integration in a Smart Grid', *In:* 2012 IEEE PES Innovative Smart Grid Technologies (ISGT), 16–20 Jan. 2012, 1–8.

Kim, H., Kim, Y.-J., Yang, K., and Thottan, M., 2011, 'Cloud-based demand response for smart grid: Architecture and distributed algorithms', Paper presented at: Smart Grid Communications (SmartGridComm), 2011 IEEE International Conference on IEEE.

Kinsy, M.A., Celanovic, I., Khan, O., and Devadas, S., 2013, 'MARTHA: architecture for control and emulation of power electronics and smart grid systems', Paper presented at: Design, Automation & Test in Europe Conference & Exhibition, 2013 (IEEE).

Kirtley, J., 2010, *Electric Power Principles: Sources, Conversion, Distribution and Use*, John Wiley & Sons, Ltd.

Kisacikoglu, M.C., Kesler, M., and Tolbert, L.M., 2015, 'Single-phase on-board bidirectional PEV charger for V2G reactive power operation', *IEEE Transactions on Smart Grid* 6, 767–775.

Kloubert, M.L., Spieker, C., Schwippe, J., Ulf, H., Rehtanz, C., and Blanco, G., 2014, 'Impact of the integration of renewable energies on the reactive power demand in the German transmission grid', IEEE PES Innovative Smart Grid Technologies, Europe, 12–15 Oct. 2014, 1–6.

Kobayashi, T., 2014, 'Power electronics technology in smart grid projects-Applications and experiences', Paper presented at: 2014 International Power Electronics Conference (IPEC-Hiroshima 2014-ECCE ASIA, IEEE).

Koh, L.H., Tan, Y.K., Wang, P., and Tseng, K.J., 2012, 'Renewable energy integration into smart grids: Problems and solutions—Singapore experience', *In*: 2012 IEEE Power and Energy Society General Meeting, 22–26 July, 2012, 1–7.

Komninos, N., Philippou, E., and Pitsillides, A., 2014, 'Survey in smart grid and smart home security: Issues, challenges and countermeasures', *IEEE Communications Surveys & Tutorials* 16(4), 1933–1954.

Kong, F., Dong, C., Liu, X., and Zeng, H., (2014), 'Quantity Versus Quality: Optimal Harvesting Wind Power for the Smart Grid', *Proceedings of the IEEE* 102(11), 1762–1776.

Kroposki, B., Margolis, R., Kuswa, G., Torres, J., Bower, W., Key, T., and Ton, D., 2008, 'Renewable systems interconnection: Executive summary', *In*: National Renewable Energy Laboratory (NREL), Golden, CO.

Kumar, N., Misra, S., Rodrigues, J., Lee, J.-H., Obaidat, M.S., and Chilamkurti, N., 2015, 'Playing the Smart Grid Game: Performance Analysis of Intelligent Energy Harvesting and Traffic Flow Forecasting for Plug-In Electric Vehicles', *IEEE Vehicular Technology Magazine* 10, 81–92.

Lakshminarayana, S., Quek, T.Q.S., and Poor, H.V., 2014, 'Combining cooperation and storage for the integration of renewable energy in smart grids', *In*: Computer Communications Workshops (INFOCOM WKSHPS), 2014 IEEE Conference on, April 2014, 622–627.

Lasseter, R.H., and Paigi, P., 2004, 'Microgrid: a conceptual solution', Paper presented at: Power Electronics Specialists Conference, 2004, PESC 04, 2004 IEEE 35th Annual.

Lee, J.Y., and Choi, S.G., 2014, 'Linear programming based hourly peak load shaving method at home area', Paper presented at: 16th International Conference on Advanced Communication Technology (IEEE).

Li, H.A., and Nair, N.-K.C., 2015, 'Multi-agent systems and demand response: A systematic review', Paper presented at: Power Engineering Conference (AUPEC), 2015 Australasian Universities (IEEE).

Liang, H., Choi, B. J., Zhuang, W., and Shen, X., (2012), 'Toward optimal energy store-carry-and-deliver for PHEVs via V2G system', Paper presented at: INFOCOM, 2012 Proceedings IEEE.

Liang, H., Tamang, A.K., Zhuang, W., and Shen, X.S., 2014, 'Stochastic information management in smart grid', *IEEE Communications Surveys & Tutorials* 16, 1746–1770.

Liserre, M., Sauter, T., and Hung, J.Y., 2010, 'Future Energy Systems: Integrating Renewable Energy Sources into the Smart Power Grid Through Industrial Electronics', *Industrial Electronics Magazine*, IEEE, 4(1), 18–37.

López, G., de Almeida, A., Moura, P., Pérez, M., Moreno, J., and Blanco, L., 2013, 'Monitoring system for the local distributed generation infrastructures of the smart grid', Paper presented at: Electricity Distribution (CIRED 2013), 22nd International Conference and Exhibition on IET.

Lund, H., and Kempton, W., 2008, 'Integration of renewable energy into the transport and electricity sectors through V2G', *Energy Policy* 36, 3578–3587.

Ma, Y., Zhou, L., Tse, N., Osman, A., and Lai, L., 2009, 'An initial study on computational intelligence for smart grid', Paper presented at: 2009 International Conference on Machine Learning and Cybernetics.

Maharjan, L., Yamagishi, T., and Akagi, H., 2012, 'Active-power control of individual converter cells for a battery energy storage system based on a multilevel cascade PWM converter', *IEEE Transactions on Power Electronics* 27, 1099–1107.

Mohamed, M.A., Eltamaly, A.M., Farh, H.M., and Alolah, A.I., 2015, 'Energy management and renewable energy integration in smart grid system', in Smart Energy Grid Engineering (SEGE), 2015 IEEE International Conference on, 1–6.

Mohd, A., Ortjohann, E., Schmelter, A., Hamsic, N., and Morton, D., (2008), 'Challenges in integrating distributed energy storage systems into future smart grid', Paper presented at: 2008 IEEE international symposium on industrial electronics (IEEE).

Molderink, A., Bakker, V., Bosman, M.G.C., Hurink, J.L., and Smit, G.J.M., 2010, 'Management and Control of Domestic Smart Grid Technology', *IEEE Transactions on Smart Grid* 1(2), 109–119.

Mukhopadhyay, S., Soonee, S.K., Joshi, R., and Rajput, A.K., 2012, 'On the progress of renewable energy integration into smart grids in India', *In*: 2012 IEEE Power and Energy Society General Meeting, 22–26 July, 2012, 1–6.

Navid, N., and Rosenwald, G., 2012, 'Market solutions for managing ramp flexibility with high penetration of renewable resource', *IEEE Transactions on Sustainable Energy* 3(4), 784–790.

Olivares, D. E., Mehrizi-Sani, A., Etemadi, A. H., Cañizares, C. A., Iravani, R., Kazerani, M., Hajimiragha, A. H., Gomis-Bellmunt, O., Saeedifard, M., and Palma-Behnke, R., (2014), 'Trends in microgrid control', *IEEE Transactions on Smart Grid* 5, 1905–1919.

Palma-Behnke, R., Benavides, C., Aranda, E., Llanos, J., and Sáez, D., 2011, 'Energy management system for a renewable based microgrid with a demand side management mechanism', Paper presented at: 2011 IEEE symposium on computational intelligence applications in smart grid (CIASG, IEEE).

Pătru, I.I., Carabaş, M., Bărbulescu, M., and Gheorghe, L., 2016, 'Smart home IoT system', 15th RoEduNet Conference: Networking in Education and Research, 1–6.

Perez, F., Custódio, J.F., de Souza, V.G., Ribeiro Filho, H.K., Motoki, E.M., and Ribeiro, P.F., 2015, 'Application of energy storage elements on a PV system in the smart grid context', Paper presented at: Innovative Smart Grid Technologies Latin America (ISGT LATAM), 2015 IEEE PES.

Peterson, S.B., Whitacre, J., and Apt, J., 2010, 'The economics of using plug-in hybrid electric vehicle battery packs for grid storage', *Journal of Power Sources* 195, 2377–2384.

Pourbeik, P., Williams, S.E., Weber, J., Sanchez-Gasca, J., Senthil, J., Huang, S., and Bolton, K., 2015, 'Modeling and Dynamic Behavior of Battery Energy Storage: A Simple Model for Large-Scale Time-Domain Stability Studies', *IEEE Electrification Magazine* 3, 47–51.

Qiu, R.C., Hu, Z., Chen, Z., Guo, N., Ranganathan, R., Hou, S., and Zheng, G., 2011, 'Cognitive radio network for the smart grid: experimental system architecture, control algorithms, security, and microgrid testbed', *IEEE Transactions on Smart Grid* 2, 724–740.

Rahbar, K., Xu, J., and Zhang, R., 2015, 'Real-time energy storage management for renewable integration in microgrid: An off-line optimization approach', *IEEE Transactions on Smart Grid* 6, 124–134.

Rahimi, F., and Ipakchi, A., 2010, 'Demand response as a market resource under the smart grid paradigm', *IEEE Transactions on Smart Grid* 1, 82–88.

Rana, M.M., and Li, L., 2015, 'Distributed generation monitoring of smart grid using accuracy dependent kalman filter with communication systems', Paper presented at: Information Technology-New Generations (ITNG), 2015 12th International Conference on IEEE.

Reddy, M.J.B., Rajesh, D.V., Gopakumar, P., and Mohanta, D.K., 2014, 'Smart fault location for smart grid operation using RTUs and computational intelligence techniques', *IEEE Systems Journal* 8, 1260–1271.

Rehmani, M.H., Kantarci, M.E., Rachedi, A., Radenkovic, M., and Reisslein, M., 2016, 'IEEE Access Special Section Editorial Smart Grids: a Hub of Interdisciplinary Research', *IEEE Access* 3, 3114–3118.

Repo, S., Maki, K., Jarventausta, P., and Samuelsson, O., 2008, 'ADINE-EU demonstration project of active distribution network', Paper presented at: SmartGrids for Distribution, 2008. IET-CIRED. CIRED Seminar (IET).

Rugthaicharoencheep, N., and Boonthienthong, M., 2012, 'Smart grid for energy management on distribution system with distributed generation', Paper presented at: Cyber Technology in Automation, Control, and Intelligent Systems (CYBER), 2012 IEEE International Conference on IEEE.

Sáez, D., Ávila, F., Olivares, D., Cañizares, C., and Marín, L., 2015, 'Fuzzy prediction interval models for forecasting renewable resources and loads in microgrids', *IEEE Transactions on Smart Grid* 6, 548–556.

Salinas, S., Li, M., Li, P., and Fu, Y., 2013, 'Dynamic energy management for the smart grid with distributed energy resources', *IEEE Transactions on Smart Grid* 4, 2139–2151.

Sardi, J., and Mithulananthan, N., 2014, 'Community energy storage, a critical element in smart grid: A review of technology, prospect, challenges and opportunity', Paper presented at: Engineering Technology and Technopreneuship (ICE2T), 2014, 4th International Conference on IEEE.

Sevilla, F.R.S., Park, C., Knazkins, V., and Korba, P., 2016, 'Model predictive control of energy systems with hybrid storage', Paper presented at: Power and Energy Society General Meeting (PESGM), 2016 (IEEE).

Shafiullah, G.M., Oo, A.M.T., Jarvis, D., Ali, A.B.M.S., and Wolfs, P., 2010, 'Potential challenges: Integrating renewable energy with the smart grid', Universities Power Engineering Conference (AUPEC), 2010 20th Australasian, 5–8 Dec., 1–6.

Sharma, H., and Kaur, G., 2016, 'Optimization and simulation of smart grid distributed generation: A case study of university campus', Paper presented at: Smart Energy Grid Engineering (SEGE), 2016 IEEE.

Shengnan, S., Pipattanasomporn, M., and Rahman, S., 2011, 'Demand Response as a Load Shaping Tool in an Intelligent Grid With Electric Vehicles', *IEEE Transactions on Smart Grid* 2(4), 624–631.

Sirisukprasert, S., 2014, 'Power electronics-based energy storages: A key component for Smart Grid technology', Paper presented at: Electrical Engineering Congress (iEECON), 2014 International (IEEE).

Sotkiewicz, P.M., and Vignolo, J.M., 2006, 'Nodal pricing for distribution networks: efficient pricing for efficiency enhancing DG', *IEEE Transactions on Power Systems PWRS* 21, 1013.

Sparacino, A.R., Reed, G.F., Kerestes, R.J., Grainger, B.M., and Smith, Z.T., 2012, 'Survey of battery energy storage systems and modeling techniques', Paper presented at: 2012 IEEE Power and Energy Society General Meeting (IEEE).

Logenthiran, T., and Srinivasan, D., 2009, 'Short Term Generation Scheduling of a Microgrid', IEEE 2009, TENCON, 1–6.

Su, H.-I., and El Gamal, A., 2011, 'Modeling and analysis of the role of fast-response energy storage in the smart grid', Paper presented at: Communication, Control, and Computing (Allerton), 2011 49th Annual Allerton Conference on IEEE.

Suslov, K., Gerasimov, D., and Solodusha, S., 2015, 'Smart grid: Algorithms for control of active-adaptive network components', Paper presented at: PowerTech, 2015 IEEE Eindhoven (IEEE).

Tang, Y., 1996, 'Power distribution system planning with reliability modeling and optimization', *IEEE Transactions on Power Systems* 11, 181–189.

Thomas, M. S., and McDonald, J. D., (2015), *Power System SCADA and Smart Grids*, CRC Press.

Turton, H., and Moura, F., 2008, 'Vehicle-to-grid systems for sustainable development: An integrated energy analysis', *Technological Forecasting and Social Change* 75, 1091–1108.

Tushar, W., Zhang, J.A., Yuen, C., Smith, D.B., and Poor, H.V., 2015, 'Smart grid energy management for a shared facility controller with renewables', Paper presented at: 2015 IEEE International Conference on Smart Grid Communications (SmartGridComm, IEEE).

Twidell, J. and Weir, T., 2006, *Renewable Energy Resources*, London: Taylor & Francis.

Vale, Z., Faria, P., Morais, H., Khodr, H., Silva, M., and Kadar, P., 2010, 'Scheduling distributed energy resources in an isolated grid—An artificial neural network approach', Paper presented at: IEEE PES General Meeting (IEEE).

Vaziri, M., Vadhva, S., Oneal, T., and Johnson, M., 2011, 'Smart grid, distributed generation, and standards', Paper presented at: 2011 IEEE Power and Energy Society General Meeting (IEEE).

Venayagamoorthy, G.K., 2011, 'Dynamic, stochastic, computational, and scalable technologies for smart grids', *IEEE Computational Intelligence Magazine* 6, 22–35.

Wan, C., Zhao, J., Song, Y., Xu, Z., Lin, J., and Hu, Z., 2015, 'Photovoltaic and solar power forecasting for smart grid energy management', *CSEE Journal of Power and Energy Systems* 1, 38–46.

Wang, G., Kowli, A., Negrete-Pincetic, M., Shafieepoorfard, E., and Meyn, S., 2011, 'A control theorist's perspective on dynamic competitive equilibria in electricity markets', *IFAC Proceedings* 44, 4933–4938.

Wang, J., Kennedy, S., and Kirtley, J., 2010, 'A new wholesale bidding mechanism for enhanced demand response in smart grids', Paper presented at: Innovative Smart Grid Technologies (ISGT), 2010 (IEEE).

Wang, X., Zhang, Y., Chen, T., and Giannakis, G.B., 2016, 'Dynamic Energy Management for Smart-Grid-Powered Coordinated Multipoint Systems', *IEEE Journal on Selected Areas in Communications* 34, 1348–1359.

Wang, Y., and Saad, W., 2015, 'On the role of utility framing in smart grid energy storage management', Paper presented at: 2015 IEEE International Conference on Communication Workshop (ICCW, IEEE).

Wang, Y., Saad, W., Mandayam, N.B., and Poor, H.V., 2014, 'Integrating energy storage into the smart grid: A prospect theoretic approach', Paper presented at: 2014 IEEE International Conference on Acoustics, Speech and Signal Processing (ICASSP, IEEE).

Wen, M.H.F., and Li, V.O.K., 2014, 'Form Follows Function: Designing Smart Grid Communication Systems Using a Framework Approach', *IEEE Power and Energy Magazine* 12(3), 37–43.

Werbos, P.J., 2011, 'Computational intelligence for the smart grid-history, challenges, and opportunities', *IEEE Computational Intelligence Magazine* 6, 14–21.

Whittingham, M.S., 2012, 'History, evolution, and future status of energy storage', *Proceedings of the IEEE* 100, 1518–1534.

Wierman, A., Liu, Z., Liu, I., and Mohsenian-Rad, H., 2014, 'Opportunities and challenges for data center demand response', Paper presented at: Green Computing Conference (IGCC), 2014 International (IEEE).

Xing, H., Fu, M., Lin, Z., and Mou, Y., 2015, 'Decentralized Optimal Scheduling for Charging and Discharging of Plug-In Electric Vehicles in Smart Grids', IEEE Transactions on Power Systems 31, 4118– 4127.

Xu, F.-Y., Leung, M., and Zhou, L., (2010), 'A RBF network for short—Term Load forecast on microgrid', Paper presented at: 2010 International Conference on Machine Learning and Cybernetics.

Yu, X., Cecati, C., Dillon, T., and Simoes, M.G., 2011, 'The new frontier of smart grids', *IEEE Industrial Electronics Magazine* 5, 49–63.

Zhang, S., Li, Q., Wu, J., Li, J., and Li, G., 2016a, 'A security mechanism for software-defined networking based communications in vehicle-to-grid', Paper presented at: Smart Energy Grid Engineering (SEGE), 2016 IEEE.

Zhang, Y., Gao, L., and Xu, Y., 2016b, 'Distributed energy management for supply-demand coordination in a smart grid', Paper presented at: Transmission and Distribution Conference and Exposition (T&D), 2016 IEEE/PES.

Zhang, Y., Rahbari-Asr, N., Duan, J., and Chow, M.Y., 2016, 'Day-ahead Smart Grid Cooperative Distributed Energy Scheduling with Renewable and Storage Integration', *IEEE Transactions on Sustainable Energy*, 1–10.

Zhao, B., Shi, Y., Dong, X., Luan, W., and Bornemann, J., 2014, 'Short-term operation scheduling in renewable-powered microgrids: A duality-based approach', *IEEE Transactions on Sustainable Energy* 5, 209–217.

Zhao, H., and Tang, Z., 2016, 'The review of demand side management and load forecasting in smart grid', Paper presented at: Intelligent Control and Automation (WCICA), 2016 12th World Congress on IEEE.

Zhu, J., Chen, M.Z., Zuo, Z., and Du, B., 2014, 'A new pricing scheme for controlling energy storage devices in smart grid', Paper presented at: 2014 American Control Conference (IEEE).

12

Environmental Monitoring for Smart Buildings

Petros Spachos[1] *and Konstantinos Plataniotis*[2]

[1] *School of Engineering, University of Guelph, Canada*
[2] *Department of Electrical and Computer Engineering, University of Toronto, Canada*

12.1 Introduction

Environmental monitoring describes the processes and activities that need to take place to characterize and monitor the quality of the environment. Indoor air quality (IAQ) refers to the quality of the air within and around buildings and structures. Environmental and IAQ monitoring is an important concern since it relates directly to the health and comfort of building occupants. Occupants of buildings with poor IAQ report a wide range of health problems, which are often called sick building syndrome (SBS) or tight building syndrome (TBS), building-related illness (BRI), and multiple chemical sensitivities (MCS) (CCOHS (2013)).

Common issues associated with IAQ include improper or inadequately maintained heating and ventilation systems as well as contamination by construction materials (glues, fiberglass, particle boards, paints, etc.) and other chemicals. At the same time, the increase in the number of building occupants and the time spent indoors directly impact the IAQ (CCOHS (2013)). Everyday, people work in their offices or sit on meeting rooms where the quality of the air can affect not only their performance but also their health.

There are many sources of indoor air contaminants:

- Carbon dioxide (CO_2), perfume, and tobacco smoke, which come from building occupants.
- Dust and gases from building materials.
- Toxic vapors and volatile organic compounds (VOCs) from workplace cleaners.

Some of these pollutants can be created also from indoor activities such as smoking and cooking. IAQ problems are more prevalent in indoor infrastructures such as houses, offices, and schools (EPA (2014)). It is necessary to develop of an accurate system for environmental and IAQ monitoring system that can be used in the current buildings as well as the smart buildings of the near future.

Traditionally, environmental monitoring includes a small number of expensive and high-precision sensor units. The units are placed in fractions of the building and monitor the quality of the air. Collected data are retrieved directly from the equipment at the end

Transportation and Power Grid in Smart Cities: Communication Networks and Services, First Edition.
Edited by Hussein T. Mouftah, Melike Erol-Kantarci and Mubashir Husain Rehmani.

of the experiment after the units have been recovered. The limited number of sensor units, due to their cost, along with the lack of correlation in real-time data collection is the main drawback of this approach.

Recent progress in wireless communication and sensor technologies can provide better, real-time monitoring of the environmental condition through smart sensors with lower cost and wireless capabilities. Every day more computer-based devices are connected to the Internet. Most of these devices have at least one sensing unit, creating opportunities for direct integration between the physical world and computer-based systems. This is the idea behind the Internet of things (IoT), a development of the Internet in which everyday objects have network connectivity, allowing them to send and receive data (Fang et al. (2014)).

In the near future, wireless sensor networks (WSNs) are expected to be integrated into the IoT and consequently to smart cities and smart buildings (Jin et al. (2014b), Bellavista et al. (2013a)). The sensing infrastructures have a major role in the IoT and great research opportunities. Sensor nodes can join the Internet dynamically and use it to collaborate and accomplish their tasks. The future Internet, designed as an IoT, is foreseen to be a world-wide network of interconnected objects uniquely addressable, based on standard communication protocols. Similar to the IoT, sensors will have a crucial role in smart cities (Jin et al. (2014a), Cardone et al. (2013)). Green cities, which span from urban wildlife initiatives to urban agriculture and green transport networks, citizen sensing, and smart cities projects are emerging that attempt to realize improved sustainability through greater urban connectivity.

At the same time, the sensor node market continues to grow rapidly, as the cost to build custom nodes decreases due to technology scaling. Moreover, the use of embedded sensors in smartphones, phablets, tablets, and mobile devices has enabled a number of popular applications (Sheng et al. (2013), Lathia et al. (2013)). Advances in technology, engineering, and materials science have opened the door for increasingly sophisticated sensors to be used in environmental monitoring research. The flexibility of sensor nodes permits fast time-to-market, which is crucial in the development of new sensor network applications. From the cost perspective, embedded sensor in mobile devices are already superior to application-specific hardware. As an example, an indoor positioning system (IPS), which based on magnetic and other sensor data or a network of devices is more flexible, secure, and inexpensive in comparison to the global positioning system (GPS) (Shin et al. (2012)).

Environmental monitoring serves a vital scientific role by revealing long-term trends that can lead to new knowledge and better understanding of the environment. WSNs are ideal systems for environmental monitoring. Inch scale sensors with low cost can be deployed over large areas to collect data. The use of WSNs for monitoring applications covers a wide area of the environmental monitoring (Oliveira and Rodrigues (2011), Othman and Shazali (2012)), from air and water quality monitoring to rain forest and biodiversity monitoring. Sensor nodes are the elementary components of any WSN, and they can provide many functionalities including signal conditioning and data acquisition, temporary storage for the data, data processing, analysis of the processed data, self monitoring (e.g., supply voltage), scheduling, receipting and transmission of data packets, and coordination and management of communications and networking.

However, they also pose a number of challenges, with survivability being one of the most crucial. The lifetime of any individual node and as a consequence, of the whole network, is solely decided by how the limited amount of energy is utilized. Energy harvesting can alleviate this problem. Self-powered WSNs provide the possibility of very long sensor node lifetimes while their deployment would have the least impact on the existing infrastructure. Moreover, with a carefully designed routing mechanism (Al-Karaki and Kamal (2004)), the total network lifetime can be extended compared with traditional battery-powered WSNs.

The objective of this chapter is to introduce an environmental monitoring system for smart buildings. It focuses on the requirements and challenges of a real-time monitoring system for CO_2 in a smart building and describes a prototype that was developed to meet the needs of the application.

The rest of the chapter is organized as follows: Section 12.2 gives a brief description of the wide area of WSN-monitoring applications. Section 12.3 introduces the design goal along with the requirements and challenges of the specific application. Section 12.4 introduces the proposed architecture for the proposed system, followed by Section 12.5 with the experimental results and discussion. Section 12.6 concludes the chapter.

12.2 Wireless Sensor Networks in Monitoring Applications

WSNs are a promising solution for a plethora of monitoring applications due to their low cost and low power characteristics (Akyildiz et al. (2002), Yick et al. (2008)). They are used from general surveillance (Ghataoura et al. (2011)) to structural health monitoring in buildings (Torfs et al. (2013)) and highway bridges (Sazonov et al. (2009)).

In recent years, environmental-monitoring applications have grown rapidly in areas such as agricultural monitoring (Kays et al. (2011), Bencini et al. (2009)), habitat monitoring (Mainwaring et al. (2002)), and climate monitoring (Martin et al. (2014)). The IoT along with the smart cities concept (Zanella et al. (2014), Spachos et al. (2015)) brings new areas for deployment for environmental monitoring. At the same time, it also brings different requirements and challenges.

Environmental-monitoring applications can be broadly categorized into outdoor and indoor monitoring (Arampatzis et al. (2005)). In outdoor-monitoring applications usually the WSN is deployed in a remote location and harsh environment with limited energy sources. Outdoor-monitoring applications include chemical hazard detection (Jan et al. (2010)) and air pollution detection (Suganya and Vijayashaarathi (2016)) with challenges such as weather condition and energy limitations. In indoor-monitoring applications, the WSN is deployed in an indoor environment along with others that might use the wireless channels. Indoor-monitoring applications include building and office monitoring (Torfs et al. (2013)) and parking space detection (Corral et al. (2012)) with challenges such as the absence of accurate location techniques and the interference from other wireless communication applications.

Environmental condition monitoring in homes have been examined in Kelly et al. (2013). The authors proposed a framework to monitor temperature, humidity, and light intensity, which is based on a combination of pervasive distributed sensing units, information system for data aggregation, and reasoning and context awareness. The reliability of the sensing information is encouraging.

Recently a number of systems have been proposed for carbon dioxide monitoring (Spachos and Hatzinakos (2016)). In Yang et al. (2013), a remote carbon dioxide concentration monitoring system is developed. The system reports geological CO_2, temperature, humidity, and light intensity of the outdoor monitoring area. Similarly, in

Mao et al. (2012) an urban CO_2 monitoring system is presented. The system operates outdoors at an urban area around 100 square kilometers.

Indoor environments can pose different challenges to a monitoring system. Indoor and outdoor air quality monitoring through a WSN is presented in Postolache et al. (2009). Each node has an array of sensors, and it is connected to the central monitoring unit either hardwired or wirelessly. In Kim et al. (2014), a real-time indoor air quality monitoring system is proposed. The system has seven sensors monitoring seven different gases. In Jelicic et al. (2013), a system with aggressive energy management at the sensor level, node level, and network level is presented. The system detects volatile organic compounds (VOCs) and CO and saves energy through context-aware adaptive sampling. A low-power ZigBee sensor network to monitor VOCs pollution levels in indoor environments is proposed in Peng et al. (2015). The network consists of end device sensors with photoionization detectors and routers.

12.3 Application Requirements and Challenges

Environmental monitoring for smart buildings has a number of requirements while it poses a few constraints and challenges. The monitoring area and the application scenario affects both the requirements and the challenges while the monitoring target, the sampling rate, and the overall deployment cost are some of the parameters that should be considered in the design decisions of the system. At the same time, wireless network coexistence and dynamic environment changes are challenges that any proposed solution should overcome.

12.3.1 Monitoring Area

Many of the design decisions are affected by the selection of the monitoring area. When the requirements and challenges of the monitoring area are well defined, the design decisions for the proposed monitoring system can be optimized. Most of the time, environmental monitoring refers to the air pollution (Boubrima et al. (2017)) and general environment conditions, so the monitoring system can be deployed either indoors or outdoors. The monitoring area can be further categorized to simple or complex based on the surrounding area.

At an outdoor-monitoring system, the nodes are deployed outside in an open area, similar to the one shown in Figure 12.1. In such monitoring area, the communication between the nodes is easier when line of sight (LoS) between the nodes is possible. A proper deployment of the nodes can minimize the interference and enhance the communication between neighboring nodes that are within LoS. Another advantage is the potential usage of renewable energy resources such as wind and solar energy. The nodes can use these resources to extend their operation lifetime and consequently extend the total network lifetime. Solar panels and wind turbines can be used for energy harvesting in order to extend the network lifetime significantly. Moreover, without the need of battery replacement, the system can work unattended in remote locations and for long periods.

An outdoor-monitoring system can also make use of the Global Positioning System (GPS) for more accurate target detection and optimal communication. Nodes can

Figure 12.1 An outdoor self-powered wireless sensor network.

have access to GPS information not only for target tracking but also for route decision making. There are a number of routing protocols in the literature that can optimize the energy consumption, if the location of the nodes in the network is known (Pantazis et al. (2013), Zorzi and Rao (2003), Maihofer (2004), Spachos and Hantzinakos (2014), Spachos et al. (2012)). However, the use of GPS will increase not only the energy requirements for each monitoring node, but it will also increase the cost per node and the total system cost.

The surrounding area can be categorized further in simple and complex. A monitoring system deployed in an isolated area can use wireless communication with minimum or no interference with other networks. The nodes deployment is easier while most of them will be in LoS. On the other hand, the deployment of a monitoring system at an urban environment is more complicated (Bellavista et al. (2013b)). The available space and location for the nodes is limited while the network has to overcome requirements such as the proper placement of the sensors in comparison with other devices in the area and challenges, such as the communication between the nodes through noise and overcrowded channels and frequencies. Moreover, these networks are characterized by multi-hop lossy links while the nodes are exposed to weather conditions.

On the other hand, an indoor-monitoring system would have its unique requirements and pose different challenges than an outdoor system. One of the main challenges of the indoor environment for wireless communication is the non-line-of-sight (NLoS) problem. The LoS path can be blocked, and the communications are conducted through reflections and diffractions (Spachos et al. (2011)). Multiple contributions between communicating nodes are constructed by reflections, transmissions and, diffractions on the building structures. At the same time, energy harvesting is more challenging (Chirap et al. (2014), Yu and Yue (2012)) while the associated cost can be high.

Regarding the location of the nodes, it should be determined through different techniques (Guo et al. (2010), Hood and Barooah (2011)), since GPS is not available. This increases the complexity of the overall system design, especially, if the nodes are mobile and they need to know their location while they are moving within the monitoring area.

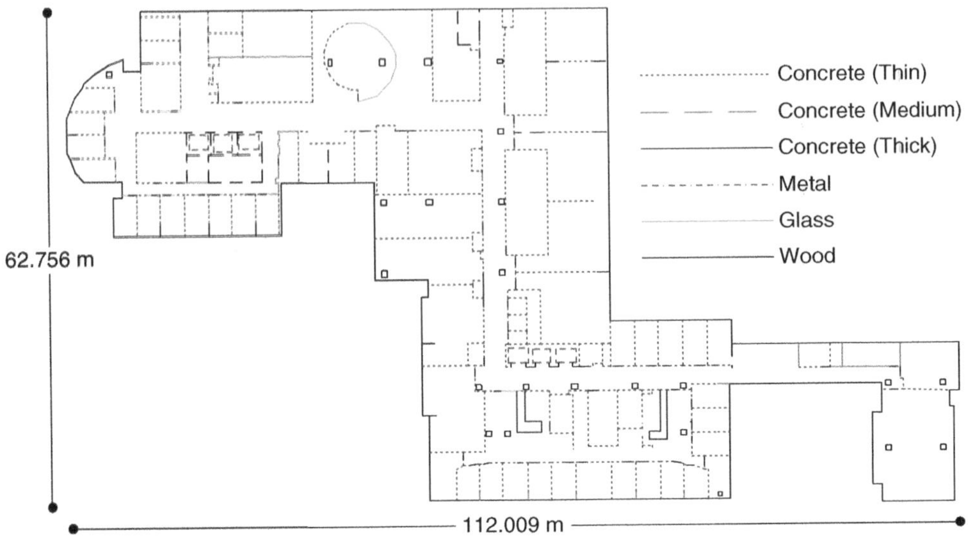

Figure 12.2 The 4th floor of Bahen Centre for Information Technology building at the University of Toronto, Canada. It is a complex indoor environment with a number of classrooms, offices, and open areas. The materials are categorized into six types: thin, medium, and thick concrete, metal, glass, and wood. Each material has a different effect on the wireless signals.

The monitoring area can be simple, inside an office with many open areas, or complex, covering the floor of a building, similar to the one shown in Figure 12.2. In this area, the materials have been categorized into six types: thin, medium, and thick concrete, metal, glass, and wood. Every material type has different characteristics that have different effects on the wireless communication between the nodes. This is a good example of a complex indoor environment and presents a lot of scatterers and obstacles such as walls, pillars, wooden doors, etc. These are of great influence when the radio links in such areas are examined. An efficient monitoring system has to take into consideration those objects to correctly predict the communication link availability in the area. Attenuation, distortion, and noise increase the complexity of the system for reliable communication.

12.3.2 Application Scenario and Design Goal

Along with the monitoring area, the application scenario is another important factor of an efficient monitoring system. In this work, a real-time monitoring system for the concentration of CO_2 at a whole floor of a crowded university building is developed. A number of wireless sensor nodes with gas leak detection capabilities and simple relay nodes are deployed in the monitoring area. The wireless sensor nodes are equipped with a gas sensor module and a wireless transmission module. The data from the sensor module are passed to the transmission module. The transmission module will forward all the necessary data to the control room, through the relay nodes. The relay nodes have only communication capabilities. Both the sensor units and the relay nodes are powered with batteries.

The floor has rooms of different sizes (offices, laboratories, meeting rooms) and number of occupants while the areas have different materials, as discussed in the previous section. Many people spend most of their time inside the building, either in their offices or in meeting rooms. The quality of the air along with the environmental conditions in the room can affect not only their health but also their general performance. General solutions that are applied to every room in the building, such as average room temperature, are not efficient while most of the time the temperature of an area of the floor is considered. The lack of real-time data of the condition inside each room decreases the efficiency of these systems while some times it ends up being more costly.

During the day, the number of people in the rooms changes; hence, the quality of the air changes. A smart heating, ventilation, and air conditioning (HVAC) system should take into consideration these parameters and take different actions in every case. An efficient environmental monitoring system should collect useful data from each room in the building, forward the data to a control room, and finally take actions toward the improvement of the environment in each room separately. For instance, the conditions in a meeting room with 10 people should be different from a similar meeting room with more occupants or with no occupants at all. The air conditioning system of an office should change based not only on the temperature of room but also on the quality in the air. Dynamic changes in the environment, such as gases from cooking, an open window, or an increase of the number of occupants in the room, need time to alter the temperature of the room, which is usually the trigger factor of the air conditioning. An environmental monitoring system for a smart building should be aware of the dynamic changes and respond as soon as possible to these changes.

The proposed system was deployed at a region of the 7th floor of Bahen Centre for Information Technology at the University of Toronto, shown in Figure 12.3. The offices that were used are for researchers, students, and visitors and a meeting room was used as well. None of the offices or the meeting room have windows and all are connected to the same ventilation system. During day time, each office has either one or two people working, while more people can come and stay in each room. The meeting room can have between 14 and 18 people.

The main focus of this work is to improve the indoor air quality of these rooms. The proposed solution will work along with the existing HVAC system and improve its efficiency. At the same time, the privacy of the occupants is also considered; hence, no cameras or any other visual sensors were used to count the number of the occupants in the room or decide the existence of people in the monitoring area.

12.3.3 Requirements

Following the described application and the monitoring area, the proposed system needs to meet the following requirements.

12.3.3.1 Sensor Type

The first requirement involves the proper selection of the sensor type. For indoor air quality monitoring there are three types of gas sensors that can be used:

1. Electrochemical
2. Optical
3. Metal Oxide Semiconductor (MOS)

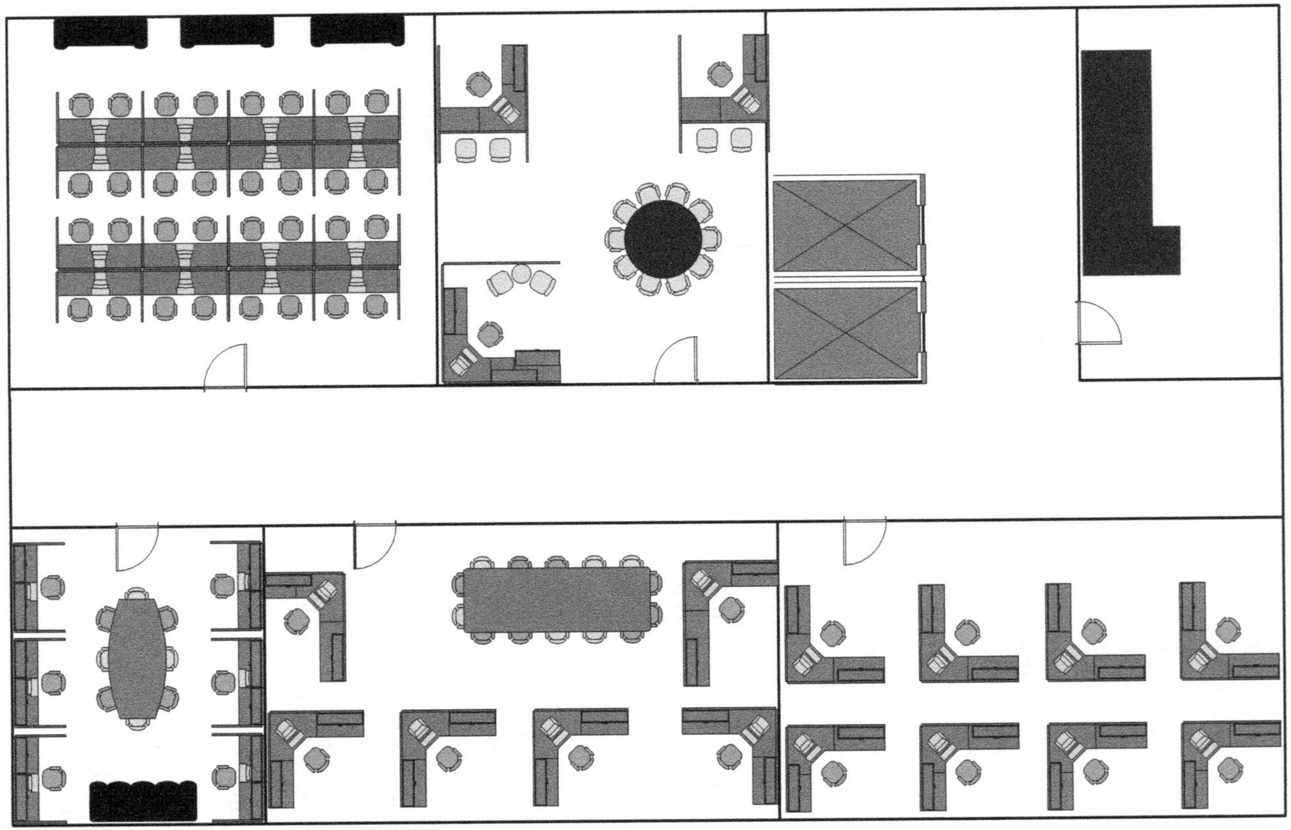

Figure 12.3 A region of the 7th floor of Bahen Centre for Information Technology building at the University of Toronto, Canada, that was used for experimentation.

Electrochemical sensors can be used to monitor several toxic gases such as carbon monoxide and oxides of nitrogen. The gas goes through a chemical reaction, producing a current. The current is directly proportional to the concentration of the gas in the environment. Electrochemical sensors have high sensitivity, with low energy requirements. However, they have relative short lifetimes, their response times can be long, while their costs and their sizes vary.

Optical sensors can be used for carbon dioxide gas determination with long lifetime, high sensitivity, and low energy consumption. Also, the warm up and the response time of optical sensors is short. However, the cost of the sensors is high.

A MOS sensor is a popular choice due to the low cost, the small size, and the long lifetime. The warm up time of these sensors is long, but the response time is short. Most of the indoor air quality sensors belong to the MOS type. The proposed system will use MOS sensors due to their advantages in comparison with the other two sensor types for the selected application and the monitoring area.

12.3.3.2 Real-Time Data Aggregation

Real-time data aggregation is an important requirement. The monitoring data should reach the control room on time and after the necessary processing, the optimal condition should be applied to each room of the smart building. There is an upper bound limit on the transmission delay of the data, after which the data are not useful any more. This requirement poses significant challenges on the selection of the routing protocol. A routing protocol should not only try to minimize the energy requirement of the wireless nodes but also deliver the data on time for further processing.

12.3.3.3 Scalability

The proposed system should cover a wide area with hundreds or even thousands of nodes. It is necessary for any proposed solution to prove that is well suited to large scale WSNs. The system scalability and node mobility are important requirements. A high number of sensors can be installed in a building while some of them might need to be placed in other parts of the building after a certain time. The framework should be scalable to include more sensors while it should also be able to provide connectivity to mobile nodes which change location over time.

At the same time, the number of sensors in the monitoring area is important. It is affected by the reliability and energy requirements of the sensor, the sensing area, and the cost. The more reliable and accurate are the sensor readings, the less sensors are needed to cover the monitoring area while the sensor calibration is faster. Similarly, when the coverage area of a sensor is large, the monitoring area can be covered with a small number of nodes. However, hardware malfunction and dynamic changes in the communication between nodes can affect the overall system performance. In general, as the number of sensors increases, the system accuracy and reliability increases as well. On the other hand, as the number of sensors increases, the cost and the network traffic increases, and more complex algorithms are necessary for reliable communication.

Another important aspect of the network scalability and the number of required nodes is the placement of them at the monitoring area. A detailed study of the monitoring area can help improve the sensor placement, minimize their number, and optimize the communication among them. For instance, the distance between the sensor and an open window or the ventilation system can affect the data. A placement of the sensor away from factors that can affect its readings would increase the reliability of the system.

12.3.3.4 Usability, Autonomy, and Reliability

The monitoring devices should be easy to use, small in size, and low cost. Each monitoring node should be able to join or leave the network at any time without the need of a special configuration. This would minimize the need for administration and maintenance. At the same time, the size of the nodes is important in order to be placed easily in a complex indoor environment. The cost of the device will affect the overall design cost of the proposed solution and many of the requirements, as described in the previous section.

The nodes should also be able to work autonomously. Batteries must be able to power the nodes during the whole deployment. Due to the wireless communication and the associated energy consumption, the network has to be efficient energy-wise, even if a renewable source of energy is used. This also affects the routing protocol. The protocol should turn off the radios periodically while it has to make sure that the remaining active nodes are able to cover all the monitoring area.

Moreover, the network should be able to perform simple operations to prevent unexpected crashes or general hardware problems in order to minimize the need for maintenance. End user maintenance should be avoided, since they may not have networking knowledge or because the areas of interest are most often remote. Achieving reliability is difficult because packet losses are more likely to happen during high network traffic, which are at the same time the most interesting episodes for data analysis.

12.3.3.5 Remote Management

The proposed system will be installed in a wide area while the nodes will be placed in locations that are hard to be visited regularly. Hence, remote access is necessary to operate, manage, and reprogram the nodes and to configure the whole system, regardless of manufacturer.

These are the minimum requirements for an efficient environmental-monitoring system. When the application changes, some other requirements come into consideration.

12.3.4 Challenges

Environmental monitoring in smart buildings faces a number of challenges. The more specific are the requirements of the system, the better estimation can be made regarding the expected challenges.

12.3.4.1 Power Management

For long-term operation of the system, an efficient power management approach should be followed. Environmental monitoring in smart buildings is used to minimize the energy requirements. Hence, the proposed system should have minimum energy requirements and be able to operate for long time. The energy footprint of the system should be small and, as a consequence, the energy consumption of the monitoring nodes should be low.

Power management also affects the routing protocol and the general design of the prototype. The protocol should minimize the communication between the nodes while it should force an energy efficient duty cycle scheme for the nodes. In this way, energy can be saved in the nodes, and the overall lifetime of the network can be extended.

The overall design of the system should take into consideration the power management. Although different sensors can improve the data accuracy, the energy requirements of the sensor is important. Since the application is specific and the requirements are clear, sensors that meet the requirements and have low energy consumption should be selected.

12.3.4.2 Wireless Network Coexistence

Another important challenge for every indoor monitoring systems is wireless network coexistence. The monitoring system is deployed in an area where there are other wireless connections. The environment of a building has a number of other wireless transmissions between laptops, desktops, smart phones, routers, etc. Especially the unlicensed band of 2.4 GHz is overcrowded. Devices that operate at the same place over the same spectrum need to have mechanisms to minimize collision, get access, and share the medium. Spectrum availability detection, interference mitigation, and spectrum sharing are requirements that every wireless system needs to properly address.

12.3.4.3 Mesh Routing

The mesh networks topologies can both provide multi-hop and path diversity. A routing protocol to support multi-hop mesh network is crucial, which must take into account the very limited features of the network. The routing protocol should also support the time and delay requirements of the application.

12.3.4.4 Robustness

The network must account for a lot of problems such as poor radio connectivity or hardware failures. The use of any protocol requiring an initialization phase to be performed synchronously by all nodes is thus inconceivable.

12.3.4.5 Dynamic Changes

Another challenge is the dynamic changes in the environment. In smart buildings, people are moving between rooms, and they can temporarily block the signals. At the same time, smart mobile devices also move between rooms, changing the interference and the link availability over time. The proposed system should be able to adopt to dynamic changes and transmit the information to the control room.

12.3.4.6 Flexibility

Nodes can join or leave the network based on the application needs for coverage in different areas. For instance, more nodes can be placed in a meeting room during a meeting while nodes can be removed from offices that are not used any more. It is crucial for the nodes to be able to automatically detect their neighborhood and minimize the need for reconfiguration.

12.3.4.7 Size and cost

The nodes should be small in size in order to be flexible in their placement in the area. Also, nodes might need to be placed in different rooms; hence, the node size is important to move the nodes around. Finally, all the requirements and challenges should be addressed with the minimum cost. The necessary hardware components and their cost should be minimum, especially for large-scale deployments.

12.4 Wireless Sensor Network Architecture

In this section, the general framework that was used in the proposed system is described, followed by a brief description of the hardware components and the data processing.

12.4.1 Framework

The system framework is shown in Figure 12.4. It has three core units: the wireless monitoring nodes, the wireless ad hoc network, and the control room. The implementation of each unit is based on the specific application; however, some minor modification can make the framework applicable to other environmental monitoring systems with similar principles. The core units have the following principles:

- **Wireless monitoring nodes:** This is the main sensor unit. A number of sensors for environmental monitoring can be connected to a transmission node. The transmission node is necessary to have a radio module in order to forward the data to the main wireless ad hoc network toward the control room. A low-cost and low-energy microcontroller can be used at the sensor unit for the initial filtering of the sensor data before the transmission. The monitoring nodes can be placed in different locations, and their location can change after the initial network configuration while more nodes can join the network or nodes can leave the network. It is important for the monitoring node to be flexible in placement and able to cope with dynamic changes. The nodes are powered with batteries while the developed prototype can also be powered through solar panels and store energy in rechargeable batteries (Spachos and Hatzinakos (2014)).
- **Relay nodes:** The relay nodes are used to provide connectivity between the sensor nodes and the control room. The relay nodes form a mesh network with a number of multi-hop connections toward the destination. They have radio modules similar to the modules at the monitoring nodes, and they implement the routing protocol. The next node selection and the spectrum selection and sharing decision is taking place in the relay nodes. If storage is available at the relay nodes, then they can also store packets.

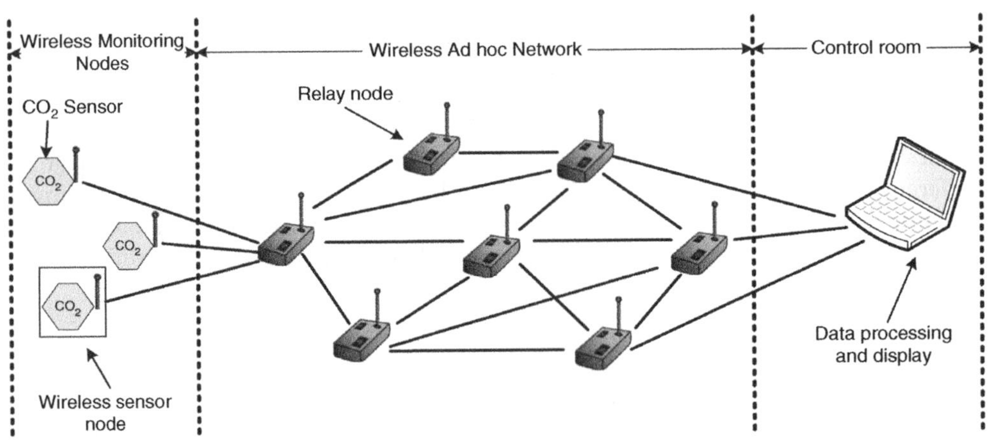

Figure 12.4 System framework of indoor CO_2 monitoring system.

However, the cost of the relay nodes should be lower than the cost of the monitoring nodes while they also need to be able to join and leave the network dynamically. The relay nodes are also powered with batteries, but their power requirements are lower than the monitoring nodes, since there is no sensor unit attached to the relay nodes.
- **Control room:** Main data processing, aggregation, and storage is taking place at the control room. The control room is at a remote location and can communicate with the monitoring nodes through the relay nodes. After collecting the monitoring data, the control room makes a decision about the environmental conditions in each room of the building. These decisions are related with the HVAC operation to improve the air quality in each room separately.

12.4.2 Hardware Infrastructure

The selection of the necessary hardware components should follow the requirements and the challenges, as described in previous sections, and also follow the main principles of the general framework. The introduced testbed has the following hardware components:

- **Sensor:** The iAQ-2000 carbon dioxide (CO_2) sensor was used. iAQ-2000 is a cost-effective solution to accurately monitor the air quality of an indoor environment (CO_2Meter). This sensor communicates using the RS-232 communication protocol and consumes 46 mW of power while active, making it ideal for long-term monitoring applications. The monitored CO_2 levels are reported in a parts-per-million (ppm) value, which can be correlated to the air quality of an indoor environment. The iAQ-2000 sensor is shown in Figure 12.5, and Table 12.1 shows the sensor specifications.
- **Microcontroller and radio module**. The OPM15 development board by RapidMesh was used as the main communication board (OMESH Networks). The OPM15 provides wireless capabilities for a network through its opportunistic wireless mesh radio. This low-power device (41 mW during active cycle) is designed for low bandwidth

Figure 12.5 iAQ-2000 sensor.

Table 12.1 iAQ-2000 specifications.

Type	MOS
Substances detected	CO, CH_4, LPG, Alcohols, Ketones, Organic acids, Amines, Aliphatic hydrocarbons, Aromatic hydrocarbons
Power supply	5 V
Power consumption	30 mA

Figure 12.6 OPM15 relay node.

applications. The microchip PIC18F26K22 programmable microcontroller enables networks to be quickly developed through open-source C-based applications. As well, the OPM15 is easily interfaced to additional components through SPI, I^2C, or RS-232 and has multiple free analog and digital ports available. The microcontroller along with the radio module is shown in Figure12.6, and Table 12.2 shows the specification of the board.

- **Power Converter**
  A prototype multifunctional power converter was used (Spachos and Hatzinakos (2015)). The assembly kit is show in Figure 12.7(a), and the circuit is shown in Figure 12.7(b). The prototype provides an all-in-one solution to power management and interfacing requirements between common hardware components. The prototype is capable of boosting and regulating an input voltage to a fixed 5V, has a built-in shutoff to protect the battery source from over-draining, is capable of recharging a lithium polymer (LiPo) battery with the aid of an energy harvesting device, and provides terminals to access the input power source's voltage. A toggle switch also enables the power to be solely drawn from an energy-harvesting device instead of a LiPo battery.

Table 12.2 OPM15 specifications.

Radio range	**20 *m***
Frequency range	2.405 – 2.483 *GHz*
Channels	3
Bandwidth (per channel)	5 *MHz*
Receiver sensitivity	−94 *dBm*
Transmitting power	−25 *dBm*
Power supply	3.3 *V*
Power cons. (Sleep)	1 *uA*
Power cons. (Work)	25 *mA*

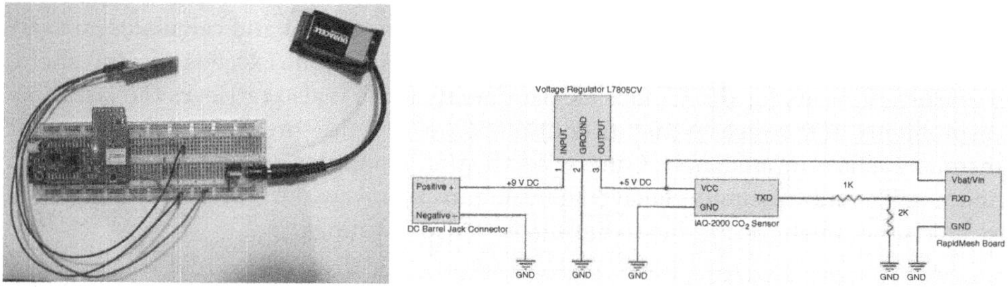

(a) Assembly kit (b) Circuit

Figure 12.7 The indoor version of the prototype with (a) the assembly kit of the monitoring board, where a sensor unit is connected to a radio module through (b) a simple electric circuit.

12.4.3 Data Processing

An important part of the monitoring system is the data processing. The collected data are not of equal importance while sometimes it might be corrupted due to unreliable readings. It is necessary that an initial processing takes place at the radio module, before data transmission. The module processes the data collected from the sensor and transmits only the necessary information to the destination. In this way both the bandwidth and energy are saved. The system minimizes useless data transmissions, minimizing the traffic in the network and the energy consumption per node. The data is transmitted into packets. The data is processed and stored at the control room.

12.4.3.1 Noise Reduction, Data Smoothing, and Calibration

One of the major problems of using low-cost MOS sensors is their unreliable readings. The sensors may report unreliable or corrupted data for a few readings. The sensor unit should decide whether the data are due to sensor malfunction or failure or there is an event in the area. This task should take place at the sensor node before the data transmission.

To reduce the sensor noise, a mobile sensor was used, similar to (Tsujita et al. (2004)). The mobile node moves in each of the monitoring rooms once every hour. The mobile node is equipped with the same sensor as the sensor units. The mobile node compares its readings with the readings reported from the different sensors in every room, for one minute. If the readings are close, the mobile sensor moves to the next sensor unit. If the sensor readings are within $\pm 4\sigma$ from the mobile sensor reading, the mobile sensor informs the control room for the necessary calibration. The calibration can take place remotely through the configuration at the control room. When the difference between the sensor unit and the mobile sensor is greater than the threshold above, there might be a need to replace the sensor node, and the system administrator needs to check the sensor on the site. The mobile node periodically updates the sensor units with the expected values according the previous levels in the area. This is important for the sensor units to detect potential outliers that can occur for a few readings.

To identify outliers and remove them from the measurements, a smoothing algorithm is also followed. When a random spike occurs within a window of time, the node ignores the data and does not transmit it. The selection of the window size follows the monitoring gas specifications. In the case of indoor CO_2, the proposed system has a window of 6 sec., 3 sec. before the possible outlier and 3 sec. after. To meet this requirement, the radio module stores the last 10 measurements of the sensor and calculates an average threshold for the monitoring area. When a measurement exceeds the threshold, the sensor unit waits for three more measurements while it also retrieves the last three measurements. If the measurements continue to exceed the threshold, the sensor unit transmits the data immediately. Otherwise, the unit computes an average value for the random spike and transmits the average value. In this way, the system minimizes the false alerts that might occur due to unreliable sensor readings.

12.4.3.2 Packet formation process

Every monitoring node forms packets and transmits data toward the destination. Packets are created every second, following the sampling rate of the sensor. The packet formation process is shown in Figure 12.8. Every packet has the following four fields:

- **Sensor identifier:** This identifier comes from the sensor module, and it is related with its product serial number. It is unique for every sensor module in the network. The identifier stores the number of the transmitter that sends the packet and allows the receiver to identify the origin of the packet. For every sensor ID there is a corresponding location stored at the control room. The location can be updated if the sensor is moved to a new location.

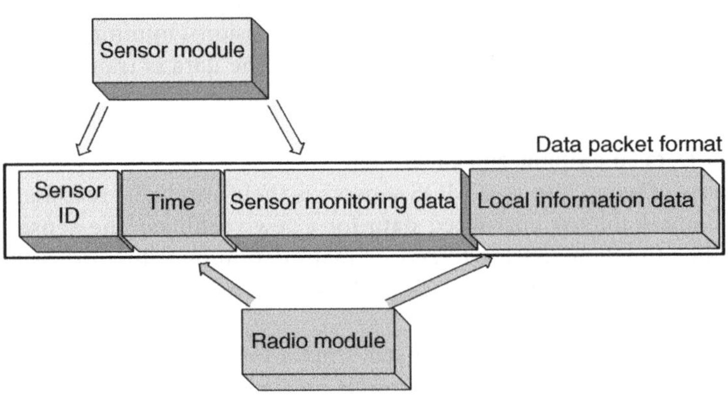

Figure 12.8 Data packet format.

- **Time:** The time comes from the radio module. This field stores the exact time when the packet was created. When a relay node receives the packets, it can place them in order based on the timestamps. For a CO_2 monitoring application, the time is of high importance since it shows the variation of the CO_2 levels over time.
- **Sensor monitoring data:** The data passed from the sensor to the radio module. The data field stores all the crucial information related with the CO_2 concentration.
- **Local information data:** This information comes from the radio module. This field includes information such as the remaining power level of a unit and average received signal strength indication (RSSI) for location information maintenance. RSSI can be used later on, when the traffic is low in the network, in order to update the location of the monitoring and relay nodes in the area.

12.4.3.3 Information Processing and Storage

The final information processing and storage takes place at the control room. Each packet is decomposed into different parts (sensor ID, time, data, etc.), and each part is plotted in real time. Further to real-time processing, each part is stored into files to keep record of the data over time. The data is available to the system administrator off-line for future reference.

The stored data is also used from the mobile node for calibration purposes. There are expected normal levels of CO_2 in the different rooms of a building. This information is given to the mobile node to help with the calibration of the sensor units as well as with the initial configuration of newly deployed sensors. The information processing, transmission, and reception are shown in Figure 12.9.

12.4.4 Indoor Monitoring System

At the proposed monitoring system application, cognitive networking (Yucek and Arslan (2009), Liang et al. (2008)) is used along with opportunistic routing (Liu et al. (2009)) in a wireless multi-hop network, for CO_2 monitoring. An overview of the different system components is shown in Figure 12.10.

The final indoor monitoring board has two types of nodes: the wireless sensor nodes (monitoring boards) and the relay nodes. The wireless sensor nodes as well as the relay nodes are deployed in the building. The wireless sensor nodes know their relative location. The relay nodes will find their location during the initialization phase through the use of the RSSI. Each wireless sensor node is able to monitor the area around it, form packets, and forward these packets to one of the neighbor relay nodes, which will then forward the packet to the destination following the principles of the routing protocol.

12.5 Experiments and Results

In this section, the experimental setup is described, followed by a brief discussion of the results.

12.5.1 Experimental Setup

For the experiments of the proposed system, the 7th floor of Bahen Centre for Information Technology at the University of Toronto was used. A simplified floor map of the monitoring area is shown in Figure 12.11.

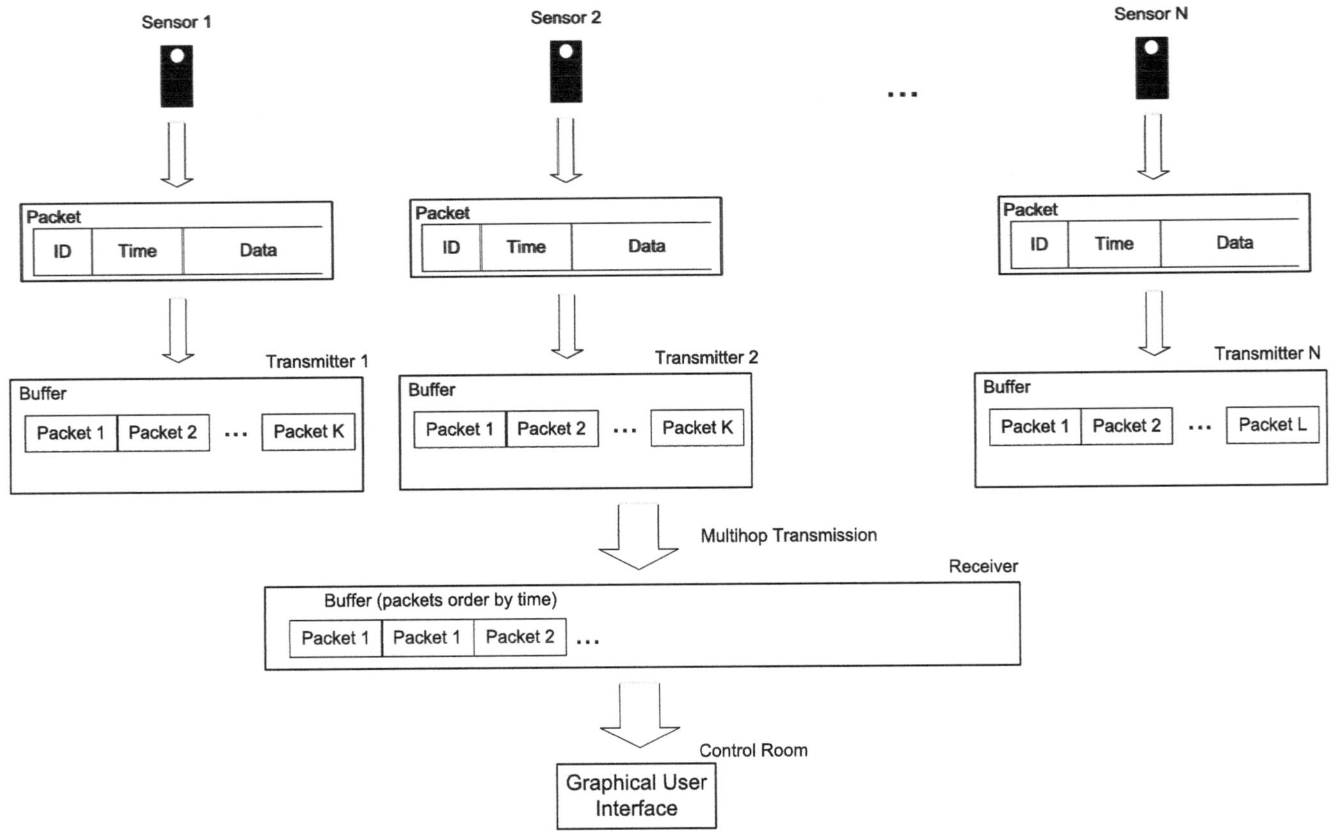

Figure 12.9 Packet process and transmission/reception.

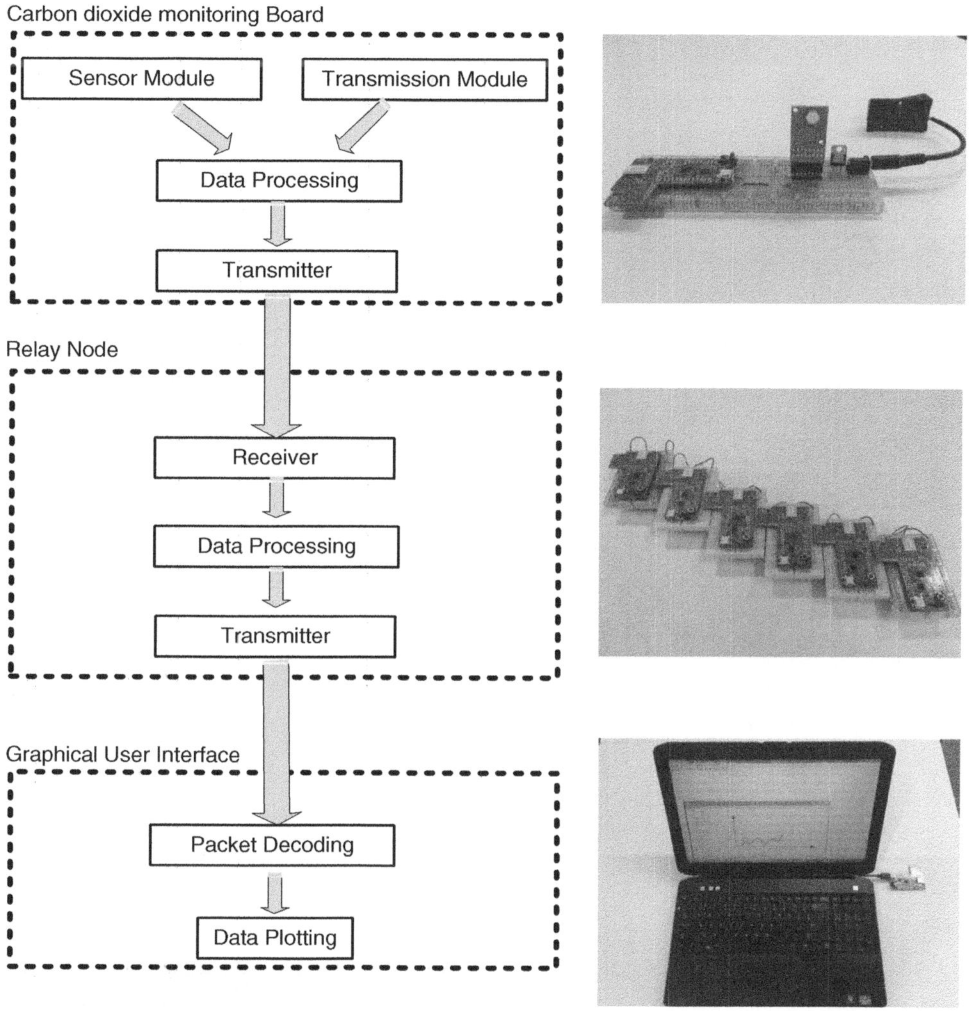

Figure 12.10 Overview of the monitoring system.

During the experiment, the CO_2 concentration of four rooms was monitored: two offices, one laboratory, and one meeting room. The rooms were selected due to the different sizes and different number of occupants they can have during the day. Specifically:

- office -1, dimension $3m \times 5m$, occupants (max) 3
- office -2, dimension $3m \times 5m$, occupants (max) 4
- laboratory, dimension $3m \times 6m$, occupants (max) 4
- meeting room, dimension $3m \times 6m$, occupants (max) 18

In all the monitoring rooms, the windows were closed during the experiment while the doors work with a mechanism that keeps them closed. The rooms are connected to the same central ventilation system. The system divides the rooms in clusters and controls the temperature of the different clusters in the floor.

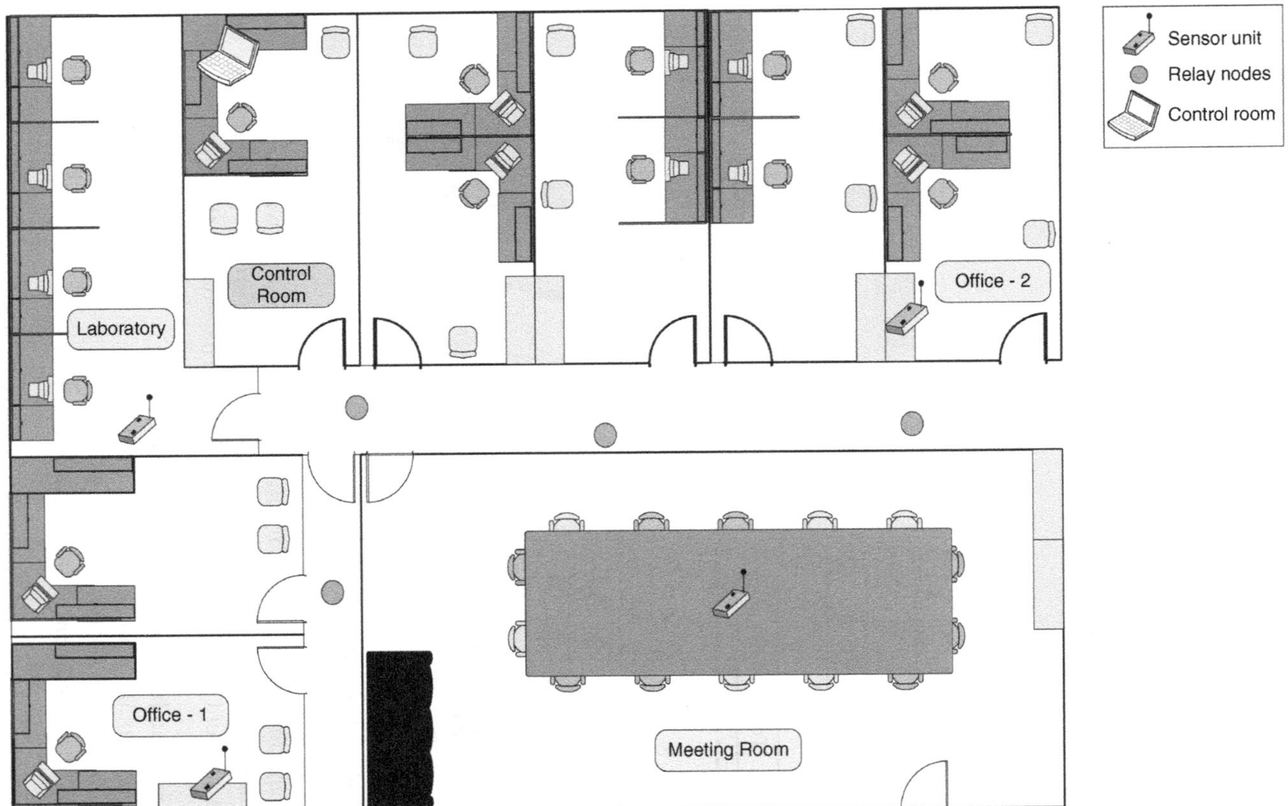

Figure 12.11 An indoor CO_2 detection and monitoring system deployed at the 7th floor of Bahen Centre for Information Technology at the University of Toronto. The sensor units have been deployed in four rooms: office -1, office -2, laboratory, and meeting room. The sensor units transmit the monitoring data to the control room through the relay nodes.

Following the specifications of the rooms and the transmission range and the sensing range of the relay nodes and the sensor units, four sensor units were deployed, one in each room, and four relay nodes were placed outside each room. In the offices and the laboratory, the sensor units were placed at equal distances from the door and the workstation while in the meeting room the sensor unit was placed in the center of the room, hanging from the ceiling. The relay nodes follow a simple routing protocol that forwards the packets to the control room. The control room has a communication unit, similar to the relay nodes and sensing units, which collects all the data. The communication unit is attached to a laptop computer that processes the data, plots them in real time, and stores them in a database.

The experiments lasted for one month, between March 1 and March 31. The mobile node visited each monitoring room once per day for any necessary calibration. After the initial deployment, the system needed 15 minutes to reach a stable level. Both the relay nodes and the sensor unit were moved throughout the experimental period, and the new positions were found with the use of the RSSI value. There was no hardware failure during the experimentation while some of the nodes needed battery replacement.

12.5.2 Results Analysis

Initially, the information processing unit was evaluated. During the experiments, some of the data were corrupted or not accurate. Outliers needed to be detected and removed to improve system reliability.

The sensor data before and after the proposed data smoothing algorithm in office- 2 and for 12 hours is shown in Figure 12.12. The sensor unit in office- 2 stored all the sensor

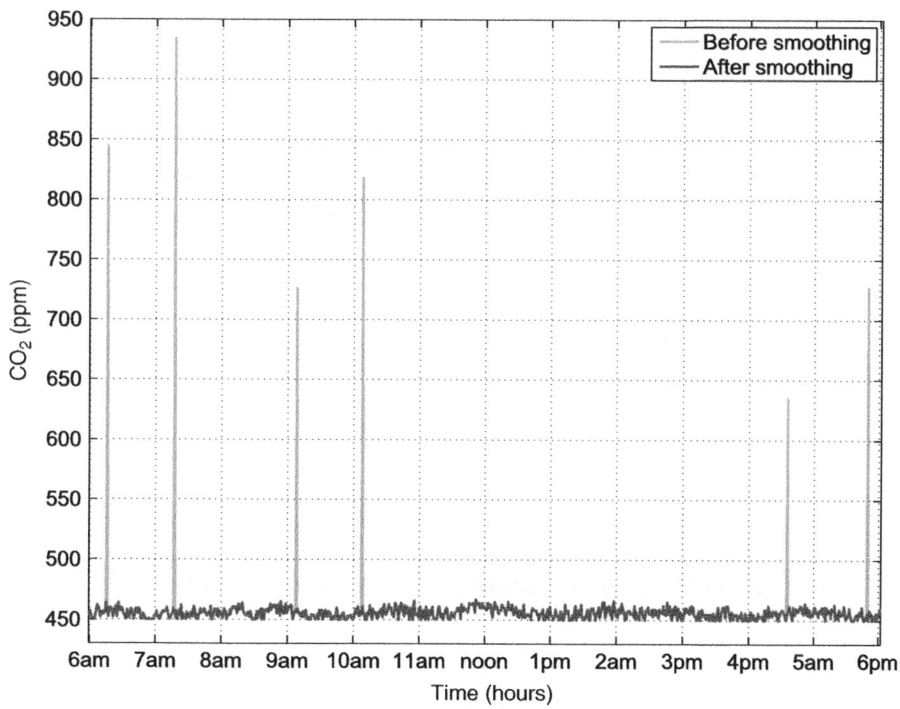

Figure 12.12 Time series representation of air quality measurements at office - 2 over 12 hours before and after smoothing.

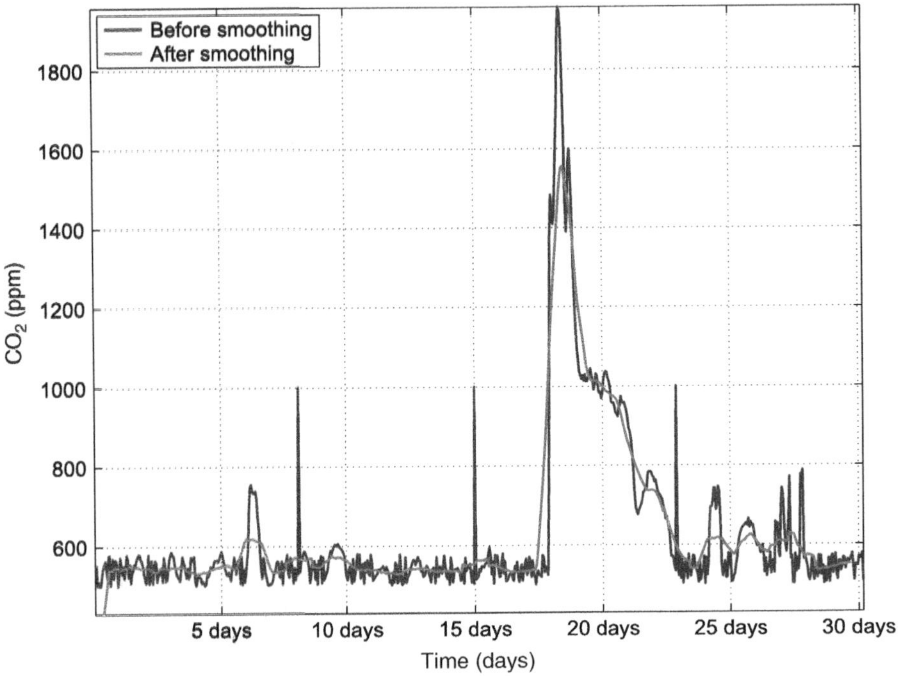

Figure 12.13 Time series representation of air quality measurements at meeting room over March 2015 before and after smoothing.

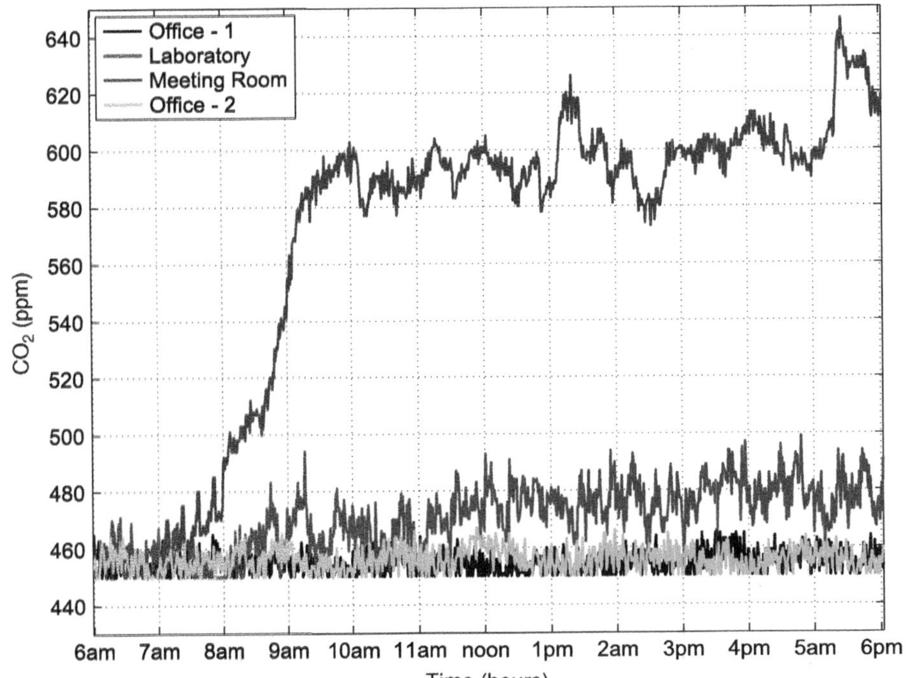

Figure 12.14 Time series representation of air quality measurements over 12 hours.

data during the experiments. These data are compared with the data the sensor transmits toward the control room, after the processing. It is clear that periodic outliers for a few seconds are cleared and ignored from the system. The decision for the potential outlier is taken at the sensor unit and is based on the outlier window of 10 sec as described in the previous section.

Similarly, the sensor data before and after smoothing in the meeting room and for a months is shown in Figure 12.13. The smoothing this time takes place at the control room. The algorithm examines the reported data in an hour period and decides the value. In a month period, the smoothing algorithm is necessary to remove noise from the data that was not removed locally inside each monitoring node.

The concentration of CO_2 for 12 hours, from the morning to the afternoon, is shown in Figure 12.14. The measurements are during the peak hours at the examined rooms. Several people are moving between the rooms, working on computers, and opening and closing the doors. Both people and computers contribute to the CO_2 levels. It is expected that when the number of people in the room increases, the CO_2 concentration increases as well.

In the early morning, between 6 *am* and 8 *am* all the rooms report similar CO_2 levels. This is expected, since the rooms are empty. Also, there is not great variation between the different room sizes since the size should not affect the CO_2 levels.

As people start coming into the examined area, there are some noticeable differences. The concentration in the two offices increases; however, because the number of the people is low, the increase is small. In the laboratory, the increase is higher than in the offices. This is due to the higher number of people who visit the laboratory during the day. It is

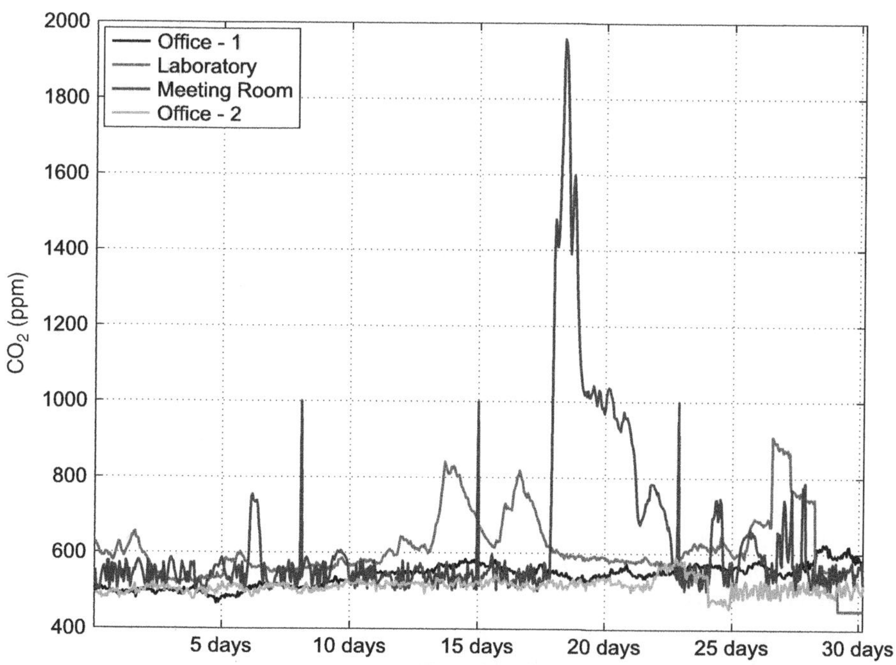

Figure 12.15 Time series representation of air quality measurements over March 2015.

also interesting to notice that when the conditions are normal (no fire detection, etc.), based on the reported levels of CO_2, it can be inferred that the number of people in the laboratory is higher than the number of the people in the offices.

The greatest increase is shown in the meeting room. As the number of people in the meeting room increases, CO_2 concentration reported increases from around 580 *ppm* to 640 *ppm*. The introduced system managed to report an increase in the level that is related to the people's presence and activity in the room (Fonollosa et al. (2014)).

Finally, the concentration for the four rooms over a month is shown in Figure 12.15. As it can be inferred, the days the monitoring rooms have a high number of people is clearly noticeable. Especially in the meeting room, when it is used, the concentration is clearly higher than the normal levels. Similarly for the laboratory, based on students activities, some days throughout the month the levels are high. The concentration of CO_2 in the offices seems stable and around normal during the monitoring month.

12.6 Conclusions

In this chapter, an environmental-monitoring system for smart buildings was proposed. The drawbacks of traditional methods for monitoring were discussed, followed by a review of some commonly used WSN-monitoring applications. Then, the requirements and challenges along with the design goal of the proposed system were discussed.

A framework was proposed, implemented, and examined for real-time environmental monitoring in a building. The necessary design decisions, hardware components, and information processing units were presented.

In the last part, experimental results over a month period from the introduced testbed were shown. A useful insight from the result is that data filtering and smoothing can help to improve the reliability of the system. Also, as it can be inferred from the results, the number of the occupants in the room affects the environmental quality in the room. In future studies, this could also be used as an estimation for the number of people in a room.

Environmental monitoring is a crucial task for smart buildings. WSNs is a promising solution due to the many advantages and the minimum energy requirements.

References

I. F. Akyildiz, Weilian Su, Y. Sankarasubramaniam, and E. Cayirci. A survey on sensor networks. *IEEE Communications Magazine*, 40(8):102–114, Aug. 2002. ISSN 0163-6804. doi: 10.1109/MCOM.2002.1024422.

J. N. Al-Karaki and A. E. Kamal. Routing techniques in wireless sensor networks: a survey. *IEEE Wireless Communications*, 11(6):6–28, Dec. 2004. ISSN 1536-1284. doi: 10.1109/MWC.2004.1368893.

T. Arampatzis, J. Lygeros, and S. Manesis. A survey of applications of wireless sensors and wireless sensor networks. In *Proceedings of the 2005 IEEE International Symposium on, Mediterrean Conference on Control and Automation Intelligent Control, 2005.*, pages 719–724, June 2005. doi: 10.1109/.2005.1467103.

P. Bellavista, G. Cardone, A. Corradi, and L. Foschini. Convergence of manet and wsn in iot urban scenarios. *IEEE Sensors Journal*, 13(10):3558–3567, Oct 2013a. ISSN 1530-437X. doi: 10.1109/JSEN.2013.2272099.

P. Bellavista, G. Cardone, A. Corradi, and L. Foschini. Convergence of manet and wsn in iot urban scenarios. *IEEE Sensors Journal*, 13(10):3558–3567, Oct. 2013b. ISSN 1530-437X. doi: 10.1109/JSEN.2013.2272099.

L. Bencini, F. Chiti, G. Collodi, D. D. Palma, R. Fantacci, A. Manes, and G. Manes. Agricultural monitoring based on wireless sensor network technology: Real long life deployments for physiology and pathogens control. In *2009 Third International Conference on Sensor Technologies and Applications*, pages 372–377, June 2009. doi: 10.1109/SENSORCOMM.2009.63.

A. Boubrima, W. Bechkit, and H. Rivano. Optimal wsn deployment models for air pollution monitoring. *IEEE Transactions on Wireless Communications*, 16(5):2723–2735, May 2017. ISSN 1536-1276. doi: 10.1109/TWC.2017.2658601.

G. Cardone, L. Foschini, P. Bellavista, A. Corradi, C. Borcea, M. Talasila, and R. Curtmola. Fostering participaction in smart cities: a geo-social crowdsensing platform. *IEEE Communications Magazine*, 51(6):112–119, June 2013. ISSN 0163-6804. doi: 10.1109/MCOM.2013.6525603.

CCOHS. Canadian Centre for Occupational Health and Safety. *Indoor Air Quality*, Jul. 2013.

A. Chirap, V. Popa, E. Coca, and D. A. Potorac. A study on light energy harvesting from indoor environment: The autonomous sensor nodes. In *2014 International Conference on Development and Application Systems (DAS)*, pages 127–131, May 2014. doi: 10.1109/DAAS.2014.6842441.

CO_2Meter. iAQ-2000 Indoor Air Quality (VOC) Sensor. http://www.co2meter.com/products/iaq-2000-indoor-air-quality-sensor.

P. Corral, J. A. Perez, A. C. De Castro Lima, and O. Ludwig. Parking spaces detection in indoor environments based on zigbee. *IEEE Latin America Transactions*, 10(1):1162–1167, Jan. 2012. ISSN 1548-0992. doi: 10.1109/TLA.2012.6142454.

EPA. Buildings and their Impact on the Environment: A statistical Summary. *U.S. Environmental Protection Agency Green Building Workgroup*, Oct. 2014.

S. Fang, L. D. Xu, Y. Zhu, J. Ahati, H. Pei, J. Yan, and Z. Liu. An integrated system for regional environmental monitoring and management based on internet of things. *IEEE Transactions on Industrial Informatics*, 10(2):1596–1605, May 2014. ISSN 1551-3203. doi: 10.1109/TII.2014.2302638.

J. Fonollosa, I. Rodriguez-Lujan, A. V. Shevade, M. L. Homer, M. A. Ryan, and R. Huerta. Human activity monitoring using gas sensor arrays. *Sensors and Actuators B: Chemical*, 199:398–402, Aug. 2014.

D.S. Ghataoura, J.E. Mitchell, and G.E. Matich. Networking and application interface technology for wireless sensor network surveillance and monitoring. *Communications Magazine, IEEE*, 49(10):90–97, Oct. 2011. ISSN 0163-6804. doi: 10.1109/MCOM.2011.6035821.

Z. Guo, Y. Guo, F. Hong, Z. Jin, Y. He, Y. Feng, and Y. Liu. Perpendicular intersection: Locating wireless sensors with mobile beacon. *IEEE Transactions on Vehicular Technology*, 59(7):3501–3509, Sept. 2010. ISSN 0018-9545. doi: 10.1109/TVT.2010.2049391.

B. N. Hood and P. Barooah. Estimating doa from radio-frequency rssi measurements using an actuated reflector. *IEEE Sensors Journal*, 11(2): 413–417, Feb. 2011. ISSN 1530-437X. doi: 10.1109/JSEN.2010.2070872.

M. Fahim Jan, Q. Habib, M. Irfan, M. Murad, K. M. Yahya, and G. M. Hassan. Carbon monoxide detection and autonomous countermeasure system for a steel mill using wireless sensor and actuator network. In *2010 6th International Conference on Emerging Technologies (ICET)*, pages 405–409, Oct. 2010. doi: 10.1109/ICET.2010.5638502.

V. Jelicic, M. Magno, D. Brunelli, G. Paci, and L. Benini. Context-Adaptive Multimodal Wireless Sensor Network for Energy-Efficient Gas Monitoring. *Sensors Journal, IEEE*, 13(1):328–338, Jan. 2013. ISSN 1530-437X. doi: 10.1109/JSEN.2012.2215733.

J. Jin, J. Gubbi, S. Marusic, and M. Palaniswami. An information framework for creating a smart city through internet of things. *IEEE Internet of Things Journal*, 1(2):112–121, April 2014a. ISSN 2327-4662. doi: 10.1109/JIOT.2013.2296516.

J. Jin, J. Gubbi, S. Marusic, and M. Palaniswami. An information framework for creating a smart city through internet of things. *IEEE Internet of Things Journal*, 1(2):112–121, April 2014b. ISSN 2327-4662. doi: 10.1109/JIOT.2013.2296516.

R. Kays, S. Tilak, M. Crofoot, T. Fountain, D. Obando, A. Ortega, F. Kuemmeth, J. Mandel, G. Swenson, T. Lambert, B. Hirsch, and M. Wikelski. Tracking animal location and activity with an automated radio telemetry system in a tropical rainforest. *The Computer Journal*, 54(12):1931, 2011. doi: 10.1093/comjnl/bxr072. URL +http://dx.doi.org/10.1093/comjnl/bxr072.

S.D.T. Kelly, N.K. Suryadevara, and S.C. Mukhopadhyay. Towards the Implementation of IoT for Environmental Condition Monitoring in Homes. *Sensors Journal, IEEE*, 13(10):3846–3853, Oct. 2013. ISSN 1530-437X. doi: 10.1109/JSEN.2013.2263379.

J. Kim, C. Chu, and S. Shin. ISSAQ: An Integrated Sensing Systems for Real-Time Indoor Air Quality Monitoring. *Sensors Journal, IEEE*, 14(12): 4230–4244, Dec. 2014. ISSN 1530-437X.

N. Lathia, V. Pejovic, K. K. Rachuri, C. Mascolo, M. Musolesi, and P. J. Rentfrow. Smartphones for large-scale behavior change interventions. *IEEE Pervasive Computing*, 12(3):66–73, July 2013. ISSN 1536-1268. doi: 10.1109/MPRV.2013.56.

Y. C. Liang, Y. Zeng, E. C. Y. Peh, and A. T. Hoang. Sensing-throughput tradeoff for cognitive radio networks. *IEEE Transactions on Wireless Communications*, 7(4):1326–1337, April 2008. ISSN 1536-1276. doi: 10.1109/TWC.2008.060869.

H. Liu, B. Zhang, H. T. Mouftah, X. Shen, and J. Ma. Opportunistic routing for wireless ad hoc and sensor networks: Present and future directions. *IEEE Communications Magazine*, 47(12):103–109, Dec 2009. ISSN 0163-6804. doi: 10.1109/MCOM.2009.5350376.

C. Maihofer. A survey of geocast routing protocols. *IEEE Communications Surveys Tutorials*, 6(2):32–42, Second 2004. ISSN 1553-877X. doi: 10.1109/COMST.2004.5342238.

A. Mainwaring, D. Culler, J. Polastre, R. Szewczyk, and J. Anderson. Wireless sensor networks for habitat monitoring. In *Proceedings of the 1st ACM International Workshop on Wireless Sensor Networks and Applications*, WSNA '02, pages 88–97, New York, NY, USA, 2002. ACM. ISBN 1-58113-589-0. doi: 10.1145/570738.570751. URL http://doi.acm.org/10.1145/570738.570751.

X. Mao, X. Miao, Y. He, X. Li, and Y. Liu. CitySee: Urban CO2 monitoring with sensors. In *INFOCOM, 2012 Proceedings IEEE*, pages 1611–1619, Mar. 2012. doi: 10.1109/INFCOM.2012.6195530.

I. Martin, T. O'Farrell, R. Aspey, S. Edwards, T. James, P. Loskot, T. Murray, I. Rutt, N. Selmes, and T. Baugé. A high-resolution sensor network for monitoring glacier dynamics. *IEEE Sensors Journal*, 14(11):3926–3931, Nov. 2014. ISSN 1530-437X. doi: 10.1109/JSEN.2014.2348534.

L. Oliveira and J. Rodrigues. Wireless sensor networks: a survey on environmental monitoring. *Journal of Communications, Special Issue: Recent Advance on Distributed Sensor Systems and Applications*, 6(2):143 –151, 2011.

OMESH Networks. OPM15. http://www.omeshnet.com/omesh/.

M. Othman and K. Shazali. Wireless sensor network applications: A study in environment monitoring system. *Procedia Engineering*, 41:1204 –1210, 2012. ISSN 1877-7058. doi: http://dx.doi.org/10.1016/j.proeng.2012.07.302. URL http://www.sciencedirect.com/science/article/pii/S1877705812027026.

N. A. Pantazis, S. A. Nikolidakis, and D. D. Vergados. Energy-efficient routing protocols in wireless sensor networks: A survey. *IEEE Communications Surveys Tutorials*, 15(2):551–591, Second 2013. ISSN 1553-877X. doi: 10.1109/SURV.2012.062612.00084.

C. Peng, K. Qian, and C. Wang. Design and Application of a VOC-Monitoring System Based on a ZigBee Wireless Sensor Network. *Sensors Journal, IEEE*, 15(4): 2255–2268, Apr. 2015. ISSN 1530-437X. doi: 10.1109/JSEN.2014.2374156.

O.A. Postolache, J.M.D. Pereira, and P.M.B.S. Girao. Smart sensors network for air quality monitoring applications. *Instrumentation and Measurement, IEEE Transactions on*, 58(9):3253–3262, Sept 2009. ISSN 0018-9456. doi: 10.1109/TIM.2009.2022372.

E. Sazonov, Haodong Li, D. Curry, and P. Pillay. Self-Powered Sensors for Monitoring of Highway Bridges. *Sensors Journal, IEEE*, 9(11):1422–1429, Nov. 2009. ISSN 1530-437X. doi: 10.1109/JSEN.2009.2019333.

X. Sheng, J. Tang, X. Xiao, and G. Xue. Sensing as a service: Challenges, solutions and future directions. *IEEE Sensors Journal*, 13(10):3733–3741, Oct. 2013. ISSN 1530-437X. doi: 10.1109/JSEN.2013.2262677.

H. Shin, Y. Chon, and H. Cha. Unsupervised construction of an indoor floor plan using a smartphone. *IEEE Transactions on Systems, Man, and Cybernetics, Part C (Applications and Reviews)*, 42(6):889–898, Nov. 2012. ISSN 1094-6977. doi: 10.1109/TSMCC.2011.2169403.

P. Spachos and D. Hantzinakos. Scalable Dynamic Routing Protocol for Cognitive Radio Sensor Networks. *Sensors Journal, IEEE*, 14(7):2257–2266, Jul. 2014. ISSN 1530-437X. doi: 10.1109/JSEN.2014.2309138.

P. Spachos and D. Hatzinakos. Poster: Cognitive networking in a self-powered wireless sensor network testbed. In *Proceedings of the 20th Annual International Conference on Mobile Computing and Networking*, MobiCom '14, pages 425–428, New York, NY, USA, 2014. ACM. ISBN 978-1-4503-2783-1. doi: 10.1145/2639108.2642909. URL http://doi.acm.org/10.1145/2639108.2642909.

P. Spachos and D. Hatzinakos. Self-powered wireless sensor network for environmental monitoring. In *2015 IEEE Globecom Workshops (GC Wkshps)*, pages 1–6, Dec. 2015. doi: 10.1109/GLOCOMW.2015.7414207.

P. Spachos and D. Hatzinakos. Real-time indoor carbon dioxide monitoring through cognitive wireless sensor networks. *IEEE Sensors Journal*, 16(2): 506–514, Jan. 2016. ISSN 1530-437X. doi: 10.1109/JSEN.2015.2479647.

P. Spachos, F. M. Bui, Liang Song, Y. Lostanlen, and D. Hatzinakos. Performance evaluation of wireless multihop communications for an indoor environment. In *2011 IEEE 22nd International Symposium on Personal, Indoor and Mobile Radio Communications*, pages 1140–1144, Sept. 2011. doi: 10.1109/PIMRC.2011.6139676.

P. Spachos, P. Chatzimisios, and D. Hatzinakos. Energy aware opportunistic routing in wireless sensor networks. In *Globecom Workshops (GC Wkshps), 2012 IEEE*, pages 405–409, Dec 2012. doi: 10.1109/GLOCOMW.2012.6477606.

P. Spachos, J. Lin, H. Bannazadeh, and A. Leon-Garcia. Smart room monitoring through wireless sensor networks in software defined infrastructures. In *Cloud Networking (CloudNet), 2015 IEEE 4th International Conference on*, pages 216–218, Oct 2015. doi: 10.1109/CloudNet.2015.7335310.

E. Suganya and S. Vijayashaarathi. Smart vehicle monitoring system for air pollution detection using wsn. In *2016 International Conference on Communication and Signal Processing (ICCSP)*, pages 0719–0722, April 2016. doi: 10.1109/ICCSP.2016.7754238.

T. Torfs, T. Sterken, S. Brebels, J. Santana, R. van den Hoven, V. Spiering, N. Bertsch, D. Trapani, and D. Zonta. Low Power Wireless Sensor Network for Building Monitoring. *Sensors Journal, IEEE*, 13(3):909–915, Mar. 2013. ISSN 1530-437X. doi: 10.1109/JSEN.2012.2218680.

W. Tsujita, H. Ishida, and T. Moriizumi. Dynamic gas sensor network for air pollution monitoring and its auto-calibration. In *Sensors, 2004. Proceedings of IEEE*, pages 56–59 vol.1, Oct. 2004. doi: 10.1109/ICSENS.2004.1426098.

H. Yang, Y. Qin, G. Feng, and H. Ci. Online Monitoring of Geological CO_2 Storage and Leakage Based on Wireless Sensor Networks. *Sensors Journal, IEEE*, 13(2):556–562, Feb. 2013. ISSN 1530-437X. doi: 10.1109/JSEN.2012.2223210.

J. Yick, B. Mukherjee, and D. Ghosal. Wireless sensor network survey. *Computer Networks*, 52(12):2292–2330, 2008. ISSN 1389-1286. doi: https://doi.org/10.1016/j.comnet.2008.04.002. URL http://www.sciencedirect.com/science/article/pii/S1389128608001254.

H. Yu and Q. Yue. Indoor light energy harvesting system for energy-aware wireless sensor node. *Energy Procedia*, 16:1027–1032, 2012. ISSN 1876-6102. doi: http://dx.doi.org/10.1016/j.egypro.2012.01.164. URL http://www.sciencedirect.com/science/article/pii/S1876610212001749.

T. Yucek and H. Arslan. A survey of spectrum sensing algorithms for cognitive radio applications. *IEEE Communications Surveys Tutorials*, 11(1):116–130, First 2009. ISSN 1553-877X. doi: 10.1109/SURV.2009.090109.

A. Zanella, N. Bui, A. Castellani, L. Vangelista, and M. Zorzi. Internet of things for smart cities. *IEEE Internet of Things Journal*, 1(1):22–32, Feb. 2014. ISSN 2327-4662. doi: 10.1109/JIOT.2014.2306328.

M. Zorzi and R. R. Rao. Geographic random forwarding (geraf) for ad hoc and sensor networks: multihop performance. *IEEE Transactions on Mobile Computing*, 2(4):337–348, Oct. 2003. ISSN 1536-1233. doi: 10.1109/TMC.2003.1255648.

13

Cooperative Energy Management in Microgrids

Ioannis Zenginis[1], John Vardakas[1], Prodromos-Vasileios Mekikis[2] and Christos Verikoukis[3]

[1] *Iquadrat Informatica, Barcelona, Spain*
[2] *Department of Informatics and Telecommunication, University of Athens, Athens, Greece*
[3] *Telecommunications Technological Centre of Catalonia (CTTC/CERCA), Barcelona, Spain*

13.1 Introduction

In the last decade, we have witnessed a significant reduction of the cost of renewable energy sources (RES) and energy storage systems (ESS), and so, these systems can be considered as a viable solution for high-efficiency off-grid energy systems (Wu et al., 2015). This environment stimulates the development of cost-effective and environment-friendly microgrids, as key building blocks of the future smart grid (Kyriakarakos et al., 2012). However, several technical challenges in control, management and operation of microgrids are posed by their implementation, which can be met by employing efficient energy management systems that are able to ensure an economical, reliable, and secure operation of microgrids, as well as to save energy especially during peak-demand periods (Olivares et al., 2014).

Several approaches for an energy management system have been proposed in the literature that target to plan, operate, and manage microgrids in an efficient way. A number of these approaches refer to the optimal allocation of the load in a single household; the interested reader may refer to Erdinc et al. (2015) for a comprehensive survey of such energy management systems. Multiple households are considered in a significant number of studies. Mohsenian-Rad et al. (2010) propose a game-theoretic approach for the management of competitive household users, where the optimal performance is achieved at a Nash equilibrium of the formulated scheduling procedure, by utilizing pricing schemes that alternate in level and time. Furthermore, Guo et al. (2013) develop a decentralized Lyapunov-based optimization algorithm in order to achieve cost minimization in a neighborhood of smart households. A decentralized collaborative coordination strategy in a residential neighborhood is proposed by Chang et al. (2013), with the objective of cost minimization of their aggregator/retailer in the real-time balancing market. The energy management approaches of Atia and Yamada (2016) and of Erdinc et al. (2015) target to provide methods for deriving the optimal sizing of a microgrid in terms of the optimal size of Photovoltaic (PV) panels and ESS that guarantees minimization of the microgrid's overall cost. In addition, the methods

Transportation and Power Grid in Smart Cities: Communication Networks and Services, First Edition.
Edited by Hussein T. Mouftah, Melike Erol-Kantarci and Mubashir Husain Rehmani.

proposed by Erol-Kantarci et al. (2011a), Yokohama et al. (2008), and Erol-Kantarci et al. (2011b) target to achieve higher service reliability in microgrids, while aiming to achieve an optimal design by considering predefined economic criteria. Nevertheless, all the above-mentioned approaches consider only a single type of buildings (i.e., residential buildings). Multiple building types are considered by Paterakis et al. (2016); however, the proposed optimization method does not target providing the optimal microgrid size. It should be noted that none of the state-of-the-art approaches considers methods for energy exchange between buildings of the same microgrid as a solution for cost minimization and peak demand reduction. It is therefore essential to develop an optimization approach that considers multiple types of buildings, which is a realistic approach especially for urban microgrid applications, and targets both the optimum size of microgrid's components and the buildings' optimum operational plan, which guarantees the overall cost minimization.

In this work, we propose a cooperative energy management scheme that is applied to a group of residential, commercial, and public buildings (e.g., offices, schools, and sport centers) as illustrated in Figure 13.1, which are able to exchange energy at diverse consuming periods based on their energy demands. Precisely, the proposed approach will be applied to generator/storage clusters installed in different buildings, which are able to work cooperatively in order to exchange energy between adjacent buildings. This cooperative energy management system makes decisions on how to optimally use the energy generated by the solar panels and/or the energy installed in the batteries, based on the building type, the different load profile, the electricity prices, and the time of the day. Therefore, the main objective of the proposed energy management scheme is to investigate the effect of energy sharing potential, in order to achieve a

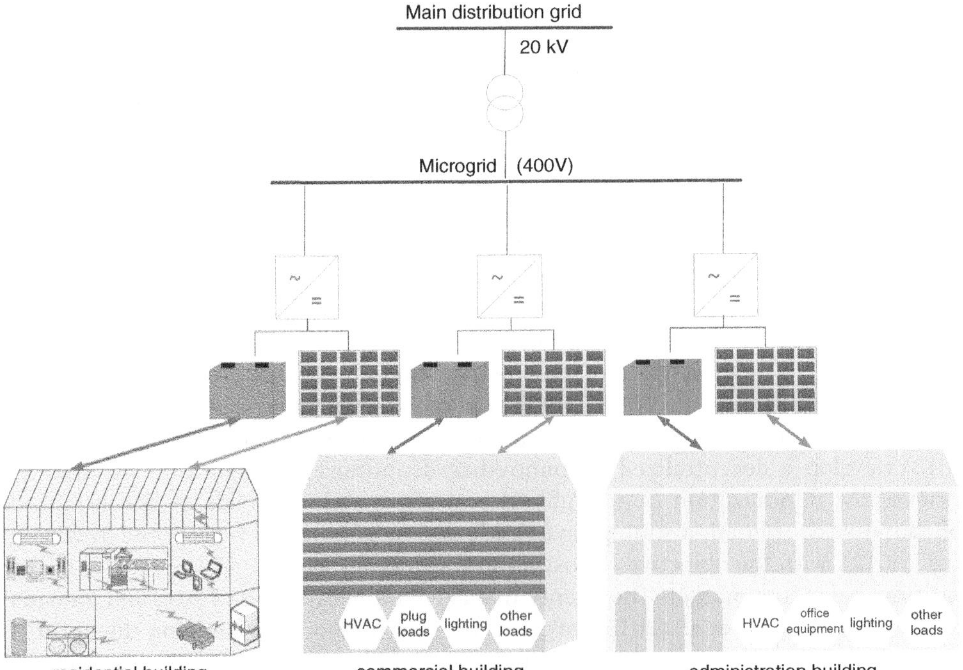

Figure 13.1 Microgrid scenario for the proposed cooperative energy management model.

significant reduction of the buildings' energy dependence on the main grid, as well as of the buildings' carbon footprint. In addition, the cooperative scheme aims to increase the buildings' self-sufficiency, through an effective energy exchange procedure, where the excess energy of buildings with energy surplus is locally consumed by buildings with energy deficit, instead of being sold back to the main grid. The proposed energy-sharing approach also contributes to a more efficient operation of the microgrid, by reducing the energy storage losses, since the excess energy is consumed by neighboring buildings, instead of being stored. Moreover, the proposed scheme can be used for the determination of the optimal microgrid sizing, in terms of the size of PV panels and energy storage devices, which guarantees the microgrid's self-management and ensures that all microgrid users benefit equally from their participation in the cooperation scheme.

The remainder of this work is organized as follows. Section 13.2 presents the cooperative energy management scheme, together with the corresponding mathematical model for the determination of the optimum size of each building's equipment (PV panels, storage devices, converters), as well as of the daily operational plan for each building, regarding the energy generated, stored, and consumed. Section 13.3 presents the evaluation scenario and the corresponding analytical results for the buildings' operational plan during different periods. Finally, Section 13.4 concludes this work.

13.2 The Cooperative Energy Management System Model

The proposed cooperative energy management system model is applied to a microgrid scenario, where buildings with diverse energy demands and energy consumption patterns exchange energy in order to reduce the energy dependence from the main grid. Precisely, we consider different building types (residential, commercial, administrative, etc.) that form a microgrid, while each building is equipped with PV panels and energy storage systems, as depicted in Figure 13.1. During PV production hours, buildings with increased energy demands, such as schools, offices or commercial buildings, are supplied with energy obtained from the neighboring residential buildings with energy surplus. On the other hand, buildings that require more energy during evening hours (e.g., residential buildings) obtain this energy by either discharging their own batteries or from the batteries of buildings with low energy demands during these periods. In addition, the proposed model incorporates the possibility that the energy produced by the PV panels is very high compared to the needs of the microgrid; in this case, a portion of the excess energy is sold to the main distribution grid, while the remaining energy is used for charging the batteries.

The proposed cooperation scheme requires that specific equipment is installed at each one of the participating buildings. Apart from the PV panels and the energy storage devices, buildings are equipped with DC/AC inverters, which are used for the connections of the storage devices with the buildings' loads, as well as with the main distribution grid. Moreover, each building is equipped with an energy management unit (EMU), which communicates with all the loads of the building, as well as with the energy storage device and the PV panels (Gelazanska and Gamage, 2014). All EMUs are coordinated by a microgrid central controller (MGCC), which is also responsible for the energy exchange between the entire microgrid and the main distribution grid.

In what follows, we present the mathematical model for the determination of: (a) the optimum size for the PV panels (kW), the storage devices (kWh), the DC/AC inverters (kW), and (b) the buildings' daily operational plan. The latter output refers to the determination of the following variables, which are calculated for different time intervals (e.g., per hour): (i) the amount of PV energy used for self-consumption, (ii) the amount of PV energy stored in the building's storage device, (iii) the amount of PV energy transferred to the neighboring buildings, (iv) the amount of energy discharged from the storage device, (v) the amount of energy that is bought from the main distribution grid, (vi) the amount of energy that is bought from the neighboring buildings, and (vii) the amount of energy that is sold to the main distribution grid. The calculation of both output data is based on specific input data: (i) annual data for the hourly load profile of each building, (ii) hourly energy generated by the PV panels, (iii) open market electricity tariffs, (iv) carbon emission tax, (v) energy price for microgrid's internal transactions (when a building sells energy to a neighboring building), and (vi) price for selling energy back to the grid. It should be noted that the price for selling energy back to the grid should be lower than the microgrid's internal energy price, which in turn should be lower than the open market electricity prices. This pricing scheme promotes the cooperation among the microgrid participants, while it also increases the self-sufficiency of the system; therefore, the users are motivated to sell their excess energy to the other microgrid users, instead of selling it back to the grid.

The optimal microgrid sizing is achieved by minimizing the total installation, equipment replacement, maintenance and operation cost of all B buildings of the microgrid:

$$\min : TC = \sum_{b} C(b), \tag{13.1}$$

where $C(b)$ is the total cost of building b (b = 1,…, B):

$$C(b) = C_I(b) + C_R(b) + C_M(b) + C_O(b) + C_C(b) - R(b), \tag{13.2}$$

with $C_I(b)$ the equipment installation cost, $C_R(b)$ the equipment replacement cost, and $C_M(b)$ the equipment maintenance cost, which are respectively given by (Erdinc et al., 2015):

$$C_I(b) = c_{PV}N_{PV}(b) + c_{ESS}N_{ESS}(b) + c_{INV}N_{INV}(b) \tag{13.3}$$

$$C_R(b) = \sum_{i=1}^{rep} \frac{c_{PV}N_{PV}(b) + c_{ESS}N_{ESS}(b) + c_{INV}N_{INV}(b)}{(1+d)^{iYrep}} \tag{13.4}$$

$$C_M(b) = \sum_{y=1}^{Y} \frac{\mu_{PV}N_{PV}(b) + \mu_{ESS}N_{ESS}(b) + \mu_{INV}N_{INV}(b)}{(1+d)^y}, \tag{13.5}$$

where $c_{PV}$is the PV of the installation cost (€/kilowatt peak (kW_p)),c_{ESS} is the ESS installation cost (€/kWh), $c_{INV}$is the inverter installation cost (€/kW), $\mu_{PV}$is the PV of maintenance cost (€/kW_p/year),μ_{ESS} is the ESS maintenance cost (€/kWh/year), $\mu_{INV}$is the inverter maintenance cost (€/kW/year), d is a discount rate, $N_{PV}(b)$, $N_{ESS}(b)$, and $N_{INV}(b)$ are the size of PV panel, ESS and inverter of building b, respectively, and y is the year in a Y total project horizon. Furthermore, $C_O(b)$ and $C_c(b)$ are the operation cost and the cost due to carbon emissions, respectively:

$$C_O(b) = \sum_{y=1}^{Y} \frac{\sum_{t=1}^{T} E_{load}^{grid}(b,t)\varphi_{grid,buy}(t)\Delta t + E_{load}^{MG}(b,t)\varphi_{MG,buy}\Delta t + E_{ESS}^{MG}(b,t)\varphi_{MG,buy}\Delta t}{(1+d)^y} \tag{13.6}$$

$$C_c(b) = \sum_{y=1}^{Y} \frac{\sum_{t=1}^{T} E_{load}^{grid}(b,t) Carb_{\text{int}} Carb_{tax} \Delta t}{(1+d)^y}, \tag{13.7}$$

where $\varphi_{grid,buy}(t)$is the price of energy bought from the main distribution grid for each time interval t, $\varphi_{MG,buy}$ is the price of energy bought from the microgrid. It should be noted that $t = 1,\ldots, T$, while T refers to the total time intervals; for instance, if hourly energy prices are considered, then the annual operation cost and the cost due to carbon emissions are calculated for $T = 8{,}760$ hours. The operational cost (Equation 13,6) is a function of the energy $E_{load}^{grid}(b,t)$ provided by the main distribution grid in order to satisfy building's b demand, the energy E_{load}^{MG} provided by the neighboring buildings to building b, and the energy $E_{ESS}^{MG}(b,t)$ provided by the neighboring buildings for charging building's b ESS. On the other hand, the cost due to carbon emissions (Equation 13.7) is a function of the energy $E_{load}^{grid}(b,t)$ provided by the main distribution grid to building b, the carbon intensity $Carb_{\text{int}}$in the country/province where the buildings are located, and the corresponding carbon tax $Carb_{tax}$.

The total cost $C(b)$ of building b is also a function of the revenue $R(b)$ of building b from selling energy to either the neighboring buildings or to the main distribution network:

$$R(b) = \sum_{y=1}^{Y} \frac{\sum_{t=1}^{T} E_{grid}^{PV}(b,t)\varphi_{grid,sell}(t)\Delta t + E_{MG}^{PV}(b,t)\varphi_{MG,sell}\Delta t + E_{MG}^{ESS}(b,t)\varphi_{MG,sell}\Delta t}{(1+d)^y}, \tag{13.8}$$

where $\varphi_{grid,sell}(t)$ is the price of selling energy back to the grid, while $\varphi_{MG,sell}$ is the price of selling energy to microgrid users. It should be noted that the aforementioned energy prices should be set so that $\varphi_{grid,buy}(t) > \varphi_{MG,buy} = \varphi_{MG,sell} > \varphi_{grid,sell}(t)$ so that cooperation between the buildings of the microgrid is promoted.

In order to solve the minimization problem of Equation 13.1, the values of the energy that is transferred between the various components of the microgrid should satisfy the following energy balance:

$$E_{load}(b,t) = E_{load}^{grid}(b,t) + E_{load}^{PV}(b,t) + E_{load}^{ESS}(b,t) + E_{load}^{MG}(b,t), \tag{13.9}$$

where $E_{load}(b,t)$ is the power demand of building b at time t. The analysis for the determination of the functions of the right-hand-side of Equation 13.9 is presented in the following paragraphs.

13.2.1 PV Panel Modeling

The proposed approach considers that each building b is equipped with a PV panel of surface $A_{PV}(b)$. The objectives of the proposed optimization model that are related to the PV panels are the determination of the optimum size $N_{PV}(b)$ installed in building b, which will guarantee the optimum buildings' daily operational plan. The size $N_{PV}(b)$ is bounded by the available surface of the building's roof:

$$N_{PV}(b) \leq A_{PV}(b)\lambda_{PV}, \tag{13.10}$$

where $\lambda_{PV}$is the amount of KW_p installed per available m^2. Based on the size of the PV panels, we are able to determine the energy $E^{PV,pro}(b,t)$produced by building's b PV panel during the time-interval t:

$$E^{PV,pro}(b,t) = N_{PV}(b)E^{PV,unit}(t), \tag{13.11}$$

where $E^{PV,unit}(t)$ is the energy produced during the time-interval t by a 1 KW_p PV panel.

Due to the fact that the energy produced by building's b PV panel during the time interval t is used by the same building ($E^{PV}_{load}(b,t)$),sold to the neighboring buildings of the microgrid ($E^{PV}_{MG}(b,t)$),sold to the main distribution grid ($E^{PV}_{grid}(b,t)$), or used for charging the buildings ESS ($E^{PV}_{ESS}(b,t)$), the following relation is valid:

$$E^{PV,pro}(b,t)n_{PV,array} = \frac{E^{PV}_{load}(b,t) + E^{PV}_{MG}(b,t) + E^{PV}_{grid}(b,t)}{n_{INV}} + E^{PV}_{ESS}(b,t), \tag{13.12}$$

where $n_{PV,array}$ is the efficiency coefficient of building's b PV panel and n_{INV} is the efficiency coefficient of the corresponding inverter.

13.2.2 Energy Storage System

As in the case of the PV panel model, the ESS model aims to find the optimum size $N_{ESS}(b)$ of ESS that is installed in building b, which guarantees the optimum daily operational plan of the building. To this end, the relation of the state of energy (SoE) of the ESS during the time interval t with the SoE of the previous time interval and the ESS charging ($E_{c,ESS}(b,t)$) and discharging ($E_{d,ESS}(b,t)$) energy is required:

$$SoE_{ESS}(b,t) = SoE_{ESS}(b,t-1) + E_{c,ESS}(b,t)\Delta t - E_{d,ESS}(b,t)\Delta t + u^{int}_{ESS}SoE_{ESS}(b,t_0), \tag{13.13}$$

where Δt is the duration of the optimization intervals and $SoE_{ESS}(b,t_0)$ is the SoE at the beginning of the optimization horizon. Since $SoE_{ESS}(b,t_0)$ is used only for the determination of $SoE_{ESS}(b,1)$, a binary variable is required:

$$u^{\text{int}}_{ESS} = \begin{cases} 1 & \text{if} \quad t = 1 \\ 0 & \text{otherwise} \end{cases} \tag{13.14}$$

The energy that is used to charge building's b ESS is either produced by the PV panel of the same building or bought from the neighboring buildings of the microgrid. Therefore, the charging energy $E_{c,ESS}(b,t)$ is given by the following equation:

$$E_{c,ESS}(b,t) = [E^{PV}_{ESS}(b,t) + E^{MG}_{ESS}(b,t)n_{INV}]n_{c,ESS}, \tag{13.15}$$

where $n_{c,ESS}$ is the ESS's charging efficiency coefficient. It should be noted that the charging energy $E_{c,ESS}(b,t)$ is bounded by the ESS's maximum charging power limit $P_{ESS,\max}(b)$; therefore, the following relation is included in the list of constraints of the proposed optimization model:

$$E_{c,ESS}(b,t) \leq Z_{ESS}N_{ESS}(b). \tag{13.16}$$

In contrast, the discharging energy $E_{d,ESS}(b,t)$of building's b ESS during time interval t is equal to the energy given by the ESS to the same building ($E_{load}^{ESS}(b,t)$) and to the neighboring buildings ($E_{MG}^{ESS}(b,t)$):

$$E_{d,ESS}(b,t)n_{d,ESS} = \frac{E_{load}^{ESS}(b,t) + E_{MG}^{ESS}(b,t)}{n_{INV}}. \tag{13.17}$$

The discharging energy $E_{d,ESS}(b,t)$is bounded by the capacity of the corresponding ESS:

$$E_{d,ESS}(b,t) \le Z_{ESS}N_{ESS}(b), \tag{13.18}$$

where $Z_{ESS}$is the power rate of the ESS. Furthermore, in order to guarantee that the ESS is not charged and discharged at the same time, the following two constraints should be considered in the optimization model:

$$E_{c,ESS}(b,t) \le \Xi u_{ESS}(b,t) \tag{13.19}$$

$$E_{d,ESS}(b,t) \le \Xi[1 - u_{ESS}(b,t)], \tag{13.20}$$

where Ξ is a constant that takes a large value, and $u_{ESS}(b,t)$ is a binary variable that is used in order to assure that the ESS is not charged and discharged simultaneously.

It should be noted that the SoE is bounded between a minimum and a maximum value, while at the end of the optimization horizon the SoE should be equal to the SoE at the beginning of the optimization horizon:

$$SoE_{ESS,\min} \le SoE_{ESS}(b,t) \le SoE_{ESS,\max} \tag{13.21}$$

$$SoE_{ESS}(b,T) = SoE_{ESS}(b,t_0). \tag{13.22}$$

13.2.3 Inverter

The size $N_{inv}(b)$ of building's b inverter is determined by considering the following energy constraints:

$$\frac{E_{load}^{PV}(b,t) + E_{MG}^{PV}(b,t) + E_{grid}^{PV}(b,t) + E_{load}^{ESS}(b,t) + E_{MG}^{ESS}(b,t)}{n_{INV}} \le N_{inv}(b) \tag{13.23}$$

and

$$E_{ESS}^{MG}(b,t) \le N_{inv}. \tag{13.24}$$

13.2.4 Microgrid Energy Exchange

Having determined the PV, ESS, and inverter models, we proceed with the description of the model for the cooperative exchange of energy. Precisely, the amount of energy sold by a group of buildings during the time interval t to another group of buildings of the same microgrid is equal to the amount of energy bought by these buildings:

$$\sum_b E_{load}^{MG}(b,t) + E_{ESS}^{MG}(b,t) = \sum_{b'} E_{MG}^{PV}(b',t) + E_{MG}^{ESS}(b',t). \tag{13.25}$$

However, in order to ensure that the buildings cannot buy and sell energy at the same time, the following relations should be included in the list of constraints of the optimization model:

$$E_{load}^{MG}(b,t) + E_{ESS}^{MG}(b,t) + E_{load}^{grid} \leq \Pi u_{ex}(b,t) \quad (13.26)$$

$$E_{MG}^{PV}(b,t) + E_{MG}^{ESS}(b,t) + E_{grid}^{PV}(b,t) \leq \Pi[1 - u_{ex}(b,t)], \quad (13.27)$$

where Π is a constant that takes a large value, and $u_{ex}(b,t)$ is a binary variable that is used in order to assure that the ESS is not charged and discharged simultaneously.

13.3 Evaluation and Discussion

The proposed cooperative energy management model is evaluated by considering a microgrid located in Barcelona, Spain, which comprises of a multi-apartment building, a school, and a municipality administrative building. The hourly power consumptions per m^2 for the considered residential buildings are provided by the municipality of Barcelona, while the total surface of each building, as well as the available surface for installing PV panels are: (i) for the residential building, 4500 (750) m^2, (ii) for the school, 3600 (900) m^2, and (iii) for the administrative building, 3000 (1000) m^2. Furthermore, the hourly PV production of a 1 kW$_p$ PV array located in Barcelona is obtained by PVWatts (2017), while we assume that 1 kW$_p$ is installed for every 7 m^2 available surface. The input parameters for the PV panels, ESS, and the inverter are summarized in Table 13.1.

Table 13.1 Case Study Input Parameters.

Parameter	Value
Discount rate (d)	3 %
Power rate of the ESS (Z_{ESS})	0.5
ESS minimum value of the SoE ($SoE_{ESS,\min}$)	0.2
ESS maximum value of the SoE ($SoE_{ESS,\max}$)	1
ESS charging efficiency coefficient ($n_{c,ESS}$)	0.95
ESS discharging efficiency coefficient ($n_{d,ESS}$)	0.95
Initial ESS acquisition cost (c_{ESS})	208 (€/kWh)
Initial ESS replacement cost (c_{ESS})	208 (€/kWh)
Annual ESS operational and management cost (μ_{ESS})	2.1 (€/kWh/yr)
PV efficiency coefficient (n_{PV})	0.95
Initial PV acquisition cost (c_{PV})	1000 (€/kW)
Annual PV operational and management cost (μ_{PV})	3.3 (€/kW/yr)
Inverter efficiency coefficient (n_{INV})	0.9
Initial inverter acquisition cost (c_{INV})	106 (€/kW)
Annual inverter operational and management cost (μ_{INV})	0.8 (€/kW/yr)

Table 13.2 Energy Prices from the Spanish Energy Retailer.

	November, December, January, February, March	April, May, June, July, August, September, October
00:00–08:00	0.079 €/kWh	0.079 €/kWh
08:00–11:00	0.109 €/kWh	0.109 €/kWh
11:00–15:00	0.109 €/kWh	0.135 €/kWh
15:00–18:00	0.109 €/kWh	0.109 €/kWh
18:00–22:00	0.135 €/kWh	0.109 €/kWh
22:00–00:00	0.109 €/kWh	0.109 €/kWh

Furthermore, Table 13.2 reports the financial data that are considered for the optimization of this case study. Specifically, the applied time-of-use (TOU) electricity rates of a Spanish retailer (Endesa, 2017) differ between winter and summer months. In both cases, the lower rate is from midnight to 8:00 a.m., while during the winter months, the higher rate is between 6:00 p.m. and 10:.00 p.m., and the peak-price period for the summer months is between 11.00 a.m. and 3:00 p.m. For the remaining hours the medium rate is activated. The rates $\varphi_{MG,sell}$ of energy transactions among the microgrid users is considered to be 40% lower than the corresponding rates provided by the energy retailer, while the price $\varphi_{grid,sell}$ of selling energy back to the main grid is 90% lower than the corresponding buying prices. Also, we consider that the carbon intensity $Carb_{int}$ in Spain is equal to 0.455 kg/kWh, while we assume a carbon tax $Carb_{tax} = 0.03$ €/Kg for the same region. Moreover, the project lifetime Y is assumed to be equal to 20 years.

In order to highlight the advantages of the proposed cooperative energy management model, we present cost results for three different scenarios: the macrogrid scenario

Table 13.3 Cost Results for the Three Scenarios.

	Residential	School	Public Building	Total
	Macrogrid scenario			
Cost	397,186	584,000	643,538	1,624,726
	No energy exchange scenario			
Cost	324,870	400,365	417,006	1,142,241
$N_{PV}(b)$ (kWp)	71.6	128.6	143	
$N_{ESS}(b)$(kWh)	28.3	0	67	
$N_{inv}(b)$ (kW)	38.5	72.5	79	
	Energy exchange scenario			
Cost	292,067	367,335	398,859	1,058,262
$N_{PV}(b)$(kWp)	107	128.6	143	
$N_{ESS}(b)$(kWh)	98	120	77.5	
$N_{inv}(b)$ (kW)	50	65	73	

assumes that all the buildings obtain the required energy exclusively from the main distribution network; therefore, no PV panels or ESS are installed. The no energy exchange scenario assumes that all buildings are equipped with PV panels and ESS; however, buildings are not able to exchange energy. Finally, the energy exchange scenario applies the proposed cooperative energy management model. As the results of Table 13.3 reveal, the application of the proposed cooperative scheme results in a significant cost reduction compared to the other two scenarios. Precisely, the application of the cooperative scheme results in 34.86% total cost savings compared to the macrogrid scenario and 7.36% compared to the no energy exchange scenario. It should be noted that the proposed scheme achieves higher savings compared to the no energy exchange scenario, despite the fact that the total size of the PV arrays, inverters, and ESS is higher under the proposed scheme.

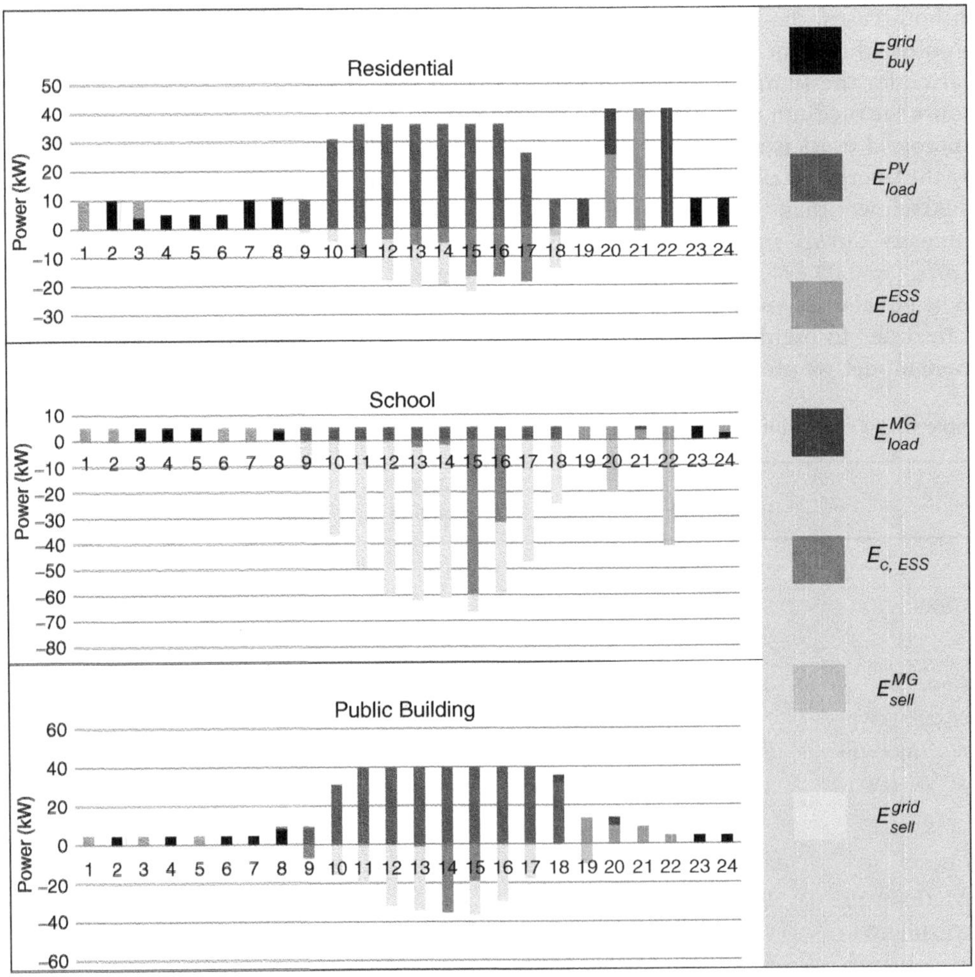

Figure 13.2 Energy exchange plan for the three buildings during a weekend day in February.

We also present the results of the case study regarding the optimal operational plan of the three considered buildings. Figure 13.2 illustrates the hourly energy exchanges that take place between the different buildings during a Sunday in February, while Figure 13.3 shows the corresponding results for the case of a Monday in the same month. In addition, Figure 13.4 and Figure 13.5 present corresponding results for a Sunday and a Monday in July, respectively. The comparison of these figures reveals that the energy exchange procedure is highly affected by the day and the season. The residential building mainly buys energy on Sundays, while it provides its excess energy on Mondays when its demand is low and the PV production is high. Furthermore, in both months the residential building uses the energy generated by its PV panel during the light-day hours, while during the high-energy consumption evening hours, the building uses the energy stored in its ESS as well as the energy bought from the other two buildings. On the other hand, the school buys energy on Mondays during their

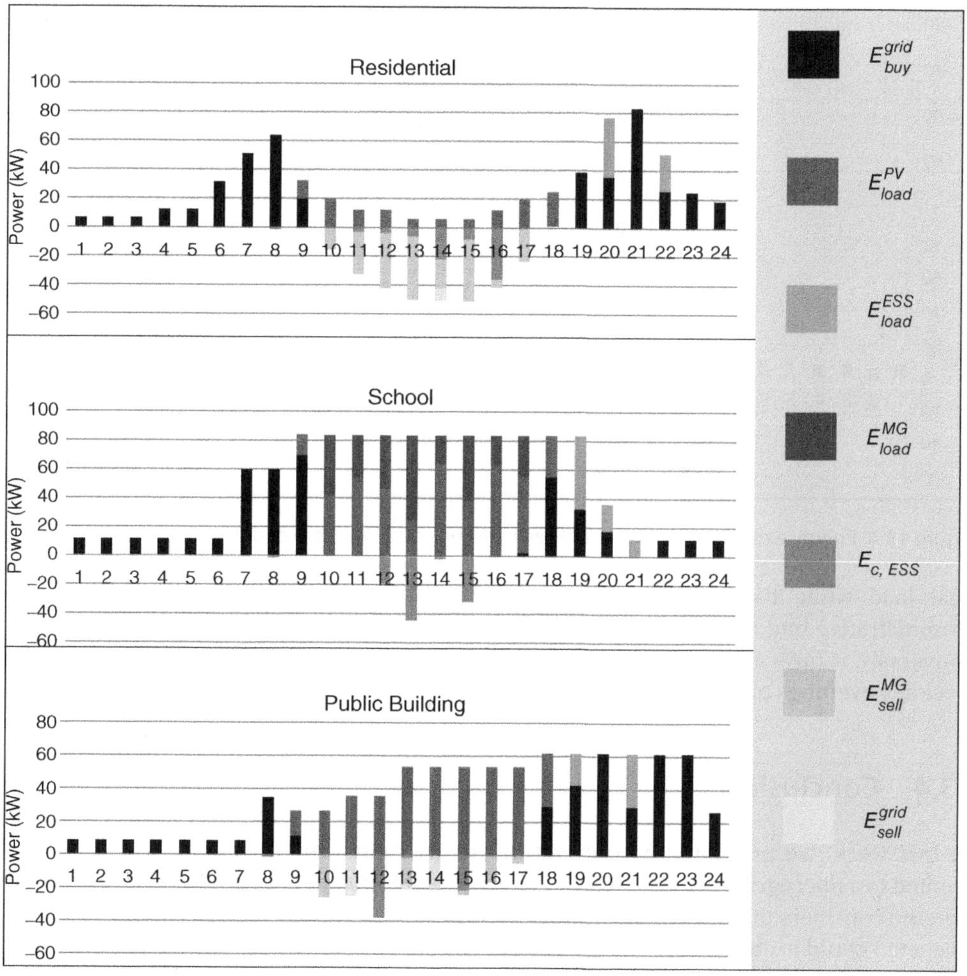

Figure 13.3 Energy exchange plan for the three buildings during a week day in February.

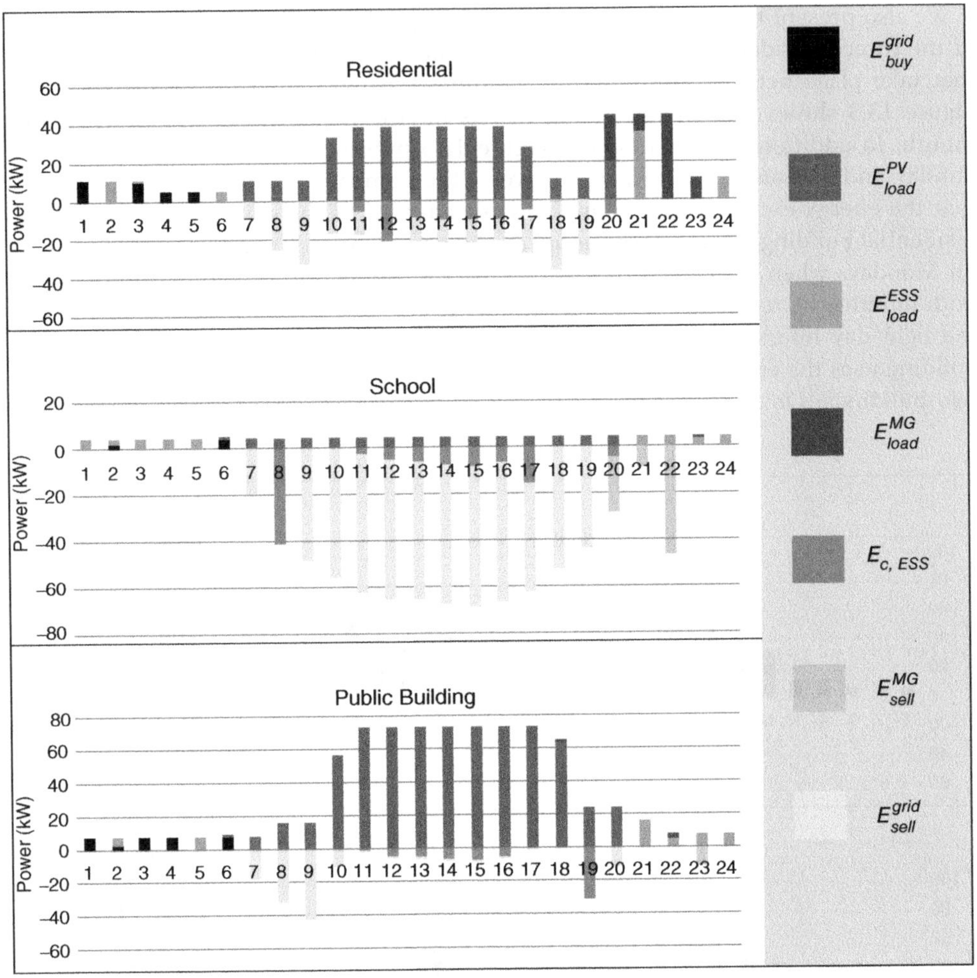

Figure 13.4 Energy exchange plan for the three buildings during a weekend day in July.

peak load, while it sells energy in any other case during low-load periods. Finally, the administrative building sells energy to the microgrid during the winter months, while conversely, it buys great amounts of energy during its peak load, which takes place on weekday evenings of summer months.

13.4 Conclusion

In this work, we presented a cooperative energy management model, which can be applied to a microgrid scenario in order to achieve significant cost savings. The proposed scheme considers that buildings of different types (residential, commercial, administrative, etc.) could exchange the energy produced by their PV panels or stored at their ESS in a smart way so that the excess energy of buildings with energy surplus is consumed by buildings of the same microgrid with energy deficit. We also propose an optimization

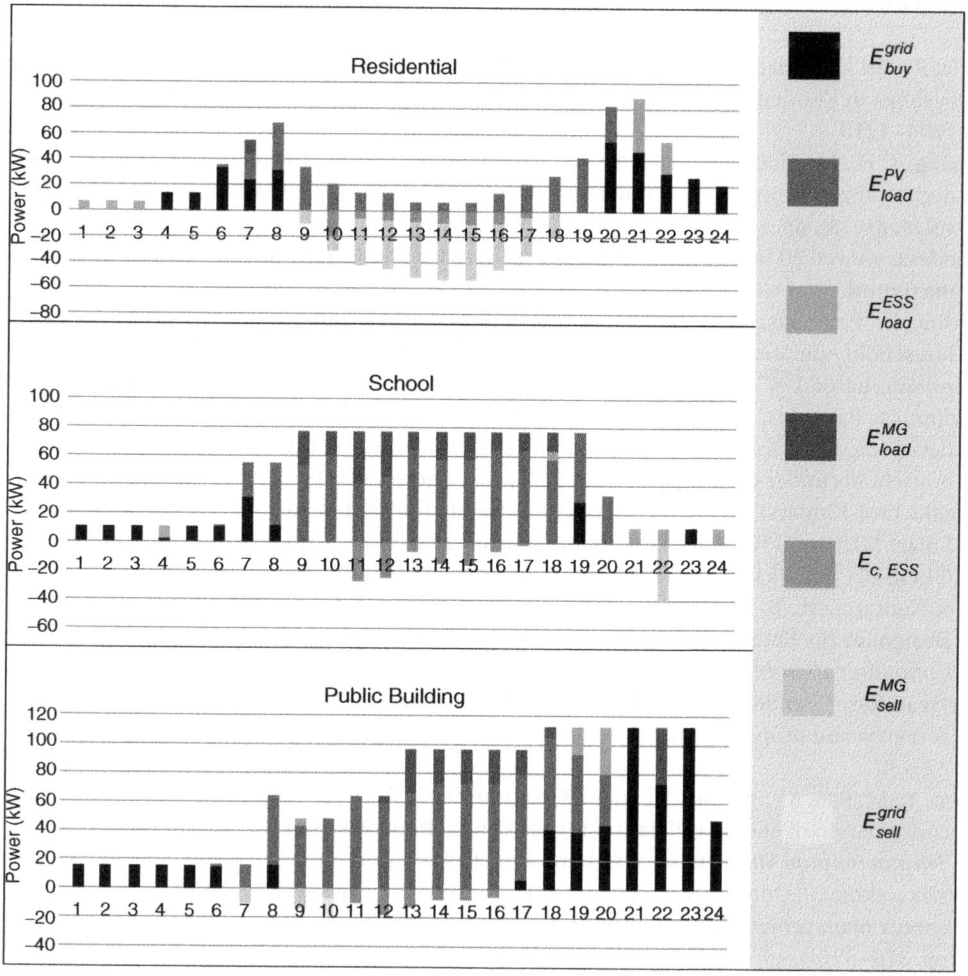

Figure 13.5 Energy exchange plan for the three buildings during a week day in July.

model for the determination of the optimum size of the various components (PV panels, ESS, inverters), as well as of the daily operational plan of each building of the microgrid. To enhance the proposed cooperative scheme, in our future work we plan to incorporate the effect of EVs to the operation of the microgrid, while also including a demand response scheme in order to achieve further peak demand reduction. Finally, we plan to improve the scalability of our cooperative scheme by incorporating the ability to sell excess energy to other microgrids. To that end, energy exchange can be achieved both with the main distribution network and between neighboring microgrids.

Acknowledgment

This work has been partially supported by the "SMART-NRG" project (grant 612294), and the "CONNECT" project, an ECSEL Joint Undertaking (grant 737434).

References

Atia, R. and N. Yamada, N., 2016, "'Sizing and Analysis of Renewable Energy and Battery Systems in Residential Microgrids'," *IEEE Transactions on Smart Grid*, vol. 7, no. (3), pp. 1204–1213.

Chang, T. H., M. Alizadeh, M., and A. Scaglione, A., 2013, "'Real-time power balancing via decentralized coordinated home energy scheduling"', *IEEE Transactions on Smart Grid*, vol. 4, no. (3), pp. 1490–1504.

eEndesa, viewed 20 July 20, 2017, URL: https://www.endesaclientes.com/one-luz-maxi.html.

Erdinc, O., Paterakis, N. G., Mendes, T. D., Bakirtzis, A. G., and Catalão, J. P., 2015, "'Smart household operation considering bi-directional EV and ESS utilization by real-time pricing-based DR"', *IEEE Transactions on Smart Grid*, vol. 6, no. (3), pp. 1281–129.

Erdinc, O., Paterakis, N. G., Pappi, I. N., Bakirtzis, A. G., and& Catalão, J. P. S., 2015, "'A new perspective for sizing of distributed generation and energy storage for smart households under demand response"', *Applied Energy*, vol. 143,(1), pp. 26–37.

Melike Erol-Kantarci, M., B. Kantarci, B., and & H. T. Mouftah, H.T., 2011, "'Cost-Aware Smart Microgrid Network Design for a Sustainable Smart Grid,'" *In:in Proc. of IEEE* GLOBECOM Wkshps, Houston, TX, USA, pp. 1223–1227.

Erol-Kantarci, M., B. Kantarci, B., and& Mouftah, H. T., 2011, "'Reliable Overlay Topology Design for the Smart Microgrid Network',", *IEEE Network, Special issue on Communication Infrastructures for Smart Grid*, vol. 25, no.(5), pp.38–43.

Gelazanskas, L., and& Gamage, K. A. A., 2014, "'Demand side management in smart grid: A review and proposals for future direction"', *Sustainable Cities and Society*, vol 11, pp. 22--30.

Guo, Y., M. Pan, M., Y. Fang, Y., and& Khargonekar, P. P., 2013, "'Decentralized coordination of energy utilization for residential households in the smart grid"', *IEEE Transactions on Smart Grid*, vol. 4, no. (3), pp. 1341–1350.

Kyriakarakos, G., Dounis, A. I., Arvanitis, K. G., and& G. Papadakis, G. 2012, "'A fuzzy logic energy management system for polygeneration micro-grids"', *Renewable Energy*, vol. 41, pp. 315–327.

Mohsenian-Rad, A., V. Wong, V.,J. Jatskevich, J., R. Schober, R., and& A. Leon-Garcia, A., 2010, "'Autonomous demand-side management based on game-theoretic energy consumption scheduling for the future smart grid"', *IEEE Transactions on Smart Grid*, vol. 1, no. (3), pp. 320–331.

Olivares D. E., et al., 2014, "'Trends in Microgrid Control"', *IEEE Transactions on Smart Grid* , vol. 5,(no. 4), pp. 1905–1919.

Paterakis, N. G., O. Erdinç, O., Pappi, I. N., Bakirtzis, A. G., and Catalão, J. P. S., 2016, "'Coordinated Operation of a Neighborhood of Smart Households Comprising Electric Vehicles, Energy Storage and Distributed Generation"', *IEEE Transactions on Smart Grid*, vol. 7(, no. 6), pp. 2736–2747.

PVWatts Calculator, viewed 20 July 20, 2017, URL: http://pvwatts.nrel.gov/.

D. Wu, D., F. Tang, F., T. Dragicevic, T., J. C. Vasquez, J.C., and J. M. Guerrero, J.M. 2015, "'A control architecture to coordinate renewable energy sources and energy storage systems in islanded microgrids'", *IEEE Transactions on Smart Grid*, vol. 6, no. 3, pp. 1156-–1166.

Yokoyama, R., T. Niimura, T., and N. Saito, N., 2008, "'Modeling and evaluation of supply reliability of microgrids including PV and wind power,"', *In*: in Proc. IEEE PESG General Meeting, Pittsburgh, PA, pp. 20–24.

14

Optimal Planning and Performance Assessment of Multi-Microgrid Systems in Future Smart Cities

Shouxiang Wang[1], Lei Wu[2], Qi Liu[1] and Shengxia Cai[3]

[1] *Key Laboratory of Smart Grid of Ministry of Education, Tianjin University, Tianjin, China*
[2] *Electrical and Computer Engineering Department, Clarkson University, Potsdam, NY, USA*
[3] *Zhou Enlai School of Government, Nankai University, China*

With the deterioration of the environment and the exhaustion of traditional energy, applications of distributed generation (DG) systems, especially distributed renewable generations, have emerged in recent years, due to their advantages of economy, environmental protection, and flexibility. However, a deeper penetration of intermittent renewable energy brings new challenges to the secure and stable operation and control of power grids.

Microgrids (MGs) present an effective way to solve the problem. An MG is a small-scale power system containing DGs, loads, energy storage systems (ESSs), and a control system. MGs have high flexibility so that they can be connected to the distribution network or work in an isolated mode when grid faults occur in the distribution network.

However, MGs have limited energy capability considering their nature of local power supply. According to the IEEE recommendations, the maximum capacity of MGs is normally limited to 10 MVA (Chowdhury et al., 2009). Hence, it is preferable to split a large amount of demand into several smaller load units and supply each unit by one MG. Furthermore, multiple MGs in close proximity can be interconnected through back-to-back (BTB) converters. In this way, each MG can operate independently under nominal situation, while during an emergent overloading in one MG, other MGs can provide support with their excessive available power.

Recently, building a resilient multi-MG system (MMGS) by interconnecting adjacent MGs through tie-lines has attracted attention among researchers. Interconnected MGs can share reserves and provide more economic dispatch as well as ancillary functionalities (e.g., black-start function) in critical conditions (e.g., loss of a major generation or a major disturbance at the utility side), which could lower the chance of system collapse, minimize emergency load shedding, and enhance the overall system reliability. As available generation and storage capacities of individual MGs can be effectively shared during emergency conditions, adequately and strategically interconnecting multiple MGs could strengthen the system against extreme weather conditions and natural disasters and in turn improve resilience of the entire grid, while still maintaining flexible operation and reliability benefits of MGs in islanded mode. The islanded mode refers to the operation

Transportation and Power Grid in Smart Cities: Communication Networks and Services, First Edition.
Edited by Hussein T. Mouftah, Melike Erol-Kantarci and Mubashir Husain Rehmani.

status that an MG is no longer connected with the distribution network, and loads of the MG are supplied by DGs and ESSs of the MGs. This chapter will comprehensively study the optimal planning and assessment of such MMGSs.

14.1 Optimal Planning of Multi-Microgrid Systems

14.1.1 Introduction

With significant technical improvements in DGs and energy efficiency in load demands in MGs, it is beneficial to build flexible and efficient MMGSs by interconnecting multiple MGs, which would further promote high penetration of distribution energy as well as enhanced reliability and resilience. There is abundant research on planning and design of MG (Basu et al., 2010; Khodaei et al., 2015; Lotfi & Khodaei, 2016; Wu et al., 2016). Khodaei et al. (2015) present an MG planning method by investigating the economic rationality of MG deployment and giving the optimal combination of distributed energy resource (DER) units for installation, taking into account multiple uncertainties that would affect the optimality of planning. Moreover, they developed approaches to determine optimal location, sizing, and type of DERs in MGs based on reliability and economic criteria to maximize benefits and satisfy technical indexes. Guo et al. (2013) and Zhang et al. (2015) presented methods for optimal planning and design of MGs for a combined utilization of cooling, heating, and power while considering economic and emission criteria. Wu et al. (2016) discussed key characteristics, challenges, and potential solution methodologies for the planning and operation of the future community MG distribution system.

Compared to literature related to MGs planning, the extent of research articles on topology and capacity planning of MMGS is relatively limited. According to the IEEE Standard 1547.4 (Standards America, 2011), large distribution systems can be clustered into a number of MGs to facilitate powerful control and operation infrastructure in future distribution systems. The reliability-based MG design, which focuses on reconfiguration of a single MG, was proposed in Arefifar et al. (2013). Shahnia et al. (2013) proposed series and parallel structures for MMGS. A preliminary study on MG network topology via integer programing was presented in Erol-Kantarci et al. (2011). Che et al. (2015) proposed a probabilistic minimal cut-set-based iterative approach for the optimal planning of interconnected MGs with variable renewable energy sources. Most of the research has been conducted on existing networks, in an attempt to develop approaches for optimally partitioning the existing network into multiple MGs. Mojtahedzadeh et al. (2016) proposed a method to design a distribution network, which is basically composed of autonomous low-voltage (LV) MGs, while considering the optimal combination of various DERs in each MG to minimize the investment cost.

An MMGS consists of various distribution generations, storage devices, and converting and control equipment, which leads to distinct operation characteristics and evaluation indexes as compared to single MGs. This section focuses on planning models of MMGS based on comprehensive energy efficiency and network structure flexibility, while considering various network structures such as nested, parallel, series, and distributed.

14.1.2 Optimal Structure Planning

14.1.2.1 Definition of Indices

According to the network structure of MMGS and characteristics of power delivery, four MMGS network structures are presented: nested MMGS, parallel MMGS, series MMGS, and dispersed MMGS. The four MMGS network structures are further evaluated via the connection index of MMGS. Finally, an improved immune genetic algorithm is used to evaluate network construction flexibility and economic optimality of the four MMGS network structures. The four MMGS network structures of MMGS are discussed as follows.

1) **Nested MMGS**
   Nested MMGSs present unique characteristics because low-voltage MGs are connected to middle-voltage (MV) MGs through the point of common coupling (PCC), which form LV- and MV-nested MMGS. In addition, components of LV MGs are DGs, energy storage assets and loads, all of which are connected to the busbar through power electronic converters, and individual MMGSs can operate independently. Different types of DGs are connected to different voltage levels. Generally, high-power DGs are connected to MV feeders, and low-power DGs are connected to LV feeders. Depending on the operation states of devices and load demand characteristics in MGs, LV and MV MMGSs can operate flexibly. Meanwhile, this structure can meet different reliability requirements of various loads, by positioning critical loads in the inner layer and providing two energy supplies. Considering the characteristics of DGs and different levels of power quality requirements, LV MMGSs can include two or more MGs. The structure is shown in Figure 14.1, where MG2 and MG3 are two low-voltage MGs that are connected to the middle-voltage MG1, forming LV- and MV-nested MMGS.
2) **Parallel MMGS**
   This structure operates multiple MGs in parallel by connecting MGs into the same LV feeders at PCC. In this way, each MG can operate either independently or in parallel. It is also convenient and beneficial to coordinately control multiple parallel MGs. The typical structure is shown in Figure 14.2.
3) **Series MMGS**
   As for long 10kV feeders in distribution networks, MGs are usually constructed at different segments according to different load levels. In this way, a series structure is formed as shown in Figure 14.3. MGs can be operated either independently or as a whole. However, downstream MGs could suffer negative impact if upstream MGs are operated in the islanded mode.

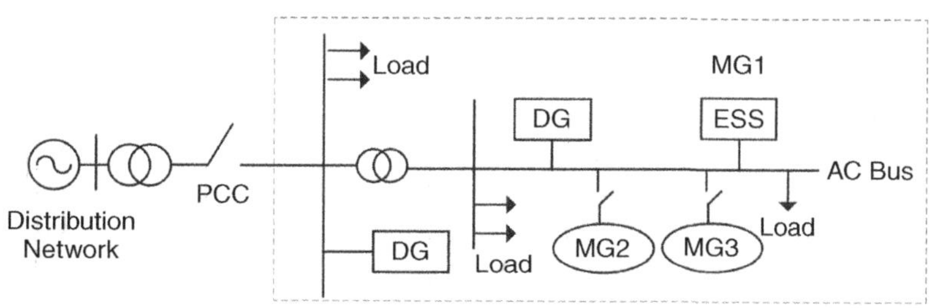

Figure 14.1 Nested MMGS.

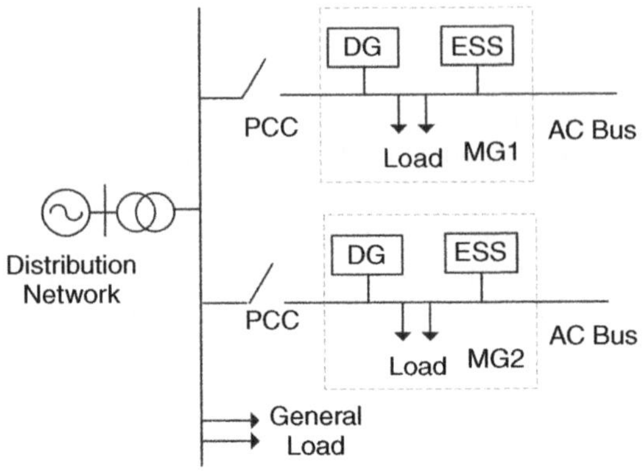

Figure 14.2 Parallel MMGS.

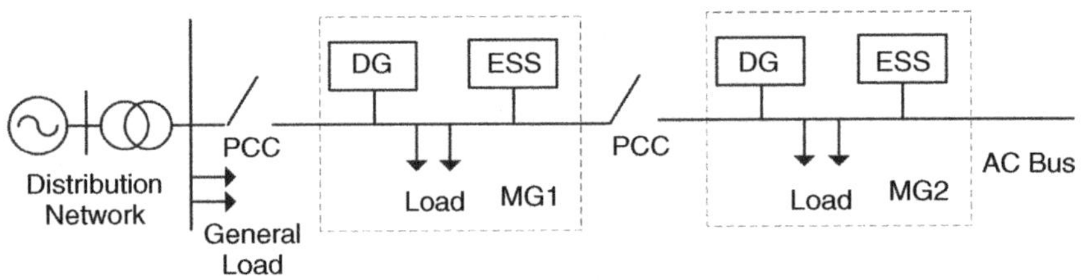

Figure 14.3 Series MMGS.

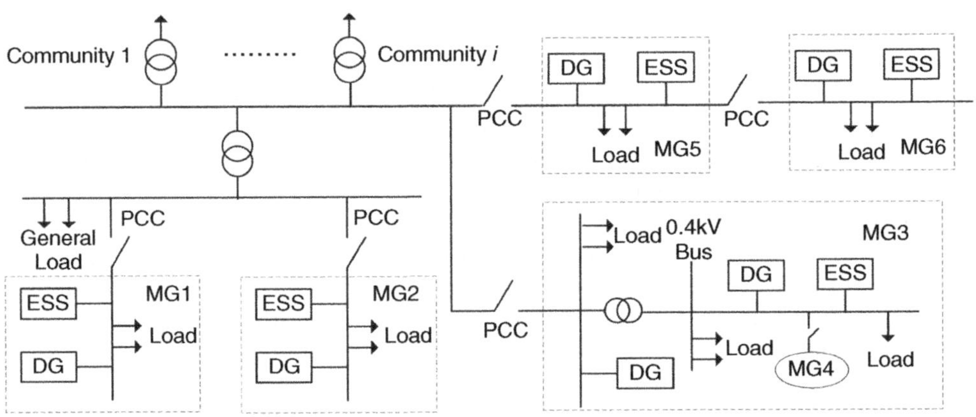

Figure 14.4 Dispersed MMGS.

4) Dispersed MMGS

According to the space as well as local distribution resources and loads, sub-MGs can form the dispersed MMGS of multi-types and multi-structures, as shown in Figure 14.4. Sub-MGs are connected to MV distribution systems. Dispersed MMGS can maximize regional advantages, improve the assimilating of distribution energy, and increase the utilization rate of distribution energy, resulting in an effective interconnected network in a wider range.

The four MMGS network structures are evaluated by the connection index, MMGS connection degree (*MCD*) (1), which is proposed to quantify the interconnection strength of MGs.

$$MCD = 2n_l / n_{mg}(n_{mg} - 1) \tag{14.1}$$

where n_l is the number of connection lines, and n_{mg} is the number of MGs. The larger the MCD, the stronger the interconnection. For instance, for two MGs connected via a single line, *MCD* is equal to 1, which shows the connection is strong and any MG can receive or deliver energy through a line to the other MGs directly.

For a series MMGS with n MGs, as only two adjacent MGs are directly connected, MCD is $2/n$. For a parallel MMGS where all MGs are directly connected to each other, MCD is 1. For a nested MMGS with $n - 1$ LV MGs in series and one MV MG, MCD is $2/n$. For a nested MMGS with $n - 1$ LV MGs in parallel and one MV MG, MCD is 1. In other scenarios, MCD of a nested MMGS with $n - 1$ LV MGs and one MV MG varies from $2/n$ to 1. A dispersed MMGS generally has no direct connections among MGs, so corresponding MCD is 0. In conclusion, a parallel MMGS has the highest MCD, followed by nested MMGS, series MMGS, and dispersed MMGS.

The structure of MMGS is planned to balance network construction flexibility and economic benefits. Network construction flexibility is to ensure the network construction is easy and flexible. A flexible MMGS would have more operation choices, which makes it more likely to meet various users' requirements. The index MMGS structure flexibility (MSF) is proposed, which quantifies the flexibility of MMGS' structure. This index is defined as:

$$MSF = (k_p * n_p + k_n * n_n + k_s * n_s + k_d * n_d) / (n_p + n_n + n_s + n_d) \tag{14.2}$$

where n_p, n_n, n_s, and n_d are numbers of parallel, nested, series, and dispersed MMGs in an MMGS, respectively. k_p, k_n, k_s, and k_d are the weights reflecting the structure flexibility of parallel, nested, series, and dispersed MMGs, respectively. An MMGS with a higher MSF is considered to be more flexible. The weights of k_p, k_n, k_s, and k_d can be set 100, 90, 80, 0, respectively, as an example according to the fact that parallel MMGS is the most flexible form, followed by nested MMGS, series MMGS, and dispersed MMGS, which has the worst flexibility.

14.1.2.2 Structure Planning Method

1) **Structure Planning Model Description**

The objective of the optimal structure planning model is to maximize flexibility of the MMGS:

$$\text{Max } F(X) = MSF(X) + G(X) \tag{14.3}$$

where *MSF(X)* is the flexibility of an MMGS under scheme *X*, and *G(X)* is the planning cost of an MMGS. The network construction flexibility is described as:

$$MSF(X) = \begin{cases} 0 & MSF(X) > MSF_{set} \\ 10^5 & MSF(X) \le MSF_{set} \end{cases} \tag{14.4}$$

where MSF_{set} is the threshold of network construction flexibility.

Cost function *G(X)* mainly includes investment as well as operation and maintenance (O&M) costs for DGs, energy storage devices, static switches, and control system.

$$G(X) = \sum_{i=1}^{N_T} N_i \cdot (C_{AI,i} + C_{M,i}) \tag{14.5}$$

where N_T is the number of types of devices, N_i is the number of the *i-th* type devices. $C_{AI,i}$ and $C_{M,i}$ are annual investment and O&M costs of the *i-th* type device.
The annual investment of a device is described as:

$$C_{AI,i} = C_i \cdot \frac{(1+\gamma)^{T_i} \cdot \gamma}{(1+\gamma)^{T_i} - 1} \tag{14.6}$$

where C_i and T_i are the *i-th* device's unit cost and operation life, respectively, and γ is the discount rate.
The annual O&M cost of a device is proportional to its annual investment cost, which is described as:

$$C_{M,i} = C_{AI,i} \cdot \eta_i \tag{14.7}$$

where η_I is ratio of the *i-th* device's annual O&M cost over its annual investment cost.
The constraint on the total installed static switches is described as:

$$0 \leq N_{\text{Switch}} \leq N_{\text{Switchmax}} \tag{14.8}$$

where N_{switch} is the number of static switches to be installed and $N_{\text{switchmax}}$ is the maximum number of static switches that can be installed.
The constraint on the location of DGs is described as:

$$0 \leq N_{de} \leq N_{\text{demax}} \tag{14.9}$$

where N_{de} is the number of DGs to be installed, and N_{demax} is the maximum number of available DGs.
The constraint on energy storage locations is described as:

$$0 \leq N_{se} \leq N_{\text{semax}} \tag{14.10}$$

where N_{se} is the number of energy storages to be installed, and N_{semax} is the maximum number of available energy storages.

2) **Structure Planning Algorithm**
Many intelligent optimization algorithms, such as particle swarm optimization, ant colony algorithm, and harmony search algorithm, can be used to solve the proposed structure planning problem. In this chapter, the immune genetic algorithm (IGA; Liao, 2006) is applied to solve the planning problem. The IGA is an improved genetic algorithm based on biological immune mechanism, which corresponds to the objective function of the problem as the antigen and to the solution of the problem as the antibody. The IGA improves the fitness of antibody population by the immune operator, guarantees the diversity of antibodies by the genetic operator, and prevents the population of degeneration phenomenon by immune selection.

The flowchart of the improved immune genetic algorithm is shown in Figure 14.5.
The planning procedure is described as follows:

Step 1: generate codes of distribution energy devices and switch locations to form the initial antibody population; set the iteration index k = 1 and the maximum generation, m;

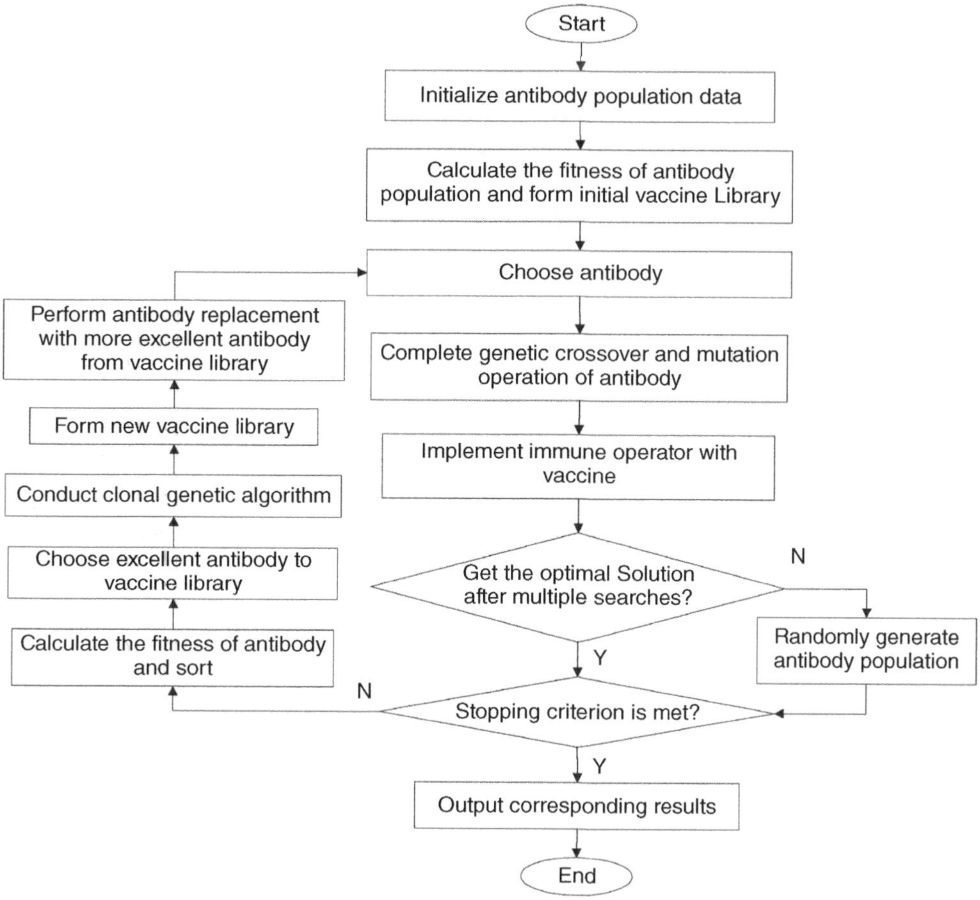

Figure 14.5 Flowchart of the improved immune genetic algorithm.

Step 2: perform the immune genetic operation to obtain new antibody population;
Step 3: calculate cost function and network construction flexibility of the current antibody population;
Step 4: build vaccine base by choosing antibodies that meet network construction flexibility and economic optimality requirements;
Step 5: Repeat Steps 2–4 until satisfying the stopping criterion.

The corresponding flowchart is shown in Figure 14.6.

14.1.3 Optimal Capacity Planning

14.1.3.1 Definition of Indexes

A variety of planning indexes are proposed for the optimal capacity planning of MMGS, focusing on distributed energy utilization, multi-energy comprehensive utilization, economy, reliability, environmental protection, and other requirements.

1) Distributed Energy Utilization

Six indexes concerning distributed energy utilization are shown as follows.

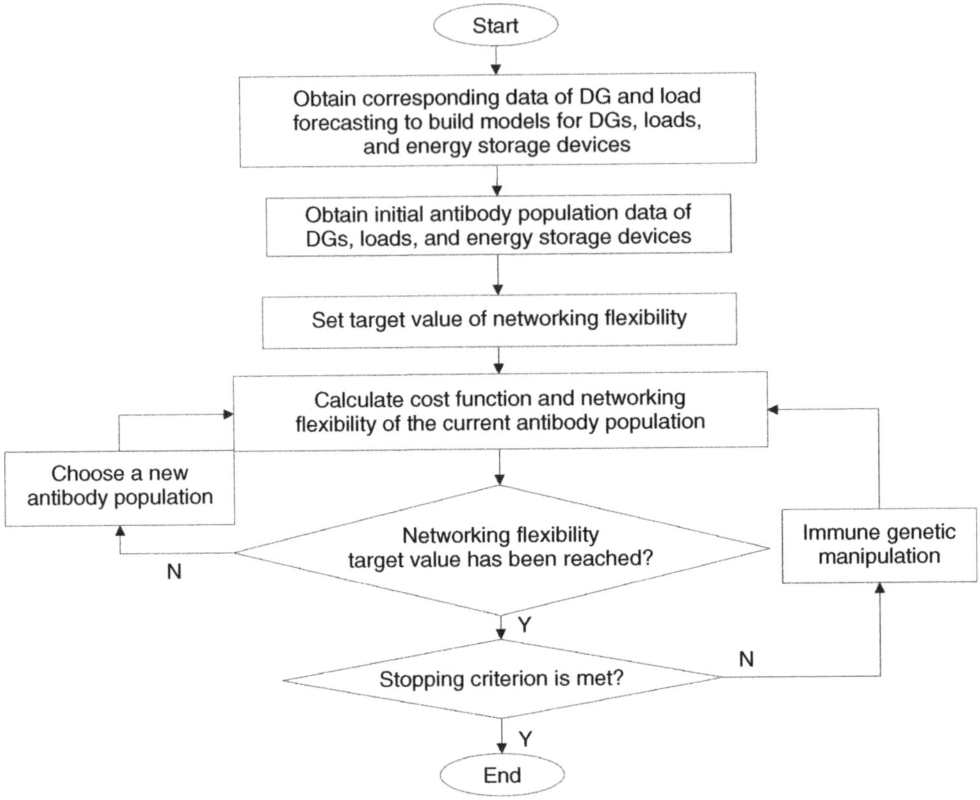

Figure 14.6 Flowchart of optimal structure planning of MMGS using the improved immune genetic algorithm.

a. Microgrid distributed energy microgrid-load consumption rate (MDEMCR) describes the ratio of consumed energy generated by DG in the MG to the total available energy generated by DG in the MG. The mathematical expression is given by:

$$MDEMCR = \left(\int_0^{8760} P_g(t)dt \text{-} \int_0^{8760} P_i(t)dt \right) / \int_0^{8760} P_g(t)dt \tag{14.11}$$

where $P_i(t)$ is the actual power dispatch of MG at time t, and $P_g(t)$ is the total available power generation of MG at time t. The total number of hours in a year is 8760. This index characterizes actual utilization of distributed energy in MG. Higher *MDEMCR* means an MG has higher degree of self-generation/self-consumption and demands less energy from other MGs.

b. Microgrid distributed generation and load capacity rate (MDGLCR) describes the ratio of rated DG capacity to the rated load in an MG. The mathematical expression is given by:

$$MDGLCR = \sum_{i=1}^{NG} P_{Gi} / \sum_{j=1}^{NL} P_{Lj} \tag{14.12}$$

where P_{Gi} and P_{Lj} are respectively rated power of DG i and load j in MG, and NG and NL are the number of DGs and loads in MG. This index characterizes the ratio of distributed energy to required energy in MG.

c. Microgrid energy storage and distributed generation capacity rate (MESDGCR) describes the ratio of rated capacity of energy storage and that of DG in an MG. The mathematical expression is given by:

$$MESDGCR = \sum_{i=1}^{NS} P_{Si} / \sum_{j=1}^{NG} P_{Gj} \tag{14.13}$$

where P_{Si} and P_{Gj} are respectively rated power of energy storage i and DG j in an MG, and NS *is* the number of energy storage assets in MG. This index characterizes the smooth degree of energy output in an MG. A higher $MESDGCR$ means the DG output can be smoother, but planning cost could increase correspondingly because of a higher penetration of energy storage devices.

d. MMGS whole distributed energy microgrid-load consumption rate (MWDEMCR) describes the ratio of energy from DGs consumed by MMGS to the total energy from DGs in MMGS. The mathematical expression is given by:

$$MWDEMCR = \left(\int_0^{8760} \sum_{i=1}^{NG} P_{ig}(t)dt \text{-} \int_0^{8760} P(t)dt \right) / \int_0^{8760} \sum_{i=1}^{NG} P_{ig}(t)dt \tag{14.14}$$

where $P_{ig}(t)$ and $P(t)$ are the total power generation of MG i and the MMGS respectively. This index characterizes actual utilization of DGs in MMGS. A higher MWDEMCR means an MMGS has higher degree of self-generation/self-consumption.

e. MMGS whole distributed generation and load capacity rate (MWDGLCR) describes the ratio of rated capacity of DGs to rated loads in MMGS. The mathematical expression is given by:

$$MWDGLCR = \frac{\sum (\sum P_{Gij})}{\sum (\sum P_{Lmn})} \tag{14.15}$$

where P_{Gij} and P_{Lmn} are the rated power of DG j in MG i and the rated power of load n in MG m, respectively. This index characterizes the ratio of distributed energy to required energy in MMGS.

f. MMGS whole energy storage and distributed generation capacity rate (MWESDGCR) describes the ratio of rated capacity of energy storage assets to that of DGs in MMGS. The mathematical expression is given by:

$$MWESDGCR = \frac{\sum (\sum P_{Sij})}{\sum (\sum P_{Gmn})} \tag{14.16}$$

where P_{Sij} and P_{Gmn} are the rated power of energy storage j in MG i and the rated power of DG n in MG m, respectively. This index characterizes the smooth degree of energy output in MMGS and reflects the consumption situation of distributed energy. A higher $MESDGCR$ means the DG output can be smoother, but planning cost could increase correspondingly because of a higher penetration of energy storage devices. However, this can be improved by effectively sharing energy storage devices among MGs.

2) **Multi-Energy Comprehensive Utilization**

Three indexes are proposed for quantifying the multi-energy comprehensive utilization.

a. Renewable energy penetration (REP) describes the ratio of average generation capacity of renewable energy to average total load demand in an MG. The mathematical expression is given by:

$$\mathrm{REP} = \frac{\int_0^T P_{R\Sigma}(t)dt}{\int_0^T P_L(t)dt} \tag{14.17}$$

where $P_{R\Sigma}(t)$ is the total output power (kW) of renewable energy DGs at time t in MG, which is restricted by DG faults and MG operation constraints. $P_L(t)$ is the total load (kW) at time t in MG. This index characterizes the ratio of actual renewable energy utilization to total energy utilization.

b. MMGS electrical/thermal capacity ratio (METCR) is used to evaluate the ratio of the installed electrical capacity to the installed thermal capacity in MMGS. The mathematical expression is given by:

$$METCR = \frac{\sum P_{ec}}{\sum P_{tc}} \tag{14.18}$$

where P_{ec} is the total rated power of all power supply units, and P_{tc} is the rated power of all heating units in MG. This index characterizes the relative size of electrical/thermal energy.

c. Regional MMGS comprehensive energy efficiency index (RMCEEI) reflects the energy efficiency in MMGS. The mathematical expression is given by:

$$RMCEEI = \frac{\sum E_L}{\sum E_G} \tag{14.19}$$

where E_L and E_G are total energy consumed by loads and all energy from generation assets, both of which include cooling, heating, and electricity energy usage. This index characterizes total energy utilization efficiency in MMGS.

3) **Reliability**

This section proposes the system reliability offset index (SROI), which can quantify the capability of system power supply and the reliability requirement of load points. The mathematical expression is given by:

$$SROI = \sum_{i=1}^{NL} w_i \left(\delta_{\lambda i} \times \max\left(0, \frac{\lambda_i - \lambda_{ti}}{\lambda_{ti}}\right) + \delta_{Ui} \times \max\left(0, \frac{U_i - U_{ti}}{U_{ti}}\right)\right) \tag{14.20}$$

where NL is the number of load points in the system, λ_i and U_i are actual outage frequency and average annual outage time of load point i, respectively, λ_{ti} and U_{ti} are relative reliability target values of load point i, respectively, $\delta_{\lambda i}$ and δ_{Ui} are weights set by load point users for outage frequency deviation and annual outage time deviation according to their own reliability preference respectively, and w_i is a weight set by system operators after comprehensively measuring the important degree of each load point.

4) **Economy and Environmental Protection**
An investment cost index of distributed energy assets while considering environmental protection benefits is proposed, which quantifies economic and environmental benefits of MMGS. Distributed energy assets include DG (such as photovoltaic, micro turbine, and energy storage) and cooling/heating equipment for combined cooling, heating, and power (CCHP) systems. The investment cost is divided into fixed cost and variable cost. The mathematical expression is given by:

$$C_{invest} = \sum_{i=1}^{N} [C_{DERE} + \alpha \cdot (V_{DERE} - V_b - C_e) - \beta \cdot R] \tag{14.21}$$

$$C_e = \sum_{i=1}^{M} (V_i \cdot Q_i) \tag{14.22}$$

where N is the number of equipment being invested, C_{DERE} is the total fixed investment cost, V_{DERE} is the annual O&M cost, V_b is the annually saved fuel cost of conventional power, C_e is the monetary environmental benefits, R is the salvage value of distributed generation, α and β are conversion factors from annual value and final value to present value respectively, V_i is the emission charge for contaminant i, Q_i is the total emission of contaminant i, and M is the total number of emission types.

14.1.3.2 Capacity Planning Method

1) **Capacity Planning Model**
The planning objective is to minimize the total investment cost of distributed energy assets, while the distributed energy consumption, multi-energy comprehensive utilization, and reliability goals are also included in the objective via a penalty function. The objective function is as follows.

$$\text{Min } F(X) = C_{invest}(X) + G(X) \tag{14.23}$$

where $C_{invest}(X)$ is the total investment cost of distributed energy assets under the planning scheme X, and $G(X)$ is the penalty function for incorporating distributed energy consumption, multi-energy comprehensive utilization, and reliability goals. The penalty function $G(X)$ is as follows.

$$G(X) = \begin{cases} 0 & other \\ 10^5 & DEC(X) < DEC_{set}, MCR(X) < MCR_{set}, SROI(X) < SROI_{set} \end{cases} \tag{14.24}$$

where *DEC(X)*, *MCR(X)*, and *SROI(X)* are actual values of distributed energy consumption, multi-energy comprehensive utilization, and reliability goals calculated under the planning scheme X, respectively. DEC_{set}, MCR_{set}, and $SROI_{set}$ are respectively target values of distributed energy consumption, multi-energy comprehensive utilization, and reliability goals that are set by system operators.

The prevailing constraints are as follows. Node power flow constraints are described as:

$$\begin{cases} P_{Gi} - U_i \sum_{j=1}^{N} U_j(G_{ij} \cos \delta_{ij} + B_{ij} \sin \delta_{ij}) = 0 \\ Q_{Gi} - U_i \sum_{j=1}^{N} U_j(G_{ij} \cos \delta_{ij} - B_{ij} \sin \delta_{ij}) = 0 \end{cases} \tag{14.25}$$

where P_{Gi} and Q_{Gi} are active and reactive power injection at node i, respectively, U_i/U_j is the voltage amplitude of node i/j, δ_i/δ_j is the voltage phase angle of node i/j, and $G_{k(i,j)}/B_{ij}$ is conductance/susceptance of branch ij.
Power balance constraints are described as:

$$\begin{cases} P_{i,CCHP,e} + P_{i,PV} + \sum_{i \neq j} P_{i,j} = \sum P_{i,l} + \sum P_{i,e,h/c} + P_{loss} \\ P_{i,CCHP,h/t} = \sum P_{i,h/c} - \sum P_{i,e,h/c} \end{cases} \tag{14.26}$$

where $P_{i,CCHP,e}$ is the electric power output of gas turbine triple-generation system connected to node i, $P_{i,PV}$ is the power output of distributed photovoltaic system connected to node i. $P_{i,j}$ is the power injection to node i from node j through branch ij, $P_{i,l}$ is non-cooling/heating electric loads at node i, $P_{i,e,h/c}$ is electricity consumption of cooling/heating loads, P_{loss} is the power losses through distribution lines, $P_{i,CCHP,h/c}$ is cooling/heating energy output of gas turbine triple-generation system, and $P_{i,h/c}$ is total energy consumption of cooling/heating loads of node i.
In order to guarantee the continuous power supply to critical loads, a minimum capacity of DG is required. That is, minimum outputs of DGs must guarantee the uninterrupted power supply to critical loads. This constraint is described as:

$$\sum_{i \in N_{DG}} P_i(t) \geq \sum_{j \in N_I} L_j(t), t = 1, 2, 3 \ldots, 8760 \tag{14.27}$$

where P_i is electricity output of DGs connected at node i *at* time t, $L_j(t)$ is critical load at node j, N_{DG} is the set of nodes connected with DGs, and N_I is the set of nodes connected with critical loads.
Due to uncertainties of DG outputs, limitation on the maximum installed capacity of DGs is required to guarantee system power supply quality. That is, the maximum installed capacity of DGs should not result in serious impact on power quality. This constraint is described as:

$$P_i \leq P_{i,\max} \tag{14.28}$$

In addition, inequality constraints also include upper and lower bounds on nodal voltages, capacity limit on branch active power, and spinning reserve. These constraints are described as:

$$\begin{cases} U_{\min} \leq U \leq U_{\max} \\ |P_l| \leq P_{l_{\max}} \\ \max\{P_{DG}\} \leq P_{s\max} - P_s \end{cases} \tag{14.29}$$

where U is the nodal voltage vector of system, P_l is the active power vector of branches, P_s and P_{smax} are the actual power and maximum power output from transmission system to distribution system, respectively.

Charging and discharging status of energy storage assets is described as follows.

$$P_{cmd}(t) = \frac{1}{24}\int_{1}^{24} P(t)dt - P(t) \tag{14.30}$$

where $P(t)$ is the net power forecast from loads and PVs at time t. If $P_{cmd}\ (t) > 0$, energy storage devices take the charging operation, and if $P_{cmd}\ (t) < 0$, energy storage devices are discharged.

The output of energy storage devices is:

$$P_{ESS}(t) = \begin{cases} 0, & SOC(t-1) = SOC\_\&P_{cmd}(t) < 0 \\ 0, & SOC(t-1) = SOC_+\&P_{cmd}(t) > 0 \\ P_{rated}, & P_{cmd}(t) > P_{rated}\&SOC_- < SOC(t-1) \le SOC_+ \\ -P_{rated}, & P_{cmd}(t) < \text{-}P_{rated}\&SOC_- \le SOC(t) < SOC_+ \\ P_{cmd}(t), & otherwise \end{cases} \tag{14.31}$$

where $SOC(t)$ is the state of charge for energy storage devices at time t, and (SOC_-, SOC_+) is the state of charge limits. In order to limit $SOC\ (t)$ within the range of (SOC_-, SOC_+), charging and discharging power is further modified as follows.

When $P_{ESS}(t) > 0$, energy storage is charging and the charging power is modified as:

$$P'_{ESS}(t) = \begin{cases} \dfrac{[SOC_+ - SOC(t-1)]\cdot E_{rated}}{\eta_c\cdot\Delta t}, & if SOC_+ - SOC(t-1) < \dfrac{\eta_c\cdot\Delta t\cdot P_{ESS}}{E_{rated}} \\ P_{ESS}(t), & otherwise \end{cases} \tag{14.32}$$

When $P_{ESS}(t)<0$, energy storage is discharging and the discharging power is modified as:

$$P'_{ESS}(t) = \begin{cases} \eta_d\dfrac{[SOC\_ - SOC(t-1)]\cdot E_{rated}}{\Delta t}, & if SOC\_ - SOC(t-1) < \dfrac{\Delta t\cdot P_{ESS}}{\eta_d\cdot E_{rated}} \\ P_{ESS}(t), & otherwise \end{cases} \tag{14.33}$$

where η_c and η_d are charging and discharging efficiencies of energy storage, respectively.

When the charging and discharging power of energy storage is determined, $SOC(t)$ will be updated as:

$$SOC(t) = SOC(t-1) + \begin{cases} \dfrac{\eta_c \cdot \Delta t \cdot P'_{ESS}(t)}{E_{rated}}, & if P'_{ESS}(t) > 0 \\ \dfrac{\Delta t \cdot P'_{ESS}(t)}{\eta_d \cdot E_{rated}}, & otherwise \end{cases} \tag{14.34}$$

2) **Capacity Planning Algorithm**
The improved immune genetic algorithm is again used to solve the optimal capacity planning model (11)-(34) of MMGS. The detailed procedure is as follows:
Step1: Generate coding for distributed energy capacities, obtain initial antibody population, set the initial iteration index $k = 1$, and the maximum generation, m.
Step2: Carry out the immune genetic manipulation of antibody population to obtain a new antibody population.
Step3: Calculate values of distributed energy consumption, multi-energy comprehensive utilization, reliability, environmental protection, and economics of the current antibody population.
Step4: Choose the antibody with a better economic objective value while meeting the goals of distributed energy consumption, multi-energy comprehensive utilization, and reliability requirements, to compose the vaccine library.
Step5: Repeat Steps 2–4 until the stop criterion is satisfied.
The corresponding flowchart is shown in Figure 14.7.

14.1.4 Conclusions

Aiming at planning and design issues for the optimal structure and capacity of MMGSs, this section establishes the planning model while considering network construction flexibility, distributed energy consumption, and multi-energy comprehensive utilization together with economic goals. An improved immune genetic algorithm is proposed to solve the problem. The proposed approaches can improve the network construction flexibility, distributed energy consumption ability, and multi-energy comprehensive utilization efficiency of MMGSs under the premise of economics, while simultaneously considering comprehensive energy efficiency, distributed energy consumption, distributed energy self-use rate of users, reliability, and environmental protection concerns.

14.2 Performance Assessment of Multi-Microgrid System

14.2.1 Introduction

The operation statuses of multiple MGs connected in an MMGS could be different from those when they are operated separately. Specifically, in an MMGS, MGs can support each other. There are articles that focus on energy scheduling, optimization and control of MMGS (Shahnia et al., 2013; Lee et al., 2015; Fathi & Bevrani, 2013; Liang et al., 2015;

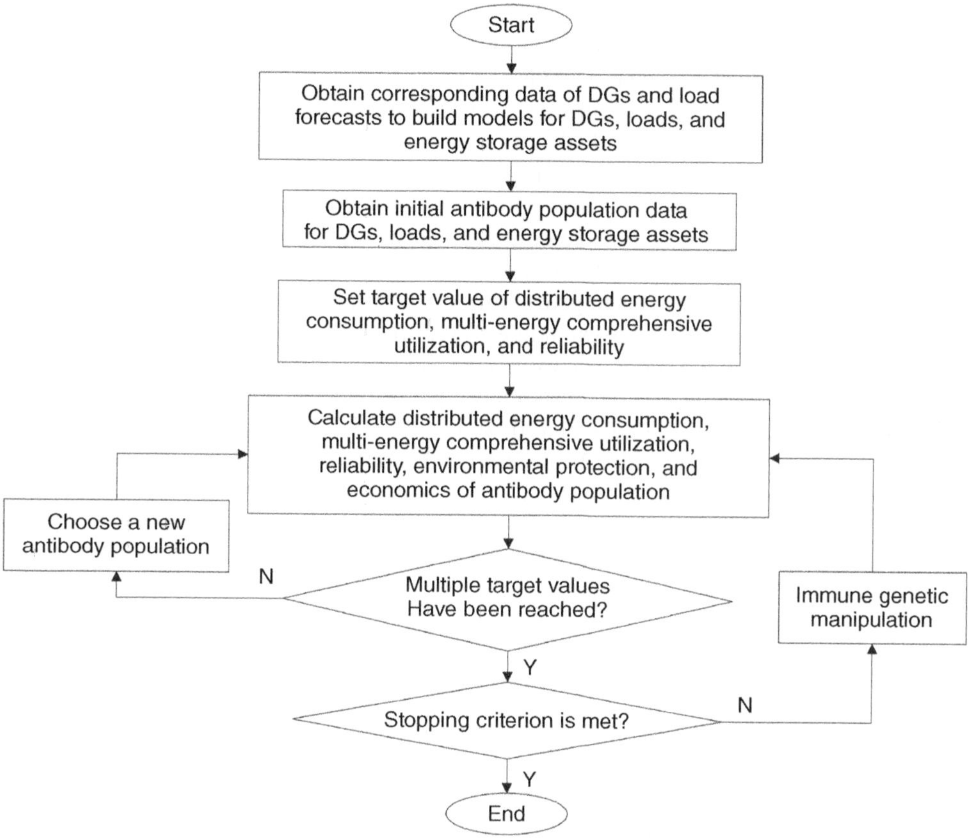

Figure 14.7 Flowchart of optimal capacity planning of MMGS using the improved immune genetic algorithm.

Nikmehr et al., 2015; Nunna et al., 2012; Wu et al., 2014; Che et al., 2015; Wu et al., 2013; Farzin et al., 2015; Farzin et al., 2016; Ouammi et al., 2014; Li et al., 2016). Shahnia et al. (2013) discussed autonomous operation of multiple interconnected MGs with self-healing capability. Lee et al. (2015) proposed a distribution mechanism for energy trading among multiple interconnected MGs in a region, while assuming that some MGs have superfluous energy for sale and other MGs need additional energy to meet local demands. A cooperative power dispatching algorithm was proposed in Fathi and Bevrani (2013) to study interactions among MGs and power sharing within the utility grid, under the assumption that demands and supplies of individual MGs are time varying. Energy exchange characteristics between MGs were investigated in Liang et al. (2015) using prospect theory. Nikmehr et al. (2015) discussed the probabilistic optimal power dispatch in an MMGS via a heuristic algorithm. An agent-based intelligent energy management system was proposed in Nunna et al. (2012), for facilitating power trading among MGs and allowing customers to participate in demand response. Wu et al. (2014) proposed a hierarchical framework for generation scheduling of MGs. A hierarchical coordination strategy was proposed in Che et al. (2015) for economic operation of an islanded community MG that contains multiple AC and DC MGs. Fathi and Bevrani (2013) addressed energy consumption schedule in a distribution network with multiple connected MGs. Wu et al. (2013) proposed a coordinated MMGS optimal control algorithm for a smart distribution management system. Farzin et al. (2015) proposed a hierarchical scheme for outage management in an MMGS. Farzin et al. (2016) presented a distributed control scheme to regulate power flow and minimize interconnections of multiple MGs. Ouammi et al. (2014) discussed a system of systems approach to study optimal control of power flows and energy local storages in a network MGs. Li et al. (2016) proposed a hybrid market-based and centralized optimal design strategy for MGs in grid-connect and islanded modes.

In the above studies, the energy exchange and monetary incomes from energy sale are often used as objective functions for validating the effectiveness of control and assessment algorithms. Wang et al. (2013) defined a set of metrics for describing operation statuses of MGs. In addition, Lee et al. (2014), Clavier et al. (2013), Salcedo et al. (2016), Appen et al. (2011), and Beer (2012) assessed the economics of different types of MGs from various aspects. However, an MMGS presents different characteristics as compared to individual MGs, and traditional metrics cannot describe the MMGS appropriately. Therefore, it is necessary to propose new metrics for rigorously assessing the performance of MMGSs.

14.2.2 Comprehensive Evaluation Indexes

Compared with traditional distribution systems, an MMGS presents more flexible and more complex operating characteristics. Although indicators for traditional power distribution systems can be applied to an MMGS for evaluating its power generation and load assets, it is necessary to define MMGS specific indexes for describing its distinct operating characteristics with multiple MGs. In this section, several new indexes are defined to describe characteristics of an MMGS, including source-charge capacity, energy interaction, reliability, multi-micro-network system economics, energy utilization efficiency, energy saving, emission reduction, and renewable energy utilization.

14.2.2.1 MMGS Source-Charge Capacity Index

MMGS, being a system that can run independently, contains an appropriate source-load capacity, which describes the basic configuration and quantifies the operating performance of an MMGS. To this end, this section discusses existing indexes and new indexes to describe an MMGS' source-charge capacity characteristics:

- MMGS total generation capacity (MSTGC) (kW): This index reflects the total generation capacity of DGs and other energy devices in an MMGS. The mathematical expression is given by:

$$MSTGC = \sum_{i=1}^{N}\left(\sum_{j=1}^{M_i} P_{i,j}\right) \tag{14.35}$$

where N is the number of MGs in an MMGS, M_i is the total number of DGs in MG i, and $P_{i,j}$ is the rated capacity of the jth DG in MG i.

- MMGS total energy generation (MSTEG) (kWh): This index reflects the total annual energy generation of DGs and other energy devices in an MMGS. The mathematical expression is given by:

$$MSTEG = \sum_{i=1}^{N} \int_0^T P_{gen,i}(t)dt \tag{14.36}$$

where T is the number of hours in a year, i.e., 8,760, and $P_{gen,i}(t)$ is the average generated power at time t in MG i.
- MMGS capacity credibility (MSCC): This index describes the ratio of the actual active power output to the rated capacity of DGs and other energy devices in an MMGS. The mathematical expression is given by:

$$MSCC = \frac{\sum_{i=1}^{N} \int_0^T P_{gen,i}(t)dt}{T \cdot \sum_{i=1}^{N}\left(\sum_{j=1}^{M} P_{i,j}\right)} \tag{14.37}$$

- MMGS total average load (MSTAL) (kW): This index reflects the total load level of all MGs in an MMGS. The mathematical expression is given by:

$$MSTAL = \sum_{i=1}^{N} \int_0^T P_{load,i}(t)dt \tag{14.38}$$

where $P_{load,i}(t)$ is the average demand level at time t in MG i.
- MMGS critical load demand (MSCLD) (kW): This index reflects the critical load level of all MGs in an MMGS. The mathematical expression is given by:

$$MSCLD = \sum_{i=1}^{N} \int_0^T P_{c\_load,i}(t)dt \tag{14.39}$$

where $P_{c\_load,i}(t)$ is the average demand level of the critical load at time t in MG i.
- MMGS noncritical load demand (MSNLD) (kW): This index reflects the demand level of noncritical loads in all MGs of an MMGS. The mathematical expression is given by:

$$MSNLD = \sum_{i=1}^{N} \int_0^T P_{nc\_load,i}(t)dt \tag{14.40}$$

where $P_{nc\_load,i}(t)$ is the average demand level of noncritical loads at time t in MG i.
- MMGS critical load ratio (MSCLR): This index reflects the proportion of critical loads in an MMGS. The mathematical expression is given by:

$$MSCLL = \frac{\sum_{i=1}^{N} \int_0^T P_{c\_load,i}(t)dt}{MSTAL} \tag{14.41}$$

14.2.2.2 MMGS Energy Interaction Index

An MMGS could contain two-way power flows among MGs, as well as mutual power flows between the MMGC/individual MGs and the distribution network. To this end, energy interaction correlation indexes are defined in this section to describe power flows between MGs in an MMGS.

1) The Energy Interaction Index between Individual MGs and the MMGS

- Micro-grid energy output ratio (MEOR): This index reflects the contribution of a single MG to other MGs within the MMGS. The mathematical expression is given by:

$$MEOR_i = \frac{\int_0^T (|p_i(t)| - p_i(t))dt}{2 \cdot MSTEG_i} = \frac{\int_0^T (|p_i(t)| - p_i(t))dt}{2 \cdot \int_0^T P_{gen,i}(t)dt} \tag{14.42}$$

  where $p_i(t)$ is the power flow through the PCC node of MG i at time t. It should be noted that the positive direction of $p_i(t)$ is defined as flowing into the PCC node of an MG. $MSTEG_i$ is the average annual generation of MG i.
- Micro-grid energy input ratio (MEIR): This index reflects the contribution of the MMGS to each MG. The mathematical expression is given by:

$$MEIR_i = \frac{\int_0^T (|p_i(t)| + p_i(t))dt}{2 \cdot MSTEG_i} = \frac{\int_0^T (|p_i(t)| + p_i(t))dt}{2 \cdot \int_0^T P_{gen,i}(t)dt} \tag{14.43}$$

- Micro-grid energy dependency index (MEDI): The ratio of power input of an MG to its load demand reflects the dependence of the MG on other MGs within the MMGS. The mathematical expression is given by:

$$MEDI_i = \frac{\int_0^T (|p_i(t)| + p_i(t))dt}{2 \cdot MSTAL_i} = \frac{\int_0^T (|p_i(t)| + p_i(t))dt}{2 \cdot \int_0^T P_{load,i}(t)dt} \tag{14.44}$$

  where $P_{load,i}(t)$ is the total load in MG i at time t.
- Micro-grid energy purchase probability (MEPP): This index reflects the frequency at which an MG will purchase electricity from the PCC-attached network or the MMGS where the MG is located. The mathematical expression is given by:

$$MEPP_i = \frac{\sum_{j=1}^{np_i} U_{ij,p}}{8760} \tag{14.45}$$

  where np_i is the number of power purchases of MG I, $U_{ij,p}$ is time duration of the j-th power purchase for MG I, and 8,760 is the total number of hours in a year.
- *Micro-grid energy sold probability* (*MESP*): This index reflects the time probability that an MG sells power to the PCC or the MMGS where the micro-network is

located. The mathematical expression is given by:

$$MESP_i = \frac{\sum_{j=1}^{ns_i} U_{ij,s}}{8760} \tag{14.46}$$

where ns_i is the number of power sales of MG i, and $U_{ij,s}$ is time duration of the j-th power sale for MG i.

- Microgrid energy local consumption ratio (MELCR): This index reflects the ratio of power consumed from local DGs to the total power generation from local DGs in each MG. The mathematical expression is given by:

$$MELCR_i = 1\text{-}\frac{\int_0^T (|p_i(t)| - p_i(t))dt}{2 \cdot \int_0^T p_{gen,i}(t)dt} \tag{14.47}$$

- Micro-grid energy local consumption time ratio (MELCTR): This index reflects the temporal frequency that power from local DGs is consumed in each MG. The mathematical expression is given by:

$$MELCTR_i = \frac{\int_0^T f(p_{load,i}(t) - p_{gen,i}(t))dt}{T} \tag{14.48}$$

The function $f(t)$ is defined as follows:

$$f(t) == \begin{cases} 1 & t > 0 \\ 0 & t \leq 0 \end{cases} \tag{14.49}$$

2) **MMGS Total Energy Interaction Index**
   - MMGS energy adequacy index (SEAI): This index reflects the degree of energy self-sufficiency in an MMGS. The mathematical expression is given by:

$$SEAI = \frac{MSTEG}{MSTAL} = \frac{\sum_{i=1}^{N} \int_0^T P_{gen,i}(t)dt}{\sum_{i=1}^{N} \int_0^T P_{load,i}(t)dt} \tag{14.50}$$

   - MMGS energy exchange capacity (SEEC) (kWh): This index reflects the annual average power flow of all tie-lines in an MMGS. The mathematical expression is given as:

$$SEEC = \sum_{i=1}^{m} \int_0^T p_{l,i}(t)dt \tag{14.51}$$

where m is the number of tie-lines in an MMGS, and $p_{l,i}(t)$ is the average power through the i-th tie-line at time t.

- MMGS energy exchange ratio (SEER): This index reflects the ratio of annual average power flow of all tie-lines to the total load in an MMGS. The mathematical expression is given by:

$$\text{SEER} = \frac{SEEC}{MSTL} = \frac{\sum_{i=1}^{m} \int_0^T p_{l,i}(t)dt}{\sum_{i=1}^{N} \int_0^T P_{load,i}(t)dt} \tag{14.52}$$

- MMGS annual output energy (SAOE) (kWh): This index reflects the total annual average power output of an MMGS to the external grid. The mathematical expression is given by:

$$SAOE = \frac{1}{2} \cdot \int_0^T \sum_{i=1}^{N} (|p_i(t)| - p_i(t))dt \tag{14.53}$$

where $p_i(t)$ is the power passing through the ith MG at time t.

- MMGS energy local consumption ratio (SELCR): This index reflects the ratio of power consumed from all local DGs to the total power generation from local DGs in an MMGS. The mathematical expression is given as:

$$SELCR = \frac{\sum_{i=1}^{N} \left[ELCR_i \cdot \int_0^T p_{gen,i}(t)dt \right]}{\sum_{i=1}^{N} \int_0^T p_{gen,i}(t)dt} \tag{14.54}$$

where $ELCR_i$ is the ELCR index of the i-th MG.

- MMGS energy local consumption time ratio (SELCTR): This index reflects the temporal frequency at which power from local DGs is consumed in an MMGS. The mathematical expression is given as:

$$SELCTR = \frac{\int_0^T \left(f\left(\sum_{i=1}^{N} p_{load,i}(t) - \sum_{i=1}^{N} p_{gen,i}(t) \right) \right) dt}{T} \tag{14.55}$$

where $f(t)$ is similarly defined as in (14.50).

14.2.2.3 MMGS Reliability Index

This section defines several reliability indexes for an MMGS from the load level, MG level, and MMGS level.

1) **MMGS Load Reliability Index**

   Loads in an MMGS, in accordance to different importance degrees, can be divided into two types of critical loads and noncritical loads. Reliable power supply requirements for critical and noncritical loads are different. An MMGS is expected to meet critical loads as much as possible, and if an MMGS has excessive power supply capacity beyond satisfying critical loads, noncritical loads could be further supplied. To this end, this section defines relevant reliability indexes for critical and noncritical loads, respectively.

Critical load reliability indexes for an MMGS include the following ones.

- Loss of critical load frequency index (LCLFI) (f/year): This index is used to assess the frequency of failing to supply critical loads. The mathematical expression is given by:

$$LCLFI = \frac{\sum_{i=1}^{NC} \lambda_i \cdot NC_i}{\sum_{i=1}^{NC} NC_i} \tag{14.56}$$

  where NC is the total number of critical loads in an MMGS, and λ_i and NC_i are the failure rate and the number of users of critical load i, respectively.
- Critical load average interrupt duration index (CLAIDI) (hours/year): This index is used to assess the power failure duration of critical loads. The mathematical expression is given by:

$$CLAIDI = \frac{\sum_{i=1}^{NC} U_i \cdot NC_i}{\sum_{i=1}^{NC} NC_i} \tag{14.57}$$

  where U_i is the annual power outage time of critical load i.
- Critical load expected energy deficiency (CLEED) (kWh/year): This index is used to estimate average load curtailment of critical loads caused per outage. The mathematical expression is given by:

$$CLEED = \frac{\sum_{i=1}^{NC} U_i \cdot P_{c\_i}}{\sum_{i=1}^{NC} N_i} \tag{14.58}$$

  where P_{c_i} is the average load power of a critical load i.
- Critical load maximum interrupt duration index (CLMIDI) (hours): This index is used to calculate the longest interrupt duration experienced by all critical loads. The significance of this index is to describe the most severe interruption to critical loads. The mathematical expression is given by:

$$CLMIDI = \max(U_{fi}) \tag{14.59}$$

  where U_{fi} is power interrupt duration of the i-th interrupt to critical loads.
- Critical load maximum energy deficiency (CLMED) (kWh/year): This index is used to calculate the most severe load curtailment experienced previously by all critical loads. The significance of this index is to describe the most severe interruption faced by critical loads. The mathematical expression is given by:

$$CLMED = \max(P_{fi} \cdot U_{fi}) \tag{14.60}$$

where P_{fi} is the average load power of the fi interruption to critical loads, and U_{fi} is the duration of the fi interruption to critical loads.

Reliability indexes for noncritical loads include:

- Loss of noncritical load probability (LNLP): This index is used to estimate the probability of failing to supply noncritical loads. The mathematical expression is given by:

$$LNLP = \frac{\sum_{i=1}^{NL} \lambda_{nc\_i} \cdot NC_i}{\sum_{i=1}^{NL} NC_i} \tag{14.61}$$

where NL is the number of noncritical loads in an MMGS, and λ_{cn_i} and NC_i are the failure rate and the number of users of noncritical load i, respectively.

- Noncritical load average interrupt duration index (NLAIDI) (hours/year): This index is used to assess the power failure duration of noncritical loads. The mathematical expression is given by:

$$NLAIDI = \frac{\sum_{i=1}^{NL} U_{n\_i} \cdot NC_i}{\sum_{i=1}^{NL} NC_i} \tag{14.62}$$

where U_{n_i} is the annual power outage time of noncritical load i.

- Noncritical load average interrupt duration index (NLAIDI) (f/year): This index is used to assess the frequency of failing to supply noncritical loads. The mathematical expression is given by:

$$NLAIDI = \frac{\sum_{i=1}^{NL} U_{n\_i} \cdot NC_i}{\sum_{i=1}^{NL} NC_i} \tag{14.63}$$

- Noncritical load expected energy deficiency (NLEED) (kWh/year): This index is used to calculate average load curtailment of noncritical loads caused per outage. The mathematical expression is given by:

$$NLAIDI = \frac{\sum_{i=1}^{NL} U_{n\_i} \cdot P_{nc\_i}}{\sum_{i=1}^{NL} NC_i} \tag{14.64}$$

2) **Microgrid Reliability Index**

Literature proposed some indices for describing reliability for the MG operation. In this section, combining with existing research results, we summarize some representative MG reliability indices. Details as follows:

- Microgrid average interruption frequency index (MAIFI) (f/year): This index is used to evaluate the frequency of load interrupt at the MG level. The mathematical expression is given by:

$$MAIFI = \frac{\sum_{i=1}^{NL} \lambda_i \cdot N_i}{\sum_{i=1}^{NL} N_i} \tag{14.65}$$

where NL is the total number of load points in an MG, and λ_i and N_i are the failure rate and the number of users at load point i, respectively.

- Microgrid average interruption duration index (MAIDI) (hours/year): The $MAIDI$ index is used to evaluate the annual average outage duration for each load point in an MG. The mathematical expression is given by:

$$MAIDI = \frac{\sum_{i=1}^{NL} U_i \cdot N_i}{\sum_{i=1}^{NL} N_i} \tag{14.66}$$

where U_i is the annual outage time of the load point i.

- Microgrid average service availability index (MASAI): The $MASAI$ index is used to estimate the temporal frequency of the available power supply for each load point. The mathematical expression is given by:

$$MGASAI = \frac{\sum_{i=1}^{NL} 8760 \cdot N_i - \sum_{i=1}^{NL} U_i \cdot N_i}{\sum_{i=1}^{NL} 8760 \cdot N_i} \tag{14.67}$$

- Microgrid customers experiencing multiple interruptions (MCEMI): The MCEMI index is used to estimate the proportion of users with more than one power outage in an MG. The mathematical expression is given by:

$$MGCEI = \frac{N_{m,o}}{\sum_{i=1}^{NL} N_i} \tag{14.68}$$

where $N_{m,o}$ is the number of users who are faced with power outages more than once.

- Microgrid islanded energy deficiency interruption (MIEDI) (kWh): The $MIEDI$ index is used to evaluate the power gap under the islanded operation mode of an MG. The mathematical expression is given by:

$$MIEDI = \sum p_i \cdot C_i \tag{14.69}$$

where p_i is the occurrence probability of state i, and C_i is the load curtailment level under condition i.

- Microgrid islanded energy deficiency probability (MIEDP): The *MIEDP* index is used to estimate the probability that an MG cannot supply sufficient power to a given load level. The mathematical expression is given by:

$$MIEDP = \Pr\{P_{gen,n} \leq P_{load,n}\} \tag{14.70}$$

where $P_{gen,n}$ is the total DG supply of MG n, and $P_{load,n}$ is the load level.

3) **MMGS Entirety Reliability Index**

An MMGS, as a high-level system composed of multiple MGs, presents different characteristics in comparison with individual MGs. To this end, this section proposes a number of MMGS reliability indexes to address operating characteristics of an MMGS. Details are as follows:

- MMGS islanded successive ratio (SISR): This index is used to evaluate the probability of success for an MMGS to switch from grid-connected mode to islanded mode. The mathematical expression is given by:

$$SISR = \frac{N_{ms}}{N_{total}} \tag{14.71}$$

where N_{ms} is the total number of success islands for an MMGS, and N_{total} is the number of annual interruptions of the distribution network to which the MMGS is connected.

- MMGS islanded time ratio (SITR): This index is used to evaluate the ratio of the total islanded operation time of an MMGS to the total interruption time of the distribution network to which the MMGS is connected. The mathematical expression is given by:

$$SITR = \frac{\sum_{i=1}^{N_{ms}} U_{ms\_i}}{\sum_{i=1}^{N_{total}} U_{ds\_i}} \tag{14.72}$$

where U_{ms_i} is the islanded operation time duration in the i-th islanding event of an MMGS, and U_{ds_i} is the interruption time duration in the i-th power failure of the distribution network where the MMGS is connected to.

- MMGS average energy not supplied (SAENS) (kWh): This index is used to estimate the annual average power shortage of an MMGS. The mathematical expression is given by:

$$SAENS = \sum_{i=1}^{N_{ms\_d}} U_{ms\_d\_i} \cdot P_{ms\_d\_i} \tag{14.73}$$

where $P_{ms\_d\_i}$ is the average interrupted power in the i-th islanding event of the MMGS, $U_{ms\_d\_i}$ is the interruption time duration of the i-th islanding event, and N_{ms_d} is the number of annual power interruptions.

- MMGS average interruption time (SAIT) (hours): This index is used to evaluate the average power interruption time of an MMGS. The mathematical expression is given by:

$$SAIT = \sum_{i=1}^{N_{ms\_d}} U_{ms\_d\_i} \tag{14.74}$$

- MMGS commination average failure duration (SCAFD) (hours): This index is used to evaluate the average communication failure time duration of the communication system in an MMGS. The mathematical expression is given by:

$$SCAFD = \frac{\sum_{i=1}^{CM} U_{c,i}}{CM} \tag{14.75}$$

where CM is the number of annual communication failures, and $U_{c,i}$ is the communication failure time duration of the i-th failure event.

- MMGS average communication available index (SACAI): This index is used to evaluate the normal validity index of the communication system for an MMGS. The mathematical expression is given by:

$$SACAI = \frac{8760 - \sum_{i=1}^{CM} U_{c,i}}{8760} \tag{14.76}$$

- MMGS communication failure affected duration (SCFAD) (hours): This index is used to evaluate the increment of system outage time duration caused by communication failure in an MMGS. The mathematical expression is given by:

$$SCFAD = \sum_{i \in (R \cup S)} U_i - \sum_{i \in R} U_i \tag{14.77}$$

where set R represents power interruption events caused by electrical device failures, and set S represents power interruption events caused by communications equipment failures.

- MMGS communication failure energy loss (SCFEL) (kWh): This index is used to evaluate power interruptions caused by communication failures in an MMGS. The mathematical expression is given by:

$$SCFEL = \sum_{i \in R} P_i \cdot U_i \tag{14.78}$$

where R is the set of power interruption events caused by communication equipment failures, U_i is the power failure time duration of power interruption event i, and P_i is the average load power corresponding to each power interruption event i.

14.2.2.4 MMGS Economics Index

Through the coordination of DGs, energy storage assets, micro-gas turbines, and other equipment, an MMGS can be smoothly transferred between grid-connected mode and islanded mode, and in turn the power supply reliability level can be significantly improved. However, power supply economics could in general be compromised at a higher reliability level. Therefore, it is of great value to quantify the power supply economics indexes, which could be used together with reliability indexes to leverage the two for the optimal operation of MMGS. Economics indexes of the MMGS are discussed as follows:

- Microgrid average energy sale income (MAESI) ($/year): This index is mainly used to assess the average annual power sale revenue of a single MG. The mathematical expression is given by:

$$MAESI_i = \frac{\int_0^T \rho_{sell,t} \cdot (|p_i(t)| - p_i(t))dt}{2} \tag{14.79}$$

  where $\rho_{sell,t}$ is the power sale price at time t.
- MMGS average energy sale income (SAESI) ($/year): This index is mainly used to evaluate the annual average revenue of power sale to the distribution network for an entire MMGS. The mathematical expression is given by:

$$SAESI = \sum_{i=1}^{N} MAESI_i \tag{14.80}$$

  where $MAESI_i$ is the average annual power sale income of MG i.
- Microgrid average energy purchase cost (MAEPC) ($/year): This index is mainly used to assess the average annual power purchase cost of a single MG. The mathematical expression is given by:

$$MAEPC_i = \frac{\int_0^T \rho_{buy,t} \cdot (|p_i(t)| + p_i(t))dt}{2} \tag{14.81}$$

  where $\rho_{buy,t}$ is the power purchase price at time t.
- MMGS average energy purchase cost (SAEPC) ($/year): This index is mainly used to assess the annual average cost of power purchase from the distribution network for an entire MMGS. The mathematical expression is given by:

$$SAEPC = \sum_{i=1}^{N} MAEPC_i \tag{14.82}$$

  where $MAEPC_i$ is the average annual power purchase cost of MG i.
- Microgrid average cash flow (MACF) ($/year): This index is mainly used to assess the difference between the average annual power purchase cost and the average annual sale income of a single MG. The mathematical expression is given by:

$$MACF_i = MAESI_i - MAEPC_i \tag{14.83}$$

  If $MACF_i$ is positive, it means that the MG will receive certain revenue each year; otherwise, the MG needs to prepare some budget for purchasing power each year.
- MMGS average cash flow (SACF) ($/year): This index is mainly used to evaluate the difference between the average annual power purchase cost and the average annual sale income for the entire MMGS. The mathematical expression is given by:

$$SACF = \sum_{i=1}^{N} MASF_i \tag{14.84}$$

- Critical load supply cost (CLSC) ($/year): This index is used to estimate the average annual power supply cost of electricity per kilowatt-hour for critical loads in an MMGS. The mathematical expression is given by:

$$CLSC = \frac{\rho_{total}}{P_{c\_load}} = \frac{\sum_{n=1}^{N}(\rho_{bld,i} + \rho_{fuel,i} + \rho_{mt,i} + MAPEC_{c,i} - SAESI_{c,i})}{\sum_{i=1}^{N}\int_{0}^{T} p_{c\_load,i}(t)dt} \tag{14.85}$$

where ρ_{total} is the total annual monetary cost of an MMGS for supplying critical loads, P_{c_load} is the average annual demand level of critical loads in the MMGS, $\rho_{bld,i}$ is the investment cost of MG i, $\rho_{fuel,i}$ is the average annual fuel cost of MG i for supplying critical loads, $\rho_{mt,i}$ is the average annual maintenance cost of MG i, $MAPEC_{c,i}$ is the power purchase cost of MG i for supplying critical loads, and $SAESI_{c,i}$ is the power sale revenue of MG i to other MGs.

- Critical load supply marginal cost (CLSMC) ($/year): This index is mainly used to evaluate the relationship between change of critical loads and change of power supply cost in an MMGS. The mathematical expression is given by:

$$\text{CLSMC} = \frac{d\rho_{total}}{dP_{c\_load}} \tag{14.86}$$

- Noncritical load supply cost (NLSC) ($/year): This index is mainly used to assess the average power supply cost of noncritical loads per kilowatt-hour in an MMGS. The mathematical expression is given by:

$$\text{NLSC} = \frac{\rho_{total,nc}}{P_{nc\_load}} = \frac{\sum_{i=1}^{N}(\rho_{fuel\_c,i} + MAPEC_{nc,i})}{\sum_{i=1}^{N}\int_{0}^{T} p_{nc\_load,i}(t)dt} \tag{14.87}$$

where $\rho_{total,nc}$ is the total annual monetary cost of an MMGS for supplying noncritical loads, P_{nc_load} is the average annual demand level of noncritical loads in an MMGS, $\rho_{fuel\_c,i}$ is the average annual fuel cost of MG i for supplying noncritical loads, and $MAPEC_{cn,i}$ is the power purchase cost by MG i for supplying noncritical loads.

- Noncritical load supply marginal cost (NLSMC) ($/year): This index is mainly used to evaluate the relationship between change of noncritical loads and change of power supply cost in an MMGS. The mathematical expression is given by:

$$MAESI = \frac{d\rho_{total,nc}}{dP_{nc\_load}} \tag{14.88}$$

- MMGS environment benefit (MSEB) ($/year): This index is mainly used to evaluate environmental benefits generated by clean energy (e.g., wind and PV) in MMGS. The mathematical expression is given by:

$$MSEB = \eta_{cl} \cdot \sum_{i=1}^{N}\int_{0}^{T} P_{gen,i}(t)dt \tag{14.89}$$

where η_{cl} is the unit environmental cost of traditional coal-fired power generation.

- MMGS reliability benefit (MSRB) ($year): This index is mainly used to assess the monetary benefits of reliability improvement for an MMGS, which is evaluated via the power interruption cost. The mathematical expression is given by:

$$MSRB = \sum_{i=1}^{N} U_i \cdot P_i \cdot IEAR_i \tag{14.90}$$

where N is the number of annual interruption events, P_i is the average load power of the system, and $IEAR_i$ is the power interruption cost of loads.

14.2.2.5 Energy Utilization Efficiency Index

- Microgrid intermittent energy capacity (MIEC): This index is used to describe the total existing intermittent energy contained in an MG. The mathematical expression is given by:

$$MIEC = \sum P_{IE} \tag{14.91}$$

where P_{IE} is the intermittent energy produced by an intermittent power source IE.

- Microgrid intermittent energy adoptability (MIEA): This index is used to describe the ratio of newly integrated intermittent energy capability into an existing MG. The mathematical expression is given by:

$$MIEA = \frac{P_{r\_i}}{MIEC} \tag{14.92}$$

where P_{r_i} is the rated power of a new integrated intermittent generation source i in an MG.

- *Renewable energy penetration* (*REP*): This index describes the percentage of the average renewable energy generation in the MG to the average total demand. The mathematical expression is given by:

$$\text{REP} = \frac{\int_0^T P_{R\Sigma}(t)dt}{\int_0^T PL(t)dt} \tag{14.93}$$

where: $P_{R\Sigma}(t)$ is the total power output of renewable DGs in the MG at time t, and $PL(t)$ is total MG loads at time t.

14.2.2.6 Energy Saving and Emission Reduction Index

- Microgrid environment benefit (MEB): This index evaluates the environmental benefit brought by clean energy in an MG. Considering wind and photovoltaic power generation as the two major renewable sources, the mathematical expression is given by:

$$MEB = \eta_{\text{coal}} \cdot (EES_{\text{PV}} + EES_{\text{WTG}}) \tag{14.94}$$

where η_{coal} is the environmental cost factor of traditional coal-fired power generation units. EES_{PV} and EES_{WTG} are respectively the expected energy supply of PVs and WTGs, in units of kWh/year.

- Microgrid carbon emission per unit energy (MCEUE): This index is used to describe the ratio of per-unit energy consumption of an MG to carbon emissions. The mathematical expression is given by:

$$MCEUE = \int_0^{8760} \frac{\mu_f \cdot F_m(t) + \mu_e E_{grid}(t)}{Q_c(t) + Q_h(t) + P_{load}(t)} dt \tag{14.95}$$

where μ_f is carbon emission of per-unit natural gas with the unit of g/MBTU, and μ_e is carbon emissions of per unit electrical energy with the unit of g/kWh.

- Renewable energy dispatch rate (REDR): This index represents the maximum ratio of the allowable renewable energy output in an MG in order to maintain stability of the MG operation.

14.2.2.7 Renewable Energy Utilization Index

- Microgrid annual output power (MAOP): This index is used to calculate the average annual power output of individual MGs in MMGS. The mathematical expression is given by:

$$MAOP = \int_0^{8760} t \cdot P_i(t)dt \tag{14.96}$$

where $P_i(t)$ is power output of the i-th MG.

- Microgrid annual generated power (MAGP): This index is used to calculate the average annual power generation of all MGs in an MMGS. The mathematical expression is given by:

$$MAGP = \int_0^{8760} t \cdot P_g(t)dt \tag{14.97}$$

- Microgrid output power/generated power ratio (MOPR): This index is used to compute the ratio of the average output power to generated power of individual MGs in an MMGS. The mathematical expression is given by:

$$MOPR = \frac{MAOP}{MAGP} \tag{14.98}$$

- MMGS annual exchange energy (MAEE): This index is used to calculate the annual average circulating power at all tie-lines of an MMGS. The mathematical expression is given by:

$$MAEE = \int_0^{8760} t \cdot P_{ex}(t)dt \tag{14.99}$$

where $P_{ex}(t)$ total power flows of all tie-lines in an MMGS.

- MMGS exchange energy and total load energy ratio (MEER): This index is used to calculate the ratio of tie-line power flow to the total load of an MMGS. The mathematical expression is given by:

$$MEER = \frac{MAEE}{\int_0^{8760} t \cdot P_{load}(t)dt} \tag{14.100}$$

where P_{load} is load level of an MMGS at time t.

- MMGS exchange energy and total output ratio (METOR): This index is used to calculate the proportion of tie-line power flows to the total generation capacity of an MMGS. The mathematical expression is given by:

$$METOR = \frac{AMIE}{\sum\left(\sum E_{ij}\right)} \tag{14.101}$$

where E_{ij} is the annual power output of the j-th unit in the i-th MG.

14.2.3 Performance Assessment

Quantitative assessment for the operation performance of an MMGS can provide a valuable guidance to the optimal planning and operation. Operation characteristics of an MMGS with multiple interconnected MGs are different from those of single MGs. Specifically, interconnected MGs can supply power and reserve for each other, which enhances reliability of power supply and economic efficiency of interconnected MGs. In addition, critical and noncritical loads have various reliability requirements and different power supply priorities, which should also be considered in the performance evaluation of an MMGS.

14.2.3.1 Performance Assessment of Grid-Connected MMGS

1) **Energy Dispatch Strategy**
   For a grid-connected MMGS, economics is the main operation objective. Assume the number of MGs in an MMGS is n, namely $MG_1, MG_2, \ldots, MG_n$; demand levels of individual MGs are $PL_1(t), PL_2(t), \ldots, PL_n(t)$; power supplies from DGs in individual MGs are $PG_1(t), PG_2(t), \ldots, PG_n(t)$. Power flow from MG i to MG j is denoted as $PT_{ij}(t)$. The following energy dispatch strategy (Fathi & Bevrani, 2013) is adopted, which minimizes the operation cost while maintaining power balance and maximum energy output constraints of individual MGs:

$$\min \sum_{i=1}^{n}\left[C\left(\sum_{i=1}^{n} PT_{ij}(t)\right) + \sum_{i=1}^{n} \mu_{ij} \cdot PT_{ij}(t)\right] \tag{14.102}$$

$$PG_j(t) + \sum_{i=1}^{n} PT_{ij}(t) - \sum_{i=1}^{n} PT_{ji}(t) = PL_j(t) \tag{14.103}$$

$$\sum_{j=1}^{N} PT_{ij}(t) \leq PT_{i,\max}(t) \tag{14.104}$$

   where C is the energy production cost, which varies with different DG types. μ_{ij} is the energy transaction cost from MG i to MG j.
2) **Performance Assessment**
   The performance assessment algorithm of a grid-connected MMGS is as follows:
   Step1: Sample DG outputs and load demands of each MG. Calculate power flows based on the energy scheduling strategy to obtain energy flowing into and out of each MG and the MMGS.
   Step 2: Calculate interactive indexes of each MG based on energy flowing into and out of that MG from Step 1.

Step 3: Calculate evaluation index of the MMGS based on energy flowing into and out of the MMGS from Step 1.

Step 4: Repeat Steps 1–3 to calculate energy flow information and evaluation indexes for multiple years. Calculate the annual evaluation index of the MMGS.

Both energy exchange data among DGs in an MMGS and those between the MMGS and the distribution network can be obtained. Meanwhile, economics indexes can also be calculated. The microgrid average energy sale income of MG i can be written as:

$$MAESI = \int_{0}^{T} \sum_{j=1,j\neq i}^{n} PT_{ij}(t) \cdot \rho_{sell,t} dt \tag{14.105}$$

where $PT_{ij}(t)$ is the power injection from MG i to MG j, $\rho_{sell,t}$ is the electricity sale price, and $\rho_{buy,t}$ is the electricity purchasing price.

The microgrid average energy purchase cost of MG j can be written as:

$$MAEPC = \int_{0}^{T} \sum_{i=1,i\neq j}^{n} PT_{ij}(t) \cdot \rho_{buy,t} dt \tag{14.106}$$

14.2.3.2 Performance Assessment of Islanded MMGS

1) **Energy Dispatch Strategy**

For an islanded MMGS, power supply priorities of critical and noncritical loads need to be rigorously considered. Critical loads have higher priority than noncritical loads. For a single islanded MG, critical loads are supplied by local energy storage assets and DGs in the same MG. On the other hand, for an islanded MMGS, critical loads of an MG can be supplied by the same MG or by other connecting MGs. The following energy scheduling plan is adopted:

a. critical loads in each MG will be supplied first;
b. critical loads that cannot be met by its own MG will be supplied by other MGs in the MMGS secondly;
c. noncritical loads of each MG are satisfied thirdly;
d. finally, noncritical loads of other MGs are considered.

When considering energy exchange in an MMGS, one should also consider connection topology and priority of MGs in the MMGS. If an MG has excessive energy, the surplus energy will flow into its nearby MGs in the MMGS. However, if priority is defined, the surplus energy will supply loads in high priority MGs.

When an MMGS is operated in the islanded mode, the operator should not only consider load demand and energy supply at the current time slot but also energy dispatch strategy for a period of future time slots that the MMGS most likely stays in the islanded mode. The key challenge is to manage uncertainties of DG outputs and to control charging and discharging behaviors of energy storage assets in the MMGS. Since energy exchange exists among MGs in an MMGS, energy storage assets is not a must in every MG. In order to satisfy critical loads in emergency conditions, the total energy storage assets in an MMGS should satisfy the following constraint:

$$(SOC - SOC_{\min}) \cdot C_{ens} \geq P_{c,j} \cdot t_{ms} \tag{14.107}$$

where $P_{c,j}$ is the maximum critical loads of MG j, t_{ms} is the start time of the micro-generator, and $SOC_{\max}$ and $SOC_{\min}$ are the maximum and minimum capacities of energy storage assets, respectively.

2) **Performance Assessment**

An MMGS presents a three-level structure including DGs, MGs, and MMGS. In addition, loads in an MMGS can be divided into two types of critical loads and noncritical loads. Therefore, the MMGS assessment algorithm should not only consider energy supply characteristics in individual MGs, but also energy exchange characteristics among MGs in an MMGS. A five-stage MMGS assessment algorithm is proposed in this section, which accurately considers operation characteristics of MMGS.

The following assumptions are applied in the proposed five-stage MMGS assessment algorithm:

a. The MMGS is owned and operated by a single entity, such as the local utility company or a third-party operator.
b. Interconnection capacity limits are not considered as most MG interconnections would be partially large.
c. Buses, cables, and power electronic devices in the MMGS are 100% reliable.
d. The AC main bus of the MMGS is in the backup state when the MMGS is operated in the grid-connected mode. It will only be switched on in the islanded mode.
e. The dynamic process from grid-connected mode to islanded mode is not considered.

The details of the proposed five-stage algorithm are shown in the following:

Stage 1: Get critical loads and total energy supplies in each MG; calculate the energy surplus or critical load shortage in each MG. Initial values of indexes for critical loads in each MG can be calculated.

Stage 2: Schedule energy in the MMGS according to energy surplus or shortage in each MG to satisfy other MGs' critical loads; update initial indices of critical loads in each MG; calculate energy surplus or shortage in each MG after all critical loads are satisfied. The final indexes of critical loads in each MG are obtained.

Stage 3: Get noncritical loads and energy surplus in each MG; update energy surplus or noncritical load shortages in each MG. Initial values of indexes for noncritical loads in each MG can be calculated.

Stage 4: Schedule energy in the MMGS according to energy surplus or shortage in each MG to further satisfy other MGs' noncritical load shortage; update initial indexes of noncritical loads in each MG. The final indexes of noncritical loads in each MG are obtained.

Stage 5: Calculate energy flows using Newton–Raphson power flow algorithm (Mumtaz et al., 2016) to obtain energy interaction indexes and economic indexes.

14.2.3.3 Annual Performance Assessment of the MMGS

Performance assessment procedures of islanded MMGS and grid-connected MMGS have been discussed above. Indeed, an MMGS is operated in the islanded mode only when the distribution network power supply is interrupted or of bad quality. From a long-term point of view, an MMGS is operated alternatively in grid-connected and islanded modes. Thus, annual performance assessment indexes, including energy

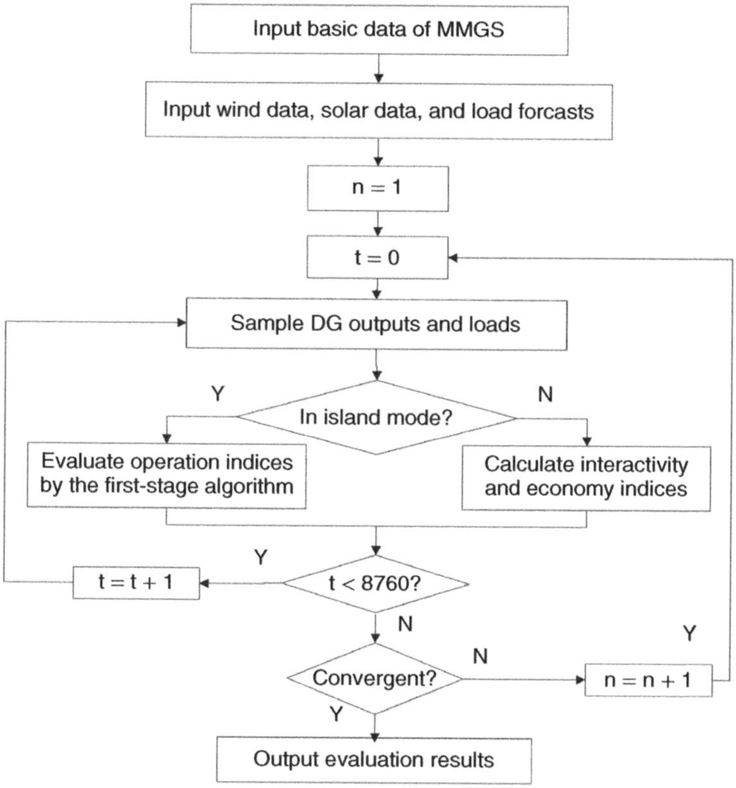

Figure 14.8 Flowchart of the annual performance assessment of an MMGS.

interaction, reliability, and economy indexes, are need to quantify an MMGS' operating state. The detailed assessment algorithm is shown in Figure 14.8.

14.2.4 Case Studies

This section provides case studies to illustrate the effectiveness of the proposed metrics.

14.2.4.1 System Description

The studied MMGS shown in Figure 14.9 includes five individual MGs. Historical wind speed and solar insolation data in 30-min intervals for one year from Tianjin, China, is used in case studies. DG data for each MG are given in Table 14.1 and Table 14.2.

A time series–based load model is used for analyzing the performance of MMGS. Hourly variations of both critical loads and noncritical loads are assumed to follow the IEEE-RTS model (Ouammi et al., 2014). Load characteristics of each MG are shown in Table 14.3.

14.2.4.2 Numerical Results

The energy exchange–related metrics of individual MGs in the MMGS are given in Table 14.4. As shown in Table 14.4, values of MEOR, MEIR, and MEDI are different in different MGs, which means that there is energy exchange among MGs in the MMGS.

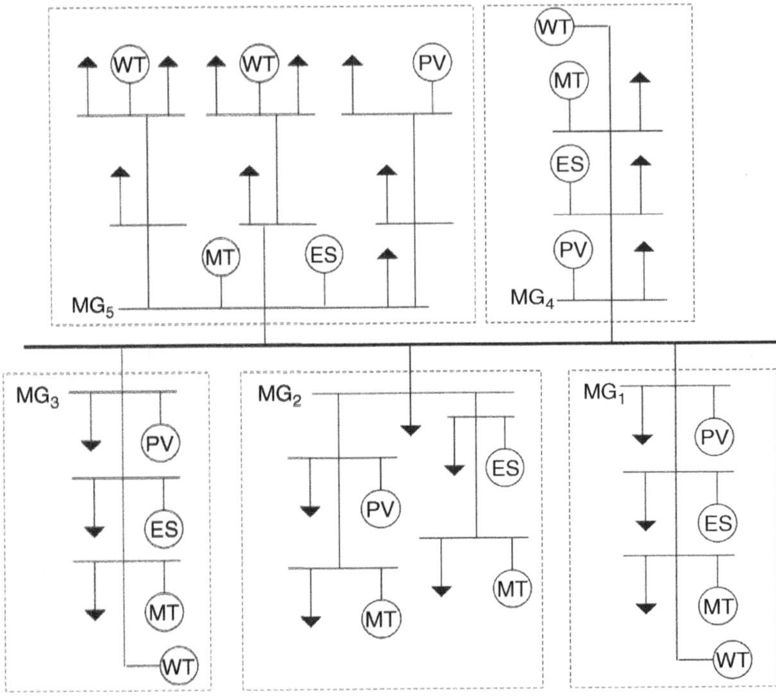

Figure 14.9 The structure of the MMGS which includes five networked MGs.

Table 14.1 DG Characteristics.

DG type	Micro turbine	Battery pack	PV array	WTG unit
Unit rating	227kW	120kWh	92kWp	200kW
MTTF (hr)	600	900	900	1900
MTTR (hr)	50	30	20	80
Other specifications	–	SOC_{min}=4%	–	Cut-in/rated/cut-out wind speeds (m/s): 3/10.5/25

Although MG_2 and MG_4 have low DG capacity, they may still provide energy to other MGs and similarly, although MG_1 and MG_5 have high DG capacity, they may buy energy from other MGs. In addition, the SEEC of the MMGS is 9.47×10^6 kWh, and *SEER* of the MMGS is 90.44%.

The reliability-related metrics of individual MGs and the MMGS are given in Table 14.5 and Table 14.6. The numbers in brackets are metrics when MGs are not interconnected with each other. As shown in Table 14.5, MG_1 and MG_5 obtain high power supply reliability ($LOLP = 4.34 \times 10^{-6}$) using micro-turbines. As shown in Table 14.6, NLEED values of MG_2, MG_3, and MG_5 are decreased when forming the MMGS. The results show that the power supply reliability of MGs that have low DG

Table 14.2 DG Installation Capacity in Each Microgrid.

MG index	Micro turbine		Battery packs		PV array		WTG unit	
	Number	Rated power (kW)	Number	Rated power (kW)	Number	Rated power (kW)	Number	Rated power (kW)
MG_1	5	1135	2	240	5	460	2	400
MG_2	0	0	4	480	5	460	5	1000
MG_3	2	450	3	360	3	276	3	600
MG_4	0	0	2	240	5	460	5	1000
MG_5	4	900	4	480	3	276	3	600

Table 14.3 Microgrid Load Characteristics.

MG index	Load size (kW)		Ratio of critical loads	User number
	Average	Peak		
MG_1	800	1300	25%	10
MG_2	700	1200	30%	9
MG_3	700	1200	30%	9
MG_4	600	1000	25%	8
MG_5	800	1300	25%	12

Table 14.4 Energy Exchange Related Metrics of MMGS.

MG index	*MEOR* (100%)	*MEIR* (100%)	*MEDI* (100%)	*EGQ* (10^6kWh)	*CLD* (10^6kWh)	*NLD* (10^6kWh)
MG_1	0.2845	0.1312	0.3610	1.01	1.85	5.55
MG_2	0.1952	0.6425	0.7924	1.56	2.05	4.78
MG_3	0.1273	0.4293	0.5352	0.93	2.05	4.78
MG_4	0.2254	0.7070	0.7738	1.56	1.42	4.27
MG_5	0.2339	0.1888	0.4261	0.93	1.85	5.55

capacity improves after being interconnected with other MGs, while reliability of MGs that have high DG capacity remains unchanged.

The economics-related metrics of individual MGs in the MMGS are given in Table 14.7. As shown in Table 14.7, although MG_2, MG_3, and MG_4 can provide energy to other MGs, they still need to prepare money to buy energy from other MGs or from the distribution network. On the other hand, MG_1 and MG_5 can fulfill their own load demands and even earn income by selling energy.

Table 14.5 Reliability Related Metrics of the MMGS.

MG Index	SAIDI (hours/year)	SAIFI (f/year)	EENS (kWh/year)	LOLP(10^{-5}) (100%)
MG_1	0.038(0.038)	0.029(0.029)	22.56(22.56)	0.434(0.434)
MG_2	0.112(0.148)	0.083(0.106)	62.18(79.96)	1.279(1.689)
MG_3	0.067(0.082)	0.051(0.062)	38.59(45.66)	0.765(0.936)
MG_4	0.170(0.227)	0.123(0.156)	90.32(117.09)	1.941(2.591)
MG_5	0.038(0.038)	0.028(0.028)	22.64(22.64)	0.434(0.434)

Table 14.6 Reliability Related Metrics of the MMGS.

MG Index	LCLFI (f/year)	CLAIDI (hours/year)	CLEED (kWh/year)	LNLP (100%)	NLEED (kWh/year)
MG_1	0.009(0.009)	0.004(0.004)	0.560(0.561)	0.434(0.434)	22.000(21.999)
MG_2	0.018(0.025)	0.011(0.020)	1.782(3.479)	1.279(1.689)	60.398(76.481)
MG_3	0.010(0.011)	0.004(0.006)	0.612(0.936)	0.765(0.936)	37.978(44.724)
MG_4	0.027(0.047)	0.021(0.049)	3.004(7.100)	1.941(2.591)	87.316(109.99)
MG_5	0.010(0.010)	0.003(0.003)	0.308(0.308)	0.434(0.434)	22.332(22.332)

Table 14.7 Economic Related Metrics of the MMGS.

MG Index	MAESI (10^5\$)	MAEPI (10^5\$)	MACF (10^5\$)	CLSC (\$/kWh)
MG_1	1.868	0.928	−0.940	0.3726
MG_2	4.007	7.016	3.009	0.3976
MG_3	0.443	2.795	2.352	0.4079
MG_4	5.328	7.720	2.392	0.3971
MG_5	1.414	1.229	−1.850	0.3774

14.3 Conclusions

While interconnecting adjacent MGs to build a multi-microgrid energy system is attracting more attention among researchers, optimal planning and operation of MMGS call for a comprehensive quantitative assessment on different aspects. This section proposes a set of new metrics to describe characteristics of MMGS, with special focus on source-charge capacity, energy interaction, reliability, economics, energy utilization efficiency, energy saving, emission reduction, and renewable energy utilization. In addition, performance assessment algorithms for MMGSs under both grid-connected and islanded modes are proposed based on their energy dispatch strategies. Case studies accurately reveal the energy interaction among MGs and the

MMGS and show that the MMGS outperforms individual MGs in terms of reliability and economics. The proposed metrics and assessment algorithms will give beneficial guidance for optimal construction and operation of future MMGS.

Acknowledgment

This work was supported in part by National High Technology Research and Development Program 863 of China (2014AA052003), and US National Science Foundation grant PFI:BIC-1534035.

References

Appen, J., Marnay, C., Stadler, M., Momber, I., Klapp, D., & Scheven, A., 2011, 'Assessment of the economic potential of microgrids for reactive power supply', IEEE, International Conference on Power Electronics and Ecce Asia, *Jeju*, 809–816.

Arefifar, S., Mohamed, A., & El-Fouly, T., 2013, 'Optimum microgrid design for enhancing reliability and supply-security', *IEEE Transactions on Smart Grid* 4(3), 1567–1575.

Basu, A., Chowdhury, S., & Chowdhury, S., 2010, 'Impact of strategic deployment of chp-based ders on microgrid reliability', *IEEE Transactions on Power Delivery* 25(3), 1697–1705.

Beer, S., Gomez, T., Dallinger, D., Momber, I., Marnay, C., & Stadler, M., 2012, 'An economic analysis of used electric vehicle batteries integrated into commercial building microgrids', *IEEE Transactions on Smart Grid* 3(1), 517–525.

Bernal-Agustín, J., Dufo-López, R., & Rivas-Ascaso, D., 2012, 'Design of isolated hybrid systems minimizing costs and pollutant emissions', *Renewable Energy* 44(14), 215–224.

Che, L., Zhang, X., Shahidehpour, M., Alabdulwahab, A., & Abusorrah, A., 2017, 'Optimal interconnection planning of community microgrids with renewable energy sources', *IEEE Transactions on Smart Grid* 8(3), 1054–1063.

Che, L., Shahidehpour, M., Alabdulwahab, A., & Al-Turki, Y., 2015, 'Hierarchical coordination of a community microgrid with ac and dc microgrids', *IEEE Transactions on Smart Grid* 6(6), 3042–3051.

Chowdhury, S., Chowdhury, S., & Crossley, P., 2009, *Microgrids and active distribution networks, The Institution of Engineering and Technology*, Stevenage Herts, SH.

Clavier, J., Joós, G., & Wong, S., 2013, 'Economic assessment of the remote community microgrid: PV-ESS-diesel study case', 26th Annual IEEE Canadian Conference on Electrical and Computer Engineering, *Regina*, 5–8.

Erol-Kantarci, M., Kantarci, B., & Mouftah, H., 2011, 'Reliable overlay topology design for the smart microgrid network', *Network IEEE* 25(5), 38–43.

Farzin, H., Fotuhi-Firuzabad, M., & Moeini-Aghtaie, M., 2015, 'Developing a hierarchical scheme for outage management in multi-microgrids', PowerTech, 2015 IEEE Eindhoven, *Eindhoven*, 1671–1676.

Farzin, H., Fotuhi-Firuzabad, M., & Moeini-Aghtaie, M., 2017, 'Enhancing power system resilience through hierarchical outage management in multi-microgrids', *IEEE Transactions on Smart Grid* 7(6), 2869–2879.

Fathi, M., & Bevrani, H., 2013, 'Adaptive energy consumption scheduling for connected microgrids under demand uncertainty', *IEEE Transactions on Power Delivery* 28(3), 1576–1583.

Fathi, M., & Bevrani, H., 2013, 'Statistical cooperative power dispatching in interconnected microgrids', *IEEE Transactions on Sustainable Energy* 4(3), 586–593.

Guo, L., Liu, W., Cai, J., Hong, B., & Wang, C., 2013, 'A two-stage optimal planning and design method for combined cooling, heat and power microgrid system', *Energy Conversion & Management* 74(10), 433–445.

Khodaei, A., Bahramirad, S., & Shahidehpour, M., 2015, 'Microgrid planning under uncertainty', *IEEE Transactions on Power Systems* 30(5), 2417–2425.

Lotfi, H., & Khodaei, A., 2016, 'AC versus DC microgrid planning', *IEEE Transactions on Smart Grid* 8(1), 296–304.

Liao, G., 2006, 'Short-term thermal generation scheduling using improved immune algorithm', *Electric Power Systems Research* 76(5), 360–373.

Lee, J., Guo, J., Choi, J., & Zukerman, M., 2015, 'Distributed energy trading in microgrids: a game-theoretic model and its equilibrium analysis', *IEEE Transactions on Industrial Electronics* 62(6), 3524–3533.

Liang, X., Mandayam, N., & Poor, H., 2014, 'Prospect theoretic analysis of energy exchange among microgrids', *IEEE Transactions on Smart Grid* 6(1), 63–72.

Li, J., Liu, Y., & Wu, L., 2016, *'Optimal operation for community based multi-party microgrid in grid-connected and islanded modes', IEEE Transactions on Smart Grid, (to appear).*

Lee, M., Soto, D., & Modi, V., 2014, 'Cost versus reliability sizing strategy for isolated photovoltaic micro-grids in the developing world', *Renewable Energy* 69(69), 16–24.

Mumtaz, F., Syed, M., Hosani, M., & Zeineldin, H., 2016, 'A novel approach to solve power flow for islanded microgrids using modified Newton Raphson with droop control of DG', *IEEE Transactions on Sustainable Energy* 7(2), 493–503.

Mojtahedzadeh, S., Najafi-Ravadanegh, S., & Haghifam, M., 2016, 'A framework for optimal clustering of a greenfield distribution network area into multiple autonomous microgrids', *Journal of Power Technologies* 96(1), 1–10.

Nikmehr, N., & Ravadanegh, S., 2015, 'Optimal power dispatch of multi-microgrids at future smart distribution grids', *IEEE Transactions on Smart Grid* 6(4), 1648–1657.

Nunna, H., & Doolla, S., 2012, 'Demand response in smart distribution system with multiple microgrids', *IEEE Transactions on Smart Grid* 3(4), 1641–1649.

Ouammi, A., Dagdougui, H., & Sacile, R., 2014, 'Optimal control of power flows and energy local storages in a network of microgrids modeled as a system of systems', *IEEE Transactions on Control Systems Technology* 23(1), 128–138.

Standards America, 2011, '*IEEE Standards Association: IEEE guide for design, operation, and integration of distributed resource island systems with electric power systems*', Standards America, New York.

Salcedo, R., Bokhari, A., Diaz-Aguilo, M., & Lin, N., 2015, 'Benefits of a non-synchronous microgrid on dense-load low-voltage secondary networks', *IEEE Transactions on Power Delivery* 31(3), 1076–1084.

Shahnia, F., Chandrasena, R., Rajakaruna, S., & Ghosh, A., 2013, 'Autonomous operation of multiple interconnected microgrids with self-healing capability', Power and Energy Society General Meeting, *Vancouver*, 1–5.

Wang, S., Li, Z., Wu, L., Shahidehpour, M., & Li, Z., 2013, 'New metrics for assessing the reliability and economics of microgrids in distribution system', *IEEE Transactions on Power Systems*, 28(3), 2852–2861.

Wu, L., Ortmeyer, T., & Li, J., 2016, 'The future community microgrid distribution system', *The Electricity Journal* 29, 16–21.

Wu, X., Wang, X., & Qu, C., 2014, 'A hierarchical framework for generation scheduling of microgrids', *IEEE Transactions on Power Delivery* 29(6), 2448–2457.

Wu, J., & Guan, X., 2013, 'Coordinated multi-microgrids optimal control algorithm for smart distribution management system', *IEEE Transactions on Smart Grid* 4(4), 2174–2181.

Zhang, D., Evangelisti, S., Lettieri, P., & Papageorgiou, L., 2015, 'Optimal design of chp-based microgrids: multiobjective optimisation and life cycle assessment', *Energy* 85, 181–193.

Section IV

Smart Cities, Intelligent Transportation System and Electric Vehicles

15

Wireless Charging for Electric Vehicles in the Smart Cities: Technology Review and Impact

Alicia Triviño-Cabrera[1] *and José A. Aguado*[2]

[1] *Associate Professor at the University of Málaga, Spain*
[2] *Full Professor at the University of Málaga, Spain*

15.1 Introduction

Electric vehicles constitute an important asset for the smart grid. In this new context, their functions are not just limited to the users' transportation, but they can also participate in the smart grid operations actively (Shareef et al., 2016). The expected inclusion of a relevant number of EVs in the near future will alter the power network performance in such a way that they will affect the load flattening, on the fast frequency control and on easing the integration of renewable energy sources.

Its influence depends on the number of EVs, the time their charges are scheduled, and their ability to deliver energy to the grid acting as mobile network storage systems. This last function is considered in the V2G (vehicle-to-grid) scenarios where the power transfer is also feasible from the vehicle to the grid. In this sense, the vehicles operate as sources when they are parked and the V2G function is enabled. The EV owners obtain some economical revenues for their participation in this kind of service (Shareef et al., 2016) while the power network also gets important benefits. The V2G procedures, when done in a coordinated way, are expected to improve power network performance and its reliability. As summarized in Shareef et al. (2016), the implementation of V2G mitigates voltage fluctuation in distribution networks, it contributes to improve the power quality, and it also helps to realize a frequency control in a decentralized power supply. In addition, considering a smart home, the EV may potentially serve as an emergency supply for home or business in case of power outages (Kalwar et al., 2015).

The scheduling and the V2G operations need to rely on the battery and on the charger features. Thus, it is important to know how the vehicles charge in order to design the procedures related to both activities. We can classify the way the EV is charged/actively discharged in two broad categories: conductive and wireless charge. In the conductive charge, the EV is connected through a wire to the grid charger. This physical connection implies some safety issues, especially when rain and snow are present (Su et al., 2012). Moreover, it is necessary that the user participate in this action.

Users' intervention is eluded in the wireless chargers as the vehicle can be charged without any physical connection. Additionally, this capability adds the autonomy to the

Transportation and Power Grid in Smart Cities: Communication Networks and Services, First Edition.
Edited by Hussein T. Mouftah, Melike Erol-Kantarci and Mubashir Husain Rehmani.

charge process so the charge can be achieved in more situations than those available with the conductive one. For instance, the EV can be charged via wireless while stopped for a short time or even while it is in motion. The range anxiety, which is a drawback conventionally associated to EVs, is alleviated with these new mechanisms (range anxiety is the fear perceived by drivers when considering that their EVs will not be able to fulfil their destinations due to the reduced battery capacities). It is said that wireless charge will allow for ubiquitous charge.

But this convenience is not the only advantage of wireless charge; it also provides electrical isolation, which makes the charge safer from an electrical point of view. In addition, since a lower number of components are manipulated by humans, the robustness of the system is increased (Mi et al., 2016).

On the other hand, the design of a wireless charge should be carefully accomplished as new issues need to be considered. Firstly, as it is supported by an electromagnetic field, it is necessary to meet the standard guidelines concerning the maximum electromagnetic emissions allowed. In this sense, the ICINRP (International Commission on Non-Ionizing Radiation Protection) guidelines should be followed, and the design of the coils, the electronics and additional modules such as shielding and ferromagnetic components need to be designed altogether for this purpose. Additionally, the interferences produced to other electronics components need to be reduced in order to prevent malfunctions on them (Song et al., 2016).

These are not the only issues to be considered. As a conventional electronic device, it has to deliver a restricted amount of reactive power to the grid and the harmonics injected to the grid must be also reduced. Both conditions are measured by two well-known electrical parameters: the power factor and the total harmonics distortion. The use of nonlinear components make a wireless charge especially productive in terms of harmonics. Moreover, as the resonant condition is not always guaranteed, the impedance offered by the wireless charge is not purely resistive, which decreases the power factor. There are some contributions in the literature to deal with these two parameters in a wireless charger for EVs, as we will study in the present chapter. We will also address how the electronics in a wireless charger is adapted to enhance power flow bidirectionality so that the EV can feed electricity back to the grid.

Electromagnetic emissions, interferences, and the electrical integration issues can be treated in every wireless charger independently. However, when considering the impact of the electric vehicle into the grid and its role as an active asset, we need to take into account a group of vehicles. In fact, to obtain evident improvements on the power network performance, we need to control the charge of a complete group of vehicles and coordinate this process following a market-based algorithm. This is the role of the aggregators.

The remainder of the chapter is structured as follows. Section 15.2 reviews the technologies to enhance wireless power transfer in an EV. It also deals with the different ways that this technology can be used. Section 15.3 focuses on the relevant electronics aspects when integrating an EV wireless charger to the grid. Section 15.4 analyzes how the coordinated process of charge/discharge a group of vehicles, that is, the scheduling, is adapted to the particularities of basing on wireless power transfer. Finally, Section 15.5 draws the main conclusions of this work and details some future guidelines that need to be addressed to make this technology progress.

15.2 Review of the Wireless Charging Methods

Wireless power transfer can be supported by multiple technologies. The suitability of this technology for its application to charge EVs depends on some physical constraints (e.g., the distance between the power emitter and the receiver), but it is also affected by the way it is going to be used (i.e., while the vehicle is moving or static). These two issues are analyzed in the following subsections.

15.2.1 Technologies Supporting Wireless Power Transfer for EVs

Wireless charging of an electric vehicle is mainly supported by an electric or magnetic field, which induces voltage in the receiver (installed in the car) when the transmitter (the power source) is electrically excited on some specific conditions.

The first approximation to wireless charging an EV could be the handheld resonant magnetic field charger proposed in Song et al. (2016). To operate, the user needs to insert the charger into the vehicle just in a similar way to fueling it. There is a coil in the handheld component that is on resonant condition with the coil in the vehicle. Since it is supported by inductive wireless power transfer, the efficiency of the power transfer greatly degrades when both terminals are not in physical contact (Musavi & Eberle, 2014). As can be observed, this approach still demands the user's intervention to put close a transmitter coil to a receiver coil, but it adds some electrical isolation in comparison with a conductive charge. This restriction is avoided in the so-called wireless power transfer systems for EV, where there is a significant gap between the transmitter coil and the receiver. This distance typically ranges from 100 to 300 mm.

The distance between the power transmitter and the receiver along with the frequency of the electromagnetic field involved in the wireless power transfer determines the technology to be used. Specifically, these two parameters decide whether the wireless power transfer is going to occur in a near-filed area, a mid-range area, or a long-distance area. Taking into account the conditions of a wireless charge for EV, near-field techniques are commonly employed. In this category, wireless power transfers applied to electric vehicles are divided into two groups: inductively coupled transfer and capacitive transfer.

In the inductively coupled wireless power transfer technology, a coil (named primary or transmitter coil) is installed in the pavement and another one (named secondary or receiver coil) is placed in the vehicle. The primary coil is excited with an alternating current, which generates a magnetic field around it according to Ampère's law. When this magnetic field traverses the secondary coil, it creates an induced voltage on this component as explained by Faraday's law. This induced voltage is used to charge the battery as the secondary coil is connected through electronic converters to the battery.

Both coils are adapted with reactive structures so that the whole system is on resonance conditions at a specific designed frequency. In this way, the battery gets the maximum real power from the grid.

Depending on the components of the reactive structures, the compensation topologies can be classified into single-resonant and multi-resonant (Villa et al., 2012). Single-resonant structures add a capacitor to each coil whereas multi-resonant topologies use multiple reactive components in the transmitter and/or the receiver coil.

Due to their robustness, single-resonant compensation topologies are frequently used. In this category, there are four compensation topologies: series-series (SS), series-parallel (SP), parallel-series (PS), and parallel-parallel (PP). The first word stands for the connection between the primary capacitor and the transmitter coil while the second word refers to the type of connection between the secondary capacitor and the receiver coil.

As observed in Villa et al. (2012), inductively coupled wireless power transfer is highly sensitive to misalignments, that is, to displacements of the primary and the secondary coils from their original designed positions. Misalignments can occur in the horizontal and vertical axis and in an angular position. The electrical consequence of misalignment is that the mutual inductance between the two coils differs from the one used in the design process. As a result, the system does not operate under resonance phenomena. To cope with misalignments, Villa et al. (2012) propose a multi-resonant compensation topology capable of offering similar power transfer efficiency when some misalignments occur. It is based on a series-parallel compensation network placed in the primary coil whereas the secondary coil is tuned with a series capacitor. Some other approaches opt for keeping single-resonant compensation networks but incorporate a control mechanism that adapts the system according to the perceived misalignment. In this sense, there are two main strategies to follow: adjust the operating frequency so that the system is on resonant conditions with the real coupling factor or adjust the angle phase of the power converters to maintain the output power.

Inductively coupled wireless power transfer technology is the most mature technology in the market by now. Nevertheless, there are alternative ways to wireless recharge the battery. Firstly, we can rely on highly coupled resonant wireless power transfer, also known as strongly coupled power transfer (Wei et al., 2014). It is perceived as the natural evolution of inductively coupled wireless power transfer, and the main difference is the use of strongly coupled coils. The primary and the secondary coil are built with high quality factors. Operating in the MHz frequency band, they can work under misalignment conditions. However, the current prototypes for EV offer a low efficiency.

Finally, the electric induction/coupling has been recently proposed to transfer power to an EV (You & Yi, 2016). Known as capacitive power transfer (Musavi & Eberle, 2014), it is based on a pair of coupling capacitors. Initially, its applications were limited to low-power transfer over short transfer distances. Chargers for tooth-brushes or cellular mobile phones were the initial candidates to benefit from this technology. However, recent progress on this type of wireless power transfer is making this technology also appropriate for the powers involved in an EV charge (Dai & Ludois, 2015). There are other non-mature proposals for wireless power transfer in EV. We can highlight microwave power transfer, which has also been applied to EVs but is still showing a reduced output power (Kalwar et al., 2015).

15.2.2 Operation Modes for Wireless Power Transfer in EVs

So far, we have detailed how the wireless power transfer from the charger to the vehicle is achieved. These technologies are also adapted to the way the chargers are used. In this sense, we can differentiate the following charging operations:

- Static, when the charge takes place in a specific position and the vehicle is expected to be turned off while a full charge is done. This is the case of home chargers or those installed in parking lots. The chargers can be enhanced with a control system that advises where to place the vehicle in order to avoid the coil misalignments. Algorithms and specific equipment to detect the presence of objects or animals between the power emitter and the receiver are also incorporated to ensure the correct operation of the charger.

- Stationary, when the vehicle is stopped but the motor is still working, and this situation holds over a short period of time that is not enough to get a full charge. This will be of utility when offering wireless charge during the stops imposed by the traffic lights or the taxi/bus/tram stops. Under these circumstances, the transmitter and the receiver could be greatly misaligned and it is not feasible to adjust their positions as that may interfere with other vehicles.
- Dynamic, which refers to the charge that is carried out when the vehicle is moving, that is, when it is conventionally circulating on a prepared road. This system promotes the roadway powered electric vehicles (RPEV; Mi et al., 2016) and it is foreseen to be applied in specific lanes for private use or for public transport such as buses or trams. In other words, it will enhance the ubiquitous charge. By now, it has already been tested in some cities in South Korea (promoted by the OLEV project) and in Spain (through the Victoria project). In order to mitigate the electromagnetic emissions and the unused power, the primary charger is divided into segments, tracks, or pads. The control techniques are responsible to detect the vehicle when it is over the primary coil and to activate the pad once the vehicle has been automatically authorized. Dynamic charging is clearly affected by misalignments, which must be handled by the control techniques. It is necessary to accomplish a study about where to place the transmitters. A work about this concern is presented in Jang, et al. (2016).

Although we have differentiated the static and the stationary charge, works in the literature name these methods indistinctly. The use of stationary and dynamic chargers make the charge available in more situations so that the battery can be charged more frequently. As a result, the battery of the EVs can be downsized, which leads to cheaper battery and lighter components. It is necessary to adopt the agreements to make the three types of charge interoperable so that a vehicle is able to recharge its battery in all the situations.

Table 15.1 summarizes the review of wireless chargers for EV done in this section.

Table 15.1 Summary of the Wireless Power Transfer Technologies

Wireless power transfer for electric vehicles	
Supporting technology	**Charge timing**
– Inductive – Inductively coupled / Magnetic-Resonant – Highly resonant coupled – Capacitive	– Static – Stationary – Dynamic, i.e., while in motion

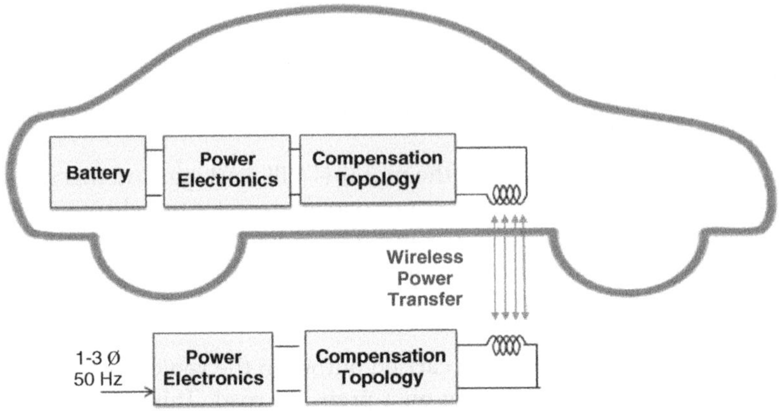

Figure 15.1 Scheme of a wireless charger for EV.

15.3 Electrical Effect of Charging Technologies on the Grid

By now, ICPT technology is the most mature technique applied to wireless charge of EVs. In this technology, a wireless charger for EVs is an electro-mechanical system composed of multiple blocks as illustrated by Figure 15.1.

The primary side is the one illustrated outside the car, which is built in the pavement. It counts on a power electronics block, which includes a rectifier, a power factor corrector (sometimes ignored in the literature), an optional DC/DC converter, and an inverter. Compensation topologies are incorporated in both the primary and the secondary side to make the coils operate on resonant conditions (Villa et al., 2012). The secondary side stands up for the electronics placed inside the car. It also requires the power electronics converter and the battery to be charged.

It is important to note that power converters are a requirement for wireless chargers in order to make the operation at high frequency feasible. These power converters usually operate in the 20–90 kHz range.

As a conventional electronic device connected to the grid, there are electrical consequences that must be taken into account when integrating a wireless charger into the grid. Concerning these integration issues, the designs proposed in the literature mainly focus on correcting the power factor and on controlling the harmonics introduced into the grid. In addition, it is relevant to analyze how the charger is designed to offer bidirectional wireless power transfer. These three aspects are studied next.

15.3.1 Harmonics Control in EV Wireless Chargers

Harmonics are mainly caused by the nonlinear components of these systems, which are localized in the power converters, and the way these components are controlled.

The mode that harmonics affect the integration of the wireless chargers is diverse. Firstly, harmonics may be responsible for the generation of undesirable magnetic near-field noise, which could affect adjacent sensitive electronic devices belonging to the wireless charger (i.e., sensors or control systems). This behavior is known as EMI (electromagnetic interference). Harmonics also decrease the efficiency of the system as losses due the harmonic currents occur too. On the other hand, the grid admits a maximum of current harmonics for those devices connected to it. Thus, harmonics provoked by the wireless chargers should be minimized.

A first step to avoid EMI consists of using shielding techniques and ferromagnetic materials to restrict the electromagnetic field in non-desirable zones (e.g., where other sensible electronic equipment may be) while guiding the electromagnetic field in the direction where the power transfer should be maximized (Kim et al., 2013). This strategy is followed in Song, et al. (2016) for a handheld wireless charger. However, the efficiency of the system decreases because of the losses occurring in the shielding. In addition, the design of the wireless power transfer system becomes quite more complex.

As an alternative, we can include lumped components. Song et al. (2014) propose the reduction of the magnetic fields by incorporating decoupled coils in both the transmitter and the secondary. By this inclusion, the impedance perceived by the harmonics is increased, especially for high frequencies. As a consequence, the harmonic currents and the current distortion are reduced.

A different strategy for reducing the harmonics currents derives in the control of the inverter. In this sense, the work in Kim et al. (2011) proposes the use of spread spectrum clock technology. In this way, the number of harmonics involved in the current of the coils increases, but their total power is reduced. Two mechanisms based on spread spectrum are also proposed in Inoue et al. (2016) to reduce harmonics in inductive chargers. According to the proposal, the implementation of the spread spectrum randomly varies the operational frequency at which the transistors in the inverter switch. The selection of the operational frequency is done using a uniform discrete probability distribution or a biased discrete probability distribution. In this last case, the probability of a frequency depends on the impedance offered by the coils and the capacitors. The main disadvantage of these approaches lies on the difficulty to implement the controller to activate the inverter transistors so that the spread spectrum is achieved.

15.3.2 Power Factor Control in EV Wireless Chargers

Reactive power should be controlled in the wireless chargers. To do so, a power factor corrector (PFC) may be incorporated in the primary side of the system. There are two main options when implementing it: a two-stage or a single-stage implementation. The first one, named two-stage PFC, consists of making it an independent block and placing it between the primary rectifier and the primary inverter. In this case, the usual implementation in wireless chargers is a boost DC/DC converter (González-Santini et al., 2016). This is a valid option when the output power does not exceed 3.5 kW. Otherwise, the ripple current in the boost capacitor makes itself so relevant that dissipation and efficiency problems appear. Nevertheless, the use of this type of scheme has been applied to a 7 kW prototype, but it requires a DC/DC converter previous to the primary inverter (Deng et al., 2014). Following a two-stage approach, the work in Shi et al. (2016) designs the PFC based on a SEPIC structure. This DC-DC (Deng et al., 2014) converter is known to force its switch to support higher voltages than boost converters for the same power outputs. Although this stress is a clear drawback, SEPIC allows for a complete regulation of the output voltage; that is, input voltage can be augmented or diminished. In order to cope with this stress feature, Shi et al. (2016) propose the use of two interleaved SEPICs as PFC. Thus, the control is more complex, and the application is still restricted to a 500-W prototype.

Including a PFC also minimizes the harmonic distortion, which has been reported in Kitamoto et al. (2016) and in Kitano et al. (2015). The authors in these works have checked this relationship by including a simple and independent PFC in their designed wireless charger. In particular, their design relies on a single-ended inverter operating with a PWM (pulse width modulation) control.

Alternatively, it is also possible to integrate the PFC in the primary rectifier leading to a single-stage PFC. In contrast to these previous works, Hsieh et al. (2016) opt for an integrated PFC based on a bridgeless structure (Siu & Ho, 2016). A single-stage PFC is also employed in Chinthavali et al. (2015).

A different approach for building a PFC is presented in González-Santini et al. (2016). Based on a Z-source network, they simultaneously control the output power and the power factor delivered by the front stage without the need of additional switches. In fact, a Z-source network is built based exclusively on reactive components with no control system applied to them. This makes the proposed PFC more reliable. Its control is indirectly related to the inverter and its duty cycles.

On the other hand, Keeling et al. (2010) address the power factor corrector from a different point of view. These researchers propose the design of a LCL compensation topology to control the reactive power delivered back on the utility. The study is theoretical, and there is no evidence about how this strategy behaves in a real prototype where the designed values are different from the values of the real components due to tolerances.

15.3.3 Implementation of Bidirectionality in EV Wireless Chargers

There is another concern on integrating wireless chargers into the grid: bidirectionality. Obviously, implementing bidirectionality has a clear effect on the grid as the vehicle is able to act as an energy storage system capable of transferring power back to the grid when needed. Next, we will review how the bidirectional chargers are implemented in the most outstanding proposals.

A symmetrical scheme is employed in Lee & Han (2015) so that the primary AC/DC and the secondary DC/AC converters are exactly the same. The symmetry is also maintained in the compensation structures and a series-series topology is followed. By these conditions, the control is eased. These structures operate at the same resonant frequency, which is kept constant, and the power output is regulated through the duty cycles of the above-mentioned power converters. Particularly, there are two current control systems based on a self-resonant PWM. The one in the primary controls the power flow from the grid to the battery whereas the second system is in charge of setting the duty cycle of the secondary DC/AC converter. The primary controller needs some measurements concerning the battery, so a wireless communication module is installed in the charger. A Bluetooth-based communication module is the usual implementation for this kind of information transfer.

As can be observed, the previous work relies on the modification of the duty cycles to control the wireless charger. The next proposal uses this parameter, but it also adjusts the operational frequency (Bojarski et al., 2014). In order to minimize the losses due to the nonlinear properties of the coils, the phase between the voltage provided by the primary inverter and the voltage of the secondary AC/DC converter is adjusted. Specifically, Bojarski et al. (2014) developed an analytical expression to maximize the system efficiency according to this phase. A previous publication applied this approach to CLCL compensation topology (Nguyen et al., 2014).

In Madawala & Thrimawithana (2010), the wireless charger relies on a parallel-parallel compensation topology as the authors argue this is the most convenient reactive structure to maintain the primary current constant independently of the number of loads. The authors derive a mathematical framework to demonstrate that the sense of the power flow can be easily controlled by the phase angle between the primary voltage (the one provided by the primary inverter) and the secondary voltage (the one set by the AC/DC converter). A matrix-based approach is followed in Thrimawithana & Madawala (2010).

15.3.4 Discussion

The implementation of a PFC, the harmonics reduction, and the bidirectionality are essential to obtain efficient V2G wireless chargers. Paying attention to the research work, we can conclude that their adequate design greatly relies on the proposal and use of appropriate control algorithms. PFC in EV wireless chargers and bidirectionality are more mature techniques than the control of harmonics.

15.4 Scheduling Considering Charging Technologies

The potential benefits of massive EVs deployments are widely recognized (Agency, 2016) (Agency, 2016). Particularly, EVs allow for a transportation sector with low carbon emissions and, due to their use patterns, they can also operate as flexible electric loads providing ancillary services to distribution grids and enabling the integration of renewable energy (Dallinger & Wietschel, 2012)(Dallinger & Wietschel, 2012). Despite these potential benefits, the EV industry is still facing several important challenges for massive EVs deployments. The first one is the limited EV driving range, which varies between 100 and 180 km for light-duty EVs. The second is related to the long charging time of EVs; full charge can take from 30 minutes for fast chargers up to 8 h for domestic chargers. The third challenge concerns the integration of EVs in the power system since it requires the development of charging infrastructure conveniently designed to comply with some integration requirements as explained in the previous sections. Moreover, uncontrolled simultaneous charging of a large number of EVs can result in voltage excursions or distribution line congestion increasing the need for capacity reinforcements both in the distribution grid and in the electricity system related to buildings and charging sites. These enforcements are expensive and will probably have small utilization factors. Thus, it is important to control when and for how long the EVs are charged. This is what scheduling is about. Scheduling defines charging coordination mechanisms, which must cope with a high degree of uncertainty as users' behavior is not predefined. Aggregators will be responsible for a fleet of vehicles and control their charge/discharge operations following a market-driven algorithm. For that, the aggregators and the electrical network must exchange information while the electrical network must manage updated information about the status of its assets.

Communication techniques specifically designed for smart grids environments are proposed in Khan et al. (2016, 2017).

Many studies suggest that adopting proper scheduling and planning strategies are necessary to mitigate the aforementioned limitations. Charging cost and the convenience level of the users are among the most relevant criteria for evaluating the adequacy of charging and scheduling algorithms. While there are several studies that address user convenience level and optimizing charging cost (Tang et al., 2014; Shao et al., 2016) separately, only a few studies consider both factors as the underlying merit factors (Tusha et al., 2014; Malhotra et al., 2016). Tang et al., (2014) try to minimize the charging cost for the parking station owner whereas a similar research in Shao et al. (2016) minimizes grid generation cost.

Scheduling and planning strategies can be classified into two main categories:

- Charging infrastructure network design strategies: EV owners prefer charging activity to be subordinated to the mobility goal and not vice versa. As a consequence, users should plug in their EVs within walking distance from their destination. Apart from domestic chargers at households and work places, numerous public charging stations will be necessary for massive EV penetration. Since massive simultaneous charging of EVs may lead to problems in distribution grids and in the electric infrastructure at charging sites, a crucial question is to determine locations and sizes of charging stations (He et al., 2013) (He et al., 2013) where up to 70% of the incremental investment costs can be avoided with optimal planning strategies. These strategies usually consider a planning expansion problem where size and location are determined subject to technical and operational constraints. Some authors have also proposed to complement the charging infrastructure with separate battery storages, potentially in combination with local generation such as solar photovoltaic panels (PV).
- EVs smart (or controlled) charging and routing strategies. One of the simplest charging strategies is the dumb charging strategy, where the vehicle owner plug in and charge once she arrives home during the evening. This behavior may add some extra load during peak demand periods. However, assuming that vehicle-to-grid (V2G) technology is available and, under the a profit maximization strategy, the EVs are to be charged when the electricity price is low and discharge back to the grid at high prices subject to mobility constraints. The idea is that since the energy price pattern often follows similar patterns as consumers demand, this strategy helps to flatten the load curve. Usually, this type of problem can be formulated as follows: given a fleet of EVs, a set of tours to be processed by vehicles, and a charging infrastructure, the problem aims to optimize the assignment of vehicles to tours and minimize the charging cost of EVs while considering several operational constraints mainly related to chargers, the electricity grid, and EVs driving range. As for the routing problem, with limited cruising range, long recharge times, and energy recuperation ability of battery-powered EVs, the problem is now to find energy efficient routes, rather than just fast or short routes. Several works that address this or similar problems can be found in López et al. (2013, 2015).

The particular case of wireless technology makes the charging and scheduling problem even more complex. Although recognized as a promising technology, it currently faces some technological barriers. This technology is available but not fully mature, and it is usually more expensive than its conductive counterpart. In terms of electrical losses, it achieves lower efficiency, particularly for the dynamic version. In order to become a relevant actor within the EV industry, it will require a relatively high critical mass of electric vehicles since installation and maintenance for the long term is still an issue that has not been well addressed.

As of today, there have been only a few scientific contributions addressing the charging and scheduling of wireless EV (Theodoropoulos et al., 2016).

From a mathematical point of view, the inclusion of a static wireless charge is nearly equivalent to a conductive one. It would only differ on the efficiency, which will be lower for a wireless charger, and it would be also associated to a low level of uncertainty due to potential misalignments. In contrast, it is a requirement to consider the particularities of the stationary and dynamic charging when including them in the scheduling algorithm. These approaches force the charge to occur over short times and frequently. Thus, both conditions must be modelled in the scheduler. Aggregators should coordinate to exchange relevant information about the incoming members and cope with the frequent changes in their fleet. More recently, the potential of demand side management during dynamic wireless charging of EVs has been studied in Theodoropoulos et al. (2016).

15.5 Conclusions and Future Guidelines

This chapter has reviewed the wireless power transfer applied to the charge of electric vehicles. When considering the operation of a wireless charge independently, we have studied it as a stand-alone electromechanical device. In particular, we have analyzed its effects on the grid in terms of harmonics and how the power factor can be adjusted so that reactive power is not delivered back to the power network. As EVs are foreseen as an important asset for future smart grids, we have also studied how they can be adapted to ensure a bidirectional wireless power transfer. Concerning these electric aspects, we can conclude that there is not a definitive solution that copes with these three functionalities. Control algorithms are foreseen as vital, as they play an important role in the implementation of these three features.

Additionally, we have studied the scheduling algorithms that control the timing of the charging process in a group of EVs. Wireless charging imposes some new challenges to the definition of scheduling algorithms. From a mathematical point of view, the static wireless charge of EVs is a mere extension of the classical conductive-based scheduling algorithms. The only difference to take into account is that the efficiency of the wireless chargers is usually lower than the conductive ones. However, wireless charge may be performed in other operation modes as in a stationary way (done in short times while the car is temporarily stopped) or dynamically (while the car is moving). Scheduling algorithms should characterize the new dynamic loads that these behaviors impose. As of today, there are still few works dealing with this issue.

After the study done in this chapter, we can conclude that there are still some open issues that need to be solved to improve the integration of wireless charging technology into the grid. We outline the following ones:

- There is no standard for the operational principles of wireless chargers. The SAE (Society of Automotive Engineers) is currently working on finalizing a standard, for which there is a strong demand, so that wireless charger operations can be performed in all types of wireless chargers in the future market and for all operation modes.

- It will be desirable to have interoperable conductive and wireless charge so that the user could use the one that is available, independently of the technology on which it is built.

  We have identified three types of charge according to the time and conditions of charge. It is necessary to adopt the agreements to make the three types of charge interoperable so that a vehicle is able to recharge its battery in all the situations.
- There is a demand to design specific aggregators to support wireless charge of EVs. In this sense, the development of these agents for static wireless charge is quite straightforward, but some new aspects need to be carefully considered for stationary and dynamic chargers. Among these aspects, aggregators should be defined for a restricted area. Aggregators should exchange data relative to vehicles when the vehicles under their control move to a different area.

References

Agency, I.E., 2016, *Global EV Outlook 2016. Beyond one million electric cars,* s.l.: s.n.

Bojarski, M., Asa, E., and Czarkowski, D., 2014, *Effect of wireless power link load resistance on the efficiency of the energy transfer*, Florence, s.n.

Bojarski, M., Czarkowski, D., and de Leon, F., 2014, *Multiphase resonant inverters for bidirectional wireless power transfer*, Florencia, s.n.

Chinthavali, M., Onar, O.C., Campbell, S.L., and Tolbert, L.M., 2015, *Isolated wired and wireless battery charger with integrated boost converter for PEV applications*, s.l., s.n.

Dai, J. and Ludois, D.C., 2015, 'A Survey of Wireless Power Transfer and a Critical Comparison of Inductive and Capacitive Coupling for Small Gap Applications', *IEEE Transactions on Power Electronics* 30, 6017–6029.

Dallinger, D. and Wietschel, M., 2012, 'Grid integration of intermittent renewable energy sources using price-responsive plug-in electric vehicles', *Renewable and Sustainable Energy Reviews* 16, 3370–3382.

Deng, J. et al., 2014, *Development of a high efficiency primary side controlled 7kW wireless power charger*, Florence, s.n.

González-Santini, N.S., Zeng, H., Yu, Y., and Peng, F.Z., 2016, 'Z-Source Resonant Converter With Power Factor Correction for Wireless Power Transfer Applications', *IEEE Transactions on Power Electronics* 31, 7691–7700.

He, F., Wu, D., Yin, Y., and Guan, Y., 2013, 'Optimal deployment of public charging stations for plug-in hybrid electric vehicles', *Transportation Research Part B: Methodological* 47, 87–101.

Hsieh, Y.-C., et al., 2016, 'High-Efficiency Wireless Power Transfer System for Electric Vehicle Applications', *IEEE Transactions on Circuits and Systems II: Express Briefs.*

Inoue, K., Kusaka, K., and Itoh, J., 2016, *Reduction on radiation noise level for inductive power transfer systems with spread spectrum focusing on combined impedance of coils and capacitors*, Milwaukee, USA, s.n.

Jang, Y. J., Suh, E.S., and Kim, J.W., 2016, 'System Architecture and Mathematical Models of Electric Transit Bus System Utilizing Wireless Power Transfer Technology', *IEEE Systems Journal* 10, 495–506.

Kalwar, K.A., Aamir, M., and Mekhilef, S., 2015, 'Inductively coupled power transfer (ICPT) for electric vehicle charging—A review', *Renewable and Sustainable Energy Reviews,* 47, 462–475.

Keeling, N.A., Covic, G.A., and Boys, J.T., 2010, 'A Unity-Power-Factor IPT Pickup for High-Power Applications', *IEEE Transactions on Industrial Electronics* 57, 744–751.

Khan, A.A., Rehmani, M.H., and Reisslan, M., 2017, 'Requirements, Design Challenges and Review of Routing and MAC Protocols for CR-based Smart Grid Systems', *IEEE Communications Magazine* 55(5), 206–215.

Khan, A.A., Rehmani, M.H., and Reisslan, M., 2016, 'Cognitive Radio for Smart Grids: Survey of Architectures, Spectrum Sensing Mechanisms, and Networking Protocols', *IEEE Communications Surveys* 18(1), 860–898.

Kim, B.-J. et al., 2013, 'Coil Design and Shielding Methods for a Magnetic Resonant Wireless Power Transfer System', *Proceeding of IEEE* 101(6), 1332–1342.

Kim, J. et al., 2011, *Spread Spectrum Technology to Reduce the EMI from the Constant Voltage Source Type Wireless Power Transfer System.* Jeju, South Korea, s.n.

Kitamoto, T. et al., 2016, *A novel type of high power-factor miniaturized wireless ev charger with optimized power receiving circuit and single-ended inverter*, s.l., s.n.

Kitano, Y. et al., 2015, *A new wireless EV charger using single switch ZVS resonant inverter with optimized power transfer and low-cost PFC*, s.l., s.n.

Lee, J.-Y. and Han, B.-M., 2015, 'A Bidirectional Wireless Power Transfer EV Charger Using Self-Resonant PWM', *IEEE Transactions on Power Electronics*, 30(4), 1784–1787.

López, M., Martín, S., Aguado, J., and Torre, S.d.l., 2013, 'V2G strategies for congestion management in microgrids with high penetration of electric vehicles', *Electric Power Systems Research* 104, 28–34.

López, M., Torre, S.d.l., Martín, S., and Aguado, J., 2015, 'Demand-side management in smart grid operation considering electric vehicles load shifting and vehicle-to-grid support', *International Journal of Electrical Power & Energy Systems* 64, 689–698.

Madawala, U. and Thrimawithana, D., 2010, 'Current sourced bi-directional inductive power transfer system', *IET Power Electronics* 4(4), 471–480.

Malhotra, A., Binetti, G., Davoudi, A., and Schizas, I.D., 2016, 'Distributed Power Profile Tracking for Heterogeneous Charging of Electric Vehicles', *IEEE Transactions on Smart Grids* 99, 1–10.

Mi, C.C., Buja, G., Choi, S.Y., and Rim, C.T., 2016, 'Modern Advances in Wireless Power Transfer Systems for Roadway Powered Electric Vehicles', *IEEE Transactions on Industrial Electronics* 63, 6533–6545.

Musavi, F., and Eberle, W., 2014, 'Overview of wireless power transfer technologies for electric vehicle battery charging', *IET Power Electronics* 7, 60–66.

Nguyen, B.X., Foo, G., Ong, A., and Vilathgamuwa, D., 2014, *Efficiency optimization for bidirectional IPT system.* Dearborn, Michigan, s.n.

Shao, C. et al., 2016, 'Hierarchical Charge Control of Large Populations of EVs', *IEEE Transactions on Smart Grids* 7(2), 1147–1155.

Shareef, H., Islam, M. and Mohamed, A., 2016, 'A review of the stage-of-the-art charging technologies, placement methodologies, and impacts of electric vehicles', *Renewable and Sustainable Energy Reviews* 64, 403–420.

Shi, C., Khaligh, A., and Wang, H., 2016, 'Interleaved SEPIC Power Factor Preregulator Using Coupled Inductors In Discontinuous Conduction Mode With Wide Output Voltage', *IEEE Transactions on Industry Applications* 52, 3461–3471.

Siu, K.K. and Ho, C.N., 2016, *A critical review of Bridgeless PFC boost rectifiers with common-mode voltage mitigation*, s.l., s.n.

Song, C. et al., 2016, 'Low EMF and EMI Design of a Tightly Coupled Handheld Resonant Magnetic Field (HH-RMF) Charger for Automotive Battery Charging', *IEEE Transactions on Electromagnetic Compatibility* 58(4), 1194–1206.

Song, C. et al., 2016, 'Low EMF and EMI Design of a Tightly Coupled Handheld Resonant Magnetic Field (HH-RMF) Charger for Automotive Battery Charging', *IEEE Transactions on Electromagnetic Compatibility* 58, 1194–1206.

Song, C. et al., 2014, *Harmonic current reduction method of hand-held resonant magnetic field charger (HH-RMFC) for electric vehicle*, Tokyo, s.n.

Su, W., Eichi, H., Zeng, W., and Chow, M.-Y., 2012, 'A Survey on the Electrification of Transportation in a Smart Grid Environment', *IEEE Transactions on Industrial Informatics* 8, 1–10.

Tang, W., Bi, S., and Zhangn, Y.J., 2014, 'Online Coordinated Charging Decision Algorithm for Electric Vehicles Without Future Information', *IEEE Transactions on Smart Grid* 5(6), 2810–2824.

Theodoropoulos, T.V., Damousis, I.G., and Amditis, A.J., 2016, 'Demand-Side Management ICT for Dynamic Wireless EV Charging', *IEEE Transactions on Industrial Electronics* 63(10), 6623–6630.

Thrimawithana, D., and Madawala, U., 2010, *A novel matrix converter based bi-directional IPT power interface for V2G applications*, Manama, s.n.

Tusha, H.K., Assi, C., Maier, M., and Uddin, M.F., 2014, 'Smart Microgrids: Optimal Joint Scheduling for Electric Vehicles and Home Appliances', *IEEE Transactions on Smart Grids* 5(1), 239–250.

Villa, J.L., Sallan, J., Osorio, J.F.S., and Llombart, A., 2012, 'High-Misalignment Tolerant Compensation Topology For ICPT Systems', *IEEE Transactions on Industrial Electronics* 59, 945–951.

Wei, X., Wang, Z., and Dai, H., 2014, 'A Critical Review of Wireless Power Transfer via Strongly Coupled Magnetic Resonances', *Energies* 7, 4316–4341.

You, Y.S. and Yi, K.H., 2016, *Capacitive coupling wireless power transfer with glass dielectric layers for electric vehicles*, Berlin, s.n.

16

Channel Access Modelling for EV Charging/Discharging Service through Vehicular ad hoc Networks (VANETs) Communications

Dhaou Said and Hussein T. Mouftah

School of Electrical Engineering and Computer Science (EECS), University of Ottawa, Ottawa, Canada

Nomenclature

ACs	access categories
ACK	acknowledgment
Cd	communication density
C_O	cut-off parameter
Cr	communication range
CW	contention window
CCH	control channel
Dch	charging/discharging rate
EDCA	enhanced distributed channel access
EV	electric vehicle
EVSE	electric vehicle supply equipment
EV-SG	EV-to-smart-grid interaction
G2V	grid-to-vehicle
ITS	intelligent transportation system
LTE	Long-Term Evolution
Mf	message frequency
PLC	power line communication
OBU	on board unit
RSU	road side unit
SAE	Society of Automotive Engineers
SCH	service channel
SG	smart grid
SoC	state of charge
TOUP	time of use pricing
Vd	vehicle density
V2G	vehicle-to-Grid
VANETs	vehicular ad hoc networks
WAVE	wireless access in vehicular environment
WSA	wave service advertisement

Transportation and Power Grid in Smart Cities: Communication Networks and Services, First Edition.
Edited by Hussein T. Mouftah, Melike Erol-Kantarci and Mubashir Husain Rehmani.

16.1 Introduction

As we see our society becomes more and more connected with an open and global living environment, the transportation technologies have to play an important and central role in cites. Moreover innovative transportation solutions are emerging as a next-generation approach for the vision of the smart city challenge. Added to the transport electrification concept, especially the electric vehicle (EV), which is beneficial in term of oil dependency and greenhouse gases (GHG) emissions reduction, the recent intelligent transportation systems (ITS) and networks development are expected to improve the electric energy management capabilities in cities while conserving clean, sustainable, economic, and safe surroundings.

In terms of electric power, the load introduced by EVs charging operations will be one of the most important challenging issues for demand response systems in the smart grid. Indeed, the efficient energy management for EVs and electric vehicle supply equipment (EVSE) will become central to achieving efficient operations of the smart grid. Moreover, the smart grid is required to enable penetration of intermittent generation sources, enable adaptive electricity price [1–3], be self-healing, optimize assets [4, 5], enable high penetration of EVs, and enable the EV grid stability participation. In addition to counting and pricing in the smart grid, the real-time control over smart grid infrastructure and energy usage promises one of the largest potentials in load management and energy saving [6–8].

For an EV, finding a vacant EVSE to plug in, especially in peak hours, is always time consuming and frustrating to drivers when the charging service is seen by the smart grid as a trade-off between consumer satisfaction and grid stability. For this end, it is necessary to design EV–smart grid communication architectures taking into account the time constraint and reliability which is expressed as a level of customer's satisfaction without stressing the grid. This communication architecture could be based on real-time information, such as EV and EVSE status, which have to be exchanged with the smart grid. As a result, the EV can be informed of the nearest available EVSE in advance since it is on the road. This can reduce the peak demand and improve EVs satisfaction with grid stability, but it requires a large area connectivity, which has to be maintained between the smart grid and EVs.

In this chapter, we discuss the technical environment of the charging/discharging problem. We introduce a survey of communication technologies used for the smart grid. By the end we present a specific case study of the channel access modelling for the EV charging service based on the IEEE802.11p/DSRC protocol.

16.2 Technical Environment of the EV Charging/Discharging Process

The EV charging/discharging process is described by the time and the place where it occurs. To reduce its potential risk to the electric power system, which exists even with a low PEV penetration [9, 10], smart charging technics based on EV scheduling algorithms can be used. Indeed, the EV charging cycle can be altered by external events, allowing for adaptive charging habits, providing the EV with the ability to integrate easily into the whole power system. To achieve those goals, a guidance model [11] has to be considered with safe, secure, reliable, and sustainable information exchangeable between different grid parts.

In this section we present an overview of electric vehicle supply equipment (EVSE) used at home and away from home (parking work place) and the fast-charging station. Second, we discuss the inductive charging technique as an innovative EV-charging scheme, and we clarify the potential that this method holds for EVs.

16.2.1 EVSE Overview

As the EV industry emerges at the beginning of the 21st century, the development of the infrastructures for recharging at home, at work, and at public locations is imperative. EVSE consists of all the equipment needed to deliver electrical energy from an electricity source to a PEV battery. The EVSE communicates with the PEV to ensure that the plug is securely connected to the vehicle receptacle before supplying a safe flow of electricity. There are three primary types of EVSE, which are classified based on charging places (at home, work places, commercial places, and supply stations), charging speed, and charging electric power (voltage, current).

- At home: the level 1 method uses a standard 120 volts AC (VAC) used for both residential and commercial buildings. Typical voltage and amp ratings can be from 110 to 120 volts AC and 15 or 20 amps, respectively.
- At parking or work place: the level 2 is described as the standard method for the EVSE for both private and publicly available facilities. It specifies a single-phase branch circuit with voltage ratings from 220 to 240 volts AC and a current as high as 80 amps AC (100 amp-rated circuit). However, current levels that high are rare, and a more typical rating would be 40 amps AC, which allows a maximum current of 32 amps. This provides approximately 7.7 kW with a 240 VAC circuit [12].
- DC fast charging station: this type of charging connection can raise the rate of charge to approximately 75–80% in as little as 20–30 minutes, depending on battery size. This type of EVSE uses an off-board charger that transforms AC power to DC and bypasses the on-board charger. Generally, 208 V three-phase or 480 V service is required for this type of charging and may not be commonly available. In many cases, a new separate service will need to be installed by the local utility.

The communication standards when an EV is being charged or plugged in at EVSE are defined by the Society of Automotive Engineers (SAE) as SAE J2293, SAE J2836/1 and J2847/1, SAE J2836/2 and J2847/2, SAE J2836/3 and J2847/3, SAE J2931/1 and SAE J2931 [12].

16.2.2 Inductive Chargers: Opportunities and Potential

As mentioned before, the most common method for EV charging is the usage of a cable plug-in system used either at home, at work, or at specific charging stations spread across some cities. An alternative to this charging method is wireless inductive charging.

The inductive charging technique focuses on the ability to charge while driving and the reduction in the space that must be dedicated to the charging infrastructure in urban areas. Moreover, with this charging method it is possible to just position the EV on a charging point and start the contactless charging process without any additional effort.

The inductive charging technique can take place statically at short stops or during short or long parking phases (the traffic light, the bus stop, or the taxi stand, while it would be impossible with the conductive charging option). It can be also possible dynamically, where the EV can charge while driving, which probably can decrease the size and cost of the battery.

This technique offers several advantages over the conductive charging method (using a cable):

- The inductive charging process is automated and easy to use.
- The system is safe against vandalism and environmental influences (e.g., humidity) because all devices are encapsulated in the vehicle or they are underground.
- There is no negative impact on the cityscape (all devices and grid connections are underground).
- The frequencies used for static inductive charging are between 80 and 500 kHz [13].
- The EV for vehicle-to-grid applications (V2G) availability is higher (e.g., for using the vehicle's battery to store excess energy in the grid).

16.3 Overview of Communication Technologies in the Smart Grid

Many technologies implemented for sensors and wired and wireless networks are already used in the smart grid. The EVs, as a new actor in the SG, need at any time the EV charging service at home or public fast charging service to fill up its battery or supplying power (V2G). As a result, a huge EV penetration, which can be translated to variable spatiotemporal power demand, can cause instability to the grid and an overloading of transformers added to the frequency regulation problem especially during rush hours [14]. However, to reduce the negative impact of the EV penetration and due to the EV dynamic mobility nature, new challenges on monitoring and communication are brought up.

In the following, we present the power line communication (PLC) technology and wireless communication, which can be used in electric mobility.

16.3.1 Power Line Communication

The key advantage of PLC is that the already existing grid infrastructure reaches every grid component and there is no need for new wires as a communication support. It uses the existing electrical lines as communication medium, and data is sent on the power lines by superposing a modulated high-frequency carrier signal on the line voltage. As a result, the PLC deployment cost is relatively low compared to other wireline or wireless technologies. Despite all those advantages, for environment regulations, the PLC cannot be allowed in some countries such as Japan. Moreover, in some cases, the PLC range can be limited by high attenuation caused by transformers.

There are three different types of PLC technologies which are classified by the frequency band and data rate used.

The oldest one is the ultra-narrow band technology. It only provides a data rate around one hundred bits per second [15]. It is used for remote communication with electrical meters (meter reading) in North America.

Table 16.1 Frequencies Reserved by the CENELECT Standard for Smart Grid Communication [17].

Band	Frequencies (kHz)	Applications
A	3–95	Utilities / smart grid; reserved for use by energy providers and customers' premises
B	95–125	Unrestricted
C	125–140	In-home networks, regulated as to channel access techniques
D	140–148.5	Alarm and security

The second one is broadband PLC, which uses the 1.8–250 MHz frequency band and the physical data rate varies between a few megabits to hundreds of megabits. It is used to bring Internet service to homes and small businesses over the electrical distribution system or as in-house networking [16].

The third one is the narrowband PLC, which operates in the 3–500 kHz band and provides lower data rates. Typically, PLC applications can include:

- automatic meter reading and management (AMR/AMM) for the smart grid;
- PLC over medium-voltage power distribution network for distribution automation application;
- PLC home area networking (PLC-HAN) for smart appliances and controls; and
- data communications over DC wires.

In the United States and Asia, there are no restrictions to use narrowband PLC frequencies. Table 16.1 summarizes some frequencies reserved by the CENELECT standard in Europe to be used for some grid applications [17].

16.3.2 Wireless Communications for EV–Smart Grid Applications

As mentioned before, in the real word the EVs power demand is dispersed in space and time. For an effective interaction between the grid and EVs, the grid has to identify and respond to any particular EV power demand nearly in real time. For this issue, advanced wireless network solutions for short or long ranges such as vehicular ad hoc networks (VANET), ZigBee, 3/4G Long Term Evolution (LTE), worldwide interoperability for microwave access (WiMAX), or wireless mesh networks (WMNs) have to be employed.

Beside the small data traffic generated by an EV charging/discharging application, the EVs interaction with the grid requires a communication coverage capability that can support the huge number of EV demands especially for mobile EVs when high data rate, wide coverage, and quality-of-service (QoS) support are needed to support a location-based application. Cellular networks are qualified to deliver connectivity to EVs and smart grid components with mobility support. Moreover the EV network application can require security support with high available and reliable communication. For example for V2G or G2V interaction, an EV scheduling process needs to receive electricity prices to control the EV charging/discharging without unwanted peaks, cable overheating, or overloading of transformers, which can disrupt the grid stability and increase the losses in the power system. The QoS needed by an effective EV–smart grid interaction is expected to gradually increase with the massive EVs penetration. Basically the communication delays and the outage probability can describe the QoS requirement for a smart grid application. However, when the EV demands exceed the network capacity, the performance of the communication system will degrade and can be unavailable in some cases.

As any network solution, the security challenge for interaction between the smart grid and EVs is very important, and all smart grid applications have to take into account the new techniques and algorithms to provide a safe and reliable solution. The security requirements can be described by three terms such as: the confidentiality level dedicated to the private EV information and the integrity aspect to protect the EV charging data and authenticity. Some attack types for the smart grid networks are defined in [18]. Indeed, for example the injection of the negative prices can increase the power consumption, which results in intense grid instability (or create virtual peak periods to destroy or paralyze the power system). Moreover, some attacks can focus on smart meter data input and output and may distort the measures in order to create enormous revenue losses.

16.4 Channel Access Model for EV Charging Service

EVs pose a great challenge for the smart grid by their effect on power grid stability, especially for high EV penetration. To maximize the benefits of using EVs, regulated and optimized charging control needs to be provided by the smart grid. This can be based on exchanged, real-time information, and large area connectivity has to be maintained between the smart grid and EVs. In this section, we propose a new channel access model for EV charging service scheme for a large number of EVs through VANET communications.

16.4.1 Overview of VANET and LTE

The proposed EV charging service is based on VANET and LTE. In the following, we present briefly the DSRC/WAVE and the LTE characteristics. The dedicated short range communication band (DSRC from 5.850 to 5.925 GHz) is a band used for short to medium range communication in vehicular networks (WAVE) [19]. This band includes safety and privates services. DSRC provides a high data rate for mobile nodes and a latency time under 100 ms. As presented in Figure 16.1, the DSRC is formed by seven

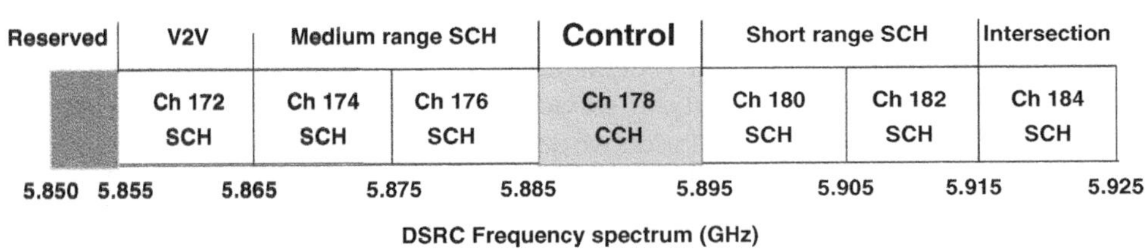

Figure 16.1 Channel allocation in WAVE (DSRC).

Table 16.2 Mobile Technologies Data Rate Ranges Comparison.

	3G	WiMax	HSPA+	LTE
Peak rate (Mbps)	3	128	168	300
Downlink rate (Mbps)	0.5–1.5	2–6	1–10	10–100
Uplink rate (Mbps)	0.2–0.5	1–2	0.5–4.5	5–50

sub bands, one in the middle, which is the control channel (CCH), and six service channels (SCHs). In WAVE, there are two types of nodes: onboard unit (OBU), located in vehicles and roadside units (RSU), which are located on the road side as service providers relaying or distributing the information. Similar to the OSI model, the WAVE model is presented as a layered architecture composed by IEEE 802.11p [19], IEEE 1609.4 [20], IEEE1609.3 [21], IEEE 1609.2 [22], IEEE1609.1 [23].

Technical references [24–26] present the 3GPP Long Term Evolution (LTE) technology as an attractive solution for mobile broadband services. Indeed LTE uses orthogonal frequency division multiple access (OFDMA) for the downlink and single carrier frequency division multiple access (SC-FDMA) for uplink, in combination with new antenna technologies such as MIMO. As presented in Table 16.2, LTE promises a downlink data rate of 100 Mbit/s and an uplink data rate of 50 Mbps over a 20 MHz bandwidth. In addition to its simplicity and reduced deployment cost, an LTE cell can support at least 200 users in the 5 MHz band and 400 users in other bands. Moreover, the LTE latency time is very small and does not require more than 10 ms. LTE supports a scalable and flexible bandwidth from 1.4 to 20 MHz. LTE also supports both modes of FDD (frequency division duplex) and TDD (time division duplex). For the mobility characteristics, LTE performance is optimized with a speed lower than 15 km/h. LTE supports speeds of 120 to 350 km/h (or 500 km/h, depending on the frequency band used). All these characteristics make the large-scale communication technology LTE favorable for V2G interaction design. We present in the next part our proposed EV charging service.

16.4.2 Case Study: Access Channel Model

We consider the system presented in Figure 16.2, where some roadsides are given out of the town highlighted by the circle; from each one derives a possible downtown way (input), and a number of EVs and EVPSSs are sharing the road infrastructure area. We suppose that each EV is equipped by an OBU to be linked to the nearest RSU. This RSU is equipped with two communication interfaces VANET and LTE. We assume that these two communication technologies are able to linking EVs, the smart grid (regional manager), and all EVPSSs. We suppose that all EVs can communicate via VANET their positions, EV state of charge (EV SoC), and priority level to the nearest RSU. All RSUs are connected via LTE to the smart grid.

The proposed EV charging service aims at exploiting the free plug-in in each EVPSS to monitor the whole EV charging system and provides, according to its initial status, an effective guidance for EVs searching for a supply station. We focused on IEEE802.11p V2I interaction between EVs and RSU, as illustrated in Figure 16.3. In the beginning,

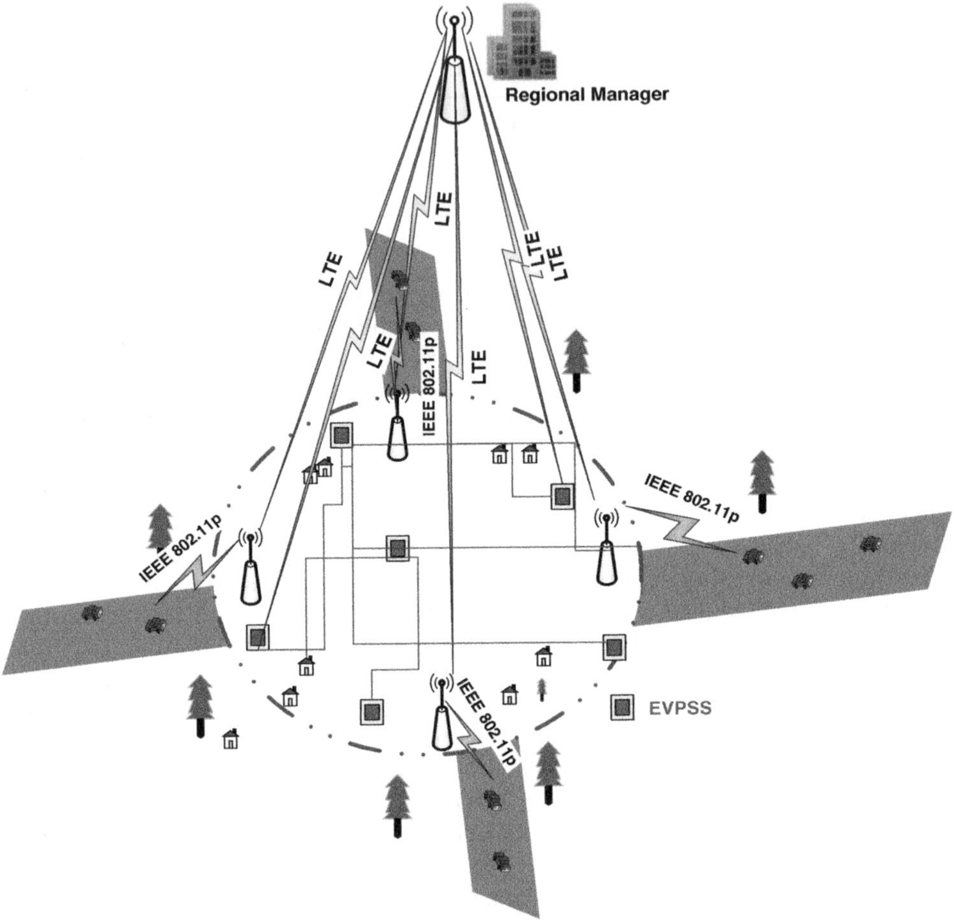

Figure 16.2 System overview.

the RSU broadcast, over CCH, the EV charge scheduling service description (num SCH, selling price, purchase price). The message format, exchanged between RSU and EV, is described in the following. There are two message types: the first one is the advertisement message sent by RSU, in the CCH channel, describing the electric charging service offered by the smart grid. The description of this message is as follows: num SCH, selling price, and purchase price. The second one is the response of any EV interested in the electric charging/discharging service.

The EV interested in sale or purchase of electricity will respond by sending its status (position, SoC, priority).

In VANET broadcast transmission context, some references [27] use the dynamic contention window (CW) adaptation technique to enhance the performance of V2V and V2I communication. Indeed, this technique is based on the network congestion level.

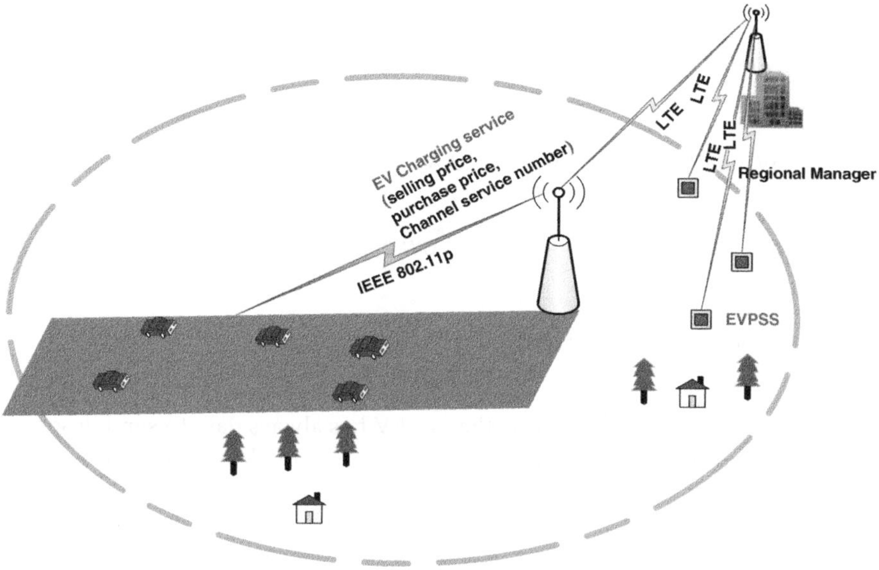

Figure 16.3 Interaction V2I between EVs and RSU.

It uses the network local state to estimate the collision rate threshold, which is used for CW adapting. In our work, for the EV charging service, the CW size fluctuates according to the EV priority level, and we prioritize the channel access for EV with high priority. We use the EV priority level parameter to dynamically regulate the CW size.

We suppose the following scenario: all RSUs are equipped with two communication interfaces, VANET and LTE. We suppose that those two communication technologies are able to link EVs, the smart grid, and all EVPSSs (see Figure 16.2 and Figure 16.3). All vehicles are assumed to be equipped with a GPS device, and they are considered to be moving along the highway in one direction (unidirectionally). We consider that the RSU periodically sends the beacon messages to announce EV charging services. According to the EV position, two cases may occur. First, if the EV is within the RSU transmission range, then the EV can connect to the RSU and exchange information related to EV charging service. Second, if the EV is out of the RSU transmission range or there is no RSU, a multi-hop communication through other EVs will be used to communicate with the nearest RSU to request the charging service announcing.

We focused on this second case. Indeed, in order to establish a connection with the RSU for the EV charging service, the EV keeps actively scanning for beacons on CCH. Once a beacon (i.e., message describing the charging/discharging service) sent by RSU is detected by an EV and if this EV is interested in the charging service, it sends an EV-charging request frame through a multi-hop path.

We are interested in the channel access issue. We present in the following a back-off counter model, based on Markov chains, related to each EV interested on the charging service. We consider that the distributed coordination function adopts an exponential back-off. This back-off is a uniform distribution integer between 0 and CW − 1, where

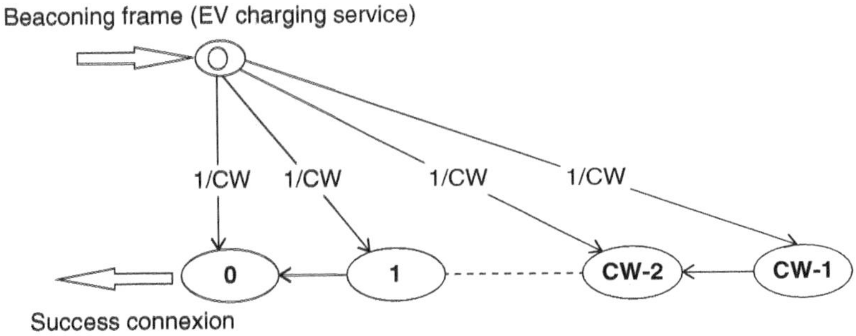

Figure 16.4 Back-off counter Markov chain model.

CW is the contention window. We assume that an EV has always data to send describing roadside traffic or other, the channel conditions are assumed to be ideal among all neighboring EVs [27, 28].

Figure 16.4 shows the system considered to represent the back-off counter process for an EV interested in the charging service. The input data is the beaconing frame describing the EV charging service, where the output data is the EV success connection status.

The computation of the back-off counter is achieved as follows:

1) select a uniform distribution integer within [0,CW];
2) multiply the chosen integer by the time slot at the physical layer;
3) if the channel is free, then the back-off is decremented; and
4) when the back-off reaches zero, the transmission is achieved.

According to Figure 16.4, the transition probability function with time-homogenous Markov chain is defined as:

$$p(l, l+k) = p_{ij}(0, k) \equiv p_{ij}^{(k)}, \quad \forall \quad l \tag{16.1}$$

where k and l are integer values. The transition matrix is defined as:

$$P = [p_{ij}] \tag{16.2}$$

The stationary condition related to our Markov chain model is given by:

$$\overline{\pi} * P = \overline{0} \tag{16.3}$$

where $\overline{\pi} = [\pi_0, \pi_1, \dots, \pi_{CW-1}]$ is the vector of stationary distribution. $CW - 1$ is the state number. Our model is completely characterized by given the transition matrix P and the initial condition π_0.

To solve this Markov chain, we have to solve the balance Equation 16.3 in each state and use the total probability to solve the initial condition π_0.

The total probability condition is:

$$\sum_{k=0}^{CW-1} \pi_k = 1 \tag{16.4}$$

The π_k and π_0 expressions can be easily derived by simplifying eq. (16.5) as:

$$\pi_k = \frac{CW - k}{CW}\pi_0 \quad \text{and} \quad \pi_0 = \frac{2}{CW + 1} \tag{16.5}$$

The probability to have a successful connection in a randomly chosen slot time is:

$$p_{succ} = \pi_0 = \frac{2}{CW + 1} \tag{16.6}$$

The probability that there is no transmission in a given slot time (collision or idle condition) is given by:

$$\rho_n = (1 - p_{succ})^n \tag{16.7}$$

where n is the number of neighboring EVs.

The probability of collision is given by:

$$p_{col} = 1 - \rho_n \tag{16.8}$$

Let P_s be the probability that exactly one EV accesses the channel, conditioned on the fact that at least one EV had access, given there are n EVs,

$$P_s = \frac{n p_{succ}(1 - p_{succ})^{n-1}}{1 - (1 - p_{succ})^n} \tag{16.9}$$

By increasing P_s or decreasing its denominator, which is given by p_{col} in Equation 16.8, we can achieve a higher throughput. This can be obtained if the value of $(1 - p_{succ})$ increases, which implies that p_{succ} needs to be small. As seen in Equation 16.6, if we increase the value of CW, we can obtain smaller p_{succ}.

The probability of collision, given by Equation 16.8, decreases when CW increases and this effect is more predominant for higher values of n.

We express TH, the normalized system throughput as:

$$TH = \frac{\rho_{col} P_s E[payload\ in\ T_{slot}]}{(1 - \rho_{col})D_{emp} + \rho_{col} P_s T_{bs} + \rho_{col}(1 - P_s)T_{bc}} \tag{16.10}$$

Given:

- $T_{st} = E[payload\ in\ T_{slot}]$ is the average time used during a successful transmission.
- T_{bs} is the average time the channel is sensed busy because of a successful transmission.
- T_{bc} is the average time the channel is sensed busy because of a collision.
- D_{emp} is the duration of an empty slot time.
- $T_{bc}{}^{*} = {}^{T_{bc}}/{}_{D_{emp}}$

Considering Equation 16.6, we can deduce the optimum CW value, which maximizes the throughput TH as:

$$CW_{opt} = n\sqrt{2T^{*}_{bc}} - 1 \tag{16.11}$$

In the following, we present the algorithm adaptive CW; it calculates CW_{opt}, which maximizes TH and assigns it to a high-priority EV.

Algorithm ***Adaptive CW***

Input: EV priority level,
Beacon message describing EV charging service
Output: Adaptive CW[ACs],CW_{opt} /* selected CW */

For each time and for each EV **do,**
If collision
If EV priority is low **then**
Maintain CW[ACs]
Else If EV priority is high **then**
Calculate n the number of neighboring EVs,
Calculate $T_{bc}{}^{*}$, /* According to network condition*/
Calculate CW_{opt} according to eq.(11)
If $CW < CW_{opt}$ **then**
Increase the corresponding CW[AC] /* CW [AC] will doubled*/
Else If $CW \geq CW_{opt}$ **then**
Maintain the corresponding CW[ACs]
End If
End If
End if
End For

The algorithm, "Adaptive CW" should be run by individual vehicles periodically to update CW [AC] values occur according to the EV priority level and the network condition (e.g., congestion, collision, etc.). In addition to the EV priority level, which is an essential parameter for our adaptive CW algorithm, each EV needs to estimate the network conditions (n and $T_{bc}{}^{*}$) to calculate CW_{opt} by analyzing the received sequence numbers at MAC layer [28].

16.4.3 Simulations Results

We evaluate the performance of our proposed channel access model using NS-2.34 and MATLAB. First, we evaluate some performance metrics such as the probability to have a successful connection in a randomly chosen slot time, the probability of collision, and the probability that exactly one EV accesses the channel, conditioned on the fact that at least one EV had access. Second, we study the efficiency of our access model for the EV charging service compared with the IEEE802.11p standard technique in terms of percentage of successful EV connections in a context of multi-hop network.

Table 16.3 summarizes the simulations parameters.

We study the efficiency of our adaptive CW algorithm compared with the standard IEEE802.11p one in terms of the percentage of successful connections for the EV-charging service.

As shown in Figure 16.5, there is an improvement in the percentage of successful connections ensured by our proposed algorithm; the latter allows more EV-charging service access per second with fewer losses. This means that, considering the total number of

Table 16.3 Simulation Parameters.

Parameter	IEEE 802.11p
Carrier frequency [GHz]	5.9
Bandwidth [MHz]	10
OFDM guard time [μs]	1.6
CWmin	3,7,15
Cwmax	1023
Tslot [μs]	13
Tsifs [μs]	32
Multi-hop	3 hops max
Transmission range	2*300 m
Vehicle density (*Vd*)	10 to 100 EV on 1000 m of road
EV speed	10 m/s, 15 m/s, 20 m/s,
Data rate	6 Mbps
Messaging frequency (*Mf*)	10 Msg/s, 20 Msg/s, and 30 Msg/s
Packet size	200 bytes
Propagation model	Nakagami with m = 2
Power limit	CCH: 44.8dBm/SCH: 33bBm
Access scheme	EDCA (supported by NS-2.34)

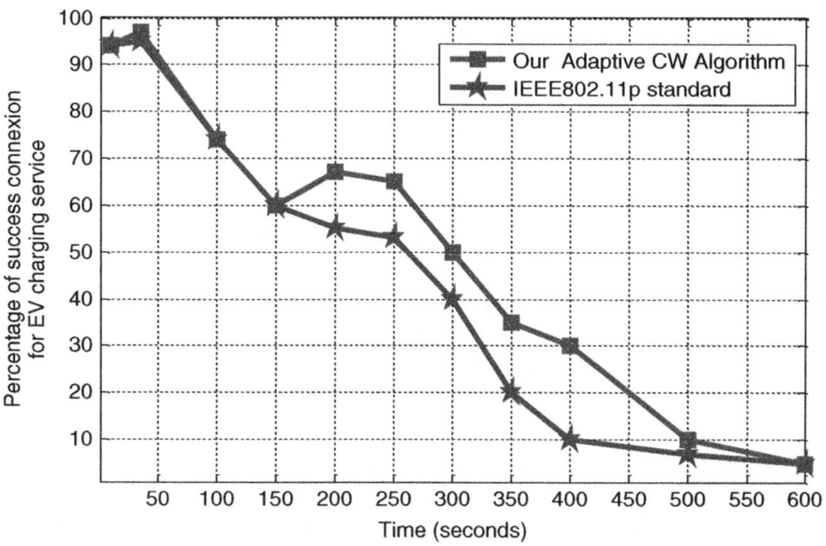

Figure 16.5 Percentage of successful EV connection for EV charging service in dense scenarios (100 EVs).

Table 16.4 Percentage of Success EV Connection for EV Charging Service Comparison between Standard IEEE802.11p and Adaptive CW.

	IEEE802.11p standard	Our algorithm: adaptive CW	Saving rates (%)
Time in [0,149]	93	95	2.1
Time in[150,600]	79	55	30.3

EV charging demands that should be received; the proposed adaptive algorithm is more successful than the standard IEEE802.11p.

Table 16.4 presents the observation results obtained from Figure 16.5, which illustrates the performance comparison between our adaptive CW algorithm and the standard IEEE802.11p one. Indeed, our adaptive approach increases the percentage of success EV connection for EV charging service with a saving rate of more than 2.1 % and 30.3%, respectively, for the two ranges of time [0,149] and [150,600]. This result proves the effectiveness of our adaptive CW algorithm mainly for a high EV number with different EV charging priority level.

16.5 Conclusions

In this chapter, the scheduling problem for electric vehicle (EV) charging/discharging in electric vehicle supply equipment is introduced. We discuss the technical environment of the charging/discharging problem. We introduce a survey of communication technologies used for the smart grid. By the end we present a specific case study of the channel access modelling for the EV charging service based on the IEEE802.11p/DSRC protocol.

References

1 Assessment of Plug-in Electric Vehicle Integration with ISO/RTO Systems, URL: http://www.iso-rto.org.
2 Kaebisch, S., Schmitt, A., Winter, M., and Heuer, J., 2010, 'Interconnections and communications of electric vehicles and smart grids', *In*: Smart Grid Communications (Smart Grid Comm), 2010 First IEEE International Conference on, 2010, 161–166.
3 IEC TC/SC 23 62196-x, Plugs, socket-outlets, vehicle couplers and vehicle inlets Conductive charging of electric vehicles, Geneva, Switzerland.
4 Ruthe, S., Schmutzler, J., Rehtanz, C., Wietfeld, C., 2011, 'Study on V2G Protocols against the Background of Demand Side Management', Conference—Interoperability in Business Information Systems IBIS, volume Issue 1, 2011, 33–44.
5 Li, N., Chen, L., and Low, S. H., 2011, 'Optimal demand response based on utility maximization in power networks', *In*: IEEE PESGM 2011, 2011.
6 Samadi, P., Mohsenian-Rad, A., Schober, R., Wong, V., and Jatskevich, J., 'Optimal real-time pricing algorithm based on utility maximization for smart grid', *In*: IEEE SmartGridComm 2010, 2010, 415–420.

7 Li. Q., and Negi, R. 'Distributed Scheduling in Cyber-physical Systems: The Case of Coordinated Electric Vehicle Charging', IEEE International Workshop on Smart Grid Communications and Networks 2011.

8 Wang, P., Rao, L., Liu, X., Qi, Y., 2011, 'Dynamic Power Management of Distributed Internet Data Centers in Smart Grid Environment', IEEE Global Telecommunications Conference GLOBECOM, 2011, 1–5.

9 Moses, P. S., Masoum, M. A. S., and Hajforoosh, S., 'Overloading of distribution transformers in smart grid due to uncoordinated charging of plug-in electric vehicles', *In*: Proc. IEEE PES Innovative Smart Grid Technologies (ISGT), 2012, 1–6.

10 Real Wireless, 'A Comparison of UNB and Spread Spectrum Wireless Technologies as used in LPWA M2M Applications', (white paper), URL: https://www.thethingsnetwork.org/forum/uploads/default/original/1X/3b1c1ae4a925e9aa897110ccde10ec61f3106b87.pdf, 2015.

11 Said, D., Cherkaoui, S., and Khoukhi, L., 2015, 'Guidance model for EV charging service', IEEE International Conference on Communications (ICC), London, June 2015, 8–12, 5765–5770.

12 'Energy Transfer System for Electric Vehicles—Part 1: Functional Requirements and System Architectures' (Stabilized Feb. 2014), URL: http://standards.sae.org/j2293/1_201402/.

13 Kesler, M., 2013, 'Highly Resonant Wireless Power Transfer: Safe, Efficient, and over Distance', WiTricity Corporation.

14 De Carne, G., Buticchi, G., Liserre, M., and Vournas, C., 'Frequencybased overload control of smart transformers', *In*: IEEE 2015 PowerTech, 1–5.

15 Mlynek, P., Misurec, J., Kolka, Z., Slacik, J., Fujdiak, R., 2015, 'Narrowband Power Line Communication for Smart Metering and Street Lighting Control', 13th IFAC and IEEE Conference on Programmable Devices and Embedded Systems—PDES 2015.

16 Oksman, V., Galli, S., 'G.hn: The New ITU-T Home Networking Standard', *IEEE Communications Magazine*, October 2009, 138–145.

17 'The European table of frequency allocations and applications in the frequency range 8.3 KHz to 3000 GHz (eca table)', report by the Electronic Communications Committee (ECC) within the European Conference of Postal and Telecommunications Administrations (CEPT; approved June 2016).

18 Baig, Z., and Amoudi, A., 2013, 'An analysis of smart grid attacks and countermeasures', *Journal of Communications* 8(8).

19 IEEE P802.11p/D3.0, 2007, 'Draft Amendment to Standard for Information Technology-Telecommunications and Information Exchange between Systems-Local and Metropolitan Area Networks-Specific Requirements—Part 11: Wireless LAN Medium Access Control (MAC) and Physical Layer (PHY) Specifications-Amendment 7: Wireless Access in Vehicular Environment'.

20 IEEE P1609.4, 2006, 'Trial-Use Standard for Wireless Access in Vehicular Environments (WAVE)—Multi-Channel Operation'.

21 IEEE Std P1609.3, 2007, 'IEEE Trial-Use Standard for Wireless Access in Vehicular Environments (WAVE)-Networking Services'.

22 IEEE P1609.2, 'Trial-Use Standard for Wireless Access in Vehicular Environments (WAVE)—Security Services for Applications and Management Messages'.

23 IEEE P1609.1, 'Trial-Use Standard for Wireless Access in Vehicular Environments (WAVE)—Resource Manager'.

24 Abdullah, M., and Yonis, A., 2012, 'Performance of LTE Release 8 and Release 10 in wireless communications', International Conference on Cyber Security, Cyber Warfare and Digital Forensic (CyberSec), 236–241.

25 Wei, Li., Jiadi, C., Hang, L., and Bin,W., 2012, 'Performance and Analysis on LTE System under Adjacent Channel Interference of Broadcasting System', IEEE 12th International Conference on Computer and Information Technology (CIT), 2012, 290–294.

26 Xu, Y., and Fischione, C., 'Real- time scheduling in LTE for Smart Grid, Proceedings of the 5th International Symposium on Communications, Control and Signal Processing, ISCCSP 2012, Rome, Italy, May 2–4, 2012.

27 Dehbi, Y., Benaboud, H., and Mikou, N., 'A geometric distribution for backoff time in IEEE 802.11 DCF:an analytical study', *International Journal of Communication Networks and Information Security (IJCNIS)* 5(3), December 2013.

28 Balon, N., and Guo, J., 'Increasing Broadcast Reliability in Vehicular Ad Hoc Networks', in VANET '06: Proceedings of the 3rd ACM international workshop on Vehicular ad hoc networks, Los Angeles, California, Sep. 2006, 104–105.

29 'Vehicle Safety Communications Project Task 3 Final Report: Identify Intelligent Vehicle Safety Applications Enabled by DSRC', Vehicle Safety Communications Consortium consisting of, BMW, DaimlerChrysler, Ford, GM, Nissan, Toyota, and VW, Mar. 2005.

17

Intelligent Parking Management in Smart Cities

Sanket Gupte and Mohamed Younis

Department of Computer Science and Electrical Engineering, University of Maryland Baltimore County, Baltimore, MD 21250

17.1 Introduction

It would be fair to say that only a small portion of urban traffic is caused by people driving through a city to get to the other side. Most people wanting to go to the other end of a city do not pass through but go around the city instead, to avoid congestion. Then it is also fair to assume that just about everyone else, even taxis, are going to have to park, maybe for a few minutes or much longer. Depending on the city, they could be spending anywhere between 30 seconds to 20 minutes on average looking for a space to park. We can hence conclude that parking has a major contribution to congestion in modern cities. As the population of a city grows, the density of vehicles on the roads consequently increases, and our society is currently on the verge of facing the great challenge of a global gridlock.

Vehicles are causing massive headaches for city planners and drivers alike, and smart cities are examining how to leverage technology to make communities more livable. The US Department of Transportation, in the first quarter of 2016, put up a challenge and pledged up to $40 million (USD) to one city that would effectively define what it means to be a smart city (US Department of Transportation). The pitch of city of Columbus, OH, which won the challenge, describes "parking" as a great impediment for the development of the city, which affects access to jobs, freight transportability and reliability, the tourism and sports industry, and of course transportation. According to ITS America (ITS-America, 2016) about 30% of urban traffic is caused by driver searching for parking, and on an average 20 minutes are spent looking for parking, which effectively translates to 55 hours every year each driver spends looking for parking. Imagine the amount of pollution and greenhouse gases emissions that could potentially be avoided.

Although cities are expanding, urban areas are so densely populated, it would not be wrong to assume that the *supply* of parking spaces is somewhat fixed. Therefore, one cannot address the supply. The only recourse is to change the demand to try and reach equilibrium. Thus, parking management in smart cities is not just about how to manage incoming and outgoing vehicles in a parking lot, but is a lot more than that. It is about how to optimize the use of available parking facilities, how to make the commuters

Transportation and Power Grid in Smart Cities: Communication Networks and Services, First Edition.
Edited by Hussein T. Mouftah, Melike Erol-Kantarci and Mubashir Husain Rehmani.

aware of the available parking locations, keep track of parking offences, speeding up the process of finding a spot, and directly and indirectly optimizing the time utilization for everyone in general.

While managing parking using RFID, WSN (Chinrungrueng, Sunantachaikul, and Triamlumlerd, 2007; Kim et al., 2010; Bachani, Qureshi, and Shaikh, 2016) and vision-based techniques (Chen et al., 2010) have been prevalent for several years, although not widespread, all the new frameworks and proposals nowadays use a combination of two or more sensing technologies (Rico et al., 2013; Stenneth et al., 2012) to achieve more precise tracking and better utilization of existing assets. Many private companies (Cisco-Systems-Inc., 2016; Siemens-AG, 2016; Streetline-Inc, 2016) are researching and developing their own hardware and software to efficiently manage parking. Siemens Mobility has designed a smart parking solution (Siemens-AG, 2016) in which they use a radar-based WSN architecture to detect parking spot occupancy and a web-based parking management system to analyze and distribute real-time status updates. Similarly, the system of Streetline (Streetline-Inc, 2016), a startup company developing smart applications and infrastructure for parking management, has successfully deployed their solution in at least three cities. They basically use in-pavement sensors and cameras to collect data and have a web-based analytics platform that in turn provides real-time parking information to users' smart phone–based applications. Meanwhile companies such as Inrix (INRIX, 2016) are developing integrated inter and intra vehicular solutions to detect whether and where a vehicle is parked and aggregating the data to a parking map. Others such as Civic-Smart (previously Duncan; Civicsmart-Inc, 2016) are using their experience over the past few decades in the parking hardware industry to modernize the existing infrastructure.

While all these systems, some proposed, some simulated, and the rest sparsely implemented, are effective ways to achieve the goal of successfully optimizing the use of parking spots and making drivers aware of the availability, there is one common pattern in each of them. All of them almost completely depend on installation of additional hardware for the detection and notification of parking availability. As shown in the glimpse above, in this chapter we provide extensive analysis, survey existing systems, and highlight cutting-edge solutions for parking management in smart cities. One such solution is our smart phone–based participatory sensing application architecture (Gupte and Younis, 2015), which is to run on the cloud in order create an efficient parking experience for citizens of smart cities, without the need for any additional hardware and infrastructure costs, making it easy to implement and quick to integrate in all the existing parking systems. We describe in depth how we can achieve real-time parking availability information and effectively regulate supply and varying demand for parking spaces. We display the most efficient ways to inform users about the parking availability and find the optimum spot based on their final destination and requirement.

With a reward-based participatory sensing architecture it is possible to incentivize the application and boost user participation with the possibility of being able to gain different kinds of rewards such as a dedicated parking spot, priority parking, etc. It is also feasible to integrate payment regulation, ticket generation, and payment processing, etc., into the smart-phone application itself. Our solution could manage street parking, which is a very complex problem given the difficulty in tracking the state of the spots. Our system can also be used to help manage the parking facilities by accessing and validating the need for having more or fewer parking spaces, etc. Higher occupancy rates and good parking turnover would mean increased revenues not just from parking fees but also from fines due to improved infringement detection. Also, such fine-tuned parking management means fewer cars on the road, and consequently less fuel consumption, and in turn lesser emissions, which is precisely one of the main visions of a true smart city.

This chapter is organized as follows. The next section summarizes the design objectives and issues for a parking management system and provides a taxonomy and general categorization. Section 17.3 surveys and classifies existing parking management solutions for smart cities. Section 17.4 describes a novel participatory sensing–based solution and discusses its applicability and performance. Finally, Section 17.5 concludes the chapter with a summary and highlights future research directions.

17.2 Design Issues and Taxonomy of Parking Solutions

17.2.1 Design Issues for Autonomous Parking Systems

A parking management system must address two fundamental issues: (1) how to track the status of parking resources, and (2) how to optimally assign a vehicle to a vacant spot. The first issue involves two key design challenges, namely, how to check the status of a parking spot, how such a status is tracked over time, and where the status of the spots is maintained. Basically, the system needs to know whether a spot is occupied or available. Doing so requires either the deployment of a sensing methodology, either hardware based or just human based. In addition, the sensing data should be either collected periodically or queried on demand. The dissemination of such data requires some communication infrastructure that may involve either wireline or wireless links, or even a mix of both. The data could be disseminated to a central unit or stay scattered in multiple units. The latter raises issues of consistency and concurrency control, which is popular in the realm of distributed databases.

Optimized assignment of spots to requesting vehicles is the core objective of the parking system. From an operator point of view, the utilization of the spots should be maximized for increased revenues and/or return on investment. For a user, finding the closest spot is desirable. The system should strive to achieve the objectives of both user and operator. When the status of the spots is closely tracked and the parking requests are deterministic, the optimization could be manageable; yet it becomes quite complicated when factoring in uncertainty about the user demands and highly error-prone state of parking spots. For example, one cannot guarantee that the requester will indeed park in the allotted spot, and the next state of the parking cannot be inferred with high fidelity. Another objective could be experienced in street parking, where the duration of occupying a parking spot is constrained in order to ensure fairness. In this case the request for parking will include a desired duration and the system will try to make the best match subject to specific regulations.

17.2.2 Taxonomy of Parking Solutions

Parking solutions have evolved by leaps over the past two decades. There are several different ways and perspectives for how these solutions are to be deployed in cities all over the world. Generally, existing solutions employ a diverse set of technologies and

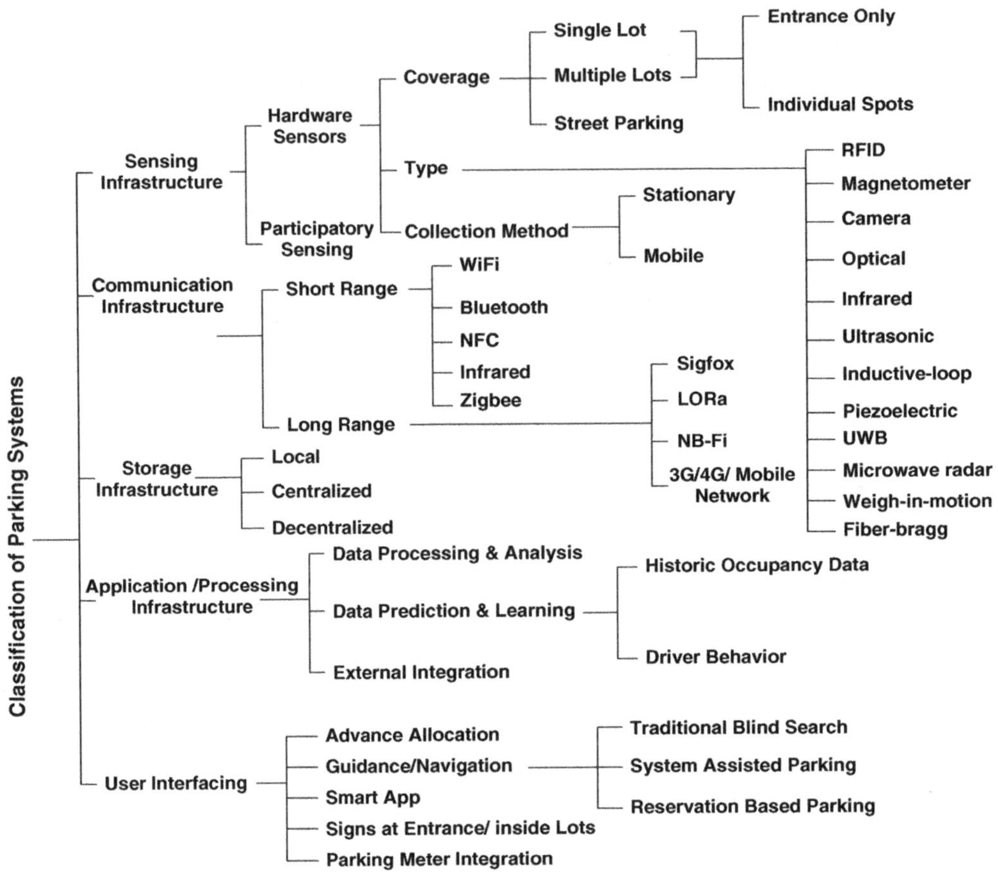

Figure 17.1 Taxonomy and general categorization of parking systems.

deployment strategies. Many variables and factors shape up the whole system and all have major roles in its success or failure. We categorize these parking systems based on five major subsystems, as explained below and illustrated in Figure 17.1.

1) **Sensing Infrastructure:** Sensing nodes are embedded systems that collect and process data into useful information and forward it to other parts of the infrastructure. They always evaluate their own accuracy, status, and health to inform the system accordingly. This poses the need for various levels of authenticity and validation checks.
2) **Communication Infrastructure:** The sensed information must be collected to be further forwarded to different parts of the system, e.g., roadside infrastructure, storage servers, parking meters, etc. Therefore, the spot sensors must be part of a network. Increasing the number of sensor devices imposes the additional burden of transmission scheduling and spectrum allocation.
3) **Storage Infrastructure:** The data received from the communication infrastructure must be stored, processed, and analyzed for dissemination as well as learning. Where the data is stored, how it is stored, and its accessibility are important aspects in its retrieval.

4) **Application Infrastructure:** The application infrastructure opts to fulfill the different requirements of the parking system and displays the spots status at the right interface, e.g., in a smart phone or signs outside parking lot, payment kiosk, etc.
5) **User Interfacing:** Parking systems use the sensed information to offer parking information to drivers. The system must efficiently cater to the interests and needs of the drivers, deliver the most relevant information, and adapt to varying service requirements.

Considering all the above factors, it would be beneficial to identify and classify existing parking solutions based on how they handle these factors. In the next section, we survey existing solutions while highlighting their advantages and shortcomings.

17.3 Classification of Existing Parking Systems

The designs of parking systems found in the literature are quite diverse and cover several aspects that range from optimizing spot allocation and parking structures to interfacing between drivers and urban resources. In this section, we survey and classify existing parking solutions based on the taxonomy provided in the previous section. We first start with how to obtain parking occupancy information from different sensors in Subsection 17.3.1 by discussing the popular sensing technologies that are employed. In Subsection 17.3.2 we cover the different ways the sensor data is transferred to information sinks or data-stores. Subsection 17.3.3 reviews the various application architectures and highlights their advantages and shortcomings. Meanwhile, Subsection 17.3.4 describes how data processing and analysis are applied in existing systems to extract useful information from the sensed data. Finally, Subsection 17.3.5 focuses on how information is displayed to the drivers to help them make an educated decision on where to park their vehicles. Finally, Subsection 17.3.6 provides a comparative summary of all existing systems.

17.3.1 Sensing Infrastructure

Assessing spot occupancy can primarily be done using hardware sensor or participatory sensing. Hardware-based sensing relies on a different modality to collect parking spot status and can be classified into *stationary* or *mobile*. Stationary sensors are usually installed in every spot to instinctively detect the presence of a vehicle and report the status. Whereas, the mobile-sensing methodology employs vehicle-mounted devices and relies on the vehicle motion in data collection. Basically, when a vehicle passes through a particular place, it reports information along the route. In essence the mobile sensing methodology minimizes the number of sensors by exploiting spatiotemporal coverage. Different sensors have distinct mechanisms to detect the presence of vehicles. Popular sensing modalities include infrared, ultrasonic, optical, magnetometer, imager, video, and RFID. In the balance of this subsection we enumerate the different types of sensors and describe them in detail.

Infrared sensors can be active or passive. The idea is to install such a sensor in the parking spot to detect whether a vehicle is present. Active ones use infrared signals to measure the distance to an obstacle in their field of view; thus, if a vehicle is present,

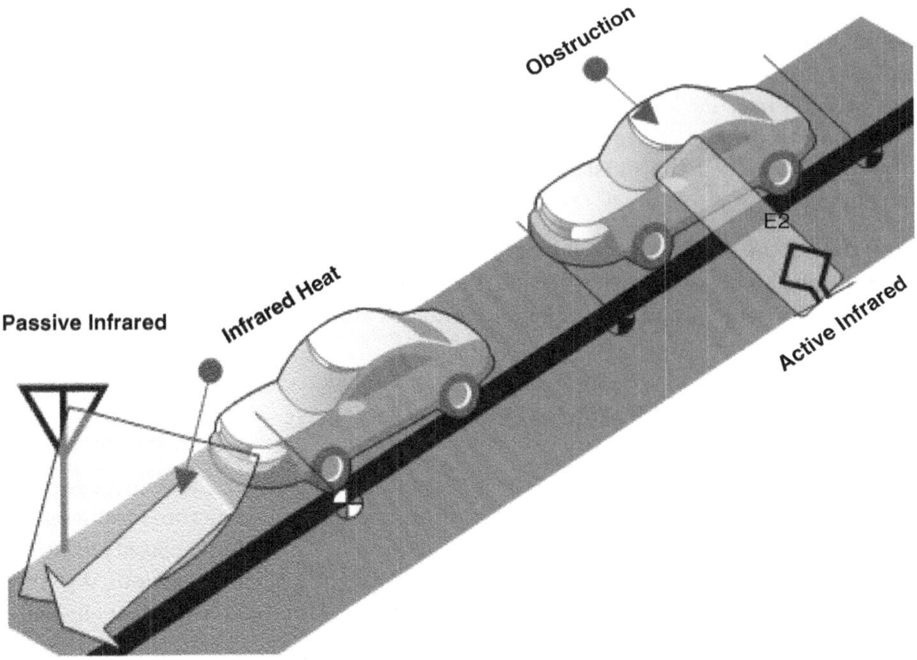

Figure 17.2 Illustrating the use of active and passive infrared sensors.

the distance is small. Active infrared sensors have the advantage of being able to detect occupancy of multiple vehicles using a single sensor. Meanwhile passive infrared sensors detect a vehicle based on the radiated heat. Examples of active and passive infrared sensors are shown in Figure 17.2. Nonetheless, like all spot-installed vehicle detection sensors, the installation and maintenance cost of infrared devices dominates. In addition, infrared sensors consume a lot of energy and are very sensitive to light and hence are not very accurate, especially in outdoor environments. Examples of systems that use infrared sensors are provided in the literature (Cao and Kong, 2011; Huang, Huang, and Sun, 2011; Mainetti et al., 2015; Pala and Inanc, 2007).

Ultrasonic sensors (Bi et al., 2006; Chen and Chang, 2011), on the other hand, detect spot occupancy by sending an ultrasonic signal and using its received echo to detect the presence of an obstacle (vehicle, in case of a parking spot), as illustrated in Figure 17.3. Together with a signal processing module, the reflected ultrasonic energy is analyzed using a roadside controller to calculate the distance between the obstacle and the sensor and infer whether a vehicle is parked. Ultrasonic sensors provide a complex signal pattern that enables multiple lane coverage and detection of vehicles that exceed certain height. Therefore, these sensors can be used for detecting occupancy in multiple spots and are consequently advantageous in outdoor environments. However, a serious shortcoming of this type of sensors is the degraded occupancy measurement on freeways with vehicles traveling at moderate to high speeds due to the large pulse repetition period. In addition, temperature changes and extreme air turbulence affect the sensor performance, although temperature compensation has been built into some of the available models in the market.

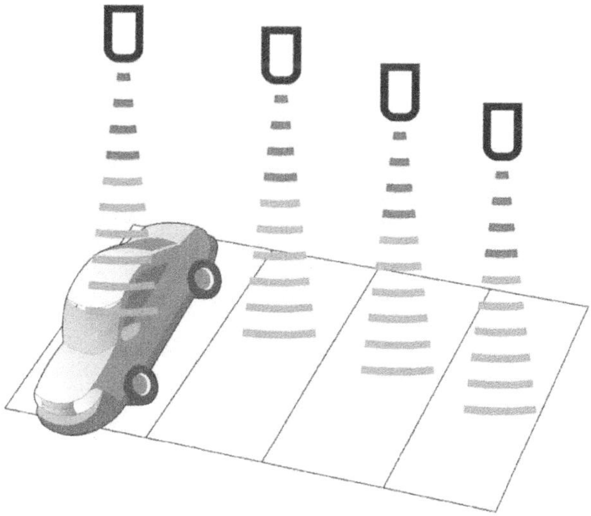

Figure 17.3 Occupancy detection by installing ultrasonic sensor above each spot to detect the presence of a vehicle.

Optical sensors detect changes in light intensity; thus they have to be installed in a spot in such a way that the incoming light will be obscured by the parked vehicle. The sensor first calculates the intensity of light of a luminous source such as the sun; the absence of this luminous intensity creates a shadow, indicating the unavailability of a spot. Hence, optical sensors are impacted by any light source and transient objects, and consequently their accuracy assessment may be questionable. Bachani, Qureshi, and Shaikh (2016) present a performance analysis of the detection accuracy of vacant parking slots under different light signals, mimicked by concentrating on light dependable resistor sensors and infrared (IR) sensors, during different times of the day and under varying weather conditions. They have concluded that IR sensors are certainly better than optical (light detecting resistor) ones since the accuracy of IR sensors is relatively unaffected by weather conditions and other sources of light. On the other hand, the accuracy of optical sensors is highly degraded by the variations in luminous intensity throughout the day and hence need different threshold values for detection during different times. Moreover, shadows also increase false detection rates for optical sensors.

Inductive loop detectors are different-sized wire loops that get excited with signals with frequencies between 10 and 50 KHz. The oscillation frequency of the inductive loop is directly controlled by the inductance of the loop, which changes with vehicle presence. This sensor technology is fairly established, mature, and well researched; compared to other commonly used techniques, it provides the best accuracy and hence is quite widely used to detect slot occupancy. The biggest disadvantages of inductive loops are its expensive installation and maintenance costs. As illustrated in Figure 17.4, the loops are placed through the pavement. In addition, multiple detectors are usually needed to monitor a location, and lots of wires are placed. The involved wires are easily susceptible to stress and temperature-based wear and tear and are also sensitive to water that seeps through cracks on the pavement surface.

Piezoelectric sensors are made from specially processed material that is able to convert kinetic energy to electrical energy when subjected to vibration or mechanical impact. The sensors not only can detect the passing of a vehicle but can also detect other physical attributes of the vehicle, making it easy to identify the category or type of vehicle.

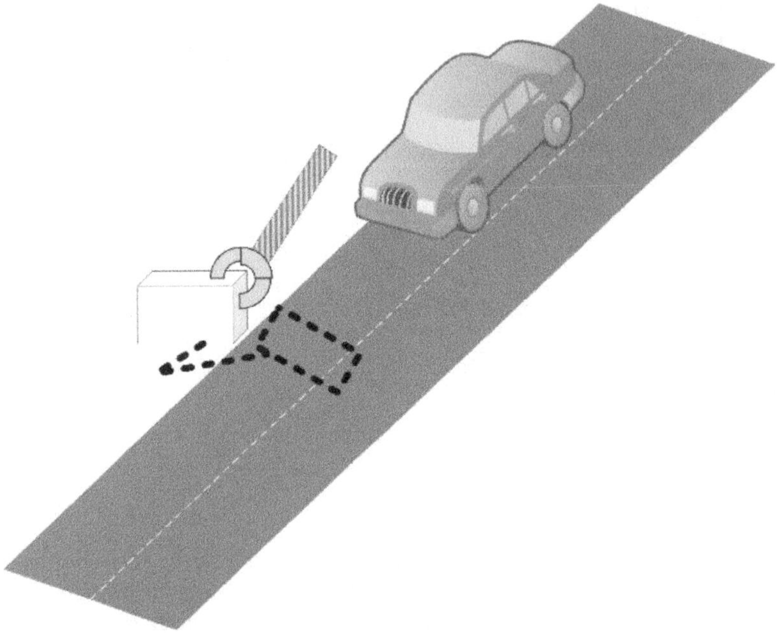

Figure 17.4 Inductive loop detector installed at the entrance of a parking lot.

In addition, it provides accurate readings on vehicle speed and classification of vehicles based on weight and axle spacing with almost the same cost as inductive loop detectors. Among the disadvantages of piezoelectric sensors is the need to use multiple detectors to instrument a location. In addition, piezoelectric sensors can also easily wear out due to high temperature and traffic stress.

Pneumatic road tube detects vehicle via the air pressure created when the vehicle passes or stops over the tube. Such air pressure closes a switch, producing signals. While a pneumatic road tube offers a low-cost solution as well as quick installation and easy maintenance, it has its disadvantages. Basically, pneumatic road sensors are temperature sensitive. Inaccuracies in axle count are also bound to happen when bus and truck numbers are high. In addition, the tubes are prone to vandalism. Furthermore, instrumenting a large parking lot with pneumatic road tubes will make it visually unappealing.

The search coil magnetometer, or induction sensor (Benson et al., 2006) measures the change in magnetic flux lines caused by heavy metal objects using the law of induction, whereas fluxgate magnetometers detect anomalies in the earth's vertical and horizontal magnetic field. Thus, by installing such a sensor in the parking spot (as illustrated in Figure 17.5), it is easy to detect if a vehicle is present. Magnetometers have an easy to read signal pattern and are insensitive to weather conditions and less susceptible to weight stresses than piezoelectric sensors or pneumatic road tubes; however, magnetometers cannot detect multiple objects. Magnetometers can be put on controller printed circuit boards and used only as sensing devices. In fact, they have become an integral part of smart phones and are primarily used to give the sense of direction, as to which direction the phone is facing relative to the geographic poles of earth. The magnetometer on board a smart phone can certainly enable the participator sensing model.

Figure 17.5 A magnetometer installed under a parking spot detects whether a large metal object, i.e., a vehicle, is present.

Using support vector classification (Villanueva et al., 2015) the data collected from the cell phone magnetometer can be used to determine whether a vehicle is parked to the left/right or on both relative to the phone. Being embedded into a smart phone, it is easy to forward the sensor data to a server. Using this data and GPS location of the phone, the server could produce a map of free parking spaces. Villanueva et al. (2015) also could successfully find strong variation in magnetometer readings when the moving test car passes by a vacant parking space. Alternatively, Boda, Nasipuri, and Howitt, (2007) compare different readings between adjacent magnetometer sensors installed on the pathway, to know how many vehicles are parked between them.

Magneto-resistive sensors measure the X, Y, and Z-axis components of the Earth's magnetic field. As vehicles come within range of the sensor, changes in the surrounding magnetic field occur and are detected. Each sensor continually monitors the background magnetic field to establish a base reference, allowing self-calibration. Magneto-resistive sensors are lightweight and small making them versatile in placement. An in-depth analysis of magneto-resistive sensors has shown that they can be used efficiently in parking spot occupancy detection (Daubaras and Zilys, 2012). It has also been concluded that their efficiency drastically degrades due to the presence of large metal objects in the vicinity and when vehicles are in motion. Although their circuit complexity is comparable to the classical fluxgate magnetometers and they are both vectorial sensors (Marathay and McCalmont, 2001), magneto-resistors are much smaller in size and also consume less power.

Microwave / millimeter wave radar detectors operate by emitting a directed signal and then measuring energy reflected from target vehicles within the field of view. The technology is fairly mature due to its past use in military applications, and its cost is not high. The internal circuit of such a radar detector is quite small in size, and the high operation frequency enables the use of small antennas. In addition, to measuring motion velocity, a single sensor can detect multiple lanes with proper placement and signal processing. On the other hand, disadvantages include unwanted vehicle detection based on reception of side-lobe radiation and false detection due to multipath. Nonetheless, these disadvantages can be overcome with proper placement, signal processing algorithms, and efficient antenna design. The most important shortcoming, however, is the limited coverage range since shorter wavelength means shorter transmission range for a given power. Sensors for parking systems need to operate at low power and consequently their transmission ranges could even be less than 10 meters. Moreover, the atmosphere absorbs millimeter waves, restricting the range, and rain, fog, and any other moisture in the air also makes signal attenuation very high.

Weight-in-motion (WIM) sensors detect the vehicle weight. The four technologies used in WIM systems are: bending plate, piezoelectric, load cell, and capacitance mat. Each of these WIM sensor types has its own advantages and disadvantages. While bending plate WIM is more accurate and costly compared to piezoelectric WIM system, it is less accurate and considerably cheaper than load cell WIM systems. Although the capacitance mat has the advantage of being portable or fixed, it is the least accurate WIM system. For accuracy, the load cell would be preferable as it is the most accurate WIM system yet with the disadvantage of being the priciest. The cheapest WIM system in use is the piezoelectric system, with a weakness of sensitivity toward weather and speed variation. Its accuracy is behind load cell WIM and the bending plate WIM system. Both the piezoelectric and capacitance mat WIM system can be used to monitor up to four lanes. All the WIM systems require replacement or refurbishment every 3 to 5 years.

Cameras provide a much more complicated signal pattern than other sensors and they require image-processing techniques to extract the required information and determine occupancy. When a vehicle is parked at a spot, it is possible to perform edge-detection on the picture/video of the spot and successfully detect the presence of a vehicle (Al-Absi et al., 2010; Banerjee, Choudekar, and Muju, 2011; Lixia and Dalin, 2012). Edge density, light intensity, and color variations are the important variables that are used (Chen et al., 2010) along with vehicle classifiers and regions of interest (Al-Absi et al., 2010) to efficiently detect spot occupancy. Due to the complexity of the signal pattern received from cameras, several different models (Deng, Jiang, and Wei, 2006) and algorithms (Lixia and Dalin, 2012) have been developed for parking spot occupancy detection. However, camera-based systems have several problems that decrease their efficiency. Basically, they have to cater to color, light, and occlusion changes of parking grounds (Chen et al., 2010) and have to resort to additional validation mechanisms, such as aerial monitoring using micro aerial vehicles (Huang et al., 2015) or to other additional hardware-based sensors in order to increase accuracy. Moreover, the data storage, processing, and communication costs of this type of occupancy detection sensors are considerably higher than those of other techniques. Despite being comparatively less efficient than alternatives, cameras can certainly provide additional functions such as license plate recognition and can thus enable vehicle identification and automated billing/charging of parking services (Du and Sun, 2012).

While discussing cameras, it is important to highlight emerging parking management solutions that leverage new driver-aiding technology found on some of the new vehicles. Basically, car manufacturers Infiniti and Nissan pioneered the "Around View Monitor"

(AVM), which is a support technology that assists drivers to park more easily by better understanding the vehicle's surroundings through a virtual bird's eye view from above the vehicle. AVM helps the driver visually confirm the vehicle's position relative to the line marks of parking spaces and adjacent objects, allowing the driver to maneuver into parking spots with more ease. The AVM processes video from four cameras, displaying the composite footage on the screen as if there were a single bird's eye view camera right above the vehicle. Video signals from four cameras feed into an image-processing unit: the side views show both left and right, the rear view showing the back of the vehicle, and a front view camera. These signals are then converted into a composite virtual bird's eye top view using image processing, as if seen from directly above the vehicle. A version of this technology is also developed by Toyota called Bird's eye View Monitor and by other big car manufacturers as well. Figure 17.6 illustrates the capabilities of AVM.

A novel parking slot detection and tracking system is proposed in Suhr and Jung, (2014), where the sensor readings of an AVM system are fused with an ultrasonic sensor–based automatic parking system to detect vacant slots. The system continuously detects parking slot markings and classifies their occupancies. Simultaneously, it presents the detection and classification results from the AVM images to help the driver identify available slots. Once a desired parking slot is selected, the system tracks the position of the selected slot, while the vehicle is moving into it. The parking maneuver is finished when the vehicle is located at the target position.

RFID is popularly being used in vehicle identification and payment at electronic toll collection points, and is hence a very viable option in parking lots as well. Vehicles installed with passive RFID tags can be identified at the entrance of parking lots that have RFID readers installed (Du and Sun, 2012; Pala and Inanc, 2007). An active RFID tag is primarily dependent on the lifetime of the battery and hence needs ultralow-power components and a design mechanism that allows to completely turn it off or put it into standby mode. Such active RFID hardware can be very helpful in multiple vehicle

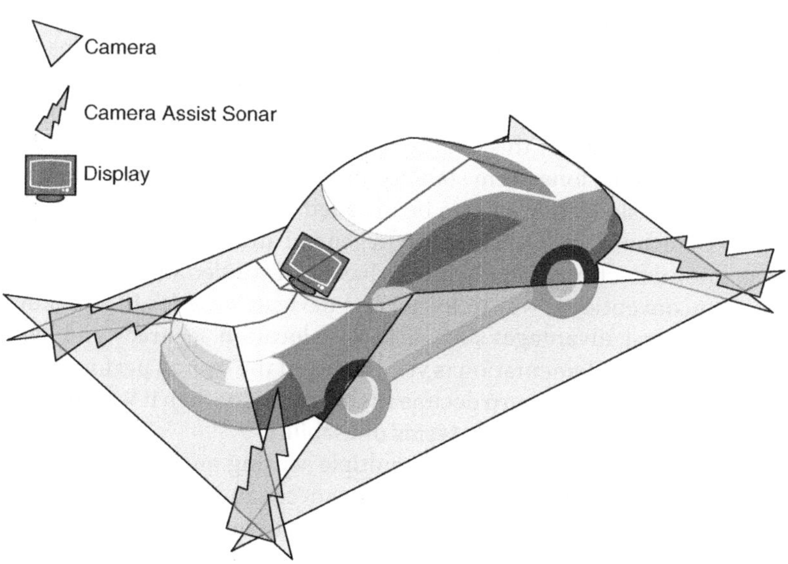

Figure 17.6 Illustrating the "around view monitor" support technology.

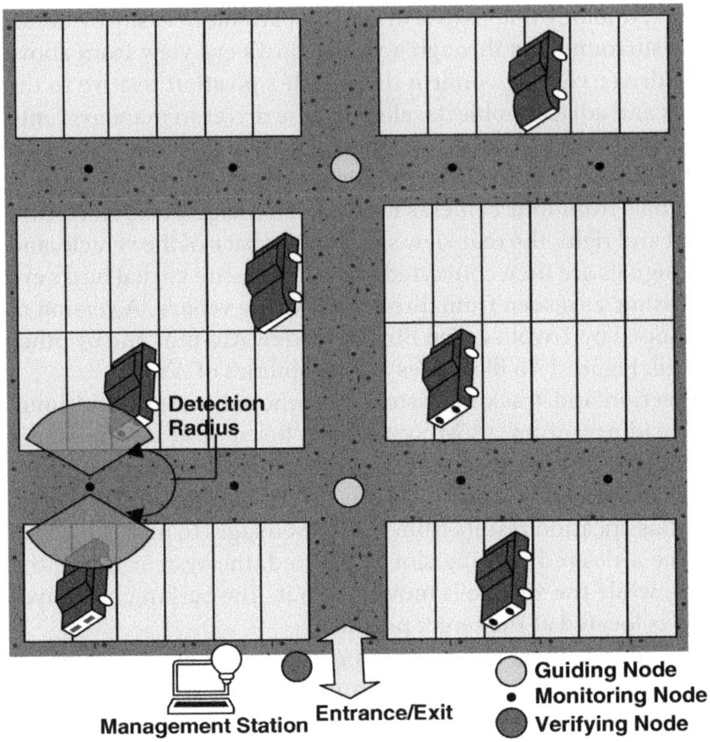

Figure 17.7 The RFID setup for parked vehicle detection, as proposed by Rahman et al. (2009).

identification (Huang, Huang, and Sun, 2011); with strategically placed receivers and an RIFD system with location detection features, it is possible to have a system with accurate occupancy detection (Rahman, Park, and Kim, 2009), as illustrated in Figure 17.7.

Fiber-bragg Sensors (FBG) have been explored in Prasad et al. (2012) as a means for detecting parking spot vacancy. These FBG sensors, which are embedded underground, determine the strain changes due to the load applied by the vehicle in the parking space. Several advantages can be noted about this technology such as electromagnetic interference immunity, remote sensing, long-term stability, miniature size, etc. Nonetheless, their deployment is expensive since they have to be installed under the pavement.

Ultra-wideband (UWB) technology is uncommon in general and absolutely rare to be used in parking occupancy detection; nonetheless, it has been studied in Zhong, Guo, and Zeng (2010) as an unconventional way to locate and navigate vehicles in a parking lot. This technology has several advantages such as precise location accuracy and high data rate, yet the initial cost of implementation is very high and the overall performance of UWB in all its applications has had a sharp decline as compared to when it was initially introduced, causing it to be near extinction in terms of usability.

It is important to note that some work employs multiple sensing modalities to effectively reduce the impact of interferences from different sources and increase accuracy. For example, Barone et al. (2014) attach RFID tags to vehicles to help in identifying the vehicles, and use an inductive magnetic loop to help detect occupancy. An illustration is provided in Figure 17.8. The RFID tags in their system can be transferred between

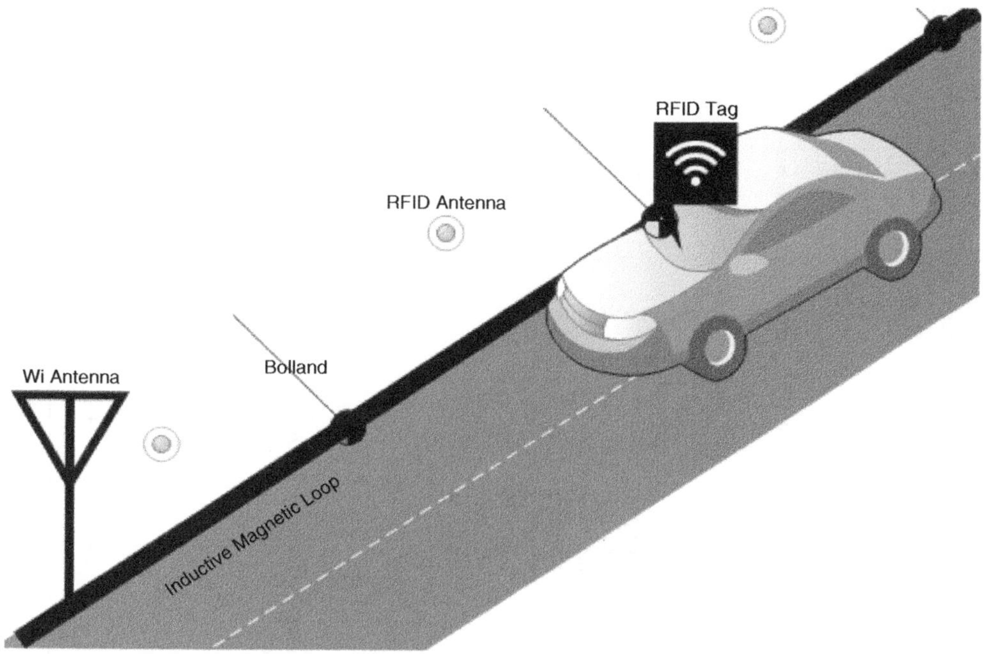

Figure 17.8 Multi-modal sensing architecture for parked vehicle detection. Picture courtesy of Barone et al. (2014).

vehicles and also help in identifying the vehicle when reserving a spot. Stenneth et al. (2012) detect the time and location of the parked vehicle by using GPS, accelerometer, and Bluetooth in conjunction with geospatial data. Basically, a change in transportation mode implies that the driver has left the vehicle, and the last reported location and time reflect where and when the vehicle is parked. To elaborate, when a person switches from a driving mode to walking mode, it becomes evident that the person parked their vehicle, and based on the GPS location and geospatial data, the parked spot status is inferred.

Huang et al. (2015) present a hybrid system where all parking spots are populated with QR codes and near-field communication (NFC) tags. QR codes are machine readable codes consisting of an array of black and white squares that have become as popular as bar codes in recent years. An onboard camera of a micro aerial vehicle (MAV) or drone is used to detect occupancy by reading the individual QR Codes or NFC tags on each spot. In the proposed system the MAV would patrol the parking lot along predefined gridlines and read the QR code at each spot and detect occupancy using the on-board camera. Since a QR code is difficult to read in low light conditions, NFC tags are installed as backup. It is not clear though why QR codes are needed if the NFC tags suffice. The authors additionally propose using low-cost web-cameras connected to Raspberry Pi controller boards on each spot for vehicle license plate recognition and spot monitoring, which seems a little overkill in terms of installation and maintenance costs for large parking lots.

Table 17.1 provides a comparative summary of the various sensors covered in this subsection.

Table 17.1 Comparative Summary of Different Sensors for Parking Management.

Sensor	Camera	Ultrasonic	RFID	Passive infrared	Passive acoustic array	Microwave radar	Weigh-in-motion	Piezo-electric cables	Pneumatic road tubes	Magneto-resistive	Magneto-meter	Inductive loop	Active infrared
Coverage	Multiple detection	Multiple Detection	–	Multiple detection	Multiple detection	Multiple detection	Multiple detection	Need multiple sensors to detect single presence	Single	Single	Need Multiple sensors to detect single	Single	Multiple spots
Most Accurate detection when	Stationary	Both	Stationary	Mobile	Mobile	Mobile	Mobile	Mobile	Mobile	Stationary	Stationary	Stationary	Stationary
Placement	Non-intrusive	Non-intrusive	Non-intrusive	Non-intrusive	Non-intrusive	Non-intrusive	Intrusive	Intrusive	Intrusive	Intrusive	Intrusive	Intrusive	Intrusive
Biggest Disadvantage	Highly sensitive to light and environment changes	Degraded occupancy measurement with vehicles traveling at moderate to high speeds due to large pulse repetition period. Temp change & extreme air turbulence affect performance	Vehicles have to equipped	Sensitive to environmental conditions	Sensitive to cold temperatures, cannot accurately detect slow moving traffic	Expensive. inefficient vehicle detection		Need multiple sensors to detect single. Extremely sensitive to high temp and traffic stress	Extremely sensitive to temperature. Prone to inaccuracies during increased traffic		Small detection zones. Need multiple sensors for single detection	Expensive and frequent maintenance. Susceptible to wear and tear due to temperature and traffic stress.	Sensitivity to environmental conditions. Need periodic lens cleaning.
Required Communication Bandwidth	Modrate to high	Low	Moderate	Low to moderate	Low to moderate	Moderate	Low to moderate	Moderate	Low to moderate		Low	Low to moderate	Low to moderate
Acquisition/ deployment cost	High	Low		Medium to high	Moderate	Moderate					Moderate	Low	Medium to high
From Factor Small size	Yes	Yes	Yes	Yes	Yes	Yes	No	No	No		Yes	No	Yes
Accuracy of detection	++	+++	+++	++	+		+++	+++	+++	+++	+++	+++	++

a) Estimated Costs per sensing unit for a single spot. Not including installation or maintenance. Low: up to \$1,000. Moderate: \$1,000 to \$8,000. High: \$8,000 to \$30,000.
b) According to data collected in (US-DOT, 2007) 2007 Study, US Dept. of Transportation.
c) +: less accurate, ++ : moderately accurate, +++ : highly accurate.
The range of communication bandwidth and cost shown for particular sensor technology reflects bandwidth usage and cost differences among specific sensor models and capabilities.

We discussed different types of sensors and sensing technologies that are currently available. Each one on its own may not be the most accurate in detecting the occupancy status at a parking spot, but using a combination of them can effectively reduce the shortcomings of each other. While installation and setup of multiple sensor devices can be fairly less complex, the cost of the entire hardware infrastructure in the system grows significantly. Moreover, the logistical complexity involving selective replacement of those sensors that run out of energy or fail is a major disadvantage. Being constrained in energy supply makes these sensors incapable of performing computation-heavy processing, and hence they become heavily dependent on the communication infrastructure to pass along the raw data to the sink nodes or servers. Intrusive sensors as the ones classified in Table 17.1 require cutting the pavement for installation and maintenance, indirectly decreasing pavement life, not to mention inconvenience and traffic obstruction.

On the flip side of having hardware sensors is using participatory sensing (Burke et al., 2006) which is a another interesting methodology for tracking the status of parking spots. Basically, smart phones these days have become such an integral part of the daily routine that it is almost impossible to imagine a day without the need of these devices. They come equipped with so many different types of sensors that if used efficiently, they can open up a world of possibilities. It is safe to assume that, even today, every person who owns a vehicle and would want to use a parking facility definitely owns a smart portable device. Thus, a parking system that uses only smart phones can successfully eliminate the need for any additional hardware for managing parking. Participatory sensing, more broadly knows as crowdsourcing, is a new sensing paradigm where a community of participants is vested in the outcome of a particular research and discovery task. Basically, participants use their smart devices to help in parking management. Although getting rid of the hardware infrastructure completely at the data collection end is certainly an advantage, starting to design a participatory sensing campaign brings out its inherent advantages and disadvantages fairly quickly. One of the biggest advantages of participatory sensing is that it is an almost negligible cost for getting the data than deploying a whole hardware sensing infrastructure. On the other hand, getting data from human participation leads to a lot of credibility questions and concerns. Recruiting participants is a major concern for such a system and consequently the design has to be such that the potential advantages of using the system are more appealing than the efforts and time spent in interpreting and providing the occupancy information to the system. It is intuitive that it is essential to have some form of user feedback, incentives, and engagement promotion mechanism. Also, without proper means of authentication and validation, there is a good chance of several security vulnerabilities. A detailed discussion of participatory sensing–based parking management will be provided in Section 17.4.

17.3.2 Communication Infrastructure

The vehicle detection infrastructure needs to be internetworked in order to communicate the sensed data between different sensors as well as servers and user interfaces. The communication support can be categorized into: (a) short range, such as low-power Wi-Fi, Bluetooth, infrared, NFC, ultraband, Zigbee, or 802.11ah, where signals can travel from several centimeters to 100 meters or so, and (b) long range, such as LTE, NB-Fi, Sigfox, or LoRa, where the signals can travel from a few kilometers to a few thousand kilometers.

Many of the commercially available sensor nodes nowadays such as magnetometers, RFID, infrared, and WSN motes are wireless and hence have to be energy efficient. The data volume and transmission power are important factors to consider while lowering energy costs from the communication point of view. WSN motes have their own network layer for communication among themselves (Bi et al., 2006; Chen and Chang, 2011; Zheng, et al., 2006). Their software support is also energy efficient and provides an event-based mechanism to collect or transmit data. The system architecture may employ multi-hop communication to disseminate data to a base station, have gateway nodes that act as intermediaries (Mainetti et al., 2015; Pham et al., 2015), or simply rely on direct links between sensors to a base station (Benson et al., 2006). As illustrated in Figure 17.9, the gateway nodes are usually shared between multiple co-located sensor nodes, whereas a base station is a single unit collecting data from all sensor nodes; in both cases they pass along sensed data to the storage and application infrastructure.

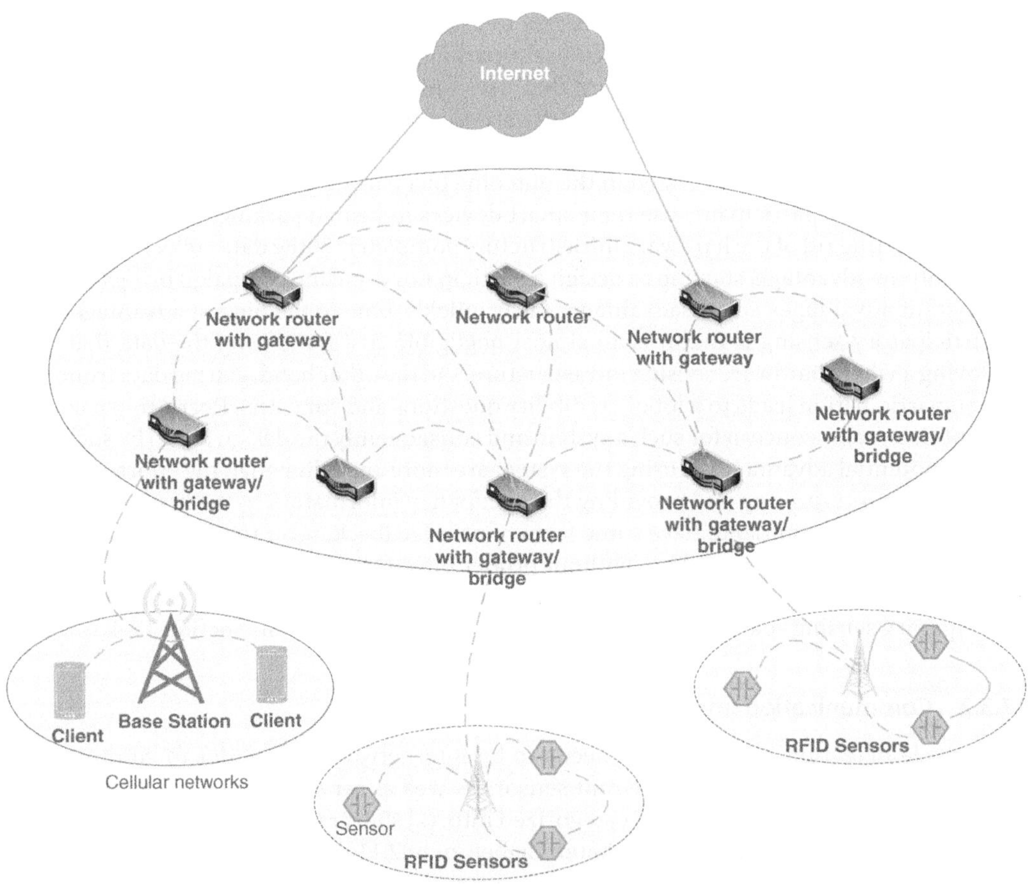

Figure 17.9 An articulation of communication infrastructure with gateways nodes from car-park-network architecture proposed in Pham et al. (2015).

Overall, internetworking sensor nodes over wireless links is subject to the conventional challenges in terms of signal interference, multipath, etc. Basically, sensor nodes often use short range RF-based communication, which is easily affected by the surroundings such as walls and other obstructions causing shadowing, reflection, scattering, fading, refraction, diffraction, distance-based signal attenuation, etc. Using long-range communication to transfer sensing information from the nodes is a viable option only if the sensing nodes are not limited on their power consumption, since data transmission over long distances consumes a significant amount of power. Hence, the option of transmitting sensing data in a sensing infrastructure over long ranges was rarely exercised. Nonetheless, this is certainly necessary and viable in participatory sensing since data is collected using mobile phones that are equipped with long-range transmission capabilities and are also rechargeable. For this reason, a dynamic infrastructure-less mechanism such as participatory sensing to determine spot occupancy is a very appealing option that gives the necessary results with negligible installation costs.

A slightly unconventional way of handling the communication aspect of the system is based on group communication in peer-to-peer (P2P) networking arrangement. Such a networking architecture suits large number of sensors that are distributed across a wide area either in a huge parking lot or in multiple parking facilities, e.g., the long-term parking lots in a major airport. For example, Cervantes et al. (2007) propose a design that combines P2P and conventional IP-based networks. The communication architecture consists of overlay group of peer nodes serving as event monitors and service connectors that act as a bridge between the P2P and conventional networks. The RFID-based parking sensors use P2P group communication, and the data is further disseminated to a data center and then to users through conventional IP-based networks. Figure 17.10 describes the system. The simulation results show an improved network performance in terms of reduced number of hops and round-trip delay. While this is a simplistic attempt at designing a hybrid networking architecture, it shows a good initiative that combines the strengths of both P2P networks and conventional IP network in the context of parking management, while also lowering the infrastructure cost significantly.

Another possible internetworking architecture is based on vehicular networking (VANET; Yousefi, Mousavi, and Fathy, 2006). Basically, vehicles are equipped with communication and processing modules, which make them a viable means for collecting, processing, and transmitting data. Most VANET systems need an on-board unit (OBU) on each vehicle to connect it with others. VANETs have been leveraged in some parking management solutions for large parking lots (Yan et al., 2011; Lu et al., 2009). In a VANET-based parking system, it is important to exchange information in such a way that each vehicle only receives relevant information. Delot et al. (2013) have developed a protocol that efficiently disseminates relevant parking information among the vehicles. They have introduced the concept of events in the context of parking management where nodes report on these events with other VANET members that might deem these events useful. For example, a vehicle leaving a parking spot could announce that a spot is becoming vacant to some vehicles driving around in the vicinity in order to aid drivers who are looking for a spot to park. By using a reservation protocol with such selective information sharing, it has been demonstrated that spots can be efficiently allocated with minimum communication bandwidth utilization.

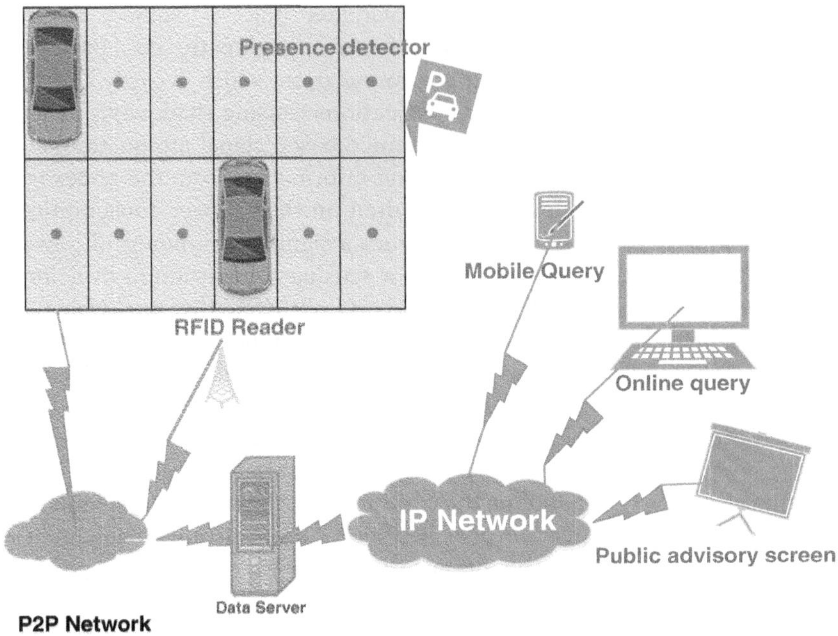

Figure 17.10 Implementation scenario of a hybrid network architecture for the parking systems of Cervantes et al. (2007).

17.3.3 Storage Infrastructure

A storage infrastructure has several important roles. It acts as the intermediary between the sensing and the application infrastructure, allowing the application infrastructure to get a consolidated view of the sensed data. We classify storage infrastructure for parking systems into three major categories: local, centralized, and distributed. In a local infrastructure, parking occupancy data and the application framework can be stored locally, close to a parking lot and isolated from other lots. This is the most inefficient system, since the occupancy information can only be obtained if one drives to the particular lot, and an interface would display the occupancy at the entrance. Figure 17.11 shows an example of such a local system.

In a centralized system multiple parking lots submit their occupancy data to a single server, and the occupancy data can be consolidated and viewed as a whole using an interface such as a smart parking app or road signs. On the other hand, local parking systems behaving as a disjoint entity that only displays occupancy at the entrances are eventually going to be obsolete since they are not as useful as those with centralized servers. It is more appealing, useful, and time saving to determine and possibly reserve a slot in advance by looking at the occupancy for all parking lots in an area than to hop around from one location to the other looking for vacancy. A centralized environment, especially cloud-based services is highly scalable and fairly cost effective in this regard; it also can support reservation quite easily.

Ideally a driver intending to park in a certain locality does not need to know the occupancy at parking lots located further away, implying all occupancy data need not be stored and accessed centrally. Nonetheless, parking systems with distributed storage

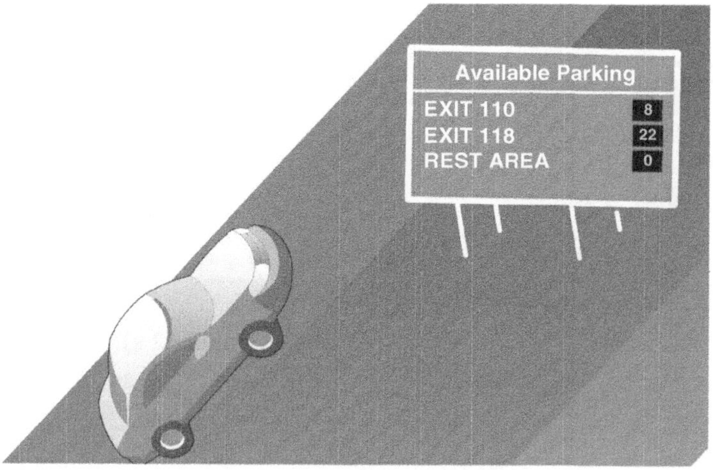

Figure 17.11 A local system may be just a road sign showing occupancy for a specific parking lot.

architecture are not very common. Ganchev, O'Droma, and Meere, (2008) propose a distributed system, which, although it employs a centralized server, also has the data distributed and cached at different locations for fast and easy access. Yamashita and Takami (2014) on the other hand propose a completely distributed and autonomous system based on VANETs by forming clusters of vehicles that store and pass along occupancy information for a parking lot. All vehicles in a parking lot are members of a VANET cluster that has an autonomously assigned cluster head, which is usually the closest vehicle to the entrance of the lot. When a vehicle comes in the vicinity, it would communicate with the cluster heads of all the neighboring parking lots, and then determine where to park. The role of a cluster head is transferred to another when the vehicle leaves the lot. Figure 17.12 illustrates the system of architecture proposed in Yamashita and Takami (2014).

17.3.4 Application Infrastructure

A comprehensive and efficient application infrastructure is essential to deal with all the parking resources and driver requests. In most parking solutions, a framework or architecture for bringing different technologies together is usually presented, and different applications or modules are proposed for spot reservation, payment, guidance, etc. Projects for large cities are often pioneered by big companies and often propose a system balancing the trade-offs between business advantages, efficiency, and reliability.

It is important to analyze the characteristics of the spatiotemporal information of parking spaces to improve the efficiency of the system. The first role of the application infrastructure is to infer a spot vacancy prediction model from collected data so as to provide drivers understandable and simple information. Although calculating spot occupancy and passing it on to the user interface is fairly straightforward (Bi et al., 2006; Chen and Chang, 2011), the difficult and more important problem to target is what occupancy information to disseminate to which driver such that the traffic flow is reduced and drivers do not have to go around to find parking or compete for the same spots. This also involves optimizing the spot location closest to a driver's final destination.

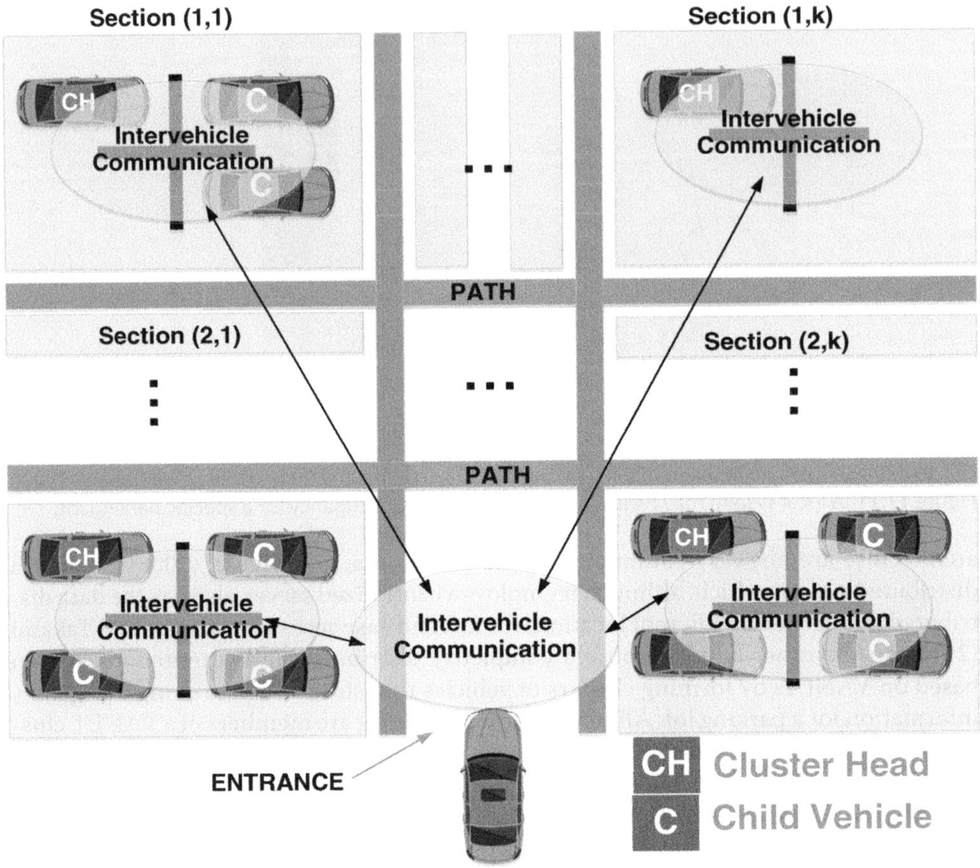

Figure 17.12 An example distributed-VANET architecture as suggested by Yamashita et al. (2014).

Moreover, parking competition as well as flow rate must also be addressed to understand how moving traffic and driver decisions can affect the performance of the system.

The parking application can also predict vehicle's arrival and departure using stochastic processes, e.g., a Markov model (Antoniou, Koutsopoulos, and Yannis, 2007), by estimating traffic on roads and tracking any associated flights (Roy, Levy, and Tomlin, 2006), i.e., when someone is to park and fly. In Klappenecker, Lee, and Welch, (2014), a system is proposed where the management infrastructure at a gated parking lot must send minimal occupancy data over a VANET so that a vehicle that is about to arrive in the vicinity and intending to park can efficiently predict future occupancy using a continuous-time Markov model and make a decision about where to park. A probabilistic traffic forecasting system such as the one presented by Horvitz et al. (2012) can also be used along with smart information dissemination (Delot and Ilarri, 2011) to help drivers make informed decisions on where to park and whether to choose a different time to travel.

Parking occupancy data can be analyzed and used in several important areas for the development of a smart city. In addition, studying historic data is essential to improve the

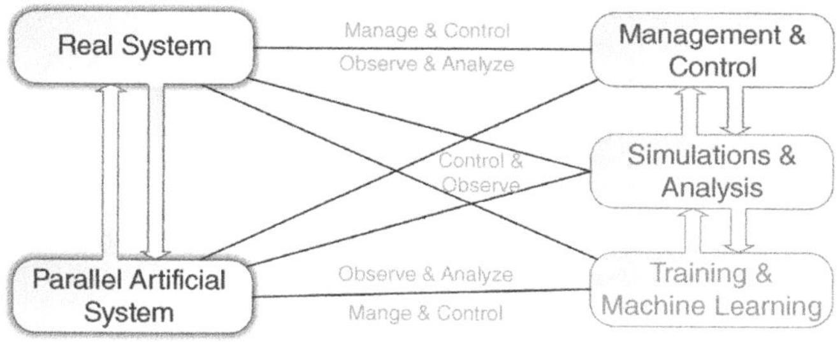

Figure 17.13 A functional diagram of the framework of parallel system of Wang et al. (2016).

system efficiency and parking policy. Parking is a problem related to land use, vehicular traffic, city's revenue, and human flow. The statistics of parking data can indicate key factors for urban development from technical, economic, and environmental perspectives. An application framework must hence be able to perform deep analysis and learning from the collected data. Wang et al. (2016) propose a parallel theory–based approach that can be fairly effective in targeting all the above-mentioned needs. According to their proposed approach, there must be an artificial system in parallel to the actual system that would constantly analyze and explore different behaviors and situations generated from the actual system and create artificial scenarios and experiments from actual data, which could help in the future for making proactive decisions. Figure 17.13 shows a functional diagram and highlights the interaction among the modules. Such a system can be a very effective mechanism to evaluate, validate, and improve the performance of complex or high-occupancy parking lots.

17.3.5 User Interfacing

Although the application models and software in existing parking solutions are quite diverse, the interfaces are kind of similar with respect to usability. Overall, the system either displays a set of parking spaces in aggregate and highlights possible available spot locations or shows the best available spot to drive to. Simpler systems with less elaborate infrastructure display a count of available empty spots at the entrance of a lot (Bi et al., 2006; Chen and Chang, 2011; Srikanth et al., 2009) for the driver to decide whether to enter the lot or go to another. Several parking systems also have an interface for fees and payments, which is either integrated with the application or as a separate payment station located in the parking area.

We can classify parking guidance approaches in parking systems into three categories: traditional blind search, system-assisted parking, and reservation-based parking. In the traditional blind search, a parking system may at the most display that there are some vacant spots in the lot, and the driver must search for a spot while driving around in the lot. Figure 17.14 shows example interfaces for mobile apps. Such a method certainly provides limited services. In the system-assisted parking approach, the driver is informed about where there is vacancy either at the entrance or through a mobile app. The location could either be a lane of parking spaces, hence narrowing down the search area, or it could be a precise set of vacant spots displayed for the driver to choose from. To enhance this further, adding a navigation module is fairly straightforward in applications

(a) (b)

Figure 17.14 User interfaces for smart phone app, (a) "Park Me" (ParkMeApp, 2016) and (b) "SF Park" (SFMTA, 2016).

for most modern smart devices, which makes it appealing and easy for the driver to get driving directions to the available spot from the interface (Du and Sun, 2012; Souissi et al., 2011; Wenzhi and Bai, 2006; Yoo et al., 2008; Zheng, Cao, et al., 2006). Although this approach is fair at guiding the users, it could lead to confusion. Basically, multiple users could potentially drive toward the same spot and find it occupied on arrival. Thus, merely displaying spot availability or even allowing the driver to navigate to the location (Wei et al., 2012), is not enough. A user hopes for a parking spot closest to her final destination, before they arrive in the vicinity of the destination.

Hence, a simple solution would be the third category, namely, a reservation-based system, where drivers are allowed to reserve a spot (Delot et al., 2013; Pham et al., 2015) from the web, a smart-app, or by a phone call, before they arrive in order to avoid the hassles of driving around looking for vacant spots. This approach has several drawbacks as well. For example, when a spot is reserved, other drivers cannot use it although it is not occupied, and stays vacant till the reserved vehicle is parked there. Given the impact of reservation on the system throughput, many concerns arise in terms of what should be the timeout of the reservation for cases such as when the driver is late to arrive or changes his plan, a better spot opens up closer to the destination, or how to handle a situation when a driver parks in a spot reserved by someone else, etc. Combining the system-assisted approach with an intelligent reservation system and smart information dissemination can potentially bring out the best from each system (Delot and Ilarri, 2011).

17.3.6 Comparison of Existing Parking Systems

A complete parking system for a smart city would be one that can incorporate all kinds of parking lots, including street parking. It must also successfully be able to inform real-time occupancy information to the drivers, who can then navigate their vehicles to an available spot without wasting time. In this subsection we present a comparison of the main features of elect parking systems that are unique by themselves and can successfully cover a majority of the aspects of a complete parking management system for a smart city.

As we can see in Table 17.2, almost all these systems can cover multiple parking lots and hence be able to not only inform the drivers if there is a vacancy in any of the incorporated lots; however, very few systems attempt to incorporate street parking as well. Moreover, of all the solutions covered in the table, only two of them can detect and disseminate occupancy information without the need of any additional hardware infrastructure. These two solutions, namely Yamashita and Takami (2014) and Delot et al. (2013), require VANET-based communication and processing, which implies the need for all vehicles to be equipped with the necessary hardware and software. Being distributed, these two solutions are also the only systems that do not need storage infrastructure to store and retrieve data. Those that do require hardware infrastructure for occupancy detection and dissemination require at least one type of sensor as well as communication support to exchange information between vehicles, infrastructure, and/or server.

Some of the systems in Table 17.2 have the additional and extremely useful facility of advance reservation of a spot; meanwhile some incorporate a navigation module to help the driver to reach to the reserved or selected spot. Analyzing and learning from the occupancy data can be very helpful to optimize allocation as well as for developing an efficient revenue model for parking fees, if needed. Some of the systems listed in Table 17.2 do attempt to incorporate that to make their system more effective.

17.4 Participatory Sensing–Based Smart Parking

In this section, we describe our own ongoing work on building a system that can overcome most of the issues mentioned in the previous sections (Gupte and Younis, 2015). Basically, we leverage smart IoT devices to accurately determine parking availability. We employ a participatory sensing paradigm to track the status of parking spots. The key advantage of this system is that it completely eliminates the need for hardware infrastructure for parking occupancy data collection. The system is also scalable without any limitations and able to handle all types of parking spots, including street parking. The participatory sensing–based framework we describe in this section displays real-time information about the parking availability and deals with the whole parking process.

Table 17.2 Comparison of Existing Parking Solutions.

Research Proposals	Focus			Interaction				Mode		Communication		Requirement				Provides navigation guidance to spot	Supports spot reservation	Data analysis and learning	
	Parking Lot		Street Parking	Vehicle-to-vehicle	Vehicle-to-infrastructure	Infrastructure-to-infrastructure	Client-Server	Push	Pull	WAN	Ad hoc	Support Infrastructure	Monitoring Sensors					Considers historic data	Considers driver behavior and decisions
	Single Lot	Multiple Lots											At entrance	1 sensor per spot	1 sensor multiple spots				
(Ganchev, O'Droma and Meere, 2008)		*				*			*		*	*		*		*	*		
(Yamashita and Takami, 2014)		*		*				*	*		*					*			
(Klappenecker, Lee and Welch, 2014)		*		*	*				*		*	*	*						
(SFMTA, 2016)		*	*			*	*			*		*		*	*	*	*	*	
(Wang et al., 2016)		*	*			*	*			*		*		*	*			*	*
(Ferreira et al., 2014)	*			*	*				*		*	*				*			
(Delot et al., 2013)		*	*	*				*			*						*		
(Mainetti et al., 2015)		*	*			*	*		*	*		*		*	*	*			
(Pham et al., 2015)		*				*	*	*		*		*		*		*			
(Barone et al., 2014)		*				*	*		*	*		*		*			*		*
(Yan et al., 2011)		*			*	*	*	*		*		*		*	*	*	*		
(Rico et al., 2013)		*	*			*	*	*		*		*		*		*	*		*
(Lu et al., 2009)		*			*	*	*	*			*	*	*		*	*			

A key challenge with a participatory model for information collection is that people could be less motivated to forward information and provide their input when the cost or effort is significant. Moreover, identifying trustworthy data is a challenge. We address the issue of recruiting participants using an incentive-based model, and we regulate data quality with a reputation-scoring module.

17.4.1 The Components

Our framework consists of four major components that seamlessly interact with each other to make up the complete parking system, as explained below.

17.4.1.1 Users

The users of the system have two primary roles, namely, requesters and reporters. Requesters look for available spots to park their vehicle. One of their important duties is to mark their parking spots as used on arrival and unused upon departure. Meanwhile, reporters inform the system about the occupancy of parking spots when they are in the vicinity of these spots. As described below, reporters can also passively participate by allowing their IoT devices to report the occupancy data. The users also allow their social media accounts to connect with the application to extract relationship data, which would be further used in building the leaderboards, as discussed later in this section.

17.4.1.2 IoT Devices

Motivating users to contribute toward a participatory sensing paradigm is a fairly big challenge. If users can provide contributions indirectly and autonomously by allowing their devices to become active reporters, that would boost participation and improve efficiency. For example the parking systems described in Stenneth et al. (2012) and Villanueva et al. (2015) use the available sensors on smart phones to report spot occupancy data without active involvement of the users themselves. In Villanueva et al. (2015) the built-in magnetometer is used to detect whether a vehicle is present in its surroundings, whereas in Stenneth et al. (2012) the accelerometer, Bluetooth, and GPS sensors on the driver's phone are employed to determine if the vehicle is parked and to report the occupancy of the used spot based on the GPS coordinates. In addition, the phone camera or the smart eyewear, e.g., Google glasses, can be utilized. Basically, image-processing techniques, accelerometer readings, and geolocation can be applied to detect the presence of a vehicle in a parking spot. The GPS, magnetometer, and accelerometer can precisely give the location and direction of the smart phone, while the images captured from the camera can be processed, as they do for camera-based sensing, and hence detect occupancy. Moreover, autonomously driven cars are expected to become a common feature on the roads; these cars are equipped with a lot of sensing capabilities, such as a magnetometer to provide a sense of direction, ultrasonic sensors for obstacle detection, and the AVM technology we discussed earlier. Suhr and Jung (2014) utilize the AVM and ultrasonic sensors that are already installed on certain modern vehicles to detect occupancy. The AVM image sequences are used to detect parking spot marks while the ultrasonic sensors assess occupancy in the detected spot area. Using theses autonomous sensing capabilities as means for detecting occupancy and as participatory devices is certainly a viable and promising option.

17.4.1.3 Server

The sensed data is collected at a server, which represents the backbone of the whole system and interfaces the users, IoT devices, and the parking resources. This server is cloud based, having a data-store and an application engine and providing a platform-as-a-service facility to perform the required computations. It may be hosted and managed by the organization owning or managing the parking area. Figure 17.15 shows the functional organization of our application hosted on the server. The server maintains the processes and communicates the information about the occupancy status of the parking spots. It triggers alerts and updates based on occupancy monitoring, manages the reservations, and assesses compliance to parking regulations. The server manages the registration of users, devices, and vehicles, as well as managing the accounting and billing. The user requests information from and reports data to the server. Similarly, the IoT devices autonomously report their data to the server. The server is also responsible for calculating the user reputation scores and rewards, as discussed later in detail. Using the received data and the reputation framework, the server calculates the spot occupancy.

17.4.1.4 Parking Spots

A parking spot is a static entity that has no active role of its own, but it is the main resource that the requester and reporter are concerned with. A parking spot has its

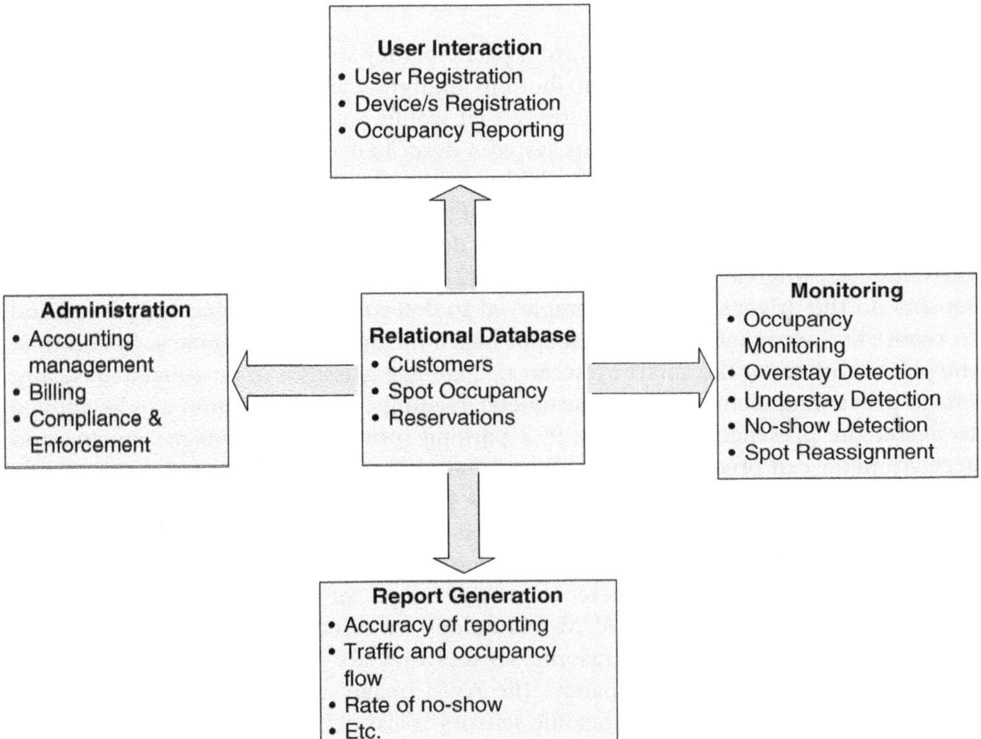

Figure 17.15 Functional organization of the parking management application.

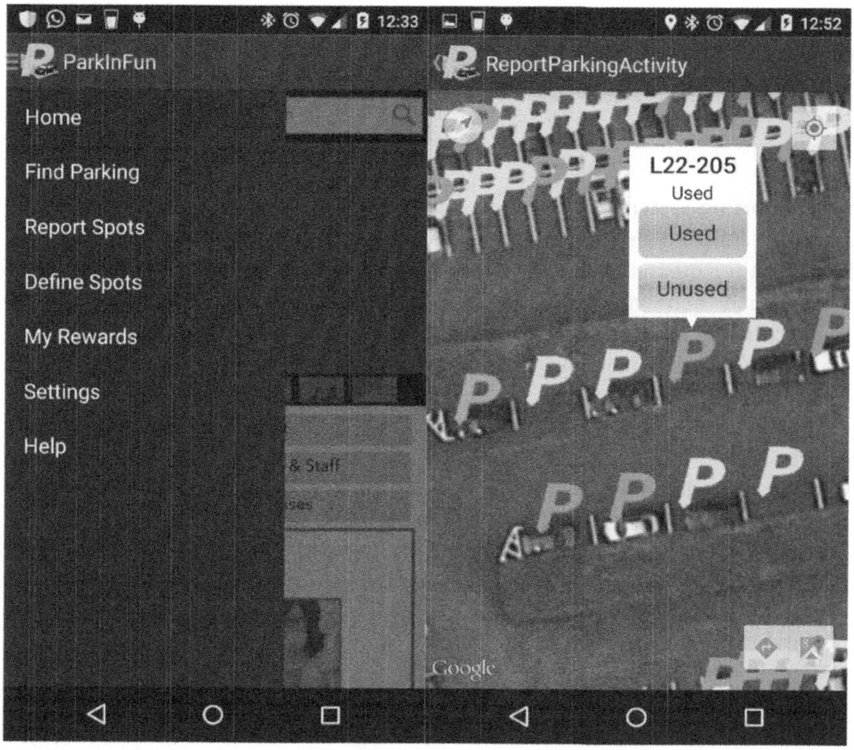

Figure 17.16 Screenshots of our initial parking Android app.

availability status, which could toggle between unknown, out of service, reserved, occupied, and available. This status can be changed by the server based on the participatory reporting received from users and other detection devices.

17.4.2 Parking Management Application

The software architecture of our parking system consists of six major interlinked modules, as explained below.

17.4.2.1 User Interface

The goal of a parking system is not only to facilitate finding a parking spot but also to make the whole user experience pleasant. We improve the user experience by proposing an easy to use smart phone–based application that is integrated with other IoT devices and has the latest advancements of payment solutions as well as convenient ticket payment options. Screenshots of our initial Android app are shown in Figure 17.16.

Upon logging into the application, a user can choose between the following two roles:

- Requester: A requester is someone who has arrived or is arriving at a location where her next move is to find a vacant parking spot. Requesters generally are looking for a spot near their current location or near a location that they are expecting to arrive at in a stipulated time. The expectation of a requester is to get different options to choose from based on the availability, proximity, cost, etc.

- Reporter: To minimize the potential of inaccurate assessment of slot occupancy, a user can report a spot status only while being in close proximity to the spot, e.g., based on the GPS sensors on the device. The user is rewarded with "bonus points" based on the reporting. Such a bonus goes toward his rewards as well as leaderboard, as explained below.

17.4.2.2 Smart Reporting System

Besides visually ascertaining a spot status and marking it occupied/available in the app, users can also use other means such as IoT devices introduced above to autonomously or semi-autonomously achieve the same goal. The rewards and reputation points received or deducted during the process of this reporting is also received by the user who owns the IoT device. The basic operation of the reporting mechanism is described in Figure 17.17.

17.4.2.3 Leaderboard

A leaderboard that would show how one is doing compared to others, provides more motivation to contribute. The app displays a leaderboard showing the user's rank in her social circle as well as in her geographic vicinity. For example, people can be ranked per locality, city, state, country, etc. The leaderboard system can be integrated with a reward store, where there would be weekly and monthly rankings, i.e., the points would be reset

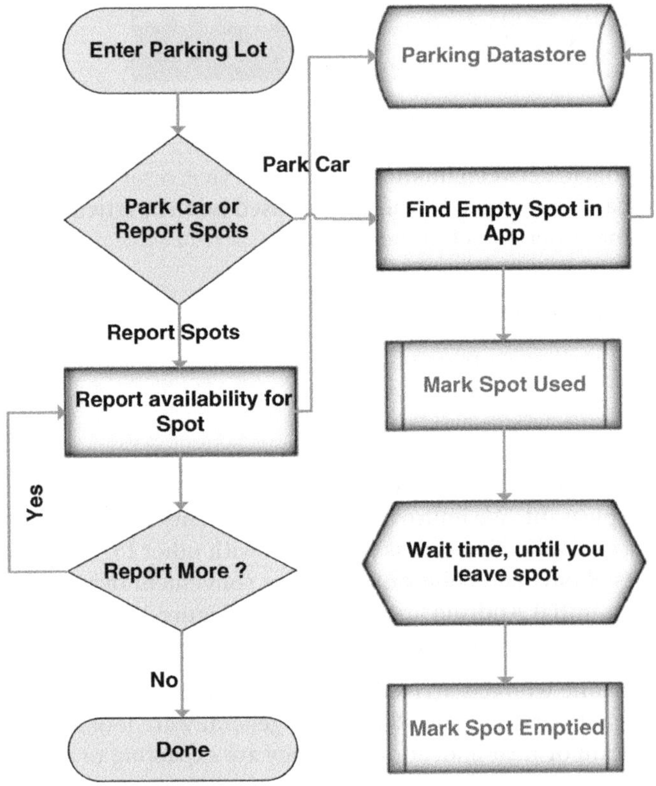

Figure 17.17 Flowchart description of the basic spot-reporting operation.

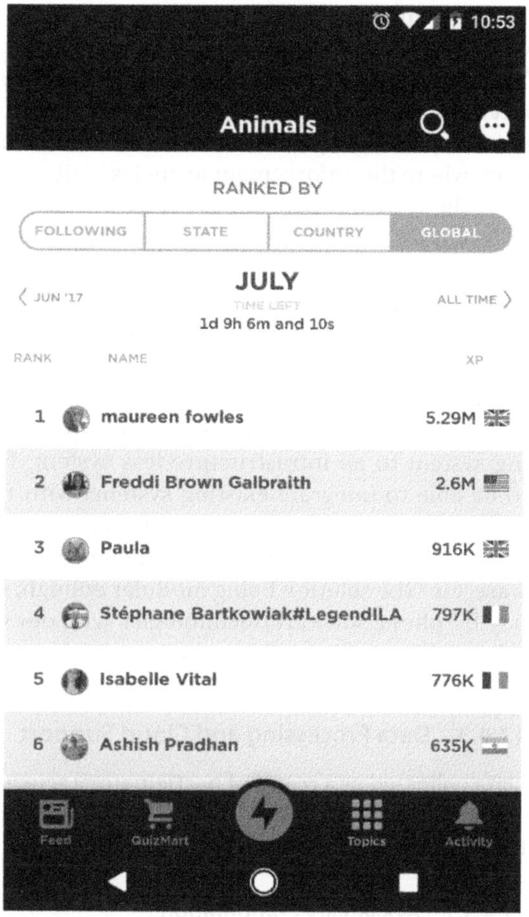

Figure 17.18 Sample screenshot of leaderboard on the QuizUp Android game.

after every week/month. The highest ranked person on the leaderboard by the end of the period could redeem the accumulated points from the rewards store. Figure 17.18 shows a simple leaderboard used in QuizUp, a competitive quiz game that shows how well one is doing compared to other people in a particular geographic vicinity. This leaderboard also shows the monthly points tally as well as all-time total score.

17.4.2.4 Rewards Store

Users collect points for accurate reporting. These points can be redeemed for various rewards such as:

1) dedicated parking spot for X amount of time (shows "reserved" on the spot in the app);
2) get monetary credit of some sort, depending on the rewards store of that parking facility;
3) reserve an available parking spot for 10 mins (before you arrive or as soon as you arrive); or
4) park without restrictions for a day, e.g., get a joker tag.

17.4.2.5 Enforcement and Compliance

Our fully integrated solution can achieve deterrence goals by collecting compliance data. Such data can be used for issuing tickets to violating vehicles. With user involvement, interaction, and feedback on availability and usage of parking spots, it is possible to enforce parking regulations and detect violations. Our system provides an interactive map, where the enforcement agencies/units would be able to view illegally parked vehicles. The system is also able to create precise queries for enforcement activities and add notes for informational use, e.g., parking is not allowed at specific time of the day.

17.4.2.6 External Integration

New technologies are being introduced almost daily to improve revenue, reduce costs, and make parking more convenient. Moreover, there is a large investment in hardware sensor-based parking systems, and they have been actively deployed and are being used all over. One cannot suddenly switch from such a hardware-based occupancy sensing system to an infrastructure-less system. For a smooth transition, it is important to be able to integrate existing systems with the futuristic framework. Therefore, the application must be modular and could be integrated with the available technology such as handheld devices, RFIDs, meters, revenue control systems, sensors, billing software, etc. The solution being modular enough, it would be possible to integrate multiple pay-by-phone software/technologies with our solution, voiding the requirement of significant investment in that area.

17.4.3 Data Processing and Cloud Support

Data collected as a result of participatory sensing is the base for parking management in our system. In this subsection we describe how the data is validated and processed and how users get rewarded for collecting the data.

17.4.3.1 Availability Computation

The parking spot availability is computed based on the reporting by multiple users using the following formula:

$$SA_x = \sum_{U_{used}} Rep_{Score} - \sum_{U_{unused}} Rep_{Score}$$

Where, SA_x reflects availability of parking spot x, U_{used} and U_{unused} indicate a user reporting "used" and "unused," respectively, and Rep_{Score} is the user reputation score. If $SA_x <= -100$, then spot is confirmed "unused," and if $SA_x >= 100$, then the spot is confirmed "used," and unless it goes beyond either of these values, its status remains as is. If reputation scores of users are low, then it would take more users to verify the availability of a spot, which automatically minimizes malicious user reporting. In the same way, users with high reputation scores help in getting accurate availability quickly since their reporting needs to be verified with a lower number of reporters. The steps are explained in the flowchart in Figure 17.19.

Each parking spot has a predefined timeout duration, after the end of which the spot status is set based on majoring vote, even if it has not been confirmed to be occupied or unoccupied. The user(s) who have contributed for most votes of occupancy get partial credit in their reputation. For example, if users A, B, and C voted "occupied" and D

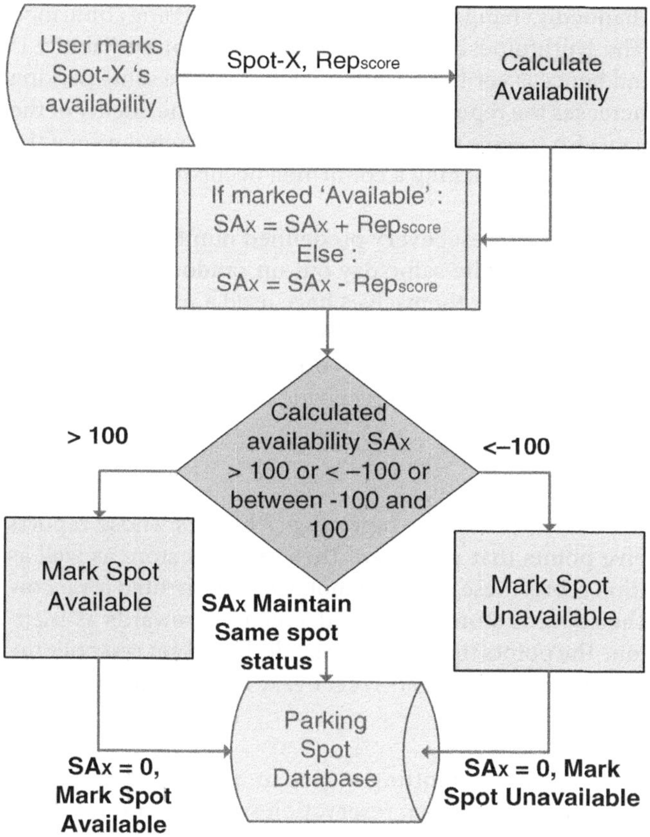

Figure 17.19 Flowchart for spot availability computation.

reported "unoccupied," at the end of the timeout, even if the total score did not exceed 100, the spot would be marked as occupied, and users A, B, and C receive partial increment in their reputation score for the reporting. The spot reporting timeout is computed based on learning and depends on the frequency of user reporting. For example, if a parking lot during peak hours receives frequent reporting and the occupancy status can be toggled and confirmed on an average of 10 minutes, then setting the timeout period to 12 minutes would be fair. Similarly, for off-peak hours, tweaking the timeout to 30 minutes, if there are not frequent changes in occupancy status, would make it more efficient. Our simulation results capture the effect of the timeout duration on the accuracy of the system.

17.4.3.2 Reputation System

Our system includes a reputation-tracking mechanism that minimizes malicious reporting and helps in giving accurate spot availability. Every user is assigned a reputation score from 0 to 99, where 99 reflects a user with the best reputation. Such a score quantifies the truthfulness of a reporter. Every time the sensed data is found to be inaccurate, the reputation score of the reporter gets decremented and vice versa. Changing the status of a spot requires a total score that exceeds 100. Since a user can never have a reputation score of 100, no one can single-handedly change the spot status without being confirmed by at least one other person. The truthfulness of a user's report on spot availability is based on the aggregate score and hence is verified by other user(s) for the same parking spot. Every correct reporting increases the reputation score and boosts the fidelity of the spot status assessment. A user can also receive partial credit toward her reputation if the vote collection for a spot times out before reaching a confirmed occupancy decision, as mentioned in the previous paragraph.

Reputation scores of individual users are reset every predefined number of days. The scores of the various users do not reset on the same day but on random days instead. Users are also encouraged to report that they themselves have used a parking spot and then report again on departure. Failure to report this information leads to decrease in the reputation score. The reputation score mechanism not only helps to weed out incorrect reporting but also prevents malicious users from receiving reporting points toward rewards and leaderboard.

17.4.3.3 Scoring System

As mentioned earlier, users also receive points for reporting. All users whose reports match the voted majority receive points that go toward their rewards store as well as leaderboard. Unlike the reputation score, these points do not get decremented for incorrect reporting. The points in the rewards store can be redeemed for rewards as mentioned in the previous subsection. The points that go on the leaderboard get reset per the rules of the leaderboard, e.g., a daily leaderboard gets reset every night at midnight, etc.

17.4.3.4 Reservation Model

Reserving a parking spot is not as trivial as allotting a spot to a user and making it unavailable to others. To have a smooth and efficient reservation system, our framework supports the enforcement of business rules. The rules could be to adopt policy on tardiness, no-show, and noncompliance. Moreover, the reservation protocol also considers the estimated time of arrival and projected availability to reserve the spot for higher efficiency and throughput. This would also maximize the probability that the spot would still be vacant when the driver arrives.

17.4.3.5 Analysis and Learning

Every parking operation generates a huge amount of data, but it is difficult to analyze parking data to improve productivity. We apply data mining and machine learning for analysis. We analyze and study the data as it comes, to provide users with an experience that gets better every day and we can evolve our solution based on the study on not just traffic trends but also ethnography.

17.4.4 Implementation and Performance Evaluation

17.4.4.1 Prototype Application

Our initial prototype application is built for Android devices. In the application, users can view the parking spot availability and navigate to it. Users can also report the availability of spots if they are in close proximity. The app calculates the spot availability based on the logic explained in the previous subsection. An approved user also can define parking lots and populate a lot with spots, which also becomes a participatory way of adding a new parking facility into our system. We use the Google cloud platform to host our app and store the data, as well as run the processing and analysis. Figure 17.16 shows a sample screenshots.

17.4.4.2 Experiment Setup

To assess the accuracy of the system, we simulate several users with smart devices that would provide participatory data. We also use some real users with real devices to provide the data. We compare the true spot availability information with the calculated spot availability by our framework. We consider a real-world scenario of the parking lot 22 at UMBC, which has 480 parking spots. We use the real parking availability information of total number of cars occupying a lot at a given time of day for a range of days, as provided by the UMBC Parking Services. We run each of the simulation for Monday to Thursday. In the simulation, the following assumptions are made. A person's reputation score is reset every 7 days, and the parking spot availability is reset to "unknown" every midnight. The average number of reporting users follows Poisson distribution with $\lambda = 60$.

Since we do not use an artificially intelligent agent to simulate human behavior in reporting the spot availability, we use a fixed truthfulness level. Say, if for a particular simulation we use truthfulness of 90, then a particular simulated user would report the true availability 90% of the times. Simulated user's truthfulness not only helps in varying their reputation score but also in avoiding frequent changes in parking spot status. A user earns reputation points for correct reports, loses points for incorrect reports, and gets partial reputation points for benefit of doubt in case the spot times out. In addition, a user earns reputation points for reporting that he parked his car on the spot and loses points on failure to report that he vacated his parking spot. For our simulation, we assume that users would earn 1 point for reporting their parking spot and 1 more when confirmed by other user(s) and similarly lose 2 points for not reporting. Unless explicitly mentioned, the initial reputation scores of a simulated user is randomly picked between 35 and 99.

17.4.4.3 Simulation Results

Figure 17.20 shows the results based on the average of four days. We observe that as the day progresses, our system reports accurate status for 83–90% of the spots, most of the remaining spots are displayed as "unknown," and very few are reported inaccurately. Since there are fewer users reporting early in the day, it is difficult to verify the availability, which results in more inaccurate reports. That drastically tapers down after 10:00 a.m., as the number of participants increases. It is worth noting that in practice the inaccurate status would not be a concern since it is observed only during times of low demand, and hence many vacant spots are available for parking.

We have studied the effect of the average number of participants. As indicated by the results in Figure 17.21, the number of users does matter but only up to a certain extent, after which the accuracy does not considerably improve with the increase in reporting user count. Such an observation is attributed to how occupancy is assessed using the reputation scores. If a smaller number of users can successfully confirm the occupancy status, then even if more users report after that, it would not matter. For example, assume that four users voted that a particular spot is occupied, and according to our calculation the spot got confirmed to be "occupied"; now having two more reports within the timeout will not contribute toward any improvement.

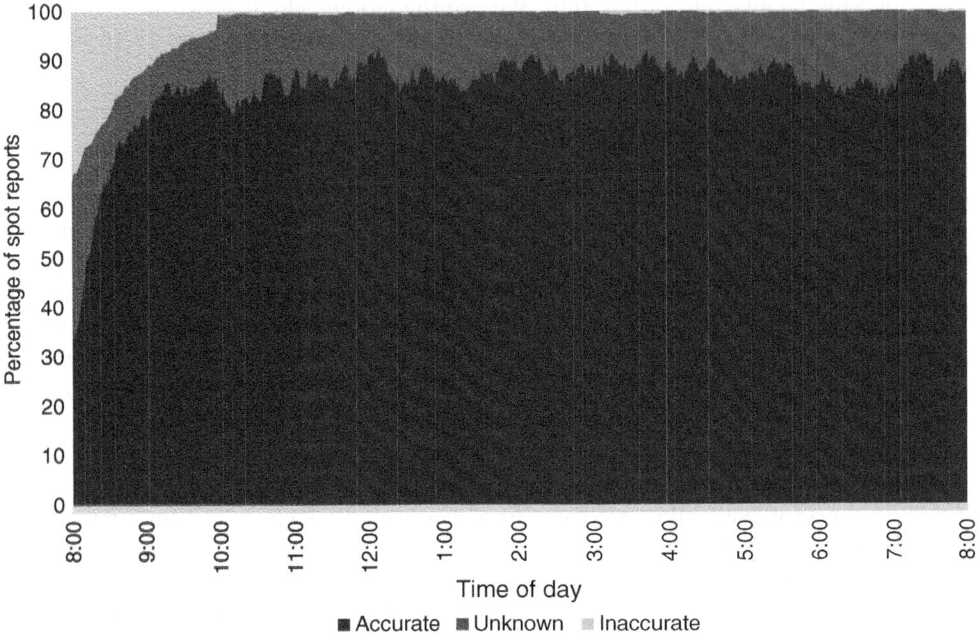

Figure 17.20 Average accuracy of system.

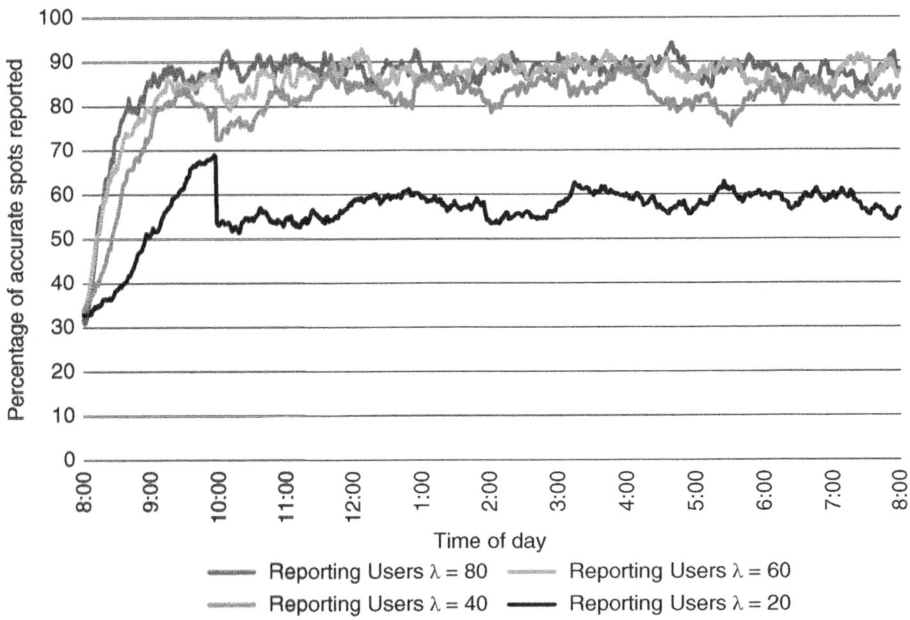

Figure 17.21 Number of accurate spot status under varying number of reports.

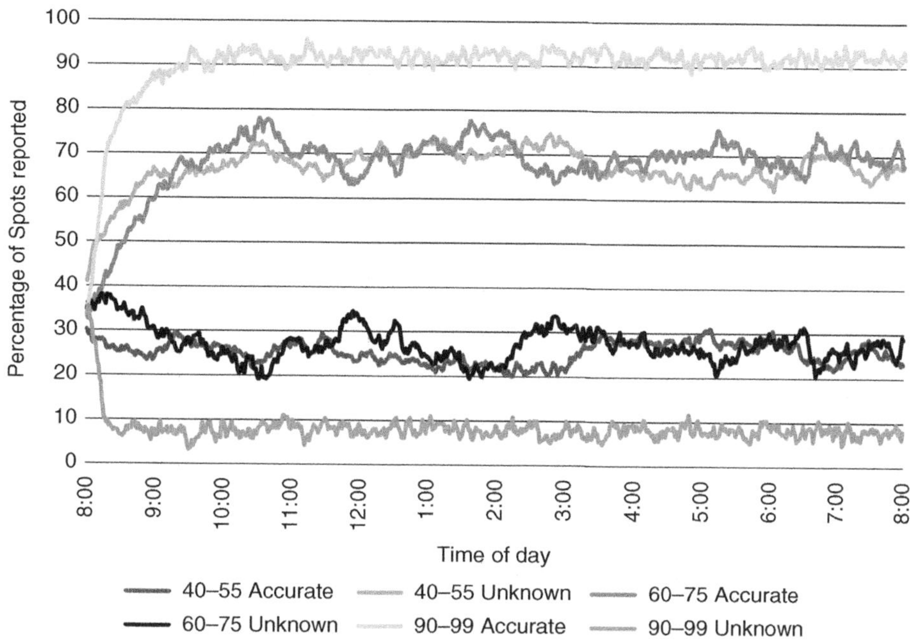

Figure 17.22 Effect of reputation scores on number of spots with accurate and unknown status.

We have also varied the average reputation score range of all the reporting users. In a real scenario, it is impossible for all users to have a low reputation score or all to have a high reputation score, but our results in Figure 17.22 show that even with all users having scores between 40 and 55, our system would still display accurate availability for 53–58% of the spots, while the rest are displayed as "unknown." Figure 17.23, on the other hand, shows the effect of varying the truthfulness range of users, starting with 40–55% of users reporting true values, to 90–99% of users reporting true values. The figure indicates that if users are reporting the spots honestly, it will lead to accurate spot availability; however, reporting false status will only result in more "unknown" values and does not lead to inaccurate spot status. This provides significant insight about how users' honesty would matter in our system. Using the reputation system and applying location constraints in reporting the spot status, we assure a high level of true participatory reporting.

17.4.5 Features and Benefits

One of the greatest benefits of our system is that it eliminates the need for any additional hardware or sensing infrastructure. This in turn makes the system setup quick, lightweight, and completely non-intrusive and non-obstructive. It does not need to disturb or change anything physically when initializing or removing the system for a parking lot, which helps in scaling easily and quickly. This also aids in supporting all kinds of parking lots and structures including street parking and multi-level parking. The spot occupancy detection accuracy does not depend on physical or environmental factors; instead it depends on human participatory engagement. As our experiments show, we can achieve near real-time spot occupancy detection, with minimal engagement. Our

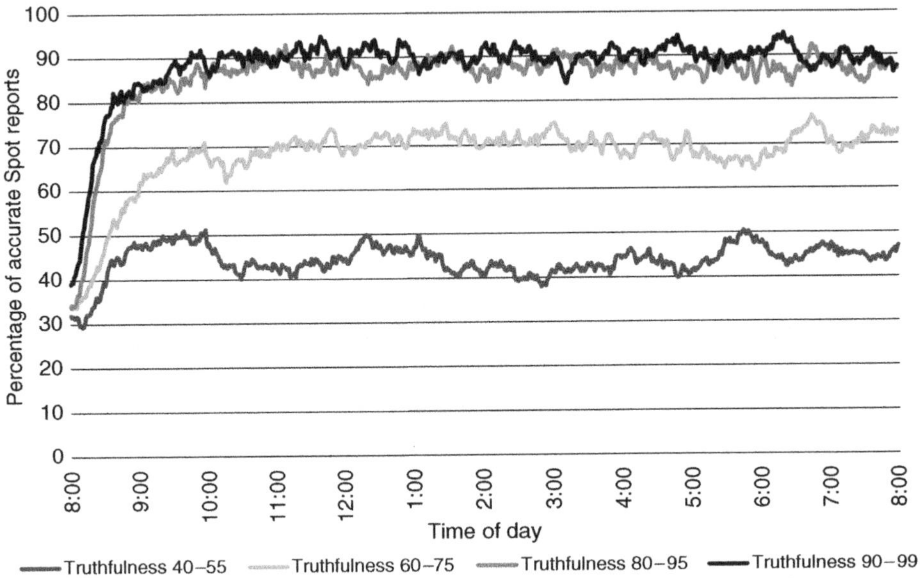

Figure 17.23 Number of accurate spot reports with varying user truthfulness.

experiments in the previous section cover all scenarios, but we assume participation from users to be proportionate to the density of occupied spots and need at the time of the hour, meaning, high-occupancy parking lots will have greater participation and hence, more accurate and precise occupancy displayed. Similarly, low-occupancy lots will have proportionately less participation but will hence have high availability of spots, so they do not need as much participation or accuracy.

Since we use smart phones for sensing as well as data transfer, the power consumption or energy efficiency does not need to be considered as a factor in the system. Moreover, the communication infrastructure for the parking system is the same as the normal network for data exchange in telecommunication networks; therefore, factors such as transmission range, transmission efficiency, and signal interference are not variables in the system. The only infrastructure needed is a storage infrastructure to store global occupancy data, which we currently envision as being centralized. This does not bring it to a disadvantage since almost all parking systems need hardware infrastructure to store occupancy data. But this too can be possibly made distributed and be eliminated completely, as we allude in the next section. Of the parking guidance approaches described in Section 17.3.5, our application interface supports system-assisted parking as well as reservation-based parking, hence providing easy navigation to unoccupied spot. During the system-assisted navigation the system temporarily changes the status of the intended spot to "reserved" hence avoiding multiple drivers racing for the same spot.

Our system can be easily integrated with other systems and modules such as payment regulation, ticket generation, etc. It can also be operated in parallel with existing sensor systems hence allowing a smooth transition from older systems, while validating the accuracy with the sensor-provided data. Hence, we optimize the use of available parking facilities, make the commuters aware of the available locations, keep track of parking offences, and speed up the process of finding a spot.

17.5 Conclusions and Future Advancements

Parking is certainly a global urban problem and one of the major concerns in a smart city. With parking information collected from sensing infrastructure or through participatory sensing, smart-parking services can provide a good platform to mitigate if not completely resolve the pain of finding a parking spot. A smart parking management system will enable drivers to shorten their travel time, and improve revenues as well as efficient utilization of the parking space. In this chapter, we have provided a general classification of parking systems. We have surveyed existing solutions and highlighted the most important aspects, advantages, and shortcomings. We have categorized regular parking systems based on the sensing infrastructure, the communication infrastructure, the storage infrastructure, application infrastructure, and user interfacing. We have also provided a brief analysis of some parking systems that we deem as unique and promising. Our survey has pointed out that deployment and maintenance of hardware sensors is a major cost burden. Moreover, to obtain a higher level of accuracy in these systems, it is also needed to employ multiple sensor modalities which require even higher expenditure. To overcome the technical issues and high costs associated with such hardware sensor-based occupancy assessment, we have promoted a participatory sensing-enabled parking management system.

We have presented a participatory sensing framework for tracking parking spot availability. Our framework employs an efficient reputation-scoring mechanism to mitigate potential inaccurate and unknown parking spot statuses. In order to boost participation, we employ an incentive- and leaderboard-based parking model. Our framework can also enable improved enforcement of parking regulation and prevent unauthorized access, as well as address any spillover problems that may occur. It is also possible to integrate fee collection, ticket generation and payment, etc. We have validated the effectiveness of our approach with real and simulated user participation. The results demonstrate that it is possible to accurately determine parking spot availability, even with nominal level of user participation. The results also show that our reputation-scoring mechanism successfully prevents malicious or careless users from undermining the system operation.

While our system is a significant step ahead of existing solutions in terms of increased capabilities and reduced cost, the current implementation relies on cloud-based storage and processing services. Realizing our framework using distributed rather than cloud-based setup and integrating VANET-based information dissemination and processing would make the system more versatile and scalable. In addition, we plan on creating leaderboards using social networks to obtain relationship information for higher user engagement. Moreover, leveraging the social ties between drivers from social networks to display who contributed toward the data would also improve the efficiency of the system while strengthening drivers' trust in the participatory data. Stored occupancy data can be used for prediction, learning and, data analysis, to improve efficiency of the system as well as guide parking policies and regulate parking rates based on traffic flow. Moreover, using smart information dissemination along with historic occupancy data can help drivers to make informed decisions on where to park or what time to travel or whether to travel to that locality. Finally, large-scale testing of the system in a populous urban setup is important for performance tuning and factoring in the behavior of drivers in real scenarios.

References

Al-Absi, H.R., Devaraj, J.D.D., Sebastian, P., and Voon, Y.V., 2010, 'Vision-based automated parking system', *In*: 10th International Conference on Information Sciences Signal Processing and their Applications (ISSPA), 2010, IEEE, 757–760.

Antoniou, C., Koutsopoulos, H.N., and Yannis, G., 2007, 'Traffic state prediction using Markov chain models', *In: Proceedings of the European Control Conference*, 2–5.

Bachani, M., Qureshi, U.M., and Shaikh, F.K., 2016, 'Performance Analysis of Proximity and Light Sensors for Smart Parking', [online] 83, 385–392. URL: http://www.sciencedirect.com/science/article/pii/S1877050916302332.

Banerjee, S., Choudekar, P., and Muju, M., 2011, 'Real time car parking system using image processing', *In*: 3rd International Conference on Electronics Computer Technology (ICECT), 2011, IEEE, 99–103.

Barone, R.E., Giuffrè, T., Siniscalchi, S.M., Morgano, M.A., and Tesoriere, G., 2014, 'Architecture for parking management in smart cities', *IET Intelligent Transport Systems*, 8(5), 445–452.

Benson, J.P., O'Donovan, T., O'Sullivan, P., Roedig, U., Sreenan, C., Barton, J., Murphy, A., and O'Flynn, B., 2006, 'Car-park management using wireless sensor networks', *In: Proceedings. 2006 31st IEEE Conference on Local Computer Networks*, IEEE, 588–595.

Bi, Y., Sun, L., Zhu, H., Yan, T., and Luo, Z., 2006, 'A parking management system based on wireless sensor network', *Acta Automatica Sinica* 32(6), 968.

Boda, V.K., Nasipuri, A., and Howitt, I., 2007, 'Design considerations for a wireless sensor network for locating parking spaces', *In: Proceedings 2007 IEEE SoutheastCon*, IEEE, 698–703.

Burke, J.A., Estrin, D., Hansen, M., Parker, A., Ramanathan, N., Reddy, S., and Srivastava, M.B., 2006, 'Participatory sensing', *Center for Embedded Network Sensing*.

Cao, Z. and Kong, Q., 2011, 'A vehicle management system of community based on radio frequency identification technology', *In*: 3rd International Conference on Communication Software and Networks (ICCSN), 2011, IEEE, 341–343.

Cervantes, L.F., Lee, Y.-S., Yang, H., and Lee, J., 2007, 'A hybrid middleware for RFID-based parking management system using group communication in overlay networks', *In*: The 2007 International Conference on Intelligent Pervasive Computing, 2007, IPC, IEEE, 521–526.

Chen, L.C., Hsieh, J.W., Lai, W.R., Wu, C.X., and Chen, S.Y., 2010, 'Vision-Based Vehicle Surveillance and Parking Lot Management Using Multiple Cameras', *In*: 2010 Sixth International Conference on Intelligent Information Hiding and Multimedia Signal Processing (IIH-MSP), 631–634.

Chen, M. and Chang, T., 2011, 'A parking guidance and information system based on wireless sensor network', *In*: 2011 IEEE International Conference on Information and Automation (ICIA), IEEE, 601–605.

Chinrungrueng, J., Sunantachaikul, U., and Triamlumlerd, S., 2007, 'Smart Parking: An Application of Optical Wireless Sensor Network', *In*: International Symposium on Applications and the Internet Workshops, 2007. SAINT Workshops 2007,.66.

Cisco-Systems-Inc., 2016, *Cisco—Smart+Connected Parking:* URL: http://www.cisco.com/c/en/us/solutions/industries/smart-connected-communities/city-parking.html, Accessed Aug 2016.

Civicsmart-Inc, 2016, URL: *http://www.civicsmart.com/, Accessed Aug 2016.*

Daubaras, A. and Zilys, M., 2012, 'Vehicle detection based on magneto-resistive magnetic field sensor,' *Elektronika ir Elektrotechnika* 118(2), 27–32.

Delot, T., Ilarri, S., Lecomte, S., and Cenerario, N., 2013, 'Sharing with Caution: Managing Parking Spaces in Vehicular Networks', *Mobile Information Systems* 9(1):69–98.

Delot, T. and Ilarri, S., 2011, 'Data gathering in vehicular networks: The VESPA experience' (invited paper), *In*: Fifth IEEE Workshop On User MObility and VEhicular Networks (LCN ON-MOVE 2011), 801–808.

Deng, H., Jiang, D., and Wei, Y., 2006, 'Parking cell detection of multiple video features with PCA-and-Bayes-based classifier', *In*: 2006 IEEE International Conference on Information Acquisition, IEEE, 655–659.

Du, S. and Sun, S., 2012, 'The research and design of intellectual parking system based on RFID', *In*: 2012 9th International Conference on Fuzzy Systems and Knowledge Discovery (FSKD), IEEE, 2427–2430.

Ferreira, M., Damas, L., Conceiçao, H., d Orey, P.M., Fernandes, R., Steenkiste, P., and Gomes, P., 2014, 'Self-automated parking lots for autonomous vehicles based on vehicular ad hoc networking', *In*: 2014 IEEE Intelligent Vehicles Symposium Proceedings, IEEE, 472–479.

Ganchev, I., O'Droma, M., and Meere, D., 2008. 'Intelligent car parking locator service', *Information Technologies and Knowledge* 2, 166–173.

Gupte, S. and Younis, M., 2015, 'Participatory-sensing-enabled efficient Parking Management in modern cities', *In*: 2015 IEEE 40th Conference on Local Computer Networks (LCN), IEEE, 241–244.

Horvitz, E.J., Apacible, J., Sarin, R., and Liao, L., 2012, 'Prediction, expectation, and surprise: Methods, designs, and study of a deployed traffic forecasting service', arXiv preprint: 1207.1352.

Huang, C.-H., Hsu, H.-S., Wang, H.-R., Yang, T.-Y., and Huang, C.-M., 2015, 'Design and management of an intelligent parking lot system by multiple camera platforms', *In*: 2015 IEEE 12th International Conference on Networking, Sensing and Control (ICNSC), IEEE, 354–359.

Huang, Y., Huang, X., and Sun, Q., 2011, 'The Design of the management system for parking lot based on 2.45 GHz active RFID', *In*: 2011 International Conference on Computer Science and Service System (CSSS), IEEE, 850–853.

INRIX, 2016, URL: http://inrix.com/, Accessed Aug 2016.

ITS-America, 2016, 'ITS America—Smart City', URL: https://itsamerica2016.org/wp-content/uploads/2016/08/Making-Cities-Smarter.Alex-Israel.pdf, Accessed Aug 2016.

Kim, J., Kim, H., Jeong, J., Seo, Y., and Mah, P., 2010, 'Field Deployment of a Large-Scale WSN for Parking Management System', *In*: 2010 7th Annual IEEE Communications Society Conference on Sensor, Mesh and Ad Hoc Communications and Networks (SECON), 1–3.

Klappenecker, A., Lee, H., and Welch, J.L., 2014, 'Finding available parking spaces made easy', *Ad Hoc Networks* 12, 243–249.

Lixia, W., and Dalin, J., 2012, 'A method of parking space detection based on image segmentation and lbp', *In*: 2012 Fourth International Conference on Multimedia Information Networking and Security. IEEE, 229–232.

Lu, R., Lin, X., Zhu, H., and Shen, X., 2009, 'SPARK: a new VANET-based smart parking scheme for large parking lots', *In*: INFOCOM 2009, IEEE, 1413–1421.

Mainetti, L., Patrono, L., Stefanizzi, M.L., and Vergallo, R., 2015, 'A Smart Parking System based on IoT protocols and emerging enabling technologies', *In*: 2015 IEEE 2nd World Forum on Internet of Things (WF-IoT), IEEE, 764–769.

Marathay, A.S. and McCalmont, J.F., 2001, 'Vector diffraction theory for electromagnetic waves', *J. Opt. Soc. Am. A* [online] 18(10), 2585–2593, URL: http://josaa.osa.org/abstract.cfm?URI=josaa-18-10-2585.

Pala, Z. and Inanc, N., 2007, 'Smart parking applications using RFID technology', *In*: 1st Annual RFID Eurasia, 2007, IEEE, pp.1–3.

ParkMeApp, 2016, 'Park Me App', URL: https://www.parkme.com/, Accessed Dec 2016.

Pham, T.N., Tsai, M.-F., Nguyen, D.B., Dow, C.-R., and Deng, D.-J., 2015, 'A Cloud-Based Smart-Parking System Based on Internet-of-Things Technologies', *IEEE Access* 3, 1581–1591.

Prasad, A.G., Sharath, U., Amith, B., Supritha, B., Asokan, S., and Hegde, G., 2012, 'Fiber Bragg Grating sensor instrumentation for parking space occupancy management', *In*: 2012 International Conference on Optical Engineering (ICOE), IEEE, 1–4.

Rahman, M.S., Park, Y., and Kim, K.-D., 2009, 'Relative location estimation of vehicles in parking management system', *In*: 2009. 11th International Conference on Advanced Communication Technology, 2009, ICACT, IEEE, 729–732.

Rico, J., Sancho, J., Cendon, B., and Camus, M., 2013, 'Parking Easier by Using Context Information of a Smart City: Enabling Fast Search and Management of Parking Resources', *In*: 2013 27th International Conference on Advanced Information Networking and Applications Workshops (WAINA),1380–1385.

Roy, K., Levy, B., and Tomlin, C.J., 2006, 'Target tracking and estimated time of arrival (ETA) prediction for arrival aircraft', *In: Proceedings of AIAA Guidance, Navigation, and Control Conference, Keystone, CO.*

Scola, N., 2014, 'What Happened When a Designer Watched the Devices That Watch New York', URL: https://nextcity.org/daily/entry/new-york-city-cameras-watch-sensors.

SFMTA, 2016, '*SF Park*', http://sfpark.org/, Accessed Dec 2016.

Siemens-AG, 2016. 'Integrated Smart Parking Solution', URL: http://www.mobility.siemens.com/mobility/global/en/urban-mobility/road-solutions/integrated-smart-parking-solution/pages/integrated-smart-parking-solution.aspx, Accessed Aug 2016.

Souissi, R., Cheikhrouhou, O., Kammoun, I., and Abid, M., 2011, 'A parking management system using wireless sensor networks', *In: ICM 2011 Proceedings*, IEEE, 1–7.

Srikanth, S., Pramod, P., Dileep, K., Tapas, S., Patil, M.U., and others, 2009, 'Design and implementation of a prototype smart parking (spark) system using wireless sensor networks', *In*: International Conference on Advanced Information Networking and Applications Workshops, 2009. WAINA'09, IEEE, 401–406.

Stenneth, L., Wolfson, O., Xu, B., and Philip, S.Y., 2012, 'PhonePark: Street Parking Using Mobile Phones', *In: MDM*, 278–279.

Streetline-Inc, 2016. 'The StreetLine Solution', URL: http://www.streetline.com/, Accessed Aug 2016.

Suhr, J.K. and Jung, H.G., 2014, 'Sensor Fusion-Based Vacant Parking Slot Detection and Tracking', *IEEE Transactions on Intelligent Transportation Systems* 15(1), 21–36.

US-Dept-of-Transportation, 2016, 'DOT—Smart City Challenge', URL: https://www.transportation.gov/smartcity, Accessed Aug 2016.

US-Dept-of-Transportation-SmartCity, 2016, 'SmartCity Vision Statement', URL: Columbushttps://www.transportation.gov/smartcity/visionstatements/columbusoh, Accessed Aug 2016.

US-DOT, 2007, 'A summary of vehicle detection and Surveillance Technologies used in Intelligent Transportation Systems', URL: http://www.fhwa.dot.gov/ohim/tvtw/vdstits.htm.

Villanueva, F.J., Villa, D., Santofimia, M.J., Barba, J., and López, J.C., 2015, 'Crowdsensing smart city parking monitoring', *In*: 2015 IEEE 2nd World Forum on Internet of Things (WF-IoT), IEEE, 751–756.

Wang, F.-Y., Yang, L.-Q., Yang, J., Zhang, Y., Han, S., and Zhao, K., 2016, 'Urban intelligent parking system based on the parallel theory', *In*: 2016 International Conference on Computing, Networking and Communications (ICNC), IEEE, 1–5.

Wei, L., Wu, Q., Yang, M., Ding, W., Li, B., and Gao, R., 2012, 'Design and implementation of smart parking management system Based on RFID and internet', *In*: 2012 International Conference on Control Engineering and Communication Technology (ICCECT), IEEE, 17–20.

Wenzhi, C. and Bai, L., 2006, 'A smart roadside parking navigation system based on sensor networks for ITS', *In*: 2006 IET International Conference on Wireless, Mobile and Multimedia Networks, IET, 1–4.

Yamashita, S. and Takami, K., 2014, 'Autonomous, Distributed Parking Lot Vacancy Management Using Intervehicle Communication', *International Journal of Vehicular Technology*.

Yan, G., Yang, W., Rawat, D.B., and Olariu, S., 2011, 'SmartParking: A secure and intelligent parking system', *IEEE Intelligent Transportation Systems Magazine*, 3(1), 18–30.

Yoo, S., Chong, P.K., Kim, T., Kang, J., Kim, D., Shin, C., Sung, K., and Jang, B., 2008, 'Pgs: Parking guidance system based on wireless sensor network', *In*: ISWPC 2008. 3rd International Symposium on Wireless Pervasive Computing, 2008, IEEE, 218–222.

Yousefi, S., Mousavi, M.S., and Fathy, M., 2006, 'Vehicular ad hoc networks (VANETs): challenges and perspectives', *In*: 2006 6th International Conference on ITS Telecommunications, IEEE, 761–766.

Zheng, Y., Cao, J., and others, 2006, 'An intelligent car park management system based on wireless sensor networks', *In*: 2006 First International Symposium on Pervasive Computing and Applications, IEEE, 65–70.

Zhong, Y., Guo, W., and Zeng, L., 2010, 'Visual navigational method of parking management based on indoor precise real-time location', *In*: 2010 International Conference on Computer and Communication Technologies in Agriculture Engineering, IEEE, 227–229.

18

Electric Vehicle Scheduling and Charging in Smart Cities

Muhammmad Amjad, Mubashir Husain Rehmani and Tariq Umer

COMSATS Institute of Information Technology, Wah Cantt, Pakistan

18.1 Introduction

Urbanization has resulted in grave concerns such as need of good governance and people's quality of living. Therefore, the need of smart solutions to mitigate the bubbling consequences of urbanization are required. For this purpose, smart cities are envisioned as the acceptable and smarter solution to minimize the drastic impacts of urbanization (Celino & Kotoulas 2013). In smart cities, the advances in information and communications technologies (ICT) have been exploited to enhance the people's standard of living. The European Initiative on smart cities by the European Commission has focused on just the four main pillars for the smooth operation of smart cities, namely, infrastructure, heating and cooling, power, and transportation (Saint 2014).

As compared to vehicles that are based on internal combustion engines (ICE), the fuel-cell electric vehicle (FCEV), the hybrid electric vehicle (HEV), and the plug-in hybrid electric vehicle (PHEV) have gained much attention for future smart cities. PHEV or electric vehicles (EVs) can now be recharged at home and other available recharging locations such as parking lots (Clement-Nyns et al. 2009). EVs are recognized as the smart solution for smart transportation in smart cities. With the inclusion of EVs, the benefits of sustainable and green application and communications support in smart cities can be leveraged. The reduction in CO_2 and good integration into the potential smart grid has also made EVs as the acceptable solution for the future smart cities (Banister 2011). This smart integration of EVs into smart cities is assumed to improve the reliability of the smart grid and smart transportation systems. From the smart grid perspective, EVs can regulate the voltage and frequency and hence not only the power is managed but also EVs can reduce the power losses. For customers, the minimization of power losses and properly scheduling the recharging duration and time enhances their confidence (Wang et al. 2016).

Smart cities also promise a smart solution for transportation management. The proper traffic control for smart cities to avoid any traffic congestion and to ensure the safety for EVs is also needed (de Andrade et al. 2016). Usually, the automation of the transportation system, traffic-control through social signals, social traffic, and ITS clouds can be employed to manage the traffic in smart cities (Zhu et al. 2016). In addition to traffic control and management, various attempts have been made from the industry and academia to well integrate EVs into smart cities.

Transportation and Power Grid in Smart Cities: Communication Networks and Services, First Edition.
Edited by Hussein T. Mouftah, Melike Erol-Kantarci and Mubashir Husain Rehmani.

18.1.1 Integration of EVs into Smart Cities

The inclusion of EVs can consume 5% of the existing electric power resources. Therefore, the recharging of EVs should be optimally scheduled to avoid any bottlenecks in the existing power structures. EVs recharging in smart cities can pose various challenges such as:

18.1.1.1 Enhancing the Existing Power Capacity

The existing power structures have been designed to support domestic and commercial needs. To exploit the power resources for the electrification of the transportation can overload the existing power grid. This requires the need of expanding the capacity of the existing power resources to support a large number of EVs into the future smart cities (Majidpour et al. 2015).

18.1.1.2 Designing the Communication Protocols to Support the Smart Recharging Structure

The smart recharging of EVs require the smart architecture for optimal and controlled recharging of EVs in smart cities. An extensive effort is required both from the industry and academia to design the wired and wireless-based communication solutions for the smart recharging architecture of EVs (Chow et al. 2014, Nolte et al. 2009).

18.1.1.3 Development of a Well-designed Recharging Architecture

A considerable amount of energy can be conserved while optimally recharging EVs in smart cities. The optimal recharging of EVs require a sophisticated EVs recharging architecture. EVs recharging in smart cities requires a smart architecture that can make the smart decisions for efficiently recharging of EVs. EVs can be recharged by employing either the centralized or distributed recharging architecture. Various energy losses due to the voltage, frequency fluctuations, aggregator unit (AU) malfunctioning, interrupted power supply, and bottlenecks in the communications architecture can only be addressed when the recharging architecture (centralized or distributed) is properly monitored or optimally controlled.

18.1.1.4 Considering the Expected Load on the Smart Grid

The unchecked integration of EVs into the future smart cities can also overload the smart grid and cause serious events such as spinning reserves. This situation can also lead to the destabilization of the smart grid. Therefore, it requires the proper consideration and energy management while considering the potential integration of EVs into the future smart cities (Huang et al. 2014).

18.1.1.5 Need for Scheduling Approaches for EVs Recharging

With the help of well-designed scheduling algorithms, various objectives such as frequency regulation, cost minimization, and estimation of the potential usage of electricity can be achieved. EVs recharging in smart cities with proper recharging and scheduling approaches can contribute in stabilizing the smart grid(He et al. 2012).

The large number of EVs integration into the future smart cities is also accompanied by various security threats and data handling issues. However, the potential benefits of EVs can only be realized with a well-controlled and optimized EVs recharging process. An extensive work has been done on the design of the scheduling approaches to achieve a controlled EVs recharging. As compared to the controlled recharging of EVs with the scheduling algorithms, the uncontrolled EVs recharging can result in:

- Uneven demand spikes with various bottlenecks.
- spinning reserves. With the help of scheduling approaches the spinning reserves can be optimized; however, without the scheduling approaches, it can not be optimized.
- Fluctuations of the frequency and voltage of the power grid.
- The inability to estimate different power losses during the recharging of EVs and from distribution network.
- Difficulty in achieving reliability in the smart grid and for customers.

18.1.2 Main Contributions

In this chapter, we have provided the state-of-the-art work on EVs inclusion into smart cities and the various scheduling approaches used for EVs recharging. However, the main contributions of the chapter can be summarized as follows:

- We have provided the main recharging approaches such as centralized and distributed for recharging EVs in smart grid.
- We have also encompassed the different scheduling algorithms involved for recharging of EVs.
- The different scheduling approaches for EVs recharging are applied to gain certain objective. We have also discussed the various objectives achieved after scheduling the EVs recharging process in smart cities.
- We have also highlighted the different communications requirements and communications development for smart transportation and recharging of smart EVs in smart cities.
- Different resource allocation schemes that employ the various optimization approaches have also been discussed in detail.
- We have also provided the open issues and challenges, as well as future research directions.

18.1.3 Organization of the Chapter

The rest of the chapter is organized as follows: Section 18.2 provides the overview of smart cities and EVs. In Section 18.3 the centralized and distributed EVs recharging approaches in smart cities are presented, while in Section 18.4 EVs scheduling approaches with various objectives have been discussed in detail. In Section 18.5, the open issues and challenges, as well as future research directions have been presented. Finally, the conclusion is presented in Section 18.6.

18.2 Smart Cities and Electric Vehicles: Motivation, Background, and Application Scenarios

18.2.1 Smart Cities: An Overview

There is no formal definition of smart cities. However, the forces that shape smart cities can be used to define them. Those forces can be the utilization of the advances in ICT for future cities (Angelidou 2015). In addition to these, various communication and business architectural models also shape the smart city dream with an innovation economy (Mulligan & Olsson 2013). The evolution of existing cities to the dynamic infrastructure with the idea of energy efficiency, sustainability, and flexibility needs to consider the above-mentioned models (Pellicer et al. 2013). Below is the description of the few pillars that shape the future smart cities:

18.2.1.1 Provision of Smart Transportation

The idea of smart transportation with innovative technology has made the concept of smart cities more attractive and applicable. To make the transportation increasingly intelligent and smart, different smart solutions in smart cities have been incorporated (Wang 2015, Bianchini & Avila 2014). In-vehicle communication with different quality-of-service (QoS) parameters can help in shaping the transportation and making it intelligent and robust (Rahmani et al. 2009). This in-vehicle communication is utilized in smart cities in shaping the traffic. To properly manage and shape the traffic of EVs in smart cities, the uncertain behavior of EVs can be predicted through in-vehicle communications or via different scenario generation (Soares et al. 2016).

In addition to shape the traffic for EVs, smart cities also produce efficient and well-managed parking for EVs. These smart parking lots not only provide a cost-efficient solution for the parking, but EVs can also be recharged at an economical rate (Barone et al. 2014). Different emergency situations such as need for an ambulance or the immediate need for charging locations have also been addressed in smart cities (Samani & Zhu 2016, Lam et al. 2014).

18.2.1.2 Energy Management in Smart cities

Conserving energy and supporting the renewable energy resources to lessen the burden on fossil fuels is considered the main contribution of smart cities. The inclusion of EVs into the existing power resources of smart cities demands the efficient management of power resources. For this purpose, EVs are optimally recharged to avoid the extra overloading of the energy reserves (He et al. 2012). In smart cities, the smart grids not only provide the smart control of power production and distribution but also efficiently control and manage the power resources (Kennel et al. 2013, Chen 2010).

18.2.1.3 Integration of the Economic and Business Model

A strategic view point regarding the development of various business and economic models is needed to fully benefit from the advances of ICT employed in smart cities. Various multi-tech and grassroot-level companies are now deciding to try their luck by initiating various businesses in smart cities (Mulligan & Olsson 2013, Dohler et al. 2013). Therefore, future smart cities need to be further explored for future opportunities regarding various businesses. Now an extensive effort has been made from both industry and academia in developing various business and economic models to facilitate future customers in smart cities (Khatoun & Zeadally 2016).

18.2.1.4 Wireless Communication Needs/Communication Architectures for Smart Cities

The traditional TCP/IP protocol has been designed for the conventional networks. To consider the implementation of the TCP/IP for smart cities would be a mistake (Chen 2010). The requirements to design the network for smart cities are totally different, and this is crucial. In smart cities, the network will be more exposed to various severe threats as compared to the conventional networks. While there are many lives at stake in smart cities, securing the network from internal and external parties is the prime goal of network designers. As compared to the conventional networks, smart cities networks have to face the other issues such as self-healing, handling of enormous data, sender authentication, and per-hop packet accounting to achieve the trace back for each packet (Vilajosana et al. 2013, Li et al. 2013, Ianuale et al. 2016).

18.2.1.5 Traffic Congestion Avoidance in Smart Cities

Smart cities provide smart transportation. That has given rise to the concept of EVs. However, the unchecked inclusion of EVs on the smart roads in smart cities needs extra measures to counter the problem of traffic congestion. For this purpose, various predictive models have been proposed to control the traffic in smart cities (Wang et al. 2015). Another solution to avoid traffic congestion in smart cities is to design automated guided vehicles (Herrero-Perez & Martinez-Barbera 2010).

18.2.1.6 Support of Heterogeneous Technologies in Smart Cities

Monitoring the traffic in real time, providing a smart solution for health, and protecting the environment are considered the main contributions of smart cities (Solanas et al. 2014). However, to support all these features, advance solutions consisting of wireless and wired networks have been proposed. This has given rise to the issue of interoperability among various networking technologies in smart cities (Avelar et al. 2015). Various innovative technologies have also been proposed to properly integrate EVs with different features such as vehicle-to-grid (V2G) (Paffumi et al. 2016). Different test-beds have also been developed to test the interoperability of different technologies of smart cities (Cardone et al. 2014).

With the concept of smart cities, the wireless sensor networks (WSNs) have gained much attention under the umbrella of the Internet of things (IoT) (Valerio 2016, Zanella et al. 2014). Now, various interoperability issues are tackled while considering the smooth operation of the IoT.

18.2.1.7 Green Applications Support in Smart Cities

Smart cities not only conserve the energy and protect the environment, but also an extensive effort has been made in developing the green applications for smart cities. Everything in smart cities has been designed to conserve and to efficiently manage energy (Wu et al. 2016). A well-defined ecosystem for smart cities based on the various architectural models has been chalked out to support the green applications (Schleicher et al. 2016).

18.2.1.8 Security and Privacy in Smart Cities

As compared to the conventional networks, resources in smart cities are more vulnerable to various threats from the external and internal parties. In smart cities, various life-supporting phenomena are being smartly controlled. Hence any threat to networks is considered as a threat to many lives of citizens (Saputro & Akkaya 2015). The citizens privacy is also a main concern of smart cities. Therefore, the privacy-aware smart cities have been envisioned to provide enough security and privacy to smart citizens (Martínez-Ballesté et al. 2013).

18.2.2 Motivation of Using EVs in Smart cities

EVs have been introduced with the concept of reduction in the carbon emissions and to support the environment. However, in smart cities EVs have been made so much smarter that they can contribute even to stabilize the power resources and to support the renewable energy resources (Adzic et al. 2013, Rodriguez et al. 2013). The decision of integrating EVs into smart cities can result into various gains such as reduction in fossil fuel consumption and stabilizing the smart grid. The main motivations for using EVs into smart cities can be summarized as follows:

- The smart integration of EVs into smart cities can improve the reliability of the smart grids. When the recharging process of EVs is optimally controlled and scheduled, it not only reduces the load on the existing power grid but also regulates the frequency and voltage of the smart grid (Bianchini & Avila 2014, Chen 2010).
- A scheduled EVs recharging process in smart cities also enhances the reliability of the customer and minimizes the prices of energy. As EVs recharging is properly controlled, the power losses and delays are countered. This all results into customer confidence and reliability.
- The aggregators' profits also increase when EVs recharging is scheduled in smart cities. By employing the centralized and distributed EVs recharging approach, EVs are recharged in an optimal way. Through this way, aggregators can recharge great numbers of EVs and enhance their profits.
- EVs integration can result into reduction of CO_2, which will ultimately have a positive impact on the environment and will reduce global warming (Comodi et al. 2016).
- EVs can also contribute in stabilizing the smart grid. For example, through V2G capability, EVs can overcome the peak load.
- Using EVs as the transportation mode will also boost the economy due to reduction in the use of fossil fuels (Shuai et al. 2016).

18.2.3 Application Scenarios

In addition to the above-mentioned main contributions of EVs in smart cities, EVs are specifically employed in various application scenarios to achieve certain goals (Ferreira et al. 2015). The following are prime objectives that are expected to be achieved from the integration of EVs into smart cities.

18.2.3.1 Avoiding Spinning Reserves

Reserved capacity to meet the unexpected peak-demands of energy is categorized as spinning reserve. To avoid the sudden peak demands and resulting spinning reserves, EVs are optimally recharged to conserve the energy (Darabi & Ferdowsi 2014).

18.2.3.2 V2G and G2V Capability

EVs can be recharged from the renewable energy resources even at home or in parking lots. Then EVs can sell the extra energy to the grid and can help in stabilizing the grid. In this way, EVs act as the transporter of the energy and integrate the renewable energy resources into the micro-grid or smart grids (Yu et al. 2016). Also, the energy from the grid to EVs can be transported to the places where there is energy scarcity (Kavousi-Fard et al. 2015).

18.2.3.3 CO_2 Minimization

Researchers from the industry and academia are focusing on the various ways to reduce carbon emissions into the environment. To give a healthy environment to the future generations, the future smart cities have been designed from every perspective to support energy-efficient applications. EVs for this purpose, have been specifically selected as the main transportation mode in smart cities.

18.2.3.4 Load Management on the Local Microgrid

The frequency and voltage fluctuations and various other irregularities in the smart grid are taken into consideration while integrating EVs into the smart grid. The optimal recharging of EVs with the help of scheduling and various optimization approaches helps in achieving the objective of reducing the load on the smart grid (Rabiee et al. 2016, Chen & Duan 2014). In addition to optimal scheduling of EVs, various other control methods such as DC/DC and DC/AC converters control for EVs to manage the load in smart cities have been proposed (Tani et al. 2013). The smart scheduling of EVs takes into consideration the optimal recharging of EVs while minimizing the peak load and helps in stabilizing the smart grid in smart cities. However, EVs scheduling in conjunction with the control methods for DC/DC and DC/AC converters not only optimize the load but also minimize the various power losses and regulates the frequency and voltage, as the EVs can operate with of 4 kWh secondary (rechargeable) batteries. Usually, the advance EVs scheduling approaches take into consideration in advance the state of charge (SOC) of the batteries to minimize any threat to the batteries from fluctuating loads. With the knowledge of SOC, the recharging of EVs are then scheduled depending on the peak or off-peak hours. This helps EVs batteries from demanding any extra requirement.

18.3 EVs Recharging Approaches in Smart Cities

To optimally recharge EVs in smart cities, the whole recharging process of EVs is categorized into the centralized and distributed approaches for recharging EVs. Depending on the recharging approach, various recharging control methods and charger types have been developed to achieve the specific recharging objectives (Wu, Gilchrist, Sealy & Bronson 2012). Figure 18.1 shows the whole EVs recharging process in smart cities.

18.3.1 Centralized EVs Recharging Approach

The centralized EVs-recharging approach as discussed in (Nguyen & Le 2014, Han et al. 2010) employs the central entity such as aggregator unit (AU) to collect all the relevant

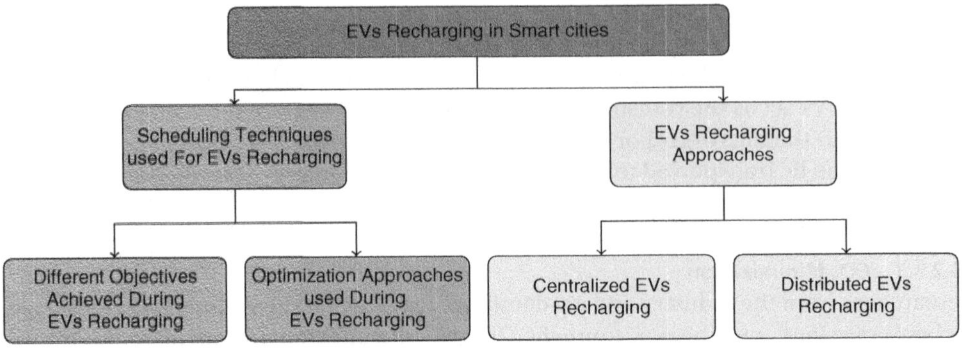

Figure 18.1 EVs in smart cities are recharged through a centralized or distributed approach using different optimization approaches.

information of EVs for recharging. An AU with the help of charging post manager (CPM) collects and controls the whole recharging process of the large number of EVs.

The AU collects the information about the time of arrival of EVs, state of charge (SoC) of batteries, duration of the recharging, prices of the recharging, and the concerned objective related to the EVs-recharging process. AUs then transfer the information to the higher unit in recharging structure named distribution system operator (DSO). The DSO then synchronizes with other DSOs and collectivity makes a decision in a centralized fashion. AUs with the help of scheduling and various optimization approaches optimally recharge EVs in a controlled way. Various conflicts related to prices and schedule are also resolved by using the scheduling algorithms for EVs recharging.

18.3.1.1 Main Contributions and Limitations of Centralized EVs-Recharging Approach

The centralized entity AU recharges EVs in a centralized way. Following are the main advantages of the centralized EVs-recharging approach.

- In a centralized EVs-recharging approach, AUs have to collect all the information of all EVs and then have to make complex recharging decisions. For this purpose, a sophisticated and smart recharging EVs architecture is employed. With the help of advanced architecture, various objectives are also achieved.
- It is estimated that, with the help of the centralized EVs-recharging approach, the network capacity is utilized up to its maximum. It avoids the waste of the network resources.
- With the help of an advanced EVs-recharging architecture, the provision of ancillary services in a smart city during EVs-recharging approach is easy to achieve as compared to the distributed EVs-recharging approach.

However, the centralized EVs-recharging approach in smart cities also faces various limitations such as:

- With the centralized point of decision, the need for backup is necessary.
- The centralized EVs approach requires an advanced recharging architecture, which proves to be a costly solution for EVs recharging in smart cities.
- The security and privacy of customer data becomes a serious concern in this mode of recharging.

18.3.2 Distributed EVs Recharging Approach

The distributed EVs-recharging approach is discussed in (He et al. 2012, Hu et al. 2014). Various limitations of the centralized EVs-recharging approach in smart cities are addressed with the introduction of distributed or decentralized EVs approach. In this recharging approach, in addition to the central entity, EVs themselves make the recharging decisions in a distributed manner. As the number of EVs increases, the computational load is also divided according to the situations and number of EVs. The distributed EVs- recharging approach in smart cities demands from EVs to be more intelligent and capable of performing the computational decisions. Hence, a simple EVs-recharging architecture is needed to perform the whole optimal scheduling of EVs recharging in a distributed manner.

18.3.2.1 Main Contributions and Limitations of the Distributed EVs-recharging Approach

The distributed EVs recharging as compared to the centralized approach adds less complexity in the whole recharging process. The main advantages achieved during the distributed EVs-recharging approach are discussed below:

- The distributed EVs recharging in smart cities requires a simpler and less complex recharging architecture.
- As the number of EVs increases, the computational load is also distributed among EVs intended for recharging.
- The security and privacy of the customer becomes less of an issue as compared to the centralized EVs-recharging approach.
- The distributed EVs-recharging approach adds more flexibility and fault tolerance in the whole recharging approach.

The distributed EVs-recharging approach addresses the various limitations of the centralized EVs recharging. However, there are also certain limitations of the distributed EVs recharging approach. Some of the limitations of the distributed EVs recharging are described below:

- As compared to the centralized EVs-recharging approach, the distributed EVs-recharging approach in smart cities do not provide enough ancillary services.
- The main limitation of the distributed EVs recharging is the avalanche effect that happens due to the larger number of EVs that are performing the computation at the same time.
- The costumer needs or future demand is difficult to predict in the distributed EVs-recharging approach. That can result into spinning reserves.

18.4 Scheduling EVs Recharging in Smart Cities

Integration of the large number of EVs into smart cities will enhance the load on the existing smart grid. It is estimated that the integration of 10% of EVs into smart cities will be a serious blow to the existing power resources. The existing smart grid was actually designed to support the domestic and commercial users. However, electrification of transportation requires extra measures to reduce the impact on the power resources (Su et al. 2012, Hadley & Tsvetkova 2009).

By taking into consideration the various potential effects of EVs integration into smart cities, it is demanded that EVs recharging is made more controlled, optimal, and flexible. For this purpose, EVs in smart cities are recharged with the help of optimal scheduling and optimization approaches (Will & Schuller 2016). With the scheduled EVs-recharging approach, various objectives such as reduction of power losses, minimization of cost of energy for customers, and V2G capability can be achieved (Singh et al. 2012).

A scheduled EVs-recharging approach with various control methods, advanced architectures, and chargers seem to conserve the energy in smart cities. (Ni et al. 2015, Rahbari-Asr & Chow 2014, Pahlevaninezhad et al. 2012). With the help of smart charging approaches (either centralized or distributed), scheduling algorithms, control methods, and smart charging locations, EVs recharging is made optimal, and EVs can contribute in stabilizing the smart grid (Mirzaei et al. 2016). To be more specific, we will discuss various scheduling methods applied to EVs recharging. Below is the detailed discussion of various scheduling approaches with respect to different objectives achieved during EVs recharging approaches in smart cities.

18.4.1 Objectives Achieved via Different Scheduling Approaches

The recharging of EVs in smart cities requires an efficient recharging process to minimize the potential impact on the smart grid. For this purpose, the optimal scheduling approach is usually employed for recharging of EVs in smart cities to achieve one or more objectives. Different objective- oriented recharging approaches are used. In this subsection, we will discuss various objectives that will be achieved by utilizing the optimal scheduling approaches in smart cities. Table 18.1 gives a description of various objectives achieved during the scheduling of EVs-recharging approaches with different conflicting objectives.

18.4.1.1 Reduction of Power Losses

It is assumed that recharging of EVs in smart cities is accompanied by various power losses. Therefore, the recharging of EVs is optimized to minimize the power losses of the smart grid. Energy in smart cities can be conserved by reducing the power losses during the EVs-recharging process. Therefore, EVs recharging in smart cities is optimally scheduled with the help of various algorithms and optimization approaches. To minimize the power losses on the smart grid while recharging EVs in the smart grid, authors in (Clement-Nyns et al. 2010) have proposed the controlled and coordinated recharging approach. With the help of stochastic programming, not only the power losses are reduced but also the load factor of the main grid is maximized. The relationship between the load factor, load variance, and the power losses has been made in (Sortomme et al. 2011). For this purpose three optimal scheduling algorithms have been proposed. The testing of the algorithms proves that this relationship between the power losses of the distribution system and load factor is independent of the system topology.

The charging of EVs power train is also considered for conserving energy (Pourabdollah et al. 2013). For this purpose, the convex optimization approach is employed, and the

Table 18.1 Scheduling approaches used for EVs recharging in smart cities with different objectives, main constraints, recharging approach, and solver tools.

Studies	Main Objective Achieved	Main Constraints	Main Solver Tools	Recharging Approach Used
(Clement-Nyns et al. 2010, Sortomme et al. 2011, Pourabdollah et al. 2013, Murgovski et al. 2013, Ko & Jang 2013)	Reduces the power losses	Power resource minimization and power cost	Matlab and JADE (Java)	Both centralized and distributed
(Wu, Aliprantis & Ying 2012, Gan et al. 2013, Bashash & Fathy 2014, Tushar et al. 2014, Yang et al. 2014, Rostami et al. 2015)	Minimizing the total cost of energy for users	Recharging availability and load	Matlab, Simulink and PowerACE	Both centralized and distributed
(Sortomme & El-Sharkawi 2012b, a, Jin et al. 2013)	Maximizing the aggregator profits	Electricity balance and rate	Matpower, PSS/E, and GAMS	In most cases the distributed EVs recharging approach is used
(Han et al. 2010, Rotering & Ilic 2011, Sortomme & El-Sharkawi 2011, Lin et al. 2014)	Frequency regulation	Minimization of the power resources	PSS/E, and PowerACE	Both centralized and distributed
(Singh et al. 2012, Richardson et al. 2012, Bai & Qiao 2015)	Voltage regulation	Minimization of the power resources	Matlab, Simulink	Both centralized and distributed
(Liu et al. 2014, Zhao et al. 2012, Huang et al. 2014, Jin et al. 2014, Zhang et al. 2014)	Support for Renewable energy	Batteries limitations and charging capacitors	Matlab, GAMS, and Matpower	Centralized but in some cases the distributed

power losses such as engine losses and the electric motor losses are also taken into consideration. Various factors such charging time, top speed, and driving patterns are also taken into consideration to optimally recharge EVs. Power train is either recharged at standstill or while driving. Both cases of recharging with an optimal recharging approach have been employed in (Murgovski et al. 2013). The problem of battery dimensioning and power split control is taken into consideration while employing dynamic programming to optimally schedule EVs recharging. The study in (Ko & Jang 2013) discusses the recharging of EVs via wireless power transfer. This require the efficient scheduling approaches to schedule the data and energy packets. For this purpose, the particle swarm optimization approach has been utilized to solve the problem of optimal recharging of EVs.

18.4.1.2 Minimizing Total Cost of Energy for Users

When EVs are optimally recharged, the total cost of energy for the customers tend to decrease. The whole recharging process is optimized in a fashion that not only minimizes the cost of energy for users, but also the profit of the AU tends to increase. A low-cost scheduling algorithm has been proposed (Wu, Aliprantis & Ying 2012) to minimize the cost for the users. Through this algorithm, the energy demand is forecast in a day-ahead manner and then applying the scheduling low-cost algorithm, the rates are negotiated in a bilateral manner. A decentralized scheduling algorithm for recharging of EVs to minimize the load and cost has been proposed in (Gan et al. 2013). The proposed algorithm fills the valley of the electric demands and takes into consideration the elasticity and controllability of EVs load and energy prices. Another scheduling algorithm that addresses the cost-optimal recharging of EVs has been proposed in (Bashash & Fathy 2014). With the help of convex optimization, not only the energy cost is considered but also the V2G support is analyzed.

Scheduling of EVs recharging in conjunction with home appliances in the smart microgrids has been proposed in (Tushar et al. 2014). The scheduling method has been formulated as mixed integer linear programming (MILP), and then it has been solved optimally to minimize the energy cost and to support the renewable energy resources. This scheduling approach has been designed for the microgrid to schedule the load between EVs recharging and home appliances. Usually, the forecast and the actual load of EVs introduces certain anomalies in the system. This issue has been resolved by introducing the risk-aware scheduling of EVs to overcome the expected and actual load mismatch and to minimize the energy cost for the users (Yang et al. 2014). This scheduling problem is formulated as a stochastic problem and then is solved with the help of the L-method. Various uncertainties regarding the price of energy for users during EVs recharging has been addressed in (Rostami et al. 2015). The EVs recharging problem is solved using the krill herd optimization technique. Here, the network topology for the recharging of EVs can be controlled from some remote places.

18.4.1.3 Maximizing Aggregator Profit

The scheduling of EVs recharging in a smart city can also result into maximizing the profit of the AU. The AU usually enhances its profit by either supporting the centralized or distributed EVs-recharging approach.

A scheduling algorithm has been proposed to schedule between the V2G capability and ancillary services while taking into consideration the aggregator and customer profits (Sortomme & El-Sharkawi 2012b). This scheduling algorithm has been specifically designed for the aggregator to address the issue of peak load shaving and to add flexibility in the system. The proposed algorithm enhances the aggregator profit while supporting the renewable energy resources. A scheduling algorithm to bid for the V2G and ancillary services has also been proposed in (Sortomme & El-Sharkawi 2012a). This algorithm also takes into consideration the unplanned departure of EVs from the pre-decided contract and maximizes the profit for the aggregator and customer. The unidirectional V2G support with spinning reserves with this algorithm significant increases the aggregator profit. EVs scheduling for recharging and enhancing the aggregator profit with energy storage capability has been addressed in (Jin et al. 2013). The scheduling problem is formulated as a MILP and then is optimally solved to enhance the aggregator profit. In this study, the aggregator with storage energy minimizes the sudden peak demands. The proposed scheduling approach is also integrated with the communication protocol to provide the efficient interaction between the aggregator, EVs, smart grid, and the energy storage resources.

18.4.1.4 Frequency Regulation

Frequency fluctuations due to unexpected loads on the smart grid can have serious effects on the smooth operations of electric appliances in smart cities. Therefore, every effort has been made to design the various applications, approaches, procedures, and algorithms to stabilize the frequency in smart cities. The recharging of EVs with various scheduling approaches also helps in regulating the frequency of the smart grid. Authors in (Han et al. 2010) propose various design paradigms for the aggregator to support the V2G capability with frequency regulation. The scheduling problem of EVs recharging in this case is optimally solved, and dynamic programming is applied to compute the charging control for each vehicle. Two other scheduling algorithms that take into consideration the frequency regulation, energy flows, charging time, and V2G have also been proposed in (Rotering & Ilic 2011). To find the optimal solution, dynamic programming has been used, and the two algorithms are designed to take into consideration the energy forecast. Unidirectional V2G capability with frequency regulation has been proposed in (Sortomme & El-Sharkawi 2011). As the rate for recharging EVs varies during the frequency regulation, this scheduling algorithm has been proposed to set the point for such variations in rate. It also takes into consideration load management and frequency regulation. The optimal scheduling approach for charging and discharging of EVs to regulate the frequency has been proposed in (Lin et al. 2014). In this study, the scheduling problem is formulated as a convex optimization problem, and then the gradient projection method is used to solve the optimization problem. In this forecast-based scheduling approach, the optimal charging and discharging of EVs with the help of a decentralized scheduling approach helps in regulating the frequency of the smart grid.

18.4.1.5 Voltage regulation

The ancillary services provided by a scheduled EVs-recharging approach also help in regulating the voltage of the smart grid in smart cities. The aggregators with the help of storage energy from EVs not only minimize the peak load but also stabilize the smart grid by regulating the voltages. The authors in (Singh et al. 2012) introduce the scheduling approach for charging and discharging of EVs into the local distribution system. Two fuzzy logic controllers named as the charging station controller and the V2G controller have been proposed to schedule the optimal recharging and discharging of EVs into the smart grid. Controlling the recharging rates can also contribute in stabilizing the smart grid (Richardson et al. 2012). Scheduling of EVs while considering the recharging rates can enhance the efficiency of the power grid. For this purpose, a scheduling approach based on the linear programming is proposed to determine the charging rates for each EVs and to regulate the voltages fluctuations. The bidirectional V2G capability has also been exploited to regulate the voltage of the smart grid during recharging or discharging of EVs (Bai & Qiao 2015). In this proposed approach, the scheduling problem of recharging and discharging of EVs is formulated as a mixed integer quadratic programming model and then solved optimally. This robust optimization model comes out to be efficient for stabilizing the power resources of smart cities.

18.4.1.6 Support for Renewable Energy Sources for Recharging of EVs

The integration of renewable energy resources into smart cities not only reduces the carbon emission but also reduces the prices of the energy for domestic and commercial users. EVs in smart cities can also be recharged while exploiting the renewable energy

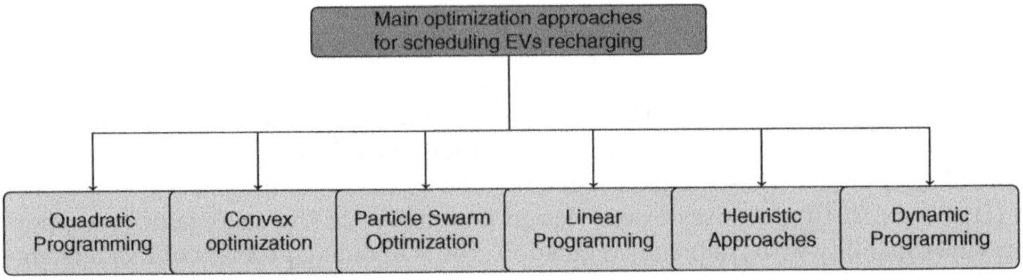

Figure 18.2 Main optimization approaches used to achieve various objectives during the scheduling of EVs-recharging approaches.

resources. For this purpose, the scheduling approaches for EVs recharging using renewable energy resources have been proposed.

Wind power has been extensively considered for recharging of EVs (Liu et al. 2014, Zhao et al. 2012). For example, the integration of wind power resources for optimal recharging of EVs is practiced in (Huang et al. 2014). The scheduling problem of EVs recharging while employing renewable energy sources is formulated as a Markov decision process (MDP) and then solved using an event-based optimization approach. The proposed scheme seems to fully support the wind energy for the recharging of EVs. The recharging of EVs from the uncertain renewable energy sources has been discussed in (Jin et al. 2014). The energy from the renewable sources is uncertain and time varying. Therefore, the scheduling of EVs recharging using the renewable energy sources is considered to be a bit complex. Lyapunov optimization is used in solving the recharging problem of EVs while supporting the fluctuating renewable energy resources. The scheduling of EVs recharging while considering the local renewable energy resources, uncertain EVs arrival, and cost minimization has been considered in (Zhang et al. 2014).

18.4.2 Resource Allocation for EVs Recharging in Smart Cities (Optimization Approaches)

Scheduling EVs recharging in smart cities is usually achieved with the help of different optimization approaches (Shen et al. 2014, Okaeme & Zanchetta 2013). To achieve the optimal recharging of EVs in smart cities, the scheduling of EVs recharging problem is solved optimally by using an optimization approach (Azizipanah-Abarghooee et al. 2016, Majidpour et al. 2015). Major optimization approaches used for EVs recharging are convex optimization, particle swarm optimization, quadratic programming, linear programming, heuristic approaches, and dynamic programming. Figure 18.2 shows the various optimization approaches used for EVs recharging with their respective reference for the work.

18.5 Open Issues, Challenges, and Future Research Directions

The integration of EVs into smart cities requires a thorough re-assessment of existing energy and other resources. The recharging of EVs in smart cities through smart recharging architectures also has to face different challenges. Regarding EVs recharging in smart cities, below is the description of the some open issues, challenges, and future research directions.

18.5.1 Support of Wireless Power Charger

Wireless power transfer has gained much attention in today's research domain. Recharging EVs through wireless power has been envisioned as the most promising solution to enhance the efficiency and flexibility of EVs recharging in smart cities (Ko & Jang 2013). To the best of our knowledge, the wireless recharging of EVs has not been explored in much detail. With the help of wireless rechargers for EVs, various renewable energy resources can be exploited and integrated in smart cities in a more reliable way.

18.5.2 Vehicle-to-Anything

Through the help of V2G capability, EVs can help in stabilizing the smart grid and provide the ancillary services. V2G or G2V capability of EVs can also be enhanced to vehicle-to-anything. Through this approach, different domestic and commercial electric appliances can be electrified (Ferreira et al. 2014). However, very limited work has been done on this domain. It requires the thorough understanding of the possibility of recharging other electric devices through EVs.

18.5.3 Energy Management for Smart Grid via EVs

The unchecked or rapid integration of EVs into the smart grid can lead to the drastic power outage and unexpected loads on the smart grid. Researchers have made a significant contribution in providing the solution for minimizing the load on the smart grid (Chen et al. 2014, Kennel et al. 2013). The impact of large penetration of EVs into the existing power resources and future smart cities has not been analyzed in detail.

18.5.4 Advance Communication Needs for Controlled EVs Recharging

The centralized and distributed EVs-recharging infrastructure requires wired or wireless communication solutions to facilitate the customers. As compared to the distributed EVs-recharging approach, the centralized EVs recharging requires an advance communication architecture for the recharging of EVs. The main need is the security and privacy of the customer data while providing EVs-recharging approach. Different solutions regarding the communications of EVs recharging architecture have been proposed (Djahel et al. 2015, Khorov et al. 2015, Alvi et al. 2016, Zhu et al. 2013, Vlacheas et al. 2013, Tal et al. 2015). However, there is still a need for an advanced wired or wireless communication architecture to securely recharge the large number of EVs in the future smart cities.

18.5.5 EVs Control Applications

Making the transportation more intelligent in smart cities requires a thorough control through different online and other applications. Some of the applications have already been proposed to control EVs drive and recharging in smart cities (Liang et al. 2013,

Bertoluzzo & Buja 2011, Steenbruggen et al. 2015, Farkas et al. 2015). However, there is a need to develop the various applications to control EVs drives and recharging in smart cities.

18.5.6 Standardization for Communication Technologies Used for EVs Recharging

Researchers from industry and academia are working on the standardization of communications protocols used for EVs recharging. So far, EVs communication architecture has not witnessed any standard to practically smooth the recharging operation. However, the standard J1772 connector (Pratt et al. 2011) has been proposed for EVs recharging architecture to support the recharging operation in the smart grid.

18.6 Conclusion

Electric vehicles (EVs) have been accepted as the smart solution for the transportation in smart cites. The integration of EVs into the smart grid requires a proper scheduling of EVs recharging to conserve the energy and to stabilize the smart grid. The controlled EVs recharging in smart cities has been studied from various perspectives. In this chapter, a detailed discussion of the scheduling of EVs recharging in smart cities have been proposed. EVs in smart cities are usually recharged with a centralized or distributed recharging approach. Different advantages and limitations regarding each recharging approach have been presented. Scheduling approaches for EVs recharging provides different objectives such as reduction in power losses, minimizing the total cost of energy for users, maximizing the aggregator profits, frequency regulation, voltage regulation, and support for renewable energy resources. To achieve the optimal scheduling of EVs recharging, a brief description of optimization techniques to gain various objectives and to allocate the resources has also been made. EVs domain still needs to be further studied from various perspectives such as wireless power recharger and EVs supportive applications. These and other perspectives have been highlighted as future research directions.

References

Adzic, E., Adzic, M., Katic, V., Marcetic, D. & Celanovic, N. (2013), 'Development of high-reliability ev and hev im propulsion drive with ultra-low latency hil environment', *IEEE Transactions on Industrial Informatics* **9** (2), 630–639.

Alvi, A., Bouk, S., Ahmed, S., Yaqub, M., Sarkar, M. & Song, H. (2016), 'Best-mac: Bitmap-assisted efficient and scalable tdma-based wsn mac protocol for smart cities', *IEEE Access* **4**, 312–322.

Angelidou, M. (2015), 'Smart cities: A conjuncture of four forces', *Cities* **47**, 95–106.

Avelar, E., Marques, L., dos Passos, D., Macedo, R., Dias, K. & Nogueira, M. (2015), 'Interoperability issues on heterogeneous wireless communication for smart cities', *Computer Communications* **58**, 4–15.

Azizipanah-Abarghooee, R., Terzija, V., Golestaneh, F. & Roosta, A. (2016), 'Multiobjective dynamic optimal power flow considering fuzzy-based smart utilization of mobile electric vehicles', *IEEE Transactions on Industrial Informatics* **12** (2), 503–514.

Bai, X. & Qiao, W. (2015), 'Robust optimization for bidirectional dispatch coordination of large-scale v2g', *IEEE Transactions on Smart Grid* **6** (4), 1944–1954.

Banister, D. (2011), 'Cities, mobility and climate change', *Journal of Transport Geography* **19** (6), 1538–1546.

Barone, R., Giuffré, T., Siniscalchi, S., Morgano, M. & Tesoriere, G. (2014), 'Architecture for parking management in smart cities', *IET Intelligent Transport Systems* **8** (5), 445–452.

Bashash, S. & Fathy, H. (2014), 'Cost-optimal charging of plug-in hybrid electric vehicles under time-varying electricity price signals', *IEEE Transactions on Intelligent Transportation Systems* **15** (5), 1958–1968.

Bertoluzzo, M. & Buja, G. (2011), 'Development of electric propulsion systems for light electric vehicles', *IEEE Transactions on Industrial Informatics* **7** (3), 428–435.

Bianchini, D. & Avila, I. (2014), 'Smart cities and their smart decisions: Ethical considerations', *IEEE Technology and Society Magazine* **33** (1), 34–40.

Cardone, G., Cirri, A., Corradi, A. & Foschini, L. (2014), 'The participact mobile crowd sensing living lab: The testbed for smart cities', *IEEE Communications Magazine* **52** (10), 78–85.

Celino, I. & Kotoulas, S. (2013), 'Smart cities.', *IEEE Internet Computing* **17** (6).

Chen, C. & Duan, S. (2014), 'Optimal integration of plug-in hybrid electric vehicles in microgrids', *IEEE Transactions on Industrial Informatics* **10** (3), 1917–1926.

Chen, T. (2010), 'Smart grids, smart cities need better networks [editor's note]', *IEEE Network* **24** (2), 2–3.

Chen, Y., Li, X., Wiet, C. & Wang, J. (2014), 'Energy management and driving strategy for in-wheel motor electric ground vehicles with terrain profile preview', *IEEE Transactions on Industrial Informatics* **10** (3), 1938–1947.

Chow, M.-Y., Baronti, F., Williamson, S. S. & Ma, C. (2014), 'Guest editorial: Special section on information and control technologies in the electrification of transportation', *IEEE Transactions on Industrial Informatics* **10** (3), 1904–1906.

Clement-Nyns, K., Haesen, E. & Driesen, J. (2009), Coordinated charging of multiple plug-in hybrid electric vehicles in residential distribution grids, *in* 'IEEE/PES Power Systems Conference and Exposition', pp. 1–7.

Clement-Nyns, K., Haesen, E. & Driesen, J. (2010), 'The impact of charging plug-in hybrid electric vehicles on a residential distribution grid', *IEEE Transactions on Power Systems* **25** (1), 371–380.

Comodi, G., Caresana, F., Salvi, D., Pelagalli, L. & Lorenzetti, M. (2016), 'Local promotion of electric mobility in cities: Guidelines and real application case in italy', *Energy* **95**, 494–503.

Darabi, Z. & Ferdowsi, M. (2014), 'An event-based simulation framework to examine the response of power grid to the charging demand of plug-in hybrid electric vehicles', *IEEE Transactions on Industrial Informatics* **10** (1), 313–322.

de Andrade, J., Ribeiro, J., Fernandez, F., Bailey, C., Barbosa, S., da Silva Neiva, S. et al. (2016), 'The adoption of strategies for sustainable cities: a comparative study between newcastle and florianópolis focused on urban mobility', *Journal of Cleaner Production* **113**, 681–694.

Djahel, S., Doolan, R., Muntean, G.-M. & Murphy, J. (2015), 'A communications-oriented perspective on traffic management systems for smart cities: challenges and innovative approaches', *IEEE Communications Surveys & Tutorials* **17** (1), 125–151.

Dohler, M., Ratti, C., Paraszczak, J. & Falconer, G. (2013), 'Smart cities [guest editorial]', *IEEE Communications Magazine* **6** (51).

Farkas, K., Feher, G., Benczur, A. & Sidlo, C. (2015), 'Crowdsending based public transport information service in smart cities', *IEEE Communications Magazine* **53** (8), 158–165.

Ferreira, J., Monteiro, V. & Afonso, J. (2014), 'Vehicle-to-anything application (v2anything app) for electric vehicles', *IEEE Transactions on Industrial Informatics* **10** (3), 1927–1937.

Ferreira, J., Monteiro, V., Afonso, J. & Afonso, J. (2015), 'Mobile cockpit system for enhanced electric bicycle use', *IEEE Transactions on Industrial Informatics* **11** (5), 1017–1027.

Gan, L., Topcu, U. & Low, S. (2013), 'Optimal decentralized protocol for electric vehicle charging', *IEEE Transactions on Power Systems* **28** (2), 940–951.

Hadley, S. & Tsvetkova, A. (2009), 'Potential impacts of plug-in hybrid electric vehicles on regional power generation', *The Electricity Journal* **22** (10), 56–68.

Han, S., Han, S. & Sezaki, K. (2010), 'Development of an optimal vehicle-to-grid aggregator for frequency regulation', *IEEE Transactions on smart grid* **1** (1), 65–72.

He, Y., Venkatesh, B. & Guan, L. (2012), 'Optimal scheduling for charging and discharging of electric vehicles', *IEEE transactions on smart grid* **3** (3), 1095–1105.

Herrero-Perez, D. & Martinez-Barbera, H. (2010), 'Modeling distributed transportation systems composed of flexible automated guided vehicles in flexible manufacturing systems', *IEEE Transactions on Industrial Informatics* **6** (2), 166–180.

Hu, J., You, S., Lind, M. & Østergaard, J. (2014), 'Coordinated charging f electric vehicles for congestion prevention in the distribution grid', *IEEE Transactions on Smart Grid* **5** (2), 703–711.

Huang, Q., Jia, Q.-S., Xia, L. & Guan, X. (2014), Event-based optimization for stochastic matching ev charging load with uncertain renewable energy, *in* '11th World Congress on Intelligent Control and Automation (WCICA)', pp. 794–799.

Ianuale, N., Schiavon, D. & Capobianco, E. (2016), 'Smart cities, big data, and communities: Reasoning from the viewpoint of attractors', *IEEE Access* **4**, 41–47.

Jin, C., Sheng, X. & Ghosh, P. (2014), 'Optimized electric vehicle charging with intermittent renewable energy sources', *IEEE Journal of Selected Topics in Signal Processing* **8** (6), 1063–1072.

Jin, C., Tang, J. & Ghosh, P. (2013), 'Optimizing electric vehicle charging with energy storage in the electricity market', *IEEE Transactions on Smart Grid* **4** (1), 311–320.

Kavousi-Fard, A., Rostami, M. & Niknam, T. (2015), 'Reliability-oriented reconfiguration of vehicle-to-grid networks', *IEEE Transactions on Industrial Informatics* **11** (3), 682–691.

Kennel, F., Görges, D. & Liu, S. (2013), 'Energy management for smart grids with electric vehicles based on hierarchical mpc', *IEEE Transactions on Industrial Informatics* **9** (3), 1528–1537.

Khatoun, R. & Zeadally, S. (2016), 'Smart cities: concepts, architectures, research opportunities', *Communications of the ACM* **59** (8), 46–57.

Khorov, E., Lyakhov, A., Krotov, A. & Guschin, A. (2015), 'A survey on ieee 802.11 ah: An enabling networking technology for smart cities', *Computer Communications* **58**, 53–69.

Ko, Y. & Jang, Y. (2013), 'The optimal system design of the online electric vehicle utilizing wireless power transmission technology', *IEEE Transactions on Intelligent Transportation Systems* **14** (3), 1255–1265.

Lam, A., Leung, Y.-W. & Chu, X. (2014), 'Electric vehicle charging station placement: Formulation, complexity, and solutions', *IEEE Transactions on Smart Grid* **5** (6), 2846–2856.

Li, F., Vögler, M., Sehic, S., Qanbari, S., Nastic, S., Truong, H. & Dustdar, S. (2013), 'Web-scale service delivery for smart cities.', *IEEE Internet Computing* **17** (4), 78–83.

Liang, Y.-W., Chen, C.-C., Liaw, D.-C. & Wei, Y.-T. (2013), 'Nonlinear reliable control with application to a vehicle antilock brake system', *IEEE Transactions on Industrial Informatics* **9** (4), 2114–2123.

Lin, J., Leung, K.-C. & Li, V. (2014), 'Optimal scheduling with vehicle-to-grid regulation service', *IEEE Internet of Things Journal* **1** (6), 556–569.

Liu, H., Guo, J. & Zeng, P. (2014), A controlled electric vehicle charging strategy considering regional wind and pv, *in* 'IEEE PES General Meeting— Conference & Exposition', pp. 1–5.

Majidpour, M., Qiu, C., Chu, P., Gadh, R. & Pota, H. (2015), 'Fast prediction for sparse time series: Demand forecast of ev charging stations for cell phone applications', *IEEE Transactions on Industrial Informatics* **11** (1), 242–250.

Martínez-Ballesté, A., Pérez-Martínez, P. & Solanas, A. (2013), 'The pursuit of citizens' privacy: a privacy-aware smart city is possible', *IEEE Communications Magazine* **51** (6), 136–141.

Mirzaei, M., Kazemi, A. & Homaee, O. (2016), 'A probabilistic approach to determine optimal capacity and location of electric vehicles parking lots in distribution networks', *IEEE Transactions on Industrial Informatics* **12** (5), 1963–1972.

Mulligan, C. & Olsson, M. (2013), 'Architectural implications of smart city business models: an evolutionary perspective', *IEEE Communications Magazine* **51** (6), 80–85.

Murgovski, N., Johannesson, L. & Sjöberg, J. (2013), 'Engine on/off control for dimensioning hybrid electric powertrains via convex optimization', *IEEE Transactions on Vehicular Technology* **62** (7), 2949–2962.

Nguyen, D. & Le, L. (2014), 'Joint optimization of electric vehicle and home energy scheduling considering user comfort preference', *IEEE Transactions on Smart Grid* **5** (1), 188–199.

Ni, W., Collings, I., Wang, X., Liu, R., Kajan, A., Hedley, M. & Abolhasan, M. (2015), 'Radio alignment for inductive charging of electric vehicles', *IEEE Transactions on Industrial Informatics* **11** (2), 427–440.

Nolte, T., Shin, I., Behnam, M. & Sjodin, M. (2009), 'A synchronization protocol for temporal isolation of software components in vehicular systems', *IEEE Transactions on Industrial Informatics* **5** (4), 375–387.

Okaeme, N. & Zanchetta, P. (2013), 'Hybrid bacterial foraging optimization strategy for automated experimental control design in electrical drives', *IEEE Transactions on Industrial Informatics* **9** (2), 668–678.

Paffumi, E., De Gennaro, M. & Martini, G. (2016), 'Innovative technologies for smart cities: towards customer driven infrastructure design for large scale deployment of electric vehicles and vehicle-to-grid applications', *Transportation Research Procedia* **14**, 4505–4514.

Pahlevaninezhad, M., Das, P., Drobnik, J., Moschopoulos, G., Jain, P. & Bakhshai, A. (2012), 'A nonlinear optimal control approach based on the control-lyapunov function for an ac/dc converter used in electric vehicles', *IEEE Transactions on Industrial Informatics* **8** (3), 596–614.

Pellicer, S., Santa, G., Bleda, A., Maestre, R., Jara, A. & Skarmeta, A. (2013), A global perspective of smart cities: a survey, *in* 'Seventh International Conference on Innovative Mobile and Internet Services in Ubiquitous Computing (IMIS)', pp. 439–444.

Pourabdollah, M., Murgovski, N., Grauers, A. & Egardt, B. (2013), 'Optimal sizing of a parallel phev powertrain', *IEEE Transactions on Vehicular Technology* **62** (6), 2469–2480.

Pratt, R., Tuffner, F. & Gowri, K. (2011), Electric vehicle communication standards testing and validation phase i: Sae j2847/1, Technical report, Pacific Northwest National Laboratory (PNNL), Richland, WA (US).

Rabiee, A., Farahani, H. F., Khalili, M., Aghaei, J. & Muttaqi, K. (2016), 'Integration of plug-in electric vehicles into microgrids as energy and reactive power providers in market environment', *IEEE Transactions on Industrial Informatics* **12** (4), 1312–1320.

Rahbari-Asr, N. & Chow, M. (2014), 'Cooperative distributed demand management for community charging of phev/pevs based on kkt conditions and consensus networks', *IEEE Transactions on Industrial Informatics* **10** (3), 1907–1916.

Rahmani, M., Tappayuthpijarn, K., Krebs, B., Steinbach, E. & Bogenberger, R. (2009), 'Traffic shaping for resource-efficient in-vehicle communication', *IEEE Transactions on Industrial Informatics* **5** (4), 414–428.

Richardson, P., Flynn, D. & Keane, A. (2012), 'Optimal charging of electric vehicles in low-voltage distribution systems', *IEEE Transactions on Power Systems* **27** (1), 268–279.

Rodriguez, J., Kazmierkowski, M., Espinoza, J., Zanchetta, P. & Rivera, M. (2013), 'Guest editorial special section on digital control systems in power electronics and electrical drives-part iii', *IEEE Transactions on Industrial Informatics* **9** (2), 587–588.

Rostami, M., Kavousi-Fard, A. & Niknam, T. (2015), 'Expected cost minimization of smart grids with plug-in hybrid electric vehicles using optimal distribution feeder reconfiguration', *IEEE Transactions on Industrial Informatics* **11** (2), 388–397.

Rotering, N. & Ilic, M. (2011), 'Optimal charge control of plug-in hybrid electric vehicles in deregulated electricity markets', *IEEE Transactions on Power Systems* **26** (3), 1021–1029.

Saint, A. (2014), 'The rise and rise of the smart city [smart cities urban britain]', *Engineering & Technology* **9** (9), 72–76.

Samani, H. & Zhu, R. (2016), 'Robotic automated external defibrillator ambulance for emergency medical service in smart cities', *IEEE Access* **4**, 268–283.

Saputro, N. & Akkaya, K. (2015), 'Parp-s: A secure piggybacking-based arp for ieee 802.11 s-based smart grid ami networks', *Computer Communications* **58**, 16–28.

Schleicher, J., Vögler, M., Dustdar, S. & Inzinger, C. (2016), 'Enabling a smart city application ecosystem: requirements and architectural aspects', *IEEE Internet Computing* **20** (2), 58–65.

Shen, J., Dusmez, S. & Khaligh, A. (2014), 'Optimization of sizing and battery cycle life in battery/ultracapacitor hybrid energy storage systems for electric vehicle applications', *IEEE Transactions on Industrial Informatics* **10** (4), 2112–2121.

Shuai, W., Maillé, P. & Pelov, A. (2016), 'Charging electric vehicles in the smart city: A survey of economy-driven approaches'.

Singh, M., Kumar, P. & Kar, I. (2012), 'Implementation of vehicle to grid infrastructure using fuzzy logic controller', *IEEE Transactions on Smart Grid* **3** (1), 565–577.

Soares, J., Borges, N., Ghazvini, M., Vale, Z. & de Moura Oliveira, P. (2016), 'Scenario generation for electric vehicles' uncertain behavior in a smart city environment', *Energy* **111**, 664–675.

Solanas, A., Patsakis, C., Conti, M., Vlachos, I., Ramos, V., Falcone, F., Postolache, O., Pérez-Martínez, P., Di Pietro, R., Perrea, D. et al. (2014), 'Smart health: a context-aware health paradigm within smart cities', *IEEE Communications Magazine* **52** (8), 74–81.

Sortomme, E. & El-Sharkawi, M. (2011), 'Optimal charging strategies for unidirectional vehicle-to-grid', *IEEE Transactions on Smart Grid* **2** (1), 131–138.

Sortomme, E. & El-Sharkawi, M. (2012a), 'Optimal combined bidding of vehicle-to-grid ancillary services', *IEEE Transactions on Smart Grid* **3** (1), 70–79.

Sortomme, E. & El-Sharkawi, M. (2012b), 'Optimal scheduling of vehicle-to-grid energy and ancillary services', *IEEE Transactions on Smart Grid* **3** (1), 351–359.

Sortomme, E., Hindi, M., MacPherson, S. & Venkata, S. (2011), 'Coordinated charging of plug-in hybrid electric vehicles to minimize distribution system losses', *IEEE transactions on smart grid* **2** (1), 198–205.

Steenbruggen, J., Tranos, E. & Nijkamp, P. (2015), 'Data from mobile phone operators: A tool for smarter cities?', *Telecommunications Policy* **39** (3), 335–346.

Su, W., Eichi, H., Zeng, W. & Chow, M.-Y. (2012), 'A survey on the electrification of transportation in a smart grid environment', *IEEE Transactions on Industrial Informatics* **8** (1), 1–10.

Tal, I., Ciubotaru, B. & Muntean, G.-M. (2015), 'Vehicular communications-based speed advisory system for electric bicycles'.

Tani, A., Camara, M., Dakyo, B. & Azzouz, Y. (2013), 'Dc/dc and dc/ac converters control for hybrid electric vehicles energy management-ultracapacitors and fuel cell', *IEEE Transactions on Industrial Informatics* **9** (2), 686–696.

Tushar, M., Assi, C., Maier, M. & Uddin, M. (2014), 'Smart microgrids: Optimal joint scheduling for electric vehicles and home appliances', *IEEE Transactions on Smart Grid* **5** (1), 239–250.

Valerio, P. (2016), 'Is the iot a tech bubble for cities?: With more cities joining the smart city revolution and investing in sensors and other iot devices, the risk of a new tech bubble is rising.', *IEEE Consumer Electronics Magazine* **5** (1), 61–62.

Vilajosana, I., Llosa, J., Martinez, B., Domingo-Prieto, M., Angles, A. & Vilajosana, X. (2013), 'Bootstrapping smart cities through a self-sustainable model based on big data flows', *IEEE Communications Magazine* **51** (6), 128–134.

Vlacheas, P., Giaffreda, R., Stavroulaki, V., Kelaidonis, D., Foteinos, V., Poulios, G., Demestichas, P., Somov, A., Biswas, A. & Moessner, K. (2013), 'Enabling smart cities through a cognitive management framework for the internet of things', *IEEE communications magazine* **51** (6), 102–111.

Wang, F., Li, S., Mei, X., Xie, W., Rodríguez, J. & Kennel, R. (2015), 'Model-based predictive direct control strategies for electrical drives: An experimental evaluation of ptc and pcc methods', *IEEE Transactions on Industrial Informatics* **11** (3), 671–681.

Wang, F.-Y. (2015), 'Scanning the issue and beyond: Transportation and mobility transformation for smart cities', *IEEE Transactions on Intelligent Transportation Systems* **16** (2), 525–533.

Wang, Q., Liu, X., Du, J. & Kong, F. (2016), 'Smart charging for electric vehicles: A survey from the algorithmic perspective', *IEEE Communications Surveys & Tutorials* **18** (2), 1500–1517.

Will, C. & Schuller, A. (2016), 'Understanding user acceptance factors of electric vehicle smart charging', *Transportation Research Part C: Emerging Technologies* **71**, 198–214.

Wu, D., Aliprantis, D. & Ying, L. (2012), 'Load scheduling and dispatch for aggregators of plug-in electric vehicles', *IEEE Transactions on Smart Grid* **3** (1), 368–376.

Wu, H., Gilchrist, A., Sealy, K. & Bronson, D. (2012), 'A high efficiency 5 kw inductive charger for evs using dual side control', *IEEE Transactions on Industrial Informatics* **8** (3), 585–595.

Wu, X., Hu, X., Moura, S., Yin, X. & Pickert, V. (2016), 'Stochastic control of smart home energy management with plug-in electric vehicle battery energy storage and photovoltaic array', *Journal of Power Sources* **333**, 203–212.

Yang, L., Zhang, J. & Poor, H. (2014), 'Risk-aware day-ahead scheduling and real-time dispatch for electric vehicle charging', *IEEE Transactions on Smart Grid* **5** (2), 693–702.

Yu, R., Zhong, W., Xie, S., Yuen, C., Gjessing, S. & Zhang, Y. (2016), 'Balancing power demand through ev mobility in vehicle-to-grid mobile energy networks', *IEEE Transactions on Industrial Informatics* **12** (1), 79–90.

Zanella, A., Bui, N., Castellani, A., Vangelista, L. & Zorzi, M. (2014), 'Internet of things for smart cities', *IEEE Internet of Things Journal* **1** (1), 22–32.

Zhang, T., Chen, W., Han, Z. & Cao, Z. (2014), 'Charging scheduling of electric vehicles with local renewable energy under uncertain electric vehicle arrival and grid power price', *IEEE Transactions on Vehicular Technology* **63** (6), 2600–2612.

Zhao, J., Wen, F., Dong, Z., Xue, Y. & Wong, K. (2012), 'Optimal dispatch of electric vehicles and wind power using enhanced particle swarm optimization', *IEEE Transactions on Industrial Informatics* **8** (4), 889–899.

Zhu, F., Li, Z., Chen, S. & Xiong, G. (2016), 'Parallel transportation management and control system and its applications in building smart cities', *IEEE Transactions on Intelligent Transportation Systems* **17** (6), 1576–1585.

Zhu, Z., Lu, P., Rodrigues, J. & Wen, Y. (2013), 'Energy-efficient wideband cable access networks in future smart cities', *IEEE Communications Magazine* **51** (6), 94–100.

Section V

Security and Privacy Issues and Big Data in Smart Cities

19

Cyber-Security and Resiliency of Transportation and Power Systems in Smart Cities

Seyedamirabbas Mousavian[1], *Melike Erol-Kantarci*[2] *and Hussein T. Mouftah*[2]

[1] *School of Business, Clarkson University, Potsdam, NY, USA*
[2] *School of Electrical Engineering and Computer Science (EECS), University of Ottawa, Ottawa, ON, Canada*

19.1 Introduction

A recent report from the University of Cambridge, Centre for Risk Studies and the Lloyd's of London insurance market discusses a severe attack scenario that is capable of leaving 9.3 million people in New York City and Washington, DC, without power, causing $1 trillion loss in the U.S. economy and in Lloyd's (2015). According to FireEye Inc.'s attack map in Fireeye (2012), globally over 25,000 attacks are detected in a day, and among these attacks, the energy sector is the third-top targeted industry. Again, according to IOActive Lab's report by Cerrudu (2015), smart cities are open to cyberattacks, as well.

A cyberattack can be launched from any component of the power and electric transportation system (PETS) including a meter, phasor measurement unit (PMU), electric vehicle, or a relay. An infectious attack can propagate further in PETS and infect critical components, utility computers, and system operator servers. Considering the vast number of connected devices in the electric power system, the attack can affect a large number of devices. Hence, numerous automation and communication devices are used to measure, monitor, and control these power grid components (DoE, 2009). Ensuring all are healthy and non-compromised is highly challenging.

It is well known that traditional boundary protection techniques are no longer effective in the energy sector (DoE, 2012). With the integration of information and communication technologies (ICT), attacks have become more permeable. In addition, with the adoption of electric vehicles, transportation is also relying on the availability of the power grid. An outage in the power grid will also incapacitate electric transportation, or malware from an electric vehicle can penetrate to the electric vehicle supply equipment (EVSE) or the so-called charging station and can make its way into utility servers and power grid components (Caryl et al., 2013).

Previous research on cybersecurity of the smart grid focuses mostly on local attacks such as jamming, denial of service (DoS), controller malfunction, and load alteration attacks that compromise only several components of the grid. Although their local impact is not to be underestimated, if those attack behaviors are replicated, and attackers infect other components, the power grid will experience failure or malfunctioning

Transportation and Power Grid in Smart Cities: Communication Networks and Services, First Edition.
Edited by Hussein T. Mouftah, Melike Erol-Kantarci and Mubashir Husain Rehmani.

at multiple components. In the interdependent PETS, the mobility of electric vehicles will contribute to the risk of cyberattack spread from one utility service area to the others. This implies that legacy protection techniques that rely on quarantining a set of components cannot cope with the complex spreading pattern of infectious cyberattacks on PETS.

Smart grid cybersecurity and resilience have received significant attention in recent years. There are several works on smart grid protection as well as EVI protection. In Mousavian et al. (2014), a malware-oriented attack toward PMUs has been investigated. The PMUs have been assumed to be connected to the independent system operator (ISO) via a multi-hop network of routers. A MILP-based response model has been proposed in order to prevent worms from spreading. This work has been extended to address attacks in EV integration in Mousavian et al. (2015).

In Li et al. (2012), attacks on dynamic electricity pricing have been considered and their effect on EV charging has been discussed. It has been shown that price information attacks affects the optimality of EV schedules, thus leading to discrepancy between real and expected costs of EV charging. Malware on EVI can easily deploy price information attacks or load alteration attacks that may cause suboptimal schedules for EVs or incorrect pricing for utilities and customers.

In Carryl et al. (2013), the authors claim that malware infections can affect upstream equipment in the smart grid. This includes EVSEs, circuit breakers, transformers, PMUs, utility computers, and so on. Security certificates are potential solutions; however, they can be stolen via a Trojan software, and malicious users can easily sign malware as legitimate software. Therefore, having a response model for such attacks is vital for the resilience of the smart grid.

Worm propagation has been studied in the context of vehicular ad hoc networks (VANETs) where vehicles communicate safety and entertainment purposes. Khayam et al. (2004) uses the so-called susceptible, infected, removed (SIR) model to analyze worm propagation among vehicles. The authors consider the effect of traffic density, channel fading conditions, and mobility models. In VANETs, interactive patching has been shown to better handle worm outbreaks in comparison to preemptive patching. In Mickens et al. (2005), contrary to Khayam et al. (2004), the SIR model has been avoided due to its inaccuracy in mobile environments, and a new model based on probabilistic queuing has been proposed.

In this chapter, we will introduce an attack-and-response model for EVI in the smart city. We estimate the threat level in the EVI based on charging duration. Then, we develop a protection scheme that responds to the attack by temporarily isolating compromised EVSEs. Our mixed integer linear programming (MILP) model computes the minimal number of EVSEs to be isolated to keep the desired level of service, while protecting the grid from attack propagation. We do not include EVSE-to-EVSE communications as most EVI applications do not require peers to communicate.

19.2 EV Infrastructure and Smart Grid Integration

The smart grid and electric vehicles are among the main drivers of the IoT, as they form a large connected network of things, such as vehicles, charging stations, smart meters, intelligent electronic devices (IED), phasor measurement units (PMU), and so on. They are also anticipated to be the drivers of green smart cities by enabling efficient integration of renewable energy and lower emissions (Northfield, 2016; Khatoun et al., 2017; Zhang et al., 2017). EVs have many advantages including reducing the nation's dependency on fossil fuels, having lower operating costs than combustion engine vehicles, and having near zero emission. They are considered as the mobile portion of the smart grid (Lopes et al., 2011; Su et al., 2012). The energy stored in EV batteries can be used as short-term or emergency power supply for homes and businesses, in addition to supplying the motor force to drive the vehicle. Using smart grid communications through utility and service provider networks, EVs can participate in more advanced applications such as providing ancillary services. Ancillary services are operated at the background in the day-to-day operation of the power grid, and they include peak power supply, spinning reserves, and regulation. Despite numerous advantages and the opportunities that come with EVs, they may pose significant threats to the grid and risk the resilience of the smart grid.

EVs and EVSE, aka charging stations, which are EV's connection points to the power grid, form the EVI. An EV connect to an EVSE via an SAE 1772 connector, which can provide AC or DC power depending on the type of EVSE. Commercially Level-1, Level-2, and fast-charging stations are available, in addition to wireless chargers in demo stage. The SAE 1772 connector is designed with a pilot communication line while most EVSE's have wireless access either using the IEEE 802.15.4 standard (Zigbee) or IEEE 802.11 (WiFi; Mouftah and Erol-Kantarci, 2012; Niyato et al., 2014). National Renewable Energy Lab (NREL) has also done tests to use power-line communications (PLC) over J1772 (Simpson et al., 2012). Communication technologies are used to exchange information between EVs and EVSEs, such as state of charge (SoC), desired charging duration, and top up amount in dollars, or distance to next destination from EVs to EVSEs, electricity price, and load control signals from EVSEs to EVs. The communication infrastructure that links EVs with EVSEs provides means to exchange such basic information. In the near future, this communication infrastructure may also be used for communicating much more information such as exchanging energy-trading bids (Mets et al., 2010; Clement et al., 2009; Sortomme, 2011; Erol-Kantarci et al., 2011).

Physical security, cybersecurity, and the resilience of the power grid may be at risk if the attackers launch attacks from EVs or EVSEs. Physical tampering, malicious software uploads, and load alteration are among the possible attacks on the EVI. The effects of such attacks can be as harsh as regional blackouts or a less severe but still undesirable disruption of the ISO/RTO market systems. More importantly, cyberattacks initiating from the EVI can propagate very fast with the help of the ubiquitous communications and mobility of EVs. Malware loaded to an EVSE can compromise an EV easily, and the worms can be carried around a city and its interconnections, as an EV travels and charges its battery from several other EVSEs on the road. Those malware can compromise other equipment in the smart grid, including PMUs, line and transformer monitoring and protection equipment, and smart meters. The cyber-component of the modernized power grid makes attack propagation much more impactful than the legacy grid (Amin et al., 2015; Fang et al., 2012). Previous studies focus on protecting the smart grid equipment from malware and other potential threats (Mousavian et al., 2014; Yan et al., 2012; Amin et al., 2012) while the most vulnerable part of the grid, i.e., EVI, has been less explored. An attack scenario has been identified in Chan et al. (2013), which uses the EVI to attack the smart grid. In the scenario, the EV passes the challenge-response authentication using the trusted computing base (TCB); however, instead of an EV, a malicious load is physically connected (plugged-in) to the grid. The malicious load does not respond to the demand-response commands and does not curtail its load, which triggers circuit breakers when multiple loads acts simultaneously. Thus, the mismatch between the physically connected device and the cyber-authenticated EV risks the resilience of the smart grid. Chan (2014) proposes using two-step challenge response to mitigate such attacks; one over the wireless communication link and another through the SEA J1772 connector. In a system where EVSEs are interconnected and EVs communicate through VANETs, EVSE2EVSE communications and EV2EV communications magnify the scale of the attacks. We consider malware spread from EV to EVSE and EVSE to EV, but the proposed response model can be easily generalized for a fully connected EVI.

19.3 System Model

19.3.1 Model Definition and Assumptions

We consider an ecosystem of EVs and EVSEs where EVs are connected to EVSEs in one hop either using wireless communications or wireline communications, which are discussed in the next section. Here, Υ denotes the set of EVSEs and Θ denotes the set of detected compromised EVSEs where $|\Theta| = M$ and $\Theta \subseteq \Upsilon$. The set of EVs that have been connected to the detected compromised EVSE is given by Γ. When an EV is connected to a compromised EVSE for charging, the worm spreads from the EVSE to the EV with probability β. We assume infections are detected with a delay of Δt. The likelihood of an EVSE or an EV being compromised is referred to as threat level. The proposed protection scheme relies on a response model that minimizes the maximum threat level of all connected EVSEs at the time of detection, which is denoted by t_d. Then, compromised EVSEs are isolated temporarily while keeping the EV network operational for a load of D_{EV} that is the forecast demand for EV charging stations. The summary of notations used throughout the next section is given in Table 19.1.

Table 19.1 Summary of notations.

Notation	Definition
Υ	Set of EVSEs
Θ	Set of detected compromised EVSEs
Γ	Set of EVs that have been connected to the detected compromised EVSE
M	Number of detected compromised EVSEs
D_{EV}	Forecast demand for EV charging stations
x_j	Binary decision variable which equals 1 if $EVSE_j$ is kept connected to the network, and 0 otherwise
$A_j(t)$	random variable which equals 1 if EV_j is compromised, and 0 otherwise
B_j	Random variable which equals 1 if EV_j infected the EVSE, and 0 otherwise
β	The probability that an attack propagates from an EVSE to an EV or vice versa
$\theta_j(t)$	The probability of an EVSE being compromised
$\lambda_j(t)$	the probability of an EV being compromised
T_j	Threshold value for an operational EVI
C_j	Number of charging stations at$EVSE_j$
Y	Maximum threat level of EVSEs

19.4 Estimating the Threat Levels in the EVSE Network

In this chapter, we study the threat levels of the EVSEs when the network is experiencing an attack. We estimate the threat levels of the uncompromised EVSEs when one or more EVSEs were detected as compromised. Let us assume at time t_d, M EVSEs are detected to be compromised while the remaining EVSEs are uncompromised. It takes time Δt for the cyberattack to be detected. Hence, all EVs connected to the detected compromised EVSEs during time interval $[t_d - \Delta t, t_d]$ are likely to be compromised. During this time, there is a chance that the attack could have been propagated to uncompromised EVSEs by the compromised EVs recharging at different stations. Moreover, these new compromised EVSEs can further infect other EVs, and so forth. We represent by the probabilities, $\theta_j(t)$ and $\lambda_j(t)$, the likelihood of an EVSE and an EV being compromised (also called threat level) at time t, respectively. The initial threat levels of the EVSEs, $\theta_j(0)$, are given as follows.

$$\theta_j(0) = 1 \qquad \forall j \in \Theta \tag{19.1}$$

$$\theta_j(0) = 0 \qquad \forall j \notin \Theta \tag{19.2}$$

First, we calculate the initial threat levels of EVs, $\lambda_j(0)$, connected to the compromised EVSE during time $[t_d - \Delta t, t_d]$, given that one of these EVs had infected the EVSE in the first place. We represent the number of EVs connected to the compromised EVSE during this time period by $N_{\Delta t}$.

$$\Pr(A_j(0) = 1) = 0 \qquad \forall j \notin \Gamma \tag{19.3}$$

$$\begin{aligned}\Pr(A_j(0) = 1) = {} & \Pr\{(A_j(0) = 1 | B_j = 1\} \\ & \times \Pr\{B_j = 1\} \\ & + \Pr\{(A_j(0) = 1 | B_j = 0\} \\ & \times \Pr\{B_j = 0)\} \qquad \forall j \in \Gamma\end{aligned} \tag{19.4}$$

Therefore, we obtain equation (19.5) where β is the probability that a cyberattack propagates from an EVSE to an EV and vice versa. Probabilistic malware spread models assume attacks propagate within the network of nodes with some probability [7]; in this paper, we adopt a similar approach.

$$\lambda_j(0) = 1 \times \frac{1}{N_{\Delta t}} + \beta \times \frac{N_{\Delta t}}{N_{\Delta t} - 1} \tag{19.5}$$

Since all parameters in equation (19.5) are constant, we represent the initial threat levels by P_0 for simplicity. The infected EVs are likely to compromise the other EVSEs. To calculate the threat levels of EVSEs at time $t = t_d$, we need to sort EVs based on their time of connection to the EVSEs in time interval $[t_d - \Delta t, t_d]$. Hence, we represent their time of connection to EVSEs by cardinal numbers for simplicity. We use equation (19.6) to calculate the threat level of an EV after getting connected to a compromised or likely compromised EVSE.

$$\lambda_j(t) = 1 - [1 - \lambda_j(t-1)] \times [1 - \beta \times \theta_j(t-1)] \tag{19.6}$$

In equation (19.6), the first bracket represents the probability that the EV has not been compromised before connection to the EVSE; the second bracket represents the probability that the attack has not been propagated to the EV after connection to the EVSE.

Similarly, the threat level of EVSEs need to be updated after each time an EV is connected to the EVSE. We use equation (19.7) to calculate the threat levels of EVSEs.

$$\theta_j(t) = 1 - [1 - \theta_j(t-1)] \times \prod (1 - \beta \times \lambda_j(t-1)) \tag{19.7}$$

In equation (19.7), the first bracket represents the probability that the EVSE has not been compromised before connection of the newly connected EVs; the second bracket represents the probability that the attack has not been propagated to the EVSE after connection of the newly connected EVs. The threat levels of EVs and EVSEs need to be calculated iteratively from time $t = t_d - \Delta t$ until time $t = t_d$. Next, we use our proposed response model to determine which EVSEs should be isolated temporarily in order to slow down the propagation of the cyberattack.

19.5 Response Model

We use mixed integer linear programming to model the problem of responding to cyberattacks in an EVSE network. The objective function of the model is the minimization of the maximum threat level of all connected EVSEs at the time of detection $t = t_d$. The threat levels at time $t = t_d$ are calculated using equation (19.7).

$$Z = \min_{\mathbf{x}} \max_{j} (\theta_j(t_d) \times x_j) \qquad \forall j \in \Upsilon \tag{19.8}$$

We reformulated the equivalent of the objective function linearly given in equations (19.9)–(19.14) by substituting $w_j = \theta_j(t_d) * x_j$.

$$Z = \min Y \tag{19.9}$$

Subject to:

$$w_j \le x_j \qquad \forall j \in \Upsilon \tag{19.10}$$

$$w_j \le \theta_j(t_d) \qquad \forall j \in \Upsilon \tag{19.11}$$

$$w_j \ge \theta_j(t_d) - (1 - x_j) \qquad \forall j \in \Upsilon \tag{19.12}$$

$$w_j \ge 0 \qquad \forall j \in \Upsilon \tag{19.13}$$

$$w_j \le Y \qquad \forall j \in \Upsilon \tag{19.14}$$

Equation (19.15) is necessary to ensure that EVSEs are kept connected to the network whenever their threat levels do not exceed the threshold value.

$$\theta_j(t_d) > T_j - x_j \qquad \forall j \in \Upsilon \tag{19.15}$$

Isolating EVSEs may affect the power supply to EVs. Therefore, we should ensure that supply meets demand.

$$\sum_{j \in \Upsilon} C_j x_j \ge D_{EV} \tag{19.16}$$

where C_j is the capacity of $EVSE_j$ and D_{EV} is the forecast demand for the EV-charging stations at $t = t_d$.

19.6 Propagation Impacts on Power System Operations

As discussed earlier, a cyberattack to PETS may further propagate to utility servers and from there, on to other power system modules such as the generator software or utility PMU networks. In this chapter, we also study propagation of cyberattacks in PMU networks for two reasons. First, since PMU networks are more protected and harder to attack, a cyberattacker, aiming for the PMU networks, may initiate the attack from PETS. More importantly, PETS components, similar to PMUs, will become connected through a communication network, which could expedite propagation of cyberattacks. Studying the propagation of cyberattacks in PMU networks sheds light on the future of the interconnected PETS and puts more emphasis on the necessity of response models against cyberattacks propagation in PETS.

19.6.1 Cyberattack Propagation in PMU Networks

PMUs measure the synchronized phasors of bus voltages and currents in real time for more accurate observability of the power grid (Dua et al., 2008). A PMU takes about 30 to 120 measurements per second and sends its measurements to a phasor data concentrator (PDC) through a wireless communication network based on the NASPInet architecture (Wang, 2013; NASPOnet, 2009). In the NASPInet architecture, PMUs are connected to an IP-based communication network such as an Intranet. Although the communication network is a dedicated Intranet and is isolated from public networks, it is not immune to cyberattacks (Lin, 2012). A technical report from CISCO proposes that PMUs should send measurements using the IP multicast routing protocol (Cisco, 2012). In this communication protocol, a PMU is directly connected to a router and sends out data packets to pre-configured destinations. The list of these predetermined destinations is a great target for cyberattackers to propagate the cyberattack to other PMUs. Furthermore, it has been reported that the communication network shows poor network security and insufficient software security (INL, 2010). Under these conditions, cyberattackers could gain access to the PMU communication network, inject false measurement data, and propagate their cyberattack to the other PMUs, jeopardize the observability of the power system, and put safe operations of the power grid at risk. The summary of notations used throughout the next section is given in Table 19.2.

19.6.2 Threat Level Estimation in PMU Networks

It is assumed that M PMUs are detected to be compromised at time $t = 0$. It takes time Δt to disable the compromised PMUs from the communication network. Naturally, their measurements will no longer be used for the observability of the system. During time Δt, there is a chance that the attack could have been propagated to uncompromised PMUs and the detection software has not detected them yet. The reason is that the detection software cannot detect at 100% efficiency (INL, 2010). The cyberattack propagates to other PMUs through a path of interconnected routers in the communication network. If the cyberattack successfully breaks into all routers between the compromised and uncompromised PMUs, it is likely to contaminate the uncompromised

Table 19.2 Summary of notations.

Notation	Definition
Υ	Set of buses
Θ	Set of buses equipped with PMUs
Γ	Set of PMUs detected as compromised
Ψ	Set of buses with conventional devices
Ω	Set of branches with conventional devices
i, j, k	Indices of buses
M	Number of detected compromised PMUs
$H_{i,j}$	Connectivity between buses i and j
T_i	Threshold threat level (between 0 and 1) of PMU_i
D_{ij}	Nodal distance between PMU_i and PMU_j
Δt	Time that a propagation attempt takes
m	Number of Δt that the system operator takes to respond to the cyberattack
x_j	Binary decision variable which equals 1 if PMU_j is kept connected to the network, and 0 otherwise
φ_i	Observability number, number of times that bus i is observed, which is ≥ 1 if bus i is observable
$y_{i,j}$	Binary variable which equals 1 if the measurement from the conventional device at bus j is assigned to compute the unknown voltage phasor of bus i, and 0 otherwise.
$\rho_{i,j}$	Binary variable which equals 1 if the measurement from the conventional device at transmission line $i-j$ is assigned to compute the unknown voltage phasor of bus i, and 0 otherwise.
$A_j(t)$	random variable which equals 1 if PMU_j is attacked, and 0 otherwise.
α_{ij}	The probability that PMU_j is attacked through PMU_i
λ	The probability that an attack propagates through a router
γ	The probability that an attack propagates to a PMU through a router
$\theta_j(t)$	The threat level of PMU_j at time t

PMUs as well. These newly compromised PMUs further contaminate other PMUs, and so forth. We represent by the probability, $\theta_j(t)$, the likelihood of a PMU being compromised (also called threat level) at time t. Notice that threat levels increase over time as long as the network still contains compromised PMUs. It takes a time period of $m\Delta t$ for the system operator to run the optimization model to obtain the optimal response and confirm that the alarm is not a false alarm. Thus, certain PMUs, determined by the optimization model, are disabled at time $(m+1)\Delta t$. At time $t = 0$, equations (19.17) and (19.18) hold.

$$\Pr(A_j(0) = 1) = 1 \qquad \forall j \in \Gamma \tag{19.17}$$

$$\Pr(A_j(0) = 1) = 0 \qquad \forall j \notin \Gamma \tag{19.18}$$

All detected compromised PMUs are disabled by time Δt. However, it is likely that the attack could have been propagated to other PMUs but not detected yet. Therefore at time Δt, equations (19.19) and (19.20) hold.

$$\Pr(A_j(\Delta t) = 1) = 0 \qquad \forall j \in \Gamma \tag{19.19}$$

$$\Pr(A_j(\Delta t) = 1) = 1 - \prod_{i \in \Gamma} (1 - \alpha_{ij}) \qquad \forall j \notin \Gamma \tag{19.20}$$

where α_{ij} is the probability that the attack propagates from compromised PMU_i to an uncompromised PMU_j during the time Δt, and it is given by equation (19.21).

$$\alpha_{ij} = \gamma \lambda^{D_{ij}} \qquad i \in \Gamma, \quad j \notin \Gamma \tag{19.21}$$

In equation (19.21), λ is the probability that an attack propagates through a router, γ represents the probability that an attack propagates to another PMU and D_{ij}, called nodal distance, is the minimum number of routers that connect PMU_i and PMU_j on the communication network.

As mentioned earlier, disabling the compromised PMUs takes time Δt. Disabled PMUs would be enabled again if, during $m\Delta t$, the system operator ensures that the alarm was false. If the alarm was true, the operator disables the PMUs as determined by the optimization model at $m\Delta t$. Compromised PMUs would be disabled by time $(m + 1)\Delta t$. Therefore, the threat levels at time $(m + 1)\Delta t$ need to be calculated. Since the threat levels are calculated in an iterative process, we need to calculate the threat levels at time $2\Delta t$, $3\Delta t$, …, $n\Delta t$. Equations (19.22) and (19.23) hold at time $2\Delta t$.

$$\Pr(A_j(2\Delta t) = 1) = 0 \qquad \forall j \in \Gamma \tag{19.22}$$

$$\begin{aligned} \Pr(A_j(2\Delta t) = 1) = {} & \Pr\{(A_j(2\Delta t) = 1 | A_j(\Delta t) = 0\} \\ & \times \Pr\{A_j(\Delta t) = 0\} \\ & + \Pr\{(A_j(2\Delta t) = 1 | A_j(\Delta t) = 1\} \\ & \times \Pr\{A_j(\Delta t) = 1)\} \qquad \forall j \notin \Gamma \end{aligned} \tag{19.23}$$

In equation (19.23), we have expressions for all terms except for $\Pr\{(A_j(2\Delta t) = 1 | A_j(\Delta t) = 0\}$, which can be obtained from equation (19.24).

$$\begin{aligned} & \Pr\{A_j(2\Delta t) = 1 | A_j(\Delta t) = 0\} \\ & = 1 - \Pr\{A_j(2\Delta t) = 0 | A_j(\Delta t) = 0\} \\ & = 1 - \prod_{\substack{k,j \notin \Gamma \\ k \neq j}} (1 - \Pr(A_k(\Delta t) = 1) \times \alpha_{kj}) \qquad j,k \in \Theta \end{aligned} \tag{19.24}$$

We denote $\Pr(A_j(t) = 1)$ by $\theta_j(t)$. Therefore, we can rewrite equation (19.23) as equation (19.25) in terms of threat levels.

$$\begin{aligned} \theta_j(2\Delta t) = {} & \left(1 - \prod_{\substack{k,j \notin \Gamma \\ k \neq j}} (1 - \theta_k(\Delta t) \times \alpha_{kj}) \right) \\ & \times (1 - \theta_j(\Delta t)) + 1 \times \theta_j(\Delta t) \qquad j,k \in \Theta \end{aligned} \tag{19.25}$$

Equation (19.26) holds when $n \geq 2$. Hence, we use equation (19.26) to estimate the threat levels for time $t \geq 2\Delta t$.

$$\begin{aligned}\theta_j(n\Delta t) &= \left(1 - \prod_{\substack{k,j\notin\Gamma\\ k\neq j}} (1-\theta_k((n-1)\Delta t)\times\alpha_{kj})\right)\\ &\quad\times(1-\theta_j((n-1)\Delta t)) + \theta_j((n-1)\Delta t)\\ &= 1 - \left(\prod_{\substack{k,j\notin\Gamma\\ k\neq j}} (1-\theta_k((n-1)\Delta t)\times\alpha_{kj})\right)\\ &\quad\times(1-\theta_j((n-1)\Delta t) \qquad j,k\in\Theta\end{aligned} \tag{19.26}$$

19.6.3 Response Model in PMU Networks

The response to cyberattacks to a PMU network is modeled using mixed integer linear programming. In this optimization model, the objective function minimizes the maximum threat level of the PMUs that remain connected to the network at time $(m+2)\Delta t$. Time $(m+2)\Delta t$ is one Δt after disabling the PMUs determined to be disabled by the proposed optimization model. We summarize the threat levels from time $t = 0$ to $t = (m+2)\Delta t$ in equations (19.27)–(19.33).

$$\theta_j(0) = 1 \qquad \forall j\in\Gamma \tag{19.27}$$

$$\theta_j(0) = 0 \qquad \forall j\notin\Gamma \tag{19.28}$$

$$\theta_j(\Delta t) = 0 \qquad \forall j\in\Gamma \tag{19.29}$$

$$\theta_j(\Delta t) = 1 - \left(\prod_{i\in\Gamma}(1-\alpha_{ij})\right) \qquad \forall j\notin\Gamma \tag{19.30}$$

$$\theta_j(n\Delta t) = 0 \qquad \forall j\in\Gamma;\quad 2\leq n\leq m+2 \tag{19.31}$$

$$\begin{aligned}\theta_j(n\Delta t) &= 1 - \left(\prod_{\substack{k,j\notin\Gamma\\ k\neq j}} (1-\theta_k((n-1)\Delta t)\alpha_{kj})\right)\\ &\quad\times(1-\theta_j((n-1)\Delta t) \qquad \forall j,k\notin\Gamma; 2\leq n\leq m+1\end{aligned} \tag{19.32}$$

$$\begin{aligned}\theta_j((m+2)\Delta t) &= 1 - \left(\prod_{\substack{k,j\notin\Gamma\\ k\neq j}} (1-\theta_k((m+1)\Delta t)\alpha_{kj}x_k)\right)\\ &\quad\times(1-\theta_j((m+1)\Delta t)) \qquad \forall j,k\notin\Gamma\end{aligned} \tag{19.33}$$

The binary decision variable x_k in equation (19.33) equals 0 if the PMU_k is disconnected from the network. The system operator cannot control the threat levels from time $t = 0$ to $t = (m+1)\Delta t$. This is due to the network constraints such as control and communication delays in disconnecting PMUs. Disabling the selected PMUs occurs at time $(m+1)\Delta t$, which decreases the threat levels from time $t = (m+2)\Delta t$.

The response optimization model disables PMUs such that the maximum threat levels at time $(m+2)\Delta t$ is minimized. Equation (19.33) is represented linearly in equation (19.34) to be solved more efficiently.

$$\begin{aligned}\ln\{1-\theta_j((m+2)\Delta t)\}\\ &= \sum_{\substack{j,k\notin\Gamma\\ k\neq j}} \ln(1-\theta_k((m+1)\Delta t)\alpha_{kj})x_k\\ &+ \ln\{1-\theta_j((m+1)\Delta t)\} \qquad \forall j\notin\Gamma\end{aligned} \tag{19.34}$$

The following equality is used to obtain equation (19.34). Notice that the parameter K is constant and x_i is a binary variable.

$$\ln(1-Kx_i) = x_i \times \ln(1-K) \tag{19.35}$$

The objective function of the response model is to minimize of the maximum threat level of all connected PMUs at time $(m+2)\Delta t$.

$$Z = \min_{\mathbf{x}} \max_{j}(\theta_j((m+2)\Delta t) \times x_j) \qquad \forall j\notin\Gamma \tag{19.36}$$

$\theta_j((m+2)\Delta t)$ and therefore the objective function is not linear. We use its equivalent function, given in equation (19.37) and reformulate it linearly using equations (19.38)–(19.43).

$$Z = \min_{\mathbf{x}} \max_{j}\{-\ln[1-\theta_j((m+2)\Delta t)] \times x_j\} \qquad \forall j\notin\Gamma \tag{19.37}$$

To find the linear equivalent of the objective function, we use equations (19.38)–(19.43) to represent equation (19.37).

$$Z = \min Y \tag{19.38}$$

Subject to:

$$w_j \le x_j \qquad \forall j\notin\Gamma \tag{19.39}$$

$$w_j \le -\ln[1-\theta_j((m+2)\Delta t)] \qquad \forall j\notin\Gamma \tag{19.40}$$

$$w_j \ge -\ln[1-\theta_j((m+2)\Delta t)] - (1-x_j) \qquad \forall j\notin\Gamma \tag{19.41}$$

$$w_j \ge 0 \qquad \forall j\notin\Gamma \tag{19.42}$$

$$w_j \le Y \qquad \forall j\notin\Gamma \tag{19.43}$$

In the response model, we used equation (19.34) in equations (19.40)–(19.41) to obtain equivalent linear equations. Hence, equations (19.40)–(19.41) can be represented as equations (19.44)–(19.45), respectively.

$$\begin{aligned}w_j \le &- \sum_{\substack{j,k\notin\Gamma\\ k\neq j}} \ln(1-\theta_k((m+1)\Delta t)\alpha_{kj})x_k\\ &- \ln\{1-\theta_j((m+1)\Delta t)\} \qquad \forall j\notin\Gamma\end{aligned} \tag{19.44}$$

$$\begin{aligned}w_j \ge &- \sum_{\substack{j,k\notin\Gamma\\ k\neq j}} \ln(1-\theta_k((m+1)\Delta t)\alpha_{kj})x_k\\ &- \ln\{1-\theta_j((m+1)\Delta t)\} - (1-x_j) \qquad \forall j\notin\Gamma\end{aligned} \tag{19.45}$$

We add equation (19.46) to avoid disabling PMUs whose threat levels do not exceed the threshold value.

$$\theta_j((m+2)\Delta t) > T_j - x_j \qquad \forall j \notin \Gamma \tag{19.46}$$

which can be represented by:

$$\ln[1-\theta_j((m+2)\Delta t)] < \ln[1-T_j+x_j] \qquad \forall j \notin \Gamma \tag{19.47}$$

The right-hand side can be reformulated as a linear equation as:

$$\ln[1-\theta_j((m+2)\Delta t)] < (1-x_j)\ln(1-T_j) + x_j\ln(2-T_j) \qquad \forall j \notin \Gamma \tag{19.48}$$

Using equation (19.34), equation (19.48) can be represented linearly as equation (19.49).

$$\sum_{\substack{j,k\notin\Gamma \\ k\neq j}} \ln(1-\theta_k((m+1)\Delta t)\alpha_{kj})x_k + \ln\{1-\theta_j((m+1)\Delta t)\} < (1-x_j)\ln(1-T_j) + x_j\ln(2-T_j) \qquad \forall j \notin \Gamma \tag{19.49}$$

Disabling PMUs may affect the observability of the power system. Equation (19.50) provides the observability function introduced in [37].

$$\varphi_i = \sum_{j\in\Theta} H_{i,j}x_j + \sum_{j\in\Psi} H_{i,j}y_{i,j} + \sum_{\{i,j\}\in\Omega} H_{i,j}\rho_{i,j} \quad \forall i \in \Upsilon \tag{19.50}$$

Where $H_{i,j}$ is the connectivity matrix. $\sum_{j\in\Theta} H_{i,j}x_j$ determines the observability of the buses through PMUs still connected to the network; $\sum_{j\in\Psi} H_{i,j}y_{i,j}$ represents the observability of the buses through conventional devices installed at buses, and $\sum_{\{i,j\}\in\Omega} H_{i,j}\rho_{i,j}$ calculates the observability of the buses through conventional devices installed on transmission lines. Equation (19.51) ensures that there is at least one measurement to make each bus observable.

$$\varphi_i \geq 1 \qquad \forall i \in \Upsilon \tag{19.51}$$

Conventional devices provides observability for a group of buses that are not directly measured by PMUs. In such cases, a system of equations needs to be solved in order to obtain the unknown state variables. Each conventional measurement needs to be assigned to one state variable, given in equations (19.52)–(19.53), to guarantee the solvability of the system of equations [37]. Equation (19.52) assigns one state variable to each conventional voltage measurement device, whereas equation (19.53) assigns one state variable to each conventional current measurement device. Equation (19.54) forces the measurements of conventional devices to observe more buses since measurements obtained by conventional devices are more reliable at the time of cyberattacks.

$$\sum_{i\in\Upsilon} H_{i,j}y_{i,j} = 1 \qquad \forall j \in \Psi \tag{19.52}$$

$$\rho_{i,j} + \rho_{j,i} = H_{i,j} \qquad \forall \{i,j\} \in \Omega \tag{19.53}$$

$$\sum_{j\in\Psi} H_{i,j}y_{i,j} + \sum_{\{i,j\}\in\Omega} H_{i,j}\rho_{i,j} \leq 1 \qquad \forall i \in \Upsilon \tag{19.54}$$

where the decision variable $\rho_{i,j}$ determines whether the state variable of bus i is obtained by a conventional device at branch i, j. The decision variables are binary and given in equation (19.55).

$$x_i,\ y_{i,j},\ \rho_{i,j} \in \{0, 1\} \tag{19.55}$$

19.6.4 PMU Networks: Experimental Results

The performance of our response model is tested on the 6-bus test system. We use the 6-bus test system, introduced in [38], to explain the problem of study and our methodology. The 6-bus test system, depicted in Figure 19.1, consists of six buses, eleven transmission lines, and three generators.

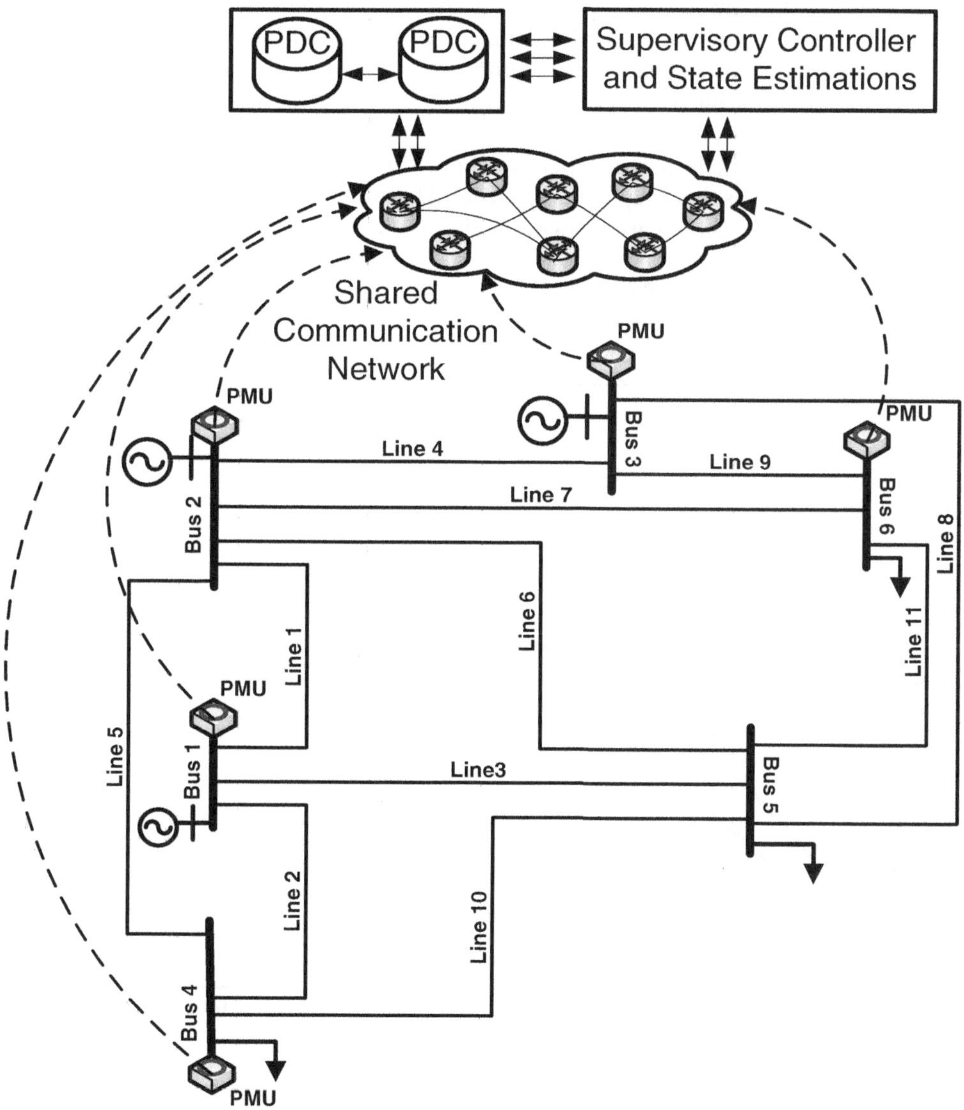

Figure 19.1 The 6-bus test system

In our experiments, we assume that an attack to a PMU propagates through a router with probability $\lambda = 0.05$, and it effectively compromises another PMU with probability $\gamma = 0.05$. For simplicity, we also assume that there is only one shortest path between PMUs. We set $N_{ij} = 1$. We set Δt to 0.1 seconds, $\Delta t = 0.1(s)$, and the threshold values to 0.005, $T_i = 0.005$.

To ensure full and redundant observability, it is assumed that PMUs are installed at buses 1, 2, 3, 4, and 6. The communication network, shown in Figure 19.1, consists of interconnected routers to send PMU measurements to PDCs. PMUs' nodal distances are given in Table 19.3.

For this case study, we assume that at time $t = 0$ the system operator is informed that PMU_1 and PMU_3 have been attacked by a cyber-intruder. In Figure 19.2, we show the threat levels of the three uncompromised PMUs over time when the system operator does not disable the compromised PMUs from the network. Notice that the threat levels increase nonlinearly until all PMUs become compromised with probability 1. PMU_4 is more at risk because it is closer to the compromised PMUs. The nodal distance from PMU_3 to PMU_4 is 1 while to PMU_2 is 3. Figure 19.3 illustrates the effect of the value of λ on the threat level of PMU_2.

To avoid propagation, the system operator should disable the compromised PMU_1 and PMU_3 and other PMUs that may be compromised because of the propagation. There are eight possible choices, which are shown in Table 19.4. The smallest threat levels can be

Table 19.3 6-bus test system – Nodal distances.

Bus	1	2	3	4	6
1	0	2	3	1	2
2	2	0	3	3	2
3	3	3	0	1	2
4	1	3	1	0	3
6	2	2	2	3	0

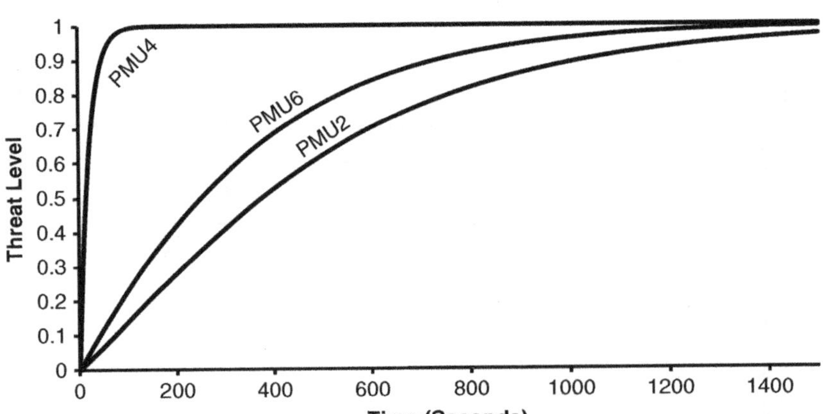

Figure 19.2 Threat levels in case of no response action

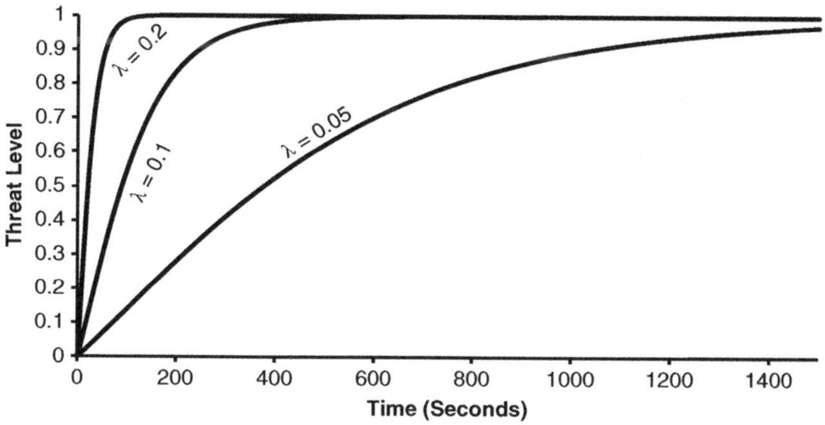

Figure 19.3 Effect of λ on threat level of PMU_2

Table 19.4 Candidate responses for the 6-bus test system.

PMU Status	Candidate Response 1	2	3	4	5	6	7	8
x_2	1	1	1	1	0	0	0	0
x_4	1	0	0	1	1	0	1	0
x_6	1	1	0	0	1	1	0	0
Threat Levels								
θ_2	0.00013	0.00013	0.00013	0.00013	0	0	0	0
θ_4	0.00500	0	0	0.00500	0.00500	0	0.00500	0
θ_6	0.00025	0.00025	0	0	0.00025	0.00025	0	0
Max(θ_j)	0.00500	0.00025	0.00013	0.00500	0.00500	0.00025	0.00500	0
Observaiblity	Yes	Yes	Yes	Yes	Yes	No	No	No

obtained when all PMUs are disabled. However, this solution is not feasible since the power system would no longer be observable. The second candidate is to disable PMU_4 and PMU_6, but the threat level of PMU_6 is less than the threshold value, $T_6 = 0.005$, and therefore it should remain connected to the network. The third candidate is to disable PMU_4. This action keeps the power network observable and minimizes the maximum threat level of all connected PMUs. In Table 19.5, we give the optimal solution, observability number of the buses obtained from equation (19.50) and the threat level of each connected PMU right after disabling PMU_4.

In Figure 19.4, the maximum threat levels of the connected PMUs are compared for two potential responses, 1) disconnecting only the compromised PMUs, and 2) disconnecting PMU_4 in addition to the compromised PMUs (optimal response). Although the threat levels still increase after the response action, the reduction in threat levels after the optimal response is considerable.

Table 19.5 6-bus system – Optimal response

PMU	x_j	Observability number (ϕ_i)	Threat level ($\theta_j(3\Delta t)$)
1	0	1	0
2	1	2	0.00013
3	0	2	0
4	0	1	0
5	NA	2	NA
6	1	2	0.00025

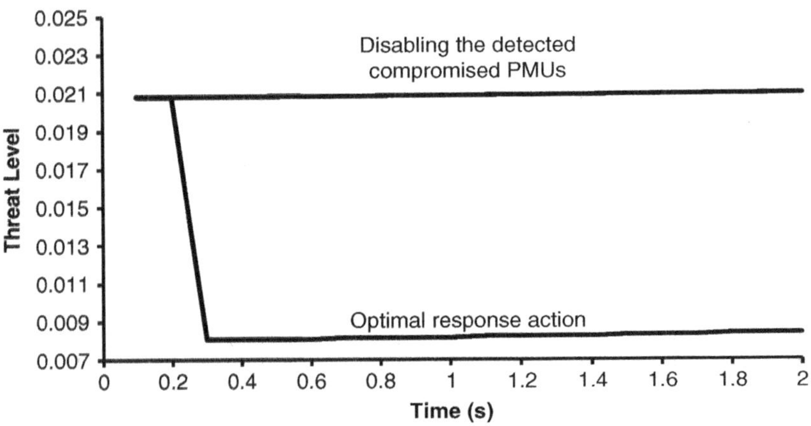

Figure 19.4 Comparison of two potential responses

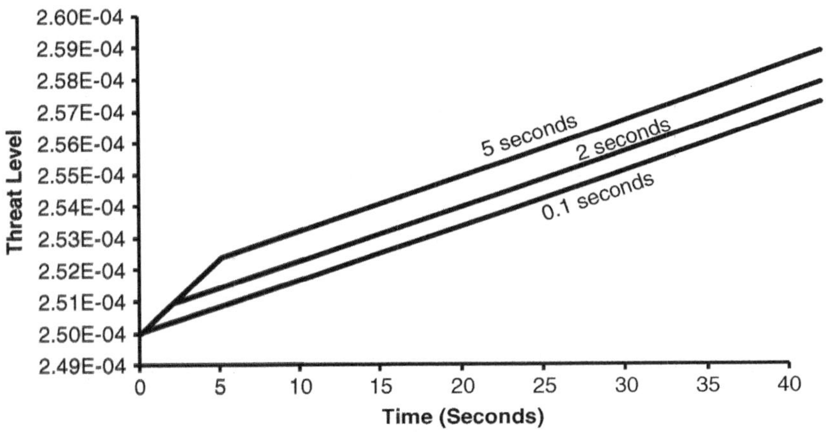

Figure 19.5 Threat level of PMU_6 for different computational times

To study how the optimization time affects the threat levels, we consider that the optimal results will be obtained in 0.1, 2, and 5 seconds. The optimal decision, in all three cases, is to disconnect PMU_4. The results are compared in Figure 19.5, which suggests that a shorter processing time is desired.

19.7 Conclusion and Open Issues

This chapter introduced the security vulnerabilities of smart cities from the smart grid and smart transportation perspective. We focused on EVs and PMUs, as they present point of access for cyber- threats; however, our approach can be easily generalized to any smart city asset. We introduced threat level computation and its corresponding response model.

Future work needs to investigate more on the response models and their implications on the service and observability side. Traditional response models isolate equipment in order to isolate attacks. Yet, in systems with inadequate redundancy this may cause disruption of service. One important open issue is to investigate the trade-off between redundancy investment and protection provided.

References

Lloyds (2015) Center for Risk Studies, University of Cambridge, "Lloyd's Emerging Risk Report, Business Blackout" [Online] www.lloyds.com

Fireeye (2017) Fireeye Inc. Cyber Threat Map. https://www.fireeye.com/cyber-map/threat-map.html

C. Cerrudu (2015) "An Emerging US (and World) Threat: Cities Wide Open to Cyber Attacks." [Online] https://securingsmartcities.org/wp-content/uploads/2015/05/CitiesWideOpenToCyberAttacks.pdf

DoE (2009) U.S. Department of Energy Office of Electricity Delivery and Energy Reliability, "Study of Security Attributes of Smart Grid Systems – Current Cyber Security Issues". [Online] http://www.inl.gov/scada/publications/d/securing_the_smart_grid_current_issues.pdf.

DoE (2012) U.S. Department of Energy Office, Electricity Subsector Cybersecurity Risk Management Process. [Online] http://energy.gov/oe/downloads/cybersecurity-risk-management-process-rmp-guideline-final-may-2012.

C. Carryl, M. Ilyas, I. Mahgoub, M. Rathod, (2013) "The PEV security challenges to the smart grid: Analysis of threats and mitigation strategies," International Conference on Connected Vehicles and Expo (ICCVE), pp. 300–305, 2-6.

S. Mousavian, J. Valenzuela, J. Wang, (2014) "A Probabilistic Risk Mitigation Model for Cyber Attacks to PMU Networks," in IEEE Transactions on Power Systems, vol. 30, no. 1, pp. 156–165.

S. Mousavian, M. Erol-Kantarci, T. Ortmeyer, (2015) "Cyber Attack Protection for a Resilient Electric Vehicle Infrastructure," IEEE Globecom-Workshop on Smart Grid Resilience, San Diego, CA, pp. 1–6.

Y. Li, R. Wang, P. Wang, D. Niyato, W. Saad, Z. Han (2012) "Resilient PHEV charging policies under price information attacks," IEEE Third International Conference on Smart Grid Communications (SmartGridComm), pp. 389–394, 5-8.

S. A. Khayam, H. Radha (2004) "Analyzing the spread of active worms over VANET," Proceedings of the 1st ACM international workshop on Vehicular ad hoc networks, pp. 86–87.

J. W. Mickens, B. D. Noble (2005) "Modeling epidemic spreading in mobile environments," Proceedings of the 4th ACM workshop on Wireless security, pp. 77–86.

R. Northfield (2016) "Greening the smart city," IEEE Engineering & Technology, vol. 11, no. 5, pp. 38–41, June 2016.

R. Khatoun and S. Zeadally (2017) "Cybersecurity and Privacy Solutions in Smart Cities," in *IEEE Communications Magazine*, vol. 55, no. 3, pp. 51–59.

K. Zhang, J. Ni, K. Yang, X. Liang, J. Ren and X. S. Shen (2017) "Security and Privacy in Smart City Applications: Challenges and Solutions," in IEEE Communications Magazine, vol. 55, no. 1, pp. 122–129.

J.A.P. Lopes, F. J. Soares, P. M. R. Almeida (2011) "Integration of Electric Vehicles in the Electric Power System," Proceedings of the IEEE, Vol. 99, No. 1, pp. 168–183.

W.C. Su, H. Rahimi-Eichi, W.T. Zeng, M.Y. Chow (2012) "A survey on the electrification of transportation in a smart grid environment," *IEEE Transactions on Industrial Informatics*, vol. 8, pp. 1–10.

H. T. Mouftah, M. Erol-Kantarci (2012) "Smart Grid Communications: Opportunities and Challenges," Handbook of Green Information and Communication Systems, Eds. M. S. Obaidat, A. Anpalagan and I. Woungang, Elsevier.

D. Niyato, N. Kayastha, E. Hossain, and Z. Han (2014) "Smart grid sensor data collection, communication, and networking: A tutorial," Wireless Communications and Mobile Computing (Wiley), vol. 14, no. 11, pp. 1055–1087.

M. Simpson, M. Jun, M. Kuss, T. Markel (2012) "Demonstrating PLC over J1772 During PEV Charging for Application at Military Microgrids." [Online] http://mydocs.epri.com/docs/PublicMeetingMaterials/0712/D2-4.pdf.

K. Mets, T. Verschueren, W. Haerick, C. Develder, F. De Turck (2010) "Optimizing smart energy control strategies for plug-in hybrid electric vehicle charging," IEEE/IFIP Network Operations and Management Symposium Workshops, pp. 293–299.

K. Clement, E. Haesen, J. Driesen (2009) "Coordinated charging of multiple plug-in hybrid electric vehicles in residential distribution grids," IEEE Power Systems Conference and Exposition, pp. 1–7.

E. Sortomme, M.M. Hindi, S. D. J. MacPherson, S. S. Venkata (2011) "Coordinated Charging of Plug-In Hybrid Electric Vehicles to Minimize Distribution System Losses," *IEEE Transactions on Smart Grid*, vol. 2, no. 1, pp. 198–205.

M. Erol-Kantarci, J. H. Sarker, H. T. Mouftah (2011) "Communication-based Plug-in Hybrid Electrical Vehicle Load Management in the Smart Grid," IEEE Symposium on Computers and Communications, pp. 404–409.

S.M. Amin, B.F. Wollenberg (2005) "Toward a smart grid: power delivery for the 21st century," *IEEE Power and Energy Magazine*, vol. 3, no. 5, pp. 34–41.

X. Fang, S. Misra, G. Xue, D. Yang (2012) "Smart Grid - The New and Improved Power Grid: A Survey," *IEEE Communications Surveys & Tutorials*, vol. 14, no. 4, pp. 944–980.

Y. Yan, Y. Qian, H. Sharif, and D. Tipper (2012) "A Survey on Cyber Security for Smart Grid Communications," *IEEE Communications Surveys and Tutorials*, Vol. 14, Issue 4, pp. 998–1010.

S. M. Amin and A. M. Giacomoni (2012) "Smart Grid- Safe, Secure, Self-Healing: Challenges and Opportunities in Power System Security, Resiliency, and Privacy," IEEE Power & Energy Magazine, pp. 33–40.

A. C-F. Chan and J. Zhou (2013) "On smart grid cybersecurity standardization: Issues of designing with NISTIR 7628," *IEEE Communication Magazine*, vol. 51, no. 1, pp. 58–65.

A. C.-F. Chan, J. Zhou (2014) "Cyber-physical device authentication for smart grid electric vehicle ecosystem," *IEEE Journal on Selected Areas in Communications*, Vol. 32, 1509–1517.

D. Dua, S. Dambhare, R. K. Gajbhiye, and S. A. Soman (2008) "Optimal multistage scheduling of PMU placement: An ILP approach," *IEEE Transactions on Power Delivery*, vol. 23, pp. 1812–1820.

W. Wang and Z. Lu (2013) "Cyber security in the smart grid: Survey and challenges," *Computer Networks*.

NASPInet (2009) Data bus technical specifications for north american syncro-phasor initiative network (NASPInet). North American Syncro-Phasor Initiative Network. [Online]. Available: https://www.naspi.org/File.aspx?fileID=587

H. Lin, Y. Deng, S. Shukla, J. Thorp, and L. Mili (2012) "Cyber security impacts on all-PMU state estimator a case study on co-simulation platform GECO," in *IEEE Third International Conference on Smart Grid Communications (SmartGridComm)*, pp. 587–592.

Cisco (2012) PMU networking with IP multicast. CISCO Public. [Online]. Available: http://www.cisco.com/

INL (2010) "NSTB assessments summary report: Common industrial control system cyber security weaknesses," Idaho National Laboratory (INL), Tech. Rep.

C. Shuguang, H. Zhu, S. Kar, T. T. Kim, H. V. Poor, and A. Tajer (2012) "Coordinated data-injection attack and detection in the smart grid: A detailed look at enriching detection solutions," *IEEE Signal Processing Magazine*, vol. 29, pp. 106–115.

S. Azizi, B. G. Gharehpetian, G. Hug-Glanzmann, and A. Dobakhshari (2012) "Optimal integration of phasor measurement units in power systems considering conventional measurements," *IEEE Transactions on Smart Grid*, vol. 4, pp. 1113–1121.

A. J. Wood and B. F. Wollenberg (1996) *Power Generation, Operation and Control*. John Wiley & Sons.

IEEE Test System (1979) Subcommittee, "IEEE reliability test system," *IEEE Transactions on Power Apparatus and Systems*, vol. 98, pp. 2047–2054.

20

Protecting the Privacy of Electricity Consumers in the Smart City

Binod Vaidya and Hussein T. Mouftah

School of Electrical Engineering and Computer Science (EECS), University of Ottawa, Ottawa, ON, Canada

20.1 Introduction

An intelligent energy infrastructure, which links together various elements of city operations, would be one of the essential features in a smart city.

A smart grid is an electric power grid that employs information and communication technologies (ICTs) to improve its efficiency, reliability, resiliency, and flexibility. Such an electric grid not only provides information about its energy usage and costs to consumers such that they can make decisions independently but also transforms electric power grid through remote monitoring and control, automation, self-healing approach as well as provides safe and reliable integration of distributed renewable energy resources (Strasser et al., 2015).

Smart meters are one of the key enablers of smart grids. An advanced metering infrastructure (AMI) allows bidirectional energy flows as well as utilizes two-way communication enabling energy service providers (ESPs) to receive energy consumption data of consumers and sends pricing or control signals back to consumers in real time. By measuring consumers' near real-time energy consumption data at high temporal resolution, AMI enables ESPs to control and optimize the supply and distribution and even offer their customers pricing schemes based on current offer and demand (Siano, 2014). Distribution system operators can monitor the electric power grid at a higher sampling rate and granularity than earlier. Electricity customers also can be benefitted from smart meter deployment by receiving timely information about consumed power and managing power consumption accordingly.

Furthermore, the smart grid allows consumers to sell electricity to the grid or other consumers by producing electricity using photovoltaic or wind turbines (Depuru, Wang, and Devabhaktuni, 2011b). Similarly, the smart grid also includes vehicle-to-grid (V2G), in which electric vehicles (EVs) can communicate with power grid operators to trade demand response services by delivering stored electricity into the electric power grid (Yu et al., 2016).

The AMI is a predominant and fundamental component in the development and deployment of the smart grid in the smart cities; nonetheless, V2G networks are also rapidly increasing in the urban cities for charging /discharging EVs.

Transportation and Power Grid in Smart Cities: Communication Networks and Services, First Edition.
Edited by Hussein T. Mouftah, Melike Erol-Kantarci and Mubashir Husain Rehmani.

Although the smart grid delivers numerous performance benefits to the electric power industry and also enables consumers to optimize their power consumption, the smart grid infrastructure (i.e., AMI and V2G) in the smart city has become increasingly susceptible to a wide range of cyber-threats.

Basically, Smart grid technologies capture customer data relating to sensitive information that is used for various purposes such as real-time pricing and demand response; however, privacy can be threatened and breached by a number of practices, which are normally considered as unacceptable (Finster and Baumgart, 2015; Li et al., 2012).

If realistic security and privacy-preserving approaches are not employed in the smart grid, new challenges to security protection, privacy, and data protection will emerge.

The objective of this chapter is to provide insights of privacy protections of the electricity consumers in the smart city. It mainly focuses on aspects of privacy principles including privacy by design (PbD). It also stipulates a basis for better understanding of the current state-of-the-art privacy engineering as well as privacy impact assessment and privacy-enhancing technologies. The rest of the chapter is organized as follows. Section 20.2 discusses privacy concern in the smart grid including AMI and V2G. Section 20.3 discusses the privacy principles while section 20.4 presents privacy engineering focusing on privacy protection goals and pertinent frameworks. Section 20.5 depicts privacy impact assessment, and section 20.6 explores privacy-enhancing technologies. Finally, section 20.7 concludes the chapter.

20.2 Privacy in the Smart Grid

The smart grid (i.e., AMI and V2G) introduces substantial benefits and opportunities to the smart city, but it also raises several challenges related to privacy. For instance, fine-grained smart metering data and control messages provide ESPs with the information about real-time electricity consumption status. However, using the fine-grained electricity consumption, not only varied information about the consumer premises can be inferred but also the probability of leaking customers' privacy including personal information, daily activities, individual behaviors, etc., increases (Lisovich, Mulligan, and Wicker, 2010; Jokar, Arianpoo, and Leung, 2016).

Indeed, security and privacy are considered as crucial components for the success of secure smart grid networks including AMI and V2G networks. Privacy emphasizes the individual's ability to control the collection, use, and dissemination of his/her personally identifying information (PII), whereas security offers mechanisms to ensure confidentiality and integrity of information and availability of ICT systems. Yet the perceptions of privacy and security do intersect. Ensuring privacy is more complicated than ensuring security (SGIPCSWG-V1, 2014; SGIPCSWG-V2, 2014; SGIPCSWG-V3, 2014).

Privacy considerations in the smart grid embrace examining the rights, values, and interests of individuals. According to NIST IR 7628v2 (SGIPCSWG-V2, 2014), four dimensions of privacy are considered: 1) privacy of personal information; 2) privacy of the person: 3) privacy of personal behavior; 4) privacy of personal communications. Though most smart grid stakeholders directly address the first dimension, since most data protection laws and regulations mainly cover privacy of personal information, the remaining three dimensions are also essential privacy considerations in the smart grid.

Not only potential risks in AMI can become ingress points for adversaries, but also potential insider risks can be exacerbates to stretch to a magnified threat level. Several serious privacy risks in AMI include eavesdropping, traffic analysis, statistical disclosure, and consumption profiling (Jiang et al., 2014; Krishna, Weaver, and Sanders, 2015). Malicious entities could use the smart metering personal data to enable malevolent uses such as identity theft, burglary, vandalism, stalking, etc.

Privacy concern is also one of the obstacles to the successful deployment of V2G networks. Privacy protection issues in V2G networks are more challenging than in AMI networks. Due to the electric mobility (E-mobility), EVs may join or depart the EV-charging network frequently, so their privacy requirements are more stringent. Privacy issues in V2G networks include location privacy of EVs and attacks of leaking privacy such as eavesdropping, man-in-the-middle attack (MiTM), impersonation attack, sybil attack, and physical attack.

Primarily, potential privacy consequences of smart grid systems (i.e. AMI and V2G) include identity theft, determine personal behavior patterns, determine specific appliances used, perform real-time surveillance, reveal activities through residual data, targeted home invasions, provide accidental invasions, activity censorship, decisions and actions based upon inaccurate data, profiling, unwanted publicity and embarrassment, tracking behavior of renters/leasers, behavior tracking (possible combination with personal behavior patterns), location tracking, and public aggregated searches revealing individual behavior (Asghar et al., 2017; Cintuglu et al., 2017).

20.2.1 Privacy Concerns over Customer Electricity Data Collected by the Utility

A primary data flow in the AMI encompasses customer electricity data from the smart meter to the utility. Smart meters can take readings of fine-grained customer electricity data and send them to the utilities. And the utilities use these data for a variety of purposes. For instance, the customer electricity data can be used for applying time-variant pricing (i.e., time-of-use rates, critical peak pricing, real-time pricing), for better understanding customer demand, or for detecting meter tampering (Depuru, Wang & Devabhaktuni, 2011a; Amin et al., 2015). Utilities also make the data available to customers for optimizing their power consumption (Hubert and Grijalva, 2012). However, this information flow in AMI raises the following privacy concerns.

1) Smart meters store meter data internally. Under existing the AMI architecture, the utility controls the meter located at the customer's premise and does not grant access to other parties without customer's consent. However, some adversaries may obtain some data about a specific electricity customer and remotely access his/her smart meter to acquire an energy consumption pattern. Such an energy consumption pattern may be key information for other malicious activities.
2) Consumer meter data moves from the smart meter to the utility, typically via a private network that the utility operates and even through public network (i.e. the Internet). Though data confidentiality is mandatory for communications between the smart meter and the utility in the existing AMI systems, user privacy is not protected, so still high privacy risks do exist in such AMI systems.
3) While storing consumer meter data at the utility premise, a customer's privacy interests should be legally protected. An adversary may be able to penetrate the AMI

network to access meter data stored at the utility premise. Privacy risks will be even higher if third parties are allowed to access energy usage information directly from the utilities.

20.2.2 Privacy Concerns on Energy Usage Information Collected by a Non-Utility-Owned Metering Device

A significant shift in the smart grid, relative to its predecessor, is that third parties will be enabled to collect metering data. In case of V2G networks, charging station operators shall collect meter data and process them. These metering data may be transferred to the utility company for third-party billing. Hence, during collecting energy usage information by a non-utility-owned metering device, the customer's privacy is a prime concern.

20.2.3 Privacy Protection

Cyber security is a critical concern owing to the growing potential of cyber-attacks and incidents against critical energy infrastructures in the smart city. The cyber-security solution for such an energy infrastructure must tackle not only intentional attacks from industrial espionage and hackers but also unintentional compromises of the electric power grid infrastructure due to user errors, equipment failures, and natural disasters.

Since private information about consumers (e.g., home energy consumption) shall be included during energy usage data exchange between multiple stakeholders of the Ssart grid, it could harm particular individuals, if the consumer data are not used with appropriate protection measures. Thus, privacy risks and challenges introduced by the smart grid (i.e., AMI and V2G) have to be properly addressed. Protecting critical energy infrastructures in the smart city has to be given great precedence, and technical measures to protect customer's privacy should be considerably prioritized.

Privacy protection comprises preventing any valuable information (i.e., private information) related to the identity of an entity to be known by other entities. Only well-protected smart grid systems in the smart city ecosystem would be considered robust and secure.

The degree of privacy protection should be well premeditated. Applying PbD in conjunction to engineering aspects is critically imperative such that privacy impact assessment (PIA) can be properly conducted, and appropriate privacy-enhancing technologies (PETs) can be deployed.

20.3 Privacy Principles

There are several privacy principles that can assist to impose privacy protection in ICT-based systems including smart grid networks.

As system design and architecture may collect PII, there is a possibility to violate privacy protection. A better methodology to protect privacy of the consumer data is to apply a PbD approach (IPCO 2009; Cavoukian 2011). With initiation of A. Cavoukian, an original Privacy-by-Design (PbD) approach was proposed in 1990. This approach identifies a set of foundational principles that should be followed when designing and

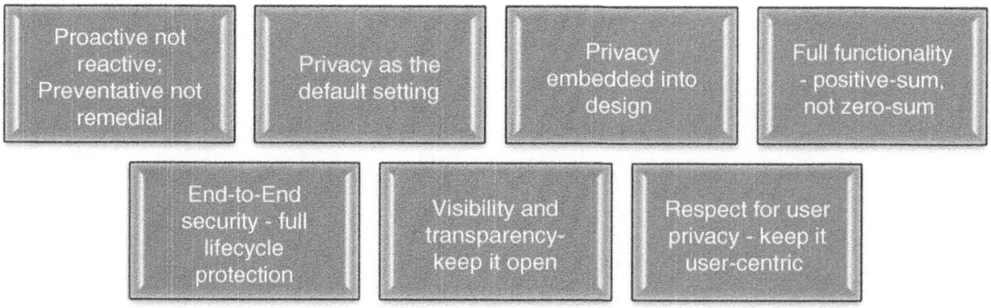

Figure 20.1 Seven foundational principles of PbD.

developing privacy-sensitive applications (IPCO, 2009). These 7 foundational principles are depicted in Figure 20.1.

With a proactive approach, PbD encompasses embedding privacy directly into design of technologies, business practices, and networked infrastructures. It compels privacy as a foundational requirement, consequently, preventing privacy-invasive occurrences before they may happen. By making privacy the default setting within an organization, its customers' privacy can be well protected (Cavoukian, 2011).

Nevertheless, as an early approach for interpretation of privacy and personal information, the Federal Trade Commission (FTC), USA, developed the Fair Information Practice Principles (FIPPs) in 1973. Its core principles of privacy in the context of information are: 1) notice/awareness; 2) choice/consent; 3) access/participation; 4) integrity/security; and 5) enforcement/redress (Landesberg et al., 1998).

Different adaptations of the FIPPs have been demarcated. For instance, inferring privacy and data protection, the Guidelines on the Protection of Privacy and Trans-border Flows of Personal Data were developed by the Organization for Economic Cooperation and Development (OECD) in 1980 and revised in 2013 as OECD Privacy Framework (OECD, 2013). The principles in the OECD documents have been widely adopted.

In order to realize a universal privacy framework, the International Organization for Standardization (ISO) and the International Electrotechnical Commission (IEC) have published ISO/IEC 29100 as an international standard in 2011 (ISO/IEC JTC 1/SC 27 IT Security techniques, 2011). Privacy principles described in the ISO/IEC 29100:2011 Privacy Framework are mainly derived from existing principles developed by various international bodies such as the FTC and OECD. This framework not only targets organizations and intends to support them in defining their privacy preservation requirements but also aims at enhancing the existing security standards by the privacy perspective whenever PII is processed. Hence, these privacy principles are used to guide the design, development, and implementation of privacy policies and privacy controls.

The American Institute of Certified Public Accounts (AICPA) and the Canadian Institute of Chartered Accountants (CICA) have developed privacy principles and applicable criteria, known as Generally Accepted Privacy Principles (GAPPs) in order to assist organizations in the design and implementation of comprehensive privacy practices and policies (Cornelius, 2009).

The relationship among three major privacy principles, namely, OECD Privacy Framework Privacy Principles, Privacy Principles of the ISO/IEC 29100:2011 Privacy

OECD Privacy Framework Privacy Principle	ISO/IEC 29100 Privacy Principle	AICPA Generally Accepted Privacy Principle
- Collection limitation	- Consent & choice - Collection limitation	- Choice & consent - Collection
- Data quality	- Accuracy & quality	- Quality
- Purpose specification	- Purpose legitimacy & specification	- Notice
- Use limitation	- Data minimization - Use, retention & disclosure limitation	- Use & retention - Disclosure to third parties
- Security safeguards	- Information security	- Security for privacy
- Openness	- Openness, transparency & notice	- Notice
- Individual participation	- Individual participation & access	- Monitoring & enforcement - Access
- Accountability	- Accountability - Privacy compliance	- Management

Figure 20.2 The relationship among OECD Privacy Framework, ISO/IEC 29100:2011 Privacy Framework, and AICPA GAPPs.

Framework, and AICPA Generally Accepted Privacy Principles (GAPPs) is shown in Figure 20.2.

By studying privacy principles from other relevant industries that handle and store sensitive information, smart grid actors, i.e. utilities, can provide utmost protection to the consumer electricity data.

The National Institute of Standards and Technology (NIST), USA, is one of the central players in promoting the growth of the smart grid by developing a framework for the smart grid that embraces interoperable standards and protocols so that all components of the smart grid shall be able to work together. In this regard, NIST published Smart Grid Cyber Security Strategy and Requirements (NIST IR 7628) in 2010 and later revised in 2014 (SGIPCSWG-V1, 2014; SGIPCSWG-V2, 2014; SGIPCSWG-V3, 2014). NIST IR 7628 Vol 2 (SGIPCSWG-V2, 2014) is dedicated to privacy in the smart grid. NIST IR 7628 applies FIPPs while deliberating privacy considerations for the smart grid and uses GAPPs as one of privacy principles.

IPCO (2010), by adapting a PbD approach with the seven fundamental foundational principles to the smart grid context, best practices for smart grid PbD have been created such that smart grid systems should not only proactively embed privacy requirements into their designs and ensure that privacy is the default but also be visible and transparent to consumers and be designed with respect for consumer privacy. A related approach that emphasizes embedding privacy into the design of electricity has been depicted in Cavoukian, Polonetsky, and Wolf (2010). Further applying privacy by design for third-party access to customer energy usage data is provided in the article Cavoukian and Polonetsky (2013).

Table 20.1 Five Privacy-Aware Principles for the Demand Response Platform.

Privacy-aware principle	Requirement
Provide full disclosure of data collection	• description requirement • enforceability requirement • irrevocability requirement • intelligibility requirement
Require consent to data collection	• acknowledgement requirement • opt-in requirement
Minimize collection of personal data	• functional requirement for collection • distributed processing requirement
Minimize identification of data with individuals	• non-attribution requirement • separate storage requirement
Minimize and secure data retention	• functional requirement for retention • security requirement • non-reusability requirement

Wicker and Thomas (2011) have advocated a framework for privacy-aware design practices for embedding privacy awareness into information networks that consists of a set of principles derived from the FIPPs. This approach that has five privacy-aware principles is specifically intended for the demand response platform (Wicker and Schrader 2011), which is depicted in Table 20.1.

20.4 Privacy Engineering

Provisioning a PbD approach is not sufficient due to the lack of holistic and systematic methodologies that address the intricacy of privacy and the absence of the translation of privacy principles into engineering activities (Alshammari and Simpson 2016). Privacy engineering deals with designing, implementing, adapting, and evaluating guidelines, protocols, and techniques to methodically apprehend and address privacy issues in the development of ICT systems (Gurses and Del Alamo, 2016; Spiekermann and Cranor, 2009). Hence, privacy engineering focuses on providing guidance that will enable organizations to make persistent decisions about resource allocation and effective deployment of controls in ICT systems in order to decrease privacy risks (Cavoukian, Shapiro, and Cronk, 2014).

There are not only several efforts to formulate concepts for privacy engineering but also approaches to define engineering, technical, and operational aspects for PbD (Hoepman, 2014; Kroener and Wright, 2014; Antignac and Le Metayer, 2015; Bringer et al., 2015).

20.4.1 Privacy Protection Goals

Protection goals are regarded as authoritative components while evaluating information security of ICT systems and choosing appropriate technical and operational safeguards in various technologies (Meis, Wirtz, and Heisel, 2015).

Classically, information security in ICT systems features three security protection goals, namely confidentiality, integrity, and availability, commonly known as CIA triad. The CIA triad is typically deemed as critical to evaluate security conditions of the ICT systems. Confidentiality guarantees preserving authorized controls on information access and disclosure including means for protecting personal privacy and sensitive information from malicious people. Integrity ensures protecting against unauthorized and improper information alteration or destruction and includes ensuring information non-repudiation and authenticity. Availability provides a guarantee of timely and reliable access to the information by authorized people.

Three privacy-specific protection goals such as unlinkability, transparency, and intervenability (i.e. UTI triad) (Hansen, Jensen, and Rost, 2015) have been identified in order to strengthen the privacy perspective by accompanying with the above security protection goals, i.e., the CIA triad.

1) Unlinkability guarantees that privacy-relevant data cannot be linked across privacy domains that are constituted by a common purpose and context. This implies that processes must function in such a way that the privacy-relevant data are not linkable to any privacy-relevant information outside of the domain. It is interrelated to not only the principles of necessity and data minimization but also the principle of purpose binding. Unlinkability refers to the property of anonymity and its different aspects of realization. The protection goal of unlinkability targets realizing unobservability and undetectability.
2) Transparency ensures that all aspects (i.e., legal, technical, and organizational) of privacy-relevant data processing should be clear and able to reassemble at any time. The information that is subjected for privacy-relevant data processing needs to be available any time. It is correlated to the principles concerning openness and a prerequisite for accountability.
3) Intervenability guarantees that intervention is likely regarding all possible privacy-relevant data processing, in particular for data owners. The purpose of intervenability is to apply corrective measures and counterbalances on demand. It is related to the principles concerning customers' rights. Furthermore, data controllers should have effectively control over data processors and their ICT systems to manipulate the data processing at any time.

Meaningful protection goals should be delivered in order to balance the requirements derived from both protection goals (i.e., CIA triad and UTI triad) concerning legal, technical, and organizational processes. Considerations on fairness, impartiality, and accountability postulate guidance for balancing the requirements as well as determining better strategies and appropriate protections.

In a similar manner, NIST has also provided three privacy engineering goals—disassociability, predictability, and manageability (i.e., DPM triad)—for developing and operating privacy-preserving ICT systems (Brooks et al., 2017). These goals are designed to enable ICT designers to build ICT systems that are capable of implementing an organization's privacy protection goals and reinforce the management of privacy risk.

1) Disassociability empowers the processing of personal information or events without any correlation with individuals or devices beyond the operational requirements

of the ICT system. It recognizes that privacy risks can result from exposure even when access is authorized or as a consequence of a transaction. Further deliberation for increasing the effectiveness of disassociability is by constructing a classification of prevailing identity-related categories, including anonymity, de-identification, unlinkability, unobservability, pseudonymity, or others. Such a classification could provision a more precise privacy risk mitigation.

2) Predictability enables reliable assumptions by individuals, owners, and operators about personal privacy and secret information and their processing by the ICT system. It supports a wide range of organizational interpretations of transparency from a value statement about the significance of open processes to a requirements-based view that particular information has to be shared. Predictability facilitates the preservation of trusted relationships between ICT systems and individuals and the capability for individuals' self-determination.
3) Manageability offers the capability for granulated administration of personal information including alteration, deletion, and selective disclosure. It is a primary property for enabling self-determination and fair treatment of individuals. When the ICT system permits fine-grained control over data, organizations can be able to implement key privacy principles including maintaining data quality and integrity and implementing individuals' privacy preferences. Manageability could provision the mapping of technical controls such as data tagging and emerging standards in identity management that relate to attribute transmission.

Combination of three privacy engineering objectives (i.e., DPM triad), complemented by the CIA triad to address unauthorized access to personal information, stipulate a chief set of information system capabilities to support the well-balanced realization of business goals and privacy goals and assist in the mapping of controls to mitigate identified privacy risks.

The UTI triad and DPM triad can be loosely associated to provide privacy protection goals for a particular privacy-preserving ICT system. A mapping between UTI and DPM triads is shown in Figure 20.3.

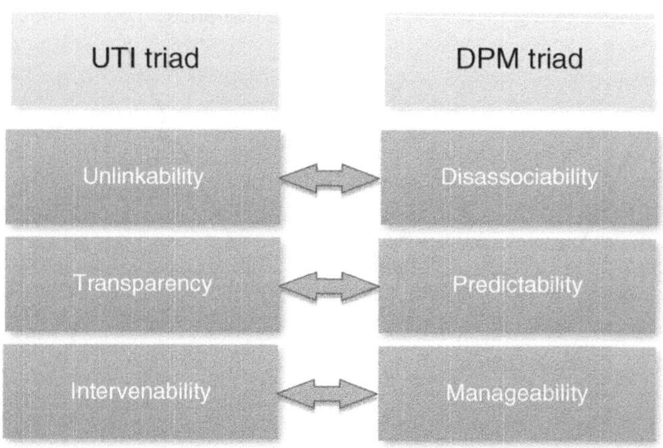

Figure 20.3 Mapping between UTI and DPM triads.

20.4.2 Privacy Engineering Framework and Guidelines

Numerous methodologies for designing privacy into modern ICT systems have been considered (Notario et al., 2014; Notario et al., 2015; Kung, 2014). Aspects of privacy engineering can be articulated by incorporating privacy requirements into the areas of the systems engineering life cycle (SELC) that could facilitate core privacy protection objectives and other organizational objectives. For some organizations, the prime motivation for privacy engineering would be for regulatory compliance purposes or reducing organizational risk.

Furthermore, MITRE Corporation has formulated a privacy engineering framework such that privacy engineering operationalizes the PbD logical framework within ICT systems (MITRE-CoP, 2014) by:

- segregating PbD into activities related with those of the SELC and adopted by particular methods that account for privacy's distinct characteristics;
- outlining and applying privacy requirements in terms of implementable system functionality and properties for focusing privacy risks within the SELC. Privacy risks are recognized and sufficiently addressed; and
- supporting deployed ICT systems by affiliating system usage and enhancement with a broader privacy platform.

Figure 20.4 demonstrates mapping of the fundamental privacy engineering activities into stages of the typical SELC. Such a mapping ensues for every SELC, including agile development.

The primary life cycle activities for privacy engineering and privacy methods are listed in Table 20.2, and a brief discussion of life cycle activities for privacy engineering framework is given as follows.

1) Privacy requirements definition. This phase shall define specification of system privacy properties in a way that supports system design and development. It includes the

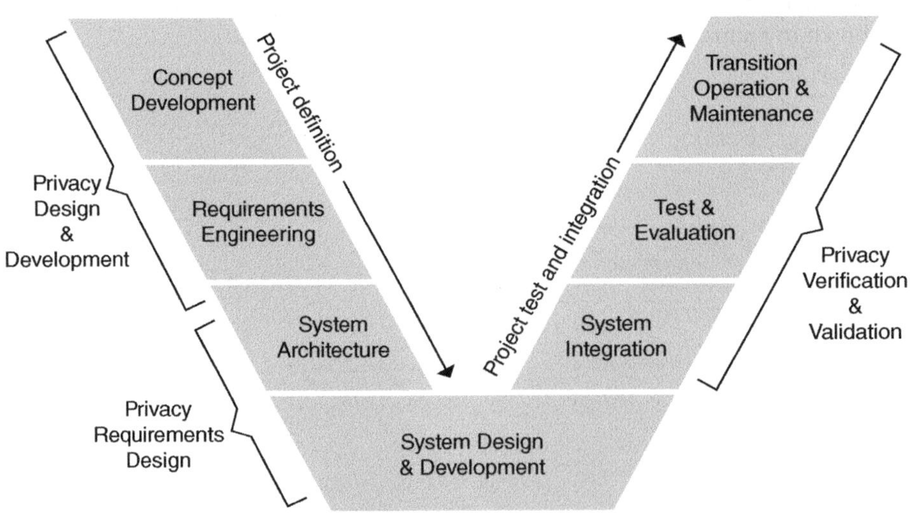

Figure 20.4 Mapping of the fundamental privacy engineering activities into stages of the typical SELC (MITRE-CoP, 2014).

Table 20.2 Privacy Engineering Activities and Methods.

Life Cycle Activity	Privacy Method
Privacy requirements definition	Baseline and custom privacy system requirements
	Privacy empirical theories and abstract concepts
Privacy design and development	Fundamental privacy design concepts
	Privacy empirical theories and abstract concepts
	Privacy design tools
	Privacy heuristics
Privacy verification and validation	Privacy testing and review
	Operational synchronization

following activities: selection and fine-tuning of baseline privacy requirements and tests, privacy risk analysis of functional requirements, and development of custom privacy requirements and tests based on results of privacy risk analysis.

2) Privacy design and development. This phase provides representation and implementation of those elements of the system that support defined privacy requirements. It includes the following activities: identification of privacy design strategies and patterns; identification of architectural, technical point, and policy privacy controls; development of data and process models reflecting identified privacy controls; alignment, integration, and implementation of privacy controls with functional elements; and privacy risk analysis of overall design.
3) Privacy verification and validation: This phase is premeditated for confirmation that defined privacy requirements has been correctly implemented and reflected stakeholder expectations. Typically, it encompasses the following activities: privacy test case development/refinement; privacy test case execution; and check of operational behavior against applicable privacy policies and procedures.

Fhom and Bayarou (2011) proposed a step-by-step approach as engineering flow to smart grid systems to address privacy concerns. An overall privacy-aware engineering flow is depicted in Figure 20.5, and the steps and guidelines of the proposed methodology are outlined as follows.

1) Identify relevant high-level privacy requirements. High-level policies that are relevant to the designed smart grid system should be identified. In general, these policies include privacy requirements that all the smart grid stakeholders must comply with while dealing with privacy sensitive data and are typically imposed by agencies and regulatory bodies. These policies may include contractual agreements as well as privacy and data protection guidelines.
2) System modeling. A detailed model of the designed smart grid system should be developed, in which every element of the model, its functionalities, and the interdependencies among relevant elements should be distinctly designated.
3) Privacy risks and impacts assessment. For an overall threat and risk assessment process, privacy impact assessment (PIA) for the smart grid should be conducted. Thus,

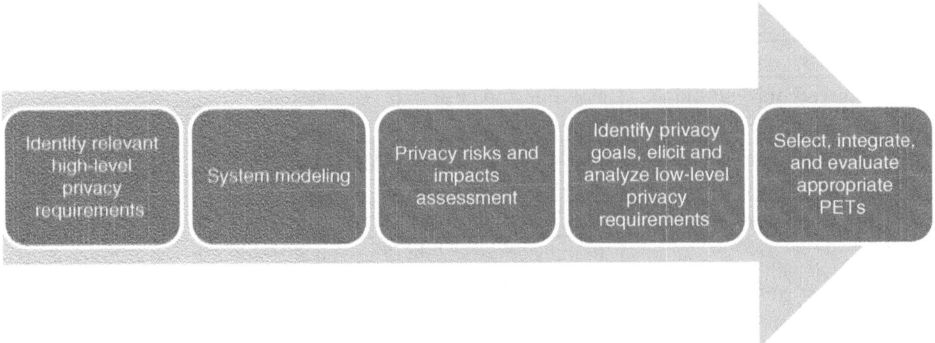

Figure 20.5 Overall privacy-aware engineering flow.

this phase ensures better perception of the privacy threats and risks associated with the accessibility of sensitive information within the designed smart grid system.

4) Identify privacy goals and elicit and analyze low-level privacy requirements. During this phase, privacy goals and requirements for the designed smart grid should be captured and analyzed. They not only perceptibly include traditional privacy properties such as confidentiality and integrity, as well as anonymity-related properties such as unlinkability and unobservability but also might explicitly target ensuring a user-centric privacy management.
5) Select, integrate, and evaluate appropriate PETs. In this phase, privacy-preserving mechanisms including PETs, which can not only be adopted as countermeasures against identified privacy risks and threats but also appropriately satisfy the related privacy requirements, are selected and incorporated into the designed smart grid system. Finally these PETs must be evaluated to fulfill privacy protection goals.

20.5 Privacy Risk and Impact Assessment

To effectively comply with the privacy protection goals, the privacy principles and other regulations need a greater understanding of the privacy risks in demand response systems (i.e., smart grid).

20.5.1 System Privacy Risk Model

Basically, a privacy risk model aims to provide a structured, repeatable, and quantifiable method for addressing privacy risk in ICT systems. The model may be defined with an equation and a series of inputs designed to enable (i) the identification of problems that may occur from the processing of personal information and (ii) the calculation of how such problems can be reflected in an organizational risk management approach that allows for prioritization and resource allocation to achieve organizational goals while minimizing overall adversative events.

In general, the system privacy risk is the product of three inputs: personal information collected or generated, data actions performed on that information, and the context

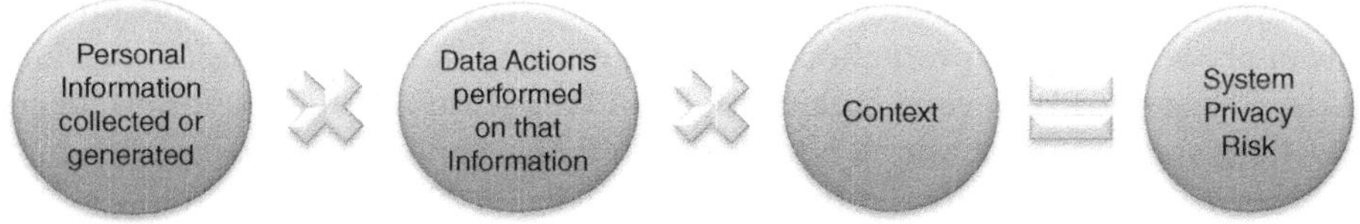

Figure 20.6 Equation expression for a system privacy risk model.

surrounding the collection, generation, and processing of this personal information (Brooks et al., 2017).

As the system privacy risk is the risk of challenging data actions occurring, its inputs can be explained as follows.

1) Data actions: They are any information system operations that process personal information. Processing of personal information may include the collection, retention, logging, generation, transformation, disclosure, transfer, and disposal of personal information.
2) Personal information and context: They are two critical inputs that modify the privacy risk of any given data action. For every data action, an organization should identify the associated personal information at a granular level. Both context and associated personal information contribute to whether a data action has the potential to cause privacy problems. Based on these pieces of information, it would be possible for an organization to illustrate initial observations about data actions.

And the equation expression for a system privacy risk model can be expressed as in Figure 20.6.

Subsequently, a system privacy risk model can assist the organization to identify possible privacy risks other than security risk. Given the emphasis on the operations of the system while processing personal information, ICT system's privacy risk, hence, can be ascribed as a function of the likelihood that a data action causes problems for individuals.

20.5.2 Privacy Impact Assessment (PIA)

A PIA is a comprehensive and methodical process for identifying and addressing privacy issues in an ICT system that processes PII. Basically, the PIA is used for determining the privacy, confidentiality, and security risks associated with the collection, use, and disclosure of PII as well as assessing the potential effects on privacy of a process and an information system. The PIA also describes the measures that can be used to mitigate and probably eliminate the identified risks.

The main goals of the PIA include: (i) to ensure that information handling obeys the pertinent legal, regulatory, and policy requirements regarding privacy; (ii) to determine the risks and effects of collecting, maintaining, and disseminating PIIs within the ICT system; and (iii) to examine and evaluate protections and alternate processes for handling information to alleviate potential privacy risks.

The PIA being integral to the process for privacy risk treatment to evaluate and manage privacy impacts and to ensure compliance with privacy protection rules and responsibilities, it is considered as an approach for making privacy engineering more specific and effectual.

In recent years, risk management focusing on PIA has started to take on a more prominent role in privacy and data protection (Wright, Finn, and Rodrigues, 2013; Kalogridis et al., 2014; Engage, 2011). Several organizations and regulatory entities have developed PIA guidelines and strategies, which often differ in scope, context, and goal, depending on the concerned organizations and their purposes (IPCO, 2015).

The ISO/IEC 29134 (ISO/IEC JTC 1/SC 27 IT Security techniques, 2017), which is currently under development, is a standard on privacy impact assessment that shall basically provide a set of guidelines for the managing PIAs. The PIA guidelines of the ISO/IEC 29134 standard can be summarized as follows: (i) Determine PIA requisite in the given system and define the information flows and other privacy impacts; (ii) distinguish privacy risks and possible solutions; (iii) formulate and implement the PIA-related recommendations; (iv) conduct third-party review and/or audit of the PIA; (v) update the PIA if revisions occur; and (vi) embed privacy awareness throughout the organization and ensure accountability. These guidelines shall assist the organization in conducting the resulting PIA.

In NIST IR 7628 v2 (SGIPCSWG-V2, 2014), the NIST has depicted a comprehensive consumer-to-utility PIA that is meant for the smart grid. Such a smart grid PIA activity delivers a structured, repeatable type of analysis aimed at determining how collected meter data can reveal personal information about individuals, and the focus of the PIA can be on a segment within the electric power grid or the power grid as a whole. Privacy principles and corresponding recommendations for smart grid high-level consumer-to-utility privacy impact assessment as depicted in NIST IR 7628 v2 is shown in Table 20.3.

Hofer et al. (2013) propose a PIA for the e-Mobility system that mainly focuses on the ISO/IEC 15118 standard. The e-Mobility PIA includes the following guidelines: (i) stipulating scope and purpose definition; (ii) identifying stakeholders; (iii) determining information assets; (iv) identifying information requirements and use; (v) determining information handling and other considerations; and (vi) conducting evaluation.

In the EU, several privacy risk assessment methodologies have been developed (Papakonstantinou and Kloza, 2015), for instance, the French *Commission Nationale de l'informatique et des Liberte* (CNIL) methodology for privacy risk management (CNIL, 2015) and the UK Information Commissioner's Office (ICO) privacy impact assessments code of practice. More prominently, the Data Protection Impact Assessment (DPIA) template for the smart grid and smart metering has been developed so that smart grid actors can conduct an assessment prior to deployment of any smart metering application (EU-SGTF 2014).

20.6 Privacy Enhancing Technologies

Privacy enhancing technologies (PET; Senicar, Jerman-Blazic, and Klobucar, 2003) can refer to particular methods that work in accordance with the data protection laws to prevent situations that might result in violation of privacy. For instance, PETs allow electricity consumers to protect privacy of their PII provided to and handled by other stakeholders in the smart grid. PETs are used to protect electricity consumers, largely against activity and behavioral analysis. Thus PETs are determined to protect privacy by minimizing personal data, subsequently preventing excessive processing of personal data,

Table 20.3 Privacy Principles and Corresponding Recommendations for Smart Grid High-Level Consumer-to-Utility Privacy Impact Assessment.

Principle	Recommendations
Management and accountability	• assign privacy responsibility • establish privacy audits • establish or amend incident response and law enforcement request policies and procedures
Notice and purpose	• provide notification for personal information collected • provide notification for new information use purposes and collection
Choice and consent	• provide notification about choices
Collection and scope	• limit the collection of data to only that necessary for smart grid operations • obtain the data by lawful and fair means
Use and retention	• review privacy policies and procedures • limit information retention
Individual access	• access to energy usage data • dispute resolution
Disclosure and limiting use	• limit information use • disclosure
Security and safeguards	• associate energy data with individuals only when and where required • de-identify information • safeguard personal information • do not use personal information for research purposes
Accuracy and quality	• keep information accurate and complete
Openness, monitoring, and challenging compliance	• policy challenge procedures • perform regular privacy impact assessments • establish breach notice practices

without the loss of the functionality of the smart grid system (Jawurek, Kerschbaum, and Danezis, 2012).

The main objective of PETs is to protect personal data as well as ensure the customers that their information remains confidential and management of data protection is a priority for the service providers who are responsible for dealing with PIIs. In this regard, PETs aim to furnish functionality and benefits to all stakeholders and to make available personal data to third parties, without disclosing any sensitive information (Kement et al., 2017; Jo, Kim, and Lee, 2016; Tonyali et al., 2017).

To address the privacy concerns in the smart grid, several privacy-preserving protocols (PPP) and PETs have been proposed and deployed (Souri et al., 2014; Ferrag et al., 2016; Han and Xiao, 2016b). The majority of PETs are focused on AMI networks (Diao et al., 2015; Birman et al., 2015; Li et al., 2015; Wang, Mu, and Chen, 2016); however, some PETs are dedicated for V2G networks (Liu et al., 2014; Wang et al., 2015; Han and Xiao, 2016a; Liu et al., 2016).

Design of innovative PETs should assure consumers' privacy and allow ESPs to monitor and control the grid securely (Abdallah and Shen, 2016; He et al., 2017; Li et al., 2017; Liao et al., 2017). PETs can be classified into several categories: anonymization, trusted computation, cryptographic computation, perturbation, and verifiable computation (Jawurek, Kerschbaum, and Danezis, 2012).

20.6.1 Anonymization

The basic notion of anonymization is that the data consumer (e.g., ESP or utility) can still perform the needed calculations although the direct association between data item (i.e., smart meter reading for consumed electricity) and the data producer (e.g., household or customer) has been isolated. In other words, an anonymization technique shall remove user-specific features from metering data before sending it to the authorized data consumer such that the ESP obtains anonymous metering data (i.e., data without any PII), which cannot be simply attributed to any specific customer. Thus, the ESP can have the statistical data processing and perform required computations. However, it is difficult or impossible for the data consumers to associate the metering data received to a specific smart meter, household, or electric vehicle.

The data consumer's incapability to attribute information determined from electricity customer data items to their producers can be viewed as a substantial privacy-enhancing effect due to the anonymization. Thus, anonymity is one fundamental form of privacy protection that can be beneficial. Several privacy-preserving mechanisms using the anonymization approach have been discussed.

Efthymiou and Kalogridis (2010) propose a mechanism for anonymizing customers' metering data using pseudonymous IDs through an escrow service by a trusted third party (TTP), so data consumers can be assured of the legality of the received metering data, but they would not be able to link the data to a specific customer. Similarly, Gong et al. (2016) propose utilizing two different IDs (i.e., anonymous ID and attributable ID) for providing privacy-preserving incentive-based demand response.

Badra and Zeadally (2014) propose a different approach to providing authorized data consumers with anonymous metering data, i.e., to use virtual ring architecture. While Finster and Baumgart (2013) propose a pseudonymous smart metering protocol without a TTP such that a smart meter uses a blinded pseudonym signed by the authorized data consumer, so the smart meter does reveal its identity.

Rottondi, Mauri, and Verticale (2015) propose a data pseudonymization protocol, which uses a secret splitting scheme to construct a unique pseudonym from different intermediate trusted nodes such that once the data consumer receives all the shares attached with the same pseudonym, it can recover the metering data associated with the pseudonym. Furthermore, Vaidya, Makrakis, and Mouftah (2014) propose ID-based partially restrictive blind signature for V2G network such that the blindness property of the e-token keeps EV's real ID anonymous to the local aggregator.

Apparently, an anonymization technique is practical only if the computation result does not have to be attributed to a specific data producer. This technique can be ineffective, since sometimes it is possible to re-identify the owner of the data. Jawurek, Johns, and Rieck (2011) find that pseudonymized consumption traces (i.e., separated from PIIs), can still be attributed to individuals using auxiliary information such as household observation correlations between power events and physical events.

20.6.2 Trusted Computation

In trusted computation approaches, the data consumer would not have direct access to the electricity usage information of the individual consumer. Instead, it only receives an aggregation of metering data, which is computed either by the data producers themselves or an additional TTP that is introduced as external aggregator. By issuing aggregation results, the data consumer cannot recover consumers' personal details (i.e., PIIs). Yet the data consumer obtains sufficiently accurate aggregated metering data. The aggregation of metering data is mostly done in either temporal (i.e., power traces of a singe user over time) or spatial (i.e., power traces of multiple users at a certain time interval) manner.

In this approach, the TTP are mainly used for provisioning unlinkability between readings and the smart meters and supporting fraud/loss detection. However, the disclosure of this individual data to the data consumer constitutes one of the major threats. The disclosure can be performed by the aggregating entities. Thus, these types of privacy-preserving protocols typically demand the strong assumptions for the trustworthiness of the aggregating entities.

Li, Luo, and Liu (2010) propose a mechanism that provides aggregation of Paillier-encrypted data while routing through a minimal-spanning-tree of smart meters toward the authorized data consumer. Similarly, Ruj, Nayak, and Stojmenovic (2011) propose a two-tier system for aggregation of smart metering data and subsequent access by authorized data consumers with a help of Paillier encryption.

Chen, Lu, and Cao (2015) propose a privacy-preserving data aggregation scheme with fault tolerance (so-called PDAFT) for smart grid communications that uses a homomorphic Paillier encryption technique to encrypt sensitive consumer data such that the data consumer can obtain the aggregated data without knowing individual ones.

20.6.3 Cryptographic Computation

In the cryptographic computation approach, either encryption schemes based on the homomorphic property or secret-sharing schemes can be deployed. Metering data items that arrive at the data consumer shall be either ciphertexts or secret shares. So the privacy-preserving protocol should ensure that the data consumer could only decrypt the aggregate of data items (i.e., ciphertexts or secret shares) but not individual data item.

In the homomorphic encryption technique, the individual power consumption items are encrypted by the data producers (i.e., smart meters) using the public key of the data consumer (Tonyali, Saputro, and Akkaya, 2015). Essentially, the encrypted data item undergoes a homomorphic operation prior to going to the data consumer. If the individual data items have to be aggregated, a homomorphic addition operation is employed. Finally, after receiving the aggregated data, the data consumer can obtain the decrypted result using its private key.

In case of a secret sharing scheme, a secret is allotted in multiple parts, and each part is given to an authorized participant. All or a subset of these participants have to contribute their shares to reconstruct the resultant secret. In the context of the smart grid, a consumer's electricity consumption reading obtained from a smart meter can be used as a share.

Garcia and Jacobs (2011) propose a privacy-friendly energy metering system that uses an aggregation protocol based on a homomorphic property of encryption and specifically targets the detection of energy theft at substations. Likewise, Nateghizad, Erkin, and Lagendijk (2016) put forward a privacy-preserving cryptographic protocol based on homomorphic encryption for smart metering that can reduce communication cost and improve efficiency of the protocol with encrypted inputs.

Tonyali et al. (2016a) investigated the feasibility and performance of fully homomorphic encryption (FHE) aggregation in the smart grid AMI networks utilizing the reliable data transport protocol, TCP, and proposed a novel packet reassembly mechanism for TCP to overcome the packet reassembly problem.

Rottondi, Verticale, and Capone (2012) describe an approach where trusted privacy-preserving nodes and a central configurator are introduced into the smart metering system and uses Shamir's secret-sharing algorithm, a secret sharing variant. Similarly, Rottondi, Fontana, and Verticale (2014) propose to employ Shamir's secret-sharing scheme in privacy-preserving mechanism for V2G networks, in which three types of EV data (i.e., plug-in time period, charge level of the battery, and amount of recharged electricity) are split into parts. Each one in a set of local aggregators holds one part of the data, and the required information is reconstructed from the parts contributed by all the local aggregators.

20.6.4 Perturbation

Privacy-enhancing protocols using a perturbation technique add certain amount of noise to individual data items or to the final aggregate such that the data consumer can still perform the required computation, but it cannot be used to derive any sensitive information of the data producer (i.e., sufficiently protect privacy of the data producer).

One of the choices of the perturbation-based PETs is a differential privacy technique, in which a differential private aggregation function appends an adequate quantity of random noise to their result so that individual input data items cannot be deduced from the function's result.

PETs that use perturbation mechanisms that add certain amount of noise to every measurement in a distributed manner such that the resultant noise values is just adequate for reaching differential privacy (Jawurek and Kerschbaum, 2012).

Acs and Castelluccia (2011) put forward a PET that guarantees differential privacy by exploiting an innovative way for adding appropriate noise to the computed aggregate. The distributed noise generation mechanism is designed such that it allows not relying on a TTP to act as an aggregator. Similarly, Shi et al. (2011) describe an aggregation protocol in which individual data producers can add random noise in such a way that a sum of the random noise of all data producers shall ensure differential privacy for the aggregate outcome. Every data producer encrypts its noisy measurement before sending it to the aggregator. The aggregator uses a final share that allows decrypting the differential private aggregate of all measurements.

Bao and Lu (2015) propose a data aggregation scheme for smart metering, named differentially private data aggregation with fault tolerance (DPAFT), which can ensure differential privacy of data aggregation along with fault tolerance by applying the Boneh–Goh–Nissim cryptosystem.

Since smart meters periodically send fine-grained power consumption data to the utility company in AMI networks, consumer privacy is one of prime concerns. In Tonyali et al. (2016b), a meter data obfuscation scheme is proposed to protect consumer privacy from eavesdroppers and the utility companies while preserving the utility companies' ability to use the data for state estimation.

20.6.5 Verifiable Computation

In the verifiable computation paradigm, the aggregator not only provides an aggregation outcome but also furnishes a proof that the computed result has been performed as claimed. Hence, such privacy-preserving protocols can be deployed in untrusted environments (i.e., having untrusted aggregators) so that the untrusted aggregators can perform the required computations while guaranteeing the integrity of the aggregation outcome. Such protocols are normally based on the zero-knowledge proof (ZKP) protocols, in which the verifier only validates the legitimacy of the statement provided by the prover to be proven, but no private information has been revealed.

The verifiable computation–based PET is suitable for deploying protocols related to billing purposes. This is due to the fact that the verifiable computation protocols using ZKP can provide integrity and accuracy of the aggregate result that can be used for billing purposes without disclosing any private information.

In the case of the smart grid, the data producer can be projected as the prover and the data consumer as the verifier. The data producer (i.e., smart meter) computes the total energy consumption for certain duration and sends it to the data consumer (i.e., utility company). Subsequently, the data consumer can verify the validity of the result without disclosing individual smart meter readings. Furthermore, the zero-knowledge proof enables the data consumer to verify cumulative price such that the total amount for the energy consumption has been calculated correctly.

Rial and Danezis (2011) have proposed an approach in which a smart meter produces metering data as well as its commitment and signature over the commitments and sends them to a user device. The user device obtains the tariff from the service provider and calculates the fee and the required proof. It implements ZKP protocols for various tariffs including cumulative tariffs, interval linear tariffs, and even cumulative polynomial tariffs.

Wan, Zhu, and Wang (2016) propose a privacy-preserving mechanism called PRAC (privacy via randomized anonymous credentials) for V2G communication, which ensures anonymous authentication and rewarding as well as guarantees unlinkable credentials and rewards. They have used zero-knowledge proof as an approach to prove that the scheme can ensure integrity and anonymous authentication. Likewise, Rahman et al. (2017) propose a solution in which a secure and privacy-preserving communication channel between a bidder and a registration manager is established and utilizes El-Gamal public key encryption and Schnorr signature scheme for ensuring zero-knowledge proof.

Acknowledgment

This work was funded by The Ontario Ministry of Energy Smart Grid Fund.

References

Abdallah, A., and Shen, X., 2016, 'A Lightweight Lattice-based Homomorphic Privacy-Preserving Data Aggregation Scheme for Smart Grid', *IEEE Trans. on Smart Grid.*

Acs, G. and Castelluccia, C., 2011, 'I Have a DREAM! (DiffeRentially privatE smArt Metering)', *In*: Filler, T., Pevny, T., Craver, S., and Ker, A. (eds.), *Information Hiding, IH 2011, Lecture Notes in Computer Science, LNCS 6958*, Berlin, Heidelberg: Springer, 118–132.

Alshammari, M. and Simpson, A., 2016, 'Towards a Principled Approach for Engineering Privacy by Design', Oxford University Research, Report No.: RR-16-05, viewed Jan 10, 2017.

Amin, S., Schwartz, G.A., Cardenas, A.A., and Sastry, S.S., 2015, 'Game-Theoretic Models of Electricity Theft Detection in Smart Utility Networks: Providing New Capabilities with Advanced Metering Infrastructure', *IEEE Control Systems* 35(1), 66–81.

Antignac, T. and Le Metayer, D., 2015, 'Trust Driven Strategies for Privacy by Design', *In*: Damsgaard Jensen, C., Marsh, S., Dimitrakos, T., Murayama, Y. (eds.), 'Trust Management IX. IFIP International Federation for Information Processing (IFIPTM 2015)', *IFIP Advances in Information and Communication Technology* 454, Springer, Cham, 60–75.

Asghar, M.R, Dan, G., Miorandi, D., and Chlamtac, I., 2017, 'Smart Meter Data Privacy: A Survey', *IEEE Communications Surveys & Tutorials.*

Badra, M. and Zeadally, S., 2014, 'Design and performance analysis of a virtual ring architecture for smart grid privacy', *IEEE Trans. on Information Forensics and Security* 9(2), 321–329.

Bao, H., and Lu, R., 2015, 'A New Differentially Private Data Aggregation With Fault Tolerance for Smart Grid Communications', *IEEE Internet of Things Journal* 2(3), 248–258.

Birman, K., Jelasity, M., Kleinberg, R., and Tremel, E., 2015, 'Building a Secure and Privacy-Preserving Smart Grid', *ACM SIGOPS Operating Systems Review—Special Issue on Repeatability and Sharing of Experimental Artifacts Archive* 49(1), 131–136.

Bringer, J., Chabanne, H., Le Metayer, D., and Lescuyer, R., 2015, 'Privacy by Design in Practice: Reasoning about Privacy Properties of Biometric System Architectures', *In*: Bjorner and de Boer (eds.), 'FM 2015: Formal Methods', *Lecture Notes in Computer Science 9109*, Springer, Cham, 90–107.

Brooks, S., Garcia, M., Lefkovitz, N., Lightman, S., and Nadeau, E., 2017, 'An Introduction to Privacy Engineering and Risk Management in Federal Systems', National Institute of Standards and Technology, NIST IR No.: 8062, viewed Jun 11, 2017.

Cavoukian A. and Polonetsky, J., 2013, 'Privacy by Design and Third Party Access to Customer Energy Usage Data', Information and Privacy Commissioner of Ontario, viewed Jan 10, 2017.

Cavoukian, A., 2011, 'Privacy by Design: The 7 Foundational Principles', Information and Privacy Commissioner of Ontario, viewed Jan 10, 2017.

Cavoukian, A., Polonetsky, J., and Wolf, C., 2010, 'SmartPrivacy for the Smart Grid: embedding privacy into the design of electricity conservation', *Identity in the Information Society* 3(2), 275–294.

Cavoukian, A., Shapiro, S., and Cronk, R.J., 2014, 'Privacy Engineering: Proactively Embedding Privacy, by Design', Information and Privacy Commissioner of Ontario, viewed Jan 10, 2017.

Chen, L., Lu, R., and Cao, Z., 2015, 'PDAFT: A privacy-preserving data aggregation scheme with fault tolerance for smart grid communications', *Peer-to-Peer Networking and Applications* 8(6), 1122–1132.

Cintuglu, M.H., Mohammed, O.A., Akkaya, K., and Uluagac, A.S., 2017, 'A Survey on Smart Grid Cyber-Physical System Testbeds', *IEEE Communications Surveys & Tutorials* 19(1), 446–464.

Commission Nationale de l'informatique et des Liberte (CNIL) 2015, 'Privacy impact assessment (PIA) Methodology—How to carry out a PIA', viewed Jan 10, 2017.

Cornelius, D 2009, *AICPA's Generally Accepted Privacy Principles*, Compliance Building, viewed 10 Jan 2017, URL: http://www.compliancebuilding.com/2009/01/09/aicpas-generally-accepted-privacy-principles/

Depuru, S.S.S.R., Wang, L., and Devabhaktuni, V., 2011a, 'Electricity theft: Overview, issues, prevention and a smart meter based approach to control theft', *Energy Policy* 39, 1007–1015.

Depuru, S.S.S.R., Wang, L., and Devabhaktuni, V., 2011b, 'Smart meters for power grid: Challenges, issues, advantages and status', *Renewable and Sustainable Energy Reviews* 15(6), 2736–2742.

Diao, F., Zhang, F., and Cheng, X., 2015, 'A Privacy-Preserving Smart Metering Scheme Using Linkable Anonymous Credential', *IEEE Trans. on Smart Grid* 6(1), 461–467.

Efthymiou, C. and Kalogridis, G., 2010, 'Smart Grid Privacy via Anonymization of Smart Metering Data', *In: Proceedings of the 1st IEEE International Conference on Smart Grid Communications (SmartGridComm 2010)*, Gaithersburg, MD, USA, Oct. 2010, IEEE, 238–243.

Engage Consulting Ltd., 2011, 'Privacy Impact Assessment: Use of Smart Metering data by Network Operators', Ref No.: ENA-CF002-007-1.0, viewed Jan 10, 2017.

EU Smart Grid Task Force (EU-SGTF) 2014, 'Data Protection Impact Assessment Template for Smart Grid and Smart Metering systems', viewed Jan 10, 2017.

Ferrag, M.A., Maglaras, L.A., Janicke, H., and Jiang, J., 2016, 'A Survey on Privacy-preserving Schemes for Smart Grid Communications', Cornell University Library, arXiv: 1611.07722, viewed Jan 20, 2017, URL: https://arxiv.org/abs/1611.07722.

Fhom, H.S. and Bayarou, K.M., 2011, 'Towards a Holistic Privacy Engineering Approach for Smart Grid Systems', *In: Proceedings of 2011 International Joint Conference of IEEE TrustCom-11/IEEE ICESS-11/FCST-11 2011*.

Finster, S. and Baumgart, I., 2013, 'Pseudonymous smart metering without a trusted third party', *In: Proceedings of 12th IEEE International Conference on Trust, Security and Privacy in Computing and Communications (TrustCom)*, Melbourne, Australia, Jul 2013, IEEE, 1723–1728.

Finster, S, and Baumgart, I., 2015, 'Privacy-Aware Smart Metering: A Survey', *IEEE Communications Surveys & Tutorial* 17(2), 1088–1101.

Garcia, F.D., and Jacobs, B., 2011, 'Privacy-Friendly Energy-Metering via Homomorphic Encryption', In: Cuellar, J., Lopez, J., Barthe, G., and Pretschner, A. (eds.), *Security and Trust Management, STM 2010, Lecture Notes in Computer Science, LNCS 6710*, Berlin, Heidelberg: Springer, 226–238.

Gong, Y., Cai, Y., Guo, Y., and Fang, Y., 2016, 'A Privacy-Preserving Scheme for Incentive-Based Demand Response in the Smart Grid', *IEEE Trans. on Smart Grid* 7(3), 1304–1313.

Gurses, S. and Del Alamo, J.M., 2016, 'Privacy Engineering: Shaping an Emerging Field of Research and Practice', *IEEE Security & Privacy* 14(2), 40–46.

Han, W. and Xiao, Y., 2016b, 'IP2DM: integrated privacy-preserving data management architecture for smart grid V2G networks', *Wireless Communications and Mobile Computing* 16(17), 2956–2974.

Han, W. and Xiao, Y., 2016a, 'Privacy preservation for V2G networks in smart grid: A survey', *Computer Communications* 91–92, 17–28.

Hansen, M., Jensen, M., and Rost, M., 2015, 'Protection Goals for Privacy Engineering', in *Proceedings of 2015 IEEE CS Security and Privacy Workshops*, San Jose, CA, USA, May 2015, IEEE, 159–166.

He, D., Kumar, N., Zeadally, S., Vinel, A., and Yang, L.T., 2017, 'Efficient and Privacy-Preserving Data Aggregation Scheme for Smart Grid against Internal Adversaries', *IEEE Trans. on Smart Grid*.

Hoepman, J.H., 2014, 'Privacy Design Strategies', in *Proceedings of IFIP International Information Security Conference SEC 2014: ICT Systems Security and Privacy Protection*, IFIPAICT, vol. 428, Marrakech, Morocco, 446–459.

Hofer, C., Petit, J., Schmidt, R., and Kargl, F., 2013, 'Popcorn: privacy-preserving charging for e-mobility', *In: Proceedings of the 2013 ACM workshop on Security, privacy & dependability for cyber vehicles*, ACM, 37–48.

Hubert, T., and Grijalva, S., 2012, 'Modeling for Residential Electricity Optimization in Dynamic Pricing Environments', *IEEE Trans. on Smart Grid* 3(4), 2224–2231.

Information and Privacy Commissioner of Ontario (IPCO) 2009, 'Privacy by Design: The 7 Foundational Principles', Information and Privacy Commissioner of Ontario, viewed Jan 10, 2017.

Information and Privacy Commissioner of Ontario (IPCO) 2010, 'Privacy by Design: Achieving the Gold Standard in Data Protection for the Smart Grid', Information and Privacy Commissioner of Ontario, viewed Jan 10, 2017.

Information and Privacy Commissioner of Ontario (IPCO) 2015, 'Planning for Success: Privacy Impact Assessment Guide', Information and Privacy Commissioner of Ontario, viewed Jan 10, 2017.

ISO/IEC JTC 1/SC 27 IT Security techniques 2011, *ISO/IEC 29100:2011 Information technology — Security techniques — Privacy framework*, 1st ed, International Organization for Standardization (ISO), viewed 10 Jan 2017.

ISO/IEC JTC 1/SC 27 IT Security techniques 2017, *ISO/IEC 29134:2017 Information technology - Security techniques - Guidelines for privacy impact assessment*, International Organization for Standardization, viewed 11 Jun 2017.

Jawurek, M. and Kerschbaum, F., 2012, 'Fault-Tolerant Privacy-Preserving Statistics', *In*: Fischer-Hubner, S., and Wright, M. (eds.), *Privacy Enhancing Technologies, PETS 2012, Lecture Notes in Computer Science*, LNCS 7384, Berlin, Heidelberg: Springer.

Jawurek, M., Johns, M., and Rieck, K., 2011, 'Smart metering depseudonymization', in *Proceedings of the 27th Annual Computer Security Applications Conference (ACSAC'11)*, New York, NY, USA, 227–236.

Jawurek, M., Kerschbaum, F., and Danezis, G., 2012, 'SoK: Privacy technologies for Smart grids—a survey of options', technical report, Microsoft Research, viewed Jan 10, 2017.

Jiang, R., Lu, R., Wang, Y., Luo, J., Shen, C., and Shen, X.S., 2014, 'Energy-theft detection issues for advanced metering infrastructure in smart grid', *Tsinghua Science and Technology* 19(2), 105–120.

Jo, H.J., Kim, I.S., and Lee, D.H., 2016, 'Efficient and Privacy-Preserving Metering Protocols for Smart Grid Systems', *IEEE Trans. on Smart Grid* 7(3), 1732–1742.

Jokar, P., Arianpoo, N., and Leung, V.C.M., 2016, 'Electricity Theft Detection in AMI Using Customers' Consumption Patterns', *IEEE Trans. on Smart Grid* 7(1), 216–226.

Kalogridis, G., Sooriyabandara, M., Fan, Z., and Mustafa, M.A., 2014, 'Toward Unified Security and Privacy Protection for Smart Meter Networks', *IEEE Systems Journal* 8(2), 641–654.

Kement, C.E., Gultekin, H., Tavli, B., Girici, T., and Uludag, S., 2017, 'Comparative Analysis of Load Shaping Based Privacy Preservation Strategies in Smart Grid', *IEEE Trans. on Industrial Informatics.*

Krishna, V.B., Weaver, G.A., and Sanders, W.H., 2015, 'PCA-Based Method for Detecting Integrity Attacks on Advanced Metering Infrastructure', *In*: Campos, J. and Haverkort, B.R. (eds.), *Proceedings of Quantitative Evaluation of Systems: 12th International Conference, QEST 2015, Lecture Notes in Computer Science (LNCS) 9259*, New York: Springer-Verlag, 70–85.

Kroener, I. and Wright, D., 2014, 'A Strategy for Operationalizing Privacy by Design', *The Information Society* 30(5), 355–365.

Kung, A., 2014, 'PEARs: Privacy Enhancing Architectures', *In*: Preneel, B., and Ikonomou, D. (eds.), *Privacy Technologies and Policy, APF 2014, Lecture Notes in Computer Science*, LNCS 8450, Springer, Cham, 18–29.

Landesberg, M.K., Levin, T.M., Curtin, C.G., and Lev, O., 1998, 'Privacy Online: A Report to Congress', Federal Trade Commission, viewed Jan 10, 2017.

Li, C., Lu, R., Li, H., Chen, L., and Chen, J., 2015, 'PDA: a privacy-preserving dual-functional aggregation scheme for smart grid communications', *Wiley Security and Communication Networks* 8, 2494–2506.

Li, F., Luo, B., and Liu, P., 2010, 'Secure information aggregation for Smart grids using homomorphic encryption', *In: Proceedings of the 1st IEEE International Conference on Smart Grid Communications (SmartGridComm 2010)*, Gaithersburg, MD, USA, Oct. 2010, IEEE, 327–332.

Li, H, Gong, S, Lai, L, Han, Z, Qiu, RC, Yang, D 2012, 'Efficient and Secure Wireless Communications for Advanced Metering Infrastructure in Smart Grids', *IEEE Trans. on Smart Grid*, vol. 3, iss. 3, pp. 1540-1551.

Li, S., Xue, K., Yang, Q., and Hong, P., 2017, 'PPMA: Privacy-Preserving Multi-Subset Aggregation in Smart Grid', *IEEE Trans. on Industrial Informatics.*

Liao, X., Srinivasan, P., Formby, D., and Beyah, A.R., 2017, 'Di-PriDA: Differentially Private Distributed Load Balancing Control for the Smart Grid', *IEEE Trans. on Dependable and Secure Computing.*

Lisovich, M.A., Mulligan, D.K., and Wicker, S.B., 2010, 'Inferring Personal Information from Demand-Response Systems', *IEEE Security & Privacy* 8(1), 11–20.

Liu, H., Ning, H., Zhang, Y., Xiong, Q., and Yang, L.T., 2014, 'Role-Dependent Privacy Preservation for Secure V2G Networks in the Smart Grid', *IEEE Trans. on Information Forensics and Security* 9(2), 208–220.

Liu, J.K., Susilo, W., Yuen, T.H., Au, M.H., Fang, J., Jiang, Z.L., and Zhou, J., 2016, 'Efficient Privacy-Preserving Charging Station Reservation System for Electric Vehicles', *The Computer Journal* 59(7), 1040–1053.

Meis, R., Wirtz, R., and Heisel, M., 2015, 'A Taxonomy of Requirements for the Privacy Goal Transparency', *In*: Fischer-Hubner, S., Lambrinoudakis, C., and Lopez, J. (eds.), 'Trust, Privacy and Security in Digital Business, TrustBus 2015', *Lecture Notes in Computer Science*, LNCS 9264, Springer International Publishing, 195–209.

MITRE Privacy Community of Practice (CoP) 2014, 'Privacy Engineering Framework', viewed Jan 10, 2017.

Nateghizad, M., Erkin, Z., and Lagendijk, R.L., 2016, 'An efficient privacy-preserving comparison protocol in smart metering systems', *EURASIP Journal on Information Security*.

Notario, N., Crespo, A., Kung, A., Kroener, I., Le Metayer, D., Troncoso, C., Del Alamo, J.M., and Martin, Y.S., 2014, 'PRIPARE: A New Vision on Engineering Privacy and Security by Design', *In*: Cleary, F., and Felici, M. (eds.), 'Cyber Security and Privacy, CSP 2014', *Communications in Computer and Information Science*, CCIS 470, Springer, Cham, 65–76.

Notario, N., Crespo, A., Martin, Y.S., Del Alamo, J.M., Le Metayer, D., Antignac, T., Kung, A., Kroener, I., and Wright, D., 2015, 'PRIPARE: Integrating Privacy Best Practices into a Privacy Engineering Methodology', *In: Proceedings of 2015 IEEE CS Security and Privacy Workshops*, San Jose, CA, USA, May 2015, IEEE, 151–158.

Organisation for Economic Co-operation and Development (OECD) 2013, 'The OECD Privacy Framework', Organisation for Economic Co-operation and Development', viewed Jan 10, 2017.

Papakonstantinou, V. and Kloza, D., 2015, 'Legal Protection of Personal Data in Smart Grid and Smart Metering Systems from the European Perspective', *In*: Goel, S., Hong, Y., Papakonstantinou, V., and Kloza, D. (eds.), *Smart Grid Security, Springer Briefs in Cybersecurity* 2193-973X, London: Springer, 41-129.

Rahman, M.S., Basu, A., Kiyomoto, S., and Bhuiyan, M.Z.A., 2017, 'Privacy-friendly secure bidding for smart grid demand-response', *Information Sciences* 379, 229–240.

Rial, A. and Danezis, G., 2011, 'Privacy-preserving smart metering', In: Proceedings of the 10th Annual ACM workshop on Privacy in the Electronic Society, WPES'11, Chicago, IL, USA, Oct 2011, ACM, 49–60.

Rottondi, C., Fontana, S., and Verticale, G., 2014, 'Enabling privacy in vehicle-to-grid inter-actions for battery recharging', *Energies* 7(5), 2780–2798.

Rottondi, C., Mauri, G., and Verticale, G., 2015, 'A protocol for metering data pseudonymization in smart grids', *IEEE Trans. on Emerging Telecommunications Technologies* 26(5), 876–892.

Rottondi, C., Verticale, G., and Capone, A., 2012, 'A security framework for smart metering with multiple data consumers', *In: Proceedings of 2012 IEEE Conference on Computer Communications Workshops (INFOCOM WKSHPS)*, Orlando, FL, USA. IEEE, 103–108.

Ruj, S., Nayak, A., and Stojmenovic, I., 2011, 'A security architecture for data aggregation and access control in smart grids', Cornell University Library, arXiv: 1111.2619, viewed Jan 20, 2017, URL: https://arxiv.org/abs/1111.2619.

Senicar, V., Jerman-Blazic, B., and Klobucar, T., 2003, 'Privacy-Enhancing Technologies—approaches and development', *Computer Standards & Interfaces* 25(2), 147–158.

Shi, E., Chan, T.H., Rieffel, E.G., Chow, R., and Song, D., 2011, 'Privacy-preserving aggregation of time-series data', *In: Proceedings of the 18th Annual Network and Distributed System Security Symposium (NDSS)*, San Diego, CA, USA, Feb 2011.

Siano, P., 2014, 'Demand response and smart grids—A survey', *Elsevier Renewable and Sustainable Energy Reviews* 30, 461–478.

Souri, H., Dhraief, A., Tlili, S., Drira, K., and Belghith, A., 2014, 'Smart Metering privacy-preserving techniques in a nutshell', *Procedia Computer Science.* 32, 1087–1094.

Spiekermann, S. and Cranor, L.F., 2009, 'Engineering Privacy', *IEEE Trans. on Software Engineering* 35(1), 67–82.

Strasser, T., Andren, F., Kathan, J., Cecati, C., Buccella, C., Siano, P., Leitao, P., Zhabelova, G., Vyatkin, V., Vrba, P., and Marik, V., 2015, 'A Review of Architectures and Concepts for Intelligence in Future Electric Energy Systems', *IEEE Trans. on Industrial Electronics* 62(4), 2424–2438.

The Smart Grid Interoperability Panel Cyber Security Working Group 2014, *Guidelines for Smart Grid Cyber security: Vol. 1,* (SGIPCSWG-V1) *Smart Grid Cyber security Strategy, Architecture, and High-Level Requirements,* National Institute of Standards and Technology, U.S. Department of Commerce, NIST IR No.: 7628 Rev 1, viewed 6 Jan 2017.

The Smart Grid Interoperability Panel Cyber Security Working Group, 2014, *Guidelines for Smart Grid Cyber security: Vol. 2,* (SGIPCSWG-V2) *Privacy and the Smart Grid,* National Institute of Standards and Technology, U.S. Department of Commerce, NIST IR No.: 7628 Rev 1, viewed 6 Jan 2017.

The Smart Grid Interoperability Panel Cyber Security Working Group, 2014, *Guidelines for Smart Grid Cyber security: Vol. 3,* (SGIPCSWG-V3) *Supportive Analyses and References,* National Institute of Standards and Technology, U.S. Department of Commerce, NIST IR No.: 7628 Rev 1, viewed 6 Jan 2017.

Tonyali, S., Akkaya, K., Saputro, N., and Uluagac, A.S., 2016a, 'A reliable data aggregation mechanism with Homomorphic Encryption in Smart Grid AMI networks', *In: Proceedings of 2016 13th IEEE Annual Consumer Communications & Networking Conference (CCNC),* Las Vegas, NV, USA, Jan 2016, IEEE.

Tonyali, S., Akkaya, K., Saputro, N., Uluagac, A.S., and Nojoumian, M., 2017, 'Privacy-preserving protocols for secure and reliable data aggregation in IoT-enabled Smart Metering systems', *Future Generation Computer Systems.*

Tonyali, S., Cakmak, O., Akkaya, K., Mahmoud, M.M.E.A., and Guvenc, I., 2016b, 'Secure Data Obfuscation Scheme to Enable Privacy-Preserving State Estimation in Smart Grid AMI Networks', *IEEE Internet of Things Journal* 3(5), 709–719.

Tonyali, S., Saputro, N., and Akkaya, K., 2015, 'Assessing the feasibility of fully homomorphic encryption for Smart Grid AMI networks', *In: Proceedings of 2015 Seventh International Conference on Ubiquitous and Future Networks (ICUFN),* Sapporo, Japan, Jul 2015, IEEE.

Vaidya, B., Makrakis, D., and Mouftah, H.T., 2014, 'Security and Privacy-Preserving Mechanism for Aggregator Based Vehicle-to-Grid Network', *In*: Mitton, N., Gallais, A., Kantarci, M., and Papavassiliou, S. (eds.), 'Ad Hoc Networks, ADHOCNETS 2014', *Lecture Notes of the Institute for Computer Sciences, Social Informatics and Telecommunications Engineering,* LNICST 140, Springer, Cham., 75–85.

Wan, Z., Zhu, W.T., and Wang, G., 2016, 'PRAC: Efficient privacy protection for vehicle-to-grid communications in the smart grid', *Computers & Security* 62, 246–256.

Wang, H., Qin, B., Wu, Q., Xu, L., and Domingo-Ferrer, J., 2015, 'TPP: Traceable Privacy-Preserving Communication and Precise Reward for Vehicle-to-Grid Networks in Smart Grids', *IEEE Trans. on Information Forensics and Security* 10(11), 2340–2351.

Wang, X.F., Mu, Y., and Chen, R.M., 2016, 'An efficient privacy-preserving aggregation and billing protocol for smart grid', *Wiley Security and Communication Networks* 9, 4536–4547.

Wicker, S. and Thomas, R., 2011, 'A Privacy-Aware Architecture For Demand Response Systems', in *Proceedings of the 44th Hawaii International Conference on System Sciences (HICSS'11)*, Hawaii, USA, Jan 2011, 1–9.

Wicker, S.B. and Schrader, D.E., 2011, 'Privacy-Aware Design Principles for Information Networks', *Proc. of the IEEE* 99(2), 330–350.

Wright, D., Finn, R., and Rodrigues, R., 2013, 'A Comparative Analysis of Privacy Impact Assessment in Six Countries', *Journal of Contemporary European Research* 9(1).

Yu, R., Zhong, W., Xie, S., Yuen, C., Gjessing, S., and Zhang, Y., 2016, 'Balancing Power Demand Through EV Mobility in Vehicle-to-Grid Mobile Energy Networks', *IEEE Trans. on Industrial Informatics*12(1), 79–90.

21

Privacy Preserving Power Charging Coordination Scheme in the Smart Grid

Ahmed Sherif[1], Muhammad Ismail[2], Marbin Pazos-Revilla[1], Mohamed Mahmoud[1], Kemal Akkaya[3], Erchin Serpedin[4] and Khalid Qaraqe[2]

[1] *Tennessee Technological University, Cookeville, TN, USA*
[2] *Texas A&M University at Qatar, Doha, Qatar*
[3] *Florida International University, Miami, FL, USA*
[4] *Texas A&M University, College Station, USA*

21.1 Introduction

Smart grid is a progressive upgrade to the existing power grid. It consists of sophisticated systems of smart electronic devices, dispersed generators, and loads (Brusco et al., 2014) and can give two-way power exchange among various components. Several functionalities of the smart grid require major dependence on an intelligent communication infrastructure. However, such a dependence makes the smart grid vulnerable to tremendous risks as well as hard challenges in protecting the smart grid from cyber-security threats (Khurana et al., 2010). Compared with the existing communication networks such as the Internet, the smart grid communication systems have different objectives, architecture, and concerns about what need to be protected (Yan et al., 2012). For example, in the smart grid, we need to guarantee the real-time performance and the features of the continuous operations. The smart grid is more concerned with the message delay than the data throughput due to the timing constraint of messages transmitted over the power networks. As a result, the existing security solutions for different communication networks cannot fit directly to the smart grid communication networks without filling the gaps where traditional communication network solutions do not work or apply. In the following, general security requirements for the smart grid are discussed. Then, we focus our attention on a specific functionality of the smart grid, discussing its security requirements, and presenting potential cyber-security solutions.

21.1.1 Smart Grid Security Requirements

The smart grid should meet rigorous security requirements such as availability, integrity, authentication, authorization, and non-repudiability. Such requirements can be briefly summarized as follows:

- Availability: denial of service (DoS) attacks are considered as the potential attacks against the availability of the smart grid networks. DoS attacks are aiming to

Transportation and Power Grid in Smart Cities: Communication Networks and Services, First Edition.
Edited by Hussein T. Mouftah, Melike Erol-Kantarci and Mubashir Husain Rehmani.

exhausting all the grid resources to prevent the legitimate users from accessing these resources. These attacks can be done in the physical layer by performing false data injection or jamming attacks. In the literature, multiple frequency channels are proposed as a defense mechanism against jamming attacks (Aravinthan et al., 2011). In addition, an intrusion detection system (IDS) can be used to defend against the DoS attacks in higher layers. There are three different categories of IDSs (Faisal et al., 2012): signature-based, specification-based, and anomaly-based. A signature-based IDS uses a black list of all known attacks to detect the intrusions, whereas specification-based IDS uses a set of rules to define the correct behavior of a program or a protocol. Finally, an anomaly-based IDS uses a model for normal behavior, and any deviation from this model is considered an anomaly.

- Integrity: it aims to prevent unauthorized persons or systems from modifying the information via message injection, device impersonation, or message replay. Different techniques have been adopted in the literature to ensure data integrity such as cryptographic hashing (Benoit, 2011), time-stamps and sequence numbers (Aravinthan et al., 2011), and nonces (numbers used once to make each message unique) (Xiao et al., 2013). Other techniques to detect data integrity attacks in the smart grid rely on secure phasor measurement units (PMUs) (Giani et al., 2011) and load profiling schemes (Aravinthan et al., 2011).
- Authentication: it is concerned with the determination of the true identity or valid account of each entity involved in the communication. Usually the mechanisms used for the authentication can provide also the integrity. Various tools exist for providing authentication and integrity, including hashes and keyed hashes such as SHA-1 or HMAC-SHA-1 and digital signatures such as RSA or ECC signatures (Stallings, 2010). In addition, other techniques rely on a physically unclonable function (PUF) for strong hardware-based authentication (Nabeel et al., 2012).
- Authorization: it is concerned with preventing access to the system by persons or systems without permission to do so. In the literature, public key certificates and zero-knowledge systems can be used for authentication, and the attribute certificates are used for authorization (Vaidya et al., 2011).
- Non-repudiability: it refers to being able to provide irrefutable proof to a third party of who initiated a certain action in the system, even if this actor is not cooperating. In the literature, such an objective can be achieved after user authentication by storing a log of transactions for a predefined number of days to resolve any repudiation by tracing back the events (Aravinthan et al., 2011).

Overall, specific smart grid functionalities require custom-made cyber-security mechanisms depending on the communication requirement of such a functionality. In what follows, we focus our attention on one functionality of the smart grid, namely charging coordination, and discuss one of its cyber-security requirements.

21.1.2 Charging Coordination Security Requirement

As per NIST (National Institute of Standards and Technology, 2010), one of the principal goals of the smart grid is the diminishment of greenhouse gas emissions by integrating more renewable energy resources (Lukic et al., 2008) and electric vehicles (EVs). Recently energy storage units (ESUs) including home batteries and EVs have received broad consideration. TESLA company recently produced a home battery product (M57) that can charge when the electricity is cheap and power the home when electricity is expensive. Numerous automotive companies have already started producing EVs from their production lines (Nissan LEAF Electric Car).

These developments imply that the smart grid will have a large number of ESUs. However, the large-scale, simultaneous charging of energy storage units (ESUs), including home batteries and EVs, will have a significant impact on the power grid, especially when the charging loads are not coordinated. This can raise technical issues including imbalance of charging demands and energy supply, more power losses, and larger voltage deviation (Sortomme et al., 2011; Hadley and Tsvetkova, 2009). These issues cannot be avoided by charging at off-peak periods, e.g., late night, since this can make another peak that cannot be handled by the power distribution network. The most ideal approach to avoid this issue and fully utilize the accessible power is via charging coordination (Deilami et al., 2011; Clement et al., 2009). However, this needs reporting a few data, for example, whether an ESU needs to charge or not, time to complete charging (TCC), the battery state of charge (SoC), and so on. Unfortunately, this data can reveal private information about the owners of ESUs, such as the location of an EV and the activities of a house's residents (P. Akula, M. Mahmoud, K. Akkaya, and M. Song, 2015; NIST, 2010). For example, the charging demands sent from an EV can reveal whether the EV's owner is at home, to what extent he/she will stay, and to what extent he/she drives. Also, if a home battery is not charging for an extended period, this can reveal that the residents do not spend time in home because they are traveling, and charging more than the normal amount can reveal that residents host guests. While one can plan solutions for particularly securing the information exchange and concealing data from the included parties, this may unfavorably influence the performance of charging coordination plans. This is on the grounds that privacy preservation often targets to hide information; however, charging coordination needs sufficient information.

Due to technical limitations, not all ESUs can charge at the same time, so each ESU ought to send a charging request to an aggregator. The aggregator should not know whether an ESU needs to charge or not, its battery state of charge, and the needed time to finish charging to preserve privacy. The aggregator forwards the requests to a charging controller (CC) that needs sufficient data to run a charging coordination scheme without linking the requests to specific ESUs to preserve privacy. In this chapter, we present a privacy-preserving power-charging coordination scheme that operates in two phases (Mahmoud et al., 2016a). The first phase aims to submit anonymous and unlinkable data to the CC, and the CC computes the charging schedules in the second phase by running a modified algorithm for the knapsack problem. The proposed charging coordination algorithm is efficient and scalable as it exhibits polynomial time complexity. In the charging coordination scheme, we define a priority function to prioritize the ESUs' requests by calculating a priority index for each request. The target of the coordination scheme is to charge in every time slot a subgroup of ESUs that maximize the total priority indices and subsequently the number of charged ESUs without surpassing the power limit. In the first phase and in order to boost the level of privacy preservation, each ESU can add a random noise to its data, i.e., TCC and SoC, to prevent the CC from linking the requests coming from a specific ESU. As adding noise to the data may influence the performance of the charging coordination scheme, we will explore the impact of noise addition on performance and the gains in privacy by utilizing information-theoretic metric referred to as entropy (Díaz et al., San Francisco, CA, USA, 2003).

The remainder of the chapter is organized as follows. Existing charging coordination and privacy preservation schemes are introduced in Section 21.2. The network and threat models and the privacy-preserving charging coordination scheme are explained in Section 21.3. Privacy analysis and simulation results are are discussed in Section 21.4. Finally, a summary is presented in Section 21.5.

21.2 Charging Coordination and Privacy Preservation

In (Ismail et al., 2015b), Ismail et al. explain that the negative impacts of vast simultaneous charging of EVs can be addressed by using optimal design of charging facilities and charging coordination of EVs. In terms of optimal design of charging facilities, Ismail et al. present a novel optimal planning algorithm for EV-charging facilities in (Ismail et al., 2015a). The planning model developed in (Ismail et al., 2015a) accounts for the stochastic nature of the EV- charging requests and determines the optimal number of chargers to be deployed within a charging station along with the queuing size within the charging facility. The proposed scheme is tested in various real-world scenarios, and the results show positive profit values when optimal planning parameters are adopted.

On the other hand, coordinated charging can prevent stressing the distribution system and avoid power outages in extreme cases (Verzijlbergh et al., 2012). The main parts of the smart-charging coordination systems include data collection and optimization (Sundstrom and Binding, 2012). The first part is responsible for collecting the data relevant to the ESUs' charging requests such as TCC and battery SoC. Typically, this part is carried out via an aggregator, which collects the data from the ESUs and sends it to the operator, and then the aggregator sends the charging decisions coming from the operator back to the ESUs. The optimization unit can make optimal coordinated charging decisions that maintain service reliability, maximize the operator's profit, satisfy system constraints, and meet the customer demands. In general, the following approaches can be distinguished for EV charging:

- Grid-to-vehicle (G2V): in this scheme, EVs are connected to the smart power grid to charge. In (Bayram et al., 2014), a pricing-based load-shifting framework is presented for G2V fast-charging stations with fixed capacity. Two different types of users are considered. The first type represents a customer who uses the charging station once. For this type, the charging coordination scheme offers incentives to the users to delay their service by one time period at a high peak load condition. So the users can take the offer and guarantee the service in the next time period or face the risk of being blocked in case of resource unavailability. On the other hand, the second type of users represents a customer who uses the same charging station in a frequent manner. This type of customers is referred to as EV subscribers, who can have a contract with the charging station that offers better service agreement. For EV subscribers, the objective of the station operator is to serve all peak-hour requests within a specific time window.
- Vehicle-to-grid (V2G): in this scheme, an EV can discharge to the grid (as it is equipped with bidirectional chargers) to balance supply and demand (Han and Xiao, 2016; Wan et al., 2016). In (Shaaban et al., 2014b), Shaaban et al. propose

a real-time system for managing the coordinated charging decisions for EVs in a smart grid. The proposed two-stage optimization unit guarantees the feasibility of the charging/discharging decisions by maximizing the satisfaction of the EV owner and minimizing the system operating costs. A smart coordination mechanism to coordinate the charging/discharging of EVs in parking lots is proposed in (Shaaban et al., 2014a). The proposed scheme maximizes the utility profit and minimizes the charging costs by taking into consideration the impacts of the future charging decisions and the normal load profile. Moreover, a sliding time window is used to deal with the dynamics associated with vehicles' arrivals and departures in a parking lot.

- Vehicle-to-vehicle (V2V): in this scheme, energy is transferred from one EV to another EV to overcome the overloading problem at the power grid. A semi-distributed V2V-charging coordination strategy for EVs is proposed in (Wang et al., 2014). Real-time information delivery for the V2V strategy is provided through a heterogeneous wireless network by utilizing both vehicular ad hoc networks (VANETs) and cellular networks. In (Wang et al., 2015), the authors incorporated the mobility of EVs into the online charging coordination and an online coordinated EV-charging strategy is proposed. The real-time vehicle information is collected via a heterogeneous network of VANETs assisted by cellular networks. By using the real-time vehicle information, the mobility-aware coordinated EV-charging strategy can be achieved jointly in a spatiotemporal framework to enhance the resource utilization while avoiding the power overload and generation-demand mismatch. In addition, a price control mechanism is proposed in (Wang et al., 2016) to enable a V2V energy-swapping strategy that maximizes the discharging revenues for supplying EVs and minimizes the charging cost for demanding EVs while avoiding power system overload.

Numerous privacy-preservation schemes are proposed in the literature for different smart grid applications such as advanced metering infrastructure (AMI) (Tonyali et al., 2016; Beussink et al., 2014; H. Mohammed and Akkaya, 2016) and power injection (Akula et al., 2015; Mahmoud et al., 2016b). AMI is a communications network that enables two-way communication between utilities and customers. Smart meters send fine-grained power consumption data in real time, e.g., every few seconds. In (Tonyali et al., 2016; Beussink et al., 2014), the authors used a data obfuscation mechanism and proposed secure and efficient algorithms to distribute obfuscation values within an AMI network. They presented a protocol that utilizes LTE-direct for exchanging of data among various gateways. An efficient privacy-preserving data collection scheme for smart grid AMI networks is proposed in (H. Mohammed and Akkaya, 2016). By using a lightweight symmetric-key-cryptography and hashing operations, the proposed scheme can collect the consumption data while preserving the customer privacy. The authors use the asymmetric-key-cryptography operations for key management that is executed every long time. In (Akula et al., 2015), the authors proposed a privacy-preserving scheme for power injection in the smart grid that is based on the idea of aggregation of sensitive information of the storage units' owners to prevent the utility from knowing individual's sensitive information. The proposed scheme can be used for the authentication of the storage units and the integrity of their data. Mahmoud et al. proposed a novel secure and privacy-preserving power injection querying scheme over AMI and LTE cellular networks in (Mahmoud et al., 2016b). The proposed scheme is based on two aggregation techniques to hide the storage units' sensitive information. They also developed a bilinear pairing-based technique to enable the utility company to ensure the integrity and authenticity of the aggregated bid without accessing the individual bids coming from the storage units. Additionally, numerous privacy-preservation schemes are proposed in the literature for different applications such as VANETs (Rabieh et al., 2016, 2015a; Mahmoud et al., 2014; Rabieh et al., 2015c, b, d), LTE networks (Haddad et al., 2016, 2015), autonomous vehicles (AVs) (Sherif et al., 2016), and mobile social network (MSN) (Oriero et al., 2016). However, these schemes cannot be applied effectively and efficiently in the smart grid due to its unique problems and requirements.

As the renewable energy will play a vital role in the smart grid, the energy storage units can be used to store the excess power at the period of strong sun or wind and inject power to the grid during peak hours. The same thing can be applied to the EVs' batteries, which can be treated as storage units. The utility companies should communicate with the storage units' owners to coordinate such a charging action. However, this type of communication may reveal some sensitive information about the owners such as their location or amount of power injected. This information can be used by the utility or by other owners to misbehave and gain more benefits. Without preserving the privacy and hiding such information, the storage unit owners or the utility company may face financial losses. However, the existing research in charging coordination in the smart grid, e.g., (Ismail et al., 2015b; Wang et al., 2014, 2015, 2016; Bayram et al., 2014; Shaaban et al., 2014a,b; Ismail et al., 2015a) do not consider the privacy preservation concerns. In addition, the works on privacy preservation in the smart grid, e.g., (Tonyali et al., 2016; Beussink et al., 2014; H. Mohammed and Akkaya, 2016; Akula et al., 2015; Mahmoud et al., 2016b) do not investigate the impact of the proposed schemes on the charging coordination efficiency. This motivates our work to study the impact of privacy preservation on charging coordination efficiency.

21.3 Privacy-Preserving Charging Coordination Scheme

21.3.1 Network and Threat Models

As outlined in Figure 21.1, the considered network model has various numbers of communities and CCs. Every community is composed of several ESUs and one aggregator. The aggregator and the CC can communicate by means of LTE or 4G (Saputro et al., 2012). The connection between the ESUs and the aggregator is based on either Wi-Fi or LTE. The CC cannot communicate directly with ESUs; however, this must be done via the aggregators. The ESUs send charging requests to the CC via the aggregators. The CC determines the appropriate charging schedules and sends them back to the ESUs. A community is associated with an electric bus with loading limit of C KWh. The normal load (residential, commercial, and industrial) signified by P_R is known to the CC. Henceforth, the accessible charging limit for ESU charging at a given time slot is given by $C - P_R$. In the event that the aggregated power requests of the ESUs are less than $C - P_R$, all ESUs can charge at the same time. Generally, the proposed privacy-preserving charging coordination scheme is utilized to satisfy the charging requests while keeping the aggregate charging power at $C - P_R$.

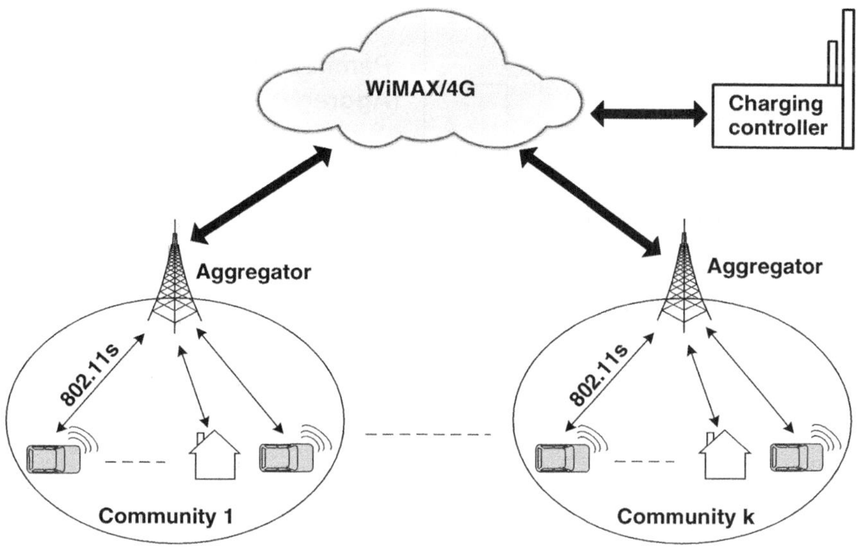

Figure 21.1 Network model under consideration.

For the threat model, an honest-but-curious model is employed, which assumes that the attackers do not aim to disturb the proper operation of the scheme, yet they are just interested in gathering private information. The attackers can be the aggregator, the CC, ESUs, and eavesdroppers. They may passively snoop on the communications to learn sensitive information. The attackers should not learn whether an EV needs to charge or not or the charging request data such as TCC and SoC. The CC needs TCC and SoC to run the charging coordination scheme, yet the information should be anonymized to preserve privacy. The CC should not have the ability to link an ESU's requests for high privacy assurance.

21.3.2 The Proposed Scheme

As outlined in Figure 21.2, the proposed scheme has two stages. In the anonymous data submission phase, ESUs send charging requests without revealing private data. In the charging coordination phase, the submitted information is utilized to generate charging schedules. First, a priority index for each ESU is calculated, and then a modified knapsack problem (Kellerer et al. 2004) algorithm is run to choose the subset of ESUs that can maximize the summation of the priority indices of scheduled ESUs without surpassing the total power limit.

21.3.2.1 Anonymous Data Submission

Charging Requests Submission The packets exchanged in the anonymous data submission phase are shown in Figure 21.3. Our goal is to empower the CC to run a charging coordination scheme without revealing private information. Time is divided into a group of slots $\mathcal{T} = \{1, 2, \dots, T\}$ of equivalent duration τ that covers the 24 hours of the day, with $T = 24/\tau$. At the start of every time slot, each ESU in a group should send a charging request packet (R_v) to the aggregator. At a given time slot, the set of ESUs with charging

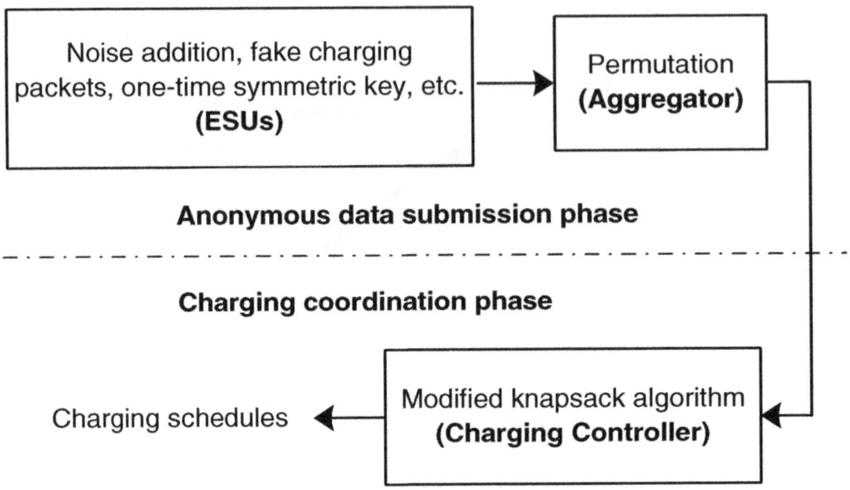

Figure 21.2 Illustration of the privacy-preserving charging coordination scheme.

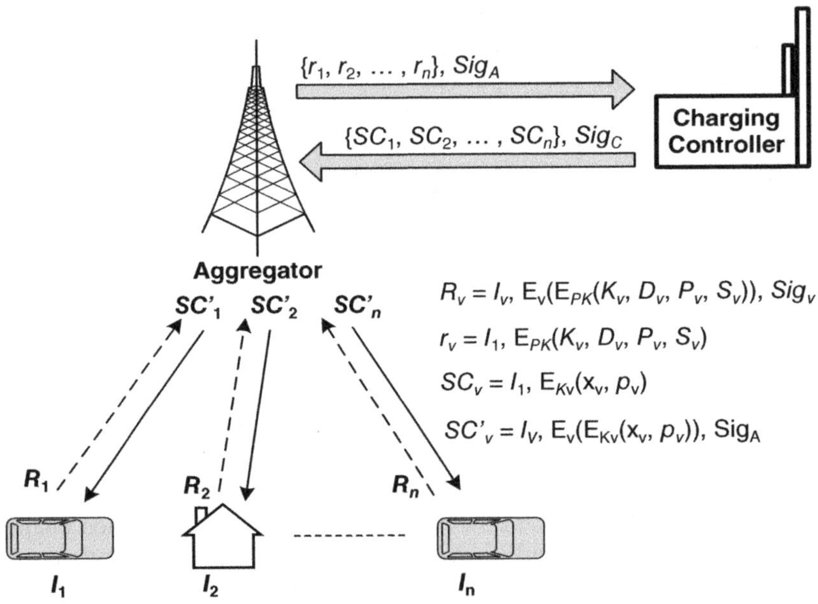

Figure 21.3 Anonymous data submission.

requests is given by $\mathcal{V} = \{1, 2, \dots, V\}$. Each ESU shares a symmetric key with the aggregator using any existing key agreement and exchange scheme such as the Diffie-Hellman (Diffie and Hellman, 1976). The charging request has the ESU's unique identity (I_v), an encryption with a symmetric key shared with the aggregator $E_v(E_{PK}(K_v, D_v, P_v, S_v))$, and its signature on the packet (Sig_v), where D_v ($D_v \geq 0$) is the TCC, S_v ($0 \leq S_v < 1$) is the battery SoC, P_v is the amount of power the ESU needs to charge, $E_{PK}(\cdot)$ is an encryption with the public key of the CC, K_v is a one-time symmetric key, and E_v is an encryption with the symmetric key shared among I_v and the aggregator. Only the CC can acquire K_v and utilize it to encrypt the ESU's charging schedule. After receiving a charging request, the aggregator first verifies the signature to guarantee that the packet is sent from I_v. Then, it decrypts the packet to acquire $E_{PK}(K_v, D_v, P_v, S_v)$. Finally, the aggregator sends the charging requests ($\{r_1, r_2, ...\}$) alongside its signature Sig_A to the CC.

The CC should not have the ability to link a charging request to an ESU from the identity sent in the packet. One way to do that is by utilizing pseudonyms rather than the real identities. Another approach is that the aggregator can utilize a secret and random permutation to shuffle the ESUs' identities and their charging requests such that the charging information of one ESU is mapped to the identity of another random ESU in the community, as demonstrated in Figure 21.4. The figure shows that the data of ESU I_n is sent under the name of ESU I_1, the data of ESU I_2 is sent under the name of ESU I_n, etc. The proposed permutation is not static, yet for each time slot, it maps charging information to various ESUs' identities and stores a mapping table. This table will be utilized to map charging schedules to their ESUs' identities. The CC should not have the ability to link an ESU's requests in various time slots by linking the charging data, i.e., TCC and SoC.

Charging Schedules Distribution The CC uses its private key to decrypt the charging requests coming from the aggregator and runs the charging coordination scheme. The CC determines the charging schedules SC_v $\forall\, v \in \mathcal{V}$, and signs and returns them back to the aggregator. As shown in Figure 21.3, a charging schedule SC has the permutated identity I_1 and the encryption of the schedule utilizing the one-time key sent by the ESU ($E_{K_v}(x_v, p_v)$). The schedule either permits ESU v to charge in the present time slot (when $x_v = 1$) and indicates its charging rate p_v, or it postpones its charging request to a future time slot ($x_v = 0$). It should be noted that $E_{K_v}(\cdot)$, $E_v(\cdot)$, and $E_{PK}(\cdot)$ indicate

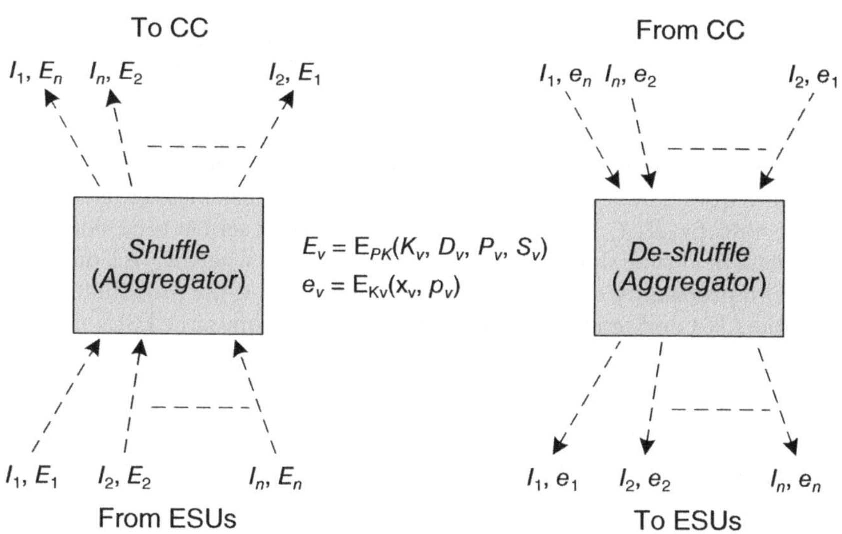

Figure 21.4 Permutation of requests.

a symmetric-key encryption with the one-time key sent in the charging request, a symmetric-key encryption with the key shared between an ESU and the aggregator, and an encryption with the public key of the CC, respectively. Upon receiving SC_v $\forall v \in \mathcal{V}$, the aggregator utilizes its permutation mapping table to undo the permutation, as shown in Figure 21.4, and transmits the charging schedules $\{SC'_1, SC'_2, ...\}$ to the ESUs, as represented in Figure 21.3.

Linking ESU's Charging Requests In light of the above discussion, the CC needs to know the charging request's data (TCC and battery SoC) to run the charging coordination scheme; however, it is hard to link the request to a particular ESU. Nonetheless, the CC can attempt to utilize the charging request data to connect an ESU's requests over various time slots since it is normal that the TCC is diminishing after some time, and if an ESU does not charge in the present time slot, its TCC and SoC in a future time slot should be near the reported values in the present time slot. Linking an ESU's requests is problematic due to the fact that it can be utilized to distinguish the ESU; furthermore, an excessive amount of data will be uncovered if the CC could recognize the ESU.

In order to prevent leaking information, ESUs should add noise to the real data in its charging requests. By adding noise to both the TCC and the battery SoC, the ESU reports its TCC as: $D_v \pm d_v$ and battery SoC as: $S_v \pm s_v$, where d_v and s_v are random noise. It should be noted that an ESU pays only for the charging power that is recorded by the smart meter. Moreover, not all the ESUs in a community need to add noise to their charging requests such as ESUs of public places, e.g., shopping malls and governmental and educational buildings. In these cases, users can hide their data in a big amount of requests coming from the same location, and the linkability of requests will be hard.

Anonymity Measurement Noise should be added to the charging requests' data (D_v and S_v) to make it hard for the charging controller to link the requests of an ESU sent at various time slots. We utilize an information-theoretic metric called entropy (Díaz et al., San Francisco, CA, USA, 2003) to quantitatively measure the privacy preservation level that can be gained by adding noise. The idea of entropy in information theory gives an estimation of the information contained in a distribution of probabilities (Cover and Thomas, 1991. ISBN 0-471-06259-6.). In (Díaz et al., San Francisco, CA, USA, 2003), the authors propose an information-theoretic model utilizing entropy to evaluate the level of anonymity given by anonymous communication schemes. The model can measure the level of information acquired by attackers about the probability that a message is sent by a specific user. The entropy ($H(X_{j,i})$) of linking a charging request sent at time slot $t+1$ to one of N requests sent at time slot t is given in Equation (21.1), where $X_{j,i}$ is a discrete random variable with probability mass function of $\chi_{j,i}$, where $\chi_{j,i}$ is the probability that request j sent at time $t+1$ and request i sent at time t are sent from same ESU.

$$H(X_{j,i}) = -\sum_{i=1}^{N} \chi_{j,i} \log_2(\chi_{j,i}) \tag{21.1}$$

The CC utilizes S_v and D_v of the packets sent in two distinct time slots to specify $\chi_{j,i}$ for every request i, where $i \in \{1, 2, .., N\}$ and N is the number of requests received at time t. The probability $\chi_{j,i}$ is higher as S_v and D_v of the two requests j and i are close.

$\chi_{j,i}$ is given in Equation (21.2), where S_j^{t+1} and S_i^t are the SoC of the request of interest j sent at time $t+1$ and the SoC of the request i sent at time t, and D_j^{t+1} and D_i^t are the TCC of the request j sent at time $t+1$ and the TCC of the request i sent at time t, and $\sum_{i=1}^{N} \chi_{j,i} = 1$.

$$\chi_{j,i} = \frac{1 - |S_j^{t+1} - S_i^t| + (1 - (\frac{|D_j^{t+1}-(D_i^t-1)|}{\sum_{i=1}^{N} |D_j^{t+1}-(D_i^t-1)|})}{\sum_{i=1}^{N}(1 - |S_j^{t+1} - S_i^t| + (1 - (\frac{|D_j^{t+1}-(D_i^t-1)|}{\sum_{i=1}^{N} |D_j^{t+1}-(D_i^t-1)|}))} \tag{21.2}$$

It is obvious that the term $1 - |S_j^{t+1} - S_i^t|$ increases as S_j^{t+1} and S_i^t are close. Likewise, the term $1 - (\frac{|D_j^{t+1}-(D_i^t-1)|}{\sum_{i=1}^{N} |D_j^{t+1}-(D_i^t-1)|})$ increases as the TCC of the two requests are close. For two requests sent from the same ESU at times t and $t+1$, TCC at time $t+1$ should be one less than the TCC at time t. To guarantee that this part always gives a value in the range [0,1], $|D_j^{t+1} - (D_i^t - 1)|$ is normalized by $\sum_{i=1}^{N} |D_j^{t+1} - (D_i^t - 1)|$.

The entropy can describe the uncertainly of the charging controller about linking various requests sent from the same ESUs. The maximum entropy (i.e., the highest anonymity level) can be achieved when the probabilities $\chi_{j,i}$ for $i \in \{1, 2, .., N\}$ follow a uniform distribution, i.e., all the N requests have a similar probability to be linked to request j. In this case, data privacy is completely ensured, and the charging controller cannot link a request to a certain ESU or even to a subset of ESUs. The maximum entropy (H_M) is given in Equation (21.3).

$$H_M = \log_2(N) \tag{21.3}$$

21.3.2.2 Charging Coordination

Once the sensitive information is concealed by using the aforementioned approaches, we next investigate how to utilize the information of the charging requests to calculate the charging schedules at every time slot. Each community is associated with an electric bus with loading limit of C. At a given time slot, the regular load capacity is given by P_R. Thus, the accessible charging limit with respect to the ESUs at a given time slot is given by $C - P_R$. Due to the limited capacity $(C - P_R)$, it is possible that not all ESUs with charging requests can charge at the present time slot. Rather, our scheme calculates a priority index for each ESU, and the ESUs with high priority will be charged at the present time slot to guarantee that $\sum_v P_v \leq C - P_R$, while other ESUs' charging requests can be postponed to future time slots. Two components play a vital role in determining the ESU's priority for charging at the present time slot, to be specific, the TCC D_v and the battery SoC S_v. In particular, an ESU with low S_v and short D_v should have higher charging priority than an ESU with high S_v and/or potentially long D_v. Subsequently, for each ESU v, we specify a priority function U_v that is given by Equation 21.4, where $F(D_v)$ is a decreasing function of D_v with a range of [0, 1] and $F(D_v) = 0$ for long TCC and equals 1 for short TCC, and SoC value $(S_v) \in [0, 1)$ with $S_v = 1$ for a completely charged ESU.

$$U_v = \alpha_1(1 - S_v) + \alpha_2 F(D_v), \tag{21.4}$$

The relative significance of S_v and D_v are given by using weights α_1 and α_2, with $\alpha_1 + \alpha_2 = 1$. An ESU v with low S_v value and short D_v will have a high priority value U_v. Henceforth, the CC's goal is to schedule the ESUs with highest priority for charging in the present time slot and postpone the charging of ESUs with lower priorities to future time slots because of the limit capacity in the present time slot. The charging schedules indicate whether a given ESU v will be charged in the present time slot (x_v) and the charging amount (P_v), where

$$\begin{aligned} \max_{x_v, p_v} \quad & \sum_{v \in \mathcal{V}} x_v U_v \\ \text{s.t.} \quad & 0 \leq p_v \leq P_v \quad \forall v \in \mathcal{V}, \\ & \sum_{v \in \mathcal{V}} x_v p_v \leq C - P_R, \\ & x_v \in \{0, 1\}. \end{aligned} \tag{21.5}$$

The charging coordination problem formulated in (21.5) is a mixed integer program (MIP) as it includes a real variable p_v and a binary variable x_v, which makes it NP-complete. For a large-scale problem (i.e., a large community with numerous ESUs), it is difficult to solve (21.5) in real time. Rather than solving the MIP in (21.5), we reformulate the charging coordination problem using an integer program (IP) that is less complex than (21.5), and is given by

$$\begin{aligned} \max_{x_v \in \{0,1\}} \quad & \sum_{v \in \mathcal{V}} x_v U_v \\ \text{s.t.} \quad & \sum_{v \in \mathcal{V}} x_v P_v \leq C - P_R. \end{aligned} \tag{21.6}$$

As indicated in (21.6), a scheduled ESU gets its charging demand (P_v) in the present time slot. However, if the difference $C - P_R - \sum_{v \in \mathcal{V}} x_v P_v$ is greater than 0 yet less than P_v of all unscheduled ESUs, inefficient resource utilization is expected, as compared to (21.5) which ensures full resource utilization. To overcome such a disadvantage, the rest of the capacity after ESUs' scheduling as indicated by (21.6), $C - P_R - \sum_{v \in \mathcal{V}} x_v P_v$, is assigned to the ESU with highest priority among all unscheduled ESUs. Such an ESU, which has not fully charged at the present time slot, will hold up to the next slot and get whatever is left of its charging request P_v. A major advantage of the formulation in (21.6) is that it can be mapped to a well-known optimization known as the knapsack problem (Kellerer et al. 2004).

In the knapsack problem, there is a knapsack with limited capacity and a set of items each with a given value (priority) and weight. The goal is to select a subset of items to be packed in the knapsack, such that the overall value of the packed items is maximized while respecting the knapsack capacity limit. The charging coordination problem at every time slot can be described as a knapsack problem as follows. The ESUs are mapped to the items, the ESU priority U_v resembles the item value, the ESU charging request P_v is equivalent to the item weight, and the charging limit constraint $C - P_R$ represents the knapsack limit. We modified a greedy algorithm designed for solving the knapsack problem in polynomial time given in (Kellerer et al. 2004) to schedule ESU charging at each time slot. Hence, the charging coordination mechanism can be described using Algorithm 1, which is implemented at the CC.

Algorithm 1 ESU Charging Coordination Mechanism

Input: $\mathcal{V}$, U_v and P_v $\forall v \in \mathcal{V}$;
Initialization: $x_v = 0\ \forall v \in \mathcal{V}$, $\mathcal{A} = \{\}$, $C_R \leftarrow C - P_R$;
Sort all ESUs in$\mathcal{V}$ such that $\frac{U_1}{P_1} \geq \frac{U_2}{P_2} \ldots \geq \frac{U_V}{P_V}$ and store the result in $\mathcal{A}$;
for $v \in \mathcal{A}$ **do**
  if $P_v \leq C_R$ **then**
    $x_v = 1$;
    $p_v = P_v$;
    $C_R = C_R - P_V$;
    $\mathcal{A} = \mathcal{A} - \backslash\{v\}$;
  end if
end for
$L = \underset{\mathcal{A}}{\text{argmax}}\ U_v$;
$x_L = 1$;
$p_L = C_R$;
Output: X and P.

21.4 Performance Evaluation

21.4.1 Privacy/Security Analysis

Neither the aggregator nor eavesdroppers can learn whether an ESU needs to charge or not or the amount of energy it needs. This is because the charging requests and schedules are encrypted by the CC's public key and one-time keys. Also, an ESU should send a charging request with zero charging demand when it does not need to charge.

In a case that an eavesdropper colludes with the CC, they cannot link the charging requests to the ESUs in light of the fact that they cannot link the public-key encryption ($E_{PK}(K_v, D_v, P_v, S_v)$) received by the CC to the symmetric-key encryption ($E_v(E_{PK}(K_v, D_v, P_v, S_v))$) sent by an ESU. They cannot likewise link a schedule to its ESU as the schedules are encrypted with the keys shared with aggregator. Without this encryption, an eavesdropper can get the schedule packet ($I_v, E_{Kv}(x_v, p_v)$) and the CC can decrypt the schedule and link it to I_v.

In a case that an ESU sends the same charging request information in consecutive time slots, the ciphertexts $E_v(E_{PK}(K_v, D_v, P_v, S_v))$ do not look the same because K_v is utilized for only one time. For a similar reason, the ciphertexts of same schedules ($E_{Kv}(x_v, p_v)$) look different.

The ESUs can guarantee that the schedules are sent from the CC as no entity can recover K_v aside from the CC. The ESUs can likewise guarantee that the schedule packets are sent from the aggregator since it appends its signature. The aggregator can ensure that the charging requests are sent from the ESUs since they are signed.

The CC gets enough information to run the charging coordination scheme, and privacy can be safeguarded due to the fact that the CC cannot link the charging requests to the ESUs. The CC cannot utilize the charging requests' identities or symmetric keys to link the requests of an ESU over consecutive time slots. This is on account of the random and secret identity permutation that is utilized by the aggregator and the one-time random key that is utilized as a part of each charging request. Furthermore, noise addition to the charging requests can make linking the requests of an ESU over consecutive time slots hard.

In addition to privacy preservation, the proposed scheme is robust against repudiation attacks. An aggregator can ensure that a particular ESU performs a specific transaction or requests a service using the ESU's signature Sig_v sent in the charging request packet (R_v). This signature is sent in the packet that contains the real identity of the ESU, and the aggregator has to verify the signature before forwarding the request to the CC to guarantee that the packet is sent from a legitimate ESU. No one other than the ESU can compute a valid signature because the private key needed to compute the signature is known only to the ESU. Furthermore, we can classify the compromised ESUs to two types, internal and external attackers. Our proposed scheme is protected against external attackers that aim to disrupt the proper operation of the charging coordination mechanism due to requesting each ESU to send a signature to the aggregator. For the internal attackers, if an attacker tries to send charging requests outside its community, the aggregator can detect the attack because it knows the identities of all ESUs in its community.

21.4.2 Experimental Study

21.4.2.1 Setup

MATLAB is utilized to evaluate the performance of the proposed privacy-preserving charging coordination scheme. We set the ESUs' battery capacity to 200 units of power, and the accessible charging limit per time slot ($C - P_R$) is 1000 units of power. We ran the simulation for 30 time slots, i.e., the slot span $\tau = 24/30$. At first, the framework has 7 ESUs that need to charge, and the arrival of new charging requests in every time slot follows a Poisson distribution with an average of λ. The SoC of each ESU battery is a random number uniformly distributed in [0, 1). Time to complete charging (D_v) is a random number that follows a geometric distribution with an average of 4. The presented results are the average of 80 runs. For the priority function, $F(D_v)$ is set as follows

$$F(D_v) = \begin{cases} 1, & D_v = 1 \\ 0.5, & D_v = 2 \\ 0.25, & D_v = 3 \\ 0, & D_v > 3 \end{cases}$$

21.4.2.2 Metrics and Baselines

To measure the performance, we use the following metrics:

- *Average satisfaction index per ESU*: This metric measures to what degree ESUs' owners are fulfilled. It is calculated by dividing the amount of power an ESU charges by the amount of power required by the ESU. The satisfaction index is a value in the range [0,1], where 1 demonstrates full satisfaction, i.e., the ESU was completely charged before its charging request expires.
- *Average number of expired requests without full charge*: This metric measures the average number of requests that expire without full charge. Performance degradation due

to the inability of the system to fully charge a large number of ESUs can be shown by the large values of this metric.

We will utilize equations explained in section 21.3.2.1 to measure the privacy level achieved by the noise addition. We will compare the privacy-aware charging coordination scheme with two benchmarks:

- *FCFS benchmark*: In this approach, the ESU that requests charging first gets charged first regardless of the request TCC and SoC.
- *Optimal coordination scheme*: This approach utilizes the modified knapsack-based ideal charging coordination mechanism without applying noise addition.

21.4.2.3 Simulation Results

For α_1 and α_2, we attempted every conceivable values, and the optimal values were observed to be 0.9 and 0.1, respectively, as they result in the largest number of served ESUs and the highest satisfaction index at various charging demand rates.

Figure 21.5 presents the average number of expired requests without full charge at various charging request rates (λ), and Figure 21.6 gives the average satisfaction index at various values of λ. The figures show the performance of the FCFS and our schemes without including noise and with adding random uniformly distributed noise to the SoC and TCC values up to $\pm$ 15% and $\pm$ 30%.

In the figures, we consider two situations where 100% and 30% of the ESUs add noise. The figures can show that our scheme has less number of expired requests before the full charge and can accomplish a higher satisfaction index unlike the FCFS. Furthermore, as λ increases, our scheme presents a significant performance improvement compared with the FCFS, which demonstrates that a charging coordination scheme is completely necessary when the charging demand is much larger than the charging limit ($C - P_R$). In this situation, our scheme can exploit the accessible resources to serve more demands before they expire by organizing the requests and serving the high priority requests first. A degradation in the performance can be observed as the number of ESUs that add

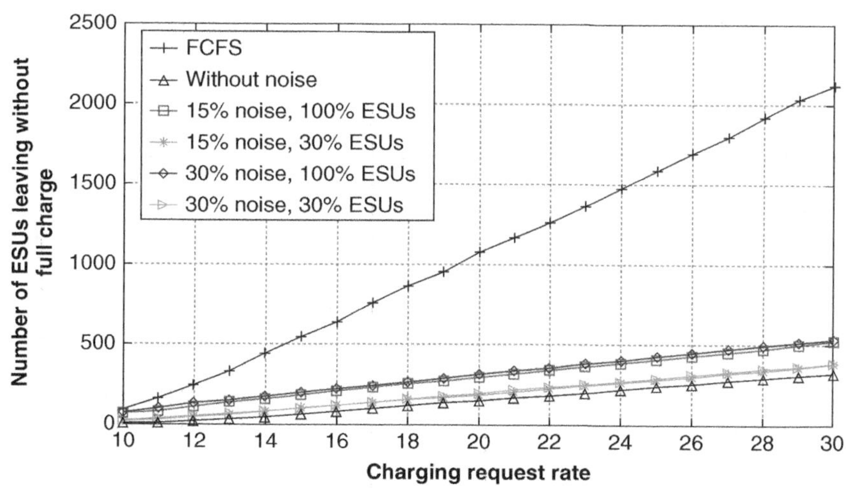

Figure 21.5 Average number of expired requests without full charge versus charging request rate (λ).

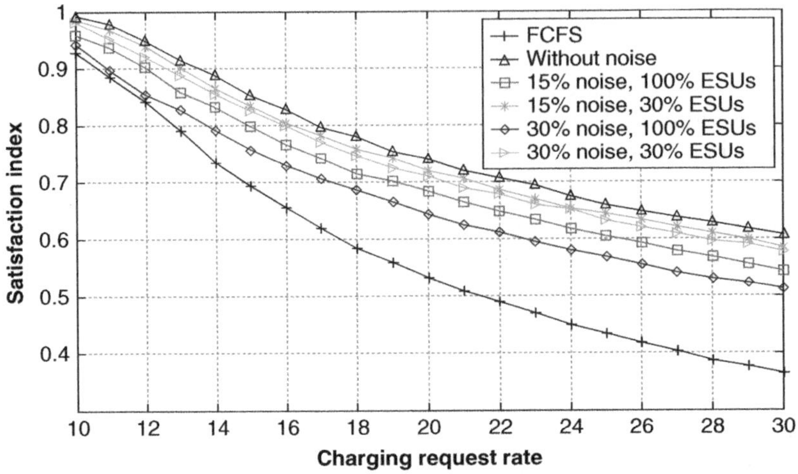

Figure 21.6 Average satisfaction index versus charging request rate.

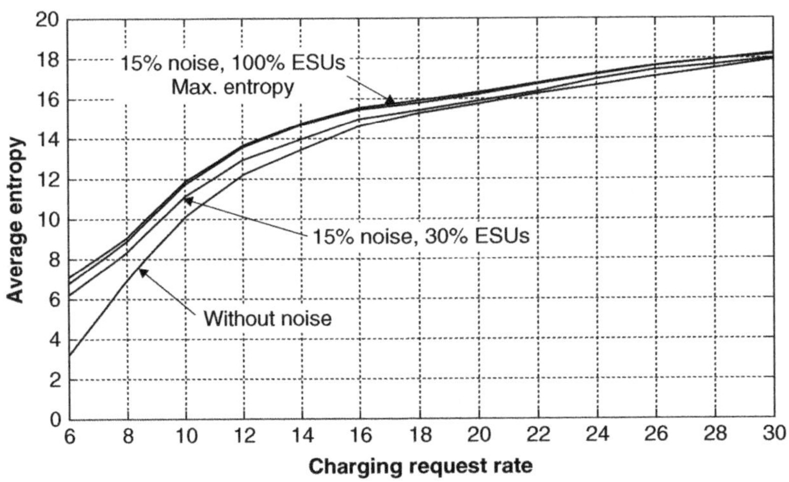

Figure 21.7 Average entropy in case of 15% noise addition versus charging request rate.

noise increases. This can be credited to the fact that as the CC gets inaccurate data, the charging coordination deviates from the ideal results.

Figures 21.7 and 21.8 present the average entropy versus the charging demand rate when adding up to $\pm$ 15% and $\pm$ 30% noise, respectively. To measure the average entropy, we utilized Equation (6) to calculate the entropy of every request at every time slot. Compared with the case that has no noise addition, it can be seen that the entropy value improves by adding noise. This shows that it is harder for the CC to link an ESU's requests since adding noise can decrease the probability of linking an ESU's request at time t to the same ESU's request at time $t + 1$. The figures also demonstrate that the entropy value improves as the noise level increases.

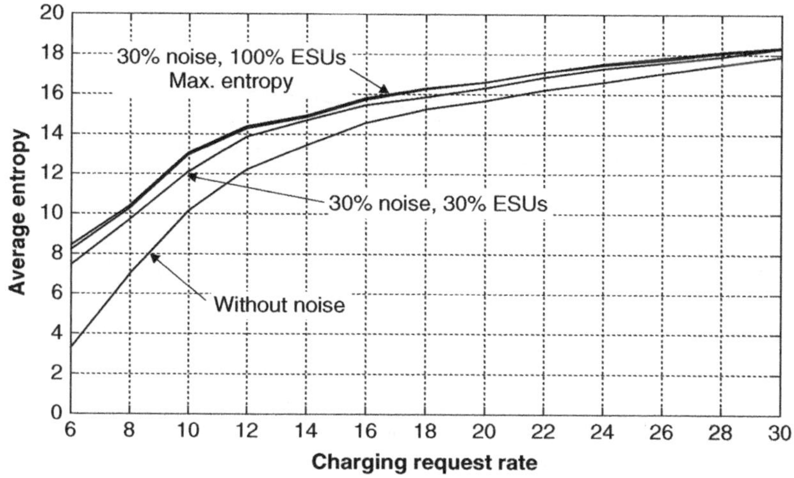

Figure 21.8 Average entropy in case of 30% noise addition versus charging request rate.

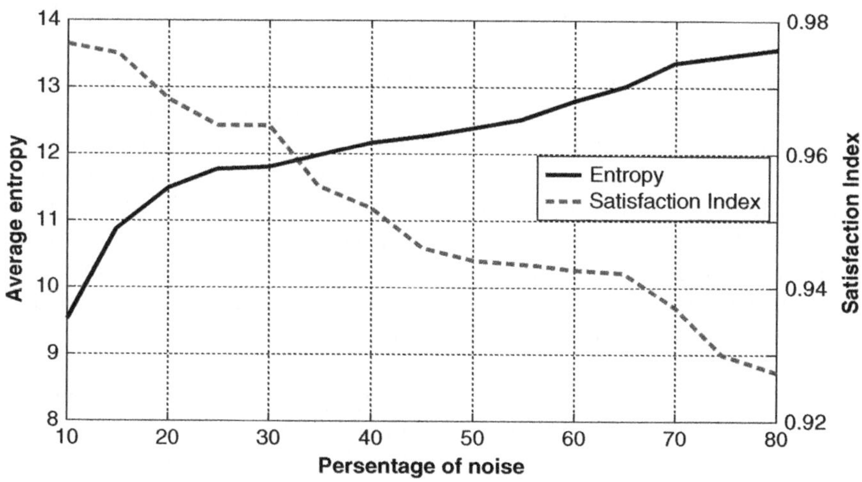

Figure 21.9 Average entropy and satisfaction versus noise with $\lambda = 10$.

The maximum entropy can be obtained when $\chi_{j,i}$ is the same for all requests. In this case, it is difficult to link requests. The figures demonstrate that the entropy achieved by adding noise is near the maximum entropy. Also, it can be seen that as λ increases, the difference in privacy level between the instances with and without noise addition reduces. This is because as the number of requests increases, several requests can present SoC and TCC values that are close to the request of interest, and thus it is harder to link requests. *This demonstrates that including noise is more helpful at low charging request rates and as the request rate increases, as it gets harder to link the requests because of the vast space of various requests.* The fewer the number of ESUs that add noise, the less the entropy, and hence more noise should be added to maintain high entropy. In all cases, the optimal and privacy-aware sub-optimal charging coordination

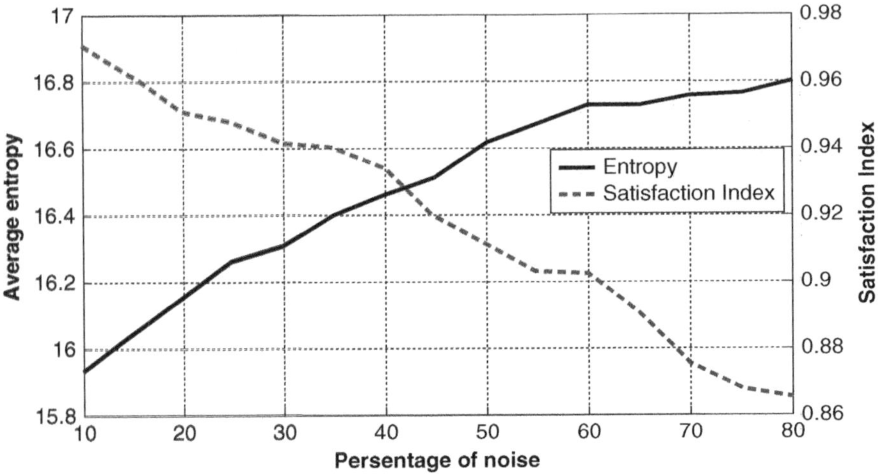

Figure 21.10 Average entropy and satisfaction versus noise with $\lambda = 20$.

schemes display better performance compared with the FCFS particularly for high charging request rates.

The outcomes additionally uncover an interesting result with respect to the trade-offs between privacy preservation and the charging coordination performance. Adding noise to SoC and D_v enhances privacy since it can make linking an ESU's requests by the CC hard as affirmed by the entropy measurements given in Figures 21.7 and 21.8. However, it somewhat degrades the charging coordination performance compared with the given results of no noise addition cases as demonstrated in Figures 21.5 and 21.6. The four figures demonstrate that the performance of the charging coordination degrades with more gains in the privacy as the noise level increases. It can be seen that the performance of our scheme is still much superior to the FCFS, and a large amount of noise may not be required in the event of extensive number of charging requests.

To investigate the impact of noise addition on the charging coordination performance, Figures 21.9 and 21.10 demonstrate the trade-off between the satisfaction index and the average entropy versus the noise level at $\lambda = 10$ and 20, respectively. As shown in the figures, increasing the noise level can increase the entropy; however, it reduces the satisfaction index. Furthermore, for the same noise level, the entropy increases as λ increases. The figures outline the trade-off between privacy preservation and charging coordination performance. Noisy data can enhance the entropy yet decreases charging coordination performance. This trade-off is not noteworthy at large λ, i.e., our scheme can accomplish high privacy preservation with superior charging coordination performance at large λ.

21.5 Summary

In this chapter, we have presented a privacy-preserving charging coordination approach that can hide sensitive information while achieving high charging coordination performance. Our goal is to maximize the number of served charging requests before they expire without revealing private information or exceeding the maximum charging capacity. Simulation results revealed that both the optimal charging coordination scheme and the privacy-aware charging coordination scheme exhibit an enhanced performance compared with FCFS approach particularly at a large charging demand situation. Also, the privacy-aware scheme offers an interesting trade-off between the charging coordination performance and privacy preservation. Our analysis has demonstrated that our approach can preserve privacy, and noise addition is an effective method to prevent linking requests. This has been qualitatively measured using entropy.

Our proposed scheme is secure against the external attackers that aim to disrupt the proper operation of the charging coordination scheme. In addition, the proposed scheme can also detect the internal attackers who send fake charging requests outside their community. In future work, we will extend the proposed charging coordination mechanism to secure it against internal attacks launched by compromised ESUs inside the same community.

Acknowledgment

This publication was made possible by NPRP grant # NPRP 9-055-2-022 from the Qatar National Research Fund (a member of Qatar Foundation). The statements made herein are solely the responsibility of the authors.

References

Tesla Home Batteries. http://money.cnn.com/2015/04/22/technology/tesla-home-battery/. Accessed: 05-06-2015.

P. Akula, M. Mahmoud, K. Akkaya, and M. Songi. Privacy-preserving and secure communication scheme for power injection in smart grid. In *2015 IEEE International Conference on Smart Grid Communications (SmartGridComm)*, pages 37–42, 2015.

V. Aravinthan, V. Namboodiri, S. Sunku, and W. Jewell. Wireless ami application and security for controlled home area networks. In *2011 IEEE Power and Energy Society General Meeting*, pages 1–8, July 2011. doi:10.1109/PES.2011.6038996.

I. Bayram, M. Ismail, M. Abdallah, K. Qaraqe, and E. Serpedin. A pricing-based load shifting framework for ev fast charging stations. In *2014 IEEE International Conference on Smart Grid Communications (SmartGridComm)*, pages 680–685, 2014.

J. Benoit. An introduction to cryptography as applied to the smart grid. 2011.

A. Beussink, K. Akkaya, I. Senturk, and M. Mahmoud. Preserving consumer privacy on ieee 802.11s-based smart grid ami networks using data obfuscation. In *2014 IEEE Conference on Computer Communications Workshops (INFOCOM WKSHPS)*, pages 658–663, 2014.

G. Brusco, A. Burgio, D. Menniti, A. Pinnarelli, and N. Sorrentino. Energy management system for an energy district with demand response availability. *IEEE Transactions on Smart Grid*, 5(5):2385–2393, Sept 2014.

K. Clement, E. Haesen, and J. Driesen. Coordinated charging of multiple plug-in hybrid electric vehicles in residential distribution grids. In *Proc. of IEEE/PES Power Systems Conference and Exposition, PSCE'09*, pages 1–7, 2009.

T. Cover and J. Thomas. *Elements of information theory. John Wiley & Sons, Inc.*, 1991. ISBN 0-471-06259-6.

S. Deilami, A. Masoum, P. Moses, and M. Masoum. Real-time coordination of plug-in electric vehicle charging in smart grids to minimize power losses and improve voltage profile. *IEEE Transactions on Smart Grid*, 2(3):456–467, 2011.

C. Díaz, S. Seys, J. Claessens, and B. Preneel. Towards measuring anonymity. *Proceedings of the 2Nd International Conference on Privacy Enhancing Technologies*, San Francisco, CA, USA, 2003.

W. Diffie and M. Hellman. New directions in cryptography. *IEEE Transactions on Information Theory*, 22(6):644–654, Nov 1976. ISSN 0018-9448. doi:10.1109/TIT.1976.1055638.

M. Faisal, Z. Aung, J. Williams, and A. Sanchez. *Securing Advanced Metering Infrastructure Using Intrusion Detection System with Data Stream Mining*, pages 96–111. Springer Berlin Heidelberg, Berlin, Heidelberg, 2012. ISBN 978-3-642-30428-6. doi:10.1007/978-3-642-30428-6_8. URL http://dx.doi.org/10.1007/978-3-642-30428-6_8.

A. Giani, E. Bitar, M. Garcia, M. McQueen, P. Khargonekar, and K. Poolla. Smart grid data integrity attacks: characterizations and countermeasures;. In *2011 IEEE International Conference on Smart Grid Communications (SmartGridComm)*, pages 232–237, Oct 2011. 10.1109/SmartGridComm.2011.6102324.

K. R. M. M. H. Mohammed, S. Tonyali and K. Akkaya. Efficient privacy-preserving data collection scheme for smart grid ami networks. In *Proc. of IEEE Globecom 2016, Washington DC, USA*, 2016.

Z. Haddad, M. Mahmoud, S. Taha, and I. Saroit. Secure and privacy-preserving ami-utility communications via lte-a networks. In *2015 IEEE 11th International Conference on Wireless and Mobile Computing, Networking and Communications (WiMob)*, pages 748–755, 2015.

Z. Haddad, M. Mahmoud, I. Saroit, and S. Taha. Secure and efficient uniform handover scheme for lte-a networks. In *2016 IEEE Wireless Communications and Networking Conference*, pages 1–6, 2016.

S. Hadley and A. Tsvetkova. Potential impacts of plug-in hybrid electric vehicles on regional power generation. *Electricity Journal*, 22(10):56–68, 2009.

W. Han and Y. Xiao. Ip2dm: integrated privacy-preserving data management architecture for smart grid v2g networks. *Wireless Communications and Mobile Computing*, 16(17):2956–2974, 2016. ISSN 1530-8677. doi:10.1002/wcm.2740. URL http://dx.doi.org/10.1002/wcm.2740. wcm.2740.

M. Ismail, I. Bayram, M. Abdallah, E. Serpedin, and K. Qaraqe. Optimal planning of fast pev charging facilities. In *2015 First Workshop on Smart Grid and Renewable Energy (SGRE)*, pages 1–6, 2015a.

M. Ismail, E. Serpedin, and K. Qaraqe. Pev charging in the future smart grid. In *2015 International Symposium on Signals, Circuits and Systems (ISSCS)*, pages 1–4, 2015b.

H. Kellerer, U. Pferschy, and D. Pisinger. *Knapsack problems*. 2004. Springer, Berlin.

H. Khurana, M. Hadley, N. Lu, and D. Frincke. Smart-grid security issues. *IEEE Security Privacy*, 8(1):81–85, Jan 2010. ISSN 1540-7993. doi:10.1109/MSP.2010.49.

S. Lukic, J. Cao, R. Bansal, F. Rodríguez, and A. Emadi. Energy storage systems for automotive applications. *IEEE Transactions on Industrial Electronics*, 55(6):2258–2267, 2008.

M. Mahmoud, S. Taha, J. Misic, and X. Shen. Lightweight privacy-preserving and secure communication protocol for hybrid ad hoc wireless networks. *IEEE Transactions on Parallel and Distributed Systems*, 2014.

M. Mahmoud, M. Ismail, P. Akula, K. Akkaya, E. Serpedin, and K. Qaraqe. Privacy-aware power charging coordination in future smart grid. In *2016 IEEE Wireless Communications and Networking Conference*, pages 1–6, April 2016a. doi:10.1109/WCNC.2016.7564967.

M. Mahmoud, N. Saputro, P. Akula, and K. Akkaya. Privacy-preserving power injection over a hybrid ami/lte smart grid network. *IEEE Internet of Things Journal*, 2016b.

M. Nabeel, S. Kerr, X. Ding, and E. Bertino. Authentication and key management for advanced metering infrastructures utilizing physically unclonable functions. In *2012 IEEE Third International Conference on Smart Grid Communications (SmartGridComm)*, pages 324–329, Nov 2012. doi:10.1109/SmartGridComm.2012.6486004.

National Institute of Standards and Technology. NIST framework and roadmap for smart grid interoperability standards, release 1.0. http://www.nist.gov/public_affairs/releases/upload/smartgrid_interoperability_final.pdf, January 2010.

Nissan LEAF Electric Car. Available: http://www.nissanusa.com/leaf-electric-car.

NIST. Guidelines for smart grid cyber security: vol. 3 supportive analyses and references. *NISTIR 7628, The Smart Grid Interoperability Panel - Cyber Security Working Group*, August 2010.

E. Oriero, K. Rabieh, M. Mahmoud, M. Ismail, E. Serpedin, and K. Qaraqe. Trust-based and privacy-preserving fine-grained data retrieval scheme for msns. In *2016 IEEE Wireless Communications and Networking Conference*, pages 1–6, 2016.

P. Akula, M. Mahmoud, K. Akkaya, and M. Song. Privacy-preserving and secure communication scheme for power injection in smart grid. *Proc. of IEEE International Conference on Smart Grid Communications*, 2015.

K. Rabieh, M. Mahmoud, M. Azer, and M. Allam. A secure and privacy-preserving event reporting scheme for vehicular ad hoc networks. *Security and Communication Networks*, 2015a.

K. Rabieh, M. Mahmoud, T. Guo, and M. Younis. Cross-layer scheme for detecting large-scale colluding sybil attack in vanets. In *2015 IEEE International Conference on Communications (ICC)*, pages 7298–7303, 2015b.

K. Rabieh, M. Mahmoud, A. Siraj, and J. Misic. Efficient privacy-preserving chatting scheme with degree of interest verification for vehicular social networks. In *2015 IEEE Global Communications Conference (GLOBECOM)*, pages 1–6, 2015c.

K. Rabieh, M. Mahmoud, and M. Younis. Privacy-preserving route reporting scheme for traffic management in vanets. In *2015 IEEE International Conference on Communications (ICC)*, pages 7286–7291, 2015d.

K. Rabieh, M. Mahmoud, and M. Younis. Privacy-preserving route reporting schemes for traffic management systems. *IEEE Transactions on Vehicular Technology*, 2016.

N. Saputro, K. Akkaya, and S. Uludag. A survey of routing protocols for smart grid communications. *Elsevier Compututer Networks*, 56(11):2742–2771, July 2012. ISSN 1389-1286.

M. Shaaban, M. Ismail, and E. El-Saadany. Pevs real-time coordination in smart grid based on moving time window. In *2014 IEEE PES General Meeting — Conference Exposition*, pages 1–5, 2014a.

M. Shaaban, M. Ismail, E. El-Saadany, and W. Zhuang. Real-time pev charging/discharging coordination in smart distribution systems. *IEEE Transactions on Smart Grid*, 2014b.

A. Sherif, K. Rabieh, M. Mahmoud, and X. Liang. Privacy-preserving ride sharing scheme for autonomous vehicles in big data era. *IEEE Internet of Things Journal*, 2016.

E. Sortomme, M. Hindi, S. MacPherson, and S. Venkata. Coordinated charging of plug-in hybrid electric vehicles to minimize distribution system losses. *IEEE Transactions on Smart Grid*, 2(1):198–205, March 2011.

W. Stallings. *Network Security Essentials: Applications and Standards*. Prentice Hall Press, Upper Saddle River, NJ, USA, 4th edition, 2010. ISBN 0136108059, 9780136108054.

O. Sundstrom and C. Binding. Flexible charging optimization for electric vehicles considering distribution grid constraints. *IEEE Transactions on Smart Grid*, 2012.

S. Tonyali, O. Cakmak, K. Akkaya, M. Mahmoud, and I. Guvenc. Secure data obfuscation scheme to enable privacy-preserving state estimation in smart grid ami networks. *IEEE Internet of Things Journal*, 2016.

B. Vaidya, D. Makrakis, and H. Mouftah. Device authentication mechanism for smart energy home area networks. In *2011 IEEE International Conference on Consumer Electronics (ICCE)*, pages 787–788, Jan 2011. doi:10.1109/ICCE.2011.5722864.

R. Verzijlbergh, M. Grond, Z. Lukszo, J. Slootweg, and M. Ilic. Network impacts and cost savings of controlled ev charging. *IEEE Transactions on Smart Grid*, 2012.

Z. Wan, W.-T. Zhu, and G. Wang. Prac: Efficient privacy protection for vehicle-to-grid communications in the smart grid. *Computers & Security*, 62: 246 –256, 2016. ISSN 0167-4048. ttps://doi.org/10.1016/j.cose.2016.07.004. URL http://www.sciencedirect.com/science/article/pii/S0167404816300761.

M. Wang, M. Ismail, R. Zhang, X. Shen, E. Serpedin, and K. Qaraqe. A semi-distributed v2v fast charging strategy based on price control. In *2014 IEEE Global Communications Conference*, pages 4550–4555, 2014.

M. Wang, M. Ismail, X. Shen, E. Serpedin, and K. Qaraqe. Spatial and temporal online charging/discharging coordination for mobile pevs. *IEEE Wireless Communications*, 2015.

M. Wang, M. Ismail, R. Zhang, X. Shen, E. Serpedin, and K. Qaraqe. Spatio-temporal coordinated v2v fast charging strategy for mobile gevs via price control. *IEEE Transactions on Smart Grid*, 2016.

Z. Xiao, Y. Xiao, and D. Du. Non-repudiation in neighborhood area networks for smart grid. *IEEE Communications Magazine*, 51(1):18–26, January 2013. ISSN 0163-6804. doi:10.1109/MCOM.2013.6400434.

Y. Yan, Y. Qian, H. Sharif, and D. Tipper. A survey on cyber security for smart grid communications. *IEEE Communications Surveys Tutorials*, 14(4): 998–1010, Fourth 2012. ISSN 1553-877X. doi:10.1109/SURV.2012.010912.00035.

22

Securing Smart Cities Systems and Services: A Risk-Based Analytics-Driven Approach

Mahmoud Gad[1] and Ibrahim Abualhaol[2]

[1] *Cognitive Labs Inc., Ottawa, Canada*
[2] *Carleton University, Ottawa, Canada*

22.1 Introduction to Cybersecurity for Smart Cities

Critical infrastructures (CI) are those assets or systems that are essential for the maintenance of vital societal functions, which include energy, utilities, and transportation. A smart city is a city that incorporates smart solutions in order to optimize its resources and improve the quality of service. Resource management includes waste management, water management, and energy management. The quality of service includes e-governance and citizen services, urban mobility, and telecommunications. This purpose is associated with a mission of driving economic growth and improves the quality of life of smart city citizens.

Smart cities services may be categorized into transportation, environment services, utility management, administration, and public safety, as shown in Figure 22.1. A more detailed taxonomy of smart cities services can be found in Lee and Lee (2014).

The concept of smart city embraces several definitions depending on the meaning of the word "smart." In Bowerman et al. (2000), it is a city that monitors and integrates all its critical infrastructures, including roads, bridges, tunnels, rails, subways, airports, seaports, communications, water, power, and even major buildings. Additionally, it can better optimize its resources, plan its preventive maintenance activities, and monitor security aspects while maximizing services to its citizens. In Harrison et al. (2010), it is a city connecting the physical infrastructure, the IT infrastructure, the social infrastructure, and the business infrastructure to leverage the collective intelligence of the city. According to Toppeta (2010), a smart city is a city that combines information and communication technology (ICT) and Web 2.0 technology with other organizational, design, and planning efforts to dematerialize and speed up bureaucratic processes and help to identify new, innovative solutions to city management complexity, in order to improve sustainability and livability. A systematic literature review of the definition can be found in Cocchia (2014). The evolution of the definition, over the years, highlights the importance of data and the integration between physical and cyber-systems in the design and operation of smart cities.

One of the challenges confronting smart cities is the volume of data collected from critical infrastructure subsystems. These data vary in type, importance, and sensitivity,

Transportation and Power Grid in Smart Cities: Communication Networks and Services, First Edition.
Edited by Hussein T. Mouftah, Melike Erol-Kantarci and Mubashir Husain Rehmani.

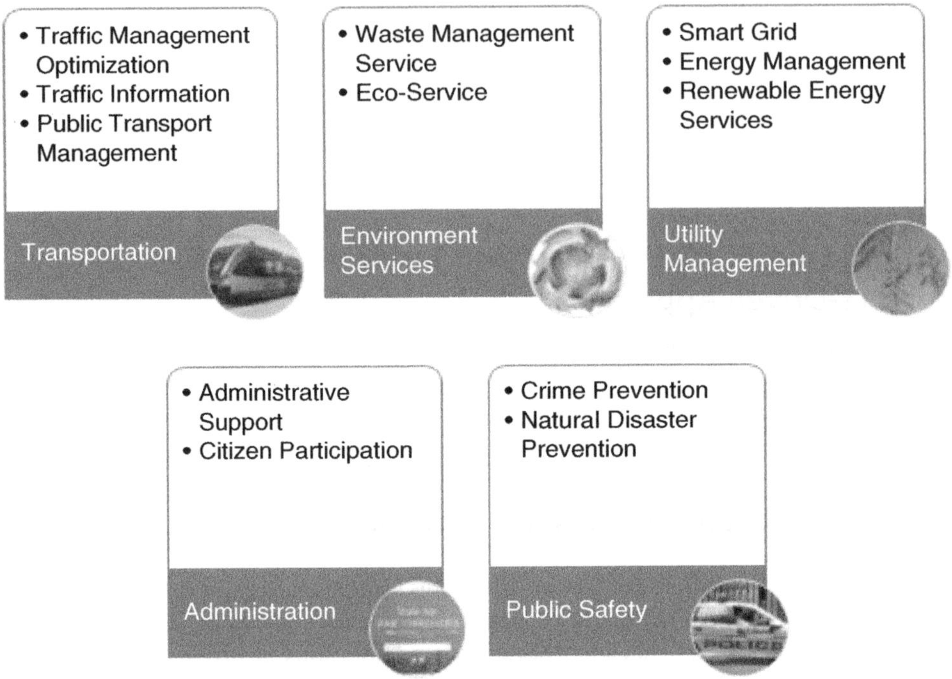

Figure 22.1 Smart city services.

which introduces a new challenge for any cybersecurity solution proposed for smart cities. In addition, any cybersecurity solution must be usable by the operators and end users. Usability is defined as the degree to which a user can complete tasks effectively and efficiently. A usable system is one that meets the needs of the user. Usability is concerned with functionality/usefulness, ease of learning, ease of use, aesthetics, user satisfaction and quality Rubin and Chisnell (2008). Usability consists of five factors: ease of learning, task efficiency, ease of remembering, understandability, and subjective satisfaction (Lauesen and Younessi, 1998). This raises the need for an innovative approach to designing cyber-fusion centers that aim at enabling the operator to make informed decisions and providing informed data-driven insights for future improvements.

Chourabi et al. (2012) identified two categories of challenges that smart cities would face: (i) managerial and organizational challenges and (ii) technological challenges. Gil-Garcia and Pardo (2005) identified the managerial and organizational challenges as well as the strategies to tackle them. From a cybersecurity point of view, the most important strategies are preplanning, setting clear and measurable deliverables, end-user involvement in the decision process by providing a feedback mechanism, and continuous improvements to the system. The technological challenges were outlined by Ebrahim and Irani (2005). They include operational challenges, organizational challenges, and IT skills. These challenges and strategies should be taken into consideration when designing any cybersecurity solution for smart cities.

The chapter is presented as follows. A typical cyberattack surface is presented in Section 22.2. Then, we present the design science approach to secure smart city systems in Section 22.3. After that, we present the risk-based NIST Cybersecurity Framework in Section 22.4. Then, we propose the use of a cybersecurity fusion center with big data analytics in Section 22.5. Finally, we conclude the chapter in Section 22.6.

22.2 Smart Cities Enablers

In October 2016, the Internet witnessed a record-breaking massive distributed denial of service (DDoS) attack of over 1.2 Tbps against Dyn, a domain name service (DNS) provider. Arbor Networks Inc. reported that the original attack was conducted by at least one Mirai IoT botnet and spread into 500,000 IoT devices, with clusters around the world, including China, Hong Kong, Taiwan, South Korea, Southeast Asia, Brazil, Spain, and elsewhere. This caused a sudden outage of popular sites and services, including Twitter, SoundCloud, Spotify, and Shopify. This incident highlights the vulnerability of one of the key enablers in smart cities.

Three key enablers of smart cities are the Internet of things (IoT), smartphones, and cloud computing. Their integration with smart city critical infrastructure should be considered with cybersecurity threats in mind.

Internet of things (IoT): The IoT is one of the major technologies that will shape the future of the digital world including smart cities and homes. It is a mesh network of physical objects that either exchange data in peer-to-peer (P2P) mode or communicate and relay information with the service provider (Atzori, Iera, and Morabito. 2010). There are many connected objects today such as electronic appliances (microwaves, cameras, refrigerators, etc.) that rely on RFID technology and state-of-the-art software and sensors (Xiaohang, 2004) in their proper operation. IoT objects can be sensed and controlled across local area networks (LANs) or wide area networks (WANs). Here are a few key applications of smart dust in the context of smart cities and homes:

- habitat monitoring;
- indoor and outdoor environmental monitoring;
- security and tracking of people and objects;
- traffic monitoring and management; and
- human health and well-being monitoring.

Smartphones: Smartphones today contain a variety of chips and sensors such as GPS, gyroscope, microphone, camera, and accelerometer, among others, that are generating a lot of raw data (Soldo, Quarto, and Di Lecce, 2012).

In the context of a smart home, for instance (Balakrishna, 2012), there are already some applications that run on smartphones to control many appliances in the home, such as TV sets, lights, garage doors, and security cameras. Moreover, users can operate their smartphones in order to interact with their city, receive live information, and connect with local authorities and public transportation systems.

Cloud Computing: Cloud computing has become a de facto platform to enable content delivery to consumers (Obaidat and Nicopolitidis, 2016). Provided pervasive computing today, largely enabled by smartphones and IoT devices, massive amounts of data need to be processed in order to transform raw data into insightful information. Current computing paradigms are no longer suitable for such endeavors. Cloud computing has three main offers:

- SaaS: software as a service;
- PaaS: platform as a service; and
- IaaS: infrastructure as a service.

A combination of the aforementioned service cloud offering leads to the emergence of other types commonly known as XaaS (everything as a service) such as storage as a service, communications as a service, network as a service, monitoring as a service, analytics as a service, data as a service, and so on (Obaidat and Nicopolitidis, 2016).

22.3 Smart Cities Attack Surface

In this section, we use the list of software weaknesses known as common attack pattern enumeration and classification (CAPEC) to describe the common patterns based on the attack domains and attack mechanisms (MITRE Corporation, 2017). This classification helps smart city security designers to understand the cybersecurity attack surface.

22.3.1 Attack Domains

22.3.1.1 Communications

In this category, attackers focus on using the weaknesses in the communication protocols. They can block, manipulate, or eavesdrop on communications in an attempt to achieve a desired negative technical impact. Attacks in this category may be classified into three subcategories: interception, protocol manipulation, and obstruction. In a typical smart city, this attack domain includes radio and wireless connections, fiber links, and cable links.

22.3.1.2 Software

Attack patterns within this category focus on the exploitation of software applications. Attacks in this category could exploit weaknesses in the design of the software, its implementation, or both. There are many subcategories of attacks including brute force, software integrity attacks, reverse engineering, and code injection. This includes applications on corporate IT networks, sensors firmware, middleware code, and others.

22.3.1.3 Hardware

Attacks in this category focus on the physical hardware used in cyber-physical systems. This includes replacement, destruction, modification, and exploitation of hardware components. Attacks against hardware components target the chips, circuit boards, device ports, or other components that comprise a cyber-physical system (e.g., SCADA and PLC systems). Sophisticated attacks may include hardware footprinting, hardware integrity attacks, and malicious logic insertion.

22.3.1.4 Social Engineering

This category of attacks focuses on the manipulation and exploitation of system users by convincing someone to perform actions or disclose confidential information, often resulting in access to computer systems or assets. Techniques in this category can range from social information gathering attacks, to information elicitation via social engineering, to target influence via social engineering.

22.3.1.5 Supply Chain

This category focuses on the disruption of the supply chain lifecycle caused by the manipulation of computer system hardware, software, or services with the purpose of espionage, data or IP theft, or the disruption of mission-critical operations. Supply chain operations are usually multinational with manufacturing and delivery occurring across multiple locations, offering an attacker multiple points for disruption. This can be done by modification during manufacturing or manipulation during distribution (MDD).

22.3.1.6 Physical Security

Attacks in this category exploit weaknesses in the physical security of a system. This includes bypassing physical security, physical theft, and physical destruction of device or component.

Table 22.1 Attack Mechanisms on Smart Cities Infrastructure Systems.

Category	Attack mechanism
Gather information	Gathering, collecting, and theft of information
Deplete resources	Depletion of a resource to affect service availability
Injection	Control or disrupt the behavior of a target through crafted input data
Deceptive interactions	Malicious interactions with a target in an attempt to deceive the target
Manipulate timing and state	Exploit weaknesses in timing or state maintaining functions to perform unauthorized actions
Abuse of functionality	Manipulate one or more functions of an application in order to achieve a malicious objective
Probabilistic techniques	Utilize probabilistic techniques to explore and overcome security properties of the target
Exploitation of authentication	Target exploitation of weaknesses, limitations, and assumptions in the mechanisms utilized to manage identity and authentication
Exploitation of authorization	Target exploitation of weaknesses, limitations, and assumptions in the mechanisms a target utilizes to manage access to its resources or authorize utilization of its functionality
Manipulate data structures	Manipulate characteristics of system data structures in order to violate their intended usage
Manipulate resources	Manipulate system resources or attributes in order to perform an attack
Analyze target	Analyze a target system, protocol, message, or application in order to overcome protections on the target or as a precursor to other attacks
Gain physical access	Gain physical access to a system or device
Execute code	Use of malicious code to achieve a desired negative technical impact
Alter system components	Alter or manipulate system components in an attempt to achieve a desired negative technical impact
Manipulate system users	Manipulate a user in an attempt to achieve a desired negative technical impact

22.3.2 Attack Mechanisms

Table 22.1 describes the different possible attack mechanisms on smart cities critical infrastructure. Each of these mechanisms must be taken into consideration when designing a security solution for smart cities.

22.4 Securing Smart Cities: A Design Science Approach

Design can be defined as the process of inventing objects that perform specific functions (Hatch, 2001). Design science has been successfully applied to many domains (Romme and Endenburg, 2006). Muegge and Craigen (2015) proposed a five-step design process to be applied to cybersecurity design problems. The five steps are presented in Figure 22.2 as follows:

1) Gather lessons learned from theory and practice: this includes published research, security reports, surveys, and input from security practitioners in the field.
2) Formulate design principles: develop a set of prescriptive propositions based on Step 1.
3) Formulate design rules: produce detailed guidelines specific to the design context.
4) Design: apply design rules to create a design representation.
5) Implementation and experimentation.

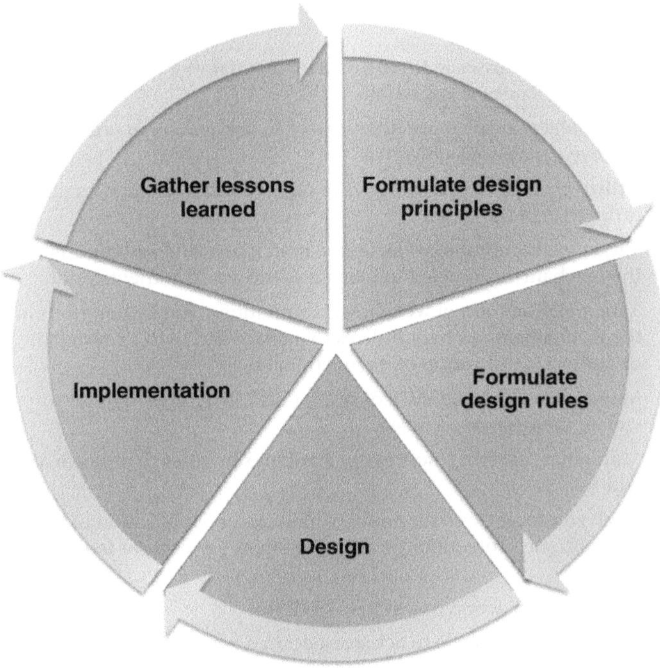

Figure 22.2 Five-step design process for securing smart cities.

In the following, we propose a set of design principles for securing smart cities infrastructure based on the literature survey in Sections 22.1 and 22.2. These principles are enabled by utilizing the NIST Framework in a cybersecurity fusion center with Big Data analytics capabilities.

- Design systems around secure design principles.
- Understand your attackers.
- Use a risk-based approach to prioritize resource allocations.
- Build your solution around continuous monitoring of system logs and components.
- Implement a threat prediction mechanism.
- Implement a feedback mechanism from end-user to report security incidents.

22.5 NIST Cybersecurity Framework

The NIST cybersecurity framework is a risk-based approach for managing cybersecurity risks. It consists of three components: *core, profile, and tiers*. The framework *core* presents the cybersecurity activities and informative references (subcategories) organized around particular outcomes (categories) to achieve five functionalities. The framework *core* functions are identify, protect, detect, respond, and recover. Identify can be thought of as the answer to the question "What are the assets that need protection?" Protect is an answer to the question "What safeguards are available?" Detect is an answer to "What techniques can identify incidents?" Respond is an answer to "What techniques can contain impacts of incidents?" And recover is an answer to "What techniques can restore capabilities?" The NIST cybersecurity framework *core* is presented in Table 22.2 (NIST, 2014).

The NIST framework *profile* aligns industry standards and best practices to the framework *core* in a particular implementation scenario. Building the *profile* is an innovation by itself, and it needs prioritization and consideration of resources and capabilities in each of the smart city critical infrastructure systems. The framework *profile* aligns functions and categories with requirements, risk tolerance, and resources of a critical infrastructure. The framework *tiers* provide context to how a critical infrastructure deals with a cybersecurity risk. A critical infrastructure can assess current security practices and use the *tier* system to prioritize improvements. A summarized list of important NIST framework definitions is provided in Table 22.3 (NIST, 2014).

The framework *profile* describes the current state or desired target state of cybersecurity activities. The *profile* enables organizations or critical infrastructures to establish a roadmap for reducing cybersecurity risks that is well aligned with organizational and sector goals and management priorities. The references recommended by the framework are standards and practices that are globally accepted, which help facilitate the operation of the framework on a global base. Industry and nonprofits are improving the framework by mapping sector's specific standards, guidelines, and best practices to the framework. They are also developing and sharing examples of how organizations are using the framework. One important example of implementing the framework in a smart city critical infrastructure component (e.g., energy sector) is given in Table 22.4.

Table 22.2 NIST Cybersecurity Framework Core.

Function	Categories	Fusion center phase
Identify	• Asset management • Business environment • Governance • Risk assessment • Risk management	Descriptive phase
Protect	• Access control • Awareness training • Data security • Information protection • Protective technology	Predictive phase
Detect	• Anomalies and events • Continuous monitoring • Detection processes	Predictive phase
Respond	• Communications • Analysis • Mitigation • Improvements	Prescriptive phase
Recover	• Recovery planning • Improvements • Communications	Prescriptive phase

Table 22.3 NIST Framework Definitions.

Term	Definition
Framework *core*	A set of cybersecurity activities and references that are common across critical infrastructure sectors and are organized around particular outcomes. The framework core comprises four types of elements: functions, categories, subcategories, and informative references.
Framework *tier*	A lens through which to view the characteristics of an organization's approach to risk—how an organization views cybersecurity risk and the processes in place to manage that risk.
Framework *profile*	A representation of the outcomes that a particular system or organization has selected from the framework categories and subcategories.
Function	One of the main components of the framework. Functions provide the highest level of structure for organizing basic cybersecurity activities into categories and subcategories. The five functions are identify, protect, detect, respond, and recover.

Table 22.4 summarizes the main highlights and recommendations of implementing the NIST cybersecurity framework at the USA energy sector. These insights align with the importance of implementing a fusion center that simplifies the operations and provides cybersecurity maturity insights, as it is proposed in the next section.

Table 22.4 USA Energy Sector Case Study (US DOE, 2015).

Highlights and recommendations
• The cybersecurity maturity model (C2M2) is an important tool to implement the NIST cybersecurity framework. • C2M2 scorecards, which highlight the level of maturity across C2M2 domains, are almost identical to the concept of framework profiles, both current and target. • The implementation of the framework should focus on the systems and assets essential to critical infrastructure functions. • The C2M2 provides *descriptive* rather than *prescriptive* guidance at a high level of abstraction. This helps organizations of all types, structures, and sizes to map C2M2 practices to framework subcategories. Also, the recommended process for using the C2M2 parallels the framework approach of setting a target, identifying gaps, and addressing gaps. • The included mapping of C2M2 practices to subcategories and tiers shows that the C2M2 adequately addresses all the objectives of the framework • The C2M2 uses maturity indicator levels that can help an organization track measurable, incremental progression in the maturity of cybersecurity practices.

22.6 Cybersecurity Fusion Center with Big Data Analytics

As described in Section 22.1, a smart city is a collection of critical infrastructures that need to be operated with minimum cybersecurity risk. One well-proven risk-based approach is the NIST cybersecurity framework that highlights the importance of minimizing the risk in three dimensions (vulnerability, threat, and damage) using five functionalities (identify, protect, detect, recover, and respond).

To enable the implementation of such framework and to provide insights into the adopted design science approach, we are proposing an analytics-driven cybersecurity fusion center, as given in Table 22.5. The fusion center consists of three phases, each of which has three characteristics inspired by the work in Delen and Demirkan (2013) and Wang et al. (2015).

Table 22.5 A Smart City Cybersecurity Fusion Center.

	Cybersecurity Fusion Center		
	Descriptive	**Predictive**	**Prescriptive**
Questions	• What happened? • What is happening?	• What will happen? • Why will it happen?	• What should we do? • Why should we do it?
Enablers	• Reports • Dash cards • Scorecards • Data warehousing	• Logs mining • Text mining • Web/media mining • Forecasting	• Optimization • Simulation • Decision model • Expert system
Outcomes	Well-defined cybersecurity incident and countermeasures	Accurate projection of future cybersecurity risks in terms of vulnerabilities, threats, and damages	Best possible operations and future design decisions

The first phase is a *descriptive* phase, where the fusion center utilizes cybersecurity reports and critical infrastructures monitoring dashboards and scorecards and data aggregated from sensors. It aims not only to answer the questions of what is happening and why it happened but also to reach to well-defined cybersecurity incidents and possible countermeasures. This phase is based on the previously articulated science design approach and provides insights into both *identify* and *detect* functionalities in the NIST Framework.

The second phase is the *predictive* phase, where the answer to questions such as what will happen and why it will happen can be reached. The patterns that are hidden in a big data (volume, velocity, variety, and veracity) collection are immensely valuable in answering those questions. Machine learning with text mining, web/media mining, and forecasting enable an accurate projection of future cybersecurity risks in terms of vulnerabilities, threats, and damages, which align with the NIST framework functionalities such as *detect* and *recover*.

The third *prescriptive* phase tackles the NIST framework's respond functionality and provides valuable insights into the previously articulated design science approach. In this phase (i.e., the *prescriptive* phase), questions such as what should we do and why should we do it in terms of operations and future designs are raised. To enable the *prescriptive* phase, optimizations, simulations, decision models, and expert systems are integrated into the fusion center to provide the best possible operations and future design decisions that would minimize the cybersecurity risk in smart cities' critical infrastructure.

The cybersecurity fusion center with big data analytics is visualized in Figure 22.3. The NIST framework functionalities (outer circle) are achieved in the three phases analytics (inner circle).

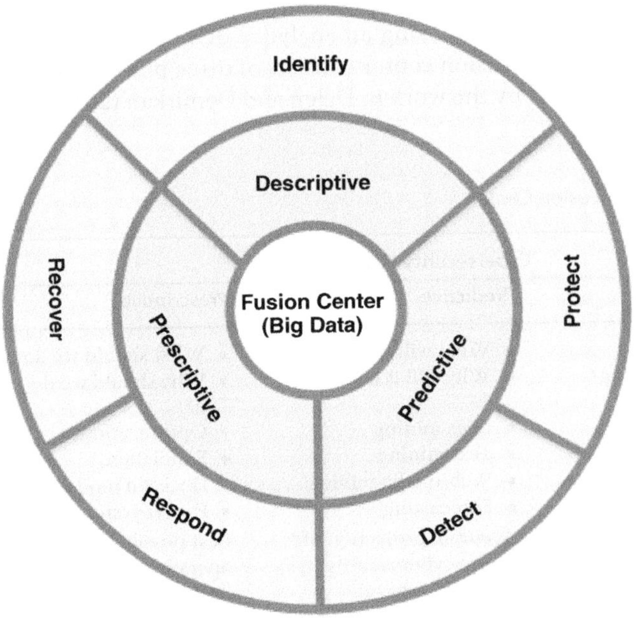

Figure 22.3 Big data fusion center with the NIST framework.

22.7 Conclusion

In this chapter, we proposed an integrated risk-based cybersecurity approach with design science in mind and enabled it with a Big Data analytics fusion center solution to help minimize the cybersecurity risk exposure. The risk-based NIST framework focuses on minimizing the risk to critical infrastructures within smart cities in five functionalities: *identify, detect, protect, recover,* and *respond.* These functionalities help improve the smart city cybersecurity in two dimensions: design and operations. The Big Data aggregated by the different smart city systems and users is utilized by the fusion center analytics engine to provide *descriptive (identify), predictive (detect and protect),* and *prescriptive (respond and recover)* functionalities that align together to minimize the cybersecurity risk during operations and provide insights for future improved designs.

22.8 Table of Abbreviations

Abbreviation	Definition
BART	Bay Area Rapid Transit
C2M2	Cybersecurity maturity model
CAPEC	Common attack pattern enumeration and classification
CI	Critical infrastructures
DDoS	Distributed denial of service
DNS	Domain name service
GPS	Global Positioning System
IaaS	Infrastructure as a service
ICT	Information and communication technology
IoT	Internet of things
LAN	Local area network
MDD	Manipulation during distribution
NHS	National Health Services
NIST	National Institute of Standards and Technology
P2P	Peer-to-peer
PaaS	Platform as a service
PLC	Programmable logic controller
RFID	Radio-frequency identification
SaaS	Software as a service
SCADA	Supervisory control and data acquisition
Tbps	Terabytes per second
WAN	Wide area network
XaaS	Everything as a service

References

Atzori, L., Iera, A., Morabito, G., 2010, 'The Internet of things: a survey', *Comput Netw.* 54(5), 2787–2805.

Balakrishna, C., 'Enabling technologies for smart city services and applications', *In*: Proceedings of the 6th IEEE international conference on next generation mobile applications, services and technologies, NGMAST'12, 223–227.

Bowerman, B., Braverman J., Taylor J., Todosow, H., von Wimmersperg, U., 2000, 'The vision of a smart city', *In*: 2nd International Life Extension Technology Workshop, Paris, Sep. 28, 2000 (Vol. 28).

Chourabi, H., Nam, T., Walker, S., Gil-Garcia, J.R., Mellouli, S., Nahon, K., Pardo, T.A., and Scholl, H.J., 'Understanding smart cities: An integrative framework', *In*: System Science (HICSS), 2012 45th Hawaii IEEE International Conference, 2289–2297.

Cocchia, A., 2014, 'Smart and digital city: A systematic literature review', In: Smart City, Springer International Publishing, 13–43

Delen, D. and Demirkan, H., 2013, "Data, information and analytics as services," *Decision Support Systems* 55(1), 359–363.

Ebrahim, Z. and Irani, Z., 2005, 'E-government adoption: architecture and barriers', *Business Process Management Journal* 11(5), 589–611.

Gil-García, J.R. and Pardo, T.A. 'E-government success factors: Mapping practical tools to theoretical foundations', *Government Information Quarterly* 2005 22(2), 187–216.

Harrison, C., Eckman, B., Hamilton, R., Hartswick, P., Kalagnanam, J., Paraszczak, J., and Williams, P, 2010, 'Foundations for smarter cities', *IBM Journal of Research and Development* 54(4), 1–6.

Hatch, N.W., 2001, 'Design rules, Volume 1: The power of modularity', *Academy of Management Review* 26(1), 130–133.

Lauesen, S. and Younessi, H., 'Six Styles for Usability Requirements', *In*: REFSQ' 1998, 98, 155–166.

Lee, J. and Lee, H., 2014, 'Developing and validating a citizen-centric typology for smart city services', *Government Information Quarterly* 2014 Jun 30(31), S93–105.

MITRE Corporation, 2017, 'Enumeration CA. Classification (CAPEC)', URL: https://capec.mitre.org.

Muegge, S. and Craigen, D., 2015, 'A Design Science Approach to Constructing Critical Infrastructure and Communicating Cybersecurity Risks', Technology *Innovation Management Review* 5(6), 6.

National Institute of Standards and Technology (NIST), 2014, 'Framework for Improving Critical Infrastructure Cybersecurity', URL: http://www.nist.gov/cyberframework/upload/cybersecurity-framework-021214.pdf.

Obaidat, M.S. and Nicopolitidis, P., 2016, *Smart Cities and Homes: Key Enabling Technologies*, Morgan Kaufmann.

Romme, A.G. and Endenburg, G., 2006, 'Construction principles and design rules in the case of circular design', *Organization Science* 17(2), 287–297.

Rubin, J. and Chisnell, D., 2008, *Handbook of Usability Testing: How to Plan, Design and Conduct Effective Tests*, John Wiley & Sons.

Soldo, D., Quarto, A., Di Lecce, V., 2012, 'M-DUST: an innovative low-cost smart PM sensor', *In*: Proceedings of the 2012 IEEE International, instrumentation and measurement technology conference, I2MTC'12, 1823–1828.

Toppeta, D., 2010, 'The smart city vision: how innovation and ICT can build smart, "livable," sustainable cities', The Innovation Knowledge Foundation. Think.

US Department of Energy (DOE), 2015, 'Energy Sector Cybersecurity Framework Implementation Guidance', URL: http://energy.gov/sites/prod/files/2015/01/f19/Energy%20Sector%20Cybersecurity%20Framework%20Implementation%20Guidance_FINAL_01-05-15.pdf.

Wang, H., Osen, O.L., Li, G., Li, W., Dai, H.N., and Zeng, W., 2015, 'Big data and industrial internet of things for the maritime industry in northwestern Norway', *In*: TENCON 2015-2015 IEEE Region 10 Conference 2015, 1–5.

Xiaohang, W., Song Dong, J., Chin, C.Y., Hettiarachchi, S.R., and Zhang, D., 2004, 'Semantic space: an infrastructure for smart spaces', *IEEE Pervasive Comput.* 3, 32–39.

23

Spatiotemporal Big Data Analysis for Smart Grids Based on Random Matrix Theory

Robert Qiu[1,2,3], Lei Chu[2,3], Xing He[2,3], Zenan Ling[2,3] and Haichun Liu[2,3]

[1] *Tennessee Technological University, Cookeville, TN 38505, USA*
[2] *Department of Electrical Engineering, Shanghai Jiaotong University, Shanghai 200240, China*
[3] *Research Center for Big Data Engineering and Technology, State Energy Smart Grid Research and Development Center, Shanghai, China*

A cornerstone of the smart grid is the advanced monitorability on its assets and operations. Increasingly pervasive installation of phasor measurement units (PMUs) allows the so-called synchrophasor measurements to be taken roughly 100 times faster than the legacy supervisory control and data acquisition (SCADA) measurements, time-stamped using the Global Positioning System (GPS) signals to capture the grid dynamics. On the other hand, the availability of low-latency two-way communication networks will pave the way to high-precision real-time grid state estimation and detection, remedial actions upon network instability, and accurate risk analysis and post-event assessment for failure prevention.

In this chapter, we firstly model spatiotemporal PMU data in large-scale grids as random matrix sequences. Secondly, some basic principles of random matrix theory (RMT), such as asymptotic spectrum laws, transforms, convergence rate, and free probability, are introduced briefly in order to improve the understanding and application of RMT technologies. Lastly, the case studies based on synthetic data and real data are developed to evaluate the performance of the RMT-based schemes in different application scenarios (i.e., state evaluation and situation awareness).

23.1 Introduction

23.1.1 Perspective on Smart Grids

The modern power grid is one of the most complex engineering systems in existence; the North American power grid is recognized as the supreme engineering achievement in the 20th century [1]. The complexity of the future's electrical grid is ever increasing: (1) the evolution of the grid network, especially the expansion in size; (2) the penetration of renewable/distributed resources, flexible/controllable electronic units, or even prosumers with dual load-generator behavior [2]; and (3) the revolution of the

Transportation and Power Grid in Smart Cities: Communication Networks and Services, First Edition.
Edited by Hussein T. Mouftah, Melike Erol-Kantarci and Mubashir Husain Rehmani.

operation mechanism, e.g., demand-side management. Also, financial, environmental, and regulatory constraints are pushing the electrical grid toward its stability limit.

Generally, power grids have experienced three ages—G1, G2, and G3 [3]. The network structures are depicted in Figure 23.1 [4]. Their data flows and energy flows, as well as corresponding data management systems and work modes, are quite different [5], which are shown in Figure 23.2 and Figure 23.3, respectively.

G1: Small-scale isolated grids

G1 was developed from the power system around 1900 to 1950, featured by small-scale isolated grids. For G1, units interchange energy and data within the isolated grid to keep generation-consumption balance. The units are mostly controlled by themselves, i.e., operating under individual-work mode. As shown in Figure 23.1(a), each apparatus collects designated data and makes corresponding decisions only with its own application. The individual-work mode works with an easy logic and little information communication. However, it means few advanced functions and inefficient utilization of resources. It is only suitable for small grids or isolated islands.

G2: Large-scale interconnected grids

G2 was developed from power grids about 1960 to 2000, featured by zone-dividing large-scale interconnected grids. For G2, units interchange energy and data with adjacent units. The units are dispatched by a control center, i.e., they are operating under team-work mode. The regional team leaders, such as local dispatching centers, substations, and microgrid control centers, aggregate their own team members (i.e., units in the region) into a standard black-box model. These standard models will be further aggregated by the global control center for control or prediction purposes. The two

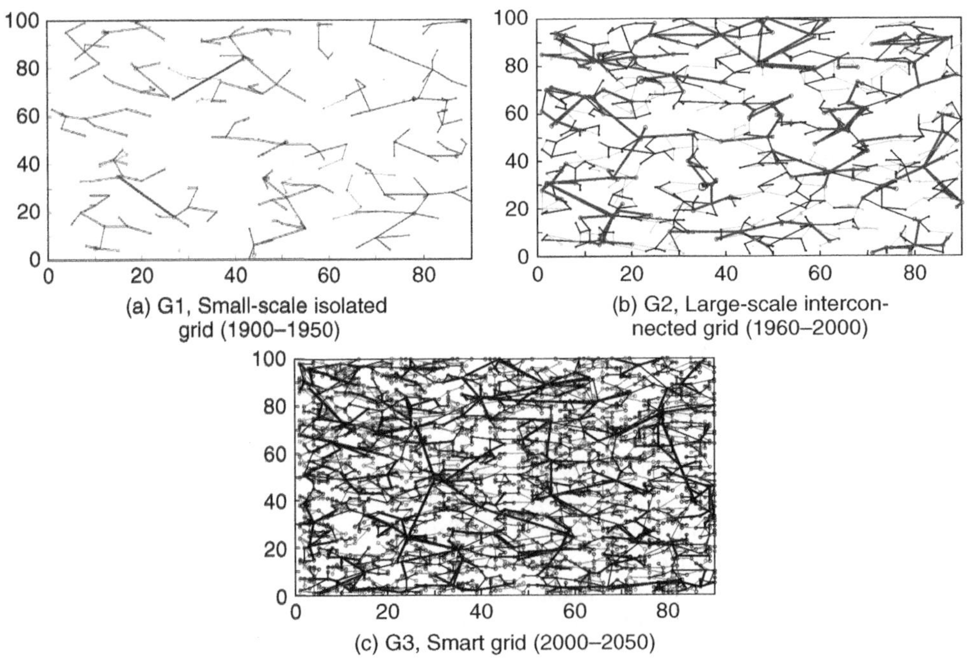

Figure 23.1 Topologies of Grid Network.

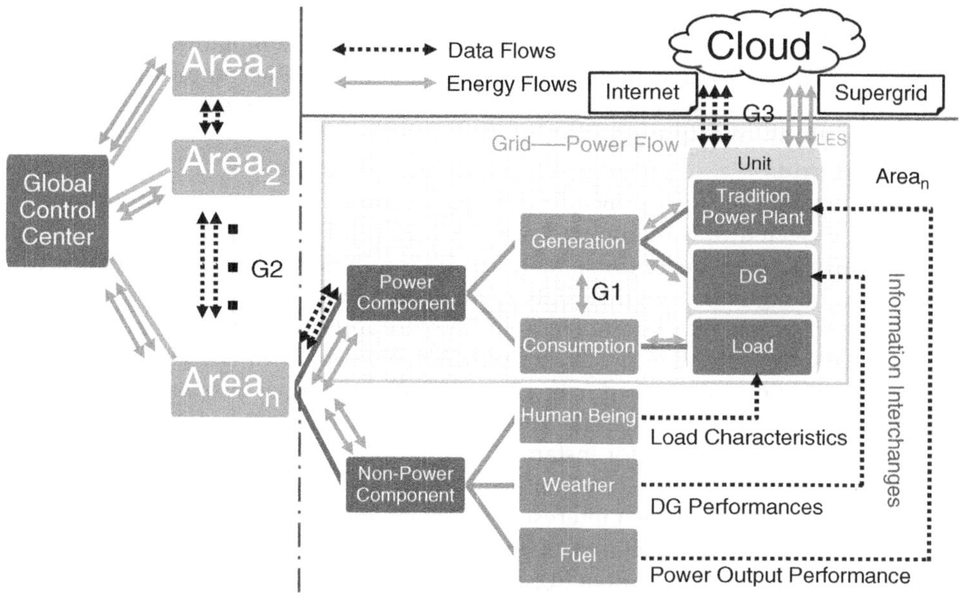

Figure 23.2 Data flows and energy flows for three generations of power systems. The single lines, double lines, and triple lines indicate the flows of G1, G2, and G3, respectively.

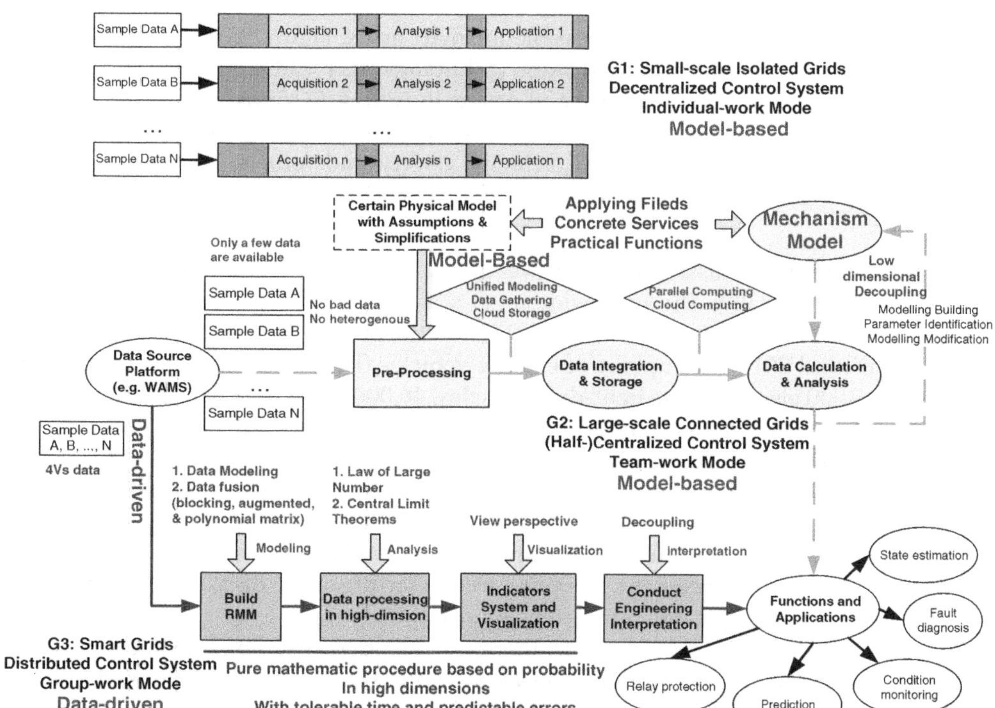

Figure 23.3 Data management systems and work modes for three ages of power systems. The above, middle, and below parts indicate the data management systems and the work modes of G1, G2, and G3, respectively. For G1, each grid works independently. For G2, global and local control centers are operating under the team-work mode. For G3, the group-work mode breaks through the regional limitation for energy.

aggregations above are achieved by four steps: data monitoring, data preprocessing, data storage, and data processing. The description above can be summarized by by dotted blue lines in Figure 23.3. In general, the team-work mode conducts model-based analysis and mainly concerns system stability rather than individual benefit; it does not work well for smart grids with 4Vs data.

G3: Smart grids

The development of G3 was launched at the beginning of the 21st century, and for China, it is expected to be completed around 2050 [3]. Figure 23.1(c) shows that the clear-cut partitioning is no longer suitable for G3, as well as the team-work mode, which is based on the regional leader. For G3, the individual units, rather than the regional center (if it still exists), play a dominant role. They are appropriately self-controlled with high intelligence, resulting in much more flexible flows for both energy exchange and data communication [6]. Accordingly, the group-work mode is proposed. Under this mode, the individuals freely operate under the supervision of the global control centers [5]. VPPs [7], MMGs [8], for instance, are typically G3 utilities. These group-work mode utilities provide a relaxed environment to benefit both individuals and the grids: the former, driven by their own interests and characteristics, are able to create or join a relatively free group to benefit mutually from sharing their own superior resources; meanwhile, these utilities are often big and controllable enough to be good customers or managers to the grids.

23.1.2 The Role of Data in the Future Power Grid

Data are more and more easily accessible in smart grids. Figure 23.4 shows numerous data sources: information communication technology (ICT), advanced metering infrastructure (AMI), supervisory control and data acquisition (SCADA), sensor technology (ST), phasor measurement units (PMUs), and intelligent electronic devices (IEDs) [9]. Hence, data with features of volume, velocity, variety, and veracity (i.e., 4Vs data) [10] are inevitably generated and daily aggregated. Particularly, the "4Vs" are elaborated as follows:

- *Volume.* There are massive data in power grids. The so-called curse of dimensionality [11] occurs inevitably. The worldwide small-scale roof-top photovoltaics (PVs) installation reached 23 GW at the end of 2013, and the growth is predicted to be 20 GW per year until 2018. The uptake of electric vehicles (EVs) also continues to grow. At least 665,000 electric-drive light-duty vehicles, 46,000 electric buses, and 235 million electric two-wheelers were in the worldwide market in early 2015 [12].
- *Velocity.* The resource costs (time, hardware, human, etc.) for big data analytics should be tolerable. To sever online decision making, massive data must be processed within a fraction of second.
- *Variety.* The data in various formats are often derived from diverse departments. In the view of data management, sampling frequency of source data, processing speed, and service objects are not completely accord.
- *Veracity.* For a massive data source, there often exist realistic bad data, e.g., incomplete, inaccurate, asynchronous, and unavailable. For system operations, decisions such as protection should be highly reliable.

As mentioned above, smart grids are always huge in size and complex in topology; big data analytics and a data-driven approach become natural solutions for the future grid

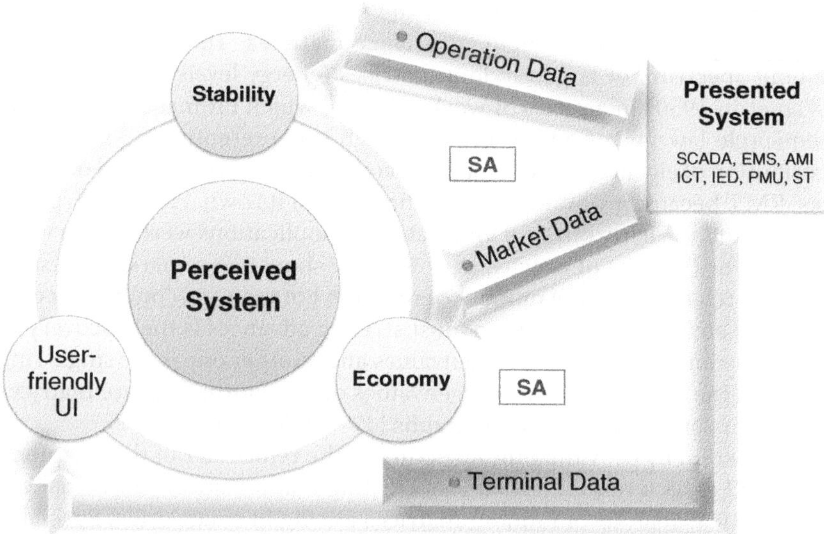

Figure 23.4 Smart grid with 4Vs data and its SA.

[13–16]. Driven by data analysis in high dimension, big data technology works out data correlations (indicated by statistical parameters) to gain insight to the inherent mechanisms. Actually, big data technology has already been successfully applied as a powerful data-driven tool for numerous phenomena, such as quantum systems [17], financial systems [18, 19], biological systems [20], as well as wireless communication networks [21–23]. For smart grids, the data-driven approach and data utilization are currently important topics, as evidenced in the special issue of "Big Data Analytics for Grid Modernization" [24]. This special issue is most relevant to our book in spirit. Several SA topics are discussed as well. We highlight anomaly detection and classification [25, 26], social media such as Twitter in [27], the estimation of active ingredients such as PV installations [28, 29], and finally the real-time data for online transient stability evaluation [30]. In addition, we point out research about the improvement in wide-area monitoring, protection and control (WAMPAC) and the utilization of PMU data [31–34], together with the fault detection and location [35–37]. Xie et al., based on principal component analysis (PCA), proposes an online application for early event detection by introducing a reduced dimensionality [38]. Lim et al. studies the quasi-steady-state operational problems relevant to the voltage instability phenomena [39]. These works provide primary exploration of the big data analysis in the smart grid. Furthermore, a brief account for random matrix theory (RMT), which can be seen as basic analysis tools for spatial-temporal grid data processing, is elaborated in the following subsection.

23.1.3 A Brief Account for RMT

The last two decades have seen the rapid growth of RMT in many science fields. Brilliant mathematical work in RMT shed light on the challenges of classical statistics. In this subsection, we present a brief introduction to the main development of RMT. The application-related account, with particular attention paid to recently rising RMT-based technology that are relevant for the smart grid, is elaborated in Section 23.3.

The research of random matrices began with the work of Wishart in 1928, which focused on the distribution of the sample covariance matrices. The first asymptotic results on the limit spectrum of large random matrices (energy levels of nuclei) were obtained by Wigner in 1950s in a series of works [40–43] which ultimately lead to the well-known semicircle law [44]. Another breakthrough was presented in [45], which studied the distribution of eigenvalues for empirical covariance matrices. Based on these excellent works, RMT became a vibrant research direction of its own. Plenty of brilliant works that branched off the early physical and statistical applications were put forward in the last decades. For the sake of brevity, here we only show two remarkable results that turned out to be related to a large number of research hotspots in economics, communications, and the smart grid. One of the most striking advances is the discovery of the Tracy Widom distribution of extreme eigenvalues and another one is the single ring law, which described the limit spectrum of eigenvalues of non-normal square matrices [46]. Interested readers are referred to monographs [47–49] for more details.

We will end this section by providing the structure of the remainder of this chapter.

Firstly, Section 23.2 gives a tutorial account of existing mathematical works that are relevant to the statistical analysis of random matrices arising in smart grids. Specially, Section 23.2.1 introduces data collected from the widely applied phasor measurement unit and data modeling using linear and nonlinear combinations of random matrices. Section 23.2.2 focuses on asymptotic spectrum laws of the major types of random matrices. Section 23.2.3 presents three dominant transforms that play key roles in describing the limit spectra of random matrices. Recent results on the convergence rate to the asymptotic limits are contained in Section 23.2.4. Section 23.2.5 is dedicated to free probability theory, which is demonstrated as a practical tool for smart grids.

Secondly, we begin with some representative problems arisen from the wide deployment of synchronous phasor measurement units that capture various features of interest in smart grids. We then show how random matrix theory has been used to characterize the data collected from synchronous phasor measurement and tackle the problems in the era of big data. In particular, Section 23.3.1 provides some basis hypothesis tests that remain fundamental to research into the behavior of the data in smart grids. Section 23.3.2 concerns stability assessment from some recently developed data-driven methods based on RMT. Section 23.3.3 focuses on situation awareness for smart grids from linear eigenvalue statistics. The early event detection problem is studied in detail using free probability in Section 23.3.4.

23.2 RMT: A Practical and Powerful Big Data Analysis Tool

In this section, we provide comprehensive existing mathematical results that are associated with the analysis of statistics of random matrices arising in smart grids. We also describe some new results on random matrices and other data-driven methods that were inspired by problems of engineering interest.

23.2.1 Modeling Grid Data using Large Dimensional Random Matrices

Before the comprehensive utilization of the RMT framework, we try to build a model for spatiotemporal PMU data using large dimensional random matrices.

It is well accepted that the transient behavior of a large electric power system can be illustrated by a set of differential and algebraic equations (DAEs) as follows [50]:

$$\dot{\mathbf{x}}^{(t)} = f(\mathbf{x}^{(t)}, \mathbf{u}^{(t)}, \mathbf{h}^{(t)}, w) \tag{23.1}$$

$$0 = g(\mathbf{x}^{(t)}, \mathbf{u}^{(t)}, \mathbf{h}^{(t)}, w) \tag{23.2}$$

where $\mathbf{x}^{(t)} \in \mathcal{C}^{m\times p}$ are the power state variables, e.g., rotor speeds and the dynamic states of loads, $\mathbf{u}^{(t)}$ represent the system input parameters, $\mathbf{h}^{(t)}$ define algebraic variables, e.g., bus voltage magnitudes, w denote the time-invariant system parameters. $t \in \mathcal{R}$, m and p are the sample time, number of system variables, and bus, respectively. The model-based stability estimators [51, 52] focus on linearization of nonlinear DAEs in (23.1) and (23.2) which gives

$$\begin{bmatrix} \Delta\dot{\underline{\mathbf{x}}} \\ \Delta\dot{\underline{\mathbf{u}}} \end{bmatrix} = \begin{bmatrix} \mathbf{A} & -\mathbf{f}_{\mathbf{u}}\mathbf{g}_{\mathbf{u}}^{-1}\mathbf{g}_{\mathbf{h}} \\ \mathbf{0} & -\mathbf{E} \end{bmatrix} \begin{bmatrix} \Delta\underline{\mathbf{x}} \\ \Delta\underline{\mathbf{u}} \end{bmatrix} + \begin{bmatrix} \mathbf{0} \\ \mathbf{C} \end{bmatrix} \xi, \tag{23.3}$$

where $\mathbf{f}_{\mathbf{x}}$, $\mathbf{f}_{\mathbf{u}}$ are the Jacobian matrices of $\mathbf{f}$ with respect to $\underline{\mathbf{x}}, \underline{\mathbf{u}}$ and $\mathbf{A} = \mathbf{f}_{\mathbf{x}} - \mathbf{f}_{\mathbf{u}}\mathbf{g}_{\mathbf{u}}^{-1}\mathbf{g}_{\mathbf{x}}$. $\mathbf{E}$ is a diagonal matrix whose diagonal entries equal t_{cor}^{-1} and t_{cor} is the correction time of the load fluctuations. $\mathbf{C}$ denotes a diagonal matrix whose diagonal entries are nominal values of the corresponding active (P) or reactive (Q) of loads; ξ is assumed to be a vector of independent Gaussian random variables.

It is noted that estimating the system stability by solving the equation (23.3) is becoming increasingly more challenging [49] as a consequence of the steady growth of the parameters, say, t, p, and m. Besides, the assumption that ξ follows a Gaussian distribution would restrict the practical applications.

On the other hand, as a novel alternative, the lately advanced data-driven estimators [38, 39, 52, 53] can assess stability without knowledge of the power network parameters or topology. However, these estimators are based on the analysis of individual window-truncated PMU data. In this chapter, we seek to provide a method with the ability of continuous learning of power system for spatiotemporal PMU data.

Firstly, we provide a novel method for modeling spatiotemporal PMU data. Figure 23.5 illustrates the conceptual representation of the structure of the spatiotemporal PMU data. More specifically, let p denote the number of the available PMUs across the whole power network, each providing c measurements. At the ith time sample, a total of $\kappa = p \times c$ measurements, say $\mathbf{z}_i$, are collected. With respect to each PMU, the c measurements could contain many categories of variables, such as voltage magnitude, power flow and frequency, etc. In this chapter, we develop PMU data analysis assuming each type of measurements is independent. That is, we assume that at each round of analysis, $\kappa := p$. Given q time periods of T seconds with K Hz sampling frequency in the kth data collection. Let $n_g = T \times K$ and $\mathbf{Z}_{ig} = \{\mathbf{z}_{i1}, \cdots, \mathbf{z}_{in_g}\}, i = 1, 2, \cdots, n$, a sequence of large random matrix

$$\left\{ \underbrace{\mathbf{Z}_{11}, \mathbf{Z}_{12}, \cdots, \mathbf{Z}_{1q}}_{q \text{ window-truncated data}}, \cdots, \underbrace{\mathbf{Z}_{n1}, \mathbf{Z}_{n2}, \cdots, \mathbf{Z}_{nq}}_{q \text{ window-truncated data}} \right\} \tag{23.4}$$

is obtained to represent the collected voltage PMU measurements.

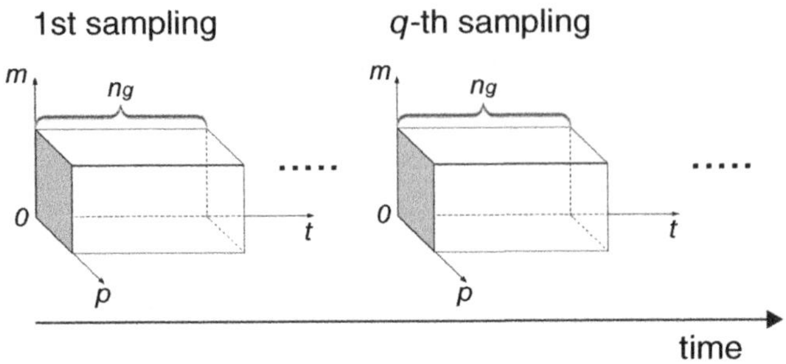

Figure 23.5 Conceptual representation of the structure of the spatiotemporal PMU data.

As illustrated in Figure 23.5, $\mathbf{Z}_i$ is a large random matrix with independent identical distributed entries. Here we also include other forms of basic random matrices that are relevant to the applications in smart grids as follows.

Gaussian unitary ensemble (GUE): Let $\mathbf{Y}$ be $N \times N$ Wigner matrix, also known as Gaussian unitary ensemble GUE, and $\mathbf{Y} = \{w_{ij}\}_{1\le i,j\le N}$. $\mathbf{Y}$ satisfies the following conditions:

1) The entries of $\mathbf{Y}$ are i.i.d Gaussian variables.
2) For $1 \le i \le j \le N$, $\mathrm{Re}(Y_{ij})$ and $\mathrm{Re}(Y_{ij})$, are i.i.d. with distribution $N\left(0, \frac{1}{2}\sigma^2\right)$.
3) For any i, j in $\{1, 2, \cdots, N\}$, $Y_{ij} = \overline{Y}_{ji}$.
4) The diagonal entries of $\mathbf{Y}$ are real random variables with distribution $N(0, \sigma^2)$.

For the convenience of analysis, we can denote GUE as $\mathbf{Y} = \frac{1}{2}(\mathbf{X} + \mathbf{X}^H)$. Besides, the joint p.d.f. of ordered eigenvalues of GUE ($\lambda_1 \ge \lambda_2 \ge \cdots \ge \lambda_N$) is [54, 55]

$$2^{-N/2}\pi^{-N^2/2}\exp\left[-\frac{\mathrm{Tr}\mathbf{Y}^2}{2}\right] \tag{23.5}$$

Laguerre unitary ensemble (LUE): Let $\{X_{ij}\}_{1\le i\le M, 1\le j\le N}$ be i.i.d. Gaussian random variables with $\mathbb{E}(X_{ij}) = 0$ and $\mathbb{E}X_{ij}^2 = \frac{1}{2}(1 + \delta_{ij})$. The so called Wishart matrix or Laguerre unitary ensemble LUE can be expressed as $\mathbf{W} = \frac{1}{N}\mathbf{X}\mathbf{X}^{\mathrm{H}}$. The p.d.f. of $\mathbf{W}$ for $N \ge M$ is [47, 55]

$$\frac{\pi^{-M(M-1)/2}}{\det \sum \prod_{i=1}^{M}(N-i)!}\exp[-\mathrm{Tr}\{\mathbf{W}\}]\det \mathbf{W}^{N-M}. \tag{23.6}$$

Large random matrix polynomials:

$$\mathbf{M} = f(\mathbf{A}) + g(\mathbf{B}),$$

where f, g are analytical functions, $\mathbf{A}$ is a GUE, and $\mathbf{B}$ is a LUE. See more details in Section 23.2.5.

23.2.2 Asymptotic Spectrum Laws

In this subsection, we provide a brief introduction to the asymptotic spectrum laws of the large basic random matrices, as shown in Section 23.2.1. There are remarkable results describing the asymptotic spectrum laws. Here special attention is paid to the limit behavior of marginal eigenvalues as the data dimensions tend to infinity.

We start with the GUE matrix $\mathbf{X} \in \mathbb{R}^N$ whose entries are independent identical distributed zero-mean (real or complex) Gaussian ensembles. As shown in [47], as $N \to \infty$, the empirical distribution of eigenvalues of $\frac{1}{\sqrt{N}}\mathbf{X}$ converges to the well-known semicircle law, whose density can be represented as

$$\rho(\lambda) = \begin{cases} \frac{1}{2\pi}\sqrt{4-\lambda^2} & \text{if } |x| \leq 2 \\ 0 & \text{if } |x| > 2 \end{cases} \tag{23.7}$$

Also shown in [40], the same result could be obtained for a symmetric $\mathbf{X}$ whose diagonal entries are 0 and whose lower-triangle entries are independent and take the values ± 1 with equal probability.

If no attempt is made to symmetrize the square matrix $\mathbf{X}$, then the eigenvalues of $\frac{1}{\sqrt{N}}\mathbf{X}$ are asymptotically uniformly distributed on the unit circle of the complex plane. This is referred to as the well-known Griko circle law, which is elaborated in the following theorem.

Theorem 23.2.1 (Circular Law [48]) *Let ξ be a complex random variable with mean zero and unit variance. For each $N \geq 1$. Let $\mathbf{X}_N$ be an iid random matrix of size N with atom variable ξ. Then, for any bounded and continuous function $f : \mathbb{C} \to \mathbb{C}$,*

$$\int_{\mathbb{C}} f(z) d\mu_{\frac{1}{\sqrt{N}}\mathbf{X}_N}(z) \to \frac{1}{\pi}\int_{\mathbb{U}} f(z) d^2 z$$

almost surely as $N \to \infty$ where $\mathbb{U}$ is the unit disk in the complex plane $|z| \leqslant 1$ and $d^2 z = dxdy$, with $z = x + iy$.

The semicircle law and circular law explain the asymptotic property of large random matrices with independent entries. However, as illustrated in Section 23.1.2, the key issues in smart grids involve the singular values of rectangular large random matrices $\mathbf{W} \in \mathbb{R}^{N\times T}$. The LUE matrices $\mathbf{W} = \frac{1}{T}\mathbf{Y}^H\mathbf{Y}$ have dependent eigenvalues of interest even if $\mathbf{Y}$ has independent entries. Let the matrix aspect ratio $c = T/N$; the asymptotic theory of singular values of $\mathbf{W}$ was presented by the landmark work [45] as follows.

As $N, T \to \infty$ and $c \leq 1$, the limit distribution of the eigenvalues of $\mathbf{W}$ converges to the so-called Marcenko-Pastur law, whose density function is

$$\rho(\lambda) = \left(1 - \frac{1}{c}\right)^+ \delta(\lambda) + \frac{\sqrt{(\lambda - a)^+(b-\lambda)^+}}{2\pi c\lambda}, \tag{23.8}$$

where $(x)^+ = \max(0, x)$ and

$$a = (1-\sqrt{c})^2, b = (1+\sqrt{c})^2.$$

Analogously, when $c \geq 1$, the limit distribution of the eigenvalues of $\mathbf{W}$ converges to

$$\tilde{\rho}(\lambda) = (1-c)^+\delta(\lambda) + \frac{\sqrt{(\lambda - a)^+(b-\lambda)^+}}{2\pi\lambda}. \tag{23.9}$$

In addition to Wigner's semicircle law above and Marchenko-Pastur law, we are also interested in the single ring law developed by Guionnet, Krishnapur, and Zeitouni

(2011) [46]. It describes the empirical distribution of the eigenvalues of a large generic matrix with prescribed singular values, i.e., an $N \times N$ matrix of the form $\mathbf{A} = \mathbf{UTV}$, with $\mathbf{U}; \mathbf{V}$ some independent Haar-distributed unitary matrices and $\mathbf{T}$ a deterministic matrix whose singular values are the ones prescribed. More precisely, under some technical hypotheses, as the dimension N tends to infinity, if the empirical distribution of the singular values of $\mathbf{A}$ converges to a compactly supported limit measure Θ on the real line, then the empirical eigenvalues distribution of $\mathbf{A}$ converges to a limit measure μ on the complex plane that depends only on Θ. The limit measure μ is rotationally invariant in $\mathbb{C}$, and its support is the annulus $S := \{z \in \mathbb{C}; a \leqslant |z| \leqslant b\}$ with $a, b \geq 0$ such that

$$a^{-2} = \int x^{-2} d\Theta(x) \text{ and } \mathrm{b}^2 = \int x^2 d\Theta(x). \tag{23.10}$$

23.2.3 Transforms

The transforms of large random matrices are especially useful to study the limit spectral properties and to tackle the problems of polynomial calculation of random matrices. In this subsection, we will review the useful transforms including Stieltjes transform, R transform, and S transform suggested by problems of interest in power grids [49, 56].

We begin with the Stieltjes transform of $\mathbf{X}$, which is defined as follows.

Definition 23.2.2 *Let $\mathbf{X}$ be a random matrix with distribution $F(\cdot)$. Its Stieltjes transform is defined as*

$$G(z) = \left\{\frac{1}{N} \operatorname{Tr}\left[(z\mathbf{I} - \mathbf{X})^{-1}\right]\right\} = \int_{\mathbb{R}} \frac{1}{x - z} dF(x), \tag{23.11}$$

where $z \in \mathbb{C}^-$ and $\mathbb{C}^- = \{z \in C : \operatorname{Im}(z) < 0\}$ and $\mathbf{I}$ represents the identity matrix of dimension N.

An important application of the Stieltjes transform is its *close* relationship with the limit spectrum density of $\mathbf{X}$.

Theorem 23.2.3 *Let $\mathbf{X}$ be a $N \times N$ random Hermitian matrix and its Stieltjes transform is $G(z)$, the corresponding eigenvalue density $\rho(x)$ can be expressed as:*

$$\rho(x) = -\frac{1}{\pi} \lim_{\operatorname{Im} z \to 0} \operatorname{Im}\{G(z)\}. \tag{23.12}$$

It is noted that the signs of the $\operatorname{Im}(z)$ and $G(z)$ coincide. This property should be emphasized in the following examples where the sign of the square root should be chosen.

For GUE and LUE matrices, the corresponding Stieltjes transforms are shown in the following examples.

Example 23.2.4 *Let $\mathbf{X}$ be an GUE matrix, and its limit spectral density is defined in (23.7), the Stieltjes transform of $\mathbf{X}$ is*

$$G(z) = \frac{1}{2\pi} \int_{-2}^{2} \frac{\sqrt{4 - x^2}}{x - z} dx = \frac{1}{2}\left[z - \sqrt{z^2 - 4}\right].$$

Example 23.2.5 *Let* $\mathbf{W}$ *be an LUE matrix, and its limit spectral density is defined in* (23.8) *and* (23.9). *The corresponding Stieltjes transform can be represented as*

$$G(z) = \frac{1}{2\pi}\int_a^b \frac{\rho(x)}{(x-z)}dx = \frac{-\beta - z + 1 \pm \sqrt{z^2 - 2(\beta+1)z + (\beta-1)^2}}{2\beta z}$$

and

$$G(z) = \frac{1}{2\pi}\int_a^b \frac{\rho'(x)}{(x-z)}dx = \frac{\beta - z - 1 \pm \sqrt{z^2 - 2(\beta+1)z + (\beta-1)^2}}{2z},$$

respectively.

Another two important transforms, which we elaborate in the following, are the R transform and the S transform. The key point of these two transforms is that the R/S transform enables the characterization of the limiting spectrum of a sum/product of random matrices from their individual limiting spectra. These properties turn out to be extremely useful in the following subsection. We start with the blue function, that is, the functional inverse of the Stieltjes transform $G(z)$, which is defined as

$$B(G(z)) = z$$

and then the R transform is simply defined by

$$R(\omega) = B(\omega) - \frac{1}{w}.$$

Two important properties of the R transform are shown in the following.

Additivity law : let $R_{\mathbf{A}}(z)$, $R_{\mathbf{B}}(z)$ and $R_{\mathbf{A+B}}(z)$ be the R transforms of matrices $\mathbf{A}$, $\mathbf{B}$ and $\mathbf{A}+\mathbf{B}$, respectively. We have

$$R_{\mathbf{A+B}}(z) = R_{\mathbf{A}}(z) + R_{\mathbf{A}}(z). \tag{23.13}$$

Scaling property : For any $\alpha > 0$,

$$R_{\alpha\mathbf{X}}(z) = \alpha R_{\mathbf{X}}(\alpha z). \tag{23.14}$$

This additivity law can be easily understood in terms of Feynman diagrams; we refer interested readers to references [49] for details. The above properties of the R transform enable us to do the linear calculation of the asymptotic spectrum of random matrices.

Another important transform of engineering significance in RMT is the S transform. The S transform is related to the R transform and is defined by

$$S(z) = \frac{1}{R(zS(z))}. \tag{23.15}$$

An interesting property of the S transform is that the S transform of the product of two independent random matrices equals the product of the S transforms:

$$S_{\mathbf{AB}}(z) = S_{\mathbf{A}}(z)S_{\mathbf{B}}(z). \tag{23.16}$$

Note that (23.16) is known as multiplication law of the S transform. For the sake of brevity, see Section 23.2.5 for more details.

23.2.4 Convergence Rate

In this section, we investigate the spectral asymptotics for GUE and LUE matrices. We are motivated by the practical problems introduced in [49]. Let $F(x)$ be the empirical spectral distribution function of GUE or LUE matrices and $G(x)$ be the distribution function of the limit law (semicircle law for GUE matrices and Marchenko-Pastur law for LUE matrices). Here, we study the convergence rate of the expected empirical distribution function $\mathbb{E}F(x)$ to $G(x)$. Especially, the bound

$$\Delta = |\mathbb{E}F(x) - G(x)|, \tag{23.17}$$

is mainly concerned in the following.

The rate of convergence for the expected spectral distribution of GUE matrices has attracted much attention due to its increasingly appreciated importance in applied mathematics and statistical physics. Wigner initially looked into the convergence of the spectral distribution of GUE matrices [57]. Bai [58] conjectured that the optimal bound for Δ in the GUE case should be of order n^{-1}. Bai and coauthors in [59] proved that $\Delta = O(N^{-1/3})$. Gotze and Tikhomirov in [60] improved the result in [59] and proved that $\Delta = O(N^{-1/2})$. Bai et al. in [61] also showed that $\Delta = O(N^{-1/2})$ on the condition that the 8th moment of $\mathbf{X}$ satisfies $\sup E|X_{ij}|^8 < \infty$. Girko in [62] stated as well that $\Delta = O(N^{-1/2})$ assuming uniform bounded 4th moment of $\mathbf{X}$. Recently, Gotze and Tikhomirov proved an optimal bound as follows.

Theorem 23.2.6 *There exists a positive constant C such that, for any $N \geq 1$,*

$$\Delta \leq CN^{-1}. \tag{23.18}$$

The convergence of the density (denoted by $g(x)$) of the standard semicircle law to the expected spectral density $p(x)$ is proved by Gotze and Tikhomirov in the following theorem.

Theorem 23.2.7 *There exists a positive constant ε and C such that, for any* $x \in \left[-2 + N^{-\frac{1}{3}}\varepsilon, 2 - N^{-\frac{1}{3}}\varepsilon\right]$,

$$|p(x) - g(x)| \leq \frac{C}{N(4 - x^2)}. \tag{23.19}$$

For LUE matrix $\mathbf{W}$ with spectral distribution function $F(x)$, let $\beta = \frac{N}{M}$ as $N, M \to \infty$; it is well known that $\mathbb{E}F(x)$ convergences to the Marchenko-Pastur law $H(x)$ with density

$$h(x) = \frac{1}{2\pi\beta x}\sqrt{(x - a)(b - x)}, \tag{23.20}$$

where $a = (1 - \sqrt{\beta})^2, b = (1 + \sqrt{\beta})^2$. The bound

$$\Delta = |\mathbb{E}F(x) - H(x)|, \tag{23.21}$$

for the convergence rate is shown in the following theorems.

Theorem 23.2.8 *For $\beta = \frac{N}{M}$, there exist some positive constants β_1 and β_2 such that $0 < \beta_1 \leq \beta \leq \beta_2 < 1$, for all $N \geq 1$. Then there exists a positive constant C depending on β_1 and β_2 and for any $N \geq 1$*

$$\Delta \leq CN^{-1}. \tag{23.22}$$

Considering the case $\beta < 1$, a similar result is shown in Theorem 23.2.9.

Theorem 23.2.9 *For $\beta = \frac{N}{M}$, there exists some positive constant β_1 and β_2 such that $0 < \beta_1 \le \beta \le \beta_2 < 1$, for all $N \ge 1$. Then there exists a positive constant C and ε depending on β_1 and β_2 and for any $N \ge 1$ and $x \in \left[a + N^{-\frac{2}{3}}\varepsilon, b - N^{-\frac{2}{3}}\varepsilon\right]$*

$$|p(x) - h(x)| \le \frac{C}{N(x-a)(b-x)}. \tag{23.23}$$

Interested readers are referred to [63] for technical details and Section 23.3.3 for applications in the smart grid.

23.2.5 Free Probability

Free probability theory, initiated in 1983 by Voiculescu in [64], together with the results published in [65] regarding asymptotic freeness of random matrices, has established a new branch of theories and tools in random matrix theory. Here, we provide some of the basic principles and then examples to enhance the understanding and application of free probability theory.

Let $x_1, \dots, x_n$ be self-adjoint elements which are freely independent. Consider a self-adjoint polynomial p in n non-commuting variables and let P be the element $P = p(x_1, \dots, x_n)$. Now we introduce the method [66, 67] to obtain the distribution of P out of the distributions of $x_1, \dots, x_n$.

Let $\mathcal{A}$ be a unital algebra and $\mathcal{B} \subset \mathcal{A}$ be a subalgebra containing the unit. A linear map

$$E : \mathcal{A} \to \mathcal{B}$$

is a conditional expectation if

$$E[b] = b \qquad \text{for all } b \in \mathcal{B}$$

and

$$E[b_1 a b_2] = b_1 E[a] b_2 \qquad \text{for all } a \in \mathcal{A} \quad \text{for all } b_1, b_2 \in \mathcal{B}$$

An operator-valued probability space consists of $\mathcal{B} \subset \mathcal{A}$ and a conditional expectation $E : \mathcal{A} \to \mathcal{B}$. Then, random variables $x_i \in \mathcal{A}(i \in I)$ are free with respect to E (or free with amalgamation over $\mathcal{B}$) if $E[a_1 \dots a_n] = 0$ whenever $a_i \in \mathcal{B} < x_{j(i)} >$ are polynomials in some $x_{j(i)}$ with coefficients from $\mathcal{B}$ and $E[a_i] = 0$ for all i and $j(1) \ne j(2) \ne \cdots \ne j(n)$. For a random variable $x \in \mathcal{A}$, we denote the operator-valued Cauchy transform:

$$G(b) := E[(b-x)^{-1}](b \in \mathcal{B})$$

whenever $(b - x)$ is invertible in $\mathcal{B}$. In order to have some nice analytic behavior, we assume that both $\mathcal{A}$ and $\mathcal{B}$ are C^*-algebras in the following; $\mathcal{B}$ will usually be of the form $\mathcal{B} = M_N(\mathbb{C})$, the $N \times N$-matrices. In such a setting and for $x = x^*$, this G is well defined and a nice analytic map on the operator-valued upper half-plane:

$$\mathbb{H}^+(B) := \{b \in B | (b - b*)/(2i) > 0\}$$

and it allows to give a nice description for the sum of two free self-adjoint elements. In the following we will use the notation

$$h(b) := \frac{1}{G(b)} - b$$

Theorem 23.2.10 ([66]) *Let x and y be self-adjoint operator-valued random variables free over $\mathcal{B}$. Then there exists a Frechet analytic map $\omega : \mathbb{H}^+(\mathcal{B}) \to \mathbb{H}^+(\mathcal{B})$ so that*

- $\Im\omega_j(b) \geq \Im b$ *for all* $b \in \mathbb{H}^+(\mathcal{B}), j \in \{1,2\}$
- $G_x(\omega_1(b)) = G_y(\omega_2(b)) = G_{x+y}(b)$

Moreover, if $b \in \mathbb{H}^+(\mathcal{B})$, then $\omega_1(b)$ is the unique fixed point of the map. $f_b : \mathbb{H}^+(\mathcal{B}) \to \mathbb{H}^+(\mathcal{B})$, $f_b(\omega) = h_y(h_x(\omega)+b)+b$, and $\omega_1(b) = \lim_{n\to\infty} f_b{}^{on}(\omega)$ for any $\omega \in \mathbb{H}^+(\mathcal{B})$, where f_b^{on} means the n-fold composition of f_b with itself. The same statements hold for $\omega_2(b)$, replaced by $\omega \to h_x(h_y(\omega)+b)+b$.

Let $\mathcal{A}$ be a complex and unital C^*-algebra and let self-adjoint elements $x_1, \dots x_n \in \mathcal{A}$. $\mathcal{A}$ is given. Then, for any non-commutative polynomial $p \in \mathbb{C} < \mathbf{X}_1, \dots, \mathbf{X}_n >$, we get an operator $P = p(x_1 \dots x_n) \in \mathcal{A}$ by evaluating p at $(x_1, \dots, x_n)$. In this situation, knowing a linearization trick [68] means to have a procedure that leads finally to an operator

$$L_p = b_0 \otimes 1 + b_1 \otimes x_1 + \cdots b_n \otimes x_n \in M_N(\mathbb{C}) \otimes \mathcal{A}$$

for some matrices $b_0, \dots, b_n \in M_N(\mathbb{C})$ of dimension N, such that $z - P$ is invertible in $\mathcal{A}$ if and only if $\Lambda(z) - L_p$ is invertible in $M_N(\mathbb{C}) \otimes \mathcal{A}$. Hereby, we put

$$\Lambda(z) = \begin{bmatrix} z & 0 & \cdots & 0 \\ 0 & 0 & \cdots & 0 \\ \vdots & \vdots & \ddots & \vdots \\ 0 & 0 & \cdots & 0 \end{bmatrix} \quad \textit{for all } z \in \mathbb{C}.$$

Let $p \in \mathbb{C} < \mathbf{X}_1, \dots, \mathbf{X}_n >$ be given. A matrix

$$L_p := \begin{bmatrix} 0 & u \\ v & \mathbf{Q} \end{bmatrix} \in M_N(\mathbb{C}) \otimes \mathbb{C} < \mathbf{X}_1, \dots, \mathbf{X}_n >,$$

where

- $N \in \mathbb{N}$ is an integer,
- $\mathbf{Q} \in M_{N-1}(\mathbb{C}) \otimes \mathbb{C} < \mathbf{X}_1, \dots, \mathbf{X}_n >$ is invertible,
- and u is a row vector and v is a column vector, both of size $N-1$ with entries in $\mathbb{C} < \mathbf{X}_1, \dots, \mathbf{X}_n >$,

is called a linearization of p, if the following conditions are satisfied:

1) There are matrices $b_0, \dots, b_n \in M_N(\mathbb{C})$, such that $L_p = b_0 \otimes 1 + b_1 \otimes \mathbf{X}_1 + \cdots b_n \otimes \mathbf{X}_n$ i.e., the polynomial entries in $\mathbf{Q}$, u and v all have degree ≤ 1.
2) It holds true that $p = -u\mathbf{Q}^{-1}v$.

To introduce the following corollary, which will enable us to shift $\Lambda_\varepsilon(z)$ for $z \in \mathbb{C}^+$ to a point

$$\Lambda_\varepsilon(z) := \begin{bmatrix} z & & & \\ & i\varepsilon & & \\ & & \ddots & \\ & & & i\varepsilon \end{bmatrix}$$

lying inside the domain $\mathbb{H}^+(M_N(\mathbb{C}))$ in order to get access to all analytic tools that are available there.

Corollary 23.2.11 *Let $(\mathcal{A}, \phi)$ be a C^*-probability space and let elements $x_1, \dots, x_n \in \mathcal{A}$ be given. For any self-adjoint $p \in \mathbb{C} < X_1, \dots, X_n >$ that has a self-adjoint linearization*

$$L_p = b_0 \otimes 1 + b_1 \otimes \mathbf{X}_1 + \cdots\ b_n \otimes \mathbf{X}_n \in M_N(\mathbb{C}) \otimes \mathbb{C} < \mathbf{X}_1, \dots, \mathbf{X}_N >$$

with matrices $b_0, \dots, b_n \in M_N(\mathbb{C})_{sa}$, we put $P = p(x_1, \dots, x_n)$ and

$$L_P = b_0 \otimes 1 + b_1 \otimes x_1 + \cdots\ b_n \otimes x_n \in M_N(\mathbb{C}) \otimes \mathcal{A}$$

Then, for each $z \in \mathbb{C}^+$ and all sufficiently small $\varepsilon > 0$, the operators $z - P \in \mathcal{A}$ and $\Lambda_\varepsilon(z) - L_p \in M_N(\mathbb{C}) \otimes \mathcal{A}$ are both invertible, and we have

$$\lim_{\varepsilon \to 0} [\mathbb{E}((\Lambda_\varepsilon(z) - L_p)^{-1})]_{1,1} = G_P(z)$$

Hereby,

$$\mathbb{E} M_N(\mathbb{C}) \otimes \mathcal{A} \to M_N(\mathbb{C})$$

denotes the conditional expectation given by $\mathbb{E}=\mathrm{id}_{M_N(\mathbb{C})} \otimes \phi$.

Let $(\mathcal{A}, \phi)$ be a non-commutative C^*-probability space, $x_1, \dots, x_n \in \mathcal{A}$ self-adjiont elements which are freely independent, and $p \in \mathbb{C} < \mathbf{X}_1, \dots, \mathbf{X}_n >$ a self-adjoint polynomial in n non-commuting variables $\mathbf{X}_1, \dots, \mathbf{X}_N$. We put $P = p(x_1, \dots, x_n)$. The following procedure leads to the distribution of P.

step 1 p has a self-adjoint linearization

$$L_p = b_0 \otimes 1 + b_1 \otimes \mathbf{X}_1 + \cdots\ b_n \otimes \mathbf{X}_n$$

with matrices $b_0, \dots, b_n \in M_N \mathbb{C}_{\mathrm{sa}}$. We put

$$L_P = b_0 \otimes 1 + b_1 \otimes x_1 + \cdots\ b_n \otimes x_n \in M_N(\mathbb{C}) \otimes \mathcal{A}.$$

step 2 The operators $b_0 \otimes 1, b_1 \otimes x_1, \dots, b_n \otimes x_n$ are freely independent elements in the operator-valued C^*-probability space $(M_N(\mathbb{C}) \otimes \mathcal{A}, \mathbb{E})$, where $\mathbb{E} := M_N(\mathbb{C}) \otimes \mathcal{A} \to M_N(\mathbb{C})$ denotes the conditional expectation given by $\mathbb{E}=\mathrm{id}_{M_N(\mathbb{C})} \otimes \phi$. Furthermore, for $j = 1, \dots, n$, the $M_N(\mathbb{C})$-valued Cauchy transform $G_{b_j \otimes x_j}(b)$ is completely determined by the scalar-valued Cauchy transforms G_{x_j} via

$$G_{b_j \otimes x_j}(b) = \lim_{\varepsilon \to 0} -\frac{1}{\pi} \int_{\mathbb{R}} (b - tb_j)^{-1} \Im(G_{x_j}(t + i\varepsilon))dt$$

for all $b \in \mathbb{H}^+(M_N(\mathbb{C}))$.

step 3 Because of Step 3, we can calculate the Cauchy transform of

$$L_p - b_0 \otimes 1 = b_1 \otimes \mathbf{X}_1 + \cdots\ b_n \otimes \mathbf{X}_n$$

by using the fixed point iteration for the operator-valued free additive convolution. The Cauchy transform of L_P is then given by

$$G_{L_P}(b) = G_{L_P - b_0 \otimes 1}(b - b_0) \quad \textit{for all } b \in \mathbb{H}^+(M_N(\mathbb{C})).$$

step 4 The corollary tells us that the scalar-valued Cauchy transform $G_P(z)$ of P is determined by

$$G_P(z) = \lim_{\varepsilon \to 0} [G_{L_P}(\Lambda_\varepsilon(z))]_{1,1} \quad \textit{for all } z \in \mathbb{C}^+$$

Finally, we obtain the desired distribution of P by applying the Stieltjes inversion formula.

Example 23.2.12 *We consider the non-commutative polynomial* $p \in \mathbb{C} < \mathbf{X}_1, \mathbf{X}_2 >$ *given by* $p(\mathbf{X}_1, \mathbf{X}_2) = \mathbf{X}_1\mathbf{X}_2 + \mathbf{X}_2\mathbf{X}_1$. *It is easy to check that*

$$L_p = \begin{bmatrix} 0 & \mathbf{X}_1 & \mathbf{X}_2 \\ \mathbf{X}_1 & 0 & -1 \\ \mathbf{X}_2 & -1 & 0 \end{bmatrix}$$

is a self-adjoint linearization of p. *Now, let* X_1, X_2 *be free semicircular or Poisson elements in a non-commutative* C^**-probability space* $(\mathcal{A}, \phi)$. *Based on the algorithm of the theorem above, we can calculate the distribution of the anticommutator* $p(\mathbf{X}_1, \mathbf{X}_2) = \mathbf{X}_1\mathbf{X}_2 + \mathbf{X}_2\mathbf{X}_1$.

Example 23.2.13 *In the same way, we can deal with the following variation of the anti-commutator:* $p(\mathbf{X}_1, \mathbf{X}_2) = \mathbf{X}_1\mathbf{X}_2 + \mathbf{X}_2\mathbf{X}_1 + \mathbf{X}_1{}^2$. *It is easy to check that*

$$L_p = \begin{bmatrix} 0 & \mathbf{X}_1 & \frac{1}{2}\mathbf{X}_1 + \mathbf{X}_2 \\ X_1 & 0 & -1 \\ \frac{1}{2}\mathbf{X}_1 + \mathbf{X}_2 & -1 & 0 \end{bmatrix}$$

is a self-adjoint linearization of p.

Then, let $\mathbf{X}_1, \mathbf{X}_2$ *be free semicircular or Poisson elements in a non-commutative* C^**-probability space* $(\mathcal{A}, \phi)$. *Based on the algorithm above, we can calculate the distribution of the polynomial* $p(\mathbf{X}_1, \mathbf{X}_2) = \mathbf{X}_1\mathbf{X}_2 + \mathbf{X}_2\mathbf{X}_1 + \mathbf{X}_1{}^2$.

For readers' convenience, experimental results obtained in various conditions are also presented. Specially, Figures 23.6, 23.7 and 23.8 illustrate the theoretical limit

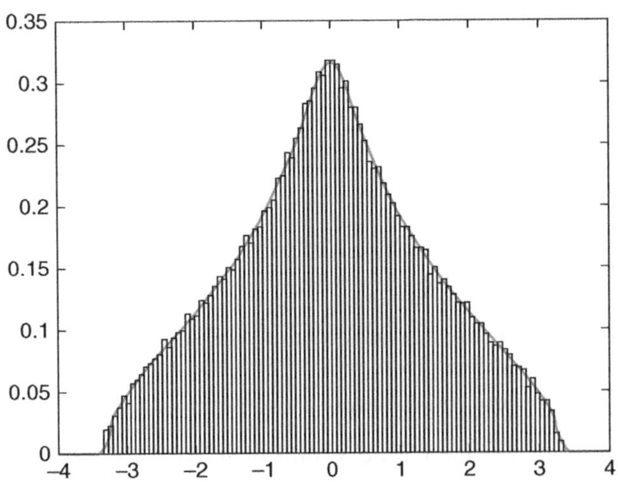

Figure 23.6 Comparison of the distribution of $p(\mathbf{X}_1, \mathbf{X}_2) = \mathbf{X}_1\mathbf{X}_2 + \mathbf{X}_2\mathbf{X}_1$ according to our algorithm, with the histogram of eigenvalues for $p(\mathbf{X}_1{}^{(n)}, \mathbf{X}_2{}^{(n)})$ according to our algorithm, with the histogram of eigenvalues for $p(\mathbf{X}_1{}^{(n)}, \mathbf{X}_2{}^{(n)})$, for $n = 1000$. $\mathbf{X}_1, \mathbf{X}_2$ are free semicircular elements and $\mathbf{X}_1{}^{(n)}, \mathbf{X}_2{}^{(n)}$ are independent standard Gaussian random matrices.

spectra and empirical one in the case of the polynomial of random matrices introduced in Example 23.2.12. Figures 23.9, 23.10 and 23.11 present simulation results for the Example 23.2.13. We see that the theoretical results agree remarkably with the numerical simulations in various conditions.

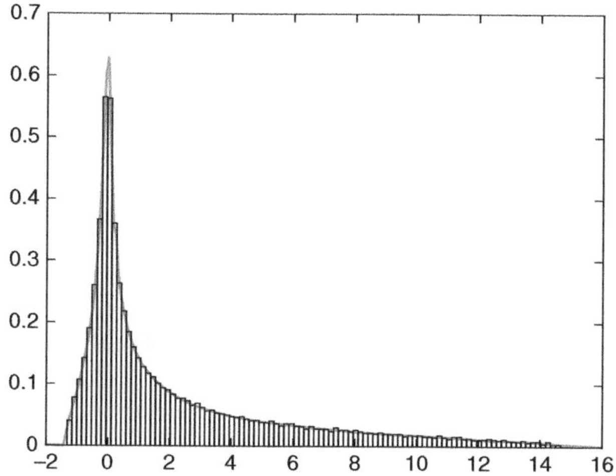

Figure 23.7 Comparison of the distribution of $p(\mathbf{X}_1, \mathbf{X}_2) = \mathbf{X}_1\mathbf{X}_2 + \mathbf{X}_2\mathbf{X}_1$ according to our algorithm, with the histogram of eigenvalues for $p(\mathbf{X}_1^{(n)}, \mathbf{X}_2^{(n)})$, for $n = 1000$. $\mathbf{X}_1, \mathbf{X}_2$ are of free Poisson elements, and $\mathbf{X}_1^{(n)}, \mathbf{X}_2^{(n)}$ are Wishart random matrices.

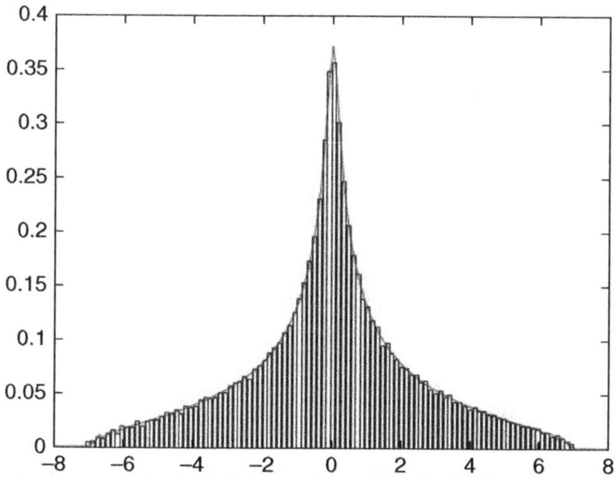

Figure 23.8 Comparison of the distribution of $p(\mathbf{X}_1, \mathbf{X}_2) = \mathbf{X}_1\mathbf{X}_2 + \mathbf{X}_2\mathbf{X}_1$ according to our algorithm, with the histogram of eigenvalues for $p(\mathbf{X}_1^{(n)}, \mathbf{X}_2^{(n)})$, for $n = 1000$. $\mathbf{X}_1$ is of free semicircular elements and $\mathbf{X}_2$ free Poisson ones. $\mathbf{X}_1^{(n)}$ is an independent standard Gaussian random matrix, and $\mathbf{X}_2^{(n)}$ is a Wishart matrix.

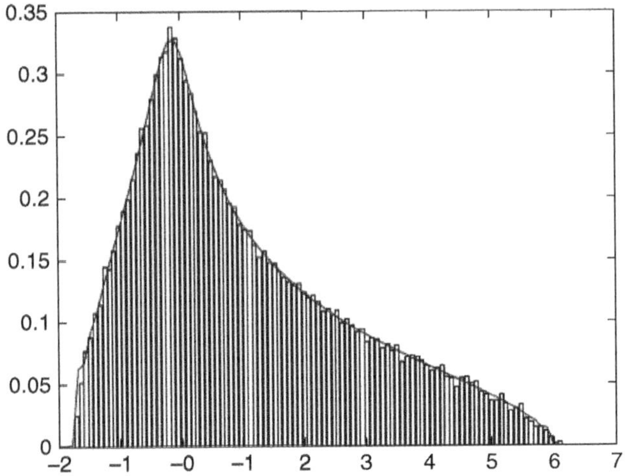

Figure 23.9 Comparison of the distribution of $p(\mathbf{X}_1, \mathbf{X}_2) = \mathbf{X}_1\mathbf{X}_2 + \mathbf{X}_2\mathbf{X}_1 + \mathbf{X}_1{}^2$ according to our algorithm, with the histogram of eigenvalues for $p(\mathbf{X}_1{}^{(n)}, \mathbf{X}_2{}^{(n)})$, for $n = 1000$. $\mathbf{X}_1, \mathbf{X}_2$ are of free Poisson elements, and $\mathbf{X}_1{}^{(n)}, \mathbf{X}_2{}^{(n)}$ are Wishart random matrices.

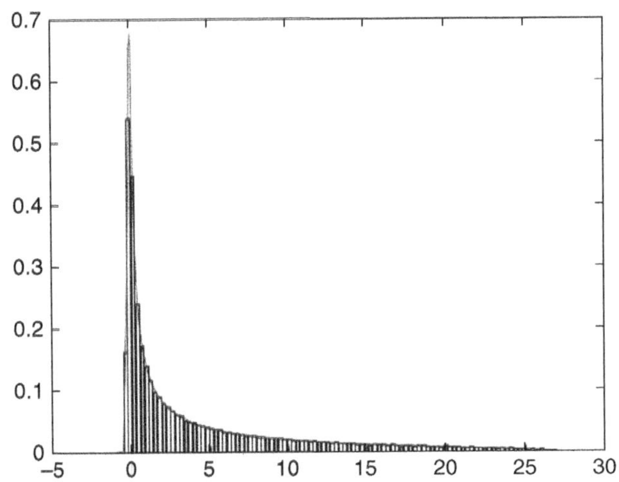

Figure 23.10 Comparison of the distribution of $p(\mathbf{X}_1, \mathbf{X}_2) = \mathbf{X}_1\mathbf{X}_2 + \mathbf{X}_2\mathbf{X}_1 + \mathbf{X}_1{}^2$ according to our algorithm, with the histogram of eigenvalues for $p(\mathbf{X}_1{}^{(n)}, \mathbf{X}_2{}^{(n)})$, for $n = 1000$. $\mathbf{X}_1, \mathbf{X}_2$ are of free Poisson elements, and $\mathbf{X}_1{}^{(n)}, \mathbf{X}_2{}^{(n)}$ are Wishart random matrices.

23.3 Applications to Smart Grids

In this section, we elaborate some of the more representative problems described in Section 23.1 that capture various features of interest in smart grids and we show how random matrix results have been used to tackle the problems that arise in the large power grid with wide deployment of PMU equipments. Besides, we also conclude with some state-of-art data driven methods for comparison.

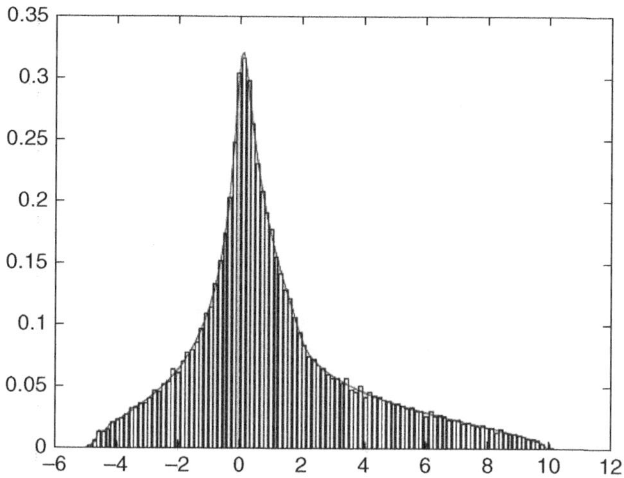

Figure 23.11 Comparison of the distribution of $p(\mathbf{X}_1, \mathbf{X}_2) = \mathbf{X}_1\mathbf{X}_2 + \mathbf{X}_2\mathbf{X}_1 + \mathbf{X}_1^2$according to our algorithm, with the histogram of eigenvalues for $p(\mathbf{X}_1^{(n)}, \mathbf{X}_2^{(n)})$, for $n = 1000$. $\mathbf{X}_1$ are of free semicircular elements and $\mathbf{X}_2$ free Poisson ones. $\mathbf{X}_1^{(n)}$ is an independent standard Gaussian random matrix, and $\mathbf{X}_2^{(n)}$ is a Wishart matrix.

23.3.1 Hypothesis Tests in Smart Grids

Considering the data model introduced in Section 23.2.1, the problem of testing hypotheses on means of populations and covariance matrices is addressed. We start with a review of traditional multivariate procedures for these tests. Then we develop adjustments of these procedures to handle high-dimensional data in smart grids.

As depicted in Section 23.2.1, a large random matrix flow $\{\mathbf{Z}_1, \mathbf{Z}_2, \cdots, \mathbf{Z}_q\}$ is adopted to represent the massive streaming PMU data in one sample period. Instead of analyzing the raw individual window-truncated PMU data $\mathbf{Z}_g$ [38, 39] or the statistic of $\mathbf{Z}_g$ [52, 53], a comprehensive analysis of the statistic of $\{\mathbf{Z}_1, \mathbf{Z}_2, \cdots, \mathbf{Z}_q\}$ is conducted in the following. More specifically, denote as $\mathbf{\Sigma}_i$ the covariance matrix of ith collected PMU measurements; we want to test the hypothesis:

$$\begin{array}{l} H_0 : \mathbf{\Sigma}_1 = \mathbf{\Sigma}_2 = \cdots = \mathbf{\Sigma}_q \\ H_1 : \exists\, j,k \ \text{ s.t. } \ \mathbf{\Sigma}_j \neq \mathbf{\Sigma}_k \end{array}. \tag{23.24}$$

It is worthy noting that the hypothesis (23.24) is a famous testing hypothesis in multivariate statistical analysis that aims to study samples share or approximately share some distribution and consider using a set of samples (data streams denoted in equation (23.4) in this paper), one from each population, to test the hypothesis that the covariance matrices of these populations are equal.

23.3.2 Data-Driven Methods for State Evaluation

The LR test [69] and CLR test [70] as introduced in Section 23.3.1 are most commonly test statistics for the hypothesis in (23.24). These tests can be understood by replacing the population covariance matrix $\mathbf{\Sigma}_g$ by its sample covariance matrix $\mathbf{Y}_g$. While direct substitution of $\mathbf{\Sigma}_g$ by $\mathbf{Y}_g$ brings invariance and good testing properties as shown in [69] for normally distributed data. The test statistic V_2 may not work for high-dimensional data as demonstrated in [71, 72]. Besides, the estimator V_3 has unnecessary terms, which slow down the convergence considerably when the dimension of PMU data is high [72, 73]. In such situations, to overcome the drawbacks, a trace criterion [72] is more suitable to the test problem. Specially, instead of estimating the population covariance matrix directly, a well-defined distance measure exploiting the difference among data flow $\{\mathbf{Z}_1, \mathbf{Z}_2, \cdots, \mathbf{Z}_q\}$ is conducted, that is, the trace-based distance measure between $\mathbf{\Sigma}_s$ and $\mathbf{\Sigma}_t$ is

$$\mathrm{tr}\{(\mathbf{\Sigma}_s - \mathbf{\Sigma}_t)^2\} = \mathrm{tr}(\mathbf{\Sigma}_s^2) + \mathrm{tr}(\mathbf{\Sigma}_t^2) - 2\mathrm{tr}(\mathbf{\Sigma}_s\mathbf{\Sigma}_t), \tag{23.25}$$

where tr(·) is the trace operator. Instead of estimating $\mathrm{tr}(\mathbf{\Sigma}_s^2)$, $\mathrm{tr}(\mathbf{\Sigma}_t^2)$ and $\mathrm{tr}(\mathbf{\Sigma}_s\mathbf{\Sigma}_t)$ by sample covariance matrix-based estimators, we adopt the merits of the U-statistics [74]. Especially, for $l = \{s, t\} \in \Omega = \{1 \leq s, t \leq q, s \neq t\}$,

$$\begin{aligned} A_l = {} & \frac{1}{n_g(n_g-1)} \sum_{i \neq j} (\mathbf{z}_{li}'\mathbf{z}_{lj})^2 \\ & - \frac{2}{n_g(n_g-1)(n_g-2)} \sum_{i,j,k}^{*} \mathbf{z}_{li}'\mathbf{z}_{lj}\mathbf{z}_{lj}'\mathbf{z}_{lk} \\ & + \frac{1}{n_g(n_g-1)(n_g-2)(n_g-3)} \sum_{i,j,k,h}^{*} \mathbf{z}_{li}'\mathbf{z}_{lj}\mathbf{z}_{lk}'\mathbf{z}_{lh} \end{aligned} \tag{23.26}$$

is proposed to estimate $\mathrm{tr}(\mathbf{\Sigma}_l^2)$. It is noted that $\sum^*$ represents summation over mutually distinct indices. For example, $\sum_{i,j,k}^*$ says summation over the set $\{(i, j, k) : i \neq j, j \neq k, k \neq i\}$. Similarly, the estimator for $\mathrm{tr}(\mathbf{\Sigma}_s\mathbf{\Sigma}_t)$ can be expressed as

$$\begin{aligned} C_{st} = {} & \frac{1}{n_g^2} \sum_i \sum_j (\mathbf{z}_{si}'\mathbf{z}_{tj})^2 \\ & - \frac{1}{(n_g-1)n_g^2} \sum_{i,h}^{*} \sum_j \mathbf{z}_{si}'\mathbf{z}_{tj}\mathbf{z}_{tj}'\mathbf{z}_{sh} \\ & - \frac{1}{(n_g-1)n_g^2} \sum_{i,l}^{*} \sum_j \mathbf{z}_{ti}'\mathbf{z}_{sj}\mathbf{z}_{sj}'\mathbf{z}_{th} \\ & + \frac{1}{(n_g-1)^2n_g^2} \sum_{i,h}^{*} \sum_{j,k}^{*} \mathbf{z}_{si}'\mathbf{z}_{tj}\mathbf{z}_{sk}'\mathbf{z}_{th}. \end{aligned} \tag{23.27}$$

The test statistic that measures the distance between $\mathbf{\Sigma}_s$ and $\mathbf{\Sigma}_t$ is

$$V_{st} = A_s + A_t - C_{st}. \tag{23.28}$$

Then the proposed test statistic can be expressed as:

$$V_1 = \frac{1}{q(q-1)} \sum_{\{s,t\} \in \Omega} T_{st}. \tag{23.29}$$

As $p, n_g \to \infty$, the asymptotic normality [73] of the test statistic (23.28) is presented in the following:

Theorem 23.3.1 *Let* $\sigma_{st}^2 = \frac{1}{n_g}(A_s + A_t)$. *Assuming the following conditions:*

1) For any k and $l \in \{s, t\}$, $\text{tr}(\boldsymbol{\Sigma}_k \boldsymbol{\Sigma}_l) \to \infty$ *and*

$$\text{tr}\{(\underset{i}{\boldsymbol{\Sigma}}\underset{j}{\boldsymbol{\Sigma}})(\underset{k}{\boldsymbol{\Sigma}}\underset{l}{\boldsymbol{\Sigma}})\} = O\{\text{tr}(\underset{i}{\boldsymbol{\Sigma}}\underset{j}{\boldsymbol{\Sigma}})\text{tr}(\underset{k}{\boldsymbol{\Sigma}}\underset{l}{\boldsymbol{\Sigma}})\}.$$

2) For $i = 1, 2, \cdots, n_g$, $\mathbf{z}^{(i)}$ *are independent and identically distributed p-dimensional vectors with finite 8th moment.*

Under the above conditions,

$$L = \frac{V_{st}}{\sigma_{st}} \xrightarrow{d} \mathcal{N}(0, 1)$$

Corollary 23.3.2 *For any* $q \geq 2$, *as* $p, n_g \to \infty$, *the proposed test statistic* V_1 *satisfies*

$$V_1 \to^d \mathcal{N}(\mu, \sigma^2), \tag{23.30}$$

where $\mu \approx 0, \sigma^2 = \sum^* \sigma_{st}^2$.

Let $R = \frac{V_1}{\sigma_{V_1}}$, the false alarm probability (FAP) for the proposed test statistic can be represented as

$$\begin{aligned} P_{FAP} &= P(R > \alpha | H_0) \\ &= \int_R^{\infty} \frac{1}{\sqrt{2\pi}} \exp\left(\frac{-t^2}{2}\right) dt \\ &= Q(R), \end{aligned} \tag{23.31}$$

where $Q(x) = \int_x^{\infty} 1/\sqrt{2\pi} \exp(-t^2/2)dt$. For a desired FAP τ, the associated threshold should be chosen such that

$$\alpha = Q^{-1}(\tau).$$

Otherwise, the detection rate (DR) can be denoted as

$$P_{DR} = P(R \geq Q(\alpha) | H_1). \tag{23.32}$$

It is noted that the computation complexity of the proposed test statistic in (23.30) is $O(\varepsilon n_g^4)$ which limits its practical application. Here, we proposed an effective approach to reduce the complexity of the proposed test statistic from $O(\varepsilon n_g^4)$ to $O(\eta n_g^2)$ by principal component calculation and redundant computation elimination. For simplicity, we briefly explained the technical details in our recent work, which is available at *https : //arxiv.org/abs/*1609.03301.

In this section, we evaluate the efficacy of the proposed test statistic for power system stability. For the experiments shown in the following, the real power flow data were of a chain-reaction fault happened in the China power grids in 2013. The PMU number, the sample rate, and the total sample time are $p = 34$, $K = 50Hz$, and $284s$, respectively. The chain-reaction fault happened from $t = 65.4s$ to $t = 73.3s$. Let $q = 5, n_g = 50$. Figure 23.12 shows that the mean and variance of λ agree well with theoretical ones. Based on the results in Figure 23.12 and event indicators (23.29), the occurrence time and the actual duration of the event can be identified as $t_0 = 65s$ and $t_dur \approx 8s$, respectively. The location of the most sensitive bus can also be identified using the data analysis above. The result shown in Figure 23.13 illustrates that $17th$ and $18th$ PMU are the most sensitive PMUs, which is in accordance with the actual accident situation.

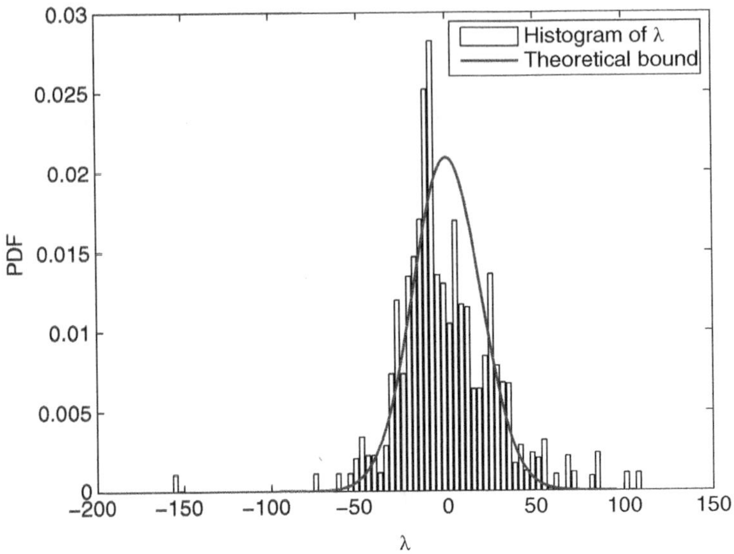

Figure 23.12 Parameter learning of the IEEE 118-bus system.

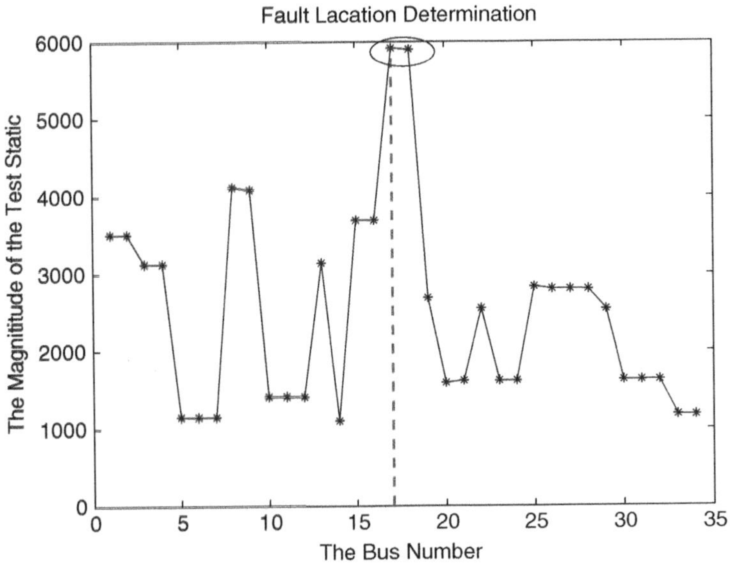

Figure 23.13 Data analysis of the realistic 34-PMU power flow around events occurrence.

23.3.3 Situation Awareness based on Linear Eigenvalue Statistics

Situation awareness (SA) is of great significance in power system operation, and a reconsideration of SA is essential for future grids [24]. These future grids are always huge in size and complex in topology. Operating under a novel regulation, their management mode is much different from previous one.

All these driving forces demand a new prominence for the term "situation awareness" (SA). The SA is essential for power grid security; inadequate SA is identified as one of the root causes for the largest blackout in history—the August 14, 2003, blackout in the United States and Canada [75].

In [76], SA is defined as the perception of the elements in an environment, the comprehension of their meaning, and the projection of their status in the near future. This chapter is aimed at the use of model-free and data-driven methodology for the comprehension of the power grid.

The massive data compose the profile of the actual grid—present state; SA aims to translate the present state into perceived state for decision making [77].

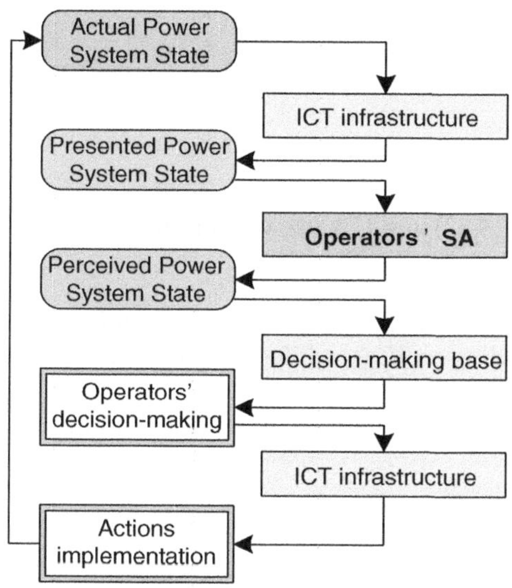

(a) SA for the operational decision-making

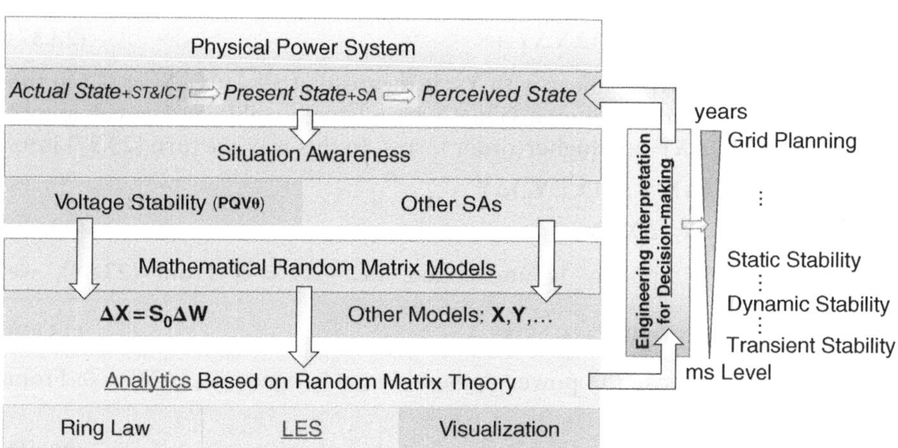

(b) SA Methodology based on RMT

Figure 23.14 SA and its methodology.

The proposed methodology consists of three essential procedures as illustrated in Figure 23.14(b): (1) big data model—to model the system using experimental data for the RMM; (2) big data analysis—to conduct high-dimensional analyses for the indicator system as the statistical solutions; (3) engineering interpretation—to visualize and interpret the statistical results for human beings for decision making.

Power grids operate in a balance situation obeying

$$\begin{cases} \Delta P_i = P_{is} - P_i(\mathbf{v}, \boldsymbol{\theta}) \\ \Delta Q_i = Q_{is} - Q_i(\mathbf{v}, \boldsymbol{\theta}) \end{cases}, \tag{23.33}$$

where P_{is} and Q_{is} are the power injections on node i, while $P_i(\mathbf{v}, \boldsymbol{\theta})$ and $Q_i(\mathbf{v}, \boldsymbol{\theta})$ are the injections of the network satisfying

$$\begin{cases} P_i = V_i \sum_{j=1}^{n} V_j(G_{ij} \cos \theta_{ij} + B_{ij} \sin \theta_{ij}) \\ Q_i = V_i \sum_{j=1}^{n} V_j(G_{ij} \sin \theta_{ij} - B_{ij} \cos \theta_{ij}) \end{cases}. \tag{23.34}$$

For simplicity, combining (23.33) and (23.34), we obtain

$$\mathbf{W}_0 = f(\mathbf{X}_0, \mathbf{Y}_0), \tag{23.35}$$

where $\mathbf{W}_0$ is the vector of power injections on nodes depending on P_{is}, Q_{is}. $\mathbf{X}_0$ is the system status variables depending on V_i, θ_i, while $\mathbf{Y}_0$ is the network topology parameters depending on B_{ij}, G_{ij}.

For a system state with certain fluctuations, and thus randomness in data sets, we formulate the system as

$$\mathbf{W}_0 + \Delta\mathbf{W} = f(\mathbf{X}_0 + \Delta\mathbf{X}, \mathbf{Y}_0 + \Delta\mathbf{Y}). \tag{23.36}$$

With a Taylor expansion, (23.36) is rewritten as

$$\begin{aligned} \mathbf{W}_0 + \Delta\mathbf{W} &= f(\mathbf{X}_0, \mathbf{Y}_0) + f'_{\mathbf{X}}(\mathbf{X}_0, \mathbf{Y}_0)\Delta\mathbf{X} + f'_{\mathbf{Y}}(\mathbf{X}_0, \mathbf{Y}_0)\Delta\mathbf{Y} \\ &\quad + \frac{1}{2}f''_{\mathbf{XX}}(\mathbf{X}_0, \mathbf{Y}_0)(\Delta\mathbf{X})^2 + \frac{1}{2}f''_{\mathbf{YY}}(\mathbf{X}_0, \mathbf{Y}_0)(\Delta\mathbf{Y})^2 \\ &\quad + f''_{\mathbf{XY}}(\mathbf{X}_0, \mathbf{Y}_0)\Delta\mathbf{X}\Delta\mathbf{Y} + \cdots . \end{aligned} \tag{23.37}$$

Equ. (23.34) shows that $\mathbf{W}_0$ is linear with $\mathbf{Y}_0$; it means that $f''_{\mathbf{YY}}(\mathbf{X}, \mathbf{Y}) = 0$. On the other hand, the values of system status variables $\mathbf{X}$ are relatively stable, and we can ignore the second-order term $(\Delta\mathbf{X})^2$ and higher-order terms. In this way, we turn (23.37) into

$$\begin{aligned} \Delta\mathbf{W} &= f'_{\mathbf{X}}(\mathbf{X}_0, \mathbf{Y}_0)\Delta\mathbf{X} + f'_{\mathbf{Y}}(\mathbf{X}_0, \mathbf{Y}_0)\Delta\mathbf{Y} \\ &\quad + f''_{\mathbf{XY}}(\mathbf{X}_0, \mathbf{Y}_0)\Delta\mathbf{X}\Delta\mathbf{Y}. \end{aligned} \tag{23.38}$$

Suppose the network topology is unchanged, i.e., $\Delta\mathbf{Y} = 0$. From (23.38), we deduce that

$$\Delta\mathbf{X} = (f'_{\mathbf{X}}(\mathbf{X}_0, \mathbf{Y}_0))^{-1}(\Delta\mathbf{W}) = \mathbf{S}_0\Delta\mathbf{W}. \tag{23.39}$$

On the other hand, suppose the power demand is unchanged, i.e., $\Delta\mathbf{W} = 0$. From (23.38), we obtain that

$$\Delta\mathbf{X} = \mathbf{S}_0\Delta\mathbf{W}_y, \tag{23.40}$$

where $\mathbf{W}_y = [\mathbf{I} + f''_{\mathbf{XY}}(\mathbf{X}_0, \mathbf{Y}_0)\Delta\mathbf{Y}\mathbf{S}_0]^{-1}[f'_{\mathbf{Y}}(\mathbf{X}_0, \mathbf{Y}_0)]$.

Note that $\mathbf{S}_0 = (f'_{\mathbf{X}}(\mathbf{X}_0, \mathbf{Y}_0))^{-1}$, i.e., the inversion of the Jacobian matrix $\mathbf{J}_0$, expressed as

$$\mathbf{J}_{ij0} = \left.\begin{bmatrix} \frac{\partial P_i}{\partial U_j} & \frac{\partial P_i}{\partial \theta_j} \\ \frac{\partial Q_i}{\partial U_j} & \frac{\partial Q_i}{\partial \theta_j} \end{bmatrix}\right|_{U_j=U_{j0},\theta_j=\theta_{j0}} . \tag{23.41}$$

Thus, we describe the power system operation using a random matrix—if there is an unexpected active power change or short circuit, the corresponding change of system status variables $\mathbf{X}_0$, i.e. V_i, θ_i, will obey (23.39) or (23.40) respectively.

For a practical system, we can always build a relationship in the form of $\mathbf{Y} = \mathbf{HX}$ with a similar procedure as (23.35) to (23.40); it is linear in high dimensions. For an equilibrium operation system in which the reactive power is almost constant or changes much more slowly than the active one, the relationship model between voltage magnitude and active power is just like the multiple input multiple output (MIMO) model in wireless communication [49, 78]. Note that most variables of vector $\mathbf{V}$ are random due to the ubiquitous noises, e.g., small random fluctuations in $\mathbf{P}$. In addition, we can add very small artificial fluctuations to make them random or replace the missing/bad data with random Gaussian variables. Furthermore, with the normalization, we can build the standard random matrix model (RMM) in the form of $\tilde{\mathbb{V}} = \tilde{\mathbf{\Xi}}\mathbf{R}$, where $\mathbf{R}$ is a standard Gaussian random matrix.

The data-driven approach conducts analysis requiring no prior knowledge of system topologies, unit operation/control mechanism, causal relationship, etc. It is able to handle massive data all at once; the large size of the data, indeed, enhances the robustness of the final decision against the bad data (errors, losses, or asynchronization). Compared with classical data-driven methodologies (e.g., PCA), the RMT-based counterpart has some unique characteristics:

- The statistical indicator is generated from all the data in the form of matrix entries. This is not true for principal components—we really do not know the rank of the covariance matrix. Thus, the RMT approach is robust against those challenges in classical data-driven methods, such as error accumulations and spurious correlations [53].
- For the statistical indicator, a theoretical or empirical value can be obtained in advance. The statistical indicator such as LES follows a Gaussian distribution, and its variance is bounded [79] and decays very fast in the order of $O(N^{-2})$ given a moderate data dimension N, say $N = 118$.
- We can flexibly handle heterogenous data to realize data fusion via matrix operations, such as the blocking [80], the sum [78], the product [78], and the concatenation [53] of matrices. Data fusion is guided by the latest mathematical research [49, Chapter 7].
- Only eigenvalues are used for further analysis, while the eigenvectors are omitted. This leads to much smaller required memory space and faster data-processing speed. Although some information is lost in this way, there is still rich information contained in the eigenvalues [81], especially those outliers [82, 83].
- Particularly, for a certain RMM, various forms of LES can be constructed by designing test functions without introducing any physical error (i.e., $\tau_{\mathrm{F}} \sum_{i=1}^{N} \varphi_F(\lambda_{\mathbf{M},i})$). Each LES, similar to a filter, provides a unique view angle. As a result, the system is systematically

understood piece by piece. Finally, with a proper LES, we can trace some specific signal.

We adopt a standard IEEE 118-node system as the grid network (Figure 23.15) and the events is shown in Table 23.1.

The power demand of nodes are assigned as

$$\tilde{y}_{\text{load\_}nt} = y_{\text{load\_}nt} \times (1 + \gamma_{\text{Mul}} \times r_1)) + \gamma_{\text{Acc}} \times r_2, \tag{23.42}$$

where r_1 and r_2 are the element of a standard Gaussian random matrix; $\gamma_{\text{Acc}} = 0.1$, $\gamma_{\text{Mul}} = 0.001$. Thus, the power demand on each node is obtained as the system injections (Figure 23.16(a)); the voltage can also be obtained (Figure 23.16(b)). Suppose we sample

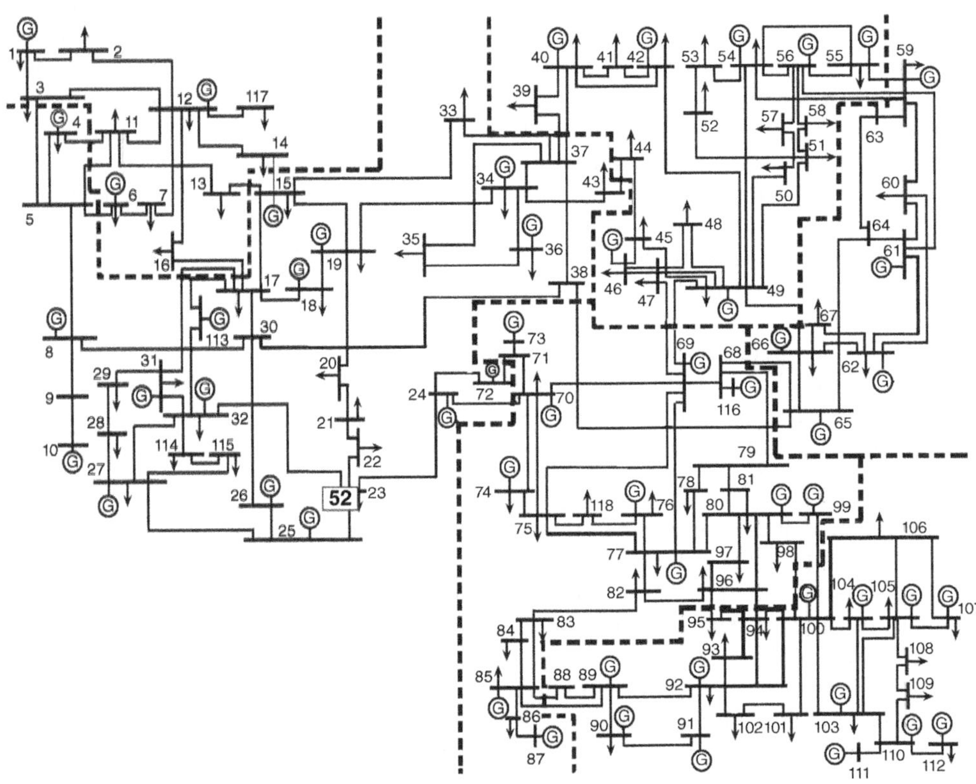

Figure 23.15 Partitioning network for the IEEE 118-node system.

Table 23.1 Series of Events.

Stage	**E1**	**E2**	**E3**	**E4**
Time (s)	1 – 500	501 – 900	901 – 1300	1301 – 2500
$P_{\text{Node-52}}$ (MW)	0	↑ 30	↑ 120	↗ $t/4 - 205$

P_{52} is the power demand of node 52.

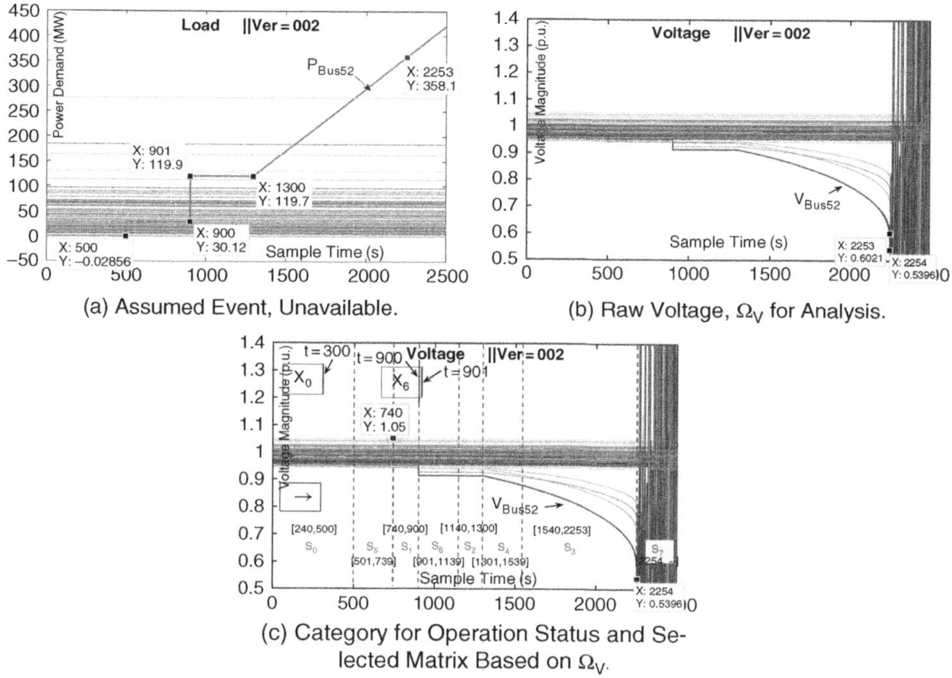

(a) Assumed Event, Unavailable.

(b) Raw Voltage, Ω_V for Analysis.

(c) Category for Operation Status and Selected Matrix Based on Ω_V.

Figure 23.16 Assumed event, data source, and category for case.

the voltage data at 1 Hz, and the data source is denoted as $\boldsymbol{\Omega}_V : \tilde{v}_{i,j} \in \mathbb{R}^{118\times 2500}$. The number of dimensions is $n = 118$ and the sampling time span is $t = 2500$.

Suppose that the power demand data (Fig. 23.16(a)) are *unknown* or unqualified for SA due to low sampling frequency or bad quality. For further analysis, we just start with data source $\boldsymbol{\Omega}_V$ (Figure 23.16(b)) and assign the analysis matrix as $\mathbf{X} \in \mathbb{R}^{118\times 240}$ (4 minutes' time span). First, we conduct category for the system operation status; the results are shown as Figure 23.16(c). In general, according to the raw data source and the analysis matrix size, we divide our system into 8 stages. Note that it is a statistical division—S4, S5, and S6 are transition stages, and their time span is right equal to the length of the analysis matrix minus one, i.e, $T - 1 = 239$. These stages are described as follows:

- For S0, S1, S2, the white noises play a dominant part. $P_{\text{Node-52}}$ is rising in turn.
- For S3, $P_{\text{Node-52}}$ maintains stable growth.
- S4, transition stage. Ramping signal exists.
- S5, S6, transition stages. Step signal exists.
- For S7, voltage collapse.

We also select two typical data cross-sections for stage S0 and S6: $X_0 \in \mathbb{R}^{118\times 240}$ during period $t = [61 : 300]$ at the sampling time $t_{\text{end}} = 300$, and 2) $X_6 \in \mathbb{R}^{118\times 240}$ during period $t = [662 : 901]$ at the sampling time $t_{\text{end}} = 901$.

Besides, as discussed in 23.2.1, we build up the RMM $\tilde{\mathbb{V}}$ from the raw voltage data. Then, τ_{MSR} is employed as a statistical indicator to conduct anomaly detection. For the

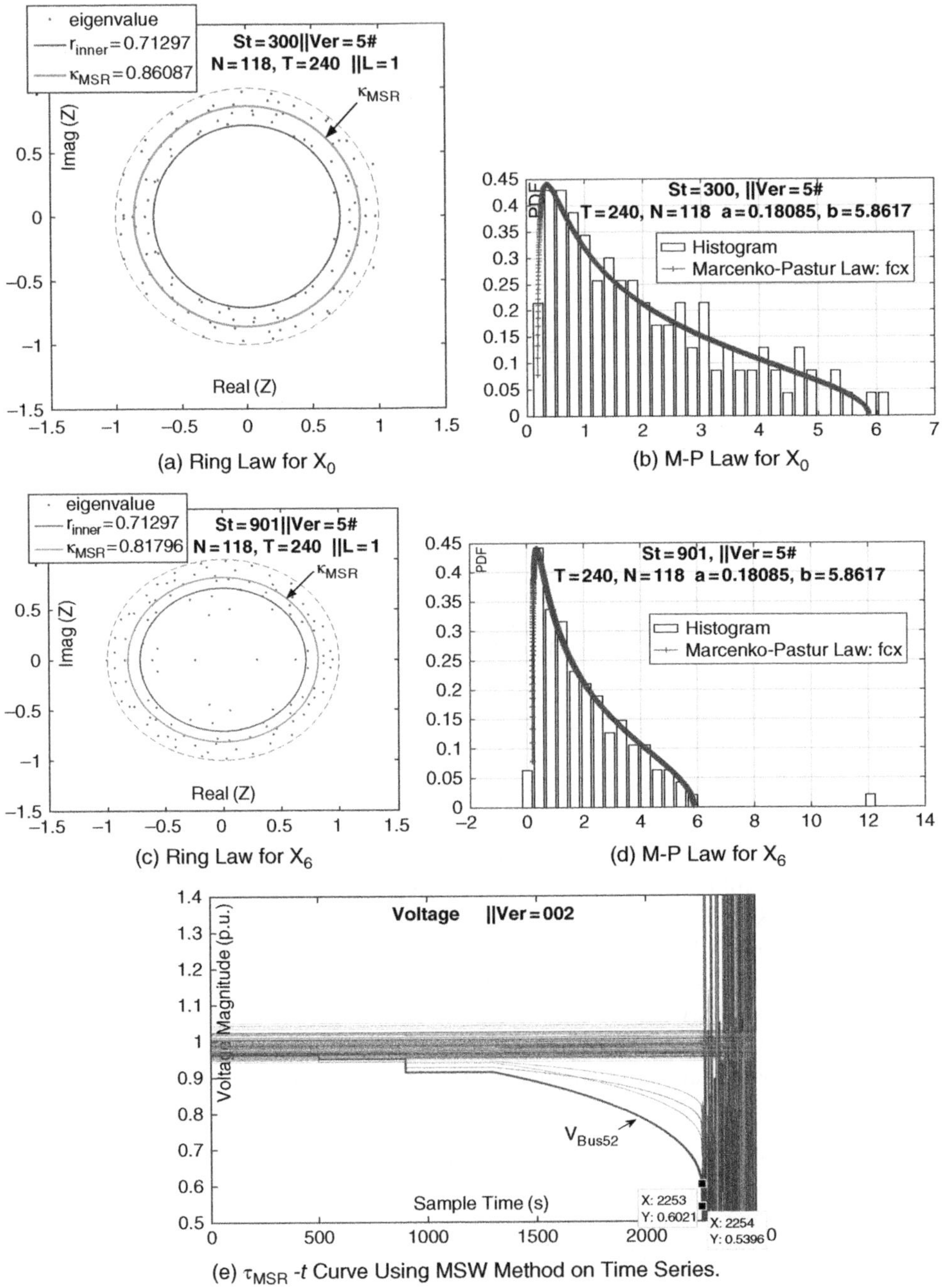

Figure 23.17 Anomaly detection result.

selected data cross section $\mathbf{X}_0$ and $\mathbf{X}_6$, their M-P law and ring law analysis are shown as Figure 23.17(a), 23.17(b), 23.17(c) and 23.17(d). With moving slide window (MSW), the $\tau_{\mathrm{MSR}} - t$ curve is obtained as Figure 23.17(e).

Fig 23.17 shows that when there is no signal in the system, the experimental RMM well matches the ring law and M-P law, and the experimental value of LES is approximately equal to the theoretical value. This validates the theoretical justification for modeling rapid fluctuation at each node with additive white Gaussian noise, as shown in Section 23.2.1. On the other hand, the ring law and M-P law are violated at the very beginning ($t_{\mathrm{end}} = 901$) of the step signal. Besides, the proposed high-dimensional indicator τ_{MSR}, is extremely sensitive to the anomaly. At $t_{\mathrm{end}} = 901$, the τ_{MSR} starts the dramatic change as shown in the $\tau_{\mathrm{MSR}} - t$ curve as Figure 23.17(e), while the raw voltage magnitudes are still in the normal range as shown in Figure 23.16(c). Moreover, we design numerous kinds of LES τ and define $\mu_0 = \tau/\mathbb{E}(\tau)$. The results are shown in Figure 23.18 and prove that different indicators have different characteristics and effectiveness; this suggests another topic to explore in the future.

Furthermore, we investigate the SA based on the high-dimensional spectrum test. The sampling time is set as $t_{\mathrm{end}} = 300$ and $t_{\mathrm{end}} = 901$. The following Lemma 23.2.7 and Lemma 23.2.9,

$\mathbf{Y}_0, Y_6 \in \mathbb{R}^{118\times240}$ (span $t = [61 : 300]$ and $t = [662 : 901]$), and $\mathbf{Z}_0, Z_6 \in \mathbb{R}^{118\times118}$ (span $t = [183 : 300]$ and $t = [784 : 901]$) are selected. The results are shown in Figure 23.19 and Fig. 23.20. These results validate that empirical spectral density test is competent to conduct anomaly detection—when the power grid is under a normal condition, the empirical spectral density $f_{\mathbf{A}}(x)$ and the ESD function $F_{\mathbf{A}}(x)$ are almost strictly bounded between the upper bound and the lower bound of their asymptotic limits. On the other hand, these results also validate that GUE and LUE are proper mathematical tools to model the power grid operation.

The $V-P$ curve (also called nose curve) and the smallest eigenvalue of the Jacobian matrix [39] are two clues for steady stability evaluation. In this case, we focus on the **E4** part during which $P_{\text{Node-52}}$ keep increasing to break down the steady stability. The related $V-P$ curve and $\lambda - P$ curve, respectively, are given in Figure 23.21(a) and

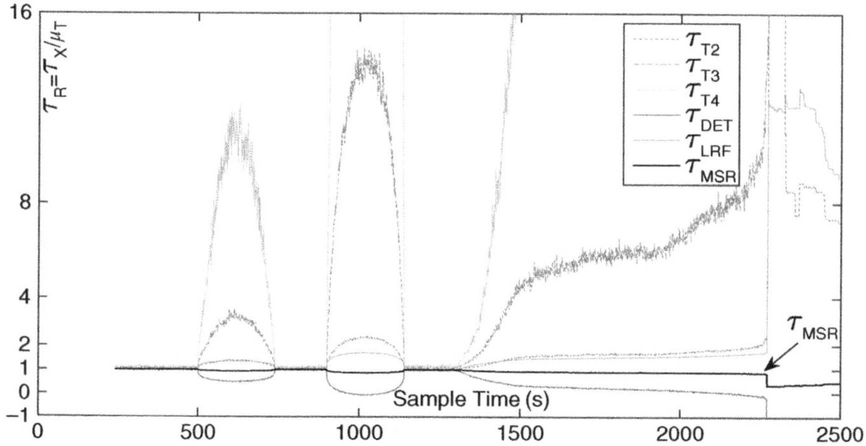

Figure 23.18 Illustration of various LES indicators.

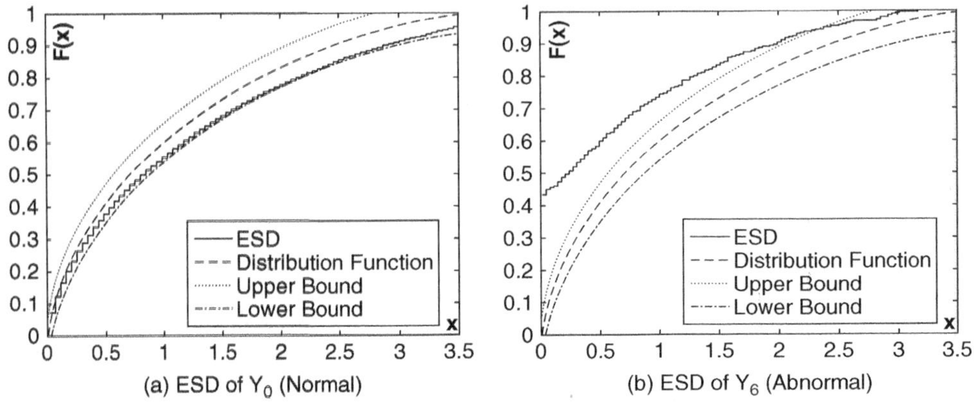

Figure 23.19 Anomaly detection using LUE matrices.

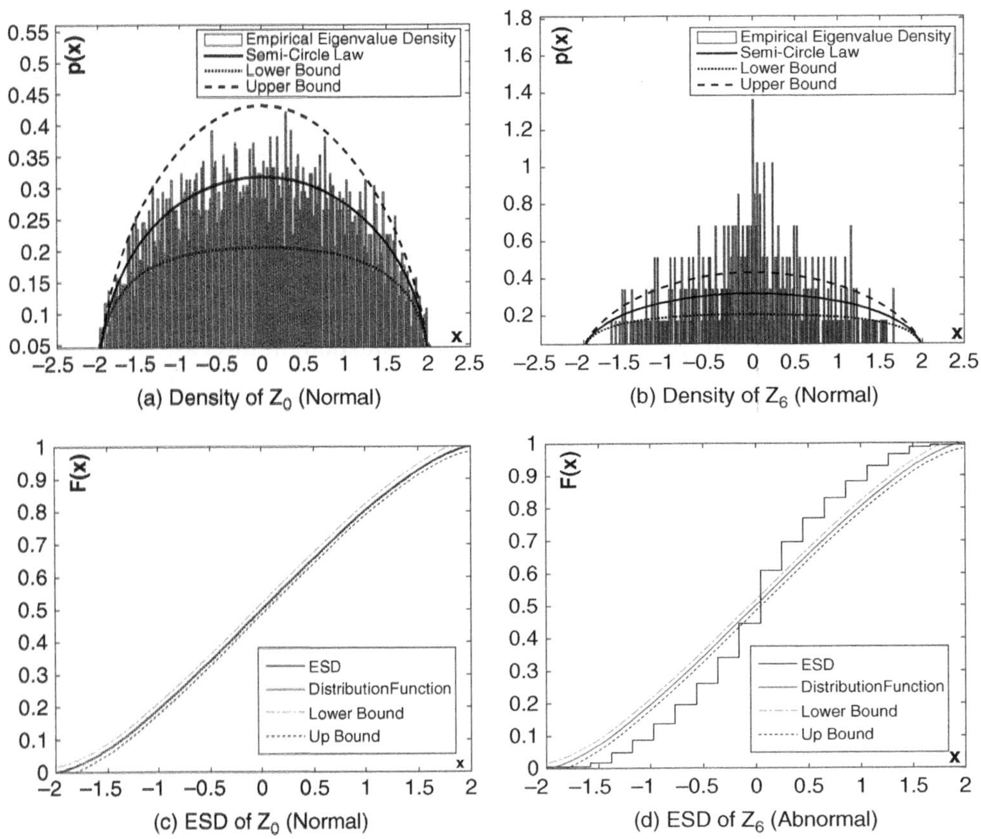

Figure 23.20 Anomaly detection using GUE matrices.

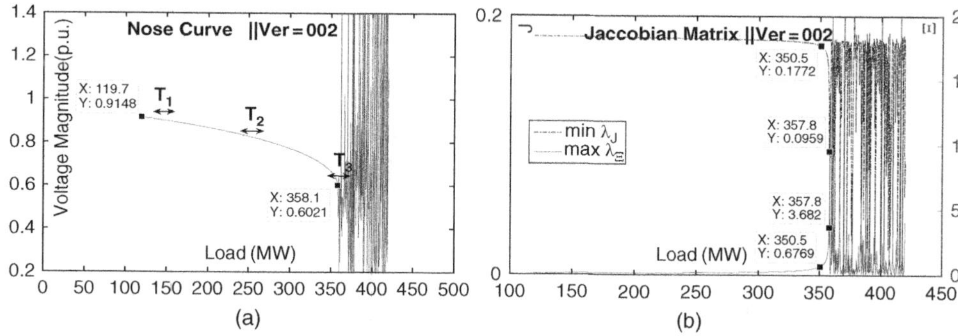

Figure 23.21 The $V - P$ curve and $\lambda - P$ curve.

Figure 23.21(b). Only using the data source $\mathbf{\Omega}_\mathrm{V}$, we choose some data cross section, $\mathbf{T}_1$: [1601 : 1840]; $\mathbf{T}_2$: [1901 : 2140]; $\mathbf{T}_3$: [2101 : 2340], as shown in Figure 23.21(a). The RMT-based results are shown as Figure 23.22. The outliers become more evident as the stability degree decreases. The statistics of the outliers are similar to the smallest eigenvalue of the Jacobian matrix, Lyapunov exponent, or the entropy in some sense.

For further analysis, we take the signal and stage division into account. Generally speaking, sorted by the stability degree, the stages are ordered as $\mathbf{S0} > \mathbf{S1} > \mathbf{S2} > \mathbf{S3} \gg \max(\mathbf{S4}, \mathbf{S5}) \gg \mathbf{S6} \gg \mathbf{S7}$. According to Figure 23.18, we make Table 23.2. The high-dimensional indicators $\overline{\tau_{\mathrm{XR}}}$ and V_R have the same trend as the stability degree order. These statistics have the potential for data-driven stability evaluation.

The key for correlation analysis is the concatenated matrix $\mathbf{A}_i$, which consists of two parts—the basic matrix $\mathbf{B}$ and a certain factor matrix $\mathbf{C}_i$, i.e., $\mathbf{A}_i = [\mathbf{B}; \mathbf{C}_i]$. For more details, see our previous work [53]. The LES of each $\mathbf{A}_i$ is computed in parallel, and Figure 23.23 shows the results.

In Figure 23.23, the blue dotted line (marked with "None") shows the LES of basic matrix $\mathbf{B}$, and the orange line (marked with "Random") shows the LES of the concatenated matrix $[\mathbf{B}; \mathbf{R}]$ ($\mathbf{R}$ is the standard Gaussian random matrix). Figure 23.23 demonstrates that: (1) node 52 is the causing factor of the anomaly; (2) sensitive nodes are 51, 53, and 58; and (3) nodes 11, 45, 46, etc., are not affected by the anomaly. Based on this algorithm, we can continue to conduct behavior analysis, e.g., detection and estimation of residential PV installations [84]. Behavior analysis is a big topic. Because of space limitations, we will not expand it here.

23.3.4 Early Event Detection Using Free Probability

Problem Modeling: Following [85], we build the statistic model for power grids. Considering T random vectors observed at time instants $i = 1, ..., T$, we form a random matrix as follows

$$[\Delta\mathbf{V}_1, \cdots, \Delta\mathbf{V}_T] = [\Xi_1 \Delta\mathbf{P}_1, \cdots, \Xi_T \Delta\mathbf{P}_T]. \tag{23.43}$$

In an equilibrium operating system, the voltage magnitude vector injections $\mathbf{V}$ with entries $V_i, i = 1, \cdots, N$ and the phase angle vector injections $\boldsymbol{\theta}$ with entries $\theta_i, i = 1, \cdots, N$ experience slight changes. Without dramatic topology changes, rich statistical empirical evidence indicates that the Jacobian matrix $\mathbf{J}$ keeps nearly constant,

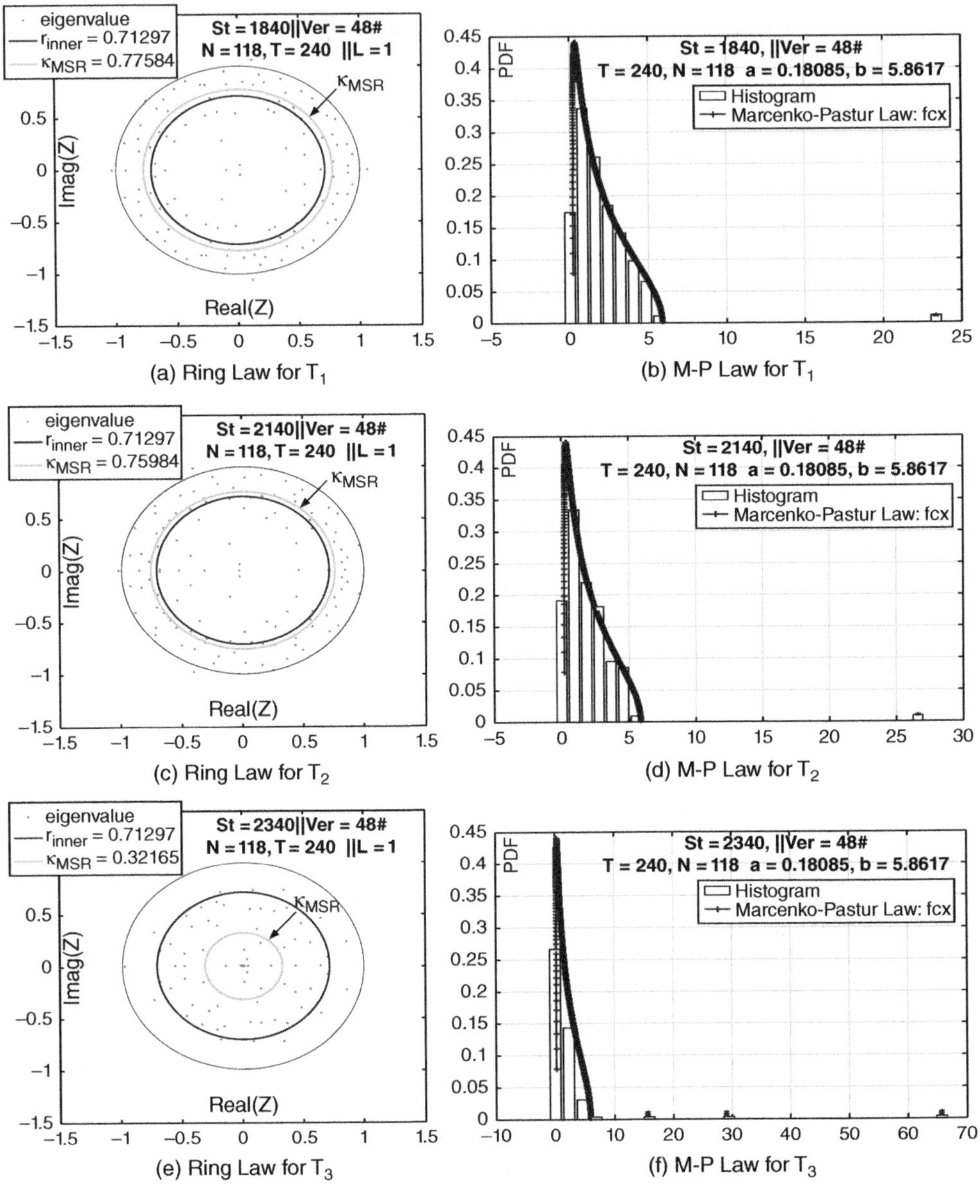

Figure 23.22 RMT-based results for voltage stability evaluation.

Table 23.2 Indicator of Various LESs at Each Stage.

	MSR	T_2	T_3	T_4	DET	LRF
E_0: Theoretical Value						
$\mathbb{E}(\tau)$	0.8645	1338.3	10069	8.35E4	48.322	73.678
$\mathbb{D}_T(\tau)$	—	665.26	93468	1.30E7	1.3532	1.4210
S0 [0240:0500, 261]: Small fluctuations around 0 MW						
$\overline{\tau_{XR}}$	0.995	1.010	1.040	1.080	0.959	1.014
V	6E −6	78.38	3.03E4	7.14E6	0.4169	0.3908
V_R	1	1	1	1	1	1
S5 [0501:0739, 239]: A step signal (0 MW↑ 30 MW) is included						
$\overline{\tau_{XR}}$	0.9331	1.280	2.565	7.661	0.5453	1.284
V_R	1.49E1	1.64E2	1.16E3	8.63E3	3.43E1	3.97E1
S1 [0740:0900, 161]: Small fluctuations around 30 MW						
$\overline{\tau_{XR}}$	0.9943	1.010	1.039	1.084	0.9568	1.015
V_R	0.8608	0.9121	0.9476	1.234	0.8972	1.101
S6 [0901:1139, 239]: A step signal (30 MW↑ 120 MW) is included						
$\overline{\tau_{XR}}$	0.8742	2.054	1.06E1	7.22E1	7E -2	1.597
V_R	5.49E1	2.06E3	3.87E4	8.54E5	1.52E2	1.62E2
S2 [1140:1300, 161]: Small fluctuations around 120 MW						
$\overline{\tau_{XR}}$	0.9930	1.019	1.067	1.135	0.9488	1.021
V_R	0.7823	1.053	1.189	1.135	0.7310	0.9255
S4 [1301:1539, 239]: A ramp signal (119.7 MW↗) is included						
$\overline{\tau_{XR}}$	0.9337	1.295	2.787	9.615	0.5316	1.294
V_R	8.50E1	7.41E2	5.63E3	5.17E4	2.14E2	2.30E2
S3 [1540:2253, 714]: Steady increase (↗ 358.1 MW)						
$\overline{\tau_{XR}}$	0.8906	1.717	6.530	3.48E1	0.1483	1.545
V_R	1.35E1	3.28E2	5.33E3	1.10E5	6.11E1	6.85E1
S7 [2254:2500, 247]: Static voltage collapse (361.9 MW↗)						
$\overline{\tau_{XR}}$	0.4259	1.02E1	2.11E2	4.65E3	−1.4E1	1.08E1
V_R	1.94E3	5.81E5	1.20E8	3.2E10	9.02E4	9.62E4

a) $\overline{\tau_{XR}} = \overline{\tau_X}/\mathbb{E}(\tau)$; $V_R(\tau_X) = V(\tau_X)/V(\tau_{X_0})$.

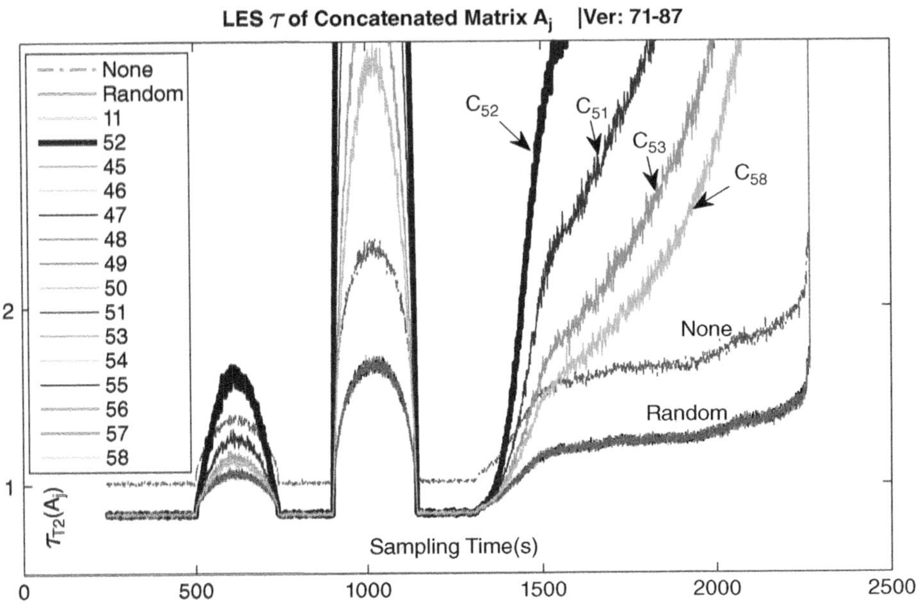

Figure 23.23 Sensitivity analysis based on concatenated matrix.

and so does Ξ. Also, we can estimate the changes of $\mathbf{V}$, $\boldsymbol{\theta}$, and Ξ only with the classical approach. Thus, we rewrite (23.43) as:

$$\mathbb{V} = \Xi_N \mathbb{P}_{N\times T} \tag{23.44}$$

where $\mathbb{V} = [\Delta\mathbf{V}_1, \cdots, \Delta\mathbf{V}_T]$, $\Xi = \Xi_1 = \cdots = \Xi_T$, and $\mathbb{P} = [\Delta\mathbf{P}_1, \cdots, \Delta\mathbf{P}_T]$. Here $\mathbb{V}$ and $\mathbb{P}$ are random matrices. In particular, $\mathbb{P}$ is a random matrix with Gaussian entries.

Model Designs: Multivariate linear or nonlinear polynomials perform a significant role in problem modeling, so we build our models on the basis of random matrix polynomials. Here, we study two typical random matrix polynomial models.

The first case is the multivariate linear polynomial:

$$P_1(S_0, S_1) = S_0 + S_1.$$

The second one is the self-adjoint multivariate nonlinear polynomial:

$$P_2(S_0, S_1) = S_0 S_1 + S_1 S_0.$$

Here, both S_0 and S_1 are the sample covariance matrices. The asymptotic eigenvalue distributions of P_1 and P_2 can be obtained via basic principles of free probability theory, as introduced above. The asymptotic eigenvalue distributions of P_i are regarded as the theoretical bounds.

Hypothesis Testing and Anomaly Detection: We formulate our problem of anomaly detection in terms of the same hypothesis testing as [85]: no outlier exists $\mathcal{H}_0$, and outlier exists $\mathcal{H}_1$.

$$\left| \begin{array}{l} \mathcal{H}_0 : \tilde{\mathbb{V}} = \tilde{\Xi} R_{N\times T} \\ \mathcal{H}_1 : \tilde{\mathbb{V}} \neq \tilde{\Xi} R_{N\times T} \end{array} \right. \tag{23.45}$$

where R is the standard Gaussian random matrix.

Generate S_0, S_1 from the sample data through the preprocess in 23.3.4. Compare the theoretical bound with the spectral distribution of raw data polynomials. If an outlier exists, $\mathcal{H}_0$ will be rejected, i.e., signals exist in the system.

The Processing of the Grid Data: The data sampled from the power grid is always non-Gaussian, so we adopt a normalization procedure in [80] to conduct data preprocessing. Meanwhile, we employ the Monte Carlo method to compute the spectral distribution of the raw data polynomial according to asymptotic property theory. See details in Algorithm 1.

Algorithm 1

Require:

The sample data matrices: V_0 and V_1;
The number of repetition times: M;
The size of V_0 and V_1: N, T;
SNR: η

for $i \leq M$ **do**
  add small white noises to sample data matrices
  $\widetilde{V_0} = V_0 + \eta \text{ randn}(N, N)$;
  $\widetilde{V_1} = V_0 + \eta \text{ randn}(N, N)$;
  standardize $\widetilde{V_0}$ and $\widetilde{V_1}$, i.e. mean=0, variance=1;
  generate the covariance matrices: $S_0 = \widetilde{V_0}\widetilde{V_0}' /N$,
  $S_0 = \widetilde{V_1}\widetilde{V_0}' /N$;
  compute the eigenvalues of $P(S_0, S_1)$;
end for
Computer the frequency of different eigenvalues and draw the spectral distribution histogram;

Ensure:

The spectral distribution histogram.

Simulation Results: Our data fusion method is tested with simulated data in the standard IEEE 118-bus system. Detailed information of the system refers to the case118.m

Table 23.3 System status and sampling data.

Cross Section (s)	Sampling (s)	Descripiton
C_0 : 118 – 900	V_0 : 100 ~ 217	Reference, no signal
C_1 : 901 – 1017	V_1 : 850 ~ 967	Existence of a step signal
C_2 : 1918 – 2600	V_2 : 2200 ~ 2317	Steady load growth for Bus 22
C_3 : 3118 – 3790	V_3 : 3300 ~ 3417	Steady load growth for Bus 52
C_4 : 3908 – 4100	V_4 : 3900 ~ 4017	Chaos due to voltage collapse
C_5 : 4118 – 5500	V_5 : 4400 ~ 4517	No signal

a) We choose the temporal end edge of the sampling matrix as the marked time for the cross section. E.g., for V_0 : 100 ~ 217, the temporal label is 217, which belongs to C_0 : 118 – 800. Thus, this method is able to be applied to conduct real-time analysis.

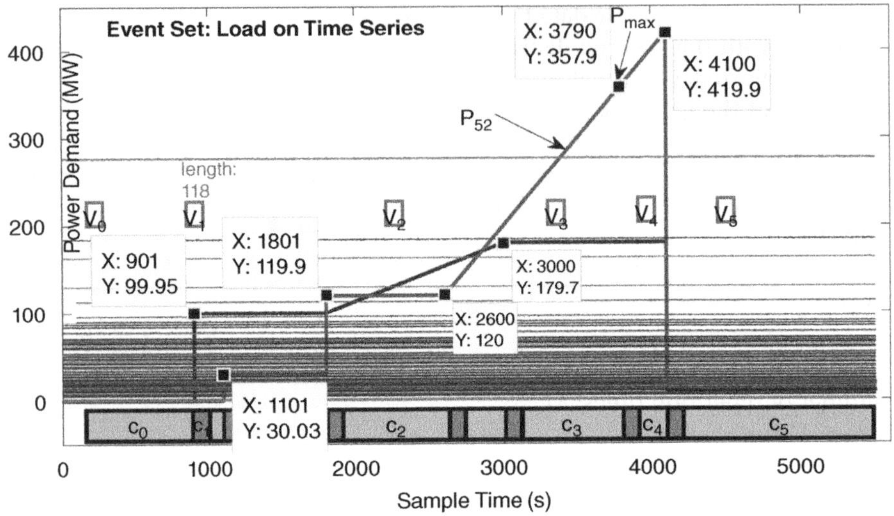

Figure 23.24 The event assumptions on time series.

in Matpower package and Matpower 4.1 User's Manual [86]. For all cases, let the sample dimension $N = 118$. In our simulations, we set the sample length equal to N, i.e., $T = 118$, $c = T/N = 1$ and select six sample voltage matrices presented in Table 23.3, as shown in Figure 23.24. The results of our simulations are presented in Figure 23.25 and Figure 23.26. The outliers existed when the system was abnormal, and its sizes become large when the anomaly becomes serious.

23.4 Conclusion and Future Directions

Motivated by the immediate demands of tackling the tricky problems raised from large-scale smart grids, this chapter introduced RMT-based schemes for spatiotemporal big data analysis. Firstly, we represent the spatiotemporal PMU data as a sequence of large random matrices. This is a crucial part for power state evaluation, as it turns the big PMU data into tiny data for practical uses. Rather than employing the raw PMU data, a comprehensive analysis of PMU data flow, namely, RMT-based techniques, is then proposed to indicate the state evaluation. The core techniques include streaming PMU data modeling, asymptotic properties analysis, and data fusion methods (based on free probability). Besides, the case studies based on synthetic data and real data are also included with the aim to bridge the technology gap between RMT and spatiotemporal data analysis in smart grids.

The current work based on RMT provides a fundamental exploration of data analysis for spatiotemporal PMU data. Much more attention is to be paid to this research direction, such as classification of power events and load forecasting. It is also noted that this work provides data-driven methods that are new substitutes for power system state estimation. The combination of power system scenario analysis, spectrum sensing mechanisms, networking protocols, and big data techniques [15, 34, 49, 87] is encouraged to be investigated for better understanding of the power system state.

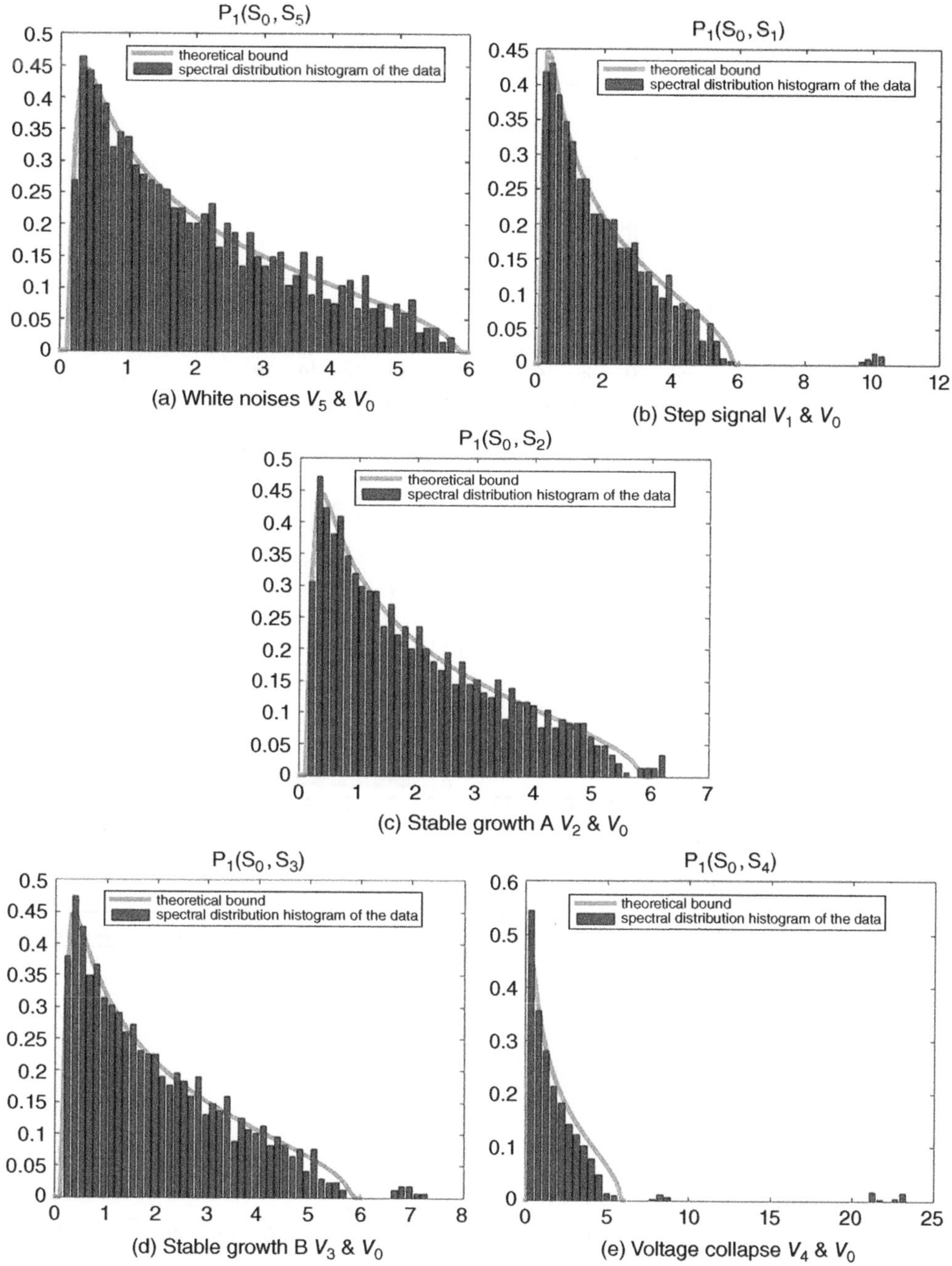

Figure 23.25 Data fusion using multivariate linear polynomial P_1.

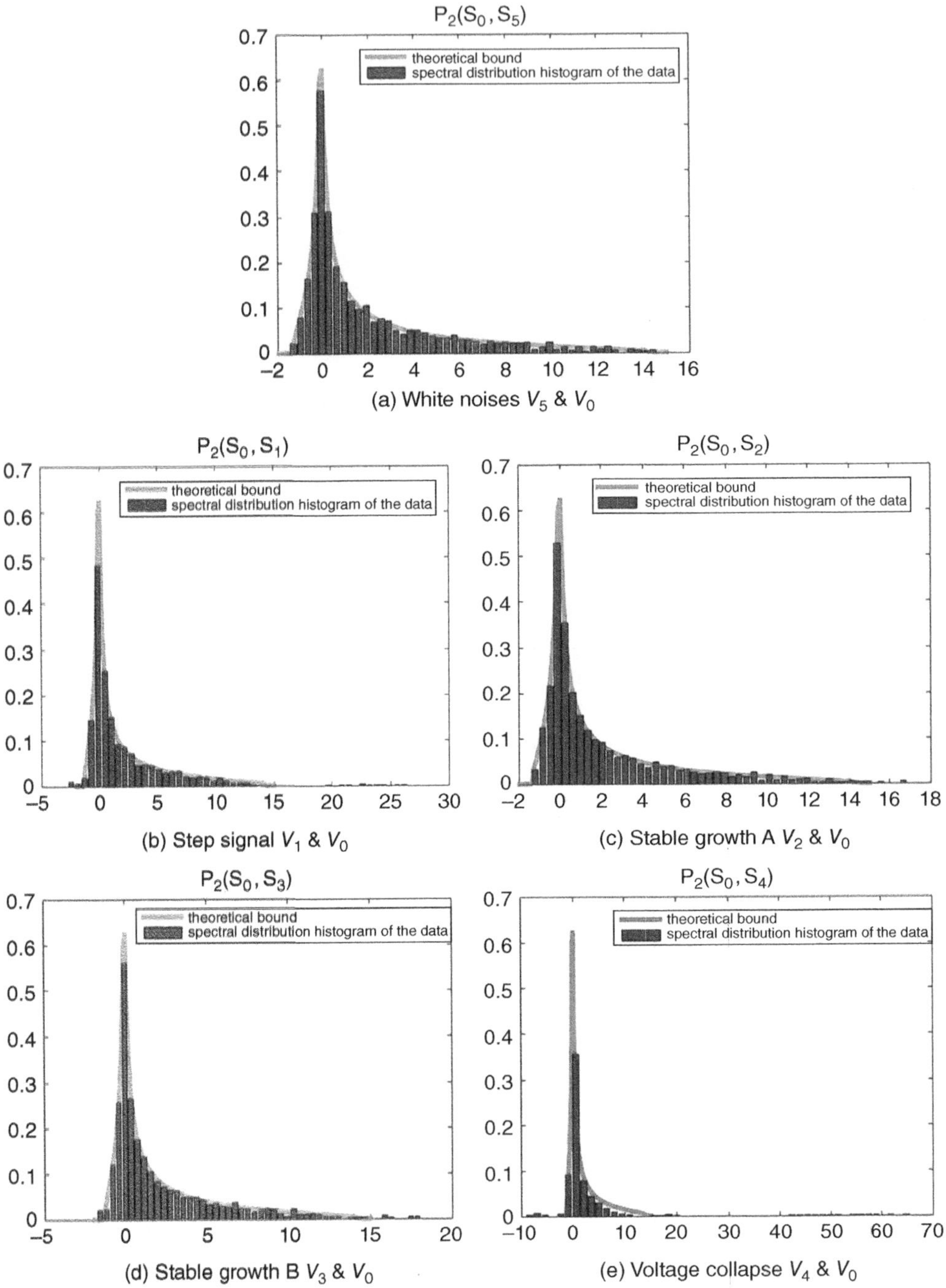

Figure 23.26 Data fusion using multivariate nonlinear polynomial P_2.

References

1 US DOE. Grid 2030: A national vision for electricity's second 100 years. *US DOE Report*, 2003.
2 Santiago Grijalva and Muhammad Umer Tariq. Prosumer-based smart grid architecture enables a flat, sustainable electricity industry. In *Innovative Smart Grid Technologies (ISGT), 2011 IEEE PES*, pages 1–6. IEEE, 2011.
3 Xiaoxin Zhou, Shuyong Chen, and Zongxiang Lu. Review and prospect for power system development and related technologies: a concept of three-generation power systems. *Proceedings of the CSEE*, 33(22):1–11, Aug. 2013.
4 Shengwei Mei, Yuan Gong, and Feng LIU. The evolution model of three generation power systems and characteristic analysis. *Proceedings of the CSEE*, 7:1003–1012, 2014.
5 Xing He, Qian Ai, Zhiwen Yu, Yiting Xu, and Jian Zhang. Power system evolution and aggregation theory under the view of power ecosystem. *Power System Protection and Control*, 42(22):100–107, Nov. 2014.
6 Zhang Hong, Dongmei Zhao, Chenghong Gu, Furong Li, and Bo Wang. Economic optimization of smart distribution networks considering real-time pricing. *Journal of Modern Power Systems and Clean Energy*, 2(4):350–356, 2014.
7 Yang Ji. Multi-agent system based control of virtual power plant and its application in smart grid. Master's thesis, School of Electronic Information and Electrical Engineering, Shanghai Jiaotong University, 2011.
8 Xing He, Qian Ai, Peng Yuan, and Xiaohong Wang. The research on coordinated operation and cluster management for multi-microgrids. In *Sustainable Power Generation and Supply (SUPERGEN 2012), International Conference on*, pages 1–3. IET, 2012.
9 Wilsun Xu and Jing Yong. Power disturbance data analytics–new application of power quality monitoring data. *Proceedings of the CSEE*, 33(19):93–101, July 2013.
10 IBM. The Four V's of Big Data. accessed date: July, 2015.
11 LS Moulin, AP Alves da Silva, MA El-Sharkawi, et al. Support vector machines for transient stability analysis of large-scale power systems. *Power Systems, IEEE Transactions on*, 19(2):818–825, 2004.
12 Yael Parag and Benjamin K Sovacool. Electricity market design for the prosumer era. *Nature Energy*, 1:16032, 2016.
13 C. Lynch. Big data: How do your data grow? *Nature*, 455(7209):28, 2008.
14 Science Staff. Dealing with data. challenges and opportunities. introduction. *Science*, 331(6018):692, 2011.
15 A. A Khan, M. H Rehmani, and M Reisslein. Cognitive radio for smart grids: Survey of architectures, spectrum sensing mechanisms, and networking protocols. *IEEE Communications Surveys and Tutorials*, 18(1):1–1, 2015.
16 Lei Chu, Robert Caiming Qiu, Xing He, Zenan Ling, and Yadong Liu. Massive streaming pmu data modeling and analytics in smart grid state evaluation based on multiple high-dimensional covariance tests. *IEEE Transactions on Big Data*, PP(99):1–1, 2017.
17 Tómas A Brody, J Flores, J Bruce French, PA Mello, A Pandey, and Samuel SM Wong. Random-matrix physics: spectrum and strength fluctuations. *Reviews of Modern Physics*, 53(3):385–480, 1981.

18 Laurent Laloux, Pierre Cizeau, Marc Potters, and Jean-Philippe Bouchaud. Random matrix theory and financial correlations. *International Journal of Theoretical and Applied Finance*, 3(3):391–397, July 2000.

19 Hsinchun Chen, Roger HL Chiang, and Veda C Storey. Business intelligence and analytics: From big data to big impact. *MIS quarterly*, 36(4):1165–1188, 2012.

20 Doug Howe, Maria Costanzo, Petra Fey, Takashi Gojobori, Linda Hannick, Winston Hide, David P Hill, Renate Kania, Mary Schaeffer, Susan St Pierre, et al. Big data: The future of biocuration. *Nature*, 455(7209):47–50, 2008.

21 Robert Qiu and Michael Wicks. *Cognitive Networked Sensing and Big Data*. Springer, 2013.

22 C. Zhang and R. C. Qiu. Data modeling with large random matrices in a cognitive radio network testbed: Initial experimental demonstrations with 70 nodes. *ArXiv e-prints*, Apr. 2014.

23 Xia Li, Feng Lin, and Robert C Qiu. Modeling massive amount of experimental data with large random matrices in a real-time UWB-MIMO system. *ArXiv e-prints*, Apr. 2014.

24 T. Hong, C. Chen, J. Huang, et al. Guest editorial big data analytics for grid modernization. *IEEE Transactions on Smart Grid*, 7(5):2395–2396, Sept 2016.

25 M. Rafferty, X. Liu, D. M. Laverty, and S. McLoone. Real-time multiple event detection and classification using moving window pca. *IEEE Transactions on Smart Grid*, 7(5):2537–2548, Sept 2016.

26 H. Jiang, X. Dai, D. W. Gao, J. J. Zhang, Y. Zhang, and E. Muljadi. Spatial-temporal synchrophasor data characterization and analytics in smart grid fault detection, identification, and impact causal analysis. *IEEE Transactions on Smart Grid*, 7(5):2525–2536, Sept 2016.

27 H. Sun, Z. Wang, J. Wang, Z. Huang, N. Carrington, and J. Liao. Data-driven power outage detection by social sensors. *IEEE Transactions on Smart Grid*, 7(5):2516–2524, Sept 2016.

28 X. Zhang and S. Grijalva. A data-driven approach for detection and estimation of residential pv installations. *IEEE Transactions on Smart Grid*, 7(5):2477–2485, Sept 2016.

29 H. Shaker, H. Zareipour, and D. Wood. A data-driven approach for estimating the power generation of invisible solar sites. *IEEE Transactions on Smart Grid*, 7(5):2466–2476, Sept 2016.

30 B. Wang, B. Fang, Y. Wang, H. Liu, and Y. Liu. Power system transient stability assessment based on big data and the core vector machine. *IEEE Transactions on Smart Grid*, 7(5):2561–2570, Sept 2016.

31 AG Phadke and Rui Menezes de Moraes. The wide world of wide-area measurement. *Power and Energy Magazine, IEEE*, 6(5):52–65, 2008.

32 Vladimir Terzija, Gustavo Valverde, Deyu Cai, Pawel Regulski, Vahid Madani, John Fitch, Srdjan Skok, Miroslav M Begovic, and Arun Phadke. Wide-area monitoring, protection, and control of future electric power networks. *Proceedings of the IEEE*, 99(1):80–93, 2011.

33 Le Xie, Yang Chen, and Huaiwei Liao. Distributed online monitoring of quasi-static voltage collapse in multi-area power systems. *Power Systems, IEEE Transactions on*, 27(4):2271–2279, 2012.

34 Athar Ali Khan, Mubashir Husain Rehmani, and Martin Reisslein. Requirements, design challenges, and review of routing and mac protocols for cr-based smart grid systems. *IEEE Communications Magazine*, 55(5):206–215, 2017.

35 Quanyuan Jiang, Xingpeng Li, Bo Wang, and Haijiao Wang. Pmu-based fault location using voltage measurements in large transmission networks. *Power Delivery, IEEE Transactions on*, 27(3):1644–1652, 2012.

36 Mahesh Venugopal and Chetan Tiwari. A novel algorithm to determine fault location in a transmission line using pmu measurements. In *Smart Instrumentation, Measurement and Applications (ICSIMA), 2013 IEEE International Conference on*, pages 1–4. IEEE, 2013.

37 Ali H Al-Mohammed and MA Abido. A fully adaptive pmu-based fault location algorithm for series-compensated lines. *Power Systems, IEEE Transactions on*, 29(5):2129–2137, 2014.

38 Le Xie, Yang Chen, and PR Kumar. Dimensionality reduction of synchrophasor data for early event detection: Linearized analysis. *IEEE Transactions on Power Systems*, 29(6):2784–2794, 2014.

39 Jong Min Lim and Christopher L DeMarco. Svd-based voltage stability assessment from phasor measurement unit data. *IEEE Transactions on Power Systems*, 31(4):2557–2565, 2016.

40 Eugene P Wigner. Lower limit for the energy derivative of the scattering phase shift. *Physical Review*, 98(1):145, 1955.

41 Eugene P. Wigner. On the distribution of the roots of certain symmetric matrices. *Annals of Mathematics*, 67(2):325–327, 1958.

42 Eugene P Wigner. The problem of measurement. *American Journal of Physics*, 31(1):6–15, 1963.

43 Eugene P. Wigner. Random matrices in physics. *Siam Review*, 9(1):1–23, 1967.

44 Eugene P. Wigner. *Characteristic Vectors of Bordered Matrices with Infinite Dimensions I*. Springer Berlin Heidelberg, 1993.

45 V. A. Marchenko and L. A. Pastur. Distribution of eigenvalues for some sets of random matrices. *Mathematics of the USSR-Sbornik*, 1(1):507–536, 1967.

46 Guionnet, Krishnapur, Manjunath, Zeitouni, and Ofer. The single ring theorem. *Annals of Mathematics*, 174(2):1189–1217, 2009.

47 Greg W. Anderson, Alice Guionnet, and Ofer Zeitouni. An introduction to random matrices. *Cambridge Studies in Advanced Mathematics*, 2009.

48 Terence Tao. *Topics in Random Matrix Theory*. American Mathematical Society, 2012.

49 R Qiu and P Antonik. *Smart Grid and Big Data*. John Wiley and Sons, 2015.

50 Math HJ Bollen. *Understanding power quality problems*, volume 3. IEEE press New York, 2000.

51 Denis Hau Aik Lee. Voltage stability assessment using equivalent nodal analysis. *IEEE Transactions on Power Systems*, 31(1):454–463, 2016.

52 Goodarz Ghanavati, Paul DH Hines, and Taras I Lakoba. Identifying useful statistical indicators of proximity to instability in stochastic power systems. *IEEE Transactions on Power Systems*, 31(2):1360–1368, 2016.

53 Xinyi Xu, Xing He, Qian Ai, and Robert Caiming Qiu. A correlation analysis method for power systems based on random matrix theory. *IEEE Transactions on Smart Grid*, PP(99):1–10, 2015.

54 Rudolf Wegmann. The asymptotic eigenvalue-distribution for a certain class of random matrices. *Journal of Mathematical Analysis and Applications*, 56(1):113–132, 1976.

55 Robert Caiming Qiu, Zhen Hu, Husheng Li, and Michael C Wicks. *Cognitive radio communication and networking: Principles and practice*. John Wiley & Sons, 2012.

56 Yingshuang Cao, Long Cai, C Qiu, Jie Gu, Xing He, Qian Ai, and Zhijian Jin. A random matrix theoretical approach to early event detection using experimental data. *arXiv preprint arXiv:1503.08445*, 2015.

57 M. L. Mehta. *Random matrices*, volume 142. Academic press, 2004.

58 Z. Bai. Convergence rate of expected spectral distributions of large random matrices. part i. wigner matrices. *The Annals of Probability*, pages 625–648, 1993.

59 Z. Bai, B. Miao, and J. Tsay. A note on the convergence rate of the spectral distributions of large random matrices. *Statistics & Probability Letters*, 34(1):95–101, 1997.

60 F. Götze and A. Tikhomirov. Rate of convergence to the semi-circular law. *Probability Theory and Related Fields*, 127(2):228–276, 2003.

61 Z. Bai, B. Miao, and J. Yao. Convergence rates of spectral distributions of large sample covariance matrices. *SIAM journal on matrix analysis and applications*, 25(1):105–127, 2003.

62 V. L. Girko. Extended proof of the statement: Convergence rate of the expected spectral functions of symmetric random matrices ξ_n is equal to o $(n^{-1/2})$ and the method of critical steepest descent. *Random Operators and Stochastic Equations*, 10(3):253–300, 2002.

63 Friedrich Götze and Alexander Tikhomirov. The rate of convergence for spectra of gue and lue matrix ensembles. *Open Mathematics*, 3(4):666–704, 2005.

64 Voiculescu Dan. *Symmetries of some reduced free product C*-algebras*. Springer Berlin Heidelberg, 1985.

65 Voiculescu Dan. Limit laws for random matrices and free products. *Inventiones mathematicae*, 104(1):201–220, 1991.

66 Serban Belinschi, Tobias Mai, and Roland Speicher. Analytic subordination theory of operator-valued free additive convolution and the solution of a general random matrix problem. *Journal Für Die Reine Und Angewandte Mathematik*, 2013.

67 Roland Speicher and Roland Speicher. Polynomials in asymptotically free random matrices. *Acta Physica Polonica*, 46(9), 2015.

68 Greg W. Anderson. Convergence of the largest singular value of a polynomial in independent wigner matrices. *Annals of Probability*, 38(1):110–C112, 2011.

69 Zhidong Bai and Hewa Saranadasa. Effect of high dimension: by an example of a two sample problem. *Statistica Sinica*, pages 311–329, 1996.

70 Zhidong Bai, Dandan Jiang, Jian-Feng Yao, and Shurong Zheng. Corrections to lrt on large-dimensional covariance matrix by rmt. *The Annals of Statistics*, pages 3822–3840, 2009.

71 Olivier Ledoit and Michael Wolf. Some hypothesis tests for the covariance matrix when the dimension is large compared to the sample size. *Annals of Statistics*, pages 1081–1102, 2002.

72 Song Xi Chen and Ying-Li Qin. A two-sample test for high-dimensional data with applications to gene-set testing. *The Annals of Statistics*, 38(2):808–835, 2010.

73 Song Xi Chen, Li-Xin Zhang, and Ping-Shou Zhong. Tests for high-dimensional covariance matrices. *Journal of the American Statistical Association*, 2012.

74 AJ Lee. U-statistics. *Theory and Practice," Marcel Dekker, New York*, 1990.

75 US-Canada Power System Outage Task Force, Spencer Abraham, Herb Dhaliwal, R John Efford, Linda J Keen, Anne McLellan, John Manley, Kenneth Vollman, Nils J Diaz, Tom Ridge, et al. *Final report on the August 14, 2003 blackout in the United states and Canada: causes and recommendations*. US-Canada Power System Outage Task Force, 2004.

76 Mica R Endsley. *Designing for situation awareness: An approach to user-centered design*. CRC Press, 2011.

77 Mathaios Panteli, Peter Crossley, Daniel S Kirschen, Dejan J Sobajic, et al. Assessing the impact of insufficient situation awareness on power system operation. *Power Systems, IEEE Transactions on*, 28(3):2967–2977, 2013.

78 Changchun Zhang and Robert C Qiu. Massive mimo as a big data system: Random matrix models and testbed. *IEEE Access*, 3:837–851, 2015.

79 Mariya Shcherbina. Central limit theorem for linear eigenvalue statistics of the wigner and sample covariance random matrices. *ArXiv e-prints*, January 2011.

80 Xing He, Qian Ai, Robert Caiming Qiu, and Wentao Huang. A big data architecture design for smart grids based on random matrix theory. *IEEE Transactions on Smart Grid*, 32(3):1, 2015.

81 Jesper R Ipsen and Mario Kieburg. Weak commutation relations and eigenvalue statistics for products of rectangular random matrices. *Physical Review E*, 89(3), 2014. Art. ID 032106.

82 Florent Benaych-Georges and Jean Rochet. Outliers in the single ring theorem. *Probability Theory and Related Fields*, pages 1–51, May 2015.

83 Terence Tao. Outliers in the spectrum of iid matrices with bounded rank perturbations. *Probability Theory and Related Fields*, 155(1-2):231–263, 2013.

84 Xiaochen Zhang and Santiago Grijalva. A data-driven approach for detection and estimation of residential pv installations. *IEEE Transactions on Smart Grid*, 7(5):2477–2485, 2016.

85 Xing He, Robert Caiming Qiu, Qian Ai, Lei Chu, Xinyi Xu, and Zenan Ling. Designing for situation awareness of future power grids: An indicator system based on linear eigenvalue statistics of large random matrices. *IEEE Access*, 4:3557–3568, 2016.

86 Ray D Zimmerman and Carlos E Murillo-Sánchez. Matpower 4.1 user's manual. *Power Systems Engineering Research Center*, 2011.

87 Robert Qiu and Michael Wicks. *Cognitive networked sensing and big data*. Springer, 2014.

Index

Transportation and Power Grid in Smart Cities: Communication Networks and Services, First Edition.
Edited by Hussein T. Mouftah, Melike Erol-Kantarci and Mubashir Husain Rehmani.

d